ECO- AND GROUND BIO-ENGINEERING: THE USE OF VEGETATION TO IMPROVE SLOPE STABILITY

Eco- and Ground Bio-Engineering: The Use of Vegetation to Improve Slope Stability

Proceedings of the First International Conference on Eco-Engineering
13–17 September 2004

Edited by

ALEXIA STOKES
IOANNIS SPANOS
JOANNE E. NORRIS

and

ERIK CAMMERAAT

Springer

A C.I.P. Catalogue record for this book is available from the Library of Congress.

ISBN-13 978-90-481-7403-4 e-ISBN-13 978-1-4020-5593-5

Published by Springer,
P.O. Box 17,3300 AA Dordrecht, The Netherlands.

www.springer.com

Cover photo:

The Finite Element Method was used to calculate the strength of tree root anchorage, with a digitized Maritime pine (*Pinus pinaster Ait.*) root system shown here as an example (data from F. Danjon, INRA). This numerical method was not only used on real root systems, but also applied to simulated schematic root patters. These theoretical investigations provided information concerning various aspects of tree anchorage mechanics, with regard to both root morphology and soil characteristics. It was shown for instance that the soil type significantly modifies the mode of failure of the root/soil plate. It was also demonstrated that, for a given total root biomass, heart-root systems are the most resistant pattern in clay-like soil and tap-root anchorage efficiency is higher in sandy-like soil.

Printed on acid-free paper

Contents

Applications at the slope level

Eco- and ground bio-engineering case studies by practitioners

Preface

In an era where climate change, natural catastrophes and land degradation are major issues, the conservation of soil and vegetation in mountainous or sloping regions has become an international priority. How to avoid substrate mass movement through landslides and erosion using sustainable and ecologically sound techniques is rapidly becoming a scientific domain where knowledge from many different fields is required. These proceedings bring together papers from geotechnical and civil engineers, biologists, ecologists and foresters, who discuss current problems in slope stability research, and how to address those problems using ground bio- and eco-engineering techniques. A selection of papers were previously published in Special Editions of Plant and Soil (2005), volume 278, 1–179, and in the Journal of Geotechnical and Geological Engineering (2006), volume 24, 427–498.

Ground bioengineering methods integrate civil engineering techniques with natural materials to obtain fast, effective and economic methods of protecting, restoring and maintaining the environment whereas eco-engineering has been defined as a long-term ecological strategy to manage a site with regard to natural or man-made hazards. Studies on slope instability, erosion, soil hydrology, mountain ecology, land use and restoration and how to mitigate these problems using vegetation are presented by both scientists and practitioners. Papers encompass many aspects of this multidisciplinary subject, including the mechanisms and modelling of root reinforcement and the development of decision support systems, areas where significant advances have been made in recent years.

Alexia Stokes
Ioannis Spanos
Joanne Norris
Erik Cammeraat

Mechanisms and modelling of root reinforcement on slopes

The influence of cellulose content on tensile strength in tree roots

Marie Genet[1,8], Alexia Stokes[1], Franck Salin[2], Slobodan B. Mickovski[1,3], Thierry Fourcaud[1,4], Jean-François Dumail[5] & Rens van Beek[6,7]

[1]*Laboratoire de Rhéologie du Bois de Bordeaux, (Mixed Unit: INRA/CNRS/Université Bordeaux I) Domaine de l'Hermitage, 69, rte d'Arcachon, 33612, Cestas Cedex, France.* [2]*INRA, Equipe de Génétique et Amélioration des Arbres Forestiers, UMR BIOGECO, 69, rte d'Arcachon, 33612, Cestas Cedex, France.* [3]*Civil Engineering Division, School of Engineering and Physical Sciences, University of Dundee, DD1 4HR, Dundee, UK.* [4]*AMAP-CIRAD AMIS, TA 40/PS2 Cedex 5, 34398, Montpellier, France.* [5]*XYLOMECA, 41, rue Michel de Montaigne, 24700, Moulin Neuf, France.* [6]*Institute for Biodiversity and Ecosystem Dynamics – Physical Geography, University of Amsterdam, Nieuwe Achtergracht 166, NL 1018 WV, Amsterdam, The Netherlands.* [7]*Department of Physical Geography, Utrecht University Heidelberglaan, 110, P.O. BOX 80.115, NL-3508, TC, Utrecht, The Netherlands.* [8]*Corresponding author**

Key words: biomechanics, *Castanea sativa* Mill., *Pinus pinaster* Ait., root reinforcement, slope stability, soil fixation

Abstract

Root tensile strength is an important factor to consider when choosing suitable species for reinforcing soil on unstable slopes. Tensile strength has been found to increase with decreasing root diameter, however, it is not known how this phenomenon occurs. We carried out tensile tests on roots 0.2–12.0 mm in diameter of three conifer and two broadleaf species, in order to determine the relationship between tensile strength and diameter. Two species, *Pinus pinaster* Ait. and *Castanea sativa* Mill., were then chosen for a quantitative analysis of root cellulose content. Cellulose is responsible for tensile strength in wood due to its microfibrillar structure. Results showed that in all species, a significant power relationship existed between tensile strength and root diameter, with a sharp increase of tensile strength in roots with a diameter <0.9 mm. In roots >1.0 mm, *Fagus sylvatica* L. was the most resistant to failure, followed by *Picea abies* L. and *C. sativa.*, *P. pinaster* and *Pinus nigra* Arnold roots were the least resistant in tension for the same diameter class. Extremely high values of strength (132–201 MPa) were found in *P. abies*, *C. sativa* and *P. pinaster*, for the smallest roots (0.4 mm in diameter). The power relationship between tensile strength and root diameter cannot only be explained by a scaling effect typical of that found in fracture mechanics. Therefore, this relationship could be due to changes in cellulose content as the percentage of cellulose was also observed to increase with decreasing root diameter and increasing tensile strength in both *P. pinaster* and *C. sativa*.

Introduction

The use of vegetation by civil engineers when dealing with unstable slopes has become increasingly popular over the last 20 years (Bischetti et al., 2005; Coppin and Richards, 1990; Gray and Sotir, 1996; Greenway, 1987; Norris, 2005; Roering et al., 2003; Schiechtl, 1980). In particular, trees and woody shrubs have been studied with regards to the soil reinforcing properties that their root systems convey to slopes subject to erosion or slippage problems (Schmidt et al., 2001; Wu, 2007). If the root system characteristics, which govern soil stabilisation, could be better identified, screening of suitable species for use on unstable slopes would be more efficient.

Vegetation has been recognised as a factor useful for increasing the shear resistance of soil on an unstable slope (Anderson and Richards, 1987; Coppin and Richards, 1990; Operstein and Frydman, 2000). The major factors which influence the shear resistance of root-permeated soil are the quantity and directional

* E-mail: genet@lrbb3.pierroton.inra.fr, stokes@liama.ia.ac.cn

This article has been previously published in the following journal—Plant and Soil, Volume 278, 1–9.
Stokes et al. (eds), Eco- and Ground Bio-Engineering: The Use of Vegetation to Improve Slope Stability, 3–11.

distribution of roots as well as their tensile strength, soil shear strength and soil–root interaction. Strength is the maximum force per unit area required to cause a material to break (Niklas, 1992). Tensile strength is considered one of the most important factors governing soil stabilisation and fixation, and has therefore been studied in great detail (Burroughs and Thomas, 1977; Hathaway and Penny, 1975; Nilaweera and Nutalaya, 1999; Operstein and Frydman, 2000; Phillips and Watson, 1994; Schiechtl, 1980). Not only is root tensile strength important when considering soil reinforcement, but can also affect plant anchorage. In herbaceous species, plants must withstand grazing pressure, whereby uprooting occurs in tension, therefore a higher root tensile strength will enable the plant to remain anchored in the soil (Ennos and Fitter, 1992). In trees, most anchorage is provided by the large structural roots (Stokes, 2002); however, the roots held in tension provide around 60% of the resistance to overturning during a storm (Coutts, 1983). Therefore, a greater root tensile strength will also be beneficial for tree anchorage.

Wide variations in root tensile strength have been reported in the literature, and appear to depend on species and site factors such as the local environment, season, root diameter and orientation (Gray and Sotir, 1996). Root resistance to failure in tension can be influenced by the mode of planting e.g. naturally regenerated Scots pine (*Pinus sylvestris* L.) had stronger roots than those of planted pines (Lindström and Rune, 1999). The time of year has also been found to affect tensile strength, roots being stronger in winter than in summer, due to the decrease in water content (Turmanina, 1965). Tensile strength usually decreases with increasing root size (Burroughs and Thomas, 1977; O'Loughlin and Watson, 1979; Operstein and Frydman, 2000; Turmanina, 1965; Wu, 1976) and this phenomenon has been attributed to differences in root structure, with smaller roots possessing more cellulose per dry mass than larger roots (Commandeur and Pyles, 1991; Hathaway and Penny, 1975; Turmanina, 1965).

The structure of cellulose has been found to be optimal for resisting failure in tension (Sjostrom, 1993). Cellulose is made up of polymer chains consisting of glucose units which are linked together by highly resistant hydrogen bonds (Delmer and Amor, 1995). These cellulose chains are then grouped together in a hemicellulose matrix and the entire structure is termed a microfibril. Each layer of the wood cell wall is made up of many microfibrils arranged in a helical structure.

In order to determine the relationship between tensile strength for a range of species and root size, mechanical tests were carried out on small roots from three conifer and two broadleaf species. To relate the root strength to the cellulose content, two species were then chosen for subsequent dosing of percentage cellulose in those roots tested mechanically. Results are discussed with regards to the structure of cellulose.

Materials and methods

Plant material

Roots with a diameter between 0.2 and 12.0 mm were collected from five tree species (Table 1). Trees were situated throughout different parts of France (Table 1). Roots of Maritime pine, Austrian pine and Sweet

Table 1. Location of the different species used in the tensile tests and parameters of the trees and the roots tested

Species common name and Latin name	Location where collected in France	Number of trees	Min.–Max. Height (m)	Min.–Max. DBH (m)	Total number of roots sampled	Number of roots successfully tested
Austrian pine (*Pinus nigra* Arnold)	Gironde	2	15.3–17.6	0.3–0.49	85	30
Maritime pine (*Pinus pinaster* Ait.)	Gironde	2	33.0–36.2	0.28–0.4	81	34
Norway spruce (*Picea abies* L.)	Isère	3	10.7–14.6	0.19–0.26	91	27
European Beech (*Fagus sylvatica* L.)	Isère	2	15.7–17.8	0.18–0.27	35	11
Sweet chestnut (*Castanea sativa* Mill.)	Gironde	2	NA	NA	202	53

NA—not available as trees were coppiced.

chestnut were collected from a sandy podzol soil in Gironde, located in SW France (Cucchi et al., 2004). Trees were growing at an altitude of 58 m in a flat region, where mean annual precipitation is 990 mm. Norway spruce and Sweet chestnut roots were sampled in the Forêt domaniale de Vaujany, Isère, in the French Alps. This forest, which is located at an altitude of 1350–1600 m, has a slope gradient of 38–42°. The soil is a crystalline soil and mean annual precipitation is 1353 mm (Stokes et al., 2005). Species were chosen in such a way as to cover a broad range of roots to test from both conifer and broadleaf trees. Roots were collected from two or three trees for each species (Table 1).

Live roots were manually excavated to a depth of about 0.6–0.7 m below the soil surface. Care was taken to avoid any damage to roots during the excavation process. Samples were collected randomly from the root system in order to have representative samples of different types of roots. Once the roots had been removed from the tree, they were put into separate bags and taken to the laboratory where they were stored at 4 °C. Mechanical testing was carried out as soon as possible, always within 1 week from sampling, to ensure that root material was still fresh.

Root tensile tests

Tensile testing was carried out on 494 root samples, using a Universal Testing machine (ADAMEL Lhomargy, France). The length of each sample was at least 15 times its central diameter. A load cell with a maximal capacity of 1.0 kN was used to measure the force required to cause failure in tension of each root. Crosshead speed was kept constant at 2.0 mm min^{-1} and both force and speed were measured constantly via a PC during each test. In order to avoid slippage of roots out of the clamps (Nilaweera and Nutalaya, 1999), thin slices of cork were inserted between the jaws and the root. The cork helped to improve the grip between the jaws and the root. Tests were considered successful only when specimens failed approximately in the middle of the root so that root rupture was due to the force applied in tension and not due to any existing damage (Table 1).

Tensile strength was calculated as the maximal force required to cause failure in the root, divided by the root cross-sectional area at the point of breakage. The diameter of each root was measured with an electronic slide gauge with 1/50 mm accuracy.

Cellulose content

Two contrasting species were chosen for consequent measurements of cellulose content: Maritime pine and Sweet chestnut. The method used to measure total cellulose content was based on that developed by Leavitt and Danzer (1993) and consisted of removing as many non-cellulosic compounds as possible from the root material. Initially, bark was removed from each root using a scalpel. The roots were then dried at 60 °C for 24 h and weighed using a balance with a precision >0.001 mg. Each root was then ground into a fine powder with a vibration mill (Retsch MM 300). This powder was poured into a Teflon sachet (no. 11803, pore size 1.2 μm), and each bag was carefully marked with the identification code of the corresponding root. Teflon sachets were used because they have a good compatibility with strong acids and solvents and are resistant to heat with inflammable temperatures around 200 °C (Lambrot and Porté, 2000).

The first compounds removed from the ground root tissue were lipids (waxes, oils and resins). Each sample was placed into a soxhlet extractor (50-mm *i.d.*, 200-mL capacity to siphon top) equipped with a flask containing a 700-mL mixture of toluene 99%–ethanol 96% (2–1; v/v) heated until boiling point. After 24 h of extraction using this method, the toluene ethanol was replaced with 700 mL of ethanol heated to the same temperature. After 24 h, the samples were removed from the soxhlet and immersed in distilled water heated to 100 °C for 6 h. This process removes hydrosoluble molecules from the sample.

The final step consisted of eliminating lignin compounds from the samples. Each sample was placed in a beaker containing 700 mL of distilled water, 7.0 g of sodium chlorite (NaClO2), and 1.0 mL of acetic acid ($C_2H_4O_2$). The samples and solution was shaken using a magnetic agitater and heated to 60–70 °C during 12 h. This procedure was repeated three times, with the solution concentrated by 100% each time. The samples were then removed and rinsed in distilled water, dried at ambient temperature during 12 h and weighed. The percentage of cellulose was evaluated by calculating the relative difference in the initial and final weight of each sample.

Statistical analyses

Linear and power regressions were carried out initially to evaluate the correlation between the different variables. A Kolmogorov–Smirnov test was used to test the normality of the data before proceeding with analyses of variance. Data were log-transformed, before analysis, to reflect the power relationship in linear regressions. To evaluate the influence of species, diameter of roots and cellulose content on tensile strength of roots, analysis of covariance (ANCOVA) and analysis of variance (ANOVA) were used. ANCOVA was used to detect differences in cellulose content of roots between species with regards to root diameter. In order to evaluate the influence of species on tensile strength only, roots were then classed into two groups according to diameter (<0.9 mm and >1.0 mm) and a Student's *t*-test was carried out to detect differences in tensile strength between the two groups. These data were then analysed with ANOVA and pair wise Tukey's Studentized Range (HSD) test in order to determine differences between species. Data were analysed with Minitab version 13 or XLstat-Pro version 7.5 software.

Results

Root tensile tests

Only 33% of the tensile tests were successful (Table 1). Failure often occurred near the jaws, or roots slipped out of the clamps. Mean root tensile strength was significantly different between species ($F_{4,152} = 15.16$, $p < 0.001$, ANCOVA) with regards to root diameter ($F_{1,155} = 113.01$, $p < 0.001$, ANCOVA). Mean root strength was 28.4 ± 2.0 MPa when all species and diameters were considered together (means are ± standard error). A power regression between tensile strength and diameter was significant for all species (Table 2, Figure 1). Tensile strength was also significantly different between root size classes ($t = 5.49$, $p < 0.001$). For roots <0.9 mm, mean tensile strength for each species was greater than for roots >1.0 mm but variability was high (Figure 1). However, when root size classes were analysed individually, no significant differences were found between species for roots <0.9 mm (ANOVA). Nevertheless, extremely high values of strength (132–201 MPa) were found in Norway spruce, Maritime pine and Sweet chestnut, for this size class of roots (Figure 1). For roots >1.0 mm, the tensile strength of roots was significantly different between species ($F = 10.17$, $p < 0.001$, ANOVA/HSD). Within this root size class, European beech was found to be the most resistant to failure in tension, followed by Norway spruce and Sweetchestnut. Maritime pine and Austrian pine roots were the least resistant in tension for the same diameter class.

Table 2. Parameters of the root tensile strength and diameter power law regressions for each tree species tested

Species	Regression Equation	R^2	p
Austrian pine	$y = 18.40x^{-0.52}$	0.23	0.010
Maritime pine	$y = 23.40x^{-0.87}$	0.51	<0.001
Norway spruce	$y = 37.86x^{-0.51}$	0.43	0.005
European Beech	$y = 63.51x^{-0.61}$	0.56	0.006
Sweet chestnut	$y = 31.92x^{-0.73}$	0.51	<0.001

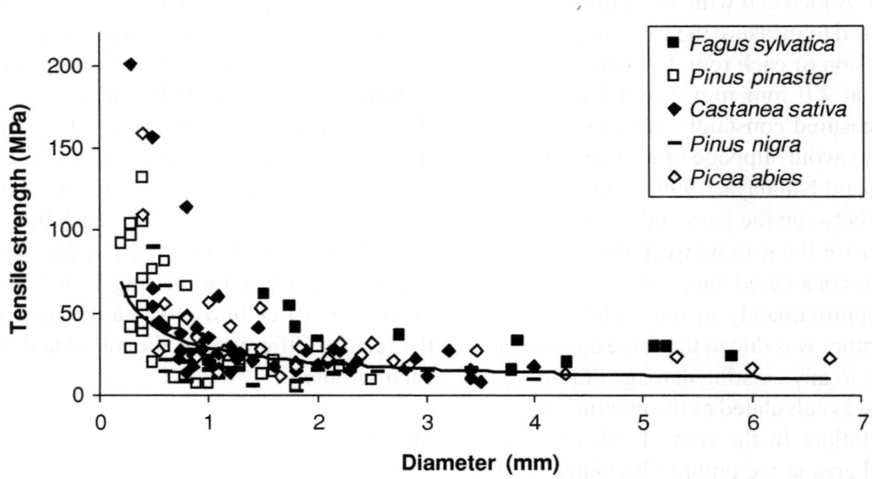

Figure 1. Tensile strength increased significantly with decreasing diameter when roots of Sweet chestnut, European beech, Maritime pine, Austrian pine and Norway spruce were considered together ($y = 28.97x^{-0.52}$, $R^2 = 0.30$, $p < 0.001$).

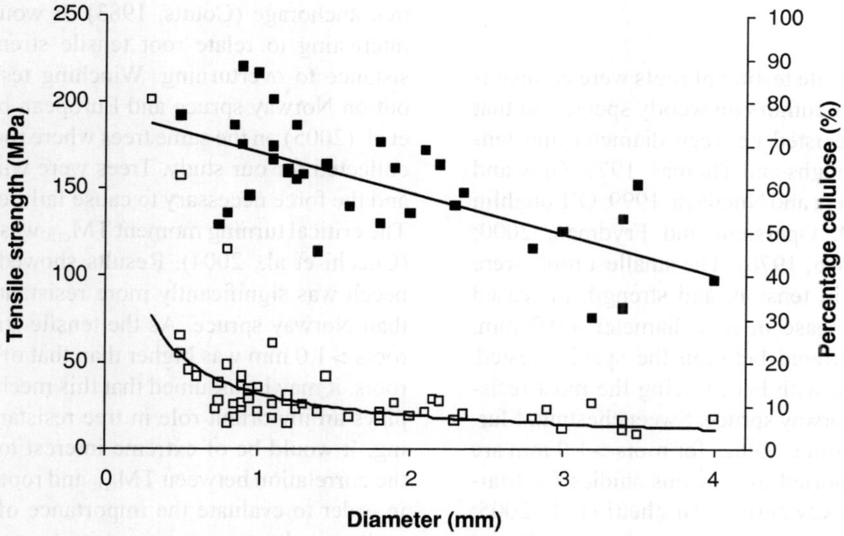

Figure 2. Tensile strength (white squares, Table 2) and cellulose content (black squares, $y = -9.44x + 77.59$, $R^2 = 0.43$, $p < 0.001$) decreased significantly with increasing root diameter in roots of Sweet chestnut.

Cellulose content

Maritime pine and Sweet chestnut roots were chosen for subsequent dosing of cellulose content, as a higher number of samples were available across the entire diameter range. The mean cellulose content was 60.0 ± 2.2% in Sweet chestnut roots and 69.9 ± 2.3% in Maritime pine roots. Cellulose content of roots was significantly different according to diameter ($F_{1,68} = 49.8$, $p < 0.001$, ANCOVA) but was not different between the two species ($F_{1,68} = 0.32$, $p = 0.58$, ANCOVA). As with tensile strength, a significant linear relationship existed between cellulose content and root diameter for both Sweet chestnut (Figure 2) and Maritime pine ($y = -13.49 + 81.87$, $R^2 = 0.34$, $p < 0.001$). Root tensile strength was also significantly related to cellulose content; however, variability was high in both Sweet chestnut (Figure 3) and Maritime pine ($y = 0.95x - 24.48$, $R^2 = 0.17$, $p = 0.026$).

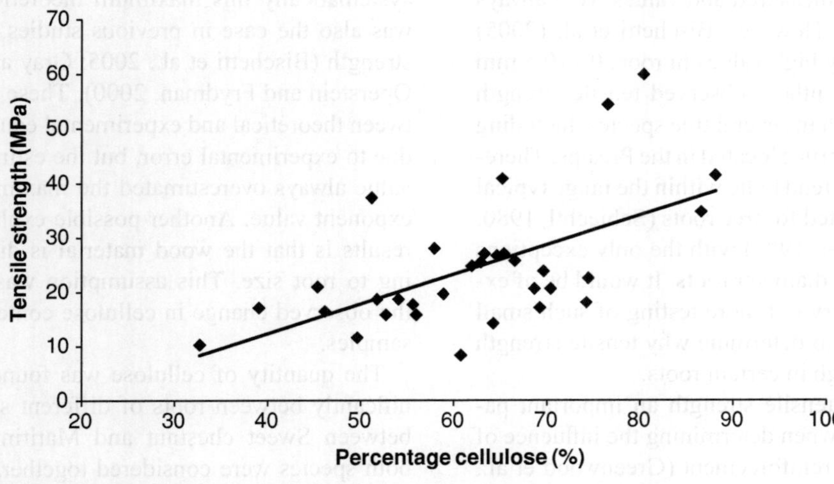

Figure 3. Tensile strength was significantly and positively related to percentage cellulose in roots of Sweet chestnut ($y = 0.56x - 9.45$, $R^2 = 0.34$, $p < 0.001$).

Discussion

Results from the tensile testing of roots were comparable to those of other authors on woody species, in that a power equation existed between diameter and tensile strength (Burroughs and Thomas, 1977; Gray and Sotir, 1996; Nilaweera and Nutalaya, 1999; O'Loughlin and Watson, 1979; Operstein and Frydman, 2000; Turmanina, 1965; Wu, 1976). The smallest roots were the most resistant in tension, and strength increased sharply with a decrease in root diameter <0.9 mm. Tensile strength differed between the species tested, for roots >1.0 mm, with beech being the most resistant, followed by Norway spruce, Sweet chestnut, Maritime and Austrian pine. Values for roots >1.0 mm are similar to those reported in previous studies for Maritime pine and Norway spruce (Bischetti et al., 2005; Turmanina, 1965). For roots <0.9 mm, no significant differences in tensile strength between species were observed, probably due to the low number of samples available. A comparison with other studies is not possible since, to our knowledge, no other studies exist concerning the tensile strength of such small roots for any of the species tested. The strength values of 132–201 MPa observed in Norway spruce, Maritime pine and Sweet chestnut were surprising, as such high values have rarely been documented in the literature. These results may be due to the fact that such small tree roots are rarely tested. To our knowledge, only Operstein and Frydman (2000) and Bischetti et al. (2005) have carried out tensile tests on small diameter roots. In the species tested by Operstein and Friedman (2000), only woody shrubs were measured and values were always lower than 80 MPa. However, Bischetti et al. (2005) also found extremely high values in roots 0.2–0.5 mm in diameter. These authors observed tensile strength values up to 750 MPa in several tree species, including beech and Norway spruce located in the Prealps. Therefore, strength values tend to lie within the range typical of that usually reported for tree roots (Schiechtl, 1980; Stokes, 2002; Ziemer, 1981) with the only exceptions being for very small diameter roots. It would be of extreme interest to carry out more testing of such small diameter roots, and to determine why tensile strength values may be so high in certain roots.

Not only is root tensile strength an important parameter to consider when determining the influence of vegetation on slope reinforcement (Greenwood et al., 2001), but is also an important factor with regards to tree anchorage (Coutts, 1983). It would therefore be interesting to relate root tensile strength to tree resistance to overturning. Winching tests were carried out on Norway spruce and European beech by Stokes et al. (2005) on the same trees where root samples were collected for our study. Trees were winched sideways and the force necessary to cause failure was measured. The critical turning moment TM_{crit} was then calculated (Cucchi et al., 2004). Results showed that European beech was significantly more resistant to overturning than Norway spruce. As the tensile strength of beech roots >1.0 mm was higher than that of Norway spruce roots, it may be assumed that this mechanical property plays an important role in tree resistance to overturning. It would be of extreme interest to study in detail the correlation between TM_{crit} and root tensile strength in order to evaluate the importance of this parameter on tree anchorage.

A power relationship, $\sigma_n \approx d^{-a}$, with $\alpha \geq 0.5$, existed between root tensile strength σ_n and diameter d. This type of relation is well known in fracture mechanics as a size effect between small and large samples (Bazant and Kazemi, 1990). The size effect is transitional between two asymptotic behaviors. There is no size effect for small dimensions of structures. For bigger dimensions a power relationship exists between the nominal strength σ_n and a characteristic dimension of the structure, e.g. the root diameter d, $\sigma_n \approx d^{-a}$ which is the size effect exhibited by Linear Elastic Fracture Mechanics (Bazant and Kazemi, 1990). Therefore the exponent term α cannot be greater than 0.5. However, our results show that this exponent exceeded systematically this maximum theoretical value. This was also the case in previous studies on root tensile strength (Bischetti et al., 2005; Gray and Sotir, 1996; Operstein and Frydman, 2000). These differences between theoretical and experimental equations could be due to experimental error, but the estimated exponent value always overestimated the maximum theoretical exponent value. Another possible explanation for our results is that the wood material is different according to root size. This assumption was confirmed by the observed change in cellulose content between the samples.

The quantity of cellulose was found to differ significantly between roots of different sizes as well as between Sweet chestnut and Maritime pine. When both species were considered together, the mean cellulose content of roots was 65%. The mean percentage cellulose in roots was therefore in the same range as other values found in the literature, e.g. Hathaway and Penny (1975) found that mean cellulose percentage in roots of six *Populus* and *Salix* species was 72%. Cellulose quantity and tensile strength of roots were significantly correlated but variability was high. In our study, cellulose content was measured using the method developed by Leavitt and Danzer (1993). In this method, hemicelluloses, which are polysaccharides linked to the cellulose present in the cell walls, were not separated from the crystalline cellulose. The quantity obtained at the end of the experiment represents therefore both cellulose and hemicelluloses. The amount of hemicelluloses of the dry weight of wood is usually around 20%. Hathaway and Penny (1975) separated hemicelluloses and crystalline cellulose. These authors found that hemicelluloses represent 17% of the dry weight of wood in roots studied. However, the hemicellulose content and composition differs between species (Sjostrom, 1993). The changes in these proportions may therefore be able to explain the high variability observed in our results. A further experiment whereby only crystalline cellulose was measured would help determine the influence of cellulose content on wood tensile strength (Akerholm et al., 2004; Andersson et al., 2003). Other chemical and anatomical parameters, which can influence tensile strength of roots, should explain the high variability observed. Lignin can also affect strength properties, especially at high moisture contents (Hathaway and Penny, 1975). The microfibril angle in root wood may also influence mechanical properties (Kerstens et al., 2001). When these microfibrils are aligned at an angle almost parallel to the cell axis, as in young wood, the combined effect of these cellulose chains is a high resistance in tension, but a low bending strength (Archer, 1986; Sjostrom, 1993). Thus, future work should concentrate on the influence of microfibril angle and lignin/cellulose ratio on tensile strength of roots.

Although cellulose content and tensile strength increases with decreasing root diameter, no measurements of annual growth rings were made in the roots studied, therefore the age of each root remains unknown. It can be imagined that cellulose content is higher in young roots, which are more resistant in tension, but this assumption should be verified through measurements of root age.

Differences in cellulose content have been proposed as the major determinant governing root tensile strength (Commandeur and Pyles, 1991; Turmanina, 1965). Nevertheless, the shape and size of a root system is influenced by its immediate environment as well being inherent to a particular species (Köstler et al., 1968). For example, trees growing on slopes may develop a specific type of root system architecture, as the mechanical function of the uphill portion of the root system is different to that downhill (Chiatante et al., 2003; Köstler et al., 1968; Shrestha et al., 2000). Root system morphology can also be modified by soil type. Nutrient supply, fertility and soil acidity all influence root growth (Fitter and Stickland, 1991; Gersani and Sachs, 1992; Gruber, 1994). Soil physical properties such as soil bulk density and strength are also important factors affecting both shoot and root growth (Campbell and Hawkins, 2003; Goodman and Ennos, 1999). In our study, samples were collected from two different habitats. As root morphology is affected by local environment and since root chemical composition also varies with root morphology, it may be possible that the local environment also influenced root cellulose content. More studies on the differences in root tensile strength of species from the same site are therefore necessary. It would also be of interest to compare the tensile strength of roots from trees growing on different types of slope or in different soil conditions, as well as testing cellulose content and tensile strength in roots around a tree, and to compare up- and down-hill roots growing on a slope (Schiechtl, 1980). Not only can cellulose content be assumed to differ between roots in a root system, but the role of this chemical compound in the overall anchorage of a root system needs to be determined, especially in young trees or woody shrubs. It has generally been assumed that root architecture is the principal component in resisting uprooting of a plant (Ennos, 2000; Dupuy et al., 2005a,b; Hamza et al., 2006; Stokes et al., 2000). However, a highly branched root system will probably not have the same percentage cellulose as a root system with fewer but thicker branches. The role each parameter plays in resisting uprooting therefore needs to be investigated.

Acknowledgements

The authors thank C. Lambrot and B. Solbes for technical assistance and F. Danjon for comments on the manuscript. This work was funded by the EU project Eco-slopes (QLK5–2001–00289).

References

Akerholm M, Hinterstoisser B and Salmen L 2004 Characterization of the crystalline structure of cellulose using static and dynamic FT-IR spectroscopy. Carbohydr. Res. 339, 569–578.

Anderson M G and Richards K S 1987 Slope Stability: Geotechnical Engineering and Geomorphology. Wiley, Chichester, 585 pp.

Andersson S, Serimaa R, Paakkari T, Saranpaa P and Pesonen E 2003 Crystallinity of wood and the size of cellulose crystallites in Norway spruce (*Picea abies*). WJ. Wood Sci. 49, 531–537.

Archer R 1986 Growth Stresses and Strains in Trees. Springer Verlag, Berlin, 240 pp.

Bazant Z P and Kazemi M T 1990 Size effect in fracture of ceramics, its use to determine fracture energy and effective process zone length. J. Am. Ceram. Soc. 73, 1841–1853.

Bischetti G B, Chiaradia E A, Simonato T, Speziali B, Vitali B, Vullo P and Zocco A 2005 Root strength and root area of forest species in Lombardy (Northern Italy). Plant Soil 278, 11–22.

Burroughs E R and Thomas B R 1977 Declining root strength in Douglas fir after felling as a factor in slope stability. USDA For. Serv. Res. Paper INT-19027pp.

Campbell K A and Hawkins C D B 2003 Paper birch and lodgepole pine root reinforcement in coarse-, medium-, and fine-, textured soils. Can. J. For. Res. 33, 1580–1586.

Chiatante D, Scippa S G, Di Iorio A and Sarnataro M 2003 The influence of steep slopes on root system development. J. Plant Growth Regul. 21, 247–260.

Commandeur P R and Pyles M R 1991 Modulus of elasticity and tensile strength of Douglas fir roots. Can. J. For. Res. 21, 48–52.

Coppin N J and Richards I G 1990 Use of Vegetation in Civil Engineering. Butterworth, London 272 pp.

Coutts M P 1983 Root architecture and tree stability. Plant Soil 71, 171–188.

Cucchi V, Meredieu C, Stokes A, Berthier S, Bert D and Najar M 2004 Root anchorage of inner and edge trees of Maritime pine (*Pinus pinaster* Ait) growing in different soil podzolic conditions. Trees-Struct. Funct. 18, 460–466.

Delmer D P and Amor Y 1995 Cellulose Biosynthesis. Plant Cell 7, 987–1000.

Dupuy L, Fourcaud T and Stokes A 2005a A numerical investigation into factors affecting the anchorage of roots in tension. Eur. J. Soil Sci. 56, 319–327.

Dupuy L, Fourcaud T, Stokes A 2005b A numerical investigation into the influence of soil type and root architecture on tree anchorage. Plant Soil 278, 119–134.

Ennos A R 2000 The mechanics of root anchorage. Adv. Bot. Res. 33, 133–157.

Ennos A R and Fitter A H 1992 Comparative functional morphology of the anchorage systems of annual dicots. Funct. Ecol. 6, 71–78.

Fitter A H and Stickland T R 1991 Architectural analysis of plant root systems 2. Influence of nutrient supply on architecture in contrasting plant species. New Phytol. 118, 383–389.

Gersani M and Sachs T 1992 Development correlations between roots in heterogeneous environments. Plant Cell. Environ. 15, 463–469.

Goodman A M and Ennos A R 1999 The effects of soil bulk density on the morphology and anchorage mechanics of the root systems of sunflower and maize. Ann. Bot. (London) 83, 293–302.

Gray D H and Sotir R D 1996 Biotechnical and Soil Bioengineering Slope Stabilization. Wiley, New York, 369 pp.

Greenway D R 1987 Vegetation and slope stability. *In* Slope Stability. Ed. M G Anderson. pp. 187–230. Wiley, New York.

Greenwood J R, Vickers A W, Morgan R P C, Coppin N J, Norris J E, 2001 Bio-engineering: The Longham Wood Cutting field trial. CIRIA Project Report 81, London. 122 pp.

Gruber F 1994 Morphology of coniferous trees: possible effects of soil acidification on morphology of Norway spruce and Silver fir. *In* DL Godbold and A Huttermann (eds), Effects of Acid Rain on Forest Processes. pp. 265–324. Wiley, New York.

Hamza O, Bengough A G, Bransby M F, Davies M C R and Hallett P D 2007 Mechanics of root-pullout from soil: a novel image and stress analysis procedure. *In* A Stokes, I Spanos, JE Norris and LH Cammeraat (eds), Eco- and Ground Bio-Engineering: The Use of Vegetation to Improve Slope Stability. Developments in Plant and Soil Sciences, vol 103, pp. 213–221. Springer, Dordrecht.

Hathaway R L and Penny D 1975 Root strength in some *Populus* and *Salix* clones. New Zeal J. Bot. 13, 333–343.

Kerstens S, Decraemer W F and Verbelen J P 2001 Cell walls at the plant surface behave mechanically like fiber reinforced composite materials. Plant Physiol. 127, 381–385.

Köstler J N, Bruckner E and Bibelriether H 1968 Die Wurzeln der Walbäume. Untersuchungen zur Morphologie der Walbäume in Mitteleuropa, 284 pp. Verlag, Hamburg and Berlin.

Lambrot C and Porté A 2000 Amélioration du protocole d'extraction de la cellulose et de l'holocellulose du bois: verification de l'absence d'un effet contaminant sur les valeurs de composition isotopique du carbone dans les cernes du bois. Cah. Techn. I.N.R.A. 45, 19–26.

Leavitt S W and Danzer S R 1993 Method for batch processing small wood samples to holocellulose for stable-carbon isotope analysis. Anal. Chem. 65, 87–89.

Lindström A and Rune G 1999 Root deformation in plantations of container-grown Scots pine trees: effects on root growth, tree stability and stem straightness. Plant Soil 217, 29–37.

Niklas K J 1992 Plant Biomechanics: An Engineering Approach to Plant Form and Function, 607 pp. The University of Chicago Press, Chicago.

Nilaweera N S and Nutalaya P 1999 Role of tree roots in slope stabilisation. Bull. Eng. Geol. Env. 57, 337–342.

Norris J E 2005 Root reinforcement by hawthorn and oak roots on a highway cut-slope in Southern England. Plant Soil 278, 43–54.

O'Loughlin C L and Watson A J 1979 Root-wood strength deterioration in radiata pine after clearfelling. N. Z. J. For. Sci. 9, 284–293.

Operstein V and Frydman S 2000 The influence of vegetation on soil strength. Ground Improvement 4, 81–89.

Phillips C J and Watson A J 1994 Structural tree root research in New Zealand: A review. Landcare Res. Sci. Ser. 7, 39–47.

Roering J J, Schmidt K M, Stock J D, Dietrich W E and Montgomery D R 2003 Shallow land sliding, root reinforcement, and the spatial distribution of trees in the Oregon Coast Range. Can. Geotech. J. 40, 237–253.

Schiechtl H M 1980 Bioengineering for Land Reclamation and Conservation, 404 pp. Edmonton Alberta, University of Alberta Press, Edmonton, Alberta.

Schmidt K M, Roering J J, Stock J D, Dietrich W E, Montgomery D R and Schaub T 2001 Root cohesion variability and shallow landslide susceptibility in the Oregon Coast Range. Can. Geotech. J. 38, 995–1024.

Shrestha M B, Horiuchi M, Yamadera Y and Miyazaki T 2000 A study on the adaptability mechanism of tree roots on steep slopes. *In* A Stokes (ed), The Supporting Roots of Trees and Woody Plants: Form, Function and Physiology. Developments in Plant and Soil Sciences, pp. 51–57. Kluwer, Dordrecht.

Sjostrom E 1993 Wood Chemistry Fundamentals and Applications, 293 pp. 2nd edition Academic Press Inc, San Diego.

Stokes A, Drexage M and Guitard D 2000 A method for predicting the possible site of failure in trees during mechanical loading. *In* A Stokes (ed), The Supporting Roots of Trees and Woody Plants: Form, Function and Physiology. Developments in Plant and Soil Sciences, pp. 279–285. Kluwer, Dordrecht.

Stokes A 2002 Biomechanics of tree root anchorage. *In* Y Waisel, A Eshel and U Kafkai (eds), Plant Roots: The Hidden Half Part, pp. 175–186. Marcel Dekker, NY.

Stokes A, Salin F, Kokutse A D, Berthier S, Jeannin H, Mochan S, Kokutse N, Dorren L, Abd.Ghani M and Fourcaud T 2005 Mechanical resistance of different tree species to rockfall in the French Alps. Plant Soil 278, 107–117.

Turmanina V 1965 On the strength of tree roots. Bull. Moscow Soc. Naturalists, Biol. Sec. 70, 36–45.

Wu T H 1976 Investigation of landslides on Prince of Wales Island, Alaska. Ohio State Univ., Dept. of Civil Eng., Geotech. Eng. Rpt. N5, 93 pp.

Wu T H 2007 Root reinforcement analyses and experiments. *In* A Stokes, I Spanos, JE Norris, LH Cammeraat (eds), Eco-and Ground Bio-Engineering: The Use of Vegetation to Improve Slope Stability. Developments in Plant and Soil Sciences. Springer, Dordrecht. Vol. 103, pp. 21–30.

Ziemer R R, 1981 Roots and the stability of forested slopes IAHS Publication 132, 343–357.

Novel biomechanical analysis of plant roots

O. Hamza[1,2], A.G. Bengough[2], M.F. Bransby[1], M.C.R. Davies[1], C. Halpin[3] & P.D. Hallett[2,4]

[1]*Division of Civil Engineering, University of Dundee, Dundee, DD1 4HN, Scotland.* [2]*Scottish Crop Research Institute, Invergowrie, Dundee, DD2 5DA, Scotland.* [3]*College of Life Sciences, University of Dundee at SCRI, Invergowrie, Dundee, DD2 5DA, Scotland.* [4]*Corresponding author**

Key words: lignin-modified, particle image velocimetry, root biomechanics, tobacco

Abstract

The mechanical behaviour of individual roots and their interaction with soil controls plant anchorage and slope stabilisation, and this is controlled by plant genotype. Tensile tests were performed on roots of tobacco (*Nicotiana tabacum* 'Samsun') plants with lignin biosynthesis pathways affected by down-regulating cinnamyl-alcohol dehydrogenase (CAD) enzyme production. Altering this pathway resulted in root stiffness <50% of the unmodified control, although failure stress was not different. Like most biological tissues, the roots had non-linear mechanical behaviour, were irregular in shape, and heterogeneous. Particle image velocimetry (PIV), applied for the first time to the tensile testing of materials, identified the localised strain fields that developed in roots under tension. PIV uses a cross correlation technique to measure localised displacements on the surface of the root between sequential digital images taken at successive strain intervals during tensile loading. Further analysis of root sections showed that non-linear mechanical behaviour is affected by cellular rupture, with a clear step-wise rupture from cortex to stele in some younger roots. This will affect slip planes that develop under pull-out at the root–soil interface. By assessing localised axial and radial strain along a root section with PIV, we have been able to determine the true stress that controls ultimate failure and the true stress–strain behaviour along the root length. The techniques used have clear potential to enhance our understanding of mechanical interactions at the root–soil interface.

Abbreviations: CAD, cinnamyl-alcohol dehydrogenase; PIV, particle image velocimetry

Introduction

Plant root systems have evolved into complex engineered structures capable of mechanically supporting a large shoot mass above ground by forming a biological anchor in soil (Niklas, 1998). The anchorage of plants is essential to understand for preventing windthrow of trees (Crook and Ennos, 1998) and reducing crop lodging in agriculture (Goodman et al., 2001). Soil stabilisation by roots has implications for the physical stability of agricultural soils (Czarnes et al., 2000), riverbank erosion, and reducing landslide risk on slopes (Sidle and Wu, 1999). The major properties of roots that control their effectiveness in either anchorage or soil stabilisation are the architecture of the root system (Stokes et al., 1996) and the biomechanics of the root tissue (Watson et al., 1999).

Root biomechanical behaviour is strongly influenced by environmental conditions and the tissue cellular structure. Niklas (1998) demonstrated that the mechanical stimulation of shoots, similar to the types of stresses induced by wind gusts or foraging, caused the biomass allocation to roots, root tensile strength and root stiffness to increase. This adaptive response allows plants to function in a wider range of environments, with more resilient species having the greatest competitive advantage (Wahl and Ryser, 2000). Considerable differences in root biomechanical behaviour have been found between species in a range of studies (Crook and

* E-mail: p.hallett@scri.sari.ac.uk

Ennos, 1994; Easson et al., 1995; Ennos, 1991; Ennos et al., 1993a,b). Crop cultivars also show differences in root biomechanical behaviour, which probably leads to differences in lodging resistance (Berry et al., 2003). Little work, however, has attempted to relate species or cultivar differences in root biomechanical behaviour to the structure of the tissue.

Modern genetic approaches can be used to change tissue structure by altering biosynthesis pathways. Hepworth and Vincent (1998) found that the genetic modification of lignin biosynthesis pathways in tobacco could reduce the tensile modulus of xylem tissue by one third. This provides an ideal model system for a more in-depth understanding of plant tissue effects on biomechanics, but it is not known if these effects would also be found in root tissue. In addition to the great value that these plants could have as a tool to improve our fundamental understanding of root biomechanics, unexpected side effects of plant modification in agricultural crops may also alter root biomechanics. Saxena and Stotzky (2001) reported higher lignin levels in Bt-maize, which is modified to express a natural insecticide. This could potentially affect lodging resistance and hence crop yields, although it is not known if the differences in lignin would affect root biomechanics, or would be expressed under field conditions.

The study of root biomechanics is complicated by the heterogeneous structure and mechanical behaviour of roots (Niklas, 1999) and soil (Bridle and Davies, 1997). As roots taper along their length, branch, bend and have defects caused by the soil environment (e.g., indentations due to stones; McCully, 1999), the stress distribution is highly spatially dependent. Many studies on root biomechanics take the diameter of the root at the point of rupture to evaluate an engineering stress (Easson et al., 1995), which provides a good estimate but does not determine the true stress distribution before ultimate failure occurs as the radial strain is unknown. Tensile tests of root biomechanics typically measure a highly nonlinear mechanical response (Ekanayake and Phillips, 1999), but the impact of radial strain verses elastic-plastic behaviour on the shape of the stress–strain relationship is impossible to discern. Many pioneering modelling studies of root biomechanics and anchorage have understandably simplified the problem by using linear elastic assumptions and simple root structures (e.g., Ennos et al., 1993a; Niklas, 1999; Stokes et al., 1996).

This paper presents data from tensile tests on individual plant roots with an aim to present new approaches that could advance our understanding of root biomechanics. In the first study we tested tobacco plants (*Nicotiana tabacum* 'Samsun') with modified lignin biosynthesis pathways to examine how genetics and cellular structure influenced root biomechanics. Altered lignin biosynthesis was hypothesised to lower the stiffness and failure stress. These tests were confounded by localised strain fields caused by the irregular shape of roots. An imaging approach was subsequently adapted from geotechnical engineering to quantify strain locally from the movement of pixels in successive digital images taken during mechanical testing. The development of this methodology and its implications for root mechanical testing will be discussed. Finally, future research opportunities using model plant systems and the novel testing procedures presented in this paper will be suggested.

Materials and methods

Root mechanical testing

A mechanical test frame was used to evaluate the mechanics of individual plant roots under tension (Model 5544, INSTRON, 100 Royall St., Canton, MA 02021-1089, USA). The cross-head displacement and force transmitted to the load cell were recorded using INSTRON Merlin software. The cross-head displacement rate was set to 0.05 mm s^{-1}, with displacement accurate to 1 μm. The load cells were accurate to 1% at 1/250 maximum load and had a range of 5, 50 N or 2 kN depending on the strength of the root tested.

Gripping the plant roots in the test frame was problematic, with slippage and surface damage potentially affecting mechanical behaviour. This was minimised by using a screw-thread clamp with rubber-faced grips to secure the older portion of the root at one end. High strength, low modulus roots such as young maize could also be clamped with the same type of grip at the younger end of the root. More mature roots were wrapped around a spindle at the younger end and affixed with rubber tape. A microscope fitted with a graticule was used to measure the diameter of the root where it failed. Other gripping approaches were evaluated including the use of super glue, medical adhesives and fast-setting araldite. These either failed to grip the root adequately or damaged the root tissue by desiccation and heat stresses.

Lignin-modified tobacco roots

Plants that differ in lignin structure and composition were produced by suppressing specific genes in tobacco (*Nicotiana tabacum* L. cv. Samsun; Halpin et al., 1994). Earlier research by Hepworth and Vincent (1998) suggested that altered lignin crosslink density changed the mechanical behaviour of shoot material. The plants used in the current study provided a model system to quantify the influence of genetics and cellular structure on biomechanical behaviour. Several lines were tested, although here we only report the findings of the unmodified wildtype, WT line and a modified line, cinnamyl-alcohol dehydrogenase (CAD), which has altered lignin structure but displays a normal phenotype including similar sized lateral roots and only slight alterations to the shape of vessels (Chabannes et al., 2001). CAD plants have been modified to down-regulate CAD, an enzyme important to lignin biosynthesis. The plants were grown in a glasshouse under natural light in 400-mm diameter × 400-mm depth pots filled with potting compost. Six plants of each line were grown. The shoots were supported with canes to reduce mechanical stresses on the root system. However, several plants were rejected because the bending of the shoot may have caused lateral loading. At 10 weeks the plants reached flowering stage and roots were harvested by washing away the potting compost. The largest adventitious roots that emerged at the highest point on the main root were selected for mechanical testing using a grip and spindle, described previously, to hold the root. There were two roots from each of four different CAD plants ($n = 8$) and two roots from each of two different WT plants tested ($n = 4$). Lateral sections of roots were imaged using a Leica SP2 confocal microscope. Intact roots of the WT line were stained with 0.1% Safranin O, which stains nuclei, chromosomes, lignified and cutinised cell walls red. Images were produced that are a 3D reconstruction using 20 slices of 0.5-μm thickness. The purpose was to investigate the potential of using confocal microscopy to investigate the strain of individual cells *in situ* caused by mechanical loading.

Determining localised strain with image analysis

Particle image velocimetry (PIV) was extended to evaluate localised strain fields in mechanically tested roots. The theory is presented in White et al. (2003) and algorithms supplied by this group were used. PIV uses a cross-correlation technique to detect the movement of pixels between sequential digital images (Adrian, 1991). It was developed first for fluid mechanics and was modified to assess deformation in soil element tests by White et al. (2003).

PIV measures the movement of patches of natural texture in an image. In soil this is provided by different colour grains and pores, but roots have little texture so it was applied artificially by dabbing graphite powder on the root surface. During mechanical testing at least ten successive digital images were taken using a Nikon D100 camera fitted with a Nikkor 60 mm f2.8 lens. The images were 6 megapixels in size and covered the length of the root and the end of the grips. During the PIV procedure, a grid of patches was placed over the root in the initial image in the sequence. In the next image, a search patch beginning at the same location as an initial patch moved around progressively to detect the new location of the patch from the peak of the auto-correlation function. The distance between patches was used to determine the localised strain fields at different stages of imposed mechanical strain.

There are several advantages of PIV over other approaches. Strain gauges placed onto roots only measure at one location, can be difficult to adhere, and may influence mechanical behaviour. Previous imaging approaches that rely on the movement of artificial targets placed on the material's surface provide a much lower resolution than PIV (White et al., 2003). We evaluated PIV first using the roots of maize (*Zea mays*) seedlings and later applied the approach to tobacco roots.

Results and discussion

Lignin-modified tobacco roots

Altered lignin biosynthesis through the genetic modification of tobacco resulted in a significantly lower modulus ($P < 0.001$), but similar maximum stress for roots of approximately the same size (Figure 1). Hepworth and Vincent (1998) found that the modulus of woody xylem tissue from the shoot of the same tobacco lines was about 10-times higher, with a 1/3 reduction in the CAD plants. The reduction between WT and CAD appears to be greater in the roots. The plants where CAD enzyme production, hence lignin monomers biosynthesis, was down-regulated appeared similar to the unmodified controls in terms of plant height, mass and root diameter ($P > 0.05$). Chabannes et al. (2001a,b) also found the plants to be phenotypically

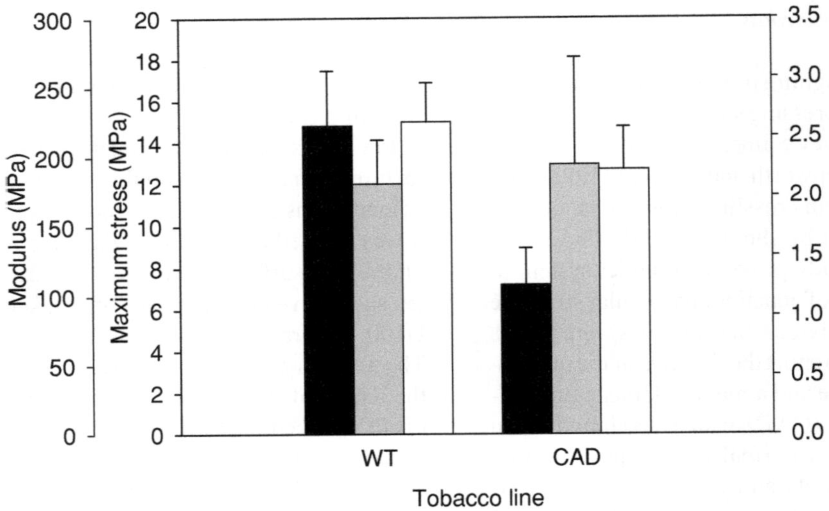

Figure 1. The Modulus (black), maximum stress (grey), and diameter (white) of tobacco that is natural, WT or modified to reduce lignin biosynthesis, CAD.

similar and reported similar lignin contents of 22% (Klason method) for both CAD and WT plants. A difference was found in lignin composition, however, with a lower ratio of syringyl (S) to guaiacyl (G) monolignal units in CAD plants. Interfascicular fibres are generally higher in S units, so this result could indicate that lignification of these structures is lower in CAD plants (Chabbanes et al., 2001b). Zhong et al. (1997) noted that elasticity reduced markedly in an interfascicular fibre mutant of *Arabidopsis*, but also reported a large reduction in strength that was not found for the tobacco plants tested here.

The genetic modification of plants to alter lignin biosynthesis pathways is being explored to improve the efficiency of paper pulping (Baucher et al., 2003). If the approach is employed to modify tree stock used for forestry, our early research on tobacco suggests that root anchorage may be influenced. Lower root stiffness may increase damage caused by wind, pull-out from soil, and ultimately the occurrence of blow-down. However, roots tend to become stiffer when repaired from mechanical damage (Niklas, 1998), so these problems may be self-correcting under mechanical damage from wind in the field. Hepworth and Vincent (1999) found that flexural stimulation increased the shoot stiffness of the same tobacco lines that we studied. In a field study, Pilate et al. (2002) found CAD modified poplar trees grew just as well as natural lines, although the biomechanics of the roots were not evaluated.

The testing approach may also need to be improved. Grip slippage and damage was problematic, particularly after the yield stress was exceeded. This would affect the maximum stress more than the modulus, as the modulus was evaluated from the linear part of the stress–strain relationship below the yield stress. Ultimate failure generally occurred near to the bottom grip, where the root was younger and smallest in diameter.

Determining localised strain with image analysis

PIV, applied for the first time to the tensile testing of biological materials, successfully measured the movement of patches placed over the surface of roots (Figure 2). The surface texture produced by graphite powder allowed for the new location of patches in successive images to be detected from the peak of the autocorrelation function. Patch movement could be used to evaluate displacement trajectories along the length of the root. These data could then be converted to radial and axial strain components.

By assessing localised axial and radial strain along a root section with PIV, the true stress that controls ultimate failure and the true stress–strain behaviour along the root length could be determined (Ashby and Jones, 1996). Figure 3 shows an image of a root at the point of cortex failure. This produces a localised zone of intensified stress and strain, where more damage occurs and

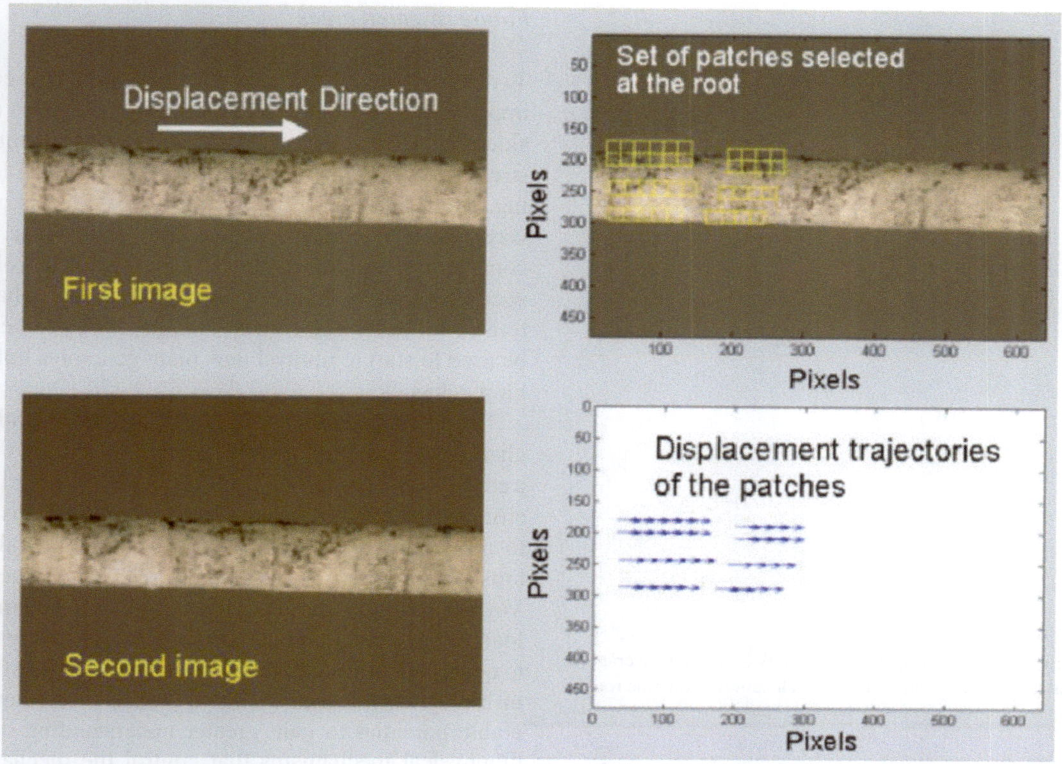

Figure 2. Calibration of PIV approach by displacing a young maize root coated in graphite powder. A set of patches were placed on the root in the first image; their movement in successive images was detected and used to evaluate displacement trajectories.

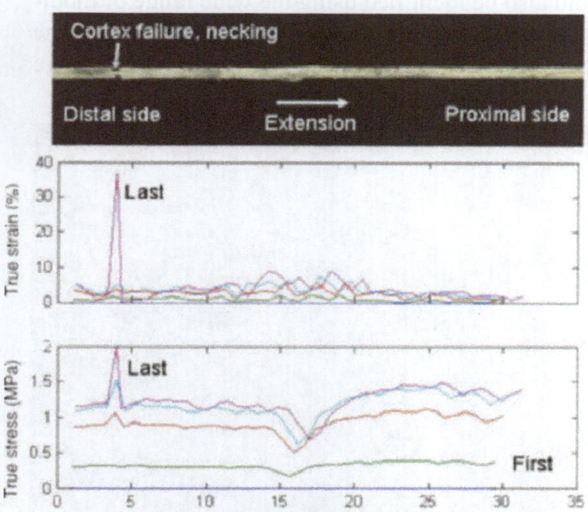

Figure 3. The 'Last' image of root undergoing tensile testing (adapted from Hamza et al., 2006). The true strain and true stress of the root were evaluated using PIV from the movement of pixels between successive images. Each line represents a step in the tensile test, starting from the 'First' image when low strain is applied and ending in the 'Last' image after cortex failure.

ultimately leads to the failure of the stele. After cortex failure, the true strain in this localised area was almost 40%, whereas the engineering strain (i.e. strain applied to the entire root) was less than 5%. The reduction in root cross-sectional area caused by cortex failure almost doubled the true stress at this point of localised failure. About half-way along the length of the root it is slightly wider, so this location has a lower true stress.

PIV and tobacco roots

PIV analysis of the tobacco root tests found a similar modulus to the values evaluated from the engineering stress–strain relationship from the loading frame. For the WT tobacco, the modulus was 358 ± 98 MPa (average ± s.e.) from the images, whereas it was 222 ± 40 MPa from the engineering stress–strain relationship ($P = 0.21$). In the CAD tobacco, the modulus was 105 ± 26 MPa from the images, compared to 108 ± 24 MPa from the engineering stress–strain relationship ($P = 0.94$). The coefficient of variation was relatively

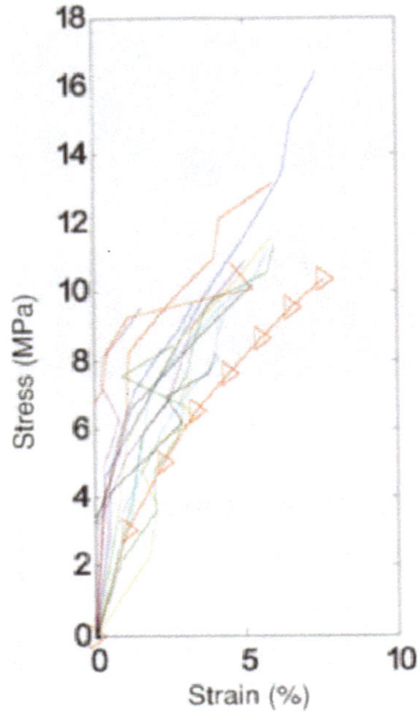

Figure 4. True stress–strain relationship for WT tobacco. Average (Δ) and local (other lines) values at various locations along the root are shown.

large (20 to 25%), which although is inherent in many biological materials, may also be due partly to experimental difficulties during testing such as grip slippage. The PIV analysis presented here looked at the average deformation along the entire root. More accurate and meaningful results could be obtained by looking at patches that are closer together, particularly near to the zones where failure ultimately occurs.

As radial strain during tensile testing reduces the area over which an applied force acts, the true tensile stress will be higher than calculated from pre-loading specimen geometries. PIV was used to convert from engineering stress–strain to true stress–strain using radial strain over the average of the root and at specific locations (Figure 4). The average true stress for a given strain was higher than the engineering stress (data not shown), as would be expected due to radial contraction of the root diameter. In the particular example, the localised stress fields determined were heterogeneously distributed along the root length. They were generally higher than the average value calculated for the entire root length because the impact of grip slippage was removed. In addition to PIV identifying localised true stress and strain fields by measuring radial and axial strain it can also be used to determine the Poisson ratio.

Future research areas

Two novel approaches have been presented that could improve our understanding of root biomechanics considerably. PIV allows for localised measurements of stress and strain that develop during mechanical loading. Model plants with controlled biosynthesis pathways provide geometrically similar plants that vary considerably in biomechanical behaviour. The role of specific genes in lignin biosynthesis is well characterised for these plant lines (Halpin, 2004) and could be used to start to unravel how plant genomics links to biomechanics.

The potential applications of PIV extend beyond mechanical tests on individual plant roots. Figure 5 shows a confocal microscope image that identifies the cellular structure of a WT tobacco root. It should be possible to mechanically test intact roots in the confocal microscope to image deformation of the tissue structure. Voytik-Harbin et al. (2003) proposed a mechanical-loading imaging technique using confocal microscopy to quantify load-induced changes to the scaffold of cells in biological materials. This approach offers considerable potential to gain greater understanding of the biophysical mechanisms that control the mechanical behaviour of roots.

The genes that shape root biomechanical behaviour could also be identified using the wide range of modified plants that are currently available. In addition to the tobacco plants used in this study, there is a wider range

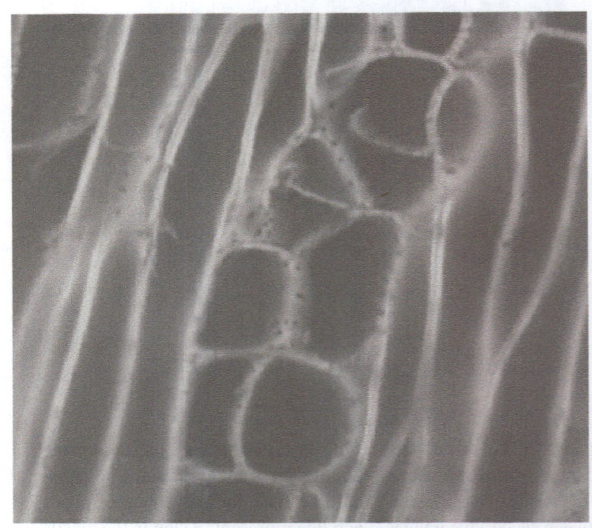

Figure 5. Confocal microscope images of WT tobacco root, 240-μm across.

of modified tobacco plants (Halpin, 2004) and poplar trees (Halpin and Boerjan, 2003) available that could help elucidate root biomechanics at the molecular level. Further advances could be achieved with *Arabidopsis* (Zhong et al., 1997), although the practicality of testing such small roots may pose a problem. Studies with *Arabidopsis* mutants have already studied the importance of root hairs to anchorage (Bailey et al., 2002).

Conclusions

As with many biological materials, roots are heterogeneous, irregularly shaped, and have a highly nonlinear mechanical behaviour. This is further confounded by genetic differences, which can have profound effects on mechanical behaviour. In model tobacco plants, suppressing a lignin biosynthesis pathway reduced stiffness by almost 2/3. The irregular shape and heterogeneous structure of roots complicates analysis using conventional mechanical testing approaches. PIV was shown to be a useful approach to help overcome this problem. It showed that localised areas of intensified stress and strain occur in the root, which will likely receive greater damage and ultimately be the locations where failure occurs. PIV also detected that cortex failure in younger roots causes a large concentration of stress during mechanical testing.

Understanding plant root biomechanics is integral to describing root anchorage and soil stabilisation by roots. New approaches from geotechnical engineering, specifically PIV and the behaviour of inclusions in soil, should help us explore this multifarious problem. Future root biomechanics research will explore how the stress–strain relationship is affected by damage at the cellular level. We are also investigating the biomechanics of Bt-Maize, a commercially grown GM crop that is reported to have lower lignin levels. This biomechanics research should ultimately increase our understanding of how plant roots stabilise slopes and anchor plants in the ground.

Acknowledgements

We thank Dr Slobodan Miskovski and Rene Sonnenberg for their help with the tobacco root tests. The confocal microscope images were prepared by Dr Trudi Gillespie. Dr David White, Cambridge University, kindly provided the PIV software and discussed its application. This work was funded by the MRC/EPSRC/BBRSC Discipline Hopper Programme. The Scottish Crop Research Institute receives grant-in-aid support from the Scottish Executive Environment and Rural Affairs Department.

References

Adrian R J 1991 Particle-imaging techniques for experimental fluid mechanics. Annu. Rev. Fluid Mech. 23, 261–304.

Ashby M F and Jones D R H 1996 Engineering Materials 1—An Introduction to their Properties and Applications. 2nd edition. Butterworth Heineman, Oxford.

Bailey P H J, Currey J D and Fitter A H 2002 The role of root system architecture and root hairs in promoting anchorage against uprooting forces in *Allium cepa* and root mutants of *Arabidopsis thaliana*. J. Exp. Bot. 53, 333–340.

Baucher M, Halpin C, Petit-Conil M and Boerjan W 2003 Lignin: genetic engineering and impact on pulping. Crit. Rev. Biochem. Mol. Biol. 38, 305–350.

Berry P M, Spink J H, Gay A P and Craigon J 2003 A comparison of root and stem lodging risks among winter wheat cultivars. J. Agr. Sci. 141, 191–202.

Bridle R J and Davies M C R 1997 Analysis of soil nailing using tension and shear: experimental observations and assessment. P. I. Civil Eng.-Geotec. 125, 155–167.

Chabannes M, Barakate A, Lapierre C, Marita J M, Ralph J, Pean M, Danoun S, Halpin C, Grima-Pettenati J and Boudet A M 2001a Strong decrease in lignin content without significant alteration of plant development is induced by simultaneous down-regulation of cinnamoyl CoA reductase (CCR) and cinnamyl alcohol dehydrogenase (CAD) in tobacco plants. Plant J. 28, 257–270.

Chabannes M, Ruel K, Yoshinaga A, Chabbert B, Jauneau A, Joseleau J P and Boudet A M 2001b In situ analysis of lignins in transgenic tobacco reveals a differential impact of individual transformations on the spatial patterns of lignin deposition at the cellular and subcellular levels. Plant J. 28, 271–282.

Crook M J and Ennos A R 1994 Stem and root characteristics associated with lodging resistance in 4 winter-wheat cultivars. J. Agr. Sci. 123, 167–174.

Crook M J and Ennos A R 1998 The increase in anchorage with tree size of the tropical tap rooted tree *Mallotus wrayi*, King (Euphorbiaceae). Ann. Bot. (London) 82, 291–296.

Czarnes S, Hallett P D, Bengough A G and Young I M 2000 Root- and microbial-derived mucilages affect soil structure and water transport. Eur. J. Soil Sci. 51, 435–443.

Easson D L, Pickles S J and White E M 1995 A study of the tensile force required to pull wheat roots from soil. Ann. Appl. Biol. 127, 363–373.

Ekanayake J C and Phillips C J 1999 A method for stability analysis of vegetated hillslopes: an energy approach. Can. Geotech. J. 36, 1172–1184.

Ennos A R 1991 The mechanics of anchorage in wheat *Triticum-aestivum* L .1. The anchorage of wheat seedlings. J. Exp. Bot. 42, 1601–1606.

Ennos A R, Crook M J and Grimshaw C 1993a A comparitive study of the anchorage systems of Himalayan Balsam *Impatiens glandulifera* and mature sunflower *Helianthus annus*. J. Exp. Bot. 44, 133–146.

Ennos A R, Crook M J and Grimshaw C 1993b The anchorage mechanics of maize, *Zea-mays*. J. Exp. Bot. 44, 147–153.
Goodman A M, Crook M J and Ennos A R 2001 Anchorage mechanics of the tap root system of winter-sown oilseed rape (*Brassica napus* L.). Ann. Bot. (London) 87, 397–404.
Halpin C 2004 Investigating and manipulating lignin biosynthesis in the postgenomic era. Adv. Bot. Res. 41, 63–106.
Halpin C and Boerjan W 2003 Stacking transgenes in forest trees. Trends Plant Sci. 8, 363–365.
Halpin C, Knight M E, Foxon G A, Campbell M M, Boudet A M, Boon J J, Chabbert B, Tollier M T and Schuch W 1994 Manipulation of lignin quality by down-regulation of cinnamyl alcohol dehydrogenase. Plant J. 6, 339–350.
Hamza O, Bengough A G, Bransby M F, Davies M C R and Hallett P D 1996 Biomechanics of plant roots: estimating localised deformation with particle image velocimetry. Biosystems Eng. 94, 119–132.
Hepworth D G and Vincent J F V 1998 The mechanical properties of xylem tissue from tobacco plants (*Nicotiana tabacum* 'Samsun'). Ann. Bot. (London) 81, 751–759.
Hepworth D G, Vincent J F V 1999 The growth response of the stems of genetically modified tobacco plants (*Nicotiana tabacum* 'Samsun') to flexural stimulation. Ann. Bot. (London) 83(1), 39–43.
McCully M E 1999 Roots in soil: Unearthing the complexities of roots and their rhizospheres. Annu. Rev. Plant Physiol. Plant Mol. Biol. 50, 695–718.
Niklas K J 1998 Effects of vibration on mechanical properties and biomass allocation pattern of *Capsella bursa*-pastoris (Cruciferae). Ann. Bot. (London) 82, 147–156.
Niklas K J 1999 A mechanical perspective on foliage leaf form and function. New Phytol. 143, 19–31.
Pilate G, Guiney E, Holt K, Petit-Conil M, Lapierre C, Leple J C, Pollet B, Mila I, Webster E A, Marstorp H G, Hopkins D W, Jouanin L, Boerjan W, Schuch W, Cornu D and Halpin C 2002 Field and pulping performances of transgenic trees with altered lignification. Nat. Biotechnol. 20(6), 607–612
Saxena D and Stotzky G 2001 Bt corn has a higher lignin content than non-Bt corn. Am. J. Bot. 88, 1704–1706.
Stokes A, Ball J, Fitter A H, Brain P and Coutts M P 1996 An experimental investigation of the resistance of model root systems to uprooting. Ann. Bot. (London) 78, 415–421.
Sidle R C and Wu W M 1999 Simulating effects of timber harvesting on the temporal and spatial distribution of shallow landslides. Z. Geomorph. 43, 185–201.
Voytik-Harbin S L, Roeder B A, Sturgis J E, Kokini K and Robinson J P 2003 Simultaneous mechanical loading and confocal reflection microscopy for three-dimensional microbiomechanical analysis of biomaterials and tissue constructs. Microsc. Microanal. 9, 74–85.
Wahl S and Ryser P 2000 Root tissue structure is linked to ecological strategies of grasses. New Phytol. 148, 459–471.
Watson A, Phillips C and Marden M 1999 Root strength, growth, and rates of decay: root reinforcement changes of two tree species and their contribution to slope stability. Plant Soil 217, 39–47.
White D J, Take W A and Bolton M D 2003 Soil deformation measurement using particle image velocimetry (PIV) and photogrammetry. Géotechnique 53, 619–631.
Zhong R Q, Taylor J J and Ye Z H 1997 Disruption of interfascicular fiber differentiation in an *Arabidopsis* mutant. Plant Cell 9, 2159–2170.

Root reinforcement: analyses and experiments

Tien H. Wu*
Ohio State University, 2070 Neil Avenue, Columbus, OH, 43210, USA.

Abstract

Simple and complex analytical models of root reinforcement and the associated requirements and limitations are reviewed. Simple models include the limiting equilibrium solution and the cable and pile solutions. The complex model is the finite element method (FEM). The simple models were used to analyze published data from laboratory and *in situ* shear tests and pullout tests on soils reinforced with synthetic materials and root systems. The models can be used for approximations when the model requirements are met. The FEM was used to simulate experiments and provided more detailed information. These results provide insight on the failure mechanisms. This forms the basis for suggestions on models to be used in stability analysis of slopes.

Introduction

The literature on the influence of vegetation on slope stability is rich, covering a broad spectrum of theoretical, experimental, and empirical studies. Comprehensive reviews are given in Coppin and Richards (1990), Gray and Sotir (1996), Morgan and Rickson (1995), and Schiechtl and Stern (1996). More specifically, the role of vegetation roots as a soil reinforcement has been the subject of many studies. Because of the wide variety of site conditions and vegetation types, the results from different studies do not always appear to be consistent. Therefore, it may be difficult to choose parameters for use in stability analysis. The overall objective of this paper is to address the question of how to evaluate the role of root reinforcement on slope stability. The first part reviews available analytical models of root reinforcement. Models may be simple or complex. Simple models are easy to use and require less data. Complex models, namely the finite element method (FEM), can provide more detailed information but require more data. In the second part, simple models are used to analyze data from a variety of laboratory and *in situ* tests. This provides a means for evaluating the failure mechanisms. The FEM is used to supplement the results from simple models. This provides insight on the failure mechanisms. Finally, suggestions are given on the use of simple models in stability analysis.

Models of root reinforcement

Bending stiffness of roots

The experiments of Shewbridge and Sitar (1989) on sand reinforced with fibers and rods clearly demonstrate the important influence of the bending stiffness on the thickness of the shear zone and the deformation of the reinforcement. A thin shear zone leads to larger extension in the reinforcement than a thick shear zone. On the other hand, flexible reinforcements require more extension to reach the failure strain than stiff reinforcements. Selection of an appropriate model requires consideration of the deformation of the reinforcement.

Simple models

For a reinforcement that is embedded at an angle of 90° to the failure surface (Figure 1a), limit equilibrium requires that

$$s_r = [T(\cos\alpha + \sin\alpha \tan\varphi)]/A = [T_y + T_z \tan\varphi]/A \quad (1a)$$

* E-mail: wu.26@osu.edu

Stokes et al. (eds), Eco- and Ground Bio-Engineering: The Use of Vegetation to Improve Slope Stability, 21–30.

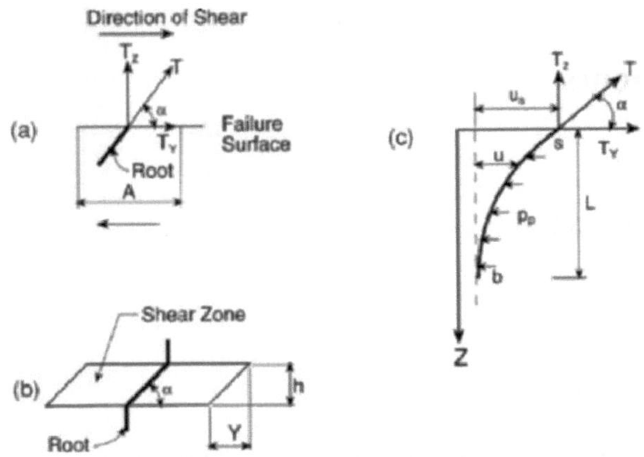

Figure 1. Simple Models. (a) Limit equilibrium, (b) flexible reinforcement, (c) cable model.

where s_r = shear strength contributed by reinforcement, T = tensile force in reinforcement, α = inclination of T, A = area of the section under consideration, φ = angle of internal friction of the soil. When written in terms of the stress, σ_r, Eq. (1a) becomes

$$s_r = [\sigma_r A_r(\cos\alpha + \sin\alpha\tan\varphi)]/A \qquad (1b)$$

where A_r = area of the reinforcement. A modified version of Eq. (1) can be used when the reinforcement is not perpendicular to the failure surface (Gray and Ohashi, 1983). For $48° < \alpha < 72°$, the quantity $(\cos\alpha + \sin\alpha\tan\varphi)$ is approximately 1.2 and Eq. (1b) may be simplified to (Wu et al., 1979)

$$s_r \approx 1.2\,\sigma_r A_r/A. \qquad (2)$$

For fibers and bars the limiting value of σ_r or T is the ultimate tension (T_u), which is the tensile strength or the friction between the soil and reinforcement. In the case of roots, the failure modes are tension failure in the main root, progressive tension failure in branch roots, and slip between root and soil. When pullout tests are used to measure the T_u, all three modes are possible. The deformation required to produce failure depends on the root properties and ranges between 0.05 and 0.15 m. This topic is considered further in a subsequent section.

The simplest application of Eq. (2) is to assume that the reinforcement deforms with the soil, or it has no influence on the shear deformation. Then α is determined by the shear strain in the soil (Figure 1b). Tension in the reinforcement is developed by extension in the shear zone and can be estimated from the tensile strain (Waldron, 1977). If the extension is insufficient to develop the ultimate tension, then σ_r would be less than the ultimate tension but Eqs. (1) and (2) remain valid provided the right value of T or σ_r is used.

The thickness of the shear zone in laboratory shear tests ranges between 5 and 50 mm for flexible fibers and depends strongly on the boundary conditions of the test (Gray, 1991; Shewbridge and Sitar, 1991). With increasing bending stiffness the thickness of the shear zone increases and the reinforcement no longer deforms with the soil (Abe and Ziemer, 1991; Jewell and Wroth, 1987). Experiments with reinforced soil walls show similar results (Jaber et al., 1987; Plumelle and Schlosser, 1991). To consider the deformation and bending resistance of the reinforcement, one can use the equation for a tie (Figure 1c), which is (Oden, 1967)

$$EI\frac{d^4u}{dz^4} - T_z\frac{d^2u}{dz^2} = q \qquad (3)$$

where E, I = Youngs modulus and moment of inertia of the reinforcement, q = soil reaction, u = displacement, L = length of tie = deformed portion of reinforcement. For the limiting case, $q = q_y$ = soil reaction or bearing pressure at yielding. For $\eta L < 1.5$, with $\eta = [T_z/EI]^{1/2}$, the tie may be represented as a beam or a laterally loaded pile. For elastic soil support, the beam solution is well known (e.g., Hetenyi, 1946). For the limiting case when the soil support is q_y, the solution for a laterally loaded pile has been summarized by Broms (1964a,b) and Jewell and Pedley (1992). The solution can be expressed in dimensionless numbers

$$N_c = T_y/cd^2,\ N_\phi = T_y/\gamma K_p d \qquad (4)$$

where γ = unit weight, d = pile diameter, K_p = coefficient of passive earth pressure. N_c and N_φ are functions of M_y = moment at yielding of the pile or beam, d, c, φ, and γ and are given in Broms (1964a,b).

Eq. (3) can be simplified to a flexible cable if $\eta L > 2.5$. The cable solution is given by (Oden, 1967)

$$T_z(0) = T(L) \qquad (5a)$$

$$T_y(0) = q_y L \qquad (5b)$$

$$u(0) = q_y L^2/2T_z(0) \qquad (5c)$$

The tension T in this case is also limited by the ultimate tension. For a root perpendicular to the slip surface, the beam or pile solution may be used for small u, which means $\alpha \to 90°$, or $T_z \to 0$. This represents the condition the initial failure at yielding in the root. If the root is ductile and does not fracture, u continues to increase, T increases and the limit is the cable solution.

Finite element method (FEM)

The FEM makes it possible to solve for the stresses and displacements in the 3-dimensional problem of a reinforcement buried in soil. It has been applied to a single root (El-Khouly, 1995; Frydman and Operstein, 2001) and to the root systems of trees (Dupuy et al., 2005). The ABAQUS software package was used for our FEM studies (El-Khouly, 1995). The root is represented by beam elements and the interface between soil and reinforcement by slide-line contact elements. The reinforcement is linear-elastic up to the yield point and the interface shear has an angle of friction δ. The soil is elastic-plastic and can be represented by either the Drucker-Prager model (Drucker et al., 1957) or the Cap model (DiMaggio and Sandler, 1971).

Evaluation of tests on reinforced soil

The simple models were used to analyze data from published laboratory and *in situ* tests to improve the understanding of failure mechanisms. Finite element analysis was used to provide more detailed information. We distinguish between tests on soils reinforced with fibers or bars from those on soils containing an entire root system of a plant. The difference is that, with the former group, the initial geometry of the reinforcement is simple and known, while with the latter, it is complex and usually not well known.

Direct shear tests

Several experimental studies that represent different reinforcement properties and measurements are reviewed to illustrate the influence of the properties and test conditions on s_r. The results are summarized in Table 1.

Tests on fibers, rods, and root members

Consider first the laboratory direct shear tests by Gray and Ohashi (1983). Eq. (2) was used to calculate s_r, with T_u calculated from friction between soil and reinforcement. The normal stress σ on the reinforcement was taken to be 1.5 times the vertical stress, σ_z, based on calculations with the FEM model. The calculated s_r's are about 1.5 times the measured values. This suggests that friction was not fully developed, although pullout was noted by the authors in some cases.

In the direct shear tests by Jewell and Wroth (1987), the rods were provided with a rough surface and the ultimate tension was equal to the friction. The displacements within the soil and the force in the reinforcement

Table 1. Direct shear tests

Author	Test	Spec.	D (cm)	Model	s_r, kPa Calc.	s_r, kPa Meas.
Gray,	direct shear	6 reed	0.18	Eq. (2)	1.5 kN	0.6 kN
Ohashi		6 copper	0.18	Eq. (2)	0.72	0.3
Jewell, Wroth	direct shear	polymer coil S9Y		Eq. (2)	$0.46\ \sigma_y$	$0.6\sigma_y$
Shewbridge	shear	14 wood	0.32	Eq. (2)	3.7 kPa	1.8 kPa
		14 para. chord	0.32	Eq. (2)	4.3	1
		14 wood	0.32	cable	2.7	1.8
		14 para. chord	0.32	cable	2.5	1
		14 wood		FEM	1.6	1.2
Nilaweera	*In situ* shear	*Hopea ordata*	0.5–0.8	Cable eq.	1.2–2.9 kPa	3 kPa
				Pile eq.	0.08	
Wu, Watson	*In situ* shear	*Pinus radiata*		cable and pile	1.7–20.8 kN	18.2 ult.
					21.5	23.2–24.2 peak
Frydman	direct shear	alfalfa		Eq. (2)	$\mu = 1.0*$	$\mu = 0.26$

$\mu = s_r/A_r$.

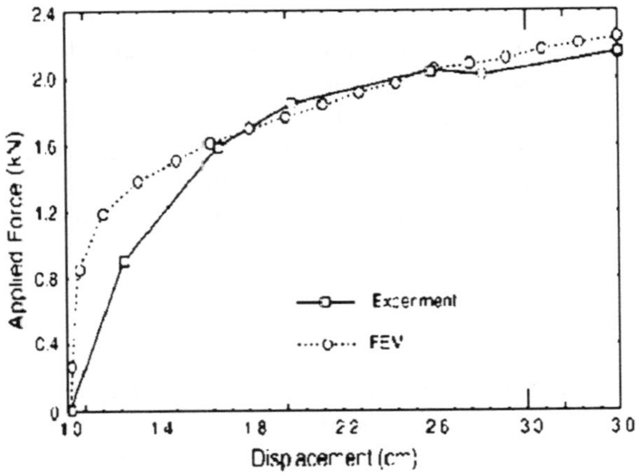

Figure 2. FEM simulation of Shewbridge and Sitar's test, load-displacement curve.

were measured. In this case the calculated value is close to the measured one and agrees with the authors' observation that friction was fully developed.

In the tests by Shewbridge and Sitar (1989), the deformations of the soil and the reinforcements were observed. The latter was used by the authors to calculate the tensile force in the reinforcement and s_r. In this study, the friction was used to calculate T_u and Eq. (2) to calculate s_r. Eq. (2) overestimates s_r. This indicates that friction was not fully developed, although Shewbridge and Sitar (1989) observed slip in the case of wooden dowels. Since the deformed shapes of the reinforcements were observed, it was possible to estimate α, L, and $u(0)$ for the cable solutions. The cable solution requires the bearing pressure q_y exerted by the soil on the reinforcement. The upper and lower limits were given by Jewell and Pedley (1992). The range between the limits is large. Palmeira and Milligan (1989) gave a modified upper limit based on experimental results. Using their relation $q_y/\sigma_z = 35$, where σ_z = overburden pressure. Broms (1964b) recommended a ratio q_y/σ_z that is 3 times Rankine's passive pressure. This gives $q_y/\sigma_z \approx 10$ and is the lower limit used here. For wooden dowels, the estimated values of α and L are 30° and 5 cm, respectively. Similar estimates were made for parachute chords. The calculated ranges of s_r are given in Table 1 and are higher than the measured values. The results of FEM simulation are shown in Figure 2. The load displacement curve for 14 wooden dowels shows that the calculated s_r is slightly larger than but close to the measured value. Figure 3 shows the calculated octahedral shear stress q. Good agreement between simulated

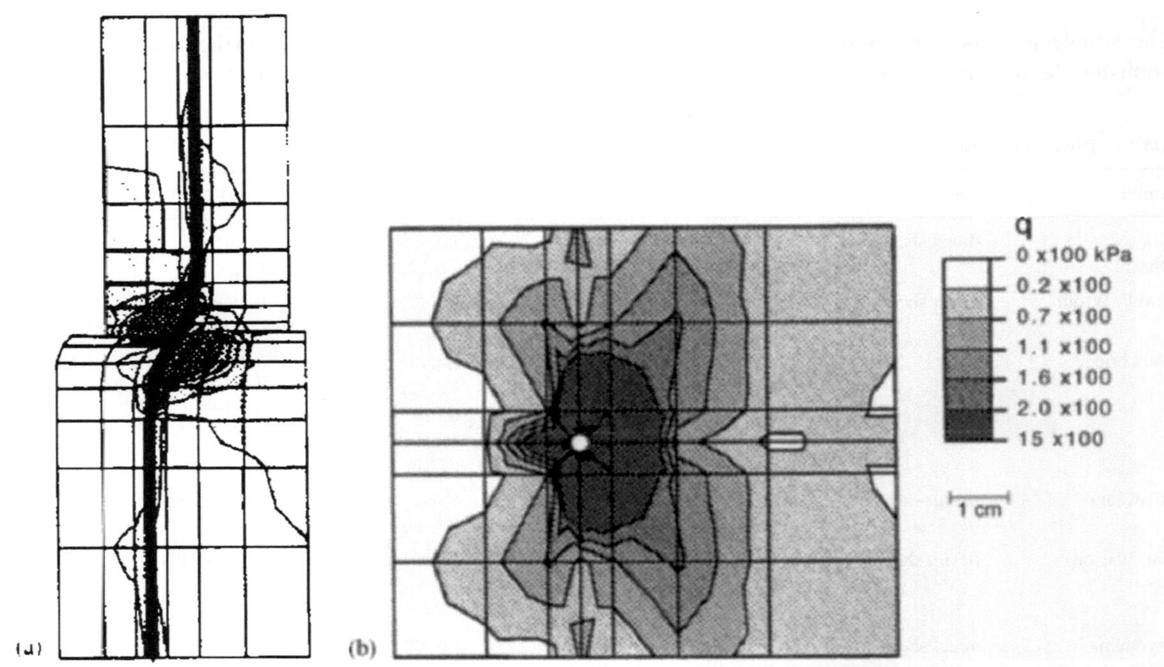

Figure 3. FEM simulation of Shewbridge and Sitar's test, stress distribution (a) on failure plane, (b) on longitudinal section.

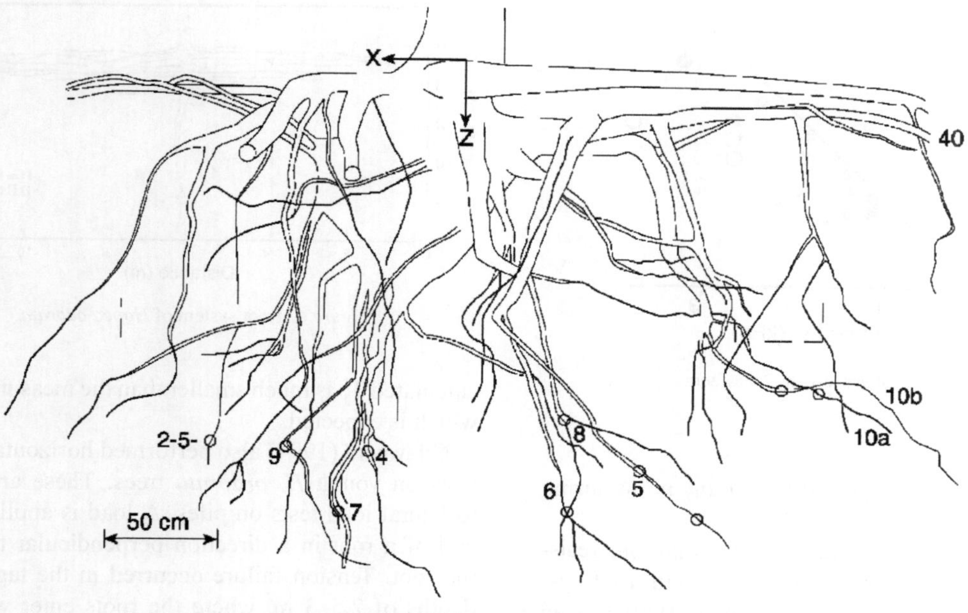

Figure 4. Root system of *Pinus radiata.*

and measured shear strengths were also obtained by Frydman and Operstein (2001). Tensile failure was not developed in any of the above experiments.

The above cases show that simple models are not accurate. The difficulty lies primarily with the evaluation of T or σ_r. One uncertainty in the above calculations is the normal stress on the rods, which controls the friction. In the cable solution, it is difficult to estimate the dimensions L and α, without detailed measurements. Furthermore, the range in the passive resistance q_y calculated by different theories is large as indicated earlier. Nevertheless, if the limitations are recognized, the simple solutions are useful for order-of-magnitude estimates and allowances can be made for inaccuracies in the estimates. On the other hand, FEM gives good results and provides more detailed information. However, it is difficult to use where the reinforcement geometry is complex.

Tests on root systems

Root system denotes a main root with branch roots and is more complex than fibers and rods. Furthermore, for *in situ* tests, the initial position of the main root and branch roots are not known and the deformed shape at failure cannot be predicted.

Wu and Watson (1998) performed *in situ* tests on soil blocks, each containing an entire root system of *Pinus radiata* D. Don. The forces in selected roots were measured. After the test, the root system was excavated so that the final positions of the roots are known. An example is shown in Figure 4. The circles on the roots denote the "bottom point," which is the point of zero displacement according to observations during excavation. No pullout failure was noted. The forces in the roots were calculated by the cable solution if the roots were in tension, with the bottom point at b in Figure 1c. Where the root is not perpendicular to the slip surface, u in Eq. (4) is the displacement perpendicular to the tangent at point b. The calculated values of T_y and T_z were used in Eq. (1a) to calculate s_r. If the roots were in tension, the cable solution was used to calculate the horizontal resistance, T_x. The details are given in the paper. Only in roots 5 and 4 are the calculated forces close to one-half of the tensile strength. The calculated and measured s_r's for the root system are given in Table 1 and the agreement is considered satisfactory. If we assume that with additional displacement, roots 10, 5, 6, 8 reach tensile strength simultaneously, s_r would increase by 0.25, which is a relatively minor increase. By comparison, the value of s_r calculated with Eq. (1b) and the tensile strength of all roots is about 3 times the measured value. In view of the root geometry, it is clear that not all the roots would fail simultaneously. A reasonable interpretation is that approximately 1/3

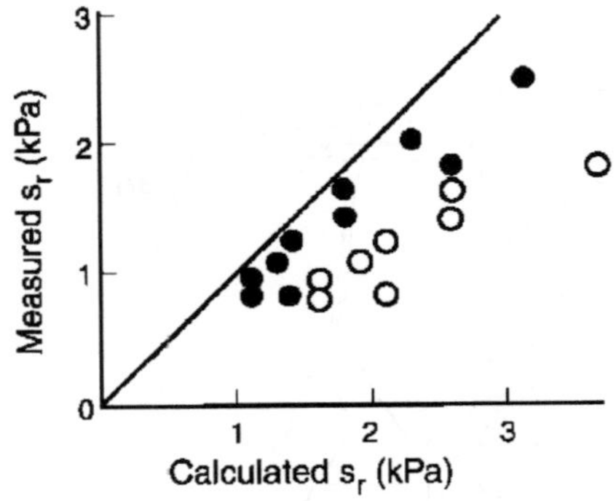

Figure 5. Shear tests by Nilaweera.

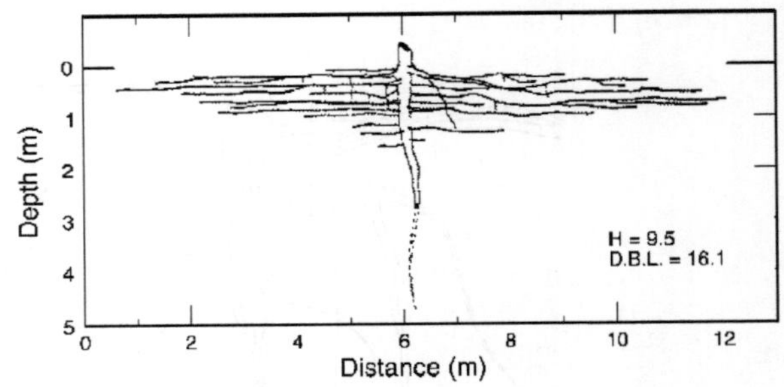

Figure 6. Root system of *Hopea odorata.*

of the roots may be expected to fail any point during shear.

Calculations with Eq. (2) and T_u = pullout resistance, for the laboratory tests on soils with root systems of alfalfa (Operstein and Frydman, 2000) gave an s_r that is 4–5 times the measured value. Both *P. radiata* and alfalfa have root systems that are plate-shaped and it is reasonable to assume that, for such root systems, only a small fraction of the roots would fail in tension.

Nilaweera (1994) performed shear tests on 2–3 year old *Hopea odorata* that have large tap roots. The roots were excavated after the tests. Tension failure was observed in the tap roots at depths of approximately 0.5 m. So the tensile strength and L were estimated and used in the cable solution, Eq. (3). Since the soil is cohesive and partially saturated, the shear strength from consolidated—undrained shear tests was used for c to calculate q_y. This assumes that at fairly rapid loading rate used in the field tests and at high moistures, failure occurs in the undrained condition. The range in calculated s_r represents the uncertainty about the root diameter at the failure point. Calculated measured s_r for 10 tests are plotted in Figure 5. During the initial stages of loading the tension in the tap root would be small and the pile solution was used to calculate T_y and the calculated T_y is much smaller than the measured value, which is expected.

Nilaweera (1994) also performed horizontal pullout tests on young *H. ordorata* trees. These are similar to lateral load tests on piles. A load is applied to the end of a root in a direction perpendicular to that of the root. Tension failure occurred in the tap roots at depths of 2.5–3 m, where the roots enter weathered rock, Figure 6. If we assume that the lateral roots and the surrounding soil in the top 0.7 m moved as a block (Wu and Watson, 1998), then the test is analogous to a shear test. The bearing pressure p_y on the tap root above the failure point was used to calculate T_y and T_z was taken to be the tensile strength. The agreement with the measured T_x is satisfactory (Table 2). The contribution of the small lateral roots was assumed to be negligible and ignored. The calculated s_r from the cable solution is close to the measured value. The s_r for initial failure, calculated with the pile solution, is smaller than the measured value.

In addition to the difficulties given above for fibers and rods, the application of simple models to root systems involve inaccuracies in estimating the dimensions and positions of roots in a root system. This is particularly true with plate-shaped roots like those of *P. radiata* and alfalfa. Therefore, the most practical approach at present may be an empirical reduction of the strength estimated from the ultimate tension, as suggested earlier.

Table 2. Pullout tests by Nilaweera

Author	Test	Spec.	d (cm)	Model	T_y Calc. (kN)	T_y Meas. (kN)
Nilaweera	pullout	*Hibiscus maccrophyllus*	6	Shear	2.3	4
				Pp on lat	1.1	
	hor. pullout	*Hopea odorata*	15	Pile eq.	13	42
				Cable eq.	40	42

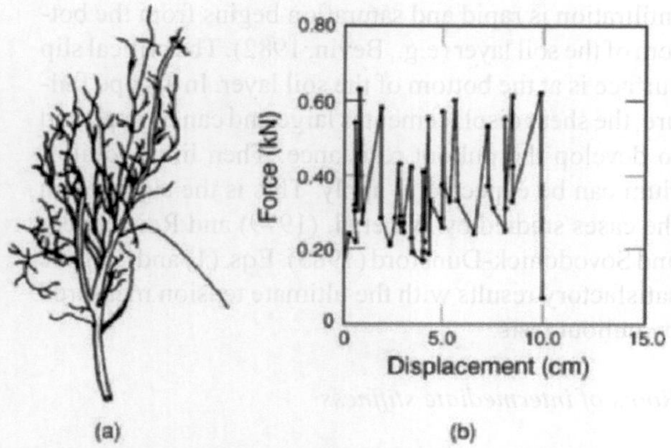

Figure 7. Pullout tests by Riestenberg. (a) Roots of *Acer sacarum* (b) Load-displacement curve.

Pullout tests

In pullout tests, a tensile force is applied to the end of a root in the direction of the root. Failure may occur by tension in the main root or progressive tension failure in the branch roots, or by slip between the root and soil. The controlling failure mode depends on the root geometry and the tensile strength of the root relative to the shear strength of the soil.

In Riestenberg's (1987) tests on sugar maple roots, Figure 7a, progressive tension failure occurred in branch roots, resulting in a load displacement with several peaks, Figure 7b, each peak representing failure of a branch root. In four tests, the branch root that failed could be identified. The tensile force was calculated for both the main root and the failed branch root. The range is shown in Figure 8. It is also known that the branch diameter ratio d_2/d_1 is approximately 0.5, in which d_1 and d_2 = diameters of the main and branch roots, respectively. A simplified model for this case is to assume that the second order roots will fail at $T \approx [\pi\ (0.5d_1)^2\ \sigma_t]/4$. The tensile strength predicted with this relation is shown as a curve in Figure 8. This curve is a good approximation for the average of the measured forces. The conclusion is that for roots with a dendritic pattern, the ultimate tension will always be smaller than the tensile strength computed from the diameter of the main root at the exposed end. The reduction depends on the ratio d_1/d_2.

Nilaweera's (1994) tests on *Hibiscus macrophyllus*. present a different picture because the root system is

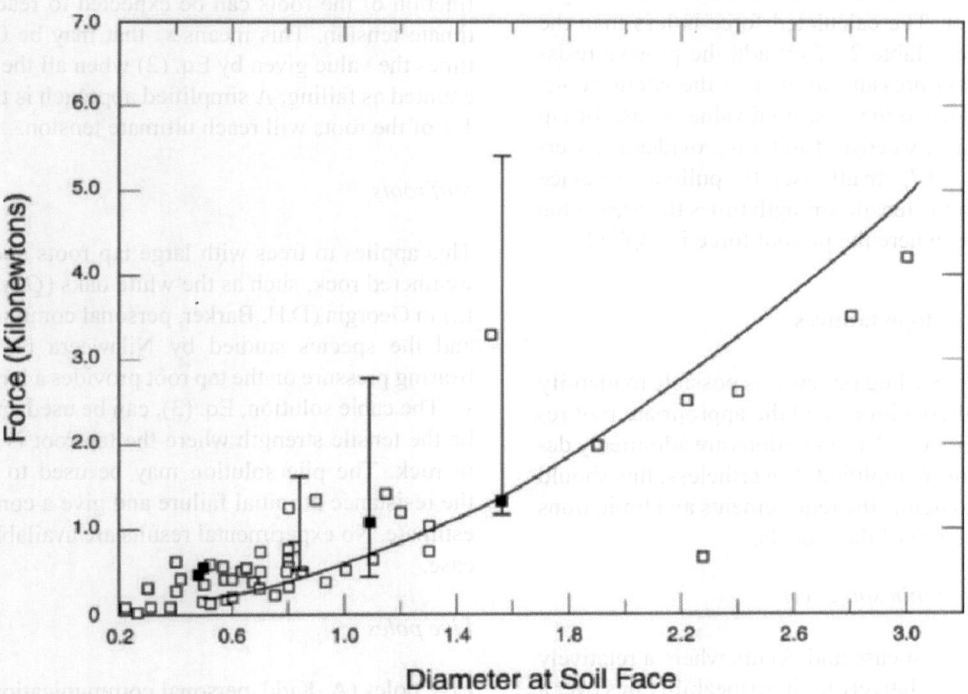

Figure 8. Pullout tests by Riestenberg, pullout resistance vs diameter.

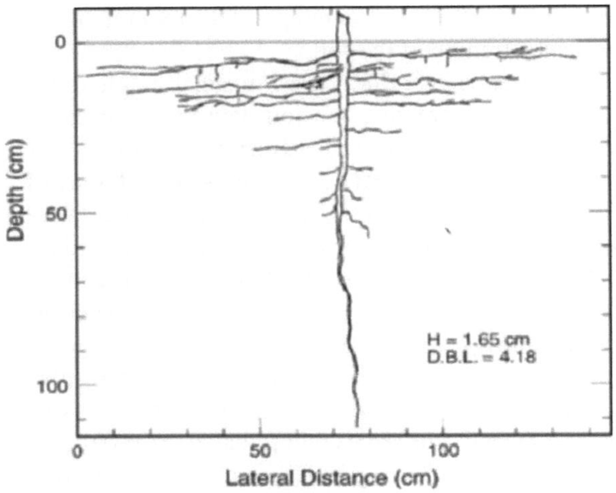

Figure 9. Root system of *Hibis macrophyllus*.

composed of a large tap root with relatively few and small lateral roots (Figure 9). No tension failure was observed. The pullout force was calculated assuming failure occurred by slip between soil and root. However, the adhesion between soil and root is unknown. The problem is simplified by assuming that the adhesion is larger than the undrained shear strength of the soil and failure occurs in the soil with a thin soil layer around the root. The calculated force is less than the measured value, Table 2. If we add the passive resistance or bearing pressure on half of the lateral roots, the result is close to the measured value. Thus, for tap roots, the shear between root and soil provides a conservative estimate of T_u. In all cases, the pullout resistance is smaller than the tensile strength times the area at the end of the root, where the pullout force is applied.

Application to slope failures

Based on the preceding review it is possible to identify several typical conditions and the appropriate root reinforcement models. The conditions are admittedly descriptive and oversimplified. Nevertheless, this should at least, help to define the requirements and limitations for appropriate use of the models.

Flexible roots in thin soil cover

This is the simplest case and occurs where a relatively thin soil layer of relatively high permeability lies over a firm base of low permeability. During large rainstorms, infiltration is rapid and saturation begins from the bottom of the soil layer (e.g., Bevin, 1982). The critical slip surface is at the bottom of the soil layer. In a slope failure, the shear displacement is large and can be expected to develop the pullout resistance. Then limit equilibrium can be expected to apply. This is the situation in the cases studied by Wu et al. (1979) and Reistenberg and Sovodonick-Dunsford (1983). Eqs. (1) and (2) gave satisfactory results with the ultimate tension measured by pullout tests.

Roots of intermediate stiffness

This is the case with a thick soil layer of relatively low permeability. During rainstorms, saturation begins at the top and the saturation front advances as the storm continues. Since root density generally decreases with depth, failure occurs where the suction is reduced to the point that the shear strength of the reinforced soil equals the shear stress. This condition is relatively common (Anderson and Pope, 1984; Ekanayake and Phillips, 1999; Greenway et al., 1984). For trees with heart-shaped or plate-shaped root systems (Dupuy et al., 2005), the roots at the failure surface would be of intermediate flexibility and their orientation, α, will range from 0 to π. Examples given earlier show that only a fraction of the roots can be expected to reach the ultimate tension. This means s_r that may be 0.20–0.33 times the value given by Eq. (2) when all the roots are counted as failing. A simplified approach is that 1/5 to 1/3 of the roots will reach ultimate tension.

Stiff roots

This applies to trees with large tap roots anchored in weathered rock, such as the white oaks (*Quercus alba* L.) in Georgia (D.H. Barker, personal communication) and the species studied by Nilaweera (1994). The bearing pressure on the tap root provides a large part of s_r. The cable solution, Eq. (3), can be used and T_z may be the tensile strength where the tap root is anchored in rock. The pile solution may be used to calculate the resistance at initial failure and give a conservative estimate. No experimental results are available for this case.

Live poles

Live poles (A. Kidd, personal communication), which consist of willow stems with diameters of 4–10 cm and lengths of about 2 m may be similar to stiff roots. However, they are of limited length. Before the roots are developed, the shear strength of the soil would be the upper limit of the "friction" between pole and soil. This friction should increase considerably as roots develop along the length of the pole. Pullout tests on 3-year old live poles have given a tensile resistance larger than 4.9 kN (D.H. Barker, personal communication). The case is similar to a reinforcement with rough surface, such as those in the tests by Jewell and Wroth, with the roots providing the shearing resistance between the pole and soil. In principle, the pullout resistance is equal to the combined pullout resistance of the roots that grow from the stem. However, we do not have quantitative data on root growth from stems. So, we can only rely on empirical data. For stability analysis, consider the example where the failure surface is near the midpoint of the pole. If we assume that the pullout resistance is uniformly distributed along the length of the pole, the friction above and below the slip surface may be taken as 1/2 of the pullout resistance. The friction before the roots are developed may be estimated from the shear strength of the soil, and after the roots are developed, it would be 2.5 kN. For either case, Eq. (1) can be used to calculate s_r with T equal to the frictional resistance.

Summary and conclusions

Simple models can be useful in understanding soil–root behavior and interpreting test results. This allows us to identify their requirements and limitations. In most cases, the tensile force may be well below the ultimate tension. The simple models can give approximate results if the tensile force can be evaluated. The analyses demonstrated the importance of root geometry, site conditions, and the nature of root displacement, which control the failure mechanism. Understanding the failure mechanism allows one to identify appropriate application of the models to stability analysis.

Acknowledgements

I wish to thank the Organizing Committee of the Conference for the opportunity to present this paper, D. H. Barker for providing unpublished information and other help, and the reviewers for their criticisms and suggestions.

References

Abe K and Ziemer R R 1991 Effect of tree roots on a shear zone: modeling reinforced shear stress. Can. J. For. Res. 21, 1012–1019.

Anderson S A and Pope R G 1984 The incorporation of soil water physics models into geotechnical studies of landslide behaviour. *In* Proceedings of the 4th International Symposium Landslides, Vol. 1, Toronto, ON, pp. 349–353.

Bevin K 1982 On subsurface storm flow. J. Sci. Hydrol. 4, 505–521.

Broms B B 1964a Lateral resistance of piles in cohesive soils. J. Soil Mech. Found. Div. 90, 27–64.

Broms B B 1964b Lateral resistance of piles in cohesionless soils. J. Soil Mech. Found. Div. 90, 123–158.

Coppin N J and Richards I G 1990 Use of Vegetation in Civil Engineering. CIRA/Buttersworth, London.

DiMaggio F L and Sandler I S 1971 Material model for granular soils. J. Soil Mech. Found. Div. 97, 935–949.

Drucker D C, Gibson R E and Henkel D J 1957 Soil mechanics and work-hardening theory of plasticity. Trans. ASCE 122, 338–346.

Dupuy L, Fourcaud T and Stokes A 2005 A numerical investigation into the influence of soil type and root architecture on tree anchorage. Plant and Soil. 278, 119–134. DOI: 10.1007/s11104-005-7577-2.

Ekanayake J C and Phillips C J 1999 A model for determining thresholds for initiation of shallow landslides, under near-saturated conditions in the East Coast region. NZ J. Hydrol. 38, 1–26.

El-Khouly M A 1995 Analysis of soil-reinforcement interaction. PhD dissertation, Ohio State Univ., Columbus, OH.

Frydman S and Operstein V 2001 Numerical simulation of direct shear of root-reinforced soil. Ground Improvement 5, 41–48.

Gray D H 1991 Discussion on deformation characteristics of reinforced sand in direct shear. J. Geotech. Eng. 117, 1810–1812.

Gray D H and Ohashi H 1983 Mechanics of fiber reinforcement in sand. J. Geotech. Eng. 109, 335–354.

Gray D H and Sotir R B 1996 Biotechnical and Soil Bioengineering Slope Stabilization. Wiley, New York.

Greenway D R, Anderson M G and Brian-Boys K C 1984 Influence of vegetation on slope stability in Hong Kong. *In* Proceedings of the 4th International Symposium Landslides, Vol. 1, Toronto, ON, pp. 399–404.

Hetenyi M 1946 Beams on Elastic Foundation. Univ. Mich. Press, Ann Arbor, MI.

Jaber M, Mitchell J K, Shen C K and Hua Z K 1987 Behavior of reinforced soil, centrifuge model tests. Final Report, Task B, Fed. Highway Admin. Contract DTFH61-84-C-00073.

Jewell R A and Pedley M J 1992 Analysis for soil reinforcement with bending stiffness. J. Geotech. Eng. 118, 1505–1528.

Jewell R A and Wroth C P 1987 Direct shear test on reinforced sand. Géotechnique 57, 53–68.

Morgan R P C and Rickson R J 1995 Slope Stabilization and Erosion Control. E and FN Spon, London.

Nilaweera N S 1994. Effects of tree roots on slope stability: the case of Khao Luang Ain area, southern Thailand. D.T.S. dissertation, Asin Inst. of Tech., Bankok, Thailand.

Oden J T 1967 Mechanics of Elastic Structures. McGraw-Hill, New York.

Operstein V and Frydman S 2000 The influence of vegetation on soil strength. Ground Improvement 4, 81–89.

Palmeira E M and Milligan G W E 1989 Scale and other factors affecting the results of pull-out tests of grids buried in sand. Géotechnique 39, 511–524.

Plumelle C and Schlosser F 1991 Three full-scale experiments of French project on soil nailing: CLOUTERRE. Transp. Res. Rec. No. 1330, 80–86.

Riestenberg M M 1987 Anchoring of thin colluvium on hillslopes in Cincinnati by roots of sugar maple and white ash. PhD dissertation, Univ. of Cincinnati, Cincinnati, OH.

Riestenberg M M and Sovodonick-Dunsford S 1983 The role of woody vegetation in stabilizing slopes in the Cincinnati area, Ohio. Geol. Soc. Am. Bull. 94, 506–518.

Scheichtl H M and Stern R 1996 Ground Bioengineering Techniques. Blackwell, Oxford.

Shewbridge S E and Sitar N 1989 Deformation characteristics of reinforced sand in direct shear. J. Geotech. Eng. 115, 1134–1147.

Shewbridge S E and Sitar N 1991 Discussion on deformation characteristics of reinforced sand in direct shear. J. Geotech. Eng. 117, 1812–1817.

Waldron L J 1977 The shear resistance of root-permeated homogeneous and stratified soil. Soil Sci. Soc. Am. Proc. 41, 843–849.

Wu T H, McKinnell W P III and Swanston D N 1979 Strength of tree roots and landslides on Prince of Wales Island, Alaska. Can. Geotech. J. 16, 19–33.

Wu T H and Watson A 1998 In-situ shear tests of soil blocks with roots. Can. Geotech. J. 35, 579–590.

Root strength and root area ratio of forest species in Lombardy (Northern Italy)

Gian Battista Bischetti[1], Enrico A. Chiaradia, Tommaso Simonato, Barbara Speziali, Barbara Vitali, Paolo Vullo & Antonio Zocco

Institute of Agricultural Hydraulics, Università degli Studi di Milano, Celoria n. 2, Milano, Italy

[1]*Corresponding author**

Key words: root area ratio, root tensile strength, soil–root interaction

Abstract

Forest vegetation is known to increase hillslope stability by reinforcing soil shear resistance and by influencing hydrologic conditions of soil. Although the importance of plant root systems for hillslope stability has received considerable attention in recent years, the quantification of such an effect needs more investigation. In this paper, we present a synthesis of the data gathered in the last 5 years for some species in different locations of the Alps and Prealps of Lombardy (Northern Italy) with the aim to increase our knowledge on root tensile strength and on root area ratio distribution within the soil. Concerning root tensile strength we developed tensile strength–diameter relationships for eight species: green alder (*Alnus viridis* (Chaix) D.C.), beech (*Fagus sylvatica* L.), red willow (*Salix purpurea* L.), goat willow (*Salix caprea* L.), hazel (*Corylus avellana* L.), European ash (*Fraxinus excelsior* L.), Norway spruce (*Picea abies* (L.) Karst.) and European larch (*Larix decidua* Mill.). Results show a great variability among the different species and also for the same species. In general, however, root strength (in terms of tension) tends to decrease with diameter according to a power law, as observed by other authors. Comparing the power law fitting curves for the considered species, it can be observed that they fall in a relatively narrow band, with the exception of hazel, which appears the most resistant. Concerning the evaluation of root distribution within the soil we estimated the root area ratio (the ratio between the area occupied by roots in a unit area of soil) according to its depth for five species (beech, Norway spruce, European larch, mixed hazel and ash) in three locations of Lombardy. Results show that there is a great variability of root density for the same species well as for different points at the same locality. The general behaviour of root density, in any case, is to decrease with depth according to a gamma function for all the studied species. The results presented in this paper contribute to expanding the knowledge on root resistance behaviour and on root density distribution within the soil. The studied location have allowed the implementation of soil–root reinforcement models and the evaluation of the vegetation contribution to soil stability.

Introduction

Vegetation affects slope stability influencing both hydrological processes and mechanical structure of the soil. The magnitude of such effects depends on root system development, which in turn is a function of genetic properties of the species and site characteristics (soil texture and structure, aeration, moisture, temperature and competition with other plants). Due to the variability of such characteristics, we observe a large spatial variability of root patterns and then a great heterogeneity in soil reinforcement.

Limiting our attention to the mechanical effects, we can recognise two main actions of roots: the small size flexible roots mobilise their tensile strength by soil–root friction increasing the compound matrix (soil–fibre) strength (Gray and Leiser, 1982), whereas the large size roots that intersect the shear plane act as individual

* E-mail: bischetti@unimi.it

This article has been previously published in the following journal—Plant and Soil, Volume 278, 11–22.
Stokes et al. (eds), Eco- and Ground Bio-Engineering: The Use of Vegetation to Improve Slope Stability, 31–41.

anchors (Coppin and Richards, 1990) and eventually tend to slip through the soil matrix without breaking, mobilising a small portion of their tensile strength (Burroughs and Thomas, 1977; O'Loughlin and Watson, 1979; Schmidt et al., 2001; Ziemer, 1981). Both the effects can be quantified by modelling (see Gray and Leiser, 1982; Wu, 1995) if appropriate parameters are provided.

Usually, the only effect considered by most of the authors is the fibre reinforcement expressed as an additional root cohesion (Abernethy and Rutherfurd, 2001; Bischetti, 2001; Bischetti et al., 2002; Burroughs and Thomas, 1977; Riestenberg and Sovonick-Dunford, 1983; Schmidt et al., 2001; Sidle 1992; Sidle et al., 1985; Wu, 1984 a,b; Wu et al., 1979; Wu and Sidle, 1995) which can be easily incorporated into spatially distributed slope stability models (Chiaradia and Bischetti, 2004; Istanbulluoglu et al., 2004; Pack et al., 1997; Wu and Sidle, 1995).

The most widespread model for root cohesion (cr) is the Wu (1976) and Waldron (1977) model:

$$C_r = K \cdot t_R, \tag{1}$$

where t_R is the mobilised root tensile strength per soil unit area; K is a factor taking into account that roots are randomly orientated with respect to the failure plane which in most of the cases varies between 1.0 and 1.3 (Waldron, 1977; Wu et al., 1979).

The mobilised root tensile strength per soil unit area (t_R) can be written as

$$t_R = T_r a_r, \tag{2}$$

where T_r is the average tensile strength per average root cross-sectional area; a_r is the *root area ratio* computed as A_r/A, where A_r is the total cross-sectional area of all roots and A is the area of soil in the sample count.

Root tensile strength is affected as much by species as by differences in size (diameter). The generally accepted form for the relationship between root tensile strength ($T_r(d)$) and diameter (d) is a simple power function (Eq. (3), Gray and Sotir, 1996; Wu, 1995):

$$T_r(d) = \alpha d^{-\beta}, \tag{3}$$

where α and β are empirical constants depending on species.

To account for the variability in root size Eq. (2) must then be rewritten as:

$$t_R = \sum_{i=1}^{N} T_{r_i} \frac{A_{r_i}}{A}, \tag{4}$$

where i indicates the diameter class and N the number of classes.

Root area ratio (RAR) provides a measure of root density within the soil and as a consequence it is strongly influenced by local soil and climate characteristics, land use management and associated vegetation communities and randomness. In general, RAR decreases with depth below the soil surface and with distance from tree trunk (Abernethy and Rutherfurd, 2001; Greenway, 1987; Nilaweera, 1994; Schmid and Kazda, 2001, 2002; Shields and Gray, 1992; Zhou et al., 1998).

On the basis of the Wu (1976) and Waldron (1977) model, the extent of root reinforcement depends on tensile strength, density and depth of roots, which vary significantly depending on species, local environmental characteristics and spatial variability of vegetation properties (density, age, fire events, erosion, trees health, etc.). Root density, in particular, shows an extremely large spatial variability, both in the vertical and in the horizontal planes. Despite the large amount of studies investigating such an issue (Böhm, 1979; Danjon et al., 1999; Glinski and Lipiec, 1990; Jackson et al., 1996; Keyes and Grier, 1981; Kramer and Boyer, 1995; Libundgut, 1981; McMinn, 1963; Paar, 1994; Sainju and Good, 1993; Watson and O'Loughlin, 1990), most of them focus on eco-physiologic behaviour of vegetation and do not provide data useful for root reinforcement estimation. Such studies, in fact, deal with nutrient and organic matter input to the soil, soil fertility maintenance and with carbon sequestration so, as a consequence, they only consider small size roots (<1–2 mm) in the upper soil layers.

Because of a renewed interest in understanding the role of vegetation on slope stability and shallow landsliding, the number of studies on such an issue is increasing (Abernethy and Ruthefurd, 2001; Bischetti et al., 2002; Roering et al., 2003; Schmidt et al., 2001; Zhou et al., 1998). Nevertheless, due to the complexity of reinforcement mechanisms, the variety of species and environments and the spatial variability of characteristics driving the processes, such work can be considered eminently site-specific and more experimental data are still needed for a whole comprehension and generalisation of the phenomenon.

The present study, focusing on some typical Alpine and Prealpine tree species in Northern Italy, aims to expand the knowledge on root strength and root density values contributing in gathering enough data to achieve such a generalisation.

Materials and methods

Study sites

Since 2000, the authors have been focusing their attention on the effect of roots on slope stability starting from a series of data and observations collected in the Central Italian Alps and Prealps within the Lombardy region territory. Most of the work was done in the areas of Valdorena, Morterone, Alpe Gigiai and Valcuvia (Figure 1); additional samples for beech were also collected in the areas of Valsassina and Val Intelvi.

Valdorena is a right-hand flank tributary of the Camonica Valley (Oglio River). Most of the slopes are dominated by superficial overconsolidated periglacial materials and till deposits, with a thickness of more than 20 m, intensively eroded by the fluvial incision. Such deposits mainly consist of Late Pleistocene and Holocene till deposits, rock glaciers (Holocene), talus, landslides and alluvial deposits, whereas gneisses are the dominant lithotype outcropping. At sampling points located between 1525 and 1575 m a.s.l., soils consist of a mixture of sand and gravel in a silty matrix. The mean annual precipitation measured at the nearest rain gauge of Edolo (690 m a.s.l.) in more than 40 years of observations is equal to about 1000 mm. Precipitation mostly occurs as snowfall from November to March and as rainfall in spring and summer, with a maximum between May and September. At this site we focused on root systems of Green alder (*Alnus viridis* (Chaix) D.C.) and Willow (*Salix spp*).

Morterone area is located in the Taleggio valley, a right hand flank tributary of Val Brembana (Brembo river). Slopes are formed by glacial and fluvio-glacial deposits with variable thickness overlying sedimentary rocks (mainly formed of fine and coarse-grained limestone with interbeds of marl and nodular chert and, occasionally, intercalation of calcareous breccias). At sample points located at 1100 m a.s.l., the soil consists of very poorly graded material, generally defined as a silt with clayey sand and fine gravel. Annual precipitation measured at the nearest raingauge of Vedeseta is equal to 1828 mm and it mostly falls as rainfall in autumn and spring, but during summer many heavy storms may occur. Mean annual air temperature is 6.1°C with an average summer temperature of 14.3°C and an average winter temperature of –1°C. At this site we focused on root strength and RAR distribution of European beech (*Fagus sylvatica* L.), which is the dominant species sometimes with intrusion of other tree species such as European white birch (*Betula pendula* Roth.), European ash (*Fraxinus excelsior* L.) and Sycamore maple (*Acer pseudoplatanus* L.).

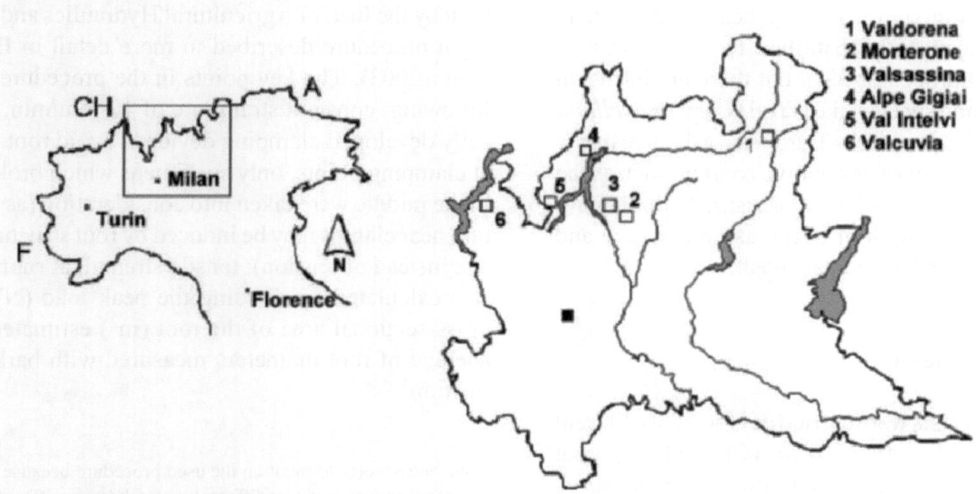

Figure 1. Location map of the study areas and species. Beech: Morterone, Valsassina, Alpe Gigiai, Val Intelvi and Valcuvia; Red willow, goat willow and green alder: Valdorena; European ash and hazel: Valcuvia; Norway spruce: Valcuvia and Alpe Gigiai; European larch: Alpe Gigiai.

Alpe Gigiai is located on the North-Western side of Como Lake, in an area called Alto Lario. The area is characterised by a steep and dissected topography, with slopes commonly ranging from 20° to 30°. Glacial and fluvio-glacial deposits with variable thickness overlie rocks in the lower part of the area, whereas the higher portion is covered by colluvial deposits. Rocks outcropping in the study area are represented by gneiss belonging to Falda Adula. At sample points soils consist of a gravel–sand mixture with silty matrix. The mean annual precipitation of the area is equal to 1604 mm. Mean annual air temperature is estimated in 9.0°C with an average summer temperature of 18.5°C and an average winter temperature of −0.1°C. At this site we focused on root strength and RAR distribution of beech, Norway spruce and European larch.

Valcuvia is a left-hand flank tributary of the Maggiore Lake (Varese, Northern Italy). The study sites are located in the St. Giulio creek catchment (about 5 km^2) which is characterised by a steep and highly dissected topography, a typical V-shaped valley and slopes commonly ranging from 25° to 45°. The outcropping rocks consist mainly of fine and coarse-grained limestone with interbeds of marl and nodular chert. The soil consists of very poorly graded granular material, generally defined as a gravel–sand mixture with silty matrix. Mean annual precipitation measured at the rain gauge of Vararo (in the upper part of the catchment) is equal to 2330 mm; this area is really one of the most rainy of the Lombardy Region and precipitation principally falls as rainfall in autumn and spring, but during summer many heavy storms may occur. In these recently disturbed terrains, forest is dominated by European beech, but there are intrusion of other hardwood species like Hazel (*Corylus avellana* L.), European white birch, European ash, Sycamore maple and some trees of non-native conifers such as the Norway spruce (*Picea abies* (L.) Karst.). At this site we focused on root strength of hazel, ash red spruce and RAR distribution of mixed hazel-ash.

Tensile strength tests

Tensile strength tests were carried out for eight different species: green alder (*Alnus viridis* (Chaix) D.C.), goat willow (*Salix caprea* L.), red willow (*Salix purpurea* L.), beech (*Fagus sylvatica* L.), hazel (*Corylus avellana* L.), European ash (*Fraxinus excelsior* L.), Norway spruce (*Picea abies* (L.) Karst.) and European larch (*Larix decidua* Mill.).

Beech roots were sampled in five different locations in the Prealps (Morterone, Valsassina, Alpe Gigiai, Val Intelvi and Valcuvia). Red willow, goat willow and green alder roots were collected in Valdorena Alpine environment, European ash and hazel roots were sampled in Valcuvia, Norway spruce roots in Valcuvia and Alpe Gigiai and European larch roots in Alpe Gigiai (Figure 1).

Live roots were collected from soil by excavating pits or trenches, taking care to avoid any root damage or stress; samples were then put in separate bags, sealed and transported to the laboratory in a refrigerated box. In most of the cases, tensile tests were carried out on fresh roots within 1 week from sampling; in some cases we have conserved roots for few weeks with three different techniques and verified that results were unaffected by the conservation method employed (Bischetti et al., 2003). The first conservation technique consisted in drying root at 60°C for 24 h and rewetting them in water for few hours (Schuurman and Goedewaagen, 1971, cited by Böhm, 1979), the second consisted in freezing roots in a plastic bag filled with water (Schuurman and Goedewaagen, 1971, cited by Böhm, 1979) while the last one by making use of alcoholic solution at 15% (Meyer and Göttsche, 1971).

Tests were carried out for roots with typical tortuousness on thread and diameter up to about 5 mm as greater size roots tended to break at clamping points or to slip out.

Testing was performed with a device designed and built by the Inst. of Agricultural Hydraulics and following a procedure described in more detail in Bischetti et al. (2003). The key points in the procedure are the following: constant strain rate of 10 mm/min, specifically developed clamping device to avoid root damage at clamping points, only specimens which broke about in the middle were taken into consideration (as the rupture near clamps may be induced by root structure damage instead of tension); tensile strength at rupture (Pa) was calculated by dividing the peak load (N) by the cross-sectional area of the root (m^2) estimated as the average of root diameters measured with bark before traction[1].

[1] We deem worth to mention the used procedure because the diversity of procedures used by different researchers can affect the results (Cofie and Koolen, 2001), with particular reference to the root diameter measure.

Table 1. Results of tensile strength test

Species	Diameter (mm)				Strength (MPa)			
	Mean	SD	Max	Min	Mean	SD	Max	Min
Fagus sylvatica	1.33	0.93	4.59	0.14	57.47	81.61	730.97	2.27
Salix purpurea	1.28	0.82	4.10	0.18	51.47	35.81	522.03	2.07
Salix caprea	1.42	1.10	5.70	0.13	47.80	60.14	408.59	6.02
Fraxinus excelsa	1.95	1.15	5.70	0.27	36.86	68.05	296.51	4.92
Alnus viridis	2.03	1.02	5.91	0.65	20.42	14.77	92.13	3.35
Corylus avellana	1.65	1.15	3.82	0.31	67.87	66.25	256.76	11.89
Picea abies	1.78	1.19	5.84	0.12	38.94	83.79	649.71	5.79
Larix decidua	1.68	1.49	5.47	0.14	66.14	99.78	427.96	6.44

Root area ratio measures

To obtain root distributions two methods are generally adopted: core-break sampling (Babu et al., 2001; Burke and Raynal, 1994; Büttner and Leuschner, 1994; Hendriks and Bianchi, 1995; Schmid and Kazda, 2002; Xu et al., 1997) and counting roots by profile trenching (Burke and Raynal, 1994; Schmid and Kazda, 2001, 2002; Vinceti et al., 1998; Xu et al., 1997). Since core-break samples provide root biomass, root number or root length, the estimation of RAR implies hypothesis about the tri-dimensional distribution of roots inside the sample (Lopez-Zamora et al., 2002). In the present work we present data obtained through root counting by image analysis, applying the trench profile wall technique photographs (Böhm, 1979; Vogt and Persson, 1991). At sampling sites in proximity of a forest roads under construction or next to landslide scarps we excavated a trench to expose a fresh profile of rooted soil down to the bedrock, applied a frame of known size and took several images that were rectified to correct geometrical deformation and roots were manually digitised. RAR values were obtained at each depth increments of 10 cm counting all roots with a diameter between 1 and 10 mm; roots less than 1 mm, in fact, are difficult to be recognised and correctly mapped, whereas big roots may strongly affect RAR values but hardly act in accordance of the reinforcement model of Eq. (1), although they are fundamental for the tree anchorage.

Results

Tensile strength

Stress–strain curves obtained by traction tests have been processed to get peak tensile strength values. Results exhibit a great variability of measured tensile strength of roots among the different species, but the variability exists also if taking only one species into consideration. The results are presented in Table 1 and in general, the value of the standard deviation for the measured tensile strength is greater than the mean value (up to twice).

Many authors (Abe and Iwamoto, 1986; Bischetti et al., 2002; Burroughs and Thomas, 1977; Gray and Sotir, 1996; Nilaweera and Nutalaya, 1999) have demonstrated that root strength is strongly influenced by root diameter. The tensile strength results must be analysed in terms of strength–diameter relationship and therefore the strength data (Tr; MPa) versus root diameter (d; mm) has been corrected by fitting the power law of Eq. (3).

Data with the corresponding fitting curve are shown in Figures 2–5, whereas the curve parameters, the number of trials and the coeficient of correlation are reported in Table 2.

Root area ratio

Root area ratio was evaluated for five species in three locations: beech in Morterone and in Alpe Gigiai, Norway spruce and European larch in Alpe Gigiai, hazel and European ash in Valcuvia.

As far as beech is concerned, both in Morterone and Alpe Gigiai, we dug three trenches along the cut slope of a forest road at a distance of about 100 m one from the other; in the case of Norway spruce and larch in Alpe Gigiai, instead, three and four trenches, respectively, were excavated at landslides scarps. Finally, in Valcuvia, due to the steep gradient of the slopes, we excavated only one trench at a landslide scarp. Results are shown in Figures 7–9.

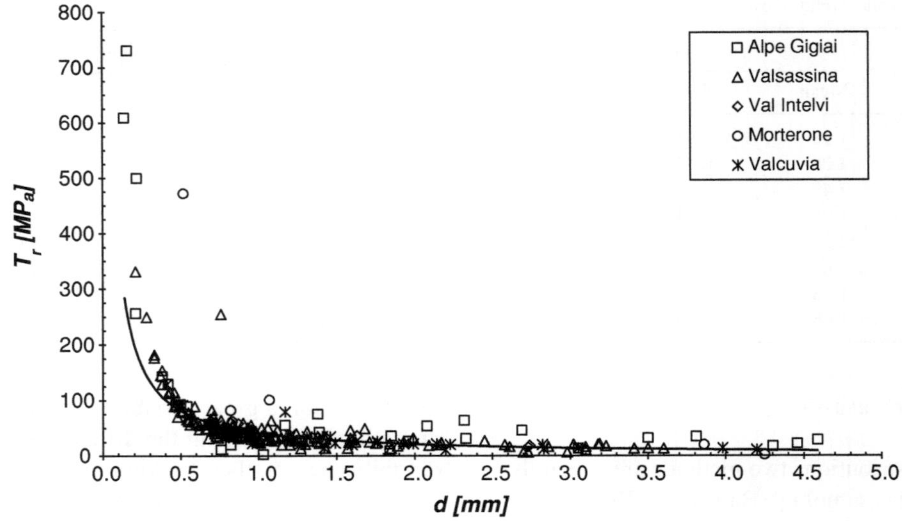

Figure 2. Strength–diameter values for beech at the study sites and the related regression line ($T_r = 41.65d^{-0.97}$, $r^2 = 0.62$).

As expected, RAR values show a great variability with depth, species and location. The general behaviour of RAR with depth is to increase its values in the first layers and then to decrease them. In most of the cases the maximum RAR values are located in the first 30 cm and the maximum depth is about 1 m. The mean values along the profiles range between 0.1% and 0.35% depending on the species, with a standard deviation of about one third of the mean for the different profiles at the same site.

Discussion

Tensile strength

The results obtained in the present study essentially confirm the power law relationship between strength and diameter for all the considered species.

Comparing root tensile strength data for beech and for Norway spruce which have been collected at different sites (Figures 2 and 5), it seems that environment

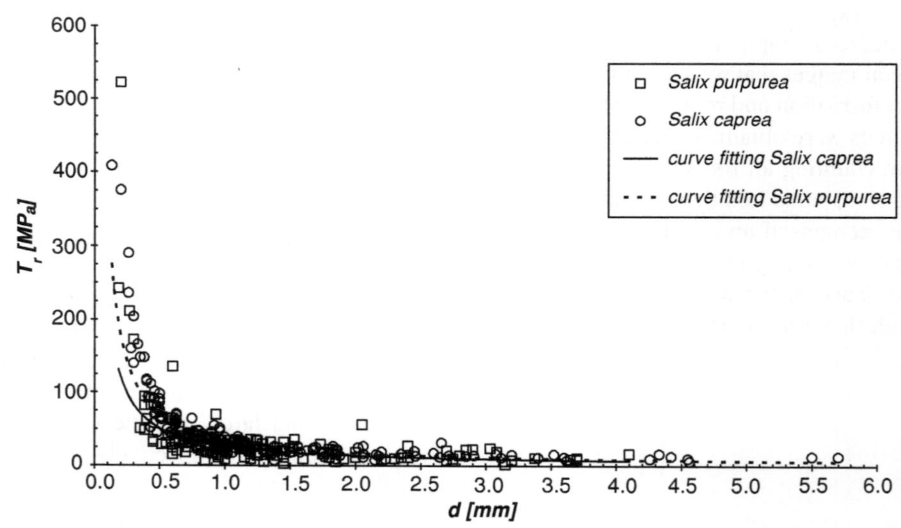

Figure 3. Strength–diameter values and the related regression lines for purple and goat willow (*Salix purpurea* $T_r = 26.33d^{-0.95}$, $r^2 = 0.55$; *Salix caprea* $T_r = 34.5d^{-1.02}$, $r^2 = 0.82$).

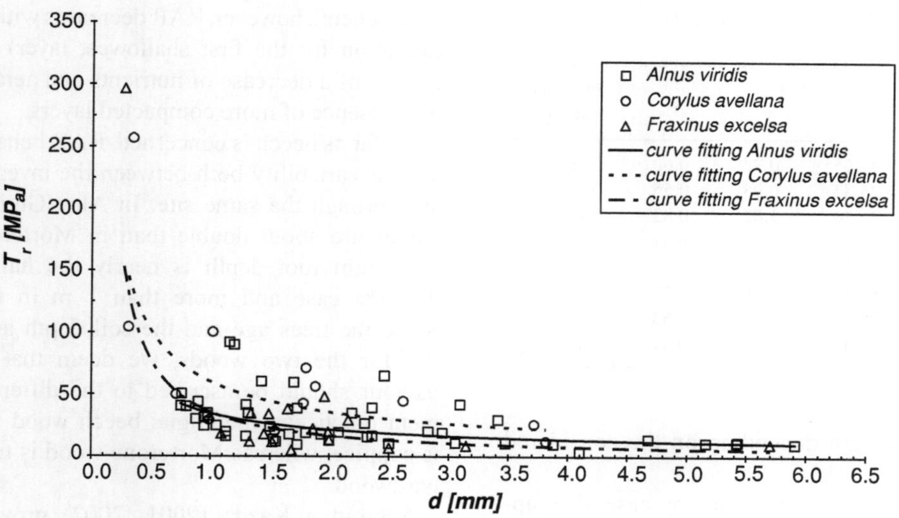

Figure 4. Strength–diameter values and the related regression lines for green alder ($T_r = 34.76d^{-0.69}$, $r^2 = 0.34$), hazel ($T_r = 60.15d^{-0.75}$, $r^2 = 0.57$) and European ash ($T_r = 35.73d^{-1.11}$, $r^2 = 0.51$).

does not significantly affect tensile strength; such an observation is in contrast with findings of other authors (Burroughs and Thomas, 1977; Schiechtl, 1980) and it must be confirmed by more analysis.

Concerning the behaviour of the different species in terms of root tensile strength, the exponent of the power law equation β controls the rate of strength decay with diameter, whereas α can be considered as a scale factor. From the results it can be recognised that there are two values of baround which the species values aggregate (Table 2): the unity for broadleaf species except hazel and alder; and 0.70 for conifers, hazel and alder; the species belonging to the same group, then, show a similar behaviour but a different scale factor. The most resistant species is hazel which has a high scale factor and a low decay rate. The less resistant species is red willow which has a little scale factor and a high exponent. The other species show an intermediate

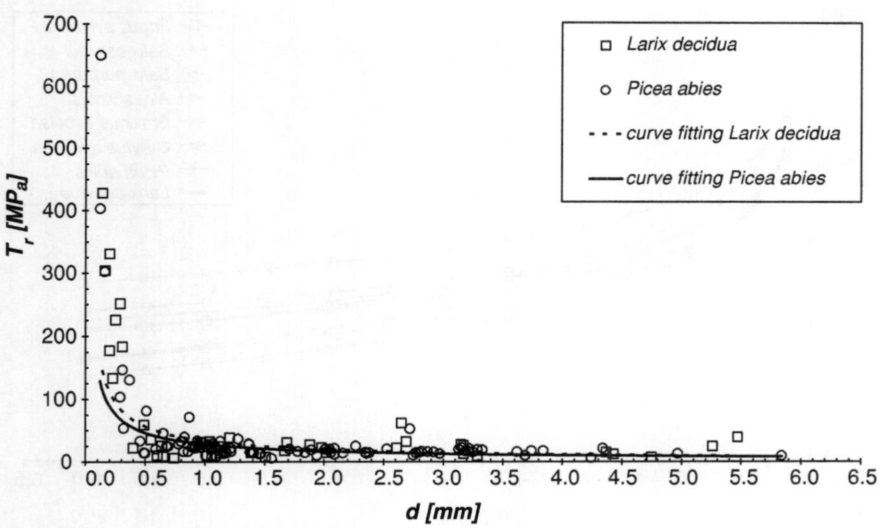

Figure 5. Strength–diameter values and the related regression lines for Norway spruce ($T_r = 28.10d^{-0.72}$, $r^2 = 0.53$) and larch ($T_r = 33.45d^{-0.75}$, $r^2 = 0.47$).

Table 2. Parameters of the root strength–diameter power law relationship

Species	α	β	r^2	Number of valid tests
Fagus sylvatica	41.65	0.97	0.62	168
Salix purpurea	26.33	0.95	0.55	150
Salix caprea	34.50	1.02	0.82	144
Fraxinus excelsa	35.73	1.11	0.51	17
Alnus viridis	34.76	0.69	0.34	49
Corylus avellana	60.15	0.75	0.57	13
Picea abies	28.10	0.72	0.53	92
Larix decidua	33.45	0.75	0.47	43

behaviour and strength depending on the considered diameter.

Comparing the fitting curves, in any case, they appear to fall in a fairly narrow band with the exception of hazel (Figure 6); such results need corroborating by additional tests, especially for those species with less strength data or a small correlation coeffcient.

Root area ratio

Observed values for RAR show a very high variability with species, location and depth. RAR is strongly influenced by genetics, by local soil and climate characteristics and by forest management; in addition, randomness must be accounted for.

In general, however, RAR decreases with depth (with exception for the first shallowest layer) as a consequence of a decrease of nutrients and aeration, and of the presence of more compacted layers.

As far as beech is concerned, RAR behaviour shows a great variability both between the investigated sites and through the same site. In Alpe Gigiai site RAR values are about double than in Morterone, but the maximum root depth is nearly the half (0.7 m in the first case and more than 1 m in the second). Since the trees age and the soil depth are very similar for the two woods, we deem that such a behaviour should be ascribed to the different management practice: Alpe Gigiai beech wood management is coppice, whereas Morterone wood is turning into a hardwood.

Schmid e Kazda (2001, 2002) provide interesting data for beech and Norway spruce root distribution; although it is not straightforward to carry out a comparison due to a different definition of root size classes, RAR values obtained in our study are consistent with those data and in general with the results reported in studies concerning other tree species in different environments dominated by hardwood forests (Wu, 1995).

In the same manner as for tensile strength, RAR distributions were approximated to an analytical function. All species had average RAR values that satisfactorily approximated a gamma function (Kottegoda and

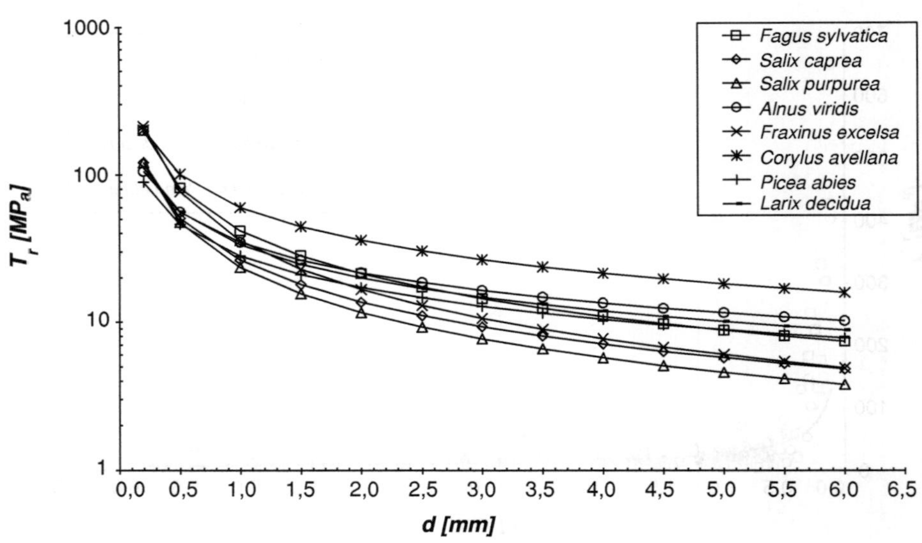

Figure 6. Strength-diameter fitting curves for the studied species (semi-logarithmic).

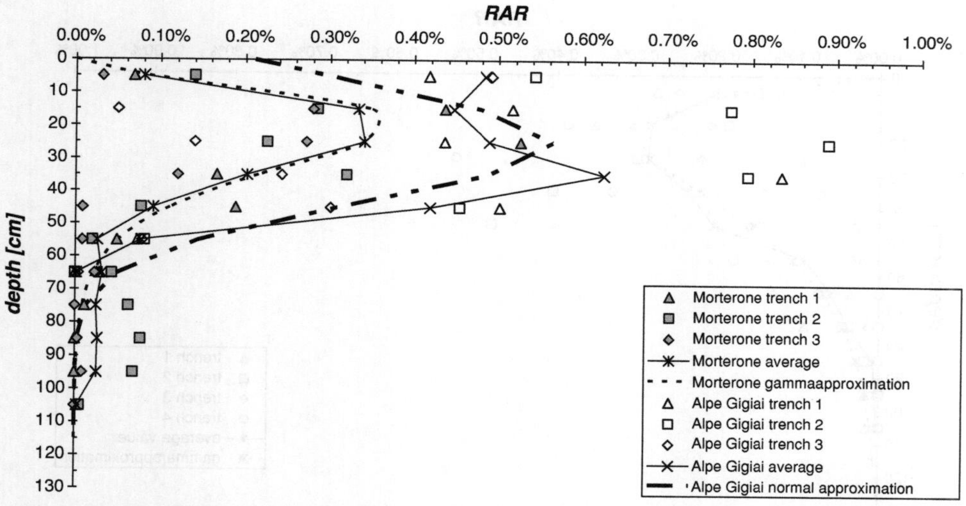

Figure 7. RAR for beech at the two study sites (values at each trench, average and analytical approximation).

Rosso, 1997), except beech at Alpe Gigiai which seem to be better approximated by a normal distribution (Figures 7–9). The results for beech agree with the results of Schmid and Kadza (2001) for root density; however, Schmid and Kadza (2001) found an exponential decrease for Norway spruce roots.

In conclusion the data presented in this study expand the knowledge on root tensile strength and root area ratio of some typical Alpine and Prealpine species in Northern Italy and are suitable to be used in hillslope stability mapping in forested areas (Chiaradia and Bischetti, 2004).

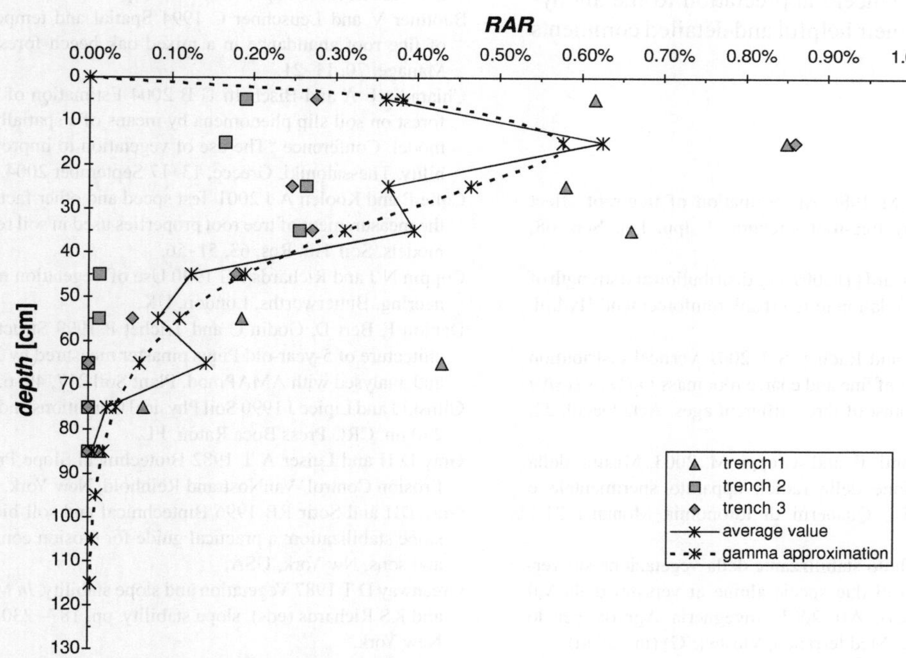

Figure 8. RAR for Norway spruce (values at each trench, average and analytical approximation).

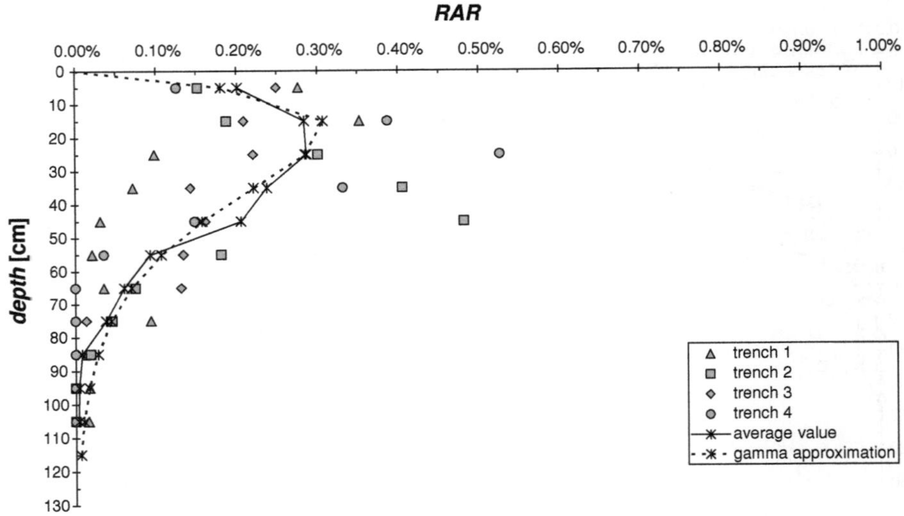

Figure 9. RAR for larch (values at each trench, average and analytical approximation).

Acknowledgements

The research was supported by MIUR-PRIN Project "*Analisi integrata dei processi erosivi per la rinaturazione del sistema alveo-versante*" and by C.M. della Valsassina, Valvarrone, Val d'Esine e Riviera. The authors express their sincere appreciation to the anonymous reviewers for their helpful and detailed comments and suggestions.

References

Abe K and Iawamoto M 1986 An evaluation of tree-root effect on slope stability by tree-root strength. J. Jpn. For. Soc. 68, 505–510.

Abernethy B and Rutherfurd I D 2001 The distribution and strength of riparian tree roots in relation to riverbank reinforcement. Hydrol. Processes 15, 63–79.

Babu J, Herendra N P and Radhey S T 2001 Vertical distribution and seasonal changes of fine and coarse root mass in *Pinus kesiya* Royle Ex. Gordon forest of three different ages. Acta Oecol. 22, 293–300.

Bischetti G B, Bonfanti F and Greppi M 2003 Misura della resistenza alla trazione delle radici: apparato sperimentale e metodologia d'analisi. Quaderni di Idronomia Montana 21/1 (in Italian).

Bischetti G B 2001 Effetto stabilizzante della vegetazione sui versanti: caso di studio di due specie alpine ai versanti della Val Dorena (Valcamonica). AII 2001: Ingegneria Agraria per lo sviluppo dei paesi del Mediterraneo, Vieste (FG) (in Italian).

Bischetti G B, Speziali B and Zocco A 2002 Effetto di un bosco di faggio sulla stabilita' dei versanti. Convegno Nazionale La Difesa della Montagna, Assisi (PG) (in Italian).

Böhm W 1979 Methods of Studying Root systems, Ecological Series No. 33. Springer-Verlag, Berlin.

Burke M K and Raynal D J 1994 Fine root growth phenology, production and turnover in a northern hardwood forest ecosystem. Plant Soil 162, 135–146.

Burroughs E R and Thomas B R 1977 Declining root strength in Douglas-fir after felling as a factor in slope stability USDA Forest Service Research Paper INT-190, 27 p.

Buöttner V and Leuschner C 1994 Spatial and temporal patterns of fine root abundance in a mixed oak beech forest. For. Ecol. Manage. 70, 11–21.

Chiaradia E A and Bischetti G B 2004 Estimation of the effect of forest on soil slip phenomena by means of a spatially distributed model. Conference "The use of vegetation to improve slope stability, Thessaloniki, Greece, 13–17 September 2004.

Cofie P and Koolen A J 2001 Test speed and other factors affecting the measurement of tree root properties used in soil reinforcement models. Soil Till. Res. 63, 51–56.

Coppin N J and Richards I G 1990 Use of Vegetation in Civil Engineering. Butterworths, London, UK.

Danjon F, Bert D, Godin C and Trichet P 1999 Structural root architecture of 5-year-old Pinus pinaster measured by 3D digitising and analysed with AMAPmod. Plant Soil 217, 49–63.

Glinski J and Lipiec J 1990 Soil Physical Conditions and Plant Roots, 240 pp. CRC Press Boca Raton, FL.

Gray D H and Leiser A T 1982 Biotechnical Slope Protection and Erosion Control. VanNostrand Reinhold, New York.

Gray DH and Sotir RB 1996 Biptechnical and soil bioengineering slope stabilization: a practical guide for erosion control. J. Wiley and sons, Nw York, USA.

Greenway D T 1987 Vegetation and slope stability. *In* MG Anderson and KS Richards (eds), slope stability. pp. 187– 230. John Wiley, New York.

Hendriks C M A and Bianchi F J J A 1995 Root density and root biomass in pure and mixed forest stands of Douglas-fir and Beech. Neth. J. Agr. Sci. 43, 321–331.

Istanbulluoglu E, Tarboton D G, Pack R T and Luce C H 2004 Modeling of the interactions between forest vegetation, disturbances, and sediment yields. J. Geophys. Res., 109.

Jackson R B, Canadell J, Ehleringer J R, Mooney H A, Sala O E and Schulze E D 1996 A global analysis of root distributions for terrestrial biomes. Oecologia 108, 389–411.

Keyes M R and Grier C C 1981 Above and below ground net production in 40-year-old Douglas-fir stands on low and high productivity sites. Can. J. For. Res. 11, 599–605.

Kottegoda N T and Rosso R 1997 Statistics, Probability and Reliability for Civil and Environmental Engineers. McGraw-Hill, New York.

Kramer J P and Boyer S J 1995 Water Relations of Plants and Soils. Academic Press, San Diego.

Libundgut H 1981 Untersuchungen uüber das verhalten von jungpflanzen einiger baumarten bei verschiedenem grundwasserstand. Schweiz. Zeitschr. Forstw. 132, 291–318.

Lopez-Zamora I, Falcão N, Comerford N B and Barros N F 2002 Root isotropy and an evaluation of a method for measuring root distribution in soil trenches. For. Ecol. Manage. 166, 303–310.

McMinn R G 1963 Characteristics of Douglas-fir root systems. Can. J. Bot. 41, 105–122.

Meyer F H and Göttsche D 1971 Distribution of root tips and tender roots of beech. *In* H Ellenberg (ed), Ecological Studies. Analysis and Synthesis, Vol. 2, pp. 47–52. Springer-Verlag, Berlin.

Nilaweera N S 1994 Effects of the roots on slope stability: the case of Khao Luang Mountain area, So. Thailand, Degree thesis, Asian Institute of Technology, Bangkok, Thailand.

Nilaweera N S and Nutalaya P 1999 Role of tree roots in slope stabilisation. Bull. Eng. Geol. Env. 57(4), 337–342.

O'Loughlin C L and Watson A 1979 Root-wood strength deterioration in Radiata Pine after clearfelling. NZ J. For. Sci. 39(3), 284–293.

Pack R T, Tarboton D G and Goodwin C N 1997 SINMAP – A Stability Index Approach to Terrain Stability Hazard Mapping – Users Manual. Available at http://www.engineering.usu.edu/dtarb.

Paar U 1994 Untersuchungen zum einfluß von ammonium unf nitrat auf wurzelphysiologische reaktionsmuster der buche. Ber. Forsch. Zentr., Waldökosysteme Göttingen A115.

Riestenberg M M and Sovonik-Dunford S 1983 The role of woody vegetation in stabilizing slopes in the Cincinnati area. Ohio. Geol. Soc. Am. Bull. 94, 506–518.

Roering J J, Schmidt K M, Stock J D, Dietrich W E and Montgomery D R 2003 Shallow landsliding, root reinforcement, and the spatial distribution of trees in the Oregon Coast Range. Can. Geotech. J. 40, 237–253.

Sainju U M and Good R E 1993 Vertical root distribution in relation to soil properties in New Jersey pinelands forests. Plant Soil 150, 87–97.

Schiechtl H M 1980 Bioengineering for Land Reclamation and Conservation, 404 pp., University of Alberta Press, Edmond, Canada.

Schmid I and Kazda M 2001 Vertical distribution and radial growth of coarse roots in pure and mixed stands of Fagus sylvatica and Picea abies. Can. J. For. Res. 31, 539–548.

Schmid I and Kazda M 2002 Root distribution of Norway spruce in monospecific and mixed stands on different soils. For. Ecol. Manage. 159, 37–47.

Schmidt K M, Roering J J, Stock J D, Dietrich W E, Montgomery D R and Schaub T 2001 The variability of root cohesion as an influence on shallow landslide susceptibility in the Oregon Coast Range. Can. Geotech. J. 38, 995–1024.

Schields F D and Gray D H 1992 Effects of woody vegetation on the structural integrity of sandy levees. Water Resour. Bull. 28(5), 917–931.

Schuurman J J and Goedewaagen M A J 1971 Methods for the Examination of Root Systems and Roots, in Methods in Use at the Insitute for Soil Fertility for Eco-morphological Root Investigations. 2nd edition. Centre for Agricultural Publishing and Documentation, Wageningen, The Netherlands, pp. 43–49.

Sidle R C 1992 A theoretical model of the effects of timber harvesting on slope stability. Water Resour. Res. 28, 1897– 1910.

Sidle R C, Pearce A J and O'Loughlin C L 1985 Hillslope stability and land use. AGU Water Resources Monography, 11, Washington D.C., USA.

Vinceti B, Paoletti E and Wolf U 1998 Analysis of soil, roots and mycorrhizae in a norway sdruce declining forest. Chemospere 36, 937–942.

Vogt K A and Persson H 1991 Measuring Growth and Development of Roots. Techniques and Approaches in Forest tree Ecophysiology. CRC Press, Boca Raton, FL.

Waldron L J 1977 The shear resistance of root-permeated homogeneous and stratified soil. Soil Sci. Soc. Am. J. 41, 843–849.

Watson A J and O'Loughlin C L 1990 Structural roots morphology and biomass of three age classes of Pinus radiate. NZ J. For. Sci. 20(1), 97–110.

Wu T H 1976 Investigation on landslides on Prince of Wales Island. Alaska Geotech. Rpt. No 5, Dpt. Of Civil Eng., Ohio State Univ., Columbus, USA, 1976.

Wu T H 1984a Effect of vegetation on slope stability. In Soil Reinforcement and Moisture Effects on Stability. Transportation Research Record 965, Transportation Research Board, Washington, USA, 37–46.

Wu T H 1984b Soil movement on permafrost slopes near Fairbanks, Alaska. Can. Geotech. J. 21, 699–709.

Wu T H 1995 Slope Stabilization. *In* RPC Morgan and RJ Rickson (eds), Slope Stabilization and Erosion Control-A Bioengineering Approach. pp. 221–264. Chapman and Hall, London.

Wu T H, McKinnel W P and Swanston D N 1979 Strength of tree roots on Prince of Wales Island, Alaska. Can. Geotech. J. 16(1), 19–33.

Wu T H and Sidle R C 1995 A distributed slope stability model for steep forested basins. Water Resour. Res. 31, 2097–2110.

Xu Y J, Röhrig E and Fölster H 1997 Reaction of root systems of grand fir (Abies grandis Lindl.) and Norway spruce (Picea abies Karst.) to seasonal waterlogging. For. Ecol. Manage. 93, 9–19.

Zhou Y, Watts D, Li Y and Cheng X 1998 A case study of effect of lateral roots of Pinus yunnanensis on shallow soil reinforcement. For. Ecol. Manage. 103, 107–120.

Ziemer R R 1981 Roots and stability of forested slopes. *In* TRH Davies and AJ Pearce (eds), Erosion and Sediment Transport in Pacific Rim Steeplands. pp. 343–361. International Association of Hydrological Sciences, Publication 132.

Biotechnical characteristics of root systems of typical Mediterranean species

Chiara Mattia[1], Gian Battista Bischetti[2,3] & Francesco Gentile[1]
[1]*Department PROGESA, Universita di Bari, via Amendola, n. 165/A, 70126, Bari, Italy,* [2]*Institute of Agricultural Hydraulics, Università degli Studi di Milano, via Celoria n. 2, 20133, Milano, Italy.* [3]*Corresponding author**

Key words: *Atriplex halimus*, biotechnical properties, hillslope stability, *Lygeum spartum*, *Pistacia lentiscus*, root systems

Abstract

Vegetation can significantly contribute to stabilise sloping terrain by adding cohesion to soil: this reinforcement depends on the morphological characteristics of the root systems and the tensile strength of single roots. The paper presents the results of research carried out in order to evaluate the biotechnical characteristics of the root system of three typical Mediterranean plant species which can affect slope stability. The species considered in the present study are *Lygeum spartum* L. (a perennial herbaceous monocotyledonous), *Atriplex halimus* L. and *Pistacia lentiscus* L. (two dicotyledonous shrub species). The plant specimens were collected in the Basilicata region (Southern Italy) by *in situ* excavation to obtain the whole root systems. Single root specimens for each species were sampled and tested for tensile strength measurement and the complete root systems were analysed to evaluate the root density distribution with depth in terms of root area ratio. The resulting data have been used to calculate the reinforcing effect in terms of increased shear strength of the soil using the model of Wu (1976, Investigation of landslides on Prince of Wales Island. Geotech. Eng. Rep. 5 Civil Eng. Dep. Ohio State Univ. Columbus, Ohio, USA) and Waldron (1977, Soil Sci. Soc. Am. J 41(3), 843–849), a simple and widespread model based on the reinforced earth theory. The results show that root reinforcement exerted by *L. spartum* is stronger than the reinforcement exerted by *P. lentiscus* and *A. halimus* in the upper layers of the soil, while *P. lentiscus* presents higher reinforcement values in deeper horizons. A. *halimus* presents lower values than either of the other species studied.

Introduction

Vegetation influences slope stability interacting with soil through hydrological and mechanical factors; mechanical factors originate from the action of root systems within the soil and result in the stabilisation of soil due to the anchorage of superficial layers to deep stable ones or into the bedrock. This reinforcement of the soil increases its shear strength and binds its particles.

The stabilising effect of vegetation is essential in preventing shallow landslides and in remediation works based on soil bioengineering techniques, which make use of vegetation as a building material (Schiechtl, 1980).

The effect of herbaceous and shrubby-arboreous associations in the control of water erosion and in slope stabilisation has been well known for centuries (e.g., restrictions on logging can be found in documents of the Republic of Venice, Italy, dating back to the 13th century). However, it was only in the second half of the last century that researchers began to quantify such effects (Greenway, 1987; Megahan and Kidd, 1972; Nilaweera, 1994; O'Loughlin, 1974; Wu, 1976), and only in the last few decades that quantification of the reinforcement due to plant root systems has been the object of particular studies (Burroughs and Thomas, 1977; Tsukamoto, 1987; Ziemer and Swanston, 1977) although the number of species studied remains fairly restricted, especially for the Mediterranean environment where the climatic conditions are not

* E-mail: bischetti@unimi.it

This article has been previously published in the following journal—Plant and Soil, Volume 278, 23–32.
Stokes et al. (eds), Eco- and Ground Bio-Engineering: The Use of Vegetation to Improve Slope Stability, 43–51.

favourable to plant growth (Gallotta et al., 2003; Schiechtl, 1980).

In this paper we analyse the characteristics of the root systems of three wild species typical of the Mediterranean region, in order to evaluate their contribution to slope stability in terms of increased soil shear strength by applying Wu's (1976) reinforcement model. This model is widely used in the evaluation of vegetated hillslope stability (Gray and Sotir, 1996; Hammond et al., 1992; Roering et al., 2003; Schmidt et al., 2001) and estimates additional cohesion due to root presence starting from root tensile strength and root cross-section per unit area of soil (RAR), two biotechnical characteristics of root systems (Gray and Sotir, 1996; Greenway, 1987; Schiechtl, 1980). Since such characteristics depend on species and on growing conditions, a high degree of variability is expected and a great deal of data collection is still needed in order to further our knowledge, especially for Mediterranean species (most of the studies carried out so far consider forest or grassland species). This paper aims to fill this gap.

Materials and methods

Reinforcement estimation

In non-rooted soil the shear strength is generally calculated by the Mohr–Coulomb equation:

$$s = c' + \sigma' \tan \Phi', \tag{1}$$

where s is the soil shear strength, c' is the soil cohesion, σ' is the effective normal stress on the shear plane and Φ' is the soil friction angle.

When the soil is permeated by fibres (synthetic or natural as in the case of roots) the displacement of soil, as a consequence of shear tension, generates friction between soil grains and fibre surfaces, causing the fibres to deform and to mobilise their tensile strength. In such a way, some of the shear tension can be transferred from soil to fibres, producing a reinforcement of the soil matrix itself.

If the soil is rooted, then the increased shear strength can be expressed as an additional cohesion:

$$s_r = s + c_r, \tag{2}$$

where s_r is the shear strength of soil reinforced by roots and c_r is the increase of shear strength due to the presence of roots (or root cohesion). Starting from concepts of reinforced earth, Wu (1976) and Waldron (1977) developed a simple theoretical model to predict the increase in shear strength due to the presence of roots. Assuming that roots are flexible, elastic and oriented perpendicularly to the slipping plane, when the soil layer moves and the roots within the shear zone bend, the tangential component of tensile strength directly counterbalances the shear force and the normal component increases the confining pressure. Assuming that the soil friction angle is not affected, the additional root cohesion can then be defined as

$$c_r = t_R(\sin\delta + \cos\delta \tan \Phi'), \tag{3}$$

where t_R is the average mobilised tensile strength of roots per unit area of soil and δ is the angle of root deformation in the shear zone.

Based on field observations and laboratory experiments, Wu et al. (1979) observed that for common values of δ and Φ', the term $(\sin\delta + \cos\delta \tan \Phi')$ varies between 1.0 and 1.3 and proposed a simplified form of Eq. (3):

$$c_r = 1.2 t_R \tag{4}$$

Even if all the above assumptions have not always been completely verified, field and laboratory direct shear tests confirmed the validity of the model (Waldron and Dakessian, 1981), which is commonly used to evaluate the contribution of roots to soil stabilisation (Abernethy and Rutherfurd, 2001; Roering et al., 2003; Schmidt et al., 2001; Sidle, 1992; Sidle et al., 1985; Wu and Sidle, 1995).

The mobilised roots tensile strength per unit area of soil (t_R) can be determined as the product of the average tensile strength of roots (T_R) and the fraction of the soil cross-section occupied by roots (A_R/A):

$$t_R = T_R(A_R/A) \tag{5}$$

The term A_R/A is called root area ratio (RAR) and it can be determined by counting roots, divided into size classes within a given soil, and by measuring their cross-section.

According to Gray and Sotir (1996) and several other authors, T_R varies with the root diameter (D) following a power law equation:

$$T_R = \alpha D^{\beta}, \tag{6}$$

where α and β are empirical constants depending on species; α can be considered as a scale factor, whereas β as the rate of strength decrease.

Study site

The area chosen to collect the plant roots used in tensile strength tests and root distribution lies in the eastern region of Basilicata, near Matera (Southern Italy) in the watershed of the Bradano River, inside the ÔFosso Scarciolla' creek basin. All of this area is in the Apennine or Bradanic Foretrough, and is characterised by the presence of Plio-Pleistocene clays and by peculiar geomorphological features such as badlands and gully erosion, which are the consequence of the prevailing formations involved in recent tectonic uplifts. The steep slopes are affected by typical badland erosion forms, while many areas are characterised by the presence of shallow translational slides. These slides involve the superficial layers of the slopes, in many cases less than 1 m in depth, where vegetation can exert a beneficial effect on stability through the reinforcing action of roots.

The climate is the typical Mediterranean semiarid climate, characterised by hot dry summers (although short periods of heavy rainfall may occur) and mild rainy winters; autumn and spring are generally very wet. The warmest month is August with a mean temperature of 24.9°C, while the coldest is January with a mean temperature of 6.3°C; the mean annual temperature is 15.1°C and the daily mean temperature is above 10°C for 8 months per year. The mean annual rainfall is 640 mm and the mean summer precipitation is 150 mm (the dry period lasts from May to September), both showing dramatic differences from 1 year to the next.

The native vegetation in the area consists of schlerophyllic evergreen plants, which are typical Mediterranean shrub associations (called 'macchia') and are able to survive both periods of drought and the difficulties of their habitat. They grow in autumn or winter whenever water is available, but are dormant in summer. This vegetation is included in the bioclimatic meso-Mediterranean zone and constitutes the climatic zone of *Oleo-Ceratonion*, with dominance of *Pistacia lentiscus* L., and with typical elements of *Quercion ilicis* class, such as *Rubia peregrina* L., *Asparagus acutifolius* L. and *Rosa sempervirens* L., while *Quercus ilex* L. is totally absent due to the presence of clay soils (Corbetta et al., 1992).

On southern slopes, grassland covers the soil near tilled areas and is constituted by *Lygeum spartum* associated with *Camphorosma monspeliaca* L., *Atriplex halimus* L. and *Polygonum tenoreanum* Nardi et Raffaelli.

Where water erosion is less evident and the soil is more stable there is a typical natural scrubland of *P. lentiscus* L. associated with *Phyllirea angustifolia* L., *Prunus spinosa* L., *R. peregrina* L., *Smilax aspera* L., *Lonicera etrusca* Santi, *A. acutifolius* L., *Rhamnus alaternus* L., *Juniperus oxycedrus* L. and also *A. halimus* L.

Species studied

Three different species, *L. spartum* L., *A. halimus* L. and *P. lentiscus* L., have been selected for the present study because of their widespread distribution in the Mediterranean environment.

Lygeum spartum is a monocotyledonous perennial grass belonging to the Graminaceae family, which is typical of clay soil; because of its resistance to drought, extreme temperature and salinity it is endemic in semi-arid environments. Its root system is thick, dense and fibrous and is composed of small-diameter roots with exclusively primary growth. Every year plants die in summer, while in autumn a new generation of buds with new roots develops from rhizomes, so at the end of the vegetative cycle the roots are at their longest. These monocotyledonous plants are anchored to the soil by the sum of all their thin roots.

Atriplex halimus and *P. lentiscus* are dicotyledonous shrubby species; the juvenile phase is characterised by a tap root and many lateral roots, whereas the lateral roots become relatively larger in the older phase (Köstler et al., 1968). *A. halimus* belongs to the Chenopodiaceae family and is a halophytic plant resistant to drought, extreme temperature and salinity. This plant is slow-growing and can reach a height of 2.0 m with well-developed foliage; it is used in the Mediterranean areas as forage. The root system of *A. halimus* consists of one main root with few branches; the tap root anchors the plant in a central position while the horizontal lateral roots act like guy ropes (Stokes, 1999). *P. lentiscus* belongs to the Anacardiaceae family, reaches a height of 6–8 m and develops a considerable volume of canopy.

Tensile strength tests

In order to obtain root specimens for tensile strength testing, plants of the selected species were dug out by hand using small tools down to a depth of about 60–70 cm below the soil surface. Plants were collected in the same period, at the end of spring, to avoid the effect of temporal changes in root reinforcement. The complete root system was obtained by gently removing the soil from the block of soil-roots by hand and washing it with jets of water. Finally the roots were left to dry in the open air for about an hour. Afterwards the roots were cut with sharp scissors and stored in airtight plastic boxes in a 15% alcoholic solution in order to prevent mould and microbial degradation (Böhm, 1979).

The roots were carefully inspected for possible damage before tensile strength tests and the root diameter was measured in three different positions along their length so as to obtain a representative value. The tensile strength tests were carried out according to the procedure described by Bischetti et al. (2003) using a device designed and built by the Institute of Agricultural Hydraulics of the University of Milan. The traction device consists of a rectified guide with a mobile bogie which is propelled by an electric motor (0.09 kW) with a speed reduction of 1:343 to ensure a constant linear speed of 10 mm $\min^{-1}$; tensile force is measured by interchangeable load cells (50 and 500 daN), while displacement is measured by a potentiometer transducer. The device has specifically developed non-serrated clamps to fix root ends; clamping of roots tips is actually a critical issue because damage to the root structure can determine a rupture of the root at the clamping points, thus affecting the measurement of tensile strength. Clamps consist of a cylinder with a groove in the centre to hold the roots, which are rolled up for three-quarters of their length and fastened by a semicircular plate screwed to the cylinder.

The tensile strength values T_R (MPa) were obtained as

$$T_R = F_{max}/\pi(D/2)^2, \tag{7}$$

where F_{max} is the maximum registered load (N) and D is the average root diameter (mm).

The inclusion or exclusion of root bark can affect evaluation of root strength from measurement of root breaking force (Bischetti et al., 2005). Nevertheless, in the present case the issue proven not to be worthy of note because the analysed roots actually had very thin bark (a few microns).

Estimation of root area ratio

In order to obtain RAR values, whole plant specimens were manually dug out in the field for collection; in the study area it was very difficult to apply the 'trench wall' method (Böhm, 1979) or 'core break' sampling (Schmid and Kazda, 2002) owing to the firmness of the clay soil and to the presence of many fine roots belonging to different plants which were indistinguishable from each other.

Collected plants were taken to the laboratory, where the whole root system was propped in the field position and architecture and spatial distribution were analysed by measuring the number of roots and their diameter at different depths (including the finest roots less than 1 mm in diameter).

The values of RAR distribution with depth in the soil were determined by counting roots for different size classes and by evaluating their cross-section with reference to horizontal planes located at different depths.

Results

Tensile strength

Twenty eight root specimens were analysed for *L. spartum*, 59 for *A. halimus* and 18 for *P. lentiscus*; in spite of the special clamping device, about 15% of root specimens were subject to anomalous rupture or slipping and the resulting data were discarded.

Lygeum spartum, as mentioned, is a monocotyledonous herbaceous plant and is characterised by a thick and fibrous root system with many thin elements. The range of the primary root diameter proved to be very limited (between 0.3 and 2.0 mm), while the secondary roots were not analysed because of their very small diameter (moreover the function of such small roots is mainly to support plant nutrition and their contribution to soil reinforcement can be considered negligible). The diameter of the roots analysed varies between 1.1 and 1.8 mm; the mean strength value is 37.8 MPa and the maximum recorded value is 58.3 MPa.

Atriplex halimus and *P. lentiscus* are dicotyledonous shrub-like plants characterised by a tap root system with one large, branching main root. Due to the characteristics of the clamping device, the maximum analysed root diameter of the two species was respectively 4.2 and 4.6 mm. The mean and the maximum strengths recorded for *A. halimus* are 57.2 and

Table 1. Mean values of root strength and root diameter, coefficient of correlation, number of samples and value of α and β of power law equation (6) of the three analysed species

Species	Mean value of root strength (MPa)	Mean value of root diameter (mm)	Coefficient of correlation (r)	Number of analysed samples	α	β
Lygeum spartum	37.8 (12.5)	1.5 (0.2)	−0.6*	28	60.7	−1.30
Atriplex halimus	57.2 (23.1)	1.9 (0.8)	−0.5*	59	73.0	−0.60
Pistacia lentiscus	55.0 (15.4)	3.4 (0.7)	−0.4 NS	18	91.2	−0.45

Numbers in parentheses represent the standard deviation of mean.
$^*P < 0.01$; NS = not significant.

116.9 MPa, whereas for *P. lentiscus* they are 55.0 and 98.3 MPa; some statistical properties of the tested roots are summarised in Table 1.

Results have also been elaborated to fit the power law of Eq. (6) and these are reported in Table 1 while the corresponding curves are represented in Figure 1.

Root distribution and root area ratio

Since *L. spartum* presents a dense and fibrous root system, whereas A. *halimus* and *P. lentiscus* present typical tap root systems with branching, as already stated (Figure 2), it is to be expected that this difference be reflected in root distribution with depth and RAR values.

Concerning RAR distribution in the soil, the resulting values for the species studied show a decrease with depth: *L. spartum* range between 0.09% at the surface layer to 0 at about 60 cm, *A. halimus* between about 0.055% to 0 at about 55 cm and *P. lentiscus* between about 0.060% to 0 at 75 cm. This decrease approximates a logarithmic law (Figure 3).

Soil reinforcement

By combining the strength-diameter relationships and the RAR distributions calculated for the three species considered, we obtained an estimate of the potential reinforcement due to vegetation using Eqs. (4) and (5) (Figure 4).

According to the RAR distribution, reinforcement decreases with depth. The strongest reinforcement effect is exerted by *L. spartum* which shows a shear strength increase of 60 kPa in upper layers and 0.3 kPa at 60 cm, where only a small number of roots are

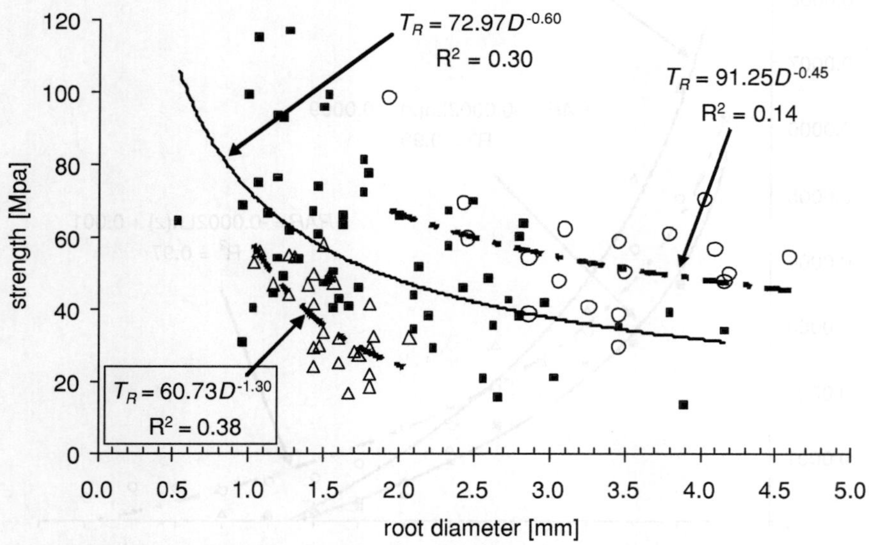

Figure 1. Relationship between root tensile strength (T_R) and root diameter (D) for *L. spartum* (triangles), *A. halimus* (squares) and *P. lentiscus* (circles). The curves reveal increasing tensile strength with decreasing root diameter. The mean value of tensile strength is 37.8 MPa for *L. spartum*, 57.2 MPa for *A. halimus* and 55.0 MPa for *P. lentiscus*.

Figure 2. Morphological differences between root systems of *L. spartum* (a), a monocotyledonous plant with a dense and smooth fibrous root system, *A. halimus* (b) and *P. lentiscus* (c), dicotyledonous plants with a tap root system consisting of a vertical main root with some branches.

present. On the contrary, *A. halimus* presents the weakest effect with values ranging approximately between 6 kPa in the upper layer and 0.2 kPa at 55 cm; *P. lentiscus* presents an intermediate effect with values varying approximately between 20 kPa in the upper layers and 3 kPa at 75 cm.

Discussion

The analysis of the root systems of the three species considered, *L. spartum, A. halimus* and *P. lentiscus*, has shown significant differences in their biotechnical characteristics.

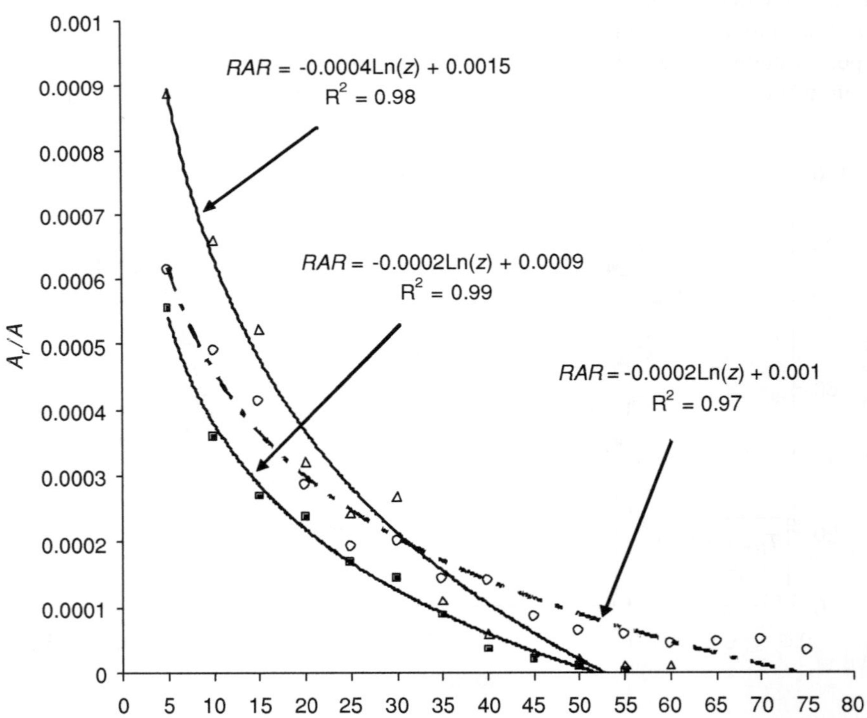

Figure 3. RAR with depth (z) of *L. spartum* (triangles), *A. halimus* (squares) and *P. lentiscus* (circles). The values of RAR were influenced by the morphological differences of the three root systems.

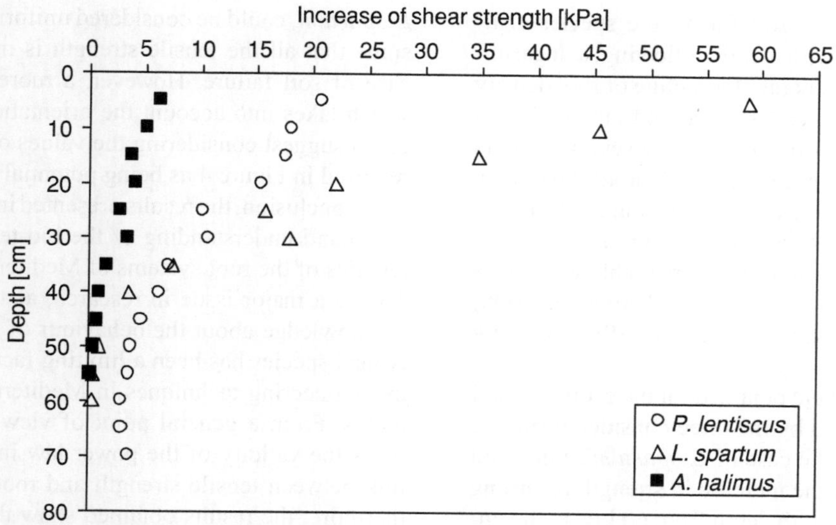

Figure 4. Shear strength increase of soil for the three plant species.

The mean tensile strength values for *A. halimus* and *P. lentiscus* (respectively 57.2 and 55.0 MPa; Table 1) are similar to those published by other authors for Alpine and Pre-alpine species (Bischetti et al., 2005; Greenway, 1987; Schiechtl, 1980) and for the Mediterranean environment (Gallotta et al. 2003). For *L. spartum* the mean tensile strength (37.8 MPa) is greater than for several herbaceous species tested by Cheng et al. (2003), greater than most of the herbaceous species reported by Schiechtl (1980) and comparable with the results of in situ shear tests carried out in rooted soil with perennial monocotyledonous grass by Tobias (1994).

The root tensile strength values for the species considered obtained by tests show that root strength decreases with diameter (Figure 1), as found by many other authors, following a power law equation (Bischetti et al., 2005; Burroughs and Thomas, 1977; Gray and Sotir, 1996; Nilaweera, 1994). The values of the parameters of the power law equation (6) α and β, obtained for the considered species, fall in the range already found for hardwood roots (between 29.1 and 87.0 for α and between –0.8 and –0.4 for β; Nilaweera, 1994) except for the decrease rate of *L. spartum* which is higher (Table 1). This behavioural difference may be ascribed to the different root anatomies of the monocotyledonous species and further study is being carried out.

From a statistical point of view the correlation coefficient is significant for *L. spartum* and *A. halimus* but not for *P. lentiscus*.

It is worth remembering how root architecture is influenced by genetic characteristics, edaphic conditions, external factors, land use management and finally by associated vegetation communities. In the study area, the clay soil is characterised by a high degree of compactness which increases with depth, and this reduces the spreading of roots down through the profile. For this reason, the roots of herbaceous species are generally confined to the faces between the large polyhedral clods and to relict channels previously occupied by annual vegetation. Root density is generally high in the top 0.5 m of the soil profile and decreases with depth; this decrease is abrupt owing to the presence of impervious layers. The limiting effect of compact clay layers on the propagation of roots is being confirmed by some preliminary data regarding the germination and growing of *L. spartum* and *A. halimus* plants in experimental plots (Mattia, 2003).

The root architecture of the *P. lentiscus* plants analysed is similar to that of *A. halimus*. It should be noted that it was necessary to extract whole root systems, this meant working with young plants whose foliage did not exceed 150 cm in height and whose root systems reached depths of 70–80 cm. Larger and older plants may have, deeper, more complex root system; some landslide scars present in the study area show how deeply penetrating these may be.

Concerning the RAR, as defined in Eq. (5), we observed a similar behavioural pattern for all the species considered: the maximum observed values are located in the upper 20 cm of soil for all the species analysed and decrease with depth following a logarithmic distribution. In general, the decline of root density with depth below the soil surface and also with distance from the stem is documented by several authors (Greenway, 1987; Nilaweera, 1994; Schmid and Kadza, 2001; Shields and Gray, 1993; Zhou et al., 1998).

The values of RAR concerning *L. spartum*, *A. halimus* and *P. lentiscus* are comparable with those determined for *Fagus sylvatica* (Bischetti et al., 2005) and other deciduous trees (Greenway, 1987; Schmidt et al., 2001).

Different results were obtained for the estimated soil reinforcement exerted by the three considered species via Eqs. (4)–(5). In the case of *L. spartum*'s thick and fibrous root system, the increase in strength regarding root cohesion varies with depth from 60 kPa in the upper layers of soil to 0 at 60 cm (Figure 4); the decrease in reinforcement with depth follows a logarithmic distribution according to RAR distribution.

P. lentiscus and *A. halimus* show a reinforcement effect which decreases with depth similarly to *L. spartum*, but with a lower rate of decrease. Maximum values are to be found in the upper layers, approximately 20 and 6 kPa, respectively, for *P. lentiscus* and *A. halimus*; the minimum observed values are approximately 3.5 kPa at 70 cm and 0 at 55 cm.

The estimated values of additional cohesion due to the presence of roots may appear high; in fact the model delineated by Eqs. (4)–(6) assumes that all the roots crossing the shear plane totally mobilise their tensile strength at rupture at the same time. Many authors agree with such a hypothesis (Roering et al., 2003; Schmidt et al., 2001), while some others showed that such a situation does not occur in reality because of the different orientation and tortuosity of roots and the possibility of uprooting (Hammond et al., 1992; Waldron and Dakessian, 1981).

We believe that the amount of potential root cohesion actually mobilised depends on the architecture of the root system considered (root size and branching). Hammond et al. (1992) considered tree vegetation which is characterised by a 'three-dimensional' branched root system with uneven sized roots and suggested a reduction factor of 0.56. In contrast, Waldron and Dakessian (1981) considered young plantations of barley with a shallow 'mono-dimensional' fine non-branched root system and uniform-sized roots and suggested a reduction factor of 0.83. As regards the plants in the present study, we think that their root systems, especially for *L. spartum*, could be considered uniform enough to assume that all the tensile strength is mobilised in the case of soil failure. However, a more realistic view, which takes into account the orientation of the roots, could suggest considering the values of root cohesion reported in Figure 4 as being potential values.

In conclusion, the results presented in the paper serve to expand understanding of the bio-technical characteristics of the root systems of Mediterranean species. This is a major issue in research, as the present lack of knowledge about the behaviour of root systems of typical species has been a limiting factor in using soil bioengineering techniques in Mediterranean environments. From a general point of view the study confirms the validity of the power law in expressing the link between tensile strength and root diameter. Furthermore, the results obtained show that the effect of *L. spartum*, *A. halimus* and *P. lentiscus* in terms of soil reinforcement are comparable with those of some trees and shrub species already studied in environments with more favourable climatic conditions (Bischetti et al., 2005; Greenway, 1987). This finding further encourages the use of Mediterranean vegetation in slope stabilisation practices.

Acknowledgement

This research was funded by the Italian Ministry of University and Research within the framework of the "Integrated analysis of soil erosion processes for the bioengineering rehabilitation of channel-slope systems" Program, Scientific Coordinator Prof. Salvatore Puglisi, University of Bari.

References

Abernethy B and Rutherfurd I D 2001 The distribution and strength of riparian tree roots in relation to riverbank reinforcement. Hydrol. Processes 15, 63–79.

Bischetti G B, Bonfanti F and Greppi M 2003 Root tensile strength measurement: testing device and protocol Quad. Idron. Mont. 21(1), 349–360 (in Italian).

Bischetti G B, Chiaradia E T, Simonato T, Speziali B, Vitali B, Vullo P and Zocco A 2005 Root strength and root area ratio of forest species in Lombardy (Northern Italy). Plant Soil 278, 11–22.

Böhm W 1979 Methods of Studying Root Systems. Ecological Series no. 33. Springer-Verlag, Berlin.

Burroughs E R and Thomas B R 1977 Declining root strength in Douglas-Fir after felling as a factor in slope stability. Res. Paper INT-190, Interm. Forest and Range Exp. Station, USFS, USA.

Cheng H, Yang X, Liu A, Fu H and Wan M 2003 A study on the performance and mechanism of soil-reinforcement by herb root system. In Proceedings of the Third International Vetiver Conference, Guangzhou, China, October 6–9 2003, pp. 384–390.

Corbetta F, Ubaldi D and Zanotti A L 1992 The *Lygeum spartum* vegetation in the badlands of Basento Valley (Basilicata Region). Arch. Bot. Ital. 67(3/4), 141–155 (in Italian).

Gallotta A, Gentile F, Puglisi S and Romano G 2003 Preliminary results of a research on mechanical resistance of root systems in hillslope stability in the Mediterranean environment. Quad. Idron. Mont. 21(1), 337–348 (in Italian).

Gray D H and Sotir R B 1996 Biotechnical and Soil Bioengineering Slope Stabilization: A Practical Guide for Erosion Control. Wiley, Chichester.

Greenway D R 1987 Vegetation and slope stability. *In* MG Anderson and KS Richards (eds.), Slope Stability Wiley, New York.

Hammond C, Hall D, Miller S and Swetik P 1992 Level I stability analysis (LISA) documentation for version 2.0, General Technical Report INT-285, USDA Forest Service Interm. Res. Station.

Köstler J N, Bruckner E and Bibelriether H 1968 Die Wurzeln der Waldbäume. Verl. Parey, Berlin (in Italian).

Mattia C 2003 Caratteristiche biotecniche degli apparati radicali di specie tipiche dell'ambiente mediterraneo. Ph.D. thesis, Agricultural University of Bari, Italy.

Megahan W F and Kidd W J 1972 Effect of logging and logging roads on erosion and sediment deposition from steep terrain. J. Forest. 70, 136–141.

Nilaweera N S 1994 Effects of tree roots on slope stability: the case of Khao Luang Mountain area, So Thailand. Dissertation. No. Gt-93-2.

O'Loughlin C L 1974 The effects of timber removal on the stability of forest soils. J. Hydrol. (N.Z.) 13, 121–134.

Roering J J, Schmidt K M, Stock J D, Dietrich W E and Montgomery D R 2003 Shallow landsliding, root reinforcement, and the spatial distribution of trees in the Oregon Coast Range. Can. Geotech. J. 40, 237–253.

Schiechtl H M 1980 Bioengineering for Land Reclamation and Conservation. University of Alberta Press, Edmonton, Alberta, Canada.

Schmid I and Kazda M 2001 Vertical distribution and radial growth of coarse roots in pure and mixed stands of Fagus sylvatica and Picea abies. Can. J. Forest Res. 31, 539–548.

Schmid I and Kazda M 2002 Root distribution of Norway spruce in monospecific and mixed stands on different soils. For. Ecol. Manage. 159, 37–47.

Schmidt K M, Roering J J, Stock J D, Dietrich W E, Montgomery D R and Schaub T 2001 The variability of root cohesion as an influence on shallow landslide susceptibility in the Oregon Coast Range. Can. Geotech. J. 38, 995–1024.

Shields F D and Gray D H 1993 Effects of woody vegetation on the structural integrity of sandy levees. Water Resour. Bull. 28(5), 917–931.

Sidle R C 1992 A theoretical model of the effects of timber harvesting on slope stability. Water Resour. Res. 28, 1897–1910.

Sidle R C, Pearce A J and O'Loughlin C L 1985 Hillslope Stability and Land Use. AGU Water Resources Monography, 11, Washington D.C., USA.

Stokes A 1999 Strain distribution during anchorage failure of Pinus pinaster Ait. at different age and tree growth response to wind-induced root movement. Plant Soil 217, 17–27.

Tobias S 1994 Shear Strength of the Soil Root Bond System. Proceedings of the International Conference on Vegetation and Slope: Stabilization, Protection and Ecology. Inst. Civil Engineers Uni. Museum Oxford, U.K.

Tsukamoto Y 1987 Evaluation of the effect of tree roots on slope stability. Bull. Exp. Forest Tokyo Univ. Agr. Tech. 23, 65–124.

Waldron L J 1977 The shear resistance of root permeated homogeneous stratified soil. Soil Sci. Soc. Am. J. 41(3), 843–849.

Waldron L J and Dakessian S 1981 Soil reinforcement by root: calculation of increased soil shear resistance from root properties. Soil Sci. 132, 427–435.

Wu T H 1976 Investigation of Landslides on Prince of Wales Island. Geotech. Eng. Rep. 5 Civil Eng. Dep. Ohio State Univ. Columbus, OH.

Wu T H, McKinell W P and Swanston D N 1979 Strength of tree roots, landslides on Prince of Wales Island, Alaska. Can. Geotech. J. 16(1), 19–33.

Wu T H and Sidle R C 1995 A distributed slope stability model for steep forested basins. Water Resour. Res. 31, 2097–2110.

Zhou Y, Watts D, Li Y and Cheng X 1998 A case study of effect of lateral roots of *Pinus yunnanensis* on shallow soil reinforcement. For. Ecol. Manage. 103, 107–120.

Ziemer R R and Swanston D N 1977 Root strength changes after logging in southeast Alaska. USDA Forest Serv. Res. Note PNW 306.

Uprooting resistance of vetiver grass (*Vetiveria zizanioides*)

S.B. Mickovski[1,3,4], L.P.H van Beek[2] and F. Salin[1]

[1]*LRBB, Domaine de l'Hermitage, 69, rte d'Arcachon, 33612, Cestas cedex, France.* [2]*IBED–Physical Geography, University of Amsterdam, Nieuwe Achtergracht 166, NL 1018, WV Amsterdam, The Netherlands.* [3]*Current address: Civil Engineering Department, University of Dundee, Dundee DD1 4HR, Scotland, UK.* [4]*Corresponding author**

Key words: pullout resistance, root system morphology, roots, uprooting, vetiver, *Vetiveria zizanioides*

Abstract

Vetiver grass (*Vetiveria zizanioides*), also known as *Chrysopogon zizanioides*, is a graminaceous plant native to tropical and subtropical India. The southern cultivar is sterile; it flowers but sets no seeds. It is a densely tufted, perennial grass that is considered sterile outside its natural habitat. It grows 0.5–1.5 m high, stiff stems in large clumps from a much branched root stock. The roots of vetiver grass are fibrous and reported to reach depths up to 3 m thus being able to stabilise the soil and its use for this purpose is promoted by the World Bank. Uprooting tests were carried out on vetiver grass in Spain in order to ascertain the resistance the root system can provide when torrential runoffs and sediments are trying to uproot the plant. Uprooting resistance of each plant was correlated to the shoot and root morphological characteristics. In order to investigate any differences between root morphology of vetiver grass in its native habitat reported in the literature, and the one planted in a sub-humid environment in Spain, excavation techniques were used to show root distribution in the soil. Results show that vetiver grass possesses the root strength to withstand torrential runoff. Planted in rows along the contours, it may act as a barrier to the movement of both water and soil. However, the establishment of the vetiver lags behind the reported rates in its native tropical environment due to adverse climatic conditions in the Mediterranean. This arrested development is the main limitation to the use of vetiver in these environments although its root strength is more than suffcient.

Introduction

Vetiver grass (*Vetiveria zizanioides*), also known as *Chrysopogon zizanioides*, is a graminaceous plant native to tropical and subtropical India. The southern cultivar is sterile; it flowers but sets no seeds. It is a densely tufted, perennial grass that is considered sterile outside its natural habitat. It is reported that vetiver grows 0.5–1.5 m high, stiff stems in large clumps from a much branched root stock (Erskine, 1992; Truong, 1999). The use of vetiver grass hedges against soil erosion increased following several key papers promoting vetiver grass planting as an effective and inexpensive erosion protection measure and the publication of World Bank's manual in 1990 (for a review see Grimshaw, 1989). Vetiver grass has wider applications due to its unique morphological, physiological and ecological characteristics that highlight its adaptability to a wide range of environmental and soil conditions. Currently used in more than 120 countries, vetiver grass applications include soil and water conservation systems in agricultural environment, slope stabilisation, rehabilitation of mines, contaminated soil and saline land, as well as wastewater treatment (Truong and Loch, 2004). In addition, vetiver has added commercial value as its roots yield aromatic compounds that are applied for domestic and cosmetic use. However, there is an argument that when the plant is harvested for this purpose it may actually increase erodibility because the process loosens the soil (van Noordwijk et al. 2000).

The most impressive characteristic of the vetiver grass is its root system that consists of fibrous roots

* E-mail: s.b.mickovski @dundee.ac.uk

This article has been previously published in the following journal—Plant and Soil, Volume 278, 33–41.
Stokes et al. (eds), Eco- and Ground Bio-Engineering: The Use of Vegetation to Improve Slope Stability, 53–60.

reported to reach depths up to 3 m (Erskine, 1992; Hellin and Haigh, 2002). Such roots extend deep enough in the soil to provide the grip and anchorage needed to prevent surficial slip in the event of heavy prolonged rainstorm (Hengchaovanich, 1999). This is the major reason why the use of vetiver grass for slope protection is promoted by the World Bank (1990) and The Vetiver Network (Paul Truong, personal communication, www.vetiver.org).

Planted in rows along slope contours, vetiver is able to quickly form a narrow but very dense hedge. Reported to tolerate adverse growing conditions (e.g. winters with ground temperatures as low as −14°C) (Truong, 1999), its stiff foliage is able to block the passage of soil and debris in cases of torrential rains (Dalton et al., 1996; Hengchaovanich 1999), in the same time allowing the trapped sediment to form a terrace upslope the hedge. Vetiver hedge is also able to slow down any surface runoff which, in turn, gives the rainfall a better chance of percolating into the soil instead of running off downslope and potentially creating rills and gullies, in the same time contributing to the increased yield of crops planted on the slope (Truong and Loch, 2004). If the sediment is not removed vetiver will continue to grow up and adjust itself in tandem with it on the newly formed terrace (Hengchaovanich, 1999; Truong, 1999) which, rises as the soil accumulates behind the hedges, thus converting highly erodible slopes into relatively more stable terraces able to support sustainable agriculture or even forestry (Meyer et al., 1995). Being a low cost, natural and environmentally friendly method for erosion control (Truong, 1999), the effciency of such contour hedges for soil and water conservation have been studied *inter alia* by Mishra et al. (1997) and Hellin and Haig (2002).

The versatility of vetiver has led to its application outside its original zones of provenance. Currently it is successfully used in Africa, Asia, Central and South America, southern Europe and Australia for stabilisation of steep batters of roads and railway embankments. For example, in China in the last 5 years it has been used for erosion and sediment control on more than 150 000 km of embankments (Truong and Loch, 2004). In principle, it would be possible to apply it also in the European Mediterranean basin although the soil and climatic conditions are harsh. Therefore, a modest field trial was set up in the Alcoy region (Spain) to evaluate its performance within the framework of the EcoSlopes project. Two plots on the riser of a cultivated bench terrace were planted with vetiver and compared to similar plots under different treatments: Spanish cane (*Arrundo donax*), natural cover and regrowth after complete stripping. In the area bench terrace risers are left unarmored and are subject to erosion and failure despite their cover with natural vegetation (mainly *Brachypodium* sp.)

Previous studies have reported on the growth and the use of vetiver grass in its natural environment (Erskine, 1992; Hellin and Haigh, 2002; Hengchaovanich, 1999, Salam et al., 1993; Truong and Loch, 2004) but the properties of vetiver root systems have not been investigated in European context. In this study the mechanical properties of vetiver roots and their architecture are studied in order to evaluate its capacity to withstand torrential rain, ponding and sediment pressure (Cheng et al., 2003; Hengchaovanich, 1999), as well as its potential for application in eco-engineering.

Materials and methods

Site characteristics

Experimental plots of vetiver grass were planted on a site near Almudaina, Spain ($X = 729275$; $Y = 4293850$ and $Z = 480$ m on UTM 30s) in the spring of 2002. The vetiver was planted on the riser of a bench terrace (Figure 1) which parts are potentially endangered by runoff and soil slippage after intense rainfall events. The local gradients on the riser ranged between

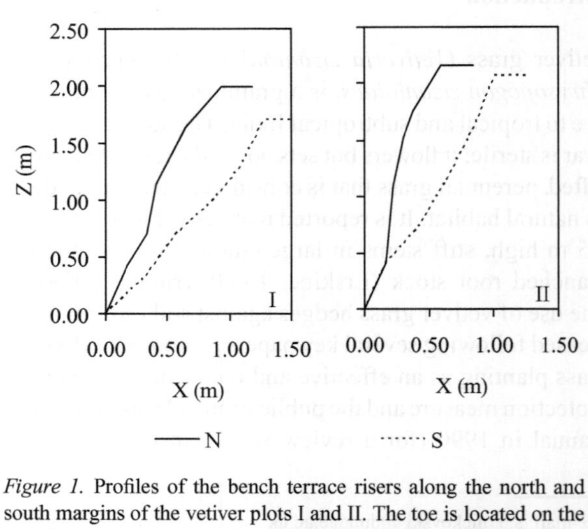

Figure 1. Profiles of the bench terrace risers along the north and south margins of the vetiver plots I and II. The toe is located on the abandoned terrace, the horizontal crest on the cultivated terrace.

35° and 60° while a nursery was established on the bench terrace. Cuttings of vetiver were planted in rows on the riser with a spacing of 10–15 cm. Rows were placed at the crest, bottom and middle of the riser that was between 1.75 and 2.25 m high (Figure 1). The vertical interval of the vetiver rows was approximately 40 cm and their length 3 m each.

The soil on the site derives from Miocene marl. The marl have a high clay content, predominantly smectites, but, due to a carbonate content of 60 per cent or more, most of the particles fall in the silt fraction. The dry bulk of the topsoil 14.6 kN m^{-3} and the porosity 0.413 m^3 m^{-3}.The soil shear strength was determined in the laboratory by means of strain-controlled, consolidated-drained direct shear tests on saturated samples (BS 1377). Sample size was 60 × 60 × 20 mm^3 and the applied strain rate 0.2 mm h^{-1}. Because of the dominance of the silt fraction, the soil has a high angle of internal friction of 34° and a cohesion of a mere 4.8 kPa ($N = 30$). These strengths have been confirmed by two *in situ* consolidated-drained direct shear tests on pristine soil with field capacity of saturation with dimensions of 32 × 32 cm in plan and 20-cm high, for which no substantial root reinforcement was found. In comparison, four tests on soil rooted with vetiver yielded a significant root reinforcement in the order of 2.7 kPa (ranging between 2.1 and 3.7 kPa) when the shearing resistance derived from the laboratory tests was subtracted.

The climate at the site is continental and Mediterranean. It shows a strong seasonality in rainfall and temperature. Most rainfall occurs in the late autumn and winter and to a lesser extent in early spring. The total annual rainfall amounts to 700 mm per year but the rainfall has a strong inter-annual variability with annual totals varying between 350 and 1050 mm. Moreover, rainfall is erratic and exceptional events occur throughout the wet season: a 24 h total of 284 mm and an event total of 553 mm have been recorded at Almudaina (van Beek, 2002).

The mean annual temperature is 16°C ranging between a mean monthly temperature of 24°C in summer and 7°C in winter. In winter, the variability in the temperature and its diurnal course are the largest with night frost occurring regularly between end December and April (van Beek, 2002). The climatic conditions at the site fall within the tolerances of vetiver (World Bank, 1990) and the conditions over the growing period of the vetiver did not, on average, deviate from them. However, the summer of 2001 was characterised by a prolonged drought that was terminated by a 90-mm storm in August. Drip irrigation was applied over this period to enable the plants to establish themselves. February 2003 experienced exceptional snowfall which cover persisted for several days. In April 2003, a 146 mm event occurred in 24 h which induced some small slips on slopes and risers. At the test site damage was restricted to one plot planted with vetiver through which the overland flow of the overlying bench terrace was routed.

Since most erosion and slippage occur in the late autumn, the investigation of the uprooting resistance of vetiver grass was carried out in November 2003 when the ambient moisture conditions were close to field capacity (observed volumetric moisture content ranged between 0.25 and 0.35 m^3 m^{-3}). At that time, the plants were well established and have proliferated multiple stems from the cuttings planted in 2002.

Preliminary tests

In order to investigate the morphological characteristics of vetiver roots, four plants were completely excavated using the block excavation method (van Noordwijk et al., 2000) (Figure 2). These plants were randomly selected from the plot, the soil surface in a radius 30 cm around each plant was carefully cleared from the litter, and the soil block with dimensions 0.3 m × 0.3 m and 0.5 m deep, containing their roots was manually excavated using a spade. Excavated

Figure 2. Morphological characteristics of a semi-excavated vetiver plant. Stiff stems grow upwards from the plant base, while primary lateral roots grow vertically down the soil. Primary lateral roots often branch into second/third, etc. order lateral roots.

plants were then transferred to the in situ root washing facilities where, to minimise root loss or damage, the plants together with their root systems were hand washed gently from the remains of the soil. Root systems were then sprinkled under a low water flow from a sprinkler. For separating the last remnants of soil on the roots, it was necessary to soak the root systems in water basins and remove the soil by gently agitating the sample after what the soil particles settled on the bottom of the basin, and the broken roots, if any, floated on the surface. After the root systems were thoroughly cleaned from the soil, they were placed on a paper mat and left to dry in the open air for half an hour and the maximum lateral spread, and maximum rooting depth was measured for each plant. All the primary lateral roots (Figure 2) were then carefully cut off from the plant base with scissors and the number of roots recorded; root diameter at its base and near its tip (d_i) was noted together with the length of the primary lateral root. The root cross-sectional area was calculated as an area of a circle with radius d_i. Observing the strong geotropical tendency in the rooting pattern of vetiver, it was assumed that all of the roots grow more or less vertically downwards and the length of each root was assumed to represent the maximum rooting depth reached by the root itself.

Pullout resistance of vetiver grass

In order to investigate the pullout resistance of vetiver grass, 22 plants were randomly chosen from the plantation and were used as a test sample. Before each pullout test the soil surface in a radius 30 cm around the plant was carefully cleared from the litter, exposing the stem base. A strong PVC rope (3-mm diameter) padded with soft tissue in order not to destroy the plant material was then tied around the stem base of the plant. The other end of the rope was connected to a hand-held portable force gauge (Alluris FMI-100) for accurate measurement of uprooting force. In order to mimic the forces applied to the plant during runoff and sediment impoundment, the pullout force was applied parallel to the slope in downslope direction. The force was applied manually with a rate of 10 mm min^{-1}, recording the change in resistance along the way. The test was terminated once the resisting force dropped sharply and the plant was uprooted. Each plant was then carefully excavated, its roots washed from the soil remnants, and left on a paper mat to air dry for an hour.

Plant morphological analysis

Aboveground characteristics such as the plant height and the average diameter at the base (Figure 2) were measured with measuring tape and callipers for each of the 22 uprooted plants. The number of stems growing from each stem base was also recorded.

Similarly as in the preliminary tests, the number of roots was recorded for each tested plant and root system characteristics including the root length, root system lateral spread and depth were measured with measuring tape (Böhm, 1979). The diameter of each root close to the stem base was measured with callipers.

Root systems were then separated from the stems using scalpel and sharp blade and placed in an oven to dry over 24 h at 70°C , after which the root:shoot ratio of each uprooted plant was calculated as the weight of dry root mass over the dry weight of shoots (Böhm, 1979).

Statistical analysis

The results of the pullout tests were analysed using the statistical package SPSS 10.0 (SPSS Inc, Chicago). A bivariate correlation analysis with Pearson's coeffcient was performed in order to investigate any underlying relation between the uprooting force for each plant and the stem and root parameters measured during the investigation (Zar, 1998). A two-tailed test of significance was used to identify the statistically significant correlations.

Results

Preliminary tests

The tests to describe the overall morphological characteristics of vetiver grass worked well in this specific plantation. Vetiver roots were shown to originate from the base of the plant that had between 8 and 10 stems on average (Figure 2). The roots were numerous, pale yellow in colour and strongly geotropic. Having diameters at the base of the plant in the range between 0.3 and 1.2 mm, the roots did not visibly taper and branched to second and third order laterals of decreasing diameters. None of the roots of the test plants had a lateral spread larger than 0.25 m from the base of the plant, nor did the depth of the excavated plants reach more than 0.3 m. These parameters justified the chosen size of

Table 1. Morphological characteristics of 22 investigated vetiver plants

Morphological characteristic	Range	Mean	Standard error
Plant height [m]	0.74–1.08	0.925	0.035
Number of stems per plant	4–23	12.5	1.25
Plant diameter at base [m]	0.030–0.092	0.062	0.005
Maximum rooting depth [m]	0.110–0.275	0.219	0.018
Lateral root spread [m]	0.151–0.292	0.229	0.015
Root diameter at base [mm]	0.30–1.45	1.02	0.04
Dry root mass [g]	4.40–37.8	22.96	3.33
Dry shoot mass [g]	36.40–114.20	70.05	8.04
Root: shoot ratio	0.121–0.636	0.353	0.059

the excavation block that provided that no mechanical damage is incurred to the root systems.

Plant morphology

Morphological characteristics of investigated vetiver plants are given in Table 1.

Pullout resistance

The plant pullout method described in the Materials and methods section was suited to the objectives of the investigation, and 19 out of 22 plants could be uprooted using this method. The other three plants were not uprooted because of the rope failure or a snap through the plant stem.

The investigated vetiver plants did not show any movement in the first several force/displacement increments. With the increase of the force applied, the plants started to rotate around a point close to the downslope end of the stem base but under the soil surface, while the upslope lateral roots were activated in tension and provided most of the resistance for the plant. In the later stages, sporadic sounds of root snapping were heard just before the plant was uprooted.

The plant pullout data showed that the plant resisting force increased with displacement until it reached the peak and then gradually started to decrease as the roots started to break or slip from the soil. A typical pullout force–displacement curve is shown on Figure 3. The slopes of the increase in the force–displacement curve to the maximum load ranged from 0.29 to 9.33, or on average 2.50 ± 0.36 (throughout this paper: mean ± SE). The maximum uprooting force ranged from 190 to 620 N, or on average 466.97 ± 31.25 N for the investigated plants.

The summary of the correlation analysis between the pullout resistance of the investigated plants and the other morphological characteristics measured during the investigation is shown in Table 2. Positive correlations were found between the uprooting force and all other measured parameters. However, only the correlation between the uprooting force and the plant height, and between the uprooting force and lateral root spread was statistically significant ($P < 0.05$).

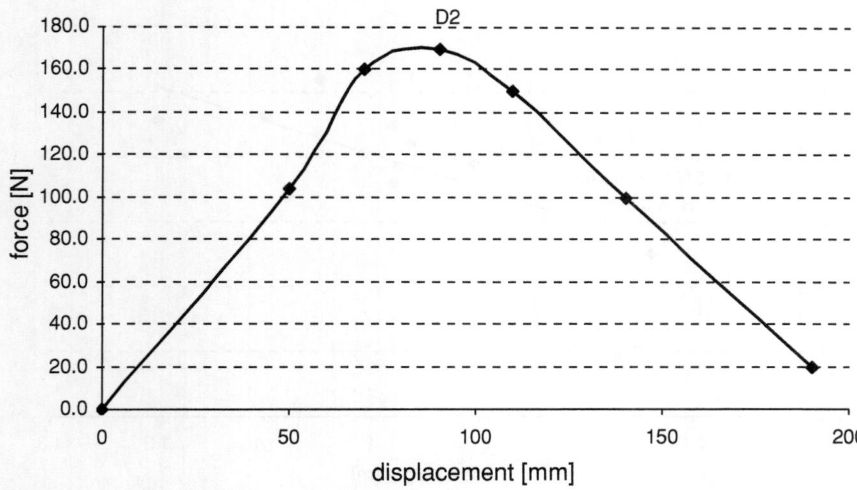

Figure 3. Typical force–displacement curve for a pullout test on vetiver grass (sample D2). The uprooting force increased with displacement to its peak value and then started to decrease due to root slippage or breakage.

Table 2. Correlation between the force necessary to uproot the plant and other morphological factors. Analysis based on $n = 22$ plants

Factor	Correlation coefficient R^2	Significance	Factor	Correlation coefficient R^2	Significance
Plant height	0.598	0.019*	Lateral root spread	0.517	0.048*
Number of stems	0.218	0.435	Root:shoot ratio	0.013	0.768
Plant diameter at base	0.130	0.644	Total CSA	0.236	0.397
Maximum rooting depth	0.201	0.472	Number of primary lateral roots	0.246	0.377
Average root length	0.378	0.165	Average root diameter	0.07	0.804

*Significant at a 0.05 level.

Figure 4 shows the dependency of the uprooting force on the plant height for the investigated plants. Taller plants show higher uprooting resistance and require higher pullout forces. Figure 5 shows the relationship between the maximum uprooting force and the maximum lateral root spread. Vetiver plants with root system that spread wider are able to resist uprooting better than the plants with root systems that do not reach far from the stem base.

Discussion

The morphology of the investigated plants did not confirm findings on the plant morphology of vetiver grass in earlier studies. While the plant height reached almost the average reported for the vetiver in its natural environment (Erskine, 1992; Hengchaovanich, 1999; Mishra et al., 1997), neither the number nor the length of the roots reached the values reported in earlier studies (Mishra et al., 1997; Salam et al., 1993; Truong, 1999). Possible causes for the 'underdevelopment' of the root system might be the soil type and the severity of climatic conditions over the growth period (Paul Truong, personal communication). The topsoil is more structured and stores most water and nutrients available to the plant as the underlying marl has a dense structure. Moreover, during the short growth season under the Mediterranean climate the soil dries out and becomes more hard. While the local availability of water due to drip irrigation also prevents root expansion. Hence, the full root development, particularly in length and abundance observed in deeper, drainable soils such as present in the Tropics have not been achieved at the time of testing.

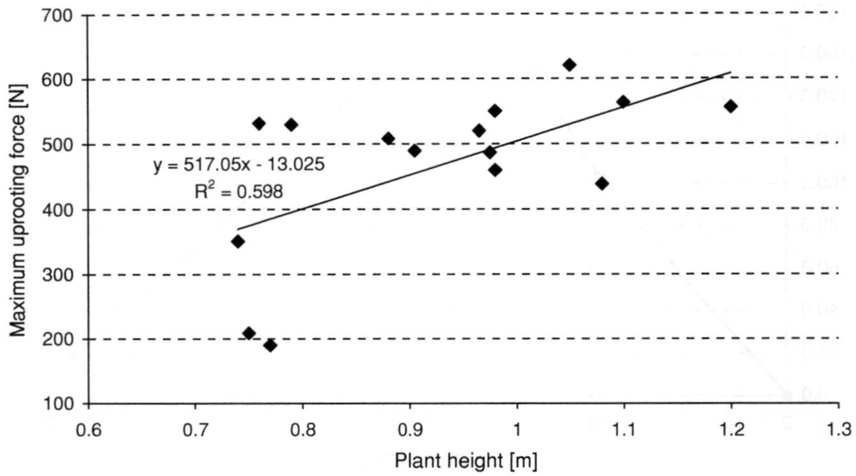

Figure 4. The relation between the maximum uprooting force and the plant height in the investigated vetiver grass (*Vetiveria zizanioides*) plants. Taller plants resist uprooting forces better than shorter plants.

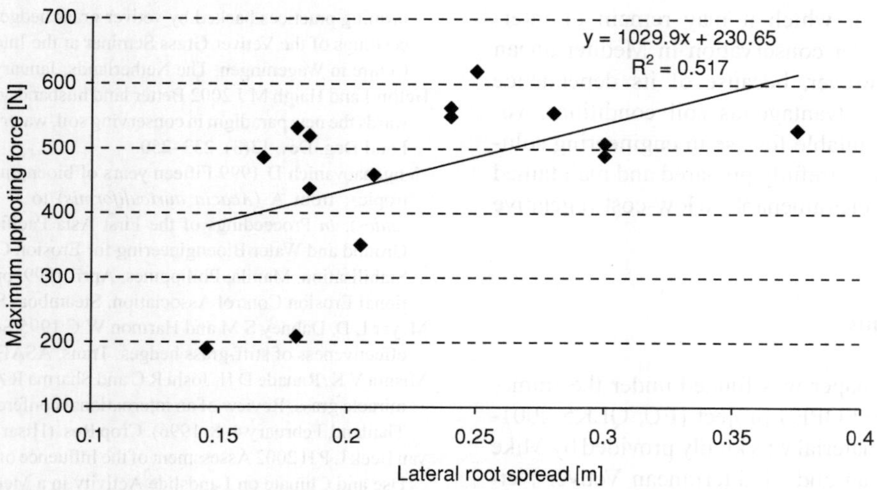

Figure 5. **The relation between the maximum uprooting force and the maximum lateral spread of root systems in investigated vetiver grass (*Vetiveria zizanioides*) plants. Plants with wider-spreading roots resist uprooting forces better than the plants with roots systems that do not reach far from the stem base.**

In Mediterranean environments therefore, where soils are shallow and water is scarce over the growing season, it would be more economical for plants to have the roots closer to the soil surface. This would explain the biased investment in aboveground biomass of vetiver observed at this site when water is available. This leads to successful growth as long as irrigation is applied over the growing season but when dependent on natural rainfall that often falls outside the growing season of vetiver, it loses the competition to endemic species that are better adapted.

Despite the poor root development, correlation and regression analysis showed that taller plants will resist uprooting better than the shorter ones, which was to be expected given the relatively constant root:shoot ratio. Plants that invest more in their above ground parts would also invest more in the proliferation of their root systems. Furthermore, the increase in uprooting resistance of plants that have root systems with extensive lateral spread can be explained by the fact that larger lateral spread also means larger anchoring length of the lateral roots. Bearing in mind that the lateral roots, especially the upslope ones (opposite of the side where the uprooting force was applied) provided most of the resistance for the plant resisting in tension, it is clear that larger anchorage length will provide better friction on the root–soil contact thus increasing the overall resistance of the root to pullout (Cheng et al., 2003). The differences in lateral root spread can be explained in terms of local differences in water and nutrient availability.

Even the limited root systems of the investigated vetiver grass proved able to withstand relatively high uprooting forces acting downslope. This high resistance shows that in a case of torrential rains and suspended runoff it can block the runoff and trap sediment behind the hedge. This function was tested during the extreme rainfall event that occurred in April 2004. One vetiver plot withstood the rain and hardly any sediment was collected. The other vetiver plot, however, received most of the overland flow generated on the overlying terrace and the riser failed as a slump with the slip plane at 30–40 cm, below the roots of the established vetiver.

The investigation of *Vetiver zizanioides* planted for soil and water conservation on a bench terrace riser in Spain showed that soil depth, water availability and to a lesser extent temperature, adversely influence root development in Mediterranean environments. Competition between native vegetation and vetiver highlights the poor adaptation of the vetiver, and shows that the dense, deep and columnar root systems cannot develop to the same extent as under its native tropical and subtropical environment (Figure 6). Rooting depth is therefore the crucial factor for the performance of vetiver on steep slopes in Mediterranean environments as the event in April 2004 showed. Still, the uprooting force of the vetiver is high and sufficient to withstand the water and sediment loads that would apply during torrential runoff for which it may remain of interest for soil and water conservation in Mediterranean environments. However, because of its dependence on irrigation and advantageous soil conditions, vetiver seems more suitable for use in engineering solutions when sites are carefully prepared and maintained rather than as a species amenable to low-cost vegetative solutions.

Acknowledgements

The work on this paper was funded under the framework of the ECOSLOPES project (EU, QLK5-2001-00289). Planting material was kindly provided by Mike Pease, the European and Mediterranean Vetiver Network Coordinator.

References

Böhm W 1979 Methods of Studying Root Systems. Springer Verlag, Berlin.

British Standard 1377 1990 Methods of Test for Soils for Civil Engineering Purposes. British Standards Institution, London.

Cheng H, Yang X, Liu A, Fu H and Wan M 2003 A study on the performance and mechanism of soil reinforcement by herb root system. *In* Proceedings of the 3rd International Vetiver Conference., Guangzhou, China, 6–9 October 2003. China Agriculture Press, Beijing, China, pp. 403–410.

Dalton P A, Smith R J and Truong P N V 1996 Vetiver grass hedges for erosion control on a cropped flood plain: hedge hydraulics. Agr. Water Manage 31(1996), 91–104.

Erskine J M 1992 Vetiver grass: its potential use in soil and moisture conservation in Southern Africa. S. Afr. J. Sci. 88(6), 298–299.

Grimshaw R G 1989 A review of existing soil conservation technologies, and a proposed method of soil conservation using contour farming practices backed by vetiver grass hedge barriers. *In* Proceedings of the Vetiver Grass Seminar at the International Agric. Centre in Wageningen, The Netherlands, January, 1989.

Hellin J and Haigh M J 2002 Better land husbandry in Honduras: towards the new paradigm in conserving soil, water and productivity. Land Deg. Dev. 13(3), 233–250.

Hengchaovanich D 1999 Fifteen years of bioengineering in the wet tropics: from A (*Acacia auriculiformis*) to V (*Vetiveria zizanioides*). *In* Proceedings of the First Asia Pacific Conference on Ground and Water Bioengineering for Erosion Control and Slope Stabilisation, Manila, Philippines, April 1999, pp 54–63. International Erosion Control Association, Steamboat Springs, USA.

Meyer L D, Dabney S M and Harmon W C 1995 Sediment-trapping effectiveness of stiff-grass hedges. Trans. ASAE 38(3), 809–815.

Mishra V K, Ranade D H, Joshi R C and Sharma R A 1997 Vetiver: A miracle grass. Review of an international conference (Chiang Rai, Thailand, February 4–8, 1996). Crop Res. (Hisar) 13(2), 507–509.

van Beek L P H 2002 Assessment of the Influence of Changes in Land Use and Climate on Landslide Activity in a Mediterranean Environment Netherlands Geographical Studies 294. KNAG, Utrecht, 363 pp.

van Noordwijk M, Brouwer G, Meijbroom F, Oliveira M R G and Bengough A G 2000 Trench profile techniques and core break methods. In AL Smit (ed), Root Methods, A Handbook. pp. 211–233. Springer, Berlin.

Salam M, Abdul Shehana R S, Mammen M K and Kunjamma P M 1993 Rooting pattern of vetiver (*Vetiveria zizanioides* (L.) Nash) in an oxisol. Trop. Agr. 70(3), 274–275.

Truong P N, 1999 Vetiver grass technology for land stabilisation, erosion and sediment control in the Asia Pacific region. *In* Proceedings of the First Asia Pacific Conference on Ground and Water Bioengineering for Erosion Control and Slope Stabilisation, Manila, Philippines, April 1999, pp. 72–84. International Erosion Control Association, Steamboat Springs, USA.

Truong P and Loch R, 2004 Vetiver system for erosion and sediment control. *In* Proceedings of the 13th International Soil Conservation Organisation Conference, Brisbane, Australia, July 2004. pp. 1–6.

World Bank 1990 Vetiver Grass—The Hedge Against Arosion, 3rd ed. Washington D.C.

Zar J H 1998 Biostatistical Analysis, 4th edition, 929 pp. Prentice Hall, New York.

Root reinforcement by hawthorn and oak roots on a highway cut-slope in Southern England

Joanne E. Norris
Nottingham Trent University, Burton Street, Nottingham, NG1 4BU, England

Key words: pull-out resistance, root morphology, root tensile strength, soil reinforcement, soil–root interaction

Abstract

Highway embankments and cutting slopes in the United Kingdom, particularly in the South East of England, are often constructed of or within stiff over-consolidated clays. These clays are prone to softening with time leading to shallow slope failures and costly repairs. Reinforcement by natural vegetation is potentially a cost-effective method of stabilising these types of slopes over the medium–long term. However, there is a lack of information on how natural vegetation reinforces and stabilises clay slopes. To investigate this problem, the potential reinforcement of selected oak (*Quercus robur* L.) and hawthorn (*Crataegus monogyna* Jacq.) roots was assessed by conducting *in situ* root pull-out experiments on a London Clay cutting in south-east England. Pull-out tests were carried out using specifically designed clamps and either a hand pull system with a spring balance and manual recording of force for oak roots or a jacking system with electronic data logging of applied force and displacement for hawthorn roots. Oak roots had a mean pull-out resistance of 7 MPa and that of hawthorn roots was 8 MPa. The electronic data logging of applied force (pull-out resistance) and displacement of the hawthorn roots provided additional data on the failure of branched roots which could be correlated with variations in root morphology. The failure of the roots can be categorised into three modes: Type A: single root failure with rapid rise in pull-out resistance until failure occurs; Type B: double peak failure of a forked or branched root and Type C: stepped failure with multiple branches failing successively. The different types of root–soil bonds are described in relation to root anchorage and soil stability.

Introduction

Many of the highway embankments and cutting slopes in the United Kingdom, particularly in the South East of England, are constructed of or within stiff over-consolidated clays which are prone to softening with time leading to shallow slope failures at depths of 1–1.5 m (Greenwood et al., 1985; Perry et al., 2003a,b). These slopes are usually seeded with grasses or planted with selected shrubs and trees in accordance with locally agreed landscaping criteria and the Highways Agency advice notes (Highways Agency, 2003, 2004). Over time, these slopes become self seeded and natural regeneration starts to take place. It is the mid-long term stability of these slopes that is critical but very little knowledge exists on how this combination of seeded grass, planted shrubs and natural vegetation are contributing to the stabilisation of these over-consolidated clay slopes.

The potential benefits of using vegetation for highway slope reinforcement (bioengineering) has been considered in recent years (e.g., Barker et al., 2004; Coppin and Richards, 1990; Gray and Sotir, 1996; Greenwood et al., 2001; MacNeil et al., 2001). However, the quantification of *in situ* root reinforcement involves a detailed appreciation of root growth, development and decay with time, the roots' interaction with the soil and the seasonal effects on the geotechnical parameters which are relevant to slope stability.

* Email: joanne.norris@ntu.ac.uk

This article has been previously published in the following journal—Plant and Soil, Volume 278, 43–53.
Stokes et al. (eds), Eco- and Ground Bio-Engineering: The Use of Vegetation to Improve Slope Stability, 61–71.

In situ root strength can be determined by conducting *in situ* root–soil tests. *In situ* root–soil strength can be obtained by carrying out *in situ* shear box tests (Endo and Tsuruta, 1969; Norris and Greenwood, 2003a,b; O'Loughlin, 1981; Van Beek et al., 2005; Waldron and Dakessian, 1981; Wu et al., 1988), whereas *in situ* root strength can be determined by *in situ* root pullout tests (e.g., Operstein and Frydman, 2000). The tensile strength or root pull-out strength is valuable information when assessing the stability of a slope and can be included in limit equilibrium stability analysis (Greenwood, 2005; Greenwood et al., 2004).

Assessing the pull-out resistance for quantification of root reinforcement and for slope stability analysis has received little attention, whereas the pull-out resistance of roots or whole plants for resistance to lodging by the wind (Bailey et al., 2002; Ennos, 1990, 2000; Ennos et al., 1993; Goodman et al., 2001), disease (Kevern and Hallauer, 1983), forest stand stability during severe gales and storms (Achim et al., 2003; Nicoll and Ray, 1996) and slope stability following clearfelling (Watson, 2000; Ziemer, 1981) is much more widely accepted. The uprooting resistance of trees by wind has been investigated through wind tunnel experiments on young trees (Stokes et al., 1995) and tree winching experiments (Cucchi et al., 2004; Stokes, 1999).

The pull-out resistance of simulated roots and their branching systems using wire models and computer-generated root systems by numerical models was attempted by Stokes et al. (1996) and Dupuy et al. (2004), respectively. Numerical models determined that the number of root branches and the diameter of roots were major components in affecting uprooting resistance.

A number of authors have carried out uprooting resistance tests on either plants or roots, but there are very few descriptions of the apparatus used to do this (Anderson et al., 1989; Denis et al., 2000; Norris and Greenwood, 2003b; Operstein and Frydman, 2000). The designs of the apparatus are based on a simple clamp, jack or pulley system to extract the roots, the ability to record resistance to pull-out or extraction force and displacement.

For the current study, the root reinforcement of natural vegetation growing on a highway cut-slope in south-east England on the M11 motorway was investigated. A bio-geotechnical site investigation was carried out to determine the ground conditions and characteristics of the vegetation. From the wide variety of naturally re-generated vegetation present, two common tree species were selected to investigate the interaction of roots and soil. Selection of the tree species was restricted by site accessibility, species abundance and time to excavate the soil from the root system. The hawthorn (*Crataegus monogyna* Jacq.) was selected for its abundance on the site and commonality on the UK's transportation infrastructure. This species also grows in a wide range of soil and climate conditions and is tolerant of all but the poorest acid soils. Hawthorn is a hardy and long lived tree (Flora for fauna, 2002). The oak (*Quercus robur* L.) was selected for its longevity. Both species chosen were tested for their potential suitablility to provide soil reinforcement on over-consolidated clay slopes.

Materials and methods

Study area

The study area is situated on a road cutting on a south-bound slip road of the M11 motorway between junctions 4 and 5, near Chigwell, Loughton, Essex, UK (coordinates Lat: 51°37′45″ N (51.6292°), Long: 0°04′14″ E (0.0704°)). The study area is a northwest facing slope having an overall slope angle of 20° and a height of 15 m. The crest of the cutting is at 40 m above sea level. The geology of the cutting is predominantly London Clay with a thin cover of superficial deposits (Boyn Hill Gravel and Boulder Clay) (BGS Sheet 257).

The soil profile consists of a surface topsoil layer of a brown sandy clay with occasional fine to medium flint gravel and a varying abundance of roots and rootlets. It varies in thickness between 0.15 and 0.25 m. A weathered soft–firm brown–grey mottled fissured (London) clay with occasional orange–brown silt partings and some roots lay beneath the topsoil layer.

The cut-slope contains a wide variety of plants, from grasses to shrubs and mature trees. Tree species present are silver birch, oak, hawthorn and pine. It was observed that natural regeneration of the vegetation was taking place as young oak trees (approximately 5 years old) were present. There seemed to be a marked change in vegetation type approximately half way up the cutting with predominantly grass, shrubs and young trees towards the lower half of the slope and the upper half of the slope consisting of mature trees. The marked difference in vegetation is probably due to reprofiling of the lower part of the slope during construction of the access road and motorway (A. Kidd, personal

Table 1. Plant, root and soil characteristics of the hawthorn and oak trees

Tree	Ref. no.	Slope angle (°)	dbh[a] (mm)	Height (m)	Biomass (kg)	Approximate age (years)	Mean soil shear strength (kPa)	Mean root diameter (at clamp) (mm)	Mean root length (m)
Oak[b]	MO1	18	5	1.45	–	8	47	5.4 ± 2.8	0.36 ± 0.08
Hawthorn[c]	H1/H2/ H3/H5	17.8 ± 0.4	84.8 ± 19.8	5.17 ± 2.6	32.8 ± 15.7	80	73 ± 17.3	21.6 ± 12.5	1.29 ± 1.49
Hawthorn[d]	HRA	14	113	6	95.5	80	61 ± 11.7	–	–

[a] dbh taken at 1.3 m above-ground level.
[b] Only one tree tested.
[c] Mean data of hawthorn trees tested for root pull-out resistance.
[d] Characteristics of the hawthorn tree excavated for root morphology observations.

communication). The original motorway was constructed in 1976.

Root pull-out tests

Four hawthorn (*Crataegus monogyna* Jacq.) trees and one oak (*Quercus robur* L.) sapling were selected for root pull-out resistance tests. Mean characteristics of each species are given in Table 1. The oak sapling and two of the hawthorn trees were tested in September 2002 and the other two hawthorn trees were tested in the Spring of the following year (May 2003).

Soil from around the base of the trunk of each tree was carefully excavated by hand trowel to a distance of 0.3 m from the trunk. Soil was removed until the main lateral roots could be clearly seen. Each main lateral root was labelled using an alphabetical labelling system and their diameters, dips and orientations recorded. Photographs and sketch drawings were taken of each root system. The tree was carefully removed in sections, so that only the stump remained. Each root was successively cut from the stump to allow the stump to be removed. The above-ground mass (biomass) of each tree was recorded.

The labelled roots were clamped and pulled out of the ground in turn. Surface roots were pulled first to cause minimal soil disturbance to roots penetrating deeper into the ground. The manual and mechanical apparatus used to pull-out the roots was designed by Nottingham Trent University (Norris and Greenwood, 2003b). The mechanical apparatus automatically recorded measurements of applied load and displacement using a 20 kN load cell and draw-wire transducer connected to a datalogger. A constant strain of 2 mm/s was applied. The hand-pull apparatus had manual recording of load and displacement using a spring balance and tape measure. The nature of the failure was recorded in both cases.

Curves of 'applied pull-out force' against 'displacement' were plotted for each root. The maximum applied force (pull-out resistance) did not necessarily correspond to the point when the root failed (broke).

Each root was sketched and/or photographed and a description of the roots sinuosity or straightness, tapering and number of branches recorded. The length of the root was determined by using a tape measure or ruler to the nearest millimetre. Root diameters at the clamp, all break points and/or root tips were measured using vernier callipers to an accuracy of 0.02 mm. Root diameter was measured by taking the average of the maximum and minimum diameter readings. The mass of the root was recorded and a portion of the root was used to determine the root moisture content by oven-drying at 80 °C for 24 h. Soil shear strength (or stifiness) was also determined as soon as possible after the tests by using a hand held shear vane (Clayton et al., 1995).

Plots were used to analyse the relationships between the maximum force (resistance) taken by the root, the failure stress and root parameters (diameter measured at the clamp, root length and number of branches). The failure stress was calculated based on the maximum applied force divided by the root diameter at the clamp (values are given as mean ± standard deviation). Regression analysis was carried out on the resultant plots. Root orientation was analysed using Geo-Orient v 9.2 (Stereographic Projections and Rose Diagram Plots) software available on the web at http://www.earth.uq.edu.au/~rodh/software. Circular statistics (Fisher, 1993; Mardia and Jupp, 1999) were applied to the data to obtain the mean root growth direction.

Results

Pull-out resistance of hawthorn and oak roots

A total of 42 roots were tested using the mechanical pull-out apparatus from the four hawthorn trees, three tests were unsuccessful as the roots were too strong for the apparatus (force to pullout exceeded the 20 kN load cell). Ten oak roots were pulled out by hand.

The maximum pull-out resistance for hawthorn roots with diameters 7.1–61.8 mm (mean 21.6 ± 12.5 mm) varied between 0.3 and 12 kN (mean 2.88 ± 2.6 kN) (Figure 1a) whereas the oak roots had maximum pull-out resistances between 0.03 and 0.44 kN (mean 0.15 ± 0.14 kN) for root diameters between 1.7 and 9.3 mm (mean 5.4 ± 2.8 mm) (Figure 1b). A positive correlation exists between maximum root pull-out resistance and root diameter for hawthorn and oak roots (Figure 1). Small root diameters have low pull-out resistance and/or breaking force whereas larger diameter roots have a high resistance to pull-out and/or high breaking forces. No significant relationship existed between root breaking force and (recovered) root length for either hawthorn or oak roots.

About 70% of hawthorn roots broke in a tensile failure along their length, 8% pulled completely out of the ground with 22% exhibiting a combined tensile slippage failure pattern, whereby the roots reached a maximum peak load and partially failed but adhesion with the soil provided a residual resistance (Greenwood et al., 2004). The oak roots had an 80% combined tensile and slippage failure pattern, 10% tensile failure and 10% pulled completely out of the ground.

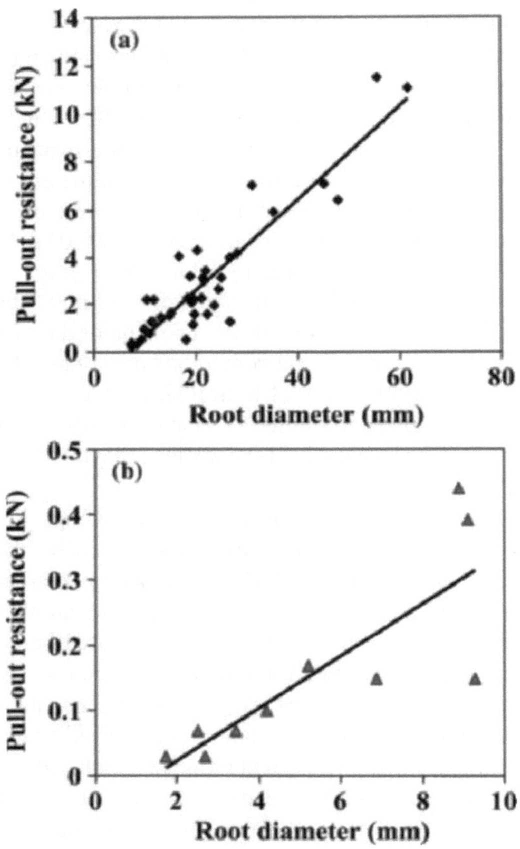

Figure 1. Root pull-out resistance was significantly correlated with root diameter in (a) hawthorn ($y = 0.1929x) - 1.2812$, $R^2 = 0.835$, $P = 0$) and (b) oak ($y = 0.0401x) - 0.0573$, $R^2 = 0.666$, $P = 0.002$).

Root pull-out failure stress

The failure stress of the hawthorn roots, based on the diameter at the clamp, ranged from approximately 5 MPa at 60-mm diameter to typically 3–15 MPa at diameters less than 30 mm (Figure 2), mean failure stress was 8.1 ± 4.6 MPa. Oak roots had failure stresses between 2 and 14 MPa, with a mean of 7.4 ± 3.5 MPa (Figure 2). When failure stress was correlated with number of branches a non-significant relationship existed.

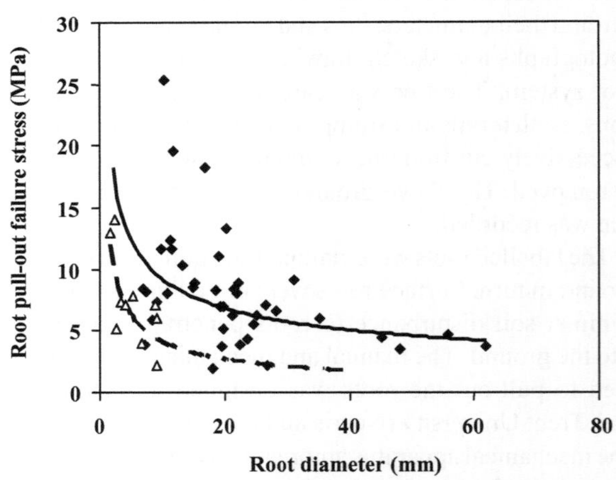

Figure 2. Root pull-out failure stress was significantly correlated with root diameter for hawthorn (solid diamonds, solid line $y = 24.919x - 0.4322$, $R^2 = 0.188$, $P = 0.004$) and oak (open triangles, dotted line $y = 16.585x - 0.6088$, $R^2 = 0.464$, $P = 0.018$).

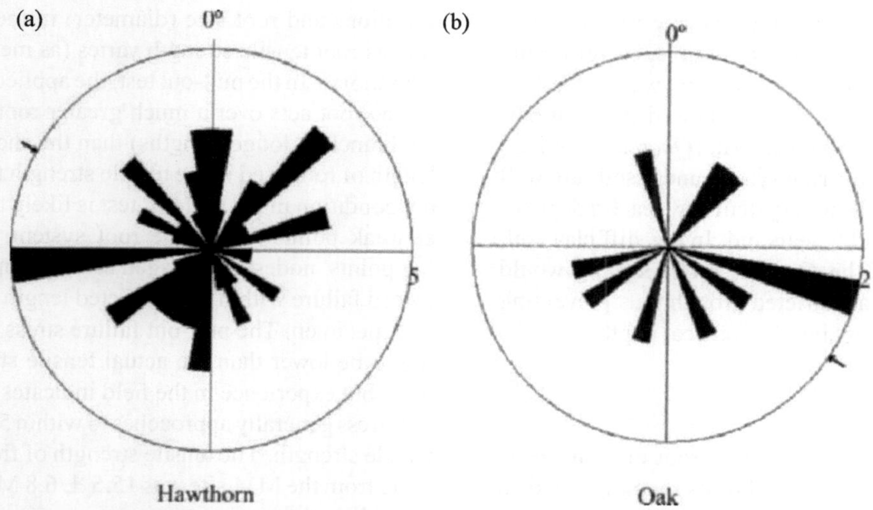

Figure 3. Root growth direction of (a) hawthorn (mean direction = 299° ± 2.2) and (b) oak (mean direction = 124° ± 1.7°). The length of each sector is equivalent to the number of roots occuring in that sector (e.g., 3 roots fall in the sector 1–10° in (a)). Each sector angle is 10°. The arrow represents mean root growth direction. Upslope direction of the cut-slope is 150°.

Root morphology and orientation

The hawthorn and oak both show an asymmetric root growth pattern (Figure 3). The hawthorn had a mean root growth direction of 299° ± 2.2° whereas the oak was 124° ± 1.7°. The most frequent number of roots occur laterally across the slope in hawthorn (Figure 3a) whereas the oak shows a greater number of roots occurring in the upslope direction (Figure 3b).

Morphology of the pulled hawthorn roots

The majority of the hawthorn roots pulled out of the ground were either short or long thick straight roots, many forking into two or more branches near the top of the root. Some of the long roots showed marked curvatures to their form. The thinner roots were sinuous in nature. Roots were ellipsoidal in cross section and showed a gradual taper along their length. The outer cortex of the roots was a reddish-brown colour, the thicker roots had prominent ridges at regular intervals along the length of the root.

During root pull-out there was no separation of the cortex (bark) and stele (inner root core) and the root generally remained intact except where lateral and forked branches had broken or snapped through tensile failure. The clay soil was observed to be smeared along many of the roots.

Hawthorn root morphology as observed from excavating the root system of one tree

To appreciate the nature of the roots in the ground and how they were resisting pull-out, a further hawthorn tree was excavated using an airspade to a distance of 1.5 m from the centre of the trunk. This hawthorn tree had a shallow rooting depth of 0.5 m below-ground level, and had other characteristics similar to the four hawthorn trees used for root pull-out tests (HRA in Table 1). The root plate showed no obvious tap-root directly below the trunk, but had many lateral roots which radiated from the base of the trunk. Roots were ellipsoidal in cross section and tapered gradually. Some lateral roots divided into multiple branches along their length.

Morphology of the pulled oak roots

The majority of the oak roots pulled out of the ground were long straight roots with many short rootlets along their length. Some roots forked into two or more branches near the root tips, others were multiple branches. Some showed right angle bends where they had obviously had to grow around an obstruction. All roots showed a gradual taper along their length.

Many of the oak roots lost the cortex during pull-out, indicating a greater adhesion between the bark and the soil than between the bark and the stele.

Excavation of a comparable oak tree for root morphology observations was not possible, therefore the actual depth of the oak root system was not established. However, Lyford (1980) showed from investigating the root system of Red Oak (*Quercus rubra* L.) that saplings have a tap root system and if soils are well drained and friable the root system may reach a depth of 0.7 m between 3 and 5 years old. In the stiff clay soil, the root system of the *Quercus robur* sapling would probably have had a restricted growth thus preventing the tap root from reaching this vertical depth.

Mode of failure

On first inspection of the plots of root pull-out resistance against displacement, all roots seemingly had an initial rapid rise in pull-out resistance (force) with relatively small displacement, to a maximum peak failure point over larger displacements. However, if root morphology is correlated to the failure curves, three different types of failure can be recognised (Figure 4).

Typically applied force (pull-out resistance) initially rises linearly with displacement to a peak point at displacements of 50–100 mm. This initial peak is either (a) followed by a rapid reduction of force until there is no resistance and the root completely pulls out of the ground (Type A failure, e.g., Root H2C) or (b) followed by a continued high resistance (force) leading to a second peak failure (Type B failure, e.g., Root H3E). In some cases, pull-out resistance increases progressively as a series of stepped peaks to a final maximum peak (Type C failure, e.g., Root H3N), these peaks correspond to the failure of lateral root branches. Type A failure generally has roots of a long length (>0.7 m) with no or few branches. Type B failures tend to have roots that are highly branched or forked. Forked roots diverge into two major branches, at angles of approximately 45°. Type C failures have roots of multibranched nature with significant lateral root branches failing before the main root. The number of branches or root divisions has more influence on the type of failure than the length of root.

Discussion

Root pull-out resistance

The pull-out resistance of the hawthorn and oak roots are affected by intra-species differences, inter-species variations and root size (diameter) in much the same way as root tensile strength varies (as measured in the laboratory). In the pull-out test, the applied force acting on the root acts over a much greater root area (multiple branches, longer lengths) than the short ~150 mm length of root used in the tensile strength test. The failure condition in the pull-out test is likely to be initiated at weak points within the root system, i.e., branching points, nodes or damaged areas, as opposed to the forced failure within the restricted length of the tensile test specimen. The pull-out failure stress is always going to be lower than the actual tensile strength of the roots but experience in the field indicates that the pull-out stress generally approaches to within 50–70% of the tensile strength. The tensile strength of fresh hawthorn roots from the M11 site was 15.5 ± 6.8 MPa (Norris, J. unpublished data).

The resistance to pull-out may be marginally affected by the stiffness of the clay but this cannot be discerned from the data. It would be expected that roots would pull more easily through a wetter softer soil than a stiff soil. Other effects may be linked to root growth around stones or roots from other trees forming barriers to pull-out.

The non-significant relationship between root pull-out resistance and root length can be explained by the fact that only the recovered root length was used and not the total length of root pulled. According to Riestenberg (1994) and Stokes et al. (1996), root length is a factor in the pull-out resistance of roots and as such a positive correlation would be expected.

No relationship was determined for root pull-out resistance and root depth because of the uncertainty as to the actual depth of the root in the ground. However, observations of the hawthorn root system showed that the roots were only shallowly rooted in a plate-like system to a maximum depth of 0.5 m. Experiments on the resistance of model root systems to uprooting concluded that the depth of roots in the soil affected the pull-out resistance (Dupuy et al., 2004; Stokes et al., 1996).

The negative correlation between pull-out failure stress and root diameter (Figure 3) is consistent with the decreasing tensile strength increasing root diameter relationship as reported by several authors, e.g., Nilaweera (1994). A decrease in root diameter (from 5 to 2 mm) can result in a doubling or even tripling of tensile strength. This phenomenon may be partially explained by considering whether or not the root bark plays a role in the root resisting pull-out. Bark has been

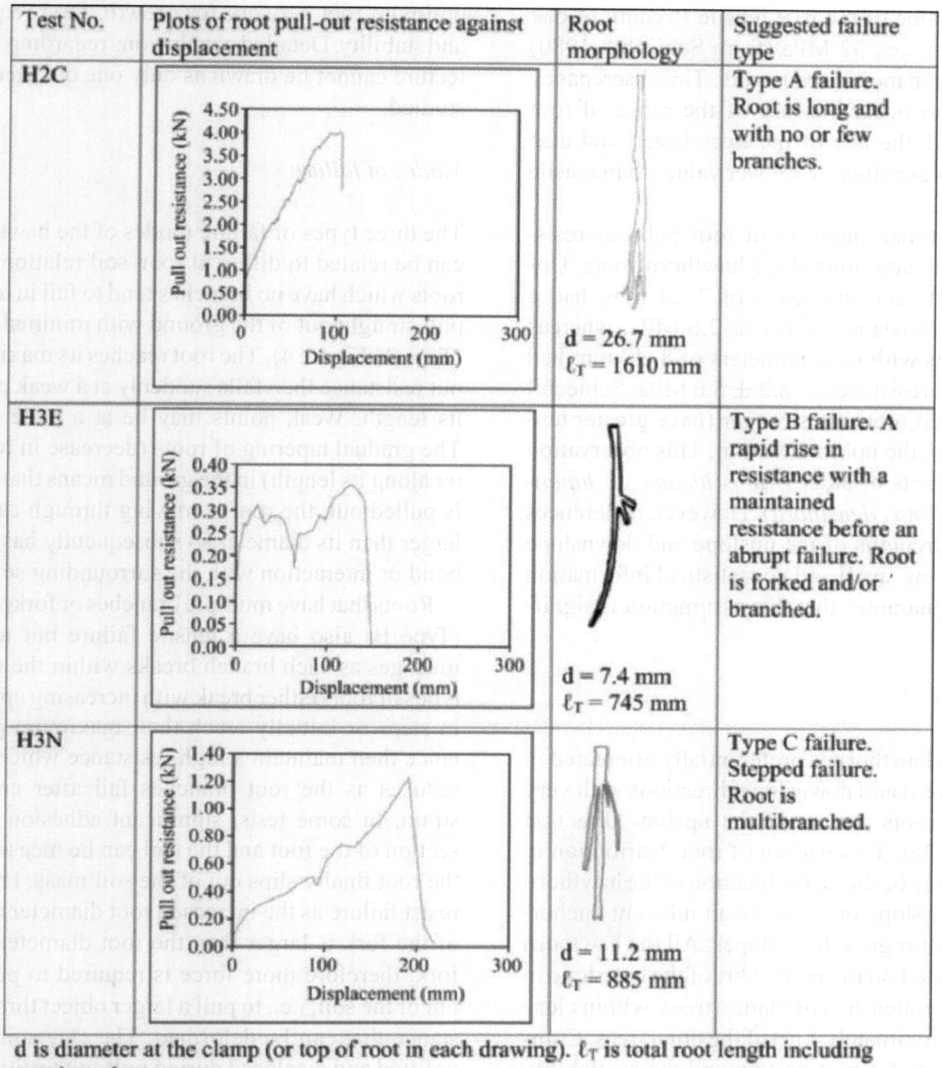

Figure 4. Examples of the three types of root failure and associated root morphologies for hawthorn roots.

shown to have minimal strength and as such should not be used as a reliable indicator of tensile strength (Hathaway and Penny, 1975). However, in the root pull-out test, the bark is in contact with the surrounding soil through root–soil adhesion, friction and mycorrhizal associations and as such affects the amount of pull-out resistance so is thus taken into account when calculating failure stress. The negative relationship has also been attributed to differences in root structure, with smaller roots possessing more cellulose than older thicker roots, cellulose being more resistant than lignin in tension (Commandeur and Pyles, 1991; Genet et al., 2005; Hathaway and Penny, 1975; Turmanina, 1965); and root straightening during tensile testing.

Both the oak and the hawthorn had similar mean failure stresses of 7.4 and 8.1 MPa, respectively. If these failure stresses are compared with published values of tree root tensile strengths of, for example, Black Poplar (*Populus nigra*) 5–12 MPa and Sallow (*Salix cinerea*) 11 MPa (both from Coppin and Richards, 1990) then the hawthorn fits within the range of this dataset. No published tensile strength data exists for hawthorn, although the unpublished value of 15 MPa (Norris, J. unpublished data) also agrees. However, the published value of the root tensile strength of oak (*Quercus robur*), i.e., 32 MPa (from Schiechtl, 1980) indicates a discrepancy in the results. This discrepancy is most likely to occur because of the range of root diameters tested, the age of the trees tested and also that pull-out failure stress has lower values than tensile strengths.

There is minimal variation in root pull-out resistance of upslope and downslope hawthorn roots. Upslope roots with root diameters of 7–24 mm, had a mean pull-out resistance of 8.1 ± 2.6 MPa, whereas downslope roots with root diameters of 8–48 mm had a mean pull-out resistance of 8.2 ± 5.6 MPa. Schiechtl (1980) suggested roots are stronger (have greater tensile strengths) in the uphill direction. This observation was based on roots of alder (*Alnus incana, A. japonica*) and pine (*Pinus densiflora*). However, differences in the tensile strengths of the upslope and downslope roots are relatively small and no statistical information is provided to guarantee that this assumption is significantly different.

Root orientation

Root growth in hawthorn is preferentially orientated in the lateral (across) and downslope directions with very few hawthorn roots present in the upslope direction (150°) (Figure 3a). This pattern of root distribution in the hawthorn may be due to the location of the hawthorn trees on the cut-slope or it maybe an inherent anchoring mechanism for growth on slopes. All the hawthorn trees were situated on the upper part of the cut-slope in the densely vegetated area of mature trees, within close proximity (approximately 1 m) of the other trees. Competition for space for root growth and the availability of nutrients and moisture would be at a premium in this environment.

The one oak sapling investigated shows an asymmetric root growth distribution, with a slight tendency for more root growth in the upslope direction, this is in partial agreement with Chiatante et al. (2003). These authors found that roots on steep slopes are preferentially orientated in the up-slope and down-slope directions so that the plant's stability is increased. The oak sapling was situated on the lower part of the cut-slope within the immature vegetation cover dominated by grasses. The sapling in this environment would have less competition for nutrients and moisture so would therefore develop a root system that would ensure its optimum root network for growth, food requirements and stability. Detailed conclusions regarding root architecture cannot be drawn as only one oak tree has been studied.

Modes of failure

The three types of failure modes of the hawthorn roots can be related to different root–soil relationships. The roots which have no branches tend to fail in tension and pull straight out of the ground with minimal resistance (Type A; Figure 4). The root reaches its maximum pull-out resistance then fails suddenly at a weak point along its length. Weak points may be at a node or branch. The gradual tapering of roots (decrease in root diameter along its length) in the ground means that as the root is pulled out, the root is moving through cavity space larger than its diameter so subsequently has no further bond or interaction with the surrounding soil.

Roots that have multiple branches or forked branches (Type B) also have a tensile failure but tend to fail in stages as each branch breaks within the soil. These types of roots either break with increasing applied force in steps or initially reach their maximum peak resistance then maintain a high resistance which gradually reduces as the root branches fail after considerable strain. In some tests, significant adhesion between a section of the root and the soil can be measured before the root finally slips out of the soil mass. Forked roots resist failure as the increased root diameter at the point of the fork is larger than the root diameter above the fork, therefore more force is required to pull the root out of the soil, i.e., to pull a larger object through a substance that can be deformed. The clay soil was often uplifted and displaced during pull-out testing of forked roots.

Multiple branched root failure (Type C; Figure 4) in the form of stepped peaks corresponded to roots of greater diameters breaking sequentially. The root gradually releases its bonds with the soil until the final tensile failure.

In some cases, when the root is of a sinusoidal nature and has many small diameter rootlets along its length. The root reaches its maximum pull-out resistance on straightening and fails at its weakest point; however in this case, it does not fail suddenly and pull straight out of the ground, it adheres and interacts with the soil producing a residual strength. If the pulling was stopped at this point, the root would provide additional strength to the soil. Since the root is pulled completely out of the ground, there is no further interaction with the soil (Greenwood et al., 2004; Norris and Greenwood, 2003b).

The oak roots, although pulled out using a manual root pull method, can mainly be classified as Type A failure, with long straight roots. Some multiple branched roots could be classified as Type B failure showing residual strength after the peak failure stress was achieved.

These modes of failures (Types A, B and C) are based on the shape of the failure curve and root morphology. In some cases, the shape of the failure curve may not be that distinct and relating branch failure points to drops in resistance is not straightforward, as proven by the non-significant relationship between number of branches and pull-out failure stress.

Dupuy et al. (2004) numerically modelled non-branching and branching root systems. These authors found that single non-branching roots have less effective resistance than branching root systems. When average pull-out resistances were determined for the three failure types (Type A (single roots): 3.3 kN, Type B (forked roots) – 2.8 kN and Type C (multiple branches) – 1.38 kN) the opposite correlation seems to apply. This difference may be due to the fixed arrangement of the root system branches in Dupuy et al.'s (2004) models which do not represent the type of morphologies and variation in root diamters as observed in the hawthorn roots. Some of the hawthorn roots classified as single branched roots had thin (approximately 1 mm in diameter) short root(-lets) occurring along their length. These rootlets would not necessarily be classed as a major subdivision or branch but would marginally affect the pull-out resistance of a single root.

The use of the hawthorn and oak for root reinforcement on highway slopes is questionable. The shallow rooting nature (0.5 m) of the hawthorn on this site does not lend itself to be used as a tree suitable for stabilising slopes that are prone to failing at depths of 1–1.5 m (Greenwood et al., 1985; Perry et al., 2003a,b). Although on other sites where root penetration to depths may be encouraged and not prevented by stiff clay or perched water tables, the hawthorn may, in conjunction with other species, form a suitable bioengineering solution. The English Oak is a slow growing tree, so would not be a suitable species for planting for immediate short term stability. However, when planted with other species that are quick growing and have only say a lifespan of 30–40 years, the oak would just be becoming established since it has a life expectancy of between 300 and 400 years (Miles, 1999).

The results presented in this paper are based on a small number of trees and on one soil type only. It is essential that more detailed investigations should be carried out to determine the relationship between root pull-out resistance and tensile strength of roots as determined by laboratory experiments. To validate the observations of the relationships between root morphology and mode of failure more experimental testing on other types of soils, trees and in other environmental settings must be carried out. The additional data obtained would increase the confidence in the value of shrubs and trees used in geotechnical engineering applications.

Acknowledgements

This work was supported by the European Commission Fifth Framework Programme, ECOSLOPES project. Technical and field assistance from Steve Goodman, Mark Flanagan, Jo Wint, David Barker and Gareth Richards are duly acknowledged. Dr R.J. Holcombe from the University of Queensland, Australia, is thanked for providing free web access to the software Geo-Orient v9.2 and for the use of this program to produce rose diagrams and circular statistics data. John Greenwood and the reviewers are thanked for their supportive comments and reviews during the preparation of this paper.

References

Achim A, Nicoll B, Mochan S and Gardiner B 2003 Wind stability of trees on slopes. In Proceedings of the International Conference 'Wind Effects on Trees'. B Ruck, C Kottmeier, C Mattheck, C Quine and G Wilhelm (eds). University of Karlsruhe, Germany, 16–18 September 2003, pp. 231–237.

Anderson C J, Coutts M P, Ritchie R M and Campbell D J 1989 Root extraction force measurements for Sitka spruce. Forestry 62, 127–137.

Bailey P H J, Currey J D and Fitter A H 2002 The role of root system architecture and root hairs in promoting anchorage against uprooting forces in *Allium cepa* and root mutants of *Arabidopsis thaliana*. J. Exp. Bot. 53, 333–340.

Barker D H, Watson A, Sombatpanit S, Northcutt B and Maglinao A R 2004 Ground and Water Bioengineering for Erosion Control. Science Publishers, New Hampshire, USA, 410 pp.

BGS Sheet 257. 1996 Geological map, Romford (Epping) 1:50000 Series, HMSO, London.

Chiatante D, Scippa S G, Di Iorio A and Sarnataro M 2003 The influence of steep slopes on root system development. J. Plant Growth Regul. 21, 247–260.

Clayton C R I, Matthews M C and Simons N E 1995 Site Investigation, 584 pp. Blackwell, Oxford.

Commandeur P R and Pyles M R 1991 Modulus of elasticity and tensile strength of Douglas for roots. Can. J. For. Res. 21, 48–52.

Coppin N J and Richards I J 1990 Use of Vegetation in Civil Engineering. Butterworths, London.

Cucchi V, Meredieu C, Stokes A, Berthier S, Bert D, Najar M, Denis A and Lastennet R 2004 Root anchorage of inner and edge trees in stands of maritime pine (*Pinus pinaster* Ait.) growing in different podzolic soil conditions. Trees Struct. Funct. 18, 460–466.

Denis S, Bremand F and Herbert Y 2000 An experimental study on the mechanical behaviour of the maize root system: analysis of the soil-root bond and determination of rigidity modulus. *In* A Stokes (ed), The Supporting Roots of Trees and Woody Plants: Form, Function and Physiology, pp. 279–285. Kluwer, Dordrecht.

Dupuy L, Fourcaud T and Stokes A 2004 A numerical investigation into factors affecting the anchorage of roots in tension. Eur. J. Soil Sci. 56(3), 319–327.

Endo T and Tsuruta T 1969 On the effect of tree roots upon the shearing strength of soil. Annual report of the Hokkaido Branch, Forest Place Experimental Station, Sapporo, Japan, 167–183.

Ennos A R 1990 The anchorage of leek seedlings: the effect of root length and soil strength. Ann. Bot. 65, 409–416.

Ennos A R 2000 The mechanics of root anchorage. Adv. Bot. Res. 33, 133–157.

Ennos A R, Crook M J and Grimshaw C 1993 The anchorage mechanics of maize, Zea mays. J. Exp. Bot. 44, 147–153.

Fisher N I 1993 Statistical Analysis of Circular Data. Cambridge Univ. Press, Cambridge.

Flora for Fauna. 2002. British native trees and shrubs. http://www.nhm.ac.uk/jubileetrees/british-native-trees/bnt-23.htm. Accessed 1/09/2003.

Genet M, Stokes A, Salin F, Mickovski S B, Fourcaud T, Dumail J and Van Beek L P H 2005 The influence of cellulose content on tensile strength in tree roots. Plant Soil 278, 1–9.

Goodman A M, Crook M J and Ennos A R 2001 Anchorage mechanics of the tap root system of winter-sown oilseed rape (*Brassica napus* L.). Ann. Bot. 87, 397–404.

Gray D H and Sotir R 1996 Biotechnical and Soil Bioengineering Slope Stabilization: A Practical Guide to Erosion Control. Wiley, New York.

Greenwood J R 2006 Slip4ex–A program for routine slope stability analysis to include the effects of vegetation, reinforcement and hydrological changes. J. Geotech. Geol. Eng., 24, 449–465.

Greenwood J R, Holt D A and Herrick G W 1985 Shallow slips in highway embankments constructed of overconsolidated clay. ICE Int. Sym on Earthworks Failures, London 6–7 March, pp. 79–92.

Greenwood J R, Norris J E and Wint J 2004 Assessing the contribution of vegetation to slope stability. J. Geotech. Eng. ASCE 157, 199–208.

Greenwood J R, Vickers A W, Morgan R P C, Coppin N J and Norris J E 2001 Bioengineering – the Longham Wood Cutting field trial, 128 pp. CIRIA PR 81, CIRIA, London.

Hathaway R L and Penny D 1975 Root Strength in some *Populus* and *Salix* cones. NZ J. Bot. 13, 333–344.

Highways Agency 2003 Design Manual for Roads and Bridges. Vol. 1. Highway structures: Approval Procedures and General Design. The Stationary Office, London. ISBN 0115524606.

Highways Agency 2004 Manual of Contract Documents for Highway Works. Vol 1. Specification for Highway Works. The Stationary Office, London. ISBN 011552634X.

Kevern T C and Hallauer A R 1983 Relation of vertical root-pull resistance and flowering in maize. Crop Sci. 23, 357–363.

Lyford W H 1980 Development of the root system of Northern Red Oak (*Quercus rubra* L.). Harvard Forest Paper. 21, 1–29.

MacNeil D J, Steele D P, McMahon W and Carder D R 2001 Vegetation for slope stability. TRL Report 515, TRL Limited, Wokingham, UK.

Mardia K V and Jupp P E 1999 Directional Statistics. Wiley, Chichester.

Miles A 1999 Silva: the Tree in Britain, 400 pp. Ebury Press, London.

Nicoll B C and Ray D 1996 Adaptive growth of tree root systems in response to wind action and site conditions. Tree Physiol. 16, 891–898.

Nilaweera N S 1994 Effects of Tree Roots on Slope Stability: the case of Khao Luang Mountain Area, Southern Thailand, Ph.D. thesis GT-93-2. Asian Institute of Technology, Bangkok, Thailand.

Norris J E and Greenwood J R 2003a Root reinforcement on unstable slopes in Northern Greece and Central Italy International Conference on Problematic Soils, July 2003, Nottingham Trent University, Nottingham, UK, pp. 411–418.

Norris J E and Greenwood J R 2003b *In situ* shear box and root pull-out apparatus for measuring the reinforcing effects of vegetation. *In* F Myrvoll (ed), Field Measurements in Geomechanics. pp. 593–597. Swets and Zeitlinger, Lisse.

O'Loughlin C L 1981 Tree roots and slope stability. What's new in forest research? Forest Research Institute Publication No. 104, Rotorua, New Zealand

Operstein V and Frydman S 2000 The influence of vegetation on soil strength. Ground Improv. 4, 81–89.

Perry J, Pedley M and Brady K 2003a Infrastructure Cuttings – Condition Appraisal and Remedial Treatment. C591 CIRIA, London

Perry J, Pedley M and Reid M 2003b Infrastructure Embankments – Condition Appraisal and Remedial Treatment. C592. 2nd edition. CIRIA, London.

Riestenberg M 1994 Anchoring of thin colluvium by roots of Sugar Maple and White Ash on hillslopes in Cincinnati. US Geological Survey Bulletin 2059-E, Washington, USA, 25 pp.

Schiechtl H M 1980 Bioengineering for land reclamation and conservation. University of Alberta Press, Alberta.

Stokes A 1999 Strain distribution during anchorage failure of Pinus pinaster Ait. at different ages and tree growth response to wind-induced root movement. Plant Soil 217, 17–27.

Stokes A, Fitter A H and Coutts M P 1995 Responses of young trees to wind and shading: effects on root architecture. J. Exp. Bot. 46, 1139–1146.

Stokes A, Ball J, Fitter A H, Brain P and Coutts M P 1996 An experimental investigation of the resistance of model root systems to uprooting. Ann. Bot. 78, 415–421.

Turmanina V I 1965 On the strength of tree roots. Bull. Moscow Soc. Nat. Biol. Sect. 70, 36–45.

Van Beek L P H, Wint J, Cammeraat L H and Edwards J P 2005 Observation and simulation of root reinforcement on abandoned Mediterranean slopes. Plant Soil 278, 55–74.

Waldron L J and Dakessian S 1981 Soil reinforcement by roots: calculation of increased soil shear resistance from root properties. Soil Sci. 132, 427–435.

Watson A 2000 Wind-induced forces in the near-surface lateral roots of radiata pine. For. Ecol. Manage. 135, 133–142.

Wu T H, Beal P E and Lan C 1988 *In situ* shear test of soil root systems. J. Geotech. Eng.-ASCE 114, 1376–1394.

Ziemer R 1981 Roots and shallow stability of forested slopes. Int. Ass. Hydrol. Sci. 132, 343–361.

Protection roles of forest and non-forest woody species on slopes in Iran

Ghassem Habibi Bibalani[1,5], Baris Majnonian[2], Ebrahim Adeli[3], and Homauon Sanii[4]

[1]*Azad University of Shabestar, Shabestar, E. Azarbayjan-5381637181, Iran.* [2]*Faculty of Natural Resource of University of Tehran, PO Box 31585-4314, Karaj, Iran.* [3]*Azad University of Tehran, Science and Research University, PO Box 14515-75, Tehran, Iran.* [4]*Tarbiat Modares University, No. 4, Bahar St., Shariati Avenue, Tehran, Iran.* [5]*Corresponding author**

Key words: factor of safety, Iran, root reinforcement, soil internal friction angle, vegetation

Abstract

The roots of trees provide an important contribution towards the stability of hill slopes. Tree roots in the soil act very similarly to steel fibers in reinforced concrete and provide resistance to shear and tensile forces induced in the soil. In addition, the roots also absorb water from the soil, which reduces moisture content, again helping to increase the stability of the slope. As Iran has a long history of landslides, our research deals with the effect of tree roots on slope stability, in particular, the following species which are of economic and environmental interest: tea (*Thea sinensis* L.), citrus (*Citrus spp.*), lilaki (*Gleditshia caspica* Dsf.) and angili (*Parrotia persica* D.C.) (Mosadegh, 1996). The study was carried out in Roudsar Township in Gilan State of Iran. Of the overall surface area of 1800 ha, 288 ha were considered suitable for the purposes of this study. A large part of the area had slopes of steep gradients on which natural vegetation was present. Other parts of the same area have been cleared and planted with tea and citrus crops. Soil samples were taken from an area of approximately 70 ha for testing in the laboratory. Direct shear tests were carried out on soil samples and the factor of safety (FOS) calculated. Results showed that the FOS was increased in soils with tree roots present. The global slope FOS was then determined using Bishop's method. We calculated the FOS in order to protect slopes where the gradient exceeds 25%. In this case study, the minimum FOS was assumed to be 1.3, which corresponds to e.g. *Parrotia sp.* vegetation with 40–60% crown cover, a soil internal friction angle of 15° and a slope angle of 21°. When soil internal friction angle equals 15° and slope angle is >31°, slope stability cannot be increased by any vegetation species.

Introduction

Due to the increase in the world population over recent years, the exploitation of renewable natural resources has increased dramatically. One of the effects of this increase has been the destruction of some of the world's forests, and this has been particularly noticeable over the last 10 years (Baher, 1994). Natural disasters, such as floods, droughts and a rise in sea levels have also had an effect on the life condition of many people. Iran has a long history of landslides, which have caused a major loss of life along with damage to infrastructure and agricultural land (Baher, 1994). It has been impossible to recover some of the damages caused by landslides, and where it has been possible, it has been at a high financial cost.

The main reason for the high number of landslides in Iran is as a result of a particular combination of geology, topography and climate. Furthermore, the cause of landslides may be due to geomorphological phenomena combined with other factors such as climate variations, vegetation cover, geology and the tectonic situation (Ahmadi, 1993). Gilan province in the north of Iran is susceptible to landslides, however, because of the unusual topographical and geological conditions in this Province, it may be possible to prevent landslides with relatively little expense and labour. Therefore, research about slope stability in this region

* E-mail: ghhabibi@iaushab.ac.ir

would be of particular interest to local stakeholders and farmers.

Vegetation has long been considered to improve slope stability (Terwilliger, 1990), therefore the influence of different species on the soil reinforcement of slopes can be studied using different methods. Measurements of soil shear strength provide an indicator of the contribution of roots to slope stability, when combined with calculations of the slope's factor of safety (FOS). The formula used to calculate the FOS of the critical landslide surface has been defined by (Terwilliger, 1990):

$$\text{Safety factor (FOS)} = \frac{\text{soil shear stress}}{\text{soil shear resistance}}. \quad (1)$$

However, the contribution of plant roots to soil shear still needs much research, and much work also needs to be carried out on the suitability of different species for stabilizing slopes.

Materials and methods

The study area is located in the east of Gilan Province in Iran and is part of the Rahimabad District of Roudsar Township. In Gilan province, the bedrock is well-bedded grey limestone and muddy limestone from the Precambrian period. Muddy limestone is a particular geological feature which causes landslides. Groundwater is seasonal from middle October to end May and is about 2.5-m deep (Geological Survey of Iran, 1975). The site is located north of Mohamad Dolagavabar, south of Slakjan, east of Goldhast and west of Bozkoyeh villages. The total surface area is about 1800 ha, of which 288 ha were selected as being suitable for this study. A large part of the study area has hillsides of steep gradients, with natural vegetation cover. In other areas, tea (*Thea sinensis* L.) plantations and citrus (*Citrus spp.*) orchards have been established. This planting of crop species has caused instability of the steep slopes and landslides occur mainly in these areas.

We divided the study area into A, B, C and D districts, in which A represents district with tea and citrus vegetation ground cover, B represents lilaki (*Gleditshia caspica* Dsf.), C represents angili (*Parrotia persica* D.C.) and D represents tea vegetation only. Twenty trial circular pits 0.9 min in diameter and 1.5 m in depth were excavated with hand on a slope where no landslides had occurred, in order to quantify the contribution of vegetation to soil reinforcement. The numbers of roots, with diameters in classes of 0.1–3.9, 4.0–7.9 and 8–10 mm were counted. Tensile testing of roots was carried out in the laboratory. Ninety soil samples with 0.2 m diameter and 0.3 m length were taken from 70 ha outside of the landslide area on which laboratory tests were carried out to determine soil mechanical characteristics (Table 1). Triaxial tests were carried out on soil samples without roots in the laboratory (Baher, 1994). Pore pressure was distinguished by triaxial tests on soil sample. Laboratory shear tests with $0.1 \times 0.1 \times 0.1$ m size box were carried out without roots using the American Standard method. The following equation was used to determine the increase in shear strength of soils (ΔSR) containing plant roots (Morgan, 1995):

$$\Delta\text{SR} = 1.15\text{TR}\frac{\text{AR}}{\text{A}} \quad (2)$$

where TR = tensile resistance (kPa) of the root, AR = total surface area (mm^2) of roots in A, A = soil surface area (m^2), and 1.15 is a coefficient (Baher, 1994).

Global slope stability was calculated using the Bishop method of analysis (Behniya, 1993; Benda and Zhang, 1990). A computer software version of this method (in Excel software) was developed by the author for calculation purposes. The FOS is calculated from:

$$\text{FOS} = \frac{L}{\sum w \sin\alpha} \times \sum\left[\{c'b + (w - ub)\tan\phi'\}\left\{\frac{\sec\alpha}{L + \frac{\tan\alpha\tan\phi'}{\text{FOS}}}\right\}\right] \quad (3)$$

where
FOS = safety factor,
α = angle of surface with horizon (°),
w = weight of soil on the slope (kN),
c' = cohesion of soil (kN/m^2),
L = length of slope (m),
u = pore pressure (kN/m^2),
b = width of area (m)
ϕ' = angle of internal friction (°).

The FOS models were run to determine the influence of different species on slope stability and soil reinforcement with two angles of internal friction, with and without roots (°). On the basis of the amount of ΔSR, it is possible to calculate the angle of internal friction of soil

Table 1. Soil properties in different slopes and areas

Slope	Area	Species	b	u	ϕ'_1	c'	α	l	h	ρ	FOS_1	d	n	TR	ϕ'_2	FOS_2	AR/A
												10	1	501			
1	C	*Parrotia sp.*	25	0.4	25.3	0.3	31.9	15	3.5	1.45	1.0	8	30	558	36.6	1.6	0.21
												4	41	591			
												10	1	501			
2	C	*Parrotia sp.*	20	0.3	26.4	0.2	29.3	15	3	1.5	1.1	8	32	558	38.9	1.8	0.22
												4	39	591			
												10	—	501			
3	B	*Gleditshia sp.*	15	0.3	22.9	0.2	20.2	15	3.5	1.55	1.3	8	15	558	25.7	1.5	0.1
												4	24	591			
												10	—	501			
4	B	*Gleditshia sp.*	12	0.4	23.8	0.3	24.8	15	2	1.88	1.9	8	16	558	28.6	1.4	0.11
												4	23	591			
												10	—	501			
5	D	Tea	70	0.4	23.5	0.3	17.1	15	3.5	1.5	1.0	8	11	558	25.1	1.2	0.08
												4	24	591			
												10	—	501			
6	D	Tea	45	0.3	21.2	0.2	26.5	15	3	1.5	0.9	8	14	558	23.3	1.1	0.1
												4	23	591			
												10	—	501			
7	D	Tea	50	0.3	18.7	0.3	23.5	20	2.5	1.5	0.9	8	13	558	21.2	1.1	0.09
												4	21	591			

b = width of area (m); u = pore pressure (kN/m^2); ϕ'_1 = angle of internal friction without vegetation (°); c' = cohesion of soil (kN/m^2); α = angle of surface (°); l = length of slope (m); h = depth of soil (m); ρ = density of soil; FOS_1 = safety factor without vegetation; d = diameter of roots (mm); n = number of roots per m^2; TR = tensile resistance of the root (kPa); ϕ'_2 = angle of internal friction with vegetation (°); FOS_2 = safety factor without vegetation; AR/A = relation between total surface area (mm^2) of roots and soil surface (m^2).

(ϕ'_2) and safety factor (FOS) with (FOS_2) or without (FOS_1) plant roots.

From ΔSR (additional soil resistance with roots) can be used to calculate the angle of internal friction with roots in soil (ϕ_2') with (Baher, 1994)

$$\tan \phi'_2 = \frac{SR + \Delta SR - c'}{hbl\rho} \tag{4}$$

where

ϕ'_2 = angle of internal friction with roots (°),
SR = soil resistance (kN/m^2),
ΔSR = additional soil resistance with roots (kN/m^2),
h = depth of soil (m),
b = width of area (m),
ℓ = length of slope (m),
ρ = density of soil (kg/m^3)
c' = cohesion of soil (kN/m^2).

Results

FOS1 (safety factor without vegetation) was calculated with soil properties (u, ϕ'_1, c' and ρ) and area characteristics (b, α, l and h) using Equation 1. Additional soil resistance with roots (ΔSR) was calculated using Equation 2, and angle of internal friction without vegetation (ϕ'_2) was then calculated. FOS2 (safety factor without vegetation) was calculated with soil properties (u, ϕ'_2, c', and ρ) and area characteristics (b, α, l and h) using Equation 1 (Table 1).

The relationship between the angle of internal friction (ϕ') and FOS of soil without roots (FOS1) was calculated for slope angles 15, 18, 23, 27, 30 and 33° (Figure 1). The relationship between angle of internal friction (ϕ') and FOS of soil with different ground covers of tea vegetation (FOS2) was calculated for a slope angle of 15° (Figure 2). The relationship between angle of internal friction (ϕ') and FOS of soil with different ground covers of *Parrotia sp.* (FOS2) was shown for a slope angle of 15° (Figure 3). The effect of different roots of 20–40% vegetation on FOS2 was compared with FOS_1 at a slope angle of 15° (Figure 4). The effect of different slopes (18, 23, 27, 30 and 33°) on FOS_2 was compared with FOS1 roots of *Parrotia sp* vegetation with ground cover above 60%

Table 2. Stability of slopes with regard to the angle of internal friction of the soil, angle of the slope and percentage vegetation cover

		α						
	Sp.	15°	18°	21°	23°	27°	30°	33°
15°	Tea	20–40%	U	U	U	U	U	U
	Tea-citrus	<20%	20–40%	U	U	U	U	U
	Parrotia sp.	<20%	<20%	40–60%	40–60%	40–60%	U	U
	Gleditshia sp.	20–40%	>60%	U	U	U	U	U
16°	Tea	S	20–40%	U	U	U	U	U
	Tea-citrus	S	<20%	40–60%	U	U	U	U
	Parrotia sp.	S	<20%	40–60%	40–60%	40–60%	>60%	U
	Gleditshia sp.	S	20–40%	U	U	U	U	U
17°	Tea	S	S	40–60%	U	U	U	U
	Tea-citrus	S	S	40–60%	U	U	U	U
	Parrotia sp.	S	S	20–40%	40–60%	40–60%	>60%	U
	Gleditshia sp.	S	S	U	U	U	U	U
18°	Tea	S	S	U	U	U	U	U
	Tea-citrus	S	S	20–40%	40–60%	U	U	U
	Parrotia sp.	S	S	<20%	20–40%	40–60%	40–60%	>60%
	Gleditshia sp.	S	S	>60%	U	U	U	U
19°	Tea	S	S	U	U	U	U	U
	Tea-citrus	S	S	20–40%	40–60%	U	U	U
	Parrotia sp.	S	S	<20%	20–40%	40–60%	40–60%	>60%
	Gleditshia sp.	S	S	40–60%	U	U	U	U
20°	Tea	S	S	40–60%	U	U	U	U
	Tea-citrus	S	S	20–40%	20–40%	>60%	U	U
	Parrotia sp.	S	S	<20%	<20%	40–60%	40–60%	>60%
	Gleditshia sp.	S	S	20–40%	40–60%	U	U	U
21°	Tea	S	S	20–40%	>60%	U	U	U
	Tea-citrus	S	S	<20%	20–40%	40–60%	U	U
	Parrotia sp.	S	S	<20%	<20%	20–40%	40–60%	40–60%
	Gleditshia sp.	S	S	20–40%	40–60%	U	U	U
22°	Tea	S	S	S	40–60%	U	U	U
	Tea-citrus	S	S	S	<20%	40–60%	U	U
	Parrotia sp.	S	S	S	<20%	20–40%	40–60%	40–60%
	Gleditshia sp.	S	S	S	20–40%	U	U	U
23°	Tea	S	S	S	20–40%	U	U	U
	Tea-citrus	S	S	S	<20%	20–40%	40–60%	U
	Parrotia sp.	S	S	S	<20%	<20%	20–40%	40–60%
	Gleditshia sp.	S	S	S	<20%	>60%	U	U
24°	Tea	S	S	S	S	>60%	U	U
	Tea-citrus	S	S	S	S	20–40%	40–60%	>60%
	Parrotia sp.	S	S	S	S	<20%	20–40%	40–60%
	Gleditshia sp.	S	S	S	S	40–60%	U	U
25°	Tea	S	S	S	S	U	U	U
	Tea-citrus	S	S	S	S	20–40%	20–40%	40–60%
	Parrotia sp.	S	S	S	S	<20%	<20%	20–40%
	Gleditshia sp.	S	S	S	S	U	U	U
26°	Tea	S	S	S	S	20–40%	>60%	U
	Tea-citrus	S	S	S	S	<20%	20–40%	40–60%
	Parrotia sp.	S	S	S	S	<20%	<20%	<20%
	Gleditshia sp.	S	S	S	S	<20%	40–60%	>60%

ϕ = angle of internal friction of soil 15–26°; Sp. = gpecies of vegetation, α = angle of slope (15, 18, 21, 23, 27, 30, or 33°); S = Stable; U = unstable with any type of vegetation.

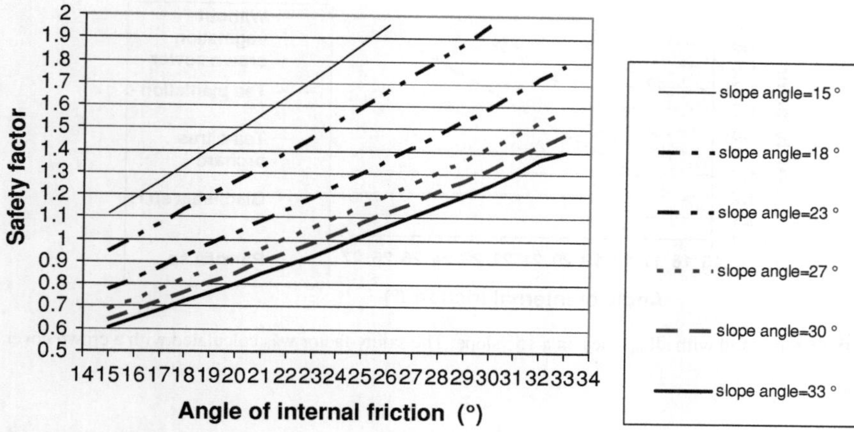

Figure 1. Calculated safety factor of soil without vegetation cover on different slope angles.

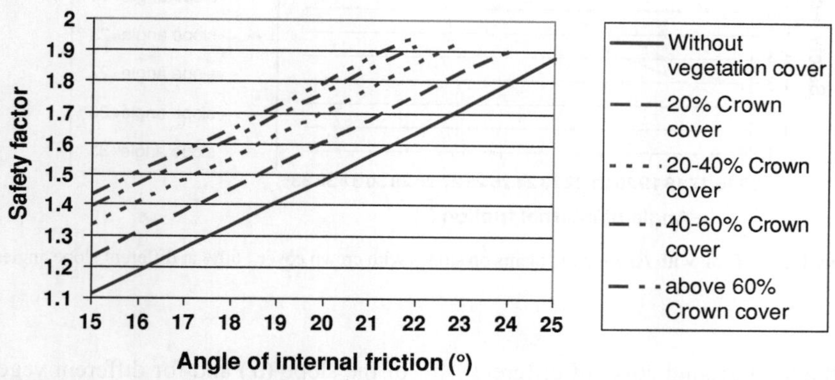

Figure 2. Calculated safety factor of soil with tea plants on a 15° slope. The safety factor was calculated with different percentages of crown covers.

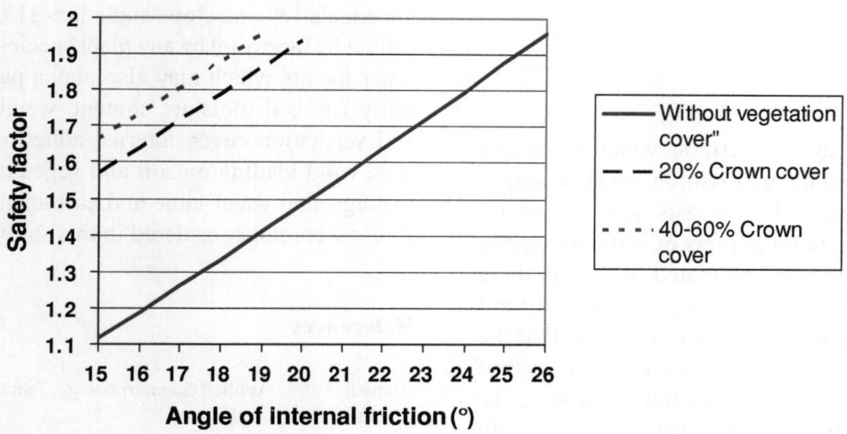

Figure 3. Calculated safety factor of soil with *Parrotia sp.* on a 15° slope. The safety factor was calculated with different percentages of crown cover.

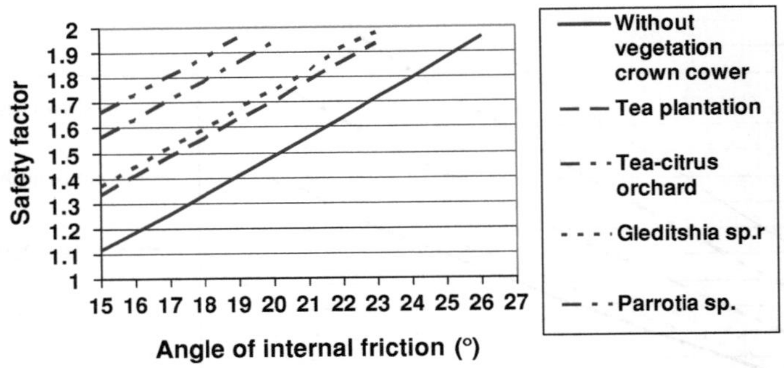

Figure 4. Calculated safety factor of soil with all species on a 15° slope. The safety factor was calculated with a crown cover of 20–40%.

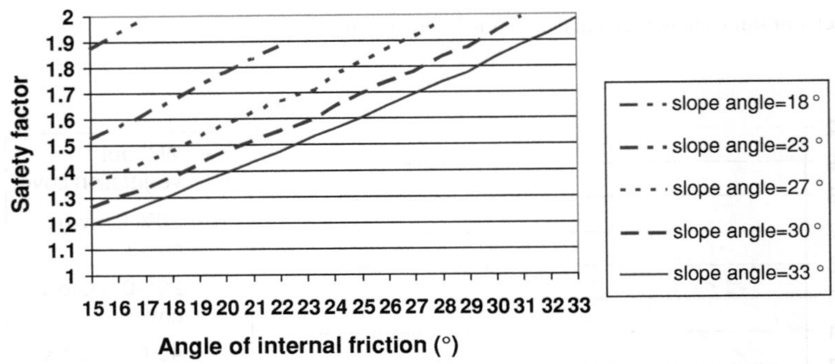

Figure 5. Calculated safety factor of soil with *Parrotia sp.* plants on slopes with crown cover >60% at different slope angles.

(Figure 5). Finally, the least ground cover of different types of vegetation that could stabilize soil on slope with different internal friction angles (ϕ') and different angles of surface from the horizon is shown in Table 2.

Conclusion

We carried out a study to determine which forest and crop species provided the best reinforcement to slopes in the Gilan province. Soil shear tests were carried out on samples with and without roots of different woody species and the slope FOS calculated. Although there are several factors which may affect slope stability and lead to landslides in the study area, it is clear that the main factor is the removal of the natural forest cover of the slopes, to create tea plantations and citrus orchards. In order to increase the FOS of slopes in this area with gradients more than 25%, the least FOS with regard to the angle of internal friction of soils (ϕ'), the gradient of the slope (α) and for different vegetation covers is given in Table 2. It was concluded that soil can be stabilized with tea, tea-citrus, *Parrotia sp.* and *Gledishia sp.* on soils with an internal friction angle >16° and a slope angle of 17–18°. When soil internal friction angle equals 15° and slope angle is >31°, slope stability cannot be increased by any plant species. Nevertheless, other factors which may also play a part in slope stability e.g. soil moisture content, weight of soil mass and vegetation cover, internal adhesion of soil particles, wind loading on soil and vegetation, location of underground water table and earthquake and tectonic forces were not considered in this study.

References

Ahmadi H 1993 Applied Geomorphology. Tehran University Press, Iran.

Baher T 1994 Research of Landslide of North Highway of Tabriz. PhD Thesis. Tarbiat Modares University, Tehran.

Bahniya K 1993 Soil Mechanic. Tehran University Press, Iran.

Benda L and Zhang W 1990 Accounting for the stochastic occurrence of landslides when predicting sediment yields. IAHSAISH 192, 115–127.
Geological Survey of Iran 1975 Explanatory Text of the Qazvin and Rasht Quadrangles Map. Ministry of Industry and Mines Press.
Morgan R 1995 Slope Stabilization and Erosion Control. Cambridge University Press, Cambridge, UK.
Mosadegh A 1996 Silviculture. Tehran University Press, Iran.
Terwilliger V 1990 Effects of Vegetation on Soil Slippage by Pore Pressure Modification. PhD Thesis. University of California.

Reinforcement of tree roots in slope stability: A case study from the Ozawa slope in Iwate Prefecture, Japan

H. Nakamura[1,3], Q.M. Nghiem[2] & N. Iwasa[2]

[1]*Graduate School of Agriculture, Tokyo University of Agriculture and Technology, Tokyo, Fuchu-shi, Saiwai cho 3-5-8, 183-0054, Japan.* [2]*Nippon Steel Metal Products Co. Ltd., Tokyo, Koto-ku 17-12-2, 135-0042, Japan.* [3]*Corresponding author**

Key words: landslide, root reinforcement, safety factor, slope stability

Abstract

The effect of root reinforcement on slope stability has been well researched through empirical studies, but to date few mechanistic studies have examined the influence of tree roots on slope stability. Furthermore, the previous research has lacked consideration of the effect of landslide displacement on root reinforcement. This paper will analyze the influence of root reinforcement on safety factors (Fs) as a function of slope displacement. A model of a root system as a cluster of straight bars inserted from unstable soil into bedrock is used to reliably estimate increases in the shear resistance of the soil. The relationship between root reinforcement and lateral displacement is analyzed under two conditions: ultimate stress and pullout resistance of root fibers. The species used in the present research was the Japanese cedar (*Cryptomeria japonica* (L.f.) D. Don.), the most common tree species in Japan. The spatial distribution of root size and root inclination was taken from field experiments performed by Japan Sabo Technical Center in 1998. The reinforcement capacity of root fibers is considered as a function of the horizontal displacement of the landslide and the depth of the slip surface. By combining the data obtained from field experiments with a calculation model of inclined roots, this paper analyzes the Ozawa slope safety factor. Thus, root reinforcement and the slope safety factor were calculated for various displacements in the process of landslide movement.

Materials and methods

Research site

Japan maintains approximately 25 million hectares of forest, 66.4% of the nation's entire surface area. The forest is concentrated in mountainous areas and hilly terrain, covering about 75% of the nation's land area (Statistical Handbook of Japan, 2004). The mountains are generally steep with intricately carved ravines. Due to unstable slopes, landslide related natural disasters annually result in serious loss of life and economic damages. Because the mountainous area was mostly forested as mentioned above, research work on landslides needs to consider the influences of root reinforcement (Abe and Iwamoto, 1990; Endo and Tsuruta, 1969; Gray and Sotir, 1996).

The properties and distribution of roots used in this paper are taken from a number of previous field experiments (Abe, 1992; Abe and Iwamoto 1986, Tsukamoto, 1987) in combination with one remarkable experiment (Sabo Technical Center, 1997) performed at the research site location in Ozawa District, Iwate Prefecture, Japan (Figure 1). In our study, the reinforcement of tree roots classified by inclination and diameter (called inclined roots) is considered as a function of both the horizontal displacement and the depth of a landslide (Nghiem et al., 2001).

Field experiment and 2-D model of tree roots

Two cedar trees (A and B in Figure 1) were cut and carefully excavated to determine the distribution of roots.

* E-mail: hnaka@cc.tuat.ac.jp

Stokes et al. (eds), Eco- and Ground Bio-Engineering: The Use of Vegetation to Improve Slope Stability, 81–90.

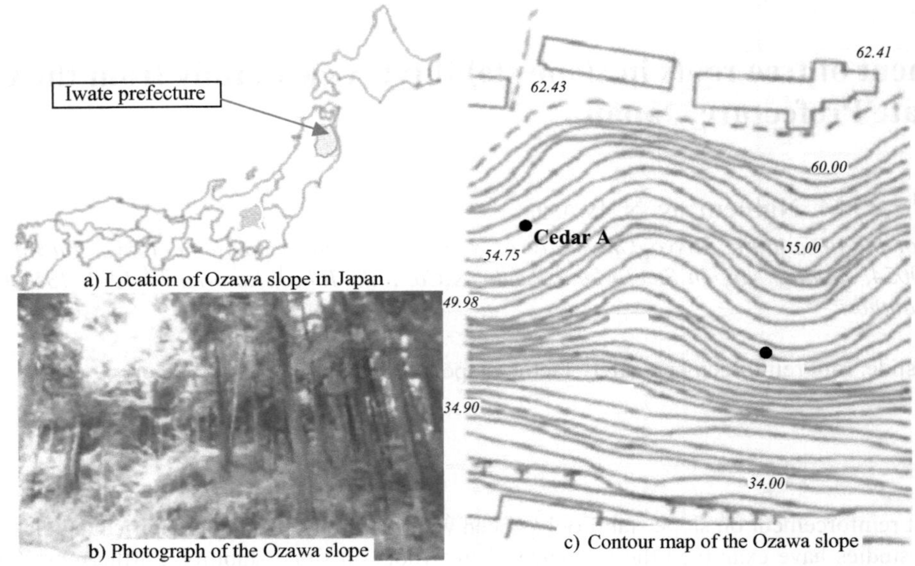

Figure 1. Cedars on Ozawa slope of research site.

Root diameter and direction were measured at each 0.1 m depths of the slope. The root distribution of cedar A is shown in Figure 2.

Considering the resistance force of an inclined root in a landslide, the component of the force perpendicular to the direction of landslide movement has a negligible influence on the landslide. Therefore, the spatial distribution of roots can be projected in a 2-D cross section of a sliding slope (Nakamura and Nghiem, 2002). The distribution of tree roots is generally very complicated and depends on many conditions such as the properties of the soil, species, age of forest, and other factors. Cedar roots usually taper with length. The diameter of each main root fiber is much larger than the diameters of the branch roots (Figure 3). The tree roots also run in random directions following various inclined angles (see Figure 2). In this simplified model, each root fiber is assumed to have a constant diameter along its length computed by taking the average diameter of its two ends. All of the tree roots are modeled as straight fibers (vertical and inclined fibers). In most cases, the length and depth of the roots in the fiber model measured from ground surface to the point at which roots were broken. The root distributions are described by two factors: size and inclination. Size of the root fibers is classified into diameters of 12.5, 15, 25, and 50 mm (Figure 4). Roots

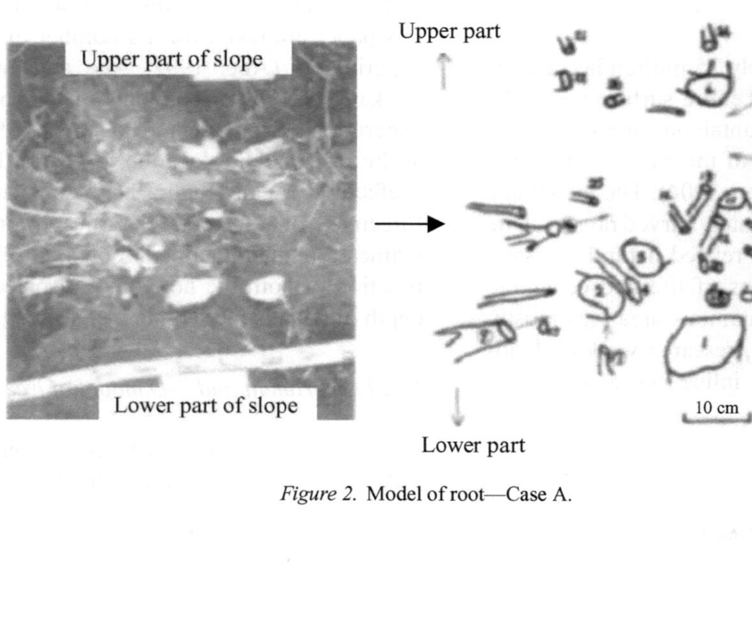

Figure 2. Model of root—Case A.

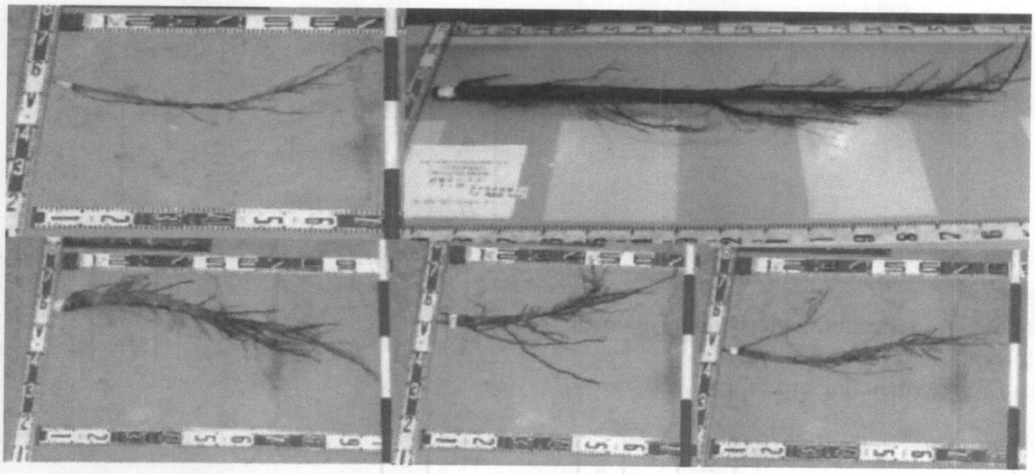

Figure 3. Root fibers—Case A.

inclined in the direction of the landslide are classified into 40°, 55°, 70°, and 80° and roots inclined against the direction of the landslide are classified into −40°, −55°, −70°, and −80°.

Root behavior in shear zone

By considering the equilibrium condition between pressure of sliding soil acting on a root body and reaction of the root in the horizontal and vertical directions (Figure 5), Eqs. (1) and (2) are formed respectively (Nghiem et al., 2003):

$$EI_i \frac{d^4y}{dx^4} + Es_i(y_i - p_i \sin\alpha) = P_{x_i} \frac{d^2y}{dx^2} \quad (1)$$

$$P_{xi} = Es_i \int_0^x (y_i \cot an\alpha - p_i \cos\alpha)\, dx \quad (2)$$

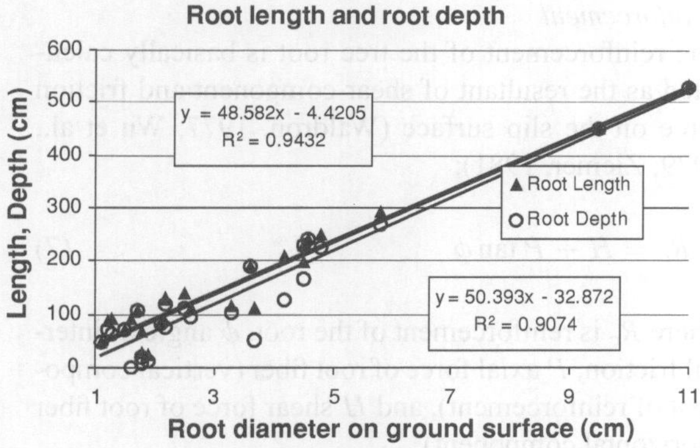

Figure 4. Distribution of roots.

where α is inclination of the root and the ith soil layer is Young's modulus Es_i, displacement p_i, deflection y_i, and axial force P_{xi}. EI_i is the stiffness of root, E is Young's modulus, and I is bending stiffness of root.

A fourth order differential Eq. (1) is the governing equation of deflected root fibers (Nakamura, 1987, 1990; Poulos, 1995). The theoretical solutions of Eq. (1) are four values of root deflection, deflection angle, bending moment, and shear force as follows:

$$\begin{aligned} yi &= e^{bx_i}[C_{1i}\cos(ax_i) + C_{2i}\sin(ax_i)] \\ &\quad + e^{-bx_i}[C_{3i}\cos(ax_i) + C_{4i}\sin(ax_i)] \\ \frac{dy_i}{dx_i} &= \theta_{xi}, \quad EI\frac{d^2y_i}{dx_i^2} = -Mx_i, \\ EI\frac{d^3y_i}{dx_i^3} &= -Hx_i \end{aligned} \quad (3)$$

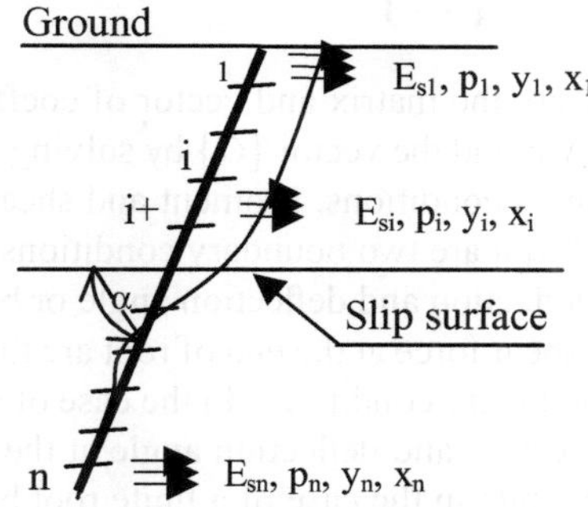

Figure 5. Deflected root.

where

$$b = \sqrt{R}\cos\left(\frac{\mu}{2}\right),\quad a = \sqrt{R}\sin\left(\frac{\mu}{2}\right),\quad R = \left(\frac{\sqrt{E}}{EI}\right),$$

and

$$\cos(\mu) = \frac{R}{2EIR}$$

θ_{xi}, $M_{xi,}$ and H_{xi} are deflection angle, bending moment and shear force at ith layer respectively.

Eq. (3) can be rewritten in matrix form.

$$\begin{Bmatrix} y_x \\ \theta_x \\ M_x \\ H_x \end{Bmatrix} = \begin{bmatrix} V_{11} & V_{12} & V_{13} & V_{14} \\ V_{21} & \cdots & \cdots & \cdots \\ V_{31} & \cdots & \cdots & \cdots \\ V_{41} & \cdots & \cdots & V_{44} \end{bmatrix} \begin{Bmatrix} C_1 \\ C_2 \\ C_3 \\ C_4 \end{Bmatrix} + \begin{Bmatrix} p \\ 0 \\ 0 \\ 0 \end{Bmatrix} \tag{4}$$

Here, matrix $[V]$ is the matrix of coefficient of functions in Eq. (3). $\{C\}$ is a vector of integral coefficients. At head ($x = 0$ or ground surface), the boundary values are

$$y_x = y_o, \theta_x = \vartheta_o, M_x = M_o, \text{ and } H_x = H_o.$$

Substituting these values into Eq. (4), the solution leads to the integral coefficients in Eq. (5).

$$\begin{Bmatrix} C_1 \\ C_2 \\ C_3 \\ C_4 \end{Bmatrix} = \begin{bmatrix} A_{11} & A_{12} & A_{13} & A_{14} \\ A_{21} & A_{22} & A_{23} & A_{24} \\ \cdots & \cdots & \cdots & \cdots \\ A_{41} & \cdots & \cdots & A_{44} \end{bmatrix} \begin{Bmatrix} y_o \\ \theta_o \\ M_o \\ H_o \end{Bmatrix} + p \begin{Bmatrix} U_1 \\ U_2 \\ U_3 \\ U_4 \end{Bmatrix} \tag{5}$$

$[A]$ and $\{U\}$ are the matrix and vector of coefficients respectively. We find the vector $\{C\}$ by solving Eq. (5) with four known conditions. Moment and shear force at the head of root are two boundary conditions. Similarly, either deflection and deflection angle or bending moment and shear force at the end of root are the other two known boundary conditions. In the case of an infinite root, deflection and deflection angle at the end of root are zero while in the case of a finite root bending moment and shear force are zero. The combination of Eqs. (4) and (5) leads to Eq. (6):

$$\begin{Bmatrix} y_x \\ \theta_x \\ M_x \\ H_x \end{Bmatrix} = \begin{bmatrix} V_{11} & V_{12} & V_{13} & V_{14} \\ V_{21} & V_{22} & V_{23} & V_{24} \\ \cdots & \cdots & \cdots & \cdots \\ V_{41} & \cdots & \cdots & V_{44} \end{bmatrix} \times \left(\begin{bmatrix} A_{11} & A_{12} & A_{13} & A_{14} \\ A_{21} & A_{22} & A_{23} & A_{24} \\ \cdots & \cdots & \cdots & \cdots \\ A_{41} & \cdots & \cdots & A_{44} \end{bmatrix} \begin{Bmatrix} y_o \\ \theta_o \\ M_o \\ H_o \end{Bmatrix} + p \begin{Bmatrix} U_1 \\ U_2 \\ U_3 \\ U_4 \end{Bmatrix} \right) + \begin{Bmatrix} p \\ 0 \\ 0 \\ 0 \end{Bmatrix} \tag{6}$$

Substituting C into Eq. (4), we have y_x, θ_x, M_x, and H_x. These values will be substituted back to Eq. (5) as new values of y_o, θ_o, $M_{o,}$ and H_o. Each segment of the root in one soil layer has four values of deflection, deflection angle, bending moment, and shear force at two ends (called deflection vector of root). When repeating the computation, the deflection vector at the lower end of a given root segment is equal to the initial deflection vector at the head of the next segment.

By using the model of inclined roots in Figure 5, we have bending moment, shear force, and axial force of tree roots (see Figure 6). The behavior of tree roots in Figure 6 contributes to reinforcement R_c, pullout resistance P_p, and normal stress σ_u. These parameters are used to determine the reinforcement capacity of tree root and are discussed in the next section of paper.

Reinforcement

The reinforcement of the tree root is basically calculated as the resultant of shear component and friction force on the slip surface (Waldron, 1977; Wu et al., 1979; Ziemer, 1981):

$$R_c = H + P\tan\phi \tag{7}$$

where R_c is reinforcement of the root, ϕ angle of internal friction, P axial force of root fiber (vertical component of reinforcement), and H shear force of root fiber (horizontal component).

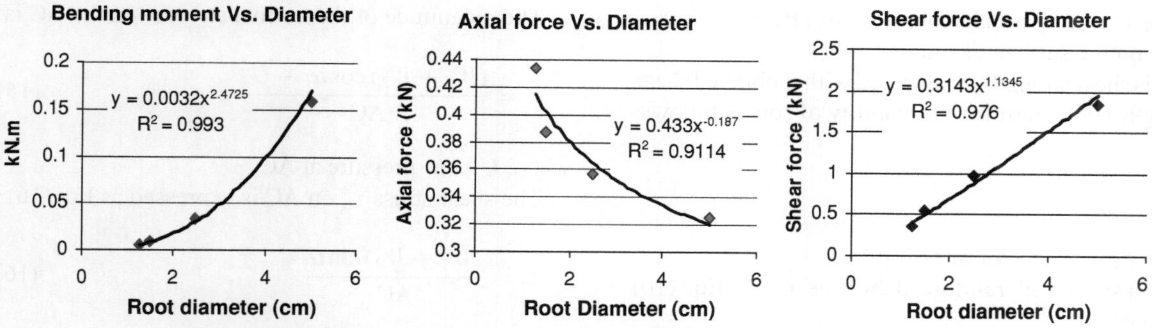

Figure 6. Behavior of cedar roots of Ozawa slope (horizontal displacement = 1 cm).

Pullout resistance
The pullout resistance of a tree root fiber is equal to the skin friction between a root fiber and soil:

$$P_p = k \cdot \pi \cdot d \cdot L_2 \tag{8}$$

where P_p pullout resistance, k coefficient of skin friction between root fiber and soil ($k = 48$ kPa for Japanese cedar roots), and L_2 length of a tree root under slip surface. The displacement of soil is called pullout displacement if the axial force of a root equals pullout resistance $P_x = P_p$.

Ultimate condition of broken roots
The normal stress of roots is calculated by the basic Eq. (9)

$$\sigma_u = \frac{P}{F} \pm \frac{M}{W} \tag{9}$$

where σ_u normal stress, P axial force, F area of cross section, M bending moment, and W bending stiffness. σ_u is compared with ultimate stress to determine the breakage of roots. The displacement of soil is called ultimate displacement if normal stress of root equals to ultimate stress (yield or permission stress).

Reinforcement capacity of tree roots
The reinforcement capacity of root fibers is determined based on the ultimate condition and pullout resistance (Nghiem et al., 2004). At the given horizontal displacement, if a tree root is broken but yet not pulled out, ultimate displacement is smaller than pullout displacement and thus the reinforcement capacity depends on ultimate displacement. Conversely, when a root fiber is pulled out while it is not broken, the reinforcement capacity depends on pullout displacement of the root fiber.

Another assumption is the reinforcement of a broken root fiber will equal zero or the reinforcement of pullout root will be constant even if displacement of the unstable soil continues to increase. Figure 7 shows the calculated results of the reinforcement of a cedar root with respect to different depths of slip surface. A displacement of 8.5 mm is the maximum for cedar at the Ozawa slope. For greater displacement, root fibers will be pulled out or broken. Thus, root reinforcement will decrease.

Safety factor of slope considering root reinforcement

The maximum shear strength of soil (S) follows Coulomb's law:

$$S = c + \sigma \tan\phi \tag{10}$$

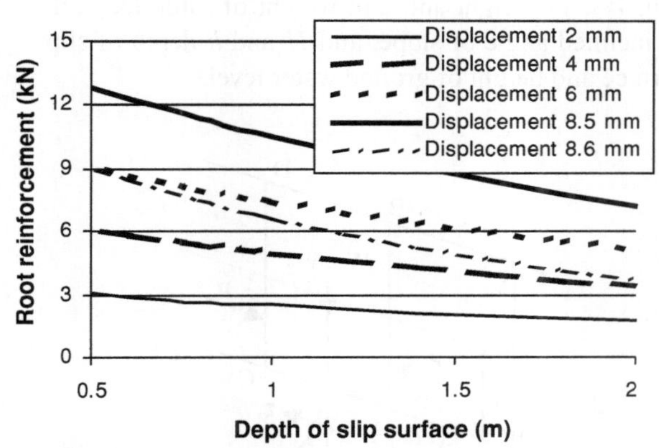

Figure 7. Reinforcement capacity as function of displacement and depth of slip surface.

where c effective unit cohesion, σ effective normal stress on the surface of rupture.

This shear strength is compared with mobilized shear strength τ to determine the stability of soil as follows

$$F_s = \frac{S}{\tau} \tag{11}$$

where F_s is the factor of safety.

In case of soil reinforced by tree roots, Eq. (10) becomes:

$$S_{(R_c)} = c + \sigma \tan\phi + \Delta S_{(R_c)} \tag{12}$$

where $\Delta S_{(R_c)}$ increment of soil cohesion by root reinforcement. This additional cohesion is a function depending on the displacement, depth of landslide, and distribution of root fibers in the soil.

The cohesion of rooted soil $c_{(R_c)} = c + \Delta S_{(R_c)}$ gives the equation:

$$S_{(R_c)} = c_{(R_c)} + \sigma \tan\phi \tag{13}$$

where : $c_{(R_c)}$ is cohesion of rooted soil, a function of horizontal displacement, depth of slip surface, and distribution of root fibers in the soil.

Figure 8 shows one slice of an infinite slope at an angle of α_l to the horizontal. The element ABDC of unit width wherein AC is parallel to BD has weight W, and has a vertical force acting on AC.

$$W = W_1 + W_2 = \gamma_1(H - h)cos\alpha_l + \gamma_2 h\, cos\alpha_l \tag{14}$$

where W_1, γ_1 weight and unit weight of unsaturated soil, W_2, γ_2 weight and unit weight of saturated soil, α_l inclined angle of slope, and H and h depth of slip surface and height of ground water level.

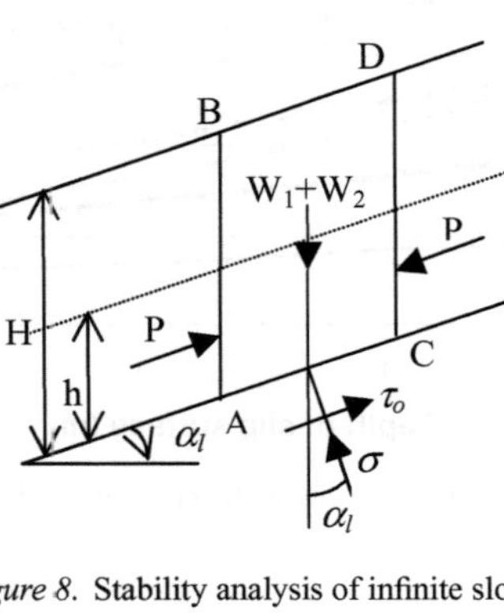

Figure 8. Stability analysis of infinite slope.

The magnitude of the normal stress σ on the AC is

$$\sigma = \frac{(W_1 + W_2)\cos\alpha_l - U}{\text{AC}} \tag{15}$$

where U pore-pressure at AC.

The shear stress τ_o on AC is expressed as Eq. (16).

$$\tau_o = \frac{(W_1 + W_2)\sin\alpha}{\text{AC}} \tag{16}$$

If resistance is mobilized on AC, substituting σ and τ_o into Eq. (11), we have

$$F_s = \frac{\sum[\{(W_1 + W_2)\cos\alpha_l - U\}\tan\phi + (c + \Delta S_{(R_c)})l]}{\sum\{(W_1 + W_2)\sin\alpha_l\}} \tag{17}$$

where l is the length of slip surface.

Rewriting Eq. (17), we have:

$$F_s = \frac{R_f + \sum S_{(R_c)}l}{D_f} \tag{18}$$

where D_f driving force and R_f resisting force. Using Figure 7 and Eq. (18) we will analyze the influence of root reinforcement on slope stability.

Results and discussion

Influence of tree roots on slope stabilization

To understand the effect of root reinforcement on slope stability, we calculated the F_s for a slope in Ozawa (Iwate prefecture, Japan). The soil properties and cross section of the Ozawa slope are shown in Table 1 and Figure 9. Cedars A and B (Figure 1) are 16–20 years old and have an average weight equivalent to 3 kN. The intervals between the cedar trees are assumed to be 4 m. The length of the slip surface is about 20 m

Table 1. Properties of slopes

γ_1	Unit weight of unsaturated soil	14 kN/m^3
γ_2	Unit weight of saturated soil	17 kN/m^3
C	Cohesion of soil	5.50 kN/m^2
ϕ	Angle of internal friction	30°

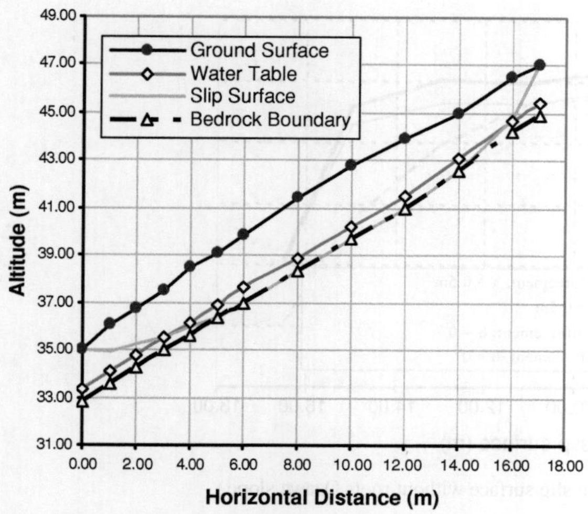

Figure 9. Cross section of Ozawa slope.

(Figure 9), thus there are five trees on the slope. The depth of cedar roots is assumed to be less than 3 m.

At the beginning of the study, the conditions in Table 1 maintain F_s of slope ≥ 1. If for some reason e.g. rainfall or erosion, F_s becomes smaller than 1, the slope is destabilized and the horizontal displacement of the landslide will increase. In case of a slope sliding shorter than the ultimate displacement 8.5 mm, a larger horizontal displacement results in greater root reinforcement (Figure 7). This reinforcement minimizes the landslide. Horizontal displacement increases until F_s of the slope returns to 1 and the slope becomes stable. At the maximum displacement *hd*, if the slope properties are restored to the initial conditions in Table 1, i.e. no rain fall or no erosion, the safety factor of the slope will be greater than the initial value. When this happens, Eq. (19) calculates the factor of safety:

$$Fs = \frac{R_f^0}{D_f^0} + \frac{\Delta S_{(hd)} l}{D_f^0} \geq 1 + \frac{\Delta S_{(hd)} l}{D_f^0} \tag{19}$$

where R_f^0 and D_f^0 resisting force and driving force of slope at beginning (horizontal displacement = 0), $\Delta S_{(hd)}$ cohesion of rooted soil at maximum horizontal displacement, hd: maximum horizontal displacement.

The steps and results of the slope stability analysis are shown in Figure 10. An increase in the ground water

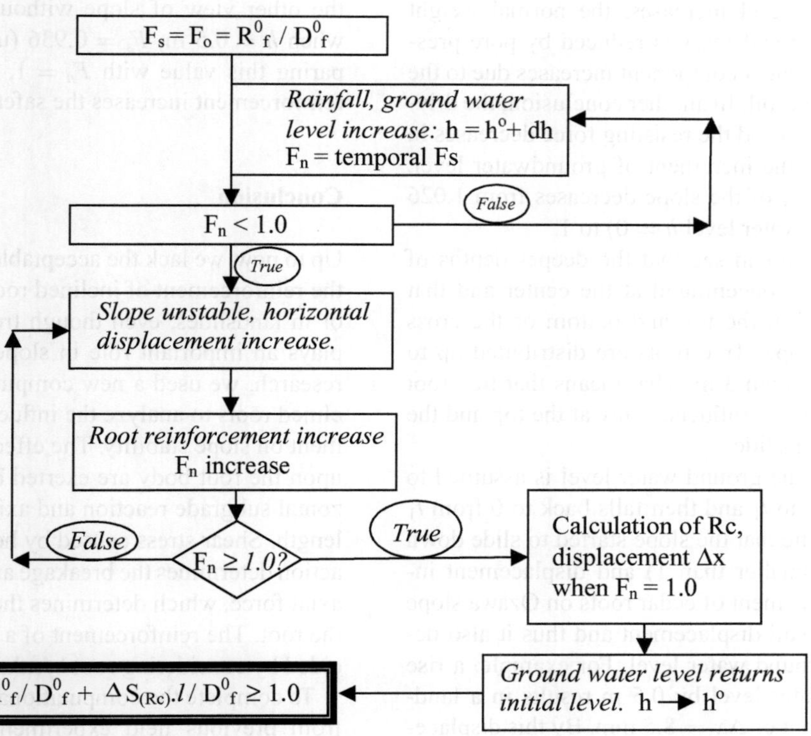

Figure 10. Landslide development of slope with root reinforcement, where h is height of ground water level, h^0 is h at start time, dh is increase of h, F_n is factor of safety when height of ground water level is $h > h^0$.

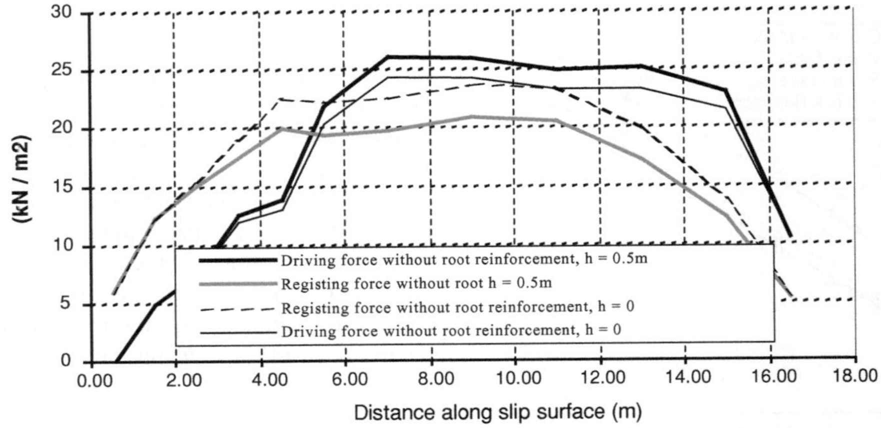

Figure 11. Distribution of driving and resisting force on slip surface without roots Ozawa slope.

level caused slope instability. Figure 11 shows the influence of groundwater level on the distribution of resisting and driving forces on the slip surface. The thin lines represent the driving force and resisting force of the slope at initial groundwater level ($h = 0$) whilst thick lines represent the driving force and resisting force of the slope at groundwater level $h = 0.5$. When the groundwater level increases, the normal weight component of the soil mass is reduced by pore pressure whilst the tangent component increases due to the effect of saturated soil. In another conclusion, the driving force increases and the resisting force decreases in accordance with the increment of groundwater level. As a result, the F_s of the slope decreases from 1.026 (at initial groundwater level $h = 0$) to 1.

In Figure 9, we can see that the deeper depths of the landslide are concentrated at the center and that they are reduced at the top and bottom of the cross section of the slope. Tree roots are distributed up to depths shallower than 3 m. That means that tree root reinforcement has an influence only at the top and the bottom of the landslide.

In Figure 12, the ground water level is assumed to rise from time 0 to t_1 and then falls back to 0 from t_1 to t_2. t_0 is the time that the slope started to slide down (F_s is slightly smaller than 1) and displacement increases. Reinforcement of cedar roots on Ozawa slope depends on the soil displacement and thus it also depends on the ground water level. For example, a rise in the groundwater level by 0.5 m results in a landslide displacement of $\Delta x = 8.5$ mm. By this displacement, reinforcement of tree roots reaches 20.3 kN, and this reinforcement function is $F_n = 1$. At this situation (displacement $= 8.5$ mm), if the groundwater level draws down to its initial value ($h = 0$), the slope displacement will still remain (it cannot return to zero). The reinforcement of tree roots (20.3 kN) also remains. At that time, $F_s = 1.092$, as calculated by Eq. (19). Comparing this value to the initial $F_s = 1.026$, we can see that tree roots increase slope stability by 6.4%. In the other view of slope without root reinforcement, when $h = 0.5$ m, $F_s = 0.936$ (unstable slope). Comparing this value with $F_n = 1$, we can see that root reinforcement increases the safety factor by 6.8%.

Conclusion

Up to now, we lack the acceptable methods to compute the reinforcement of inclined roots in mass movement or in landslides, even though tree root reinforcement plays an important role in slope stabilization. In our research, we used a new computational model for inclined roots to analyze the influence of root reinforcement on slope stability. The effects of mass movement upon the root body are exerted by coefficient of horizontal subgrade reaction and axial reaction along root length. Shear stress caused by horizontal subgrade reaction determines the breakage and axial stress causing axial force, which determines the pullout resistance of the root. The reinforcement of a tree root is finally decided by considering root breakage and pullout.

To complete the computational model, we used data from previous field experiments in Ozawa—Iwate, Japan. Properties of an actual slope were also used with the model as an illustration. In the case study, the

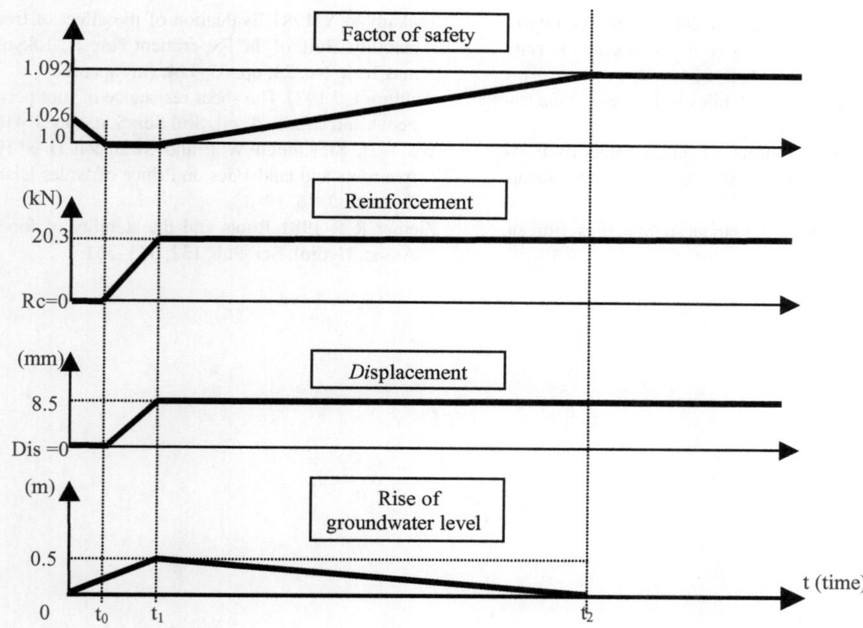

Figure 12. Influence of ground water level on root reinforcement and slope stability.

calculation shows that the F_s increased up to 6.8% through the influence of tree root reinforcement. The important conclusion is the tree roots kept Ozawa slope stable even though ground water increased up to 0.5 m. A slope without root reinforcement will become unstable when groundwater level increases by only 0.15 m.

This paper analyzes slope stability by considering the safety factor of the slope as a function of displacement of soil, depth of landslide and root distribution. To complete the calculation, data for tree roots classified by size, inclination, species, and age of trees are necessary. The distribution and properties of the roots are needed. Also needed are surveys of experiments on the properties of various vegetated soils in different seasons and weather conditions. Having a full database of roots and soil will help us predict how tree roots stabilize vegetated soils better.

Acknowledgments

We thank Prof. Gregory and Dr. Alexia Stokes for valuable suggestions and useful advice. Without their help we may have needed much more time to finish this paper.

References

Abe K 1992 Studies on Evaluation Methods for Effect of Tree Roots on Preventing Landslide. Ph.D. Dissertation. Forestry and Forest Product Research Institute. Tsukuba, Japan. (in Japanese)

Abe K and Iwamoto M 1986 An evaluation of tree root effect on slope stability by tree root strength. J. Jpn. Forest Soc. 68, 505–510.

Abe K and Iwamoto M 1990 Simulation model for the distribution of tree roots. J. Jpn. Forest Soc. 72, 375–387.

Endo T and Tsuruta T 1969 The Effect of the Tree's Roots Upon the Shear Strength of Soil. Annual Report of the Hokkaido Branch Forest Experiment Station, pp. 167–182.

Gray D H and Sotir R B 1996 Biotechnical and Soil Bioengineering Slope Stabilization: A Practical Guide for Erosion Control. Wiley, New York.

Nakamura H 1987 Fundamental and practical studies on landslide prevention works. Ph.D. Dissertation. Tokyo University. Tokyo, Japan. (in Japanese)

Nakamura H 1990 Design of Steel Pile in Landslides. Japan Landslide Prevention Technical Society, Tokyo, Japan. (in Japanese)

Nakamura H and Nghiem M Q 2002 Evaluation of tree root reinforcement in slope stability. International Association of Geology and The Environment, Proc. 9, pp. 1072–1080.

Nghiem M Q, Nakamura H and Shiraki K 2001 Analysis of tree root reinforcement in slope stability. The Japan Landslide Society, Proc. 40, pp. 187–191.

Nghiem M Q, Nakamura H and Shiraki K 2003 Analysis of root reinforcement at slip surface. J. Jpn. Landslide Soc. 40(4), 44–52.

Nghiem M Q, Nakamura H and Shiraki K 2004 Slope stability of forested slopes considering effect of tree root and steel bar reinforcement. J. Jpn. Landslide Soc. 41(3), 264–272.

Poulos H G 1995 Design of reinforcing piles to increase slope stability. Can. Geot. J. 32, 808–818.

Sabo Technical Center 1997 Reports of pullout-tree root experiment in fields. Sabo Technical Center, Tokyo, Japan. (in japanese)

Statistical Handbook of Japan 2002 Japanese Statistics Bureau, Tokyo, Japan.

Tsukamoto Y 1987 Evaluation of the effect of tree roots on slope stability. Bull. of the Experiment Forests, Tokyo Univ. of Agric. and Tech. No. 23, pp. 65–124. (in Japanese)

Waldron L J 1977 The shear resistance of root permeated homogeneous and stratified soil. Soil Sci. Soc. Am. J. 41(3), 843–849.

Wu T H, McKinnell W P and Swanston D N 1979 Strength of tree roots and landslides on Prince of Wales Island, Alaska. Can. Geotech. J. 16, 19–33.

Ziemer R R 1981 Roots and the stability of forested slopes. Int. Assoc. Hydrol. Sci. Pub. 132, 343–361.

Observation and simulation of root reinforcement on abandoned Mediterranean slopes

L.P.H. van Beek[1,5], J. Wint[2], L.H. Cammeraat[3] & J.P. Edwards[4]

[1]*Department of Physical Geography, Utrecht Centre of Geosciences, Utrecht University, P.O. Box 80115, NL 3508 TC, Utrecht, The Netherlands.* [2]*Department of Civil Engineering, Nottingham trent University, UK.* [3]*IBED-Physical Geography, University of Amsterdam, The Netherlands.* [4]*Scott Wilson Pavement Engineering Ltd., Nottingham, UK.* [5]*Corresponding author**

Key words: FLAC 2D, *in-situ* direct shear tests, root pull-outs, root reinforcement, slope stability modelling, vegetation

Abstract

The mechanics of root reinforcement have been described satisfactorily for a single root or several roots passing a potential slip plane and verified by field experiments. Yet, precious little attempts have been made to apply these models to the hillslope scale pertinent to landsliding at which variations in soil and vegetation become important. On natural slopes positive pore pressures occur often at the weathering depth of the soil profile. At this critical depth root reinforcement is crucial to avert slope instability. This is particularly relevant for the abandoned slopes in the European part of the Mediterranean basin where root development has to balance the increasing infiltration capacity during re-vegetation. Detailed investigations related to root reinforcement were made at two abandoned slopes susceptible to landsliding located in the Alcoy basin (SE Spain). On these slopes semi-natural vegetation, consisting of a patchy herbaceous cover and dispersed Aleppo pine trees, has established itself. Soil and vegetation conditions were mapped in detail and large-scale, *in-situ* direct shear tests on the topsoil and pull-out tests performed in order to quantify root reinforcement under different vegetation conditions. These tests showed that root reinforcement was present but limited. Under herbaceous cover, the typical reinforcement was in the order of 0.6 kPa while values up to 18 kPa were observed under dense pine cover. The tests indicate that fine root content and vegetation conditions are important factors that explain the root reinforcement of the topsoil. These findings were confirmed by the simulation of the direct shear tests by means of an advanced root reinforcement model developed in FLAC 2D. Inclusion of the root distribution for the observed vegetation cover mimics root failure realistically but returns over-optimistic estimates of the root reinforcement. When the root reinforcement is applied with this information at the hillslope scale under fully saturated and critical hydrological conditions, root pull-out becomes the dominant root failure mechanism and the slip plane is located at the weathering depth of the soil profile where root reinforcement is negligible. The safety factors increase only slightly when roots are present but the changes in the surface velocity at failure are more substantial. Root reinforcement on these natural slopes therefore appears to be limited to a small range of critical hydrological conditions and its mitigating effect occurs mainly after failure.

Introduction

Roots can contribute significantly to the stability of shallow soils on slopes (e.g., O'Loughlin, 1974; Schmidt et al., 2001). The mechanics of root reinforcement have been described satisfactorily for a single root or several roots passing a potential slip plane (e.g., Abe and Ziemer, 1991; Waldron, 1977; Wu, 1984, 1995) and these models have been corroborated by field and laboratory measurements (e.g., Riestenberg, 1994; Wu

* E-mail: r.vanbeek@geog.uu.nl

This article has been previously published in the following journal—Plant and Soil, Volume 278, 55–74.
Stokes et al. (eds), Eco- and Ground Bio-Engineering: The Use of Vegetation to Improve Slope Stability, 91–109.

et al., 1988). Yet, precious little attempts have been made to apply these models to the hillslope scale pertinent to landsliding. This omission should be attributed to the limited size at which root reinforcement can be tested in the field and the emergence of the variability in soil and vegetation properties as a major factor at the hills-lope scale. Consequently, the strain and strength at failure may be very different for actual landslides when compared to those under the controlled and idealised conditions of *in-situ* tests.

This paper therefore aims to translate local measurements of root reinforcement to the hills-lope scale. Field evidence of root–soil interaction is combined with a geomechanical model by which root reinforcement and ultimately root failure can be simulated under different conditions. Such a model is needed because root behaviour under strain is essentially different from that of soils.

The model that considers the root reinforcing mechanisms in detail has been developed as a routine in FLAC 2D, a commercial finite difference code with widespread application in geo-engineering (Itasca, 2002). It simulates the stress–strain behaviour numerically so the strain-dependent effect of reinforcement can be simulated more realistically with fewer simplifying assumptions. Moreover, the root reinforcement model in FLAC offers the user to specify varying root and soil properties along the slope and the influence of the hydrology on the effective stress can be evaluated rigorously. This is highly advantageous since root reinforcement is influenced by the type and nature of the vegetation and local variations in soil conditions.

Descriptions of *in-situ* measurements and the root reinforcement model are given prior to the presentation of the case study that provides the field evidence for the application of the model. The field evidence stems from two slope transects affected by slope instability in the region near Alcoy in SE Spain. Here, landsliding is rife on steep slopes in weathered Miocene marl (see for a detailed description: La Roca and Calvo-Cases, 1988; La Roca, 1991; Van Beek, 2002). This region conforms to the general trend of the abandonment of marginal agricultural fields in the European part of the Mediterranean basin (MacDonald et al., 2000). Subsequent re-vegetation of these fields increases the infiltration and storage capacity of the topsoil. On abandoned slopes this leads to elevated pore pressures during prolonged or intense rainstorms and induce landsliding (Cammeraat et al., 2005). However, these negative aspects may be partly counter-acted by increased root reinforcement that would define the ultimate stability of such abandoned slopes.

In-situ measurement of root reinforcement

Root reinforcement was studied by *in-situ* direct shear tests and root–soil interaction by pull-out tests. Additional descriptions and measurements provide background information to these tests and are presented briefly as part of the case studies where appropriate.

Pull-out tests were performed *in-situ* using a frame that allowed steady straining of the root until failure occurred. During the tests, the soil remained at the ambient moisture content. The maximum mobilised force was measured by means of a spring balance and recorded against the applied displacement (Cammeraat et al., 2002). In addition, species, diameter, orientation and inclination were noted for every root tested.

The shearing resistance of the rooted topsoil was tested *in-situ* by large-scale direct shear tests, as small samples cannot truthfully represent the effect of root reinforcement. The dimensions of the shear box used were $0.6 \times 0.6\ m^2$ in plan and 0.4-m deep to encompass a sufficiently large volume of rooted soil. The box was sunk vertically into the soil and the base excavated to provide a level surface over which the encased block could travel. Shear was applied by means of a jack and the generated shearing resistance measured with a proving ring as a function of the displacement. A normal load was applied by means of a dead weight of concrete blocks. Two loads were used, resulting in normal stresses of 3.3 and 4.1 kPa. The soil was wetted thoroughly prior to testing to eliminate any suction-derived resistance and the shear rate kept low (4 mm min^{-1} on average) to avoid the build-up of excess pore pressures. Tests were carried out in two modes. In the first mode, a four-sided shear box was used, in the second mode the soil block was bounded by plates only perpendicular to the direction of shearing so that roots extending through the soil block were not truncated. In the second mode, 10 metal rods were installed on either side of the box to measure strain in the adjacent soil. After testing, root counts were made (see below). Root reinforcement was calculated by subtracting the theoretical shearing resistance of the non-rooted soil, which was determined in the laboratory on undisturbed samples.

From the direct shear tests and the pull-outs roots were collected to be tested in the laboratory in order to determine root elasticity and tensile strength.

Root reinforcement model

The root reinforcement model is based on the theory of reinforced soil (Vidal, 1966) and an extension of existing two-dimensional analytical models (e.g., Waldron, 1977; Wu, 1984). This process description applies to a 2D case in the vertical X–Y plane on which all roots are projected. According to the concept of reinforced earth root reinforcement is the result of the elongation of roots across a potential slip plane which generates a root force F_r that is transferred to the soil by the cohesive and frictional contacts between the root and the soil (Figure 1). Roots have been shown to deform elastically to imposed stresses (Waldron, 1977; Wu et al., 1979) and the root stress can therefore be calculated by means of Hooke's Law:

$$\sigma_r = E_r \frac{\Delta L}{L} \tag{1}$$

where σ_r is the resultant root stress [Pa],
E_r is the modulus of elasticity (Young's modulus) of a root [Pa],
$\Delta L/L$ is the elongation of a root per unit length of a root [m m^{-1}].
The actual reinforcement that can be mobilised is limited by two failure modes. Dependent on the loading, fully anchored roots will snap when the root stress exceeds the tensile or compressive strength. Alternatively, roots may fail prematurely by pull-out if during the loading of the root the resistance along the root–soil interface is overcome (Waldron, 1977). For a root extending across the slip plane the longitudinal stress, σ_a, along the root before failure by is given by (Waldron, 1977):

$$\frac{d\sigma_a(x)}{dx} = 4^B/d_r, \tag{2}$$

where d_r is the root diameter [m], and,
B is the bond strength, i.e. the shearing resistance at the root–soil interface for a unit length of an embedded root [Pa] and x is the distance along the root.
The bond strength is assumed to be independent of the root stress and the Mohr–Coulomb failure criterion is adopted here to describe it:

$$B = \mathrm{CO} \cdot (c_r + \sigma_n \tan\phi_r) = \mathrm{CO} \cdot R(c + \sigma_n \tan\phi), \tag{3}$$

where CO is the effective contact length along a root [m m^{-1}],
σ_n is the normal stress acting on a root [Pa], c_r and c, are respectively the cohesion along the root–soil interface and of the soil itself [Pa],
ϕ_r and ϕ, are respectively the friction angle along the root–soil interface and of the soil itself [°],
and R is a reduction parameter [–] that relates the strength properties along the interface to those of the soil.

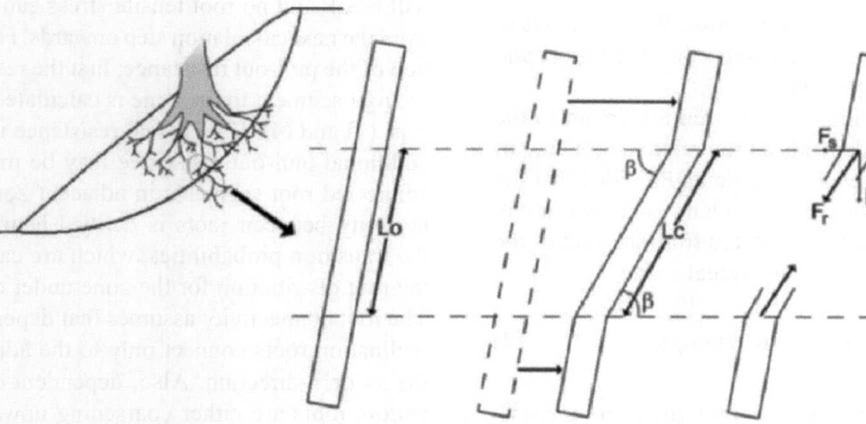

Figure 1. Schematisation of root reinforcement. A root passing a shear zone—indicated by the dashed horizontal lines—is extended from its original length L_0–L_c. This generates the root force $F_r = \sigma_r \cdot A_r$ that can be resolved by the angle of root inclination β into components normal and parallel to the shear plane, respectively, F_n and F_s.

At the slip plane $x = 0(x_0)$ the root stress is at a maximum, $\sigma_a(0) = \sigma_r$ and it decreases towards its end (Waldron, 1977). To avoid failure by pull-out, the equilibrium along the soil–root interface before failure is given by the integration of Eq. (2) over the root segment x_0x_1:

$$\sigma_a(x_1) - \sigma_a(x_0) = 4B(x_1 - x_0)/d_r, \tag{4}$$

assuming that the diameter d_r is constant along this segment.

Since $\sigma_a(x_1)$ would be zero under equilibrium conditions when the bond strength is fully mobilised, the maximum shearing resistance along the soil–root interface is used to determine whether the root would fail by pull-out, i.e., $\sigma_r \geq 4B(x_1 - x_0)/d_r$, or in tension $\sigma_r < 4B(x_1 - x_0)/d_r$.

The mobilised root stress can be resolved for the enclosed angle β into two components of root reinforcement, working respectively, normal and parallel to the slip plane (Figure 1):

$$F_n/A_r = \sigma_r \sin\beta, \tag{5}$$

$$F_s/A_r = \sigma_r \cos\beta, \tag{6}$$

where A_r is the root cross-sectional area.

The component normal to the potential slip plane exerts a confining stress and adds to the frictional component of the shearing resistance along the slip plane. The parallel component is aligned with the shear stress along the slip plane. When a root is in compression, in which case the enclosed angle β is obtuse, the root stress component works in the direction of the shear stress and the component is negative. When a root is in tension the component counteracts the shear stress and the contribution is positive.

The total contribution of root reinforcement to the shear strength, S_r, depends on the relative root area of every root that passes the slip plane. Provided that the root distribution along the slip plane is known or can be estimated, S_r can be calculated from the sum of the root reinforcement of the individual roots:

$$S_r = \sum \sigma_r \frac{A_r}{A}(\cos\beta + \sin\beta\tan\phi), \tag{7}$$

where A_r/A is the relative root area (root cross-sectional area over total area).

The root reinforcement model based on Eqs. (1)–(7) has been coded in FLAC 2D (Itasca, 2002). The main differences between the original analytical formulations of the model and its implementation in FLAC concern the deformation of the soil mass and the description of distorting roots therein. In FLAC the soil mass is described by a grid of contiguous zones that connect at nodes. FLAC does not employ predefined shear planes but describes the stress–strain behaviour of the soil numerically leading to elastic or plastic deformation under the imposed loads. Thus, some assumptions on the deformation of the shear plane after failure can be relaxed and root reinforcement can be determined at any location and at any moment during the deformation of the slope.

For the analysis a plane strain configuration is used with the 2D plane coinciding with that in which the major and minor principal stresses are acting. On this plane, roots are projected with an inclination i to the positive x-axis (0°–180°). Roots are classified according to this inclination and their diameter to give a root distribution, which specifies the number of roots passing a horizontal plane of one square meter per class.

The roots are treated implicitly in the model and an average root passing through the midpoint of a zone is taken to represent each class. On deformation, the inclination and length of the root segment change as a result of the normal strain, the shear strain and rotation experienced by each zone. For the deforming root, the root stress can be calculated from Eq. (1) if not broken already. The root stress of elongating roots is limited by their tensile strength, that of shortening roots by their compressive strength. If the root stress can be matched by the pull-out resistance and is limited by the tensile or compressive strength of the root, the root will break and no root tensile stress can be mobilised from the next calculation step onwards. For the calculation of the pull-out resistance, first the resistance along the root segment in the zone is calculated by means of Eqs. (3) and (4). If this local resistance is insufficient, additional pull-out resistance may be mobilised from connected root segments in adjacent zones. The connectivity between roots is derived heuristically from the transition probabilities, which are calculated from the root distribution for the zone under consideration. The root connectivity assumes that depending on their inclination roots connect only to the adjacent zone in the x- or y-direction. Also, dependent on their inclination, roots are either coarsening upwards or fining downwards. This results in an equivalent root length of which the corresponding resistance is added to the pull-out resistance. Different layers or root types can be

Table 1. Parameterisation of the root reinforcement model in FLAC 2D

Root mechanical properties	Generation of root stress and root failure by breakage
Root elasticity, E_r [Pa]	
Root tensile strength, T_r [Pa]	
Ratio between compressive and tensile strength, RCS [–]	
Root–soil interaction	Failure by pull-out
Cohesion*, c' [Pa]	
Friction angle*, ϕ' [°]	
Reduction factor, R [–]	
Effective contact length, CO [–]	
Root distribution	Classified root content linked to vegetation type and depth; used to describe root deformation and cross-sectional area
Root count	
Inclination in X–Y plane	
Diameter	

*Also used by FLAC.

used to capture the distribution of roots with depth or along the slope. For each root class, the parameters CO and R of Eq. (4), root elasticity (Eq. (1)) and the root tensile strength must be specified. The latter is related by means of a global parameter to the root compressive strength (Table 1).

The root reinforcement of Eq. (7) is calculated in FLAC as the reinforcement per class and summed. The term A_r/A of Eq. (7) is merely the root cross-sectional area times the root count, N_r, over 1 m^2. The resulting root reinforcement is treated as an additional cohesion for the slope normal component or an additional tension for the slope parallel component. Both material properties cannot be negative in FLAC. Moreover, FLAC constrains the tensile strength to the tension cut-off. In the unlikely event that the root reinforcement violates the physical or theoretical limits of the tensile strength and cohesion, the following procedure is invoked and an error message issued: (1) if the tension cut-off is exceeded, the remainder is added to the cohesion; (2) if the tensile component is smaller than zero, it is subtracted from the cohesion, provided that the overall value, soil cohesion included, does not become negative.

The root reinforcement calculations are invoked at the start of every calculation cycle in FLAC and change the shear resistance of the soil on the basis of the stresses and deformations from the previous time step. The resulting alterations in the strength have consequences for the deformations and stresses that are calculated in main program of FLAC for the current step, and this process is reiterated during the solving process.

Case studies

General site description

Detailed studies were made regarding morphology, vegetation and soil conditions at the two slope transects. Each transect was set out with three survey lines, 10 m apart, thus delineating a 20 m wide area. Along each survey line points were marked at every 10 m so a regular sampling network was created. From the elevation of these sampling points a profile was generated. This survey provided the basis for the mapping of the morphology and vegetation cover and the positioning of additional sample points.

The transects are located along a ravine (*barranco*) that dissects a pediment developed in Miocene marl (Transects A and B, see Figure 2). The ephemeral stream in the barranco forms the base of both slopes and signs of fluvial erosion are present. The transects receive a similar amount of insolation although their expositions differ (West and East, respectively). The slope transects have similar dimensions but slope B is steeper than A (Table 2). The relatively flat area of the pediment is presently cultivated but the bench terraces on the slopes have been abandoned and fallen into disrepair. In addition to the old terraces, landslide scars, gullies and deposition areas were identified along the slope. These morphological units were subject to diminishing degrees of erosion and secondary mass movement activity. The activity of these processes on the flatter old terraces was negligible or absent.

Sampling in soil pits and auger holes along the survey lines was used to describe the soil and to determine the porosity, dry bulk density and shear strength. Soil descriptions according to the FAO classification (FAO, 1990) also included determination of the particle size distribution (by dry sieving for the fraction >63 μm, by hydrometer tests for the fraction <63 μm), carbon content (cf. Wesemael, 1955) and organic carbon content (cf. Allison, 1935) for the soil pits. For the determination of the porosity and dry bulk density, undisturbed samples of 10^{-4} m^3 were taken in fivefold for every

Figure 2. Overview of the slope transects A and B.

Table 2. Morphometry of the slope transects A and B

		Slope angle [°]		
Transect	Length [m]	Average	Min.	Max.
A	110	20.7	3.6	32
B	100	23.7	7.8	39.3

horizon of the soil pits and at 0.25-, 0.75- and 1.25-m depth in the auger holes, if not restricted by the presence of bedrock. Undisturbed block samples were also used to determine the shear strength of the non-rooted soil on which anchorage and overall slope stability depend (see below).

In total, five soil pits and 80 auger holes were described and sampled. The observed soil depth was 0.95 m on average for both slopes but tended to be more variable on slope B, where the presence of buried topsoils and Pleistocene soils was attested. Notwithstanding these differences, the soil properties were relatively uniform. The sampled profiles can be classified as calcaric cambisols and the carbonate content is high (>55% by weight). Partly due to the cementation of finer particles by carbonate, the silt fraction dominates the texture. Only in the topsoil a variation in organic carbon content was found and this and the porosity decreased with depth (Cammeraat et al., 2005). No obvious differences were attested between the slope transects, hence the average bulk density and porosity, determined from the mass difference between a saturated and oven-dry sample, are respectively, 14 kN m^{-3} and 0.43 m^3 m^{-3} (246 samples).

Characterisation of vegetation and root distribution

Tree location and the extent of the tree canopy as well as that of the herbaceous cover were mapped. Ground cover of the herbaceous cover was estimated visually and classified: <10, 10–25, 25–50, 50–75 and >75%. For each polygon a tally of common species was made over a grid of 1 × 1 m at 0.1 m intervals. At selected places this inventory was combined with a determination of the dry above-ground biomass over the grid (cf. Cammeraat et al., 2002).

Of trees taller than 2 m, species and location were noted, as well as its height, diameter at breast height (DBH, 1.3 m) and canopy extent, shape and volume (Cammeraat et al., 2002). Foliage density was measured with a LI-COR LAI2000 (LI-COR, 1990).

The inventory of common species revealed that areas with near complete cover (>90% on average) were dominated by *Brachypodium* sp. (over 75%), those with a sparse cover (<50% on average) were characterised by succulent and aromatic species such as *Sedum* sp., *Sempervivum* sp. and *Thymus* sp. This finding was confirmed by the dry biomass determination of, respectively, 1.08 and 0.58 kg m^{-2} (respectively, 3 and 6 samples). Consequently, the vegetation units were lumped into two classes, representing a denser and a sparser vegetation type (respectively, type I and II). Vegetation type II prevails on scars and other less stable surfaces. Overall, slope A had a sparser cover than slope B (48.5 and 80.3% classified as vegetation type I),

An equal number of trees >2 m was present on slope A and B (respectively, 44 and 41). The majority of these trees were Aleppo pine (*Pinus halepensis* (Mill.)) with insubordinate numbers of olives (*Olea europaea* (L.)), almonds (*Prunus dulcis* (Mill. D.A. Webb)) and hawthorns (*Crataegus monogyna* (Jacq.)). The former crop trees especially abounded on slope B (37% compared to 7% on slope A). Mature pine trees varied in height between 7.5 and 15 m and had a DBH of 0.2–0.5 m and pine trees were generally better developed on slope A than on slope B (Table 3). A good linear relationship exists between height and DBH ($R^2 = 0.85$).

Root counts were made after the *in-situ* direct shear tests and in the soil pits, predominantly to serve as input for the 2D model. Therefore, the roots were projected on the X–Y plane, classified on the basis of their diameter, inclination and depth, and expressed as a number for a given soil volume, in this case 1 m^2 of basal area times the zone height.

For the direct shear tests, all roots were counted over the basal area of the shear box at every 0.1 m depth up to the depth of the imposed slip plane (0.4 m). Roots were counted as totals over four root diameter classes: <1, 1–3, 3–6 and >6 mm.

The five soil pits provide information on the overall root content under the two vegetation types (respectively 3, and 2 pits). A 0.1 × 0.1 m grid and 0.5 m wide was placed in the pit and all roots larger than 1 mm in diameter described by its diameter, inclination, orientation and position. Fine root content was determined gravimetrically. At every 0.1-m depth, three undisturbed cores of 10^{-4} m^3 were taken from which the roots were extracted by wet-sieving in the laboratory. These roots were dried and weighed. Thus, the volume of roots for the known volume of soil could be calculated by assuming the specific gravity of the roots. By taking the average length of the sampling rings (50 mm) and an average diameter of 0.5 mm the number of fine roots could be estimated.

The root counts from the pits were deemed insufficient in number and size to represent the structural roots of the trees present on the slopes. As an alternative, the three-dimensional information from a digitised root system was used (cf. Danjon et al., 1999). The digitised root was transformed into a two-dimensional representation by means of pole coordinates and broken down into 1-m-wide concentric rings in the X–Y plane.

Figure 3 summarises the root distribution data for the vegetation types. The overall root content showed that vegetation type I is more rooted than type II. In both cases, however, coarse roots penetrate over 1.2 m into the soil. This applies both for the coarse and fine root content although the conversion factor for the latter had to be tuned significantly in order to bring the root numbers in agreement with the counts from the direct-shear tests. The data on tree roots derive from a digitised root system of maritime pine (*Pinus pinaster* Ait.; Fourcaud et al., 2003). This system had a maximum lateral extent of 3 m and reached a depth of 1.2 m. Since the outer ring only contained 11% of the total root count and 56% of the area of 28.3 m^2 its influence on root reinforcement will be small and its roots have been redistributed over the inner two rings. Compared to the overall root count, the number of structural tree roots is small (Figure 3). All roots are fairly evenly

Table 3. Characteristics of the pine trees at the slope transects

		Tree height		DBH		Drip line		Foliage density	
Transect	Number	Average	SD	Average	SD	Average	SD	Average	SD
A	41	6.53	3.12	0.14	0.12	1.92	1.15	1.47	0.99
B	26	5.13	2.72	0.10	0.08	1.48	0.72	0.97	0.49

All values in [m] except foliage density [m^2 m^{-3}]. Shown are the average and standard deviation (SD).

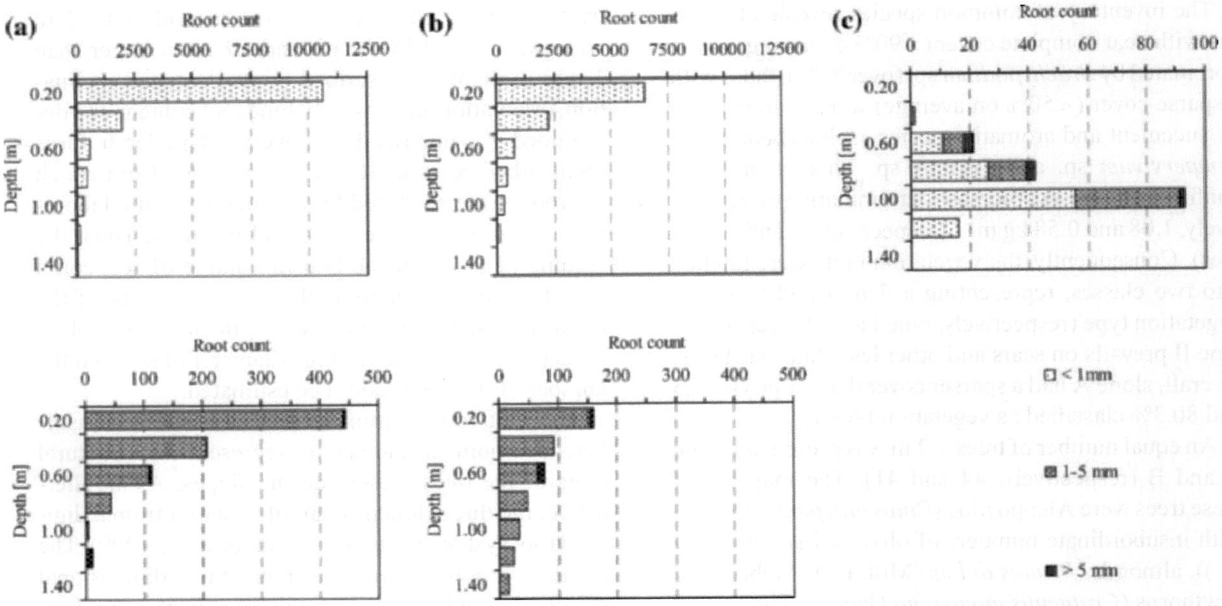

Figure 3. Root distributions for the different vegetation types: (a) Type I (dense), (b) Type II (sparse), (c) pine trees. Note the different scales used to represent the root contents.

distributed over the different inclination classes (not shown).

Root mechanical properties

Root elasticity and tensile strength were determined by Xylomecá in the laboratory on behalf of the Laboratoire du Rhéologie du Bois de Bordeaux (LRBB). The roots were kept in cold storage and soaked for 24 h prior to testing in tension using a load cell of 1 kN maximum capacity and a constant deformation rate of 2 mm min^{-1} (Genet, 2004). The maximum force at failure and the cross-sectional area were used to calculate the root tensile strength. Root elasticity (Young's modulus) was calculated from the slope of the stress–strain curve during the first, recoverable part of the root deformation.

A total of 52 root samples from the direct shear tests and pull-outs were tested of which 39 were *Pinus halepensis* and the remainder *Olea Europaea* and *Crataegus monogyna*. Of these, 17 had a diameter between 1 and 5 mm with bark, the remaining roots were larger (maximum of 14 mm). No roots smaller than 1 mm were tested.

No apparent differences in root elasticity or tensile strength were found and all roots have been lumped in the analysis. Overall, root elasticity showed a decrease with increasing root diameter. This relationship can be described by a power-function but the variability is large ($R^2 = 0.22$):

$$E_r = 1.25 \cdot 10^7 d_r^{-0.76}, \qquad (8)$$

where d_r is the root diameter [m] and E is the root elasticity [Pa].

The observed values range from 0.1 to 2.9 GPa. For the diameter classes of 1–5 and >5 mm the mean and standard deviation are respectively, 1.24 ± 0.81 GPa and 0.66 ± 0.52 GPa.

A constant but highly variable root tensile strength was found for the diameters tested, ranging from 3 to 43 MPa with an average of 13 MPa and a standard deviation of 6.8 MPa.

After rejection of the *in-situ* pull-out tests in which the root snapped at the attached clamp, 28 successful tests were available with diameters between 1 and 11 mm, the majority on pine. Most roots snapped and only few were pulled clear from the soil. The pull-out resistance measured *in-situ* thus largely coincides with the tensile strength determined in the laboratory with a range between 3 and 24 MPa an average of 9.3 MPa. Again, the tensile strength is independent of the root diameter.

Shear strength of non-rooted and rooted soil

The shear strength of the non-rooted soil was determined on saturated, undisturbed samples from the slope transects. Because of its simple and speedy execution, the tor vane test was used to obtain shear strength measurements from the 232 samples gathered at the 80 sample points (USACE, 1983). These data were spatially interpolated by block-kriging at the respective sample depths of 0.25, 0.75 and 1.25 m. Due to the rapid deformation of the saturated material excess pore pressures are generated and the results should be interpreted in terms of undrained strength (Yarbrough, 2000). Since the onset of failure of the natural slopes in the area generally occurs under drained conditions, the drained shear strength was determined in the laboratory by means of consolidated-drained strain-controlled tests (BS, 1990). The tests were performed on block samples from the soil pits with dimensions of 60 × 60 mm wide and 20-mm high. Of five representative horizons six samples were tested at three imposed normal stresses within the range from 50 to 125 kPa and a strain rate of 0.2 mm h^{-1}.

The undrained shear strength showed a weak increase with depth. The same tendency was observed for the drained shear strength (Table 4). The close similarity between the undrained shear strength and the drained cohesion points towards the complete absence of the frictional component in the former. The variation in the undrained shear strength and the drained cohesion are large and the latter does not significantly differ from zero. The drained friction angle ϕ' varies between 31° and 36° for the different horizons. A linear regression resulted in an overall friction angle of 33.6° when the cohesion was set to zero. The peak of the drained shear strength was generally achieved at 12% strain or 7-mm strain.

Table 4. Shear strength properties (six tests per horizon at three imposed normal loads between 59 and 117 kPa)

Horizon		Depth [m]*	c' [kPa]**	c_u [kPa]†	ϕ' [°]**
Ah	Top soil	0.00–0.17	4.8	4.5	34.4
Bw	Weathered soil	0.17–0.26	1.9	4.8	35.3
C1g*	Gleyic horizon	0.26–0.59	10.2	8.1	31.2
C12*	Colluvium	0.59–0.69	9.8	8.2	31.7
C2	Regolith	0.69–0.92	4.1	6.5	36.4

*The horizons C1g and C12 are not necessarily present and the given depths are indicative only. The depth is the average of the observed layers and indicative only.
**c' and ϕ' are the drained shear strength parameters.
†c_u is the average undrained shear strength (cohesion) based on field measurements ($N = 232$ in total).

At slope B eight large-scale *in-situ* direct shear tests were carried out on the rooted topsoil under varying vegetation conditions (see test description above). Of these, two were of the four-sided design (Tests 1 and 8), six were of the two-sided design. The strain at failure is more variable than in the laboratory, ranging between 6 and 25% (34 and 150 mm, respectively) and is weakly correlated to the root reinforcement of the soil (Figure 4). During the tests roots could be heard snapping and corresponding drops in the stress–strain curve were observed (Figure 5c). Some arching into the soil was observed in the two-sided design but the movement was small compared to the displacement of the enclosed block and few roots crossed through the lateral sides with the exception of Test 2 under dense forest cover of *Pinus halepensis*. This test returned the highest root reinforcement of 18.2 kPa while only in one test no reinforcement was observed (−0.4 kPa; Figure 4). The four-sided tests near pine trees yielded reinforcements of respectively 3.3 kPa and 7.8 kPa, the latter made on an 11-year-old sapling. Over all tests, the average root reinforcement is 3.9 kPa. When the largest reinforcement of Test 2 is excluded, a good correlation between the fine root content (<1 mm) of the slip plane and the root reinforcement is found ($R^2 = 0.96$).

Model applications

General

The root reinforcement model in FLAC has been applied to model the *in-situ* direct shear tests as a test of its validity and to assess the slope stability of the two slope transects.

For the modelling, the average soil properties for the friction angle, bulk density and porosity have been adopted (Tables 5 and 6). For unsaturated conditions an average degree of saturation of 65% at field capacity has been used to calculate the bulk density of the material (Van Beek, 2002). The applied cohesion varied per application (see below). The bulk modulus, K, and shear modulus, G, are needed to calculate the deformations in FLAC and initially literature values were assigned ($K = 5.0$ MPa, $G = 2.3$ MPa).

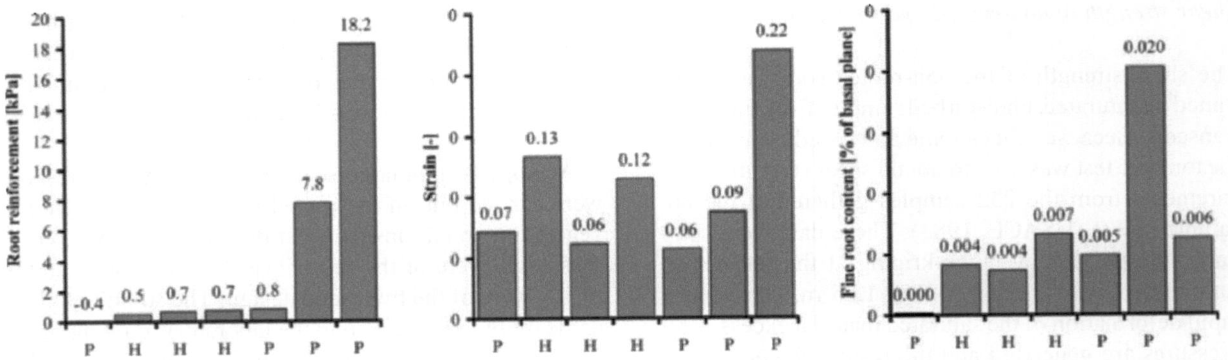

Figure 4. Reinforcement, strain and fine root content (<1 mm) for seven out of the eight *in-situ* direct shear tests. H and P refer respectively, to herbaceous cover and the presence of pine trees.

For each root diameter root elasticity was calculated from the power function of Eq. (8). The mean tensile strength was used for all roots. Without any reliable data to suggest otherwise, the ratio between the root compressive strength and the tensile strength, the root contact length and reduction parameter were kept at unity (Eq. (4)).

The surcharge due to the self-weight of the vegetation has been ignored due to the want of data and the patchy nature of the vegetation cover.

In-situ *direct shear tests*

Model settings

In the simulation of the direct shear tests, the X–Y plane of the problem was aligned parallel to the imposed shear displacement of the soil block (Table 5). The shear box was located in the centre of the block, on which the corresponding normal load was imposed, and the adjacent areas excavated (Figure 5). The shear plane was modelled as a detachable interface to allow for the observed horizontal displacement. The simulation was carried out in two stages (Table 5). After the initial stress distribution was obtained, the actual test was simulated. A horizontal strain rate was applied to the soil block contained by the shear box. The imposed strain rate was slow enough to ensure consolidated-drained conditions and pore pressure effects were not considered.

Along the shear plane, both the average normal and shear displacements and stresses were monitored in order to compare them with the observed stress–strain curve. The shearing of the soil block was simulated until the calculated shear displacement equalled the observed displacement in the field.

Parameterisation

All *in-situ* direct shear tests were modelled with the corresponding normal load and the actual root distribution of each test, the root distribution of vegetation type I and without any roots. The actual root distributions were available for the first 0.4 m at a vertical resolution of 0.1 m. Below this depth, root counts were extrapolated to 1-m depth. The actual root distributions were summarised into four diameter classes, as used for the counts in the field (<1, 1–3, 3–6 and >6 mm). Except for some structural roots, the inclination and orientation of the roots was not recorded, so the root numbers were divided equally over three inclination classes (0°–60°, 60°–120°, 120°–180°). The same inclination classes were used for the root distribution of vegetation type I for which a distinction in three diameter classes was made (<1, 1–5 and >5 mm).

As the imposed shear plane was most times located in the Bw horizon of low cohesion, no cohesion or tensile strength were attributed to the soil.

Results

The result shows that the model simulates the strain-dependent nature of root reinforcement and the failure of roots by breakage or pull-out (Figure 5c). However, the assumed plane strain conditions implied that only the four-sided direct shear tests can be simulated directly and therefore the relative root reinforcement is presented (Figure 6). The simulated root reinforcement for the actual root distributions is approximately 0.3 kPa with the exception of Test 8 for which the reinforcement is 1.5 kPa. Inclusion of the root distribution of vegetation type I gives a simulated root reinforcement of 1.6–1.9 kPa.

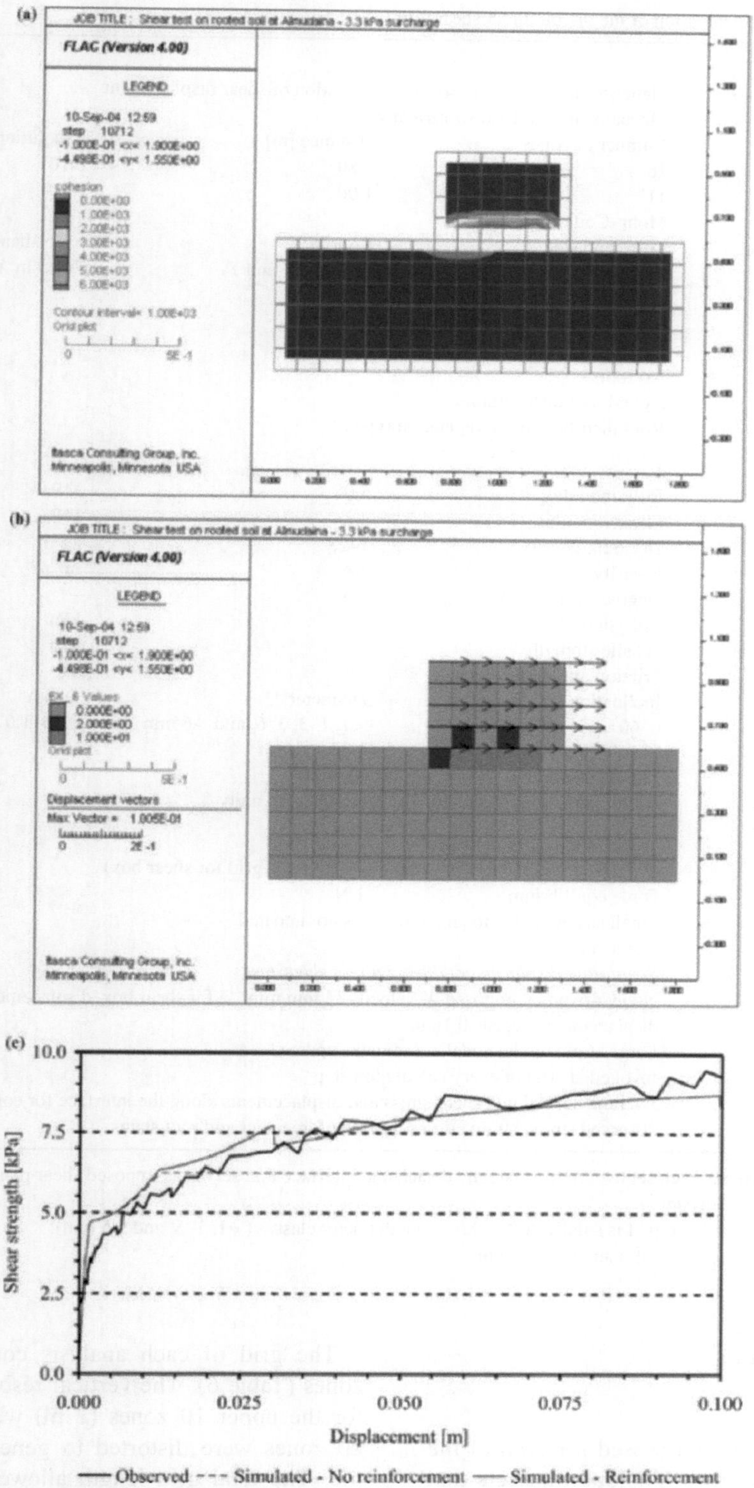

Figure 5. Simulated direct shear tests. (a) Mobilised root reinforcement [Pa], (b) zones containing broken roots (zones with values larger than 0), (c) observed and simulated stress–strain curve.

Table 5. Model settings for the simulation of the *in-situ* direct shear tests

Model settings			
Analysis	Plane strain, X–Y plane aligned to direction of shear displacement		
	Mechanical only, total stress analysis		
Grid dimensions	Number of zones	Distance [m]	Resolution in area of interest [m]
Horizontal (X)	18	1.80	0.10
Vertical (Y)	11*	1.00	0.10
Model	Mohr–Coulomb		
Mechanical boundary conditions	Left-hand side	Bottom	Right-hand side
	Fixed in X	Fixed in X and Y	Fixed in X
Loads	Gravity, 9.81 m s^{-2}		
	Surcharge (3.3 or 4.1 kPa) at top shear box		
Imposed groundwater conditions	NA		
Imposed root conditions	No roots		
	Actual root distributions		
	Root distribution for vegetation type I		
Parameterisation			
Soil properties**	Bulk modulus, K	5.0	MPa
	Shear modulus, G	2.3	MPa
	Dry bulk density	1425	kg m^{-3}
	Porosity	NA	m^3 m^{-3}
	Degree of saturation	NA	–
	Cohesion, c'	0	kPa
	Tensile strength	0	kPa
	Friction angle, ϕ'	33.6	°
Root distribution	Inclination	Diameter***	Depth
	0–60°, 60–120°, 120–180°	<1, 1–3, 3–6, and >6 mm	0.10–1.0 m depth
Root properties	Elasticity†	$1.25 \times 10^7\ d_r^{-0.76}$	Pa
	Tensile strength	13×10^9	Pa
	All remaining parameters (CO, R, RCS) set to unity		
Modelling stages			
Initial	*Obtaining initial geostatic stresses* (with fixed grid for shear box)		
	Force equilibrium‡	1 N	
Strain mode	Small strain mode, nodal coordinates not updated		
Root reinforcement	Not invoked		
Main	*Simulating displacement during direct shear test*		
Convergence	None, driven by imposed X-velocity (4 mm min^{-1}) for shear box, displacement halted when total displacement exceeds 0.10 m		
Strain mode	Large strain mode, nodal coordinates updated		
Root reinforcement	Invoked at start of every calculation step		
Additionally reported variables	Average normal and shear stress and displacements along the interface for comparison with the observed stress–strain curves, root reinforcement and root status		

*One zone (7th in Y-direction) has been set to zero to create the detachable interface that serves as imposed shear plane.
**Properties for the main modelling stage.
***The root distribution of vegetation type I is subdivided in three root diameter classes (<1, 1–5, and >5 mm).
†Root elasticity is given as function of root diameter, d_r, in m.
‡Maximum unbalanced force.

Slope stability assessments

Model settings

The slope stability has been assessed for each of the three survey lines of the two hillslope transects that were aligned to the maximum slope (6 in total, i.e., 3 profiles × 2 transects).

The grid of each analysis contained 220 by 30 zones (Table 6). The vertical resolution was constant for the upper 10 zones (2 m) whilst the underlying 20 zones were distorted to generate the slope profile. The total grid length allowed for the inclusion of two flat areas in the order of 5–10 m wide at the toe and crest of the slope. All analyses were made

Table 6. Model settings for the slope stability assessments

Model settings			
Analysis	Plane strain, X–Y plane aligned in the direction of the survey line		
	Mechanical only, effective stress analysis		
Grid dimensions	Number of zones	Distance [m]	Resolution in area of interest [m]
Horizontal (X)	220	110	0.50
Vertical (Y)	30	Var.	0.20
Model	Mohr–Coulomb		
Mechanical boundary conditions	Left-hand side	Bottom	Right-hand side
	Fixed in X	Fixed in X and Y	Fixed in X
Loads	Gravity, 9.81 m s^{-2}		
Imposed groundwater conditions	Fully saturated		
	Critical groundwater level*		
Imposed root conditions	No roots		
	Actual root distribution in connection with vegetation density		
	Fully rooted (complete tree and vegetation cover)		
Parameterisation			
Soil properties**	Bulk modulus, K	5.0	MPa
	Shear modulus, G	2.3	MPa
	Dry bulk density	1425	kg m^{-3}
	Porosity	0.43	m^3 m^{-3}
	Degree of saturation	0.65	–
	Cohesion, c'	0–29***	kPa
	Tensile strength	0–43***	kPa
	Friction angle, ϕ'	33.6	°
Root distribution	Inclination	Diameter***	Depth
	0–60°, 60–120°, 120–180°	<1, 1–5, >5 mm	0.20–1.6 m depth
Root properties	Elasticity†	$1.25 \times 10^7\ d^{-0.76}$	Pa
	Tensile strength	13×10^9	Pa
	All remaining parameters (CO, R, RCS) set to unity		
Modelling stages			
Initial	Obtaining initial geostatic stresses (under increased strength)		
Convergence	Force equilibrium‡		100 N
Strain mode	Small strain mode, nodal coordinates not updated		
Root reinforcement	Not invoked		
Main	Calculating stability without and with root reinforcement		
Convergence	Factor of safety ($\Delta F < 0.005$) and critical groundwater levels ($\Delta W < 0.01$ m)		
Strain mode	Small strain mode, nodal coordinates not updated		
Root reinforcement	Invoked at start of every calculation step		
Additionally reported variables	Safety factor, critical groundwater depth, surface velocities, root reinforcement and root status		

*Obtained from back analysis for the non-rooted case.
**Properties for the main modelling stage.
***Cohesion for the first 1.5 m derived from interpolated undrained shear strength measurements, below this depth an exponential increase to a maximum of 29 kPa at 2.5 m. No tensile strength was assigned to the first 1.5 m, thereafter it was taken equal to the tension cut-off of the Mohr–Coulomb failure envelope.
†Root elasticity is given as function of root diameter, d_r, in m.
‡Maximum unbalanced force.

in terms of effective stress assuming fully drained conditions.

After an initial stress distribution was obtained, the safety factor was calculated for each survey line by means of the parameter reduction method (Dawson et al., 1999). In addition the surface velocity field after 1000 calculation steps was evaluated. These velocities are the hypothetical deformation rates by which the model attempts to accommodate the unbalanced body forces that act on each zone of the grid. Since the model

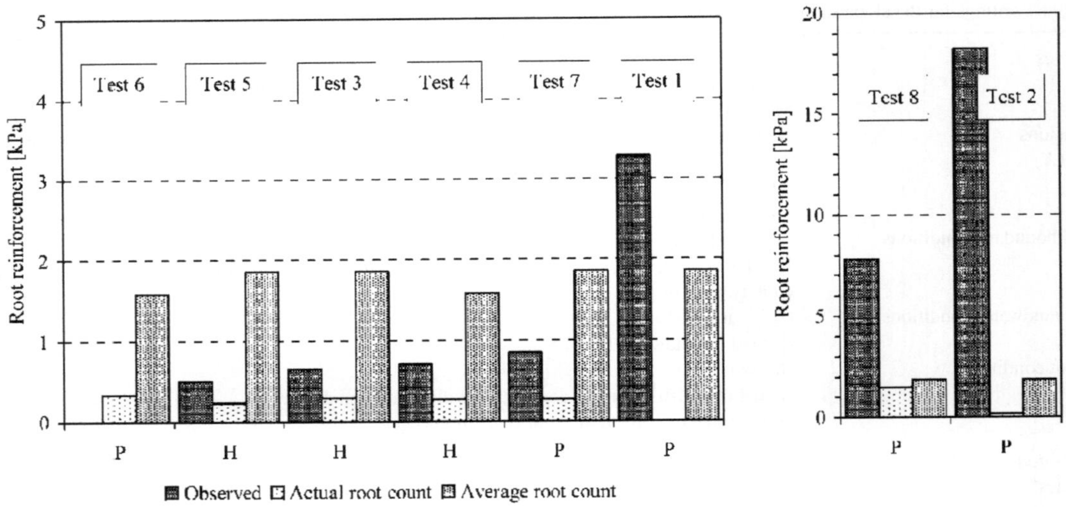

Figure 6. Comparison of the observed and modelled root reinforcement for the *in-situ* direct shear tests. Note the difference in scale between the graphs.

was run in small-strain mode for the safety factor calculations (i.e., coordinates of the nodes are not updated), these velocities may be large when they represent plastic strain at the onset of failure. Still, the velocity offers valuable information on local slope stability and the values along the profiles have been interpolated by means of inverse distance interpolation so that the surface velocity field can be compared to the morphology and vegetation along the slope transects.

Root conditions that were evaluated included a completely non-rooted case, the actual rooted case, and fully rooted case for the central survey line. For each root condition, the safety factor was calculated for a worst-case scenario in which the slope was completely saturated and for a more realistic case in which a constant piezometric line was imposed for which the slope was critical (factor of safety of unity) for the non-rooted case.

Parameterisation

The slope stability assessments differ from the simulation of the *in-situ* direct shear tests by the fact that the root distributions are less accurately known, yet have to be specified along the entire slope. In this case root distributions were assigned according to the presence of the two vegetation types along the slope. Where trees were present, the distribution of the digitised tree roots was added. The extent of the tree roots was determined from the ratio between the drip line and the extent of the digitised root system (3 m). This approach seems justified by the good relationship between the drip line and the DBH, which is a good estimator of the dry root mass for conifers (Drexhage and Gruber, 1999). In order to take the opposed inclination of roots on either side of the stem into account the number of root classes was doubled and the inclination of the roots mirrored along the *Y*-axis. This results in four root types that in combination with the two vegetation types and the presence or absence of trees result in 10 root types. For the modelling, all trees intersecting the profiles were included and all trees, including the deciduous ones, were treated as pines. Since no inclination was known for the fine roots (<1 mm) the root numbers were equally divided over the three inclination classes used (Table 6).

For the analysis under fully rooted conditions trees with a canopy extent of 3 m on either side of the stem were positioned continuously along the slope and the dense vegetation type I was used for the undergrowth.

For the upper 1.5 m of the soil, the interpolated tor vane readings were substituted for the local cohesion. Below this depth, the cohesion increased exponentially until it reached the average cohesion for sound bedrock (Van Beek, 2002). No tensile strength was attributed for the upper 1.5 m of the soil and the theoretical maximum of $c' \times \cot \phi'$ was used below this depth.

Results

Both slopes are potentially unstable as the safety factors under full saturation indicate (Table 7). On both slopes the areas of simulated failure coincide with the

Table 7. Safety factors for the survey lines under different conditions for root density and hydrology

Transect	Hydrological conditions	Root conditions	Survey line 1	2	3
A	Saturated	None	0.98	0.68	0.92
		Actual	1.06	0.70	0.93
		Full		0.70	
	Critical	None	1.00	1.00	1.00
		Actual	1.09	1.05	0.99
		Full		1.05	
B	Saturated	None	0.74	0.59	0.67
		Actual	0.76	0.63	0.70
		Full		0.64	
	Critical	None	1.00	1.00	1.00
		Actual	0.98	1.05	0.99
		Full		1.05	

Table 8. Average and standard deviation of surface velocities (see also Figure 7) for the two slope transects under different conditions for root density and hydrology

Transect	Hydrological conditions	Root conditions	Surface velocity at failure [mm h^{-1}] Average	Standard deviation
A	Saturated	None	−117.4	171.4
		Actual	−63.0	103.6
	Critical	None	−2.7	8.6
		Actual	−29.3	43.4
B	Saturated	None	−567.5	442.5
		Actual	−492.0	395.3
	Critical	None	−18.2	23.3
		Actual	−28.7	41.6

actual scars and deposition areas. For slope A, the stability of the central survey line becomes critical when the groundwater is at 0.2 m below the surface. The adjacent survey lines are marginally stable when the groundwater is close to the surface. Slope B is generally less stable and critical groundwater levels vary between 0.25 and 0.45 m below the surface. In the case of full saturation, the presence of roots generally increases slope stability but this increase is small and irrespective of the number of roots since the factor of safety does not increase appreciably whether the actual root content or the fully rooted soil is considered. For the critical groundwater depth, the increase in stability is less marked and sometimes marginally negative. In all cases the soil fails in the C-horizon at approximately 1-m depth. In the unstable soil mass, the simulated root reinforcement is found at the scarp, toe and base. The mobilised reinforcement at the base is less than 1 kPa due to the low root content.

The surface velocities associated with the plastic deformation field are high (Table 8). The velocities generally show a significant decrease in displacement when the hydrological conditions change from fully saturated to the critical water level regardless the root content. Under saturation, root content has a positive influence on the velocities. Again, the velocities do not change along the central profile when fully rooted conditions are imposed (not shown). Under critical conditions, the marginal destabilising influence due to compressing roots is found here again but the associated deformations are generally small (in the order of a few centimetres when FLAC is run in 'large-strain' mode, i.e., when the nodal coordinates are updated).

When the velocity patterns along the slope are compared, it becomes apparent that for both slopes the scars, gullies and deposition areas are the least stable. Slope A clearly shows the largest velocities in the category with less than 10% vegetation cover. Due to the dense cover on slope B, the picture is less clear with the largest velocities occurring in the categories between 25 and 75%. The influence of vegetation on slope stability remains, however, confused. Only in a few locations a clear decrease in the surface velocity is simulated, such as in the middle section of Slope A (Figure 7).

Discussion

In the study area abandonment has led to the emergence of a semi-natural vegetation cover that is characterised by a patchy ground cover, remnants of crops and isolated pine trees of different age and size. Since abandonment slope instability has occurred at both transects. At slope B the buried horizons indicate a higher activity of mass wasting in the past, which is consistent with the difference in slope angle but contradicts the sparser vegetation cover on slope A. Possibly the instability is of a younger date or re-vegetation is slower at this site.

Notwithstanding the differences in process activity, local variability obscures any larger trends in soil properties along the slopes. The uniform composition of the Miocene marl and the associated high carbonate content are overriding factors that limit soil development.

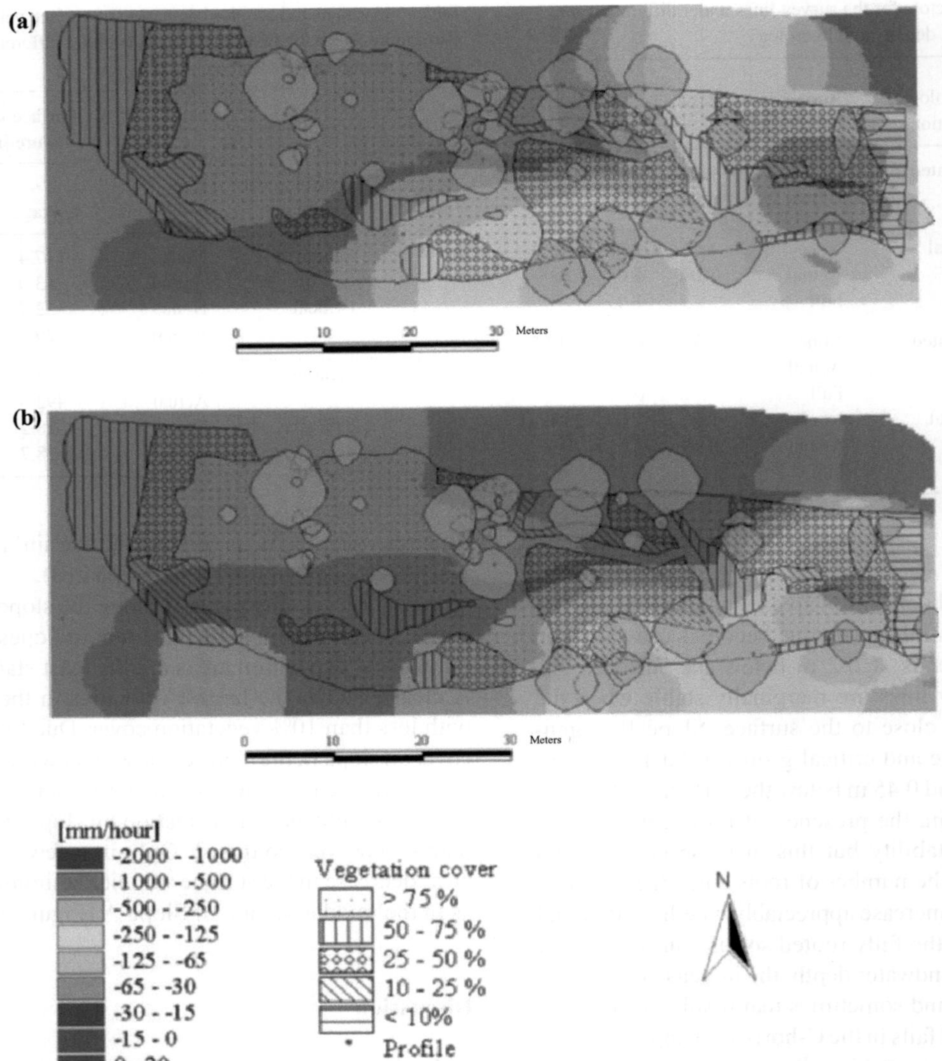

Figure 7. Surface velocities along slope A. Shown are the non-rooted (a) and rooted case (b) under fully saturated conditions. The crest of the slope is to the right. Negative velocities point downhill. Superimposed on the surface velocities, the location of the sampling profiles and the vegetation cover and extent of the tree canopy are shown.

Cementation by carbonates decreases the activity of clays, thus reducing cohesion, and increases the particle size, thus increasing friction (Lamas et al., 2002; Skempton, 1985).

Since the slopes are prone to instability, root reinforcement may be important to maintain stability under critical hydrological conditions. In the present case root reinforcement was attested in eight *in-situ* direct shear tests. The observed values fall well within the range of values listed in literature for a wide range of vegetation types (op. cit. Sidle et al., 1985). Reinforcement derives mainly from the loading of fine roots as the correlation between root reinforcement and fine root content suggests. Vegetation type appears to be of secondary importance to the actual reinforcement, which should be attributed to the fact that few coarse roots permeated the imposed shear planes in those cases and that there is little variation in root mechanical properties between species and root sizes. Where abundant, coarse roots contribute significantly to the soil shear strength as shown by the high values obtained for two tests made under pine.

The significance of fine roots is confirmed by the simulated direct shear tests. The root reinforcement model approximates the observed stress–strain behaviour with inclusion of the observed mechanisms of root failure during the tests. Yet, the actual root counts returned simulated root reinforcements that underestimate the observed values while the average root count overestimate them. This difference should be attributed to the amount of fine roots included. For the two tests under pine (Tests 2 and 8), neither the actual nor the average root distribution yielded satisfactory results. For Test 2, this shortcoming is partly explained by the absence of lateral roots crossing the soil block. For both tests, also the vertical extension of coarse roots may be underestimated by the vertical extrapolation of the actual root count.

The overestimation of the simulated root reinforcement for the average root distribution implies that the results of the slope stability assessments will be optimistic for the areas under herbaceous cover. The poor representation of the effect of coarse roots, however, called for the explicit inclusion of the root distribution under pine trees. Although these roots add considerably to the coarse root content, the total is small compared to the root counts established under the two vegetation types and the root number dwindles rapidly with depth (Figure 3).

Despite its optimistic nature, the influence of the simulated root reinforcement on the slope stability is small. Under saturated conditions, a small increase in the safety factor and a substantial decrease in the surface velocity at failure are observed. Under critical conditions, these results are more ambiguous due to the strain-dependent nature of root reinforcement. Under small deformations, the model simulates that roots remain in compression and this affects the shearing resistance negatively. Hence the stabilising influence of root reinforcement on the near-critical slopes studied here is confined to a small range of critical hydrological conditions. It can be argued whether roots, and especially the finer ones, are not too flexible to be loaded in compression and this is an obvious limitation of the root reinforcement model. Certainly, compression of roots will affect the bond strength along the root–soil interface and cap the pull-out resistance that can be mobilised in later stages of slope deformation when these roots will be loaded in tension and the soil fabric may fail before the root reinforcement can be mobilised (Mulder, 1991). It is possible to describe such modifications in bond strength conceptually but at the moment the required data from field and laboratory tests to parameterise such routines are lacking.

Since the rooting depth is limited due to the presence of hard regolith or bedrock at the slope transects, root anchorage is deficient (Tsakumoto and Kusakabe, 1984). Failure will occur by root pull-out at low loads due to a loss of effective strength along the root–soil interface. Root properties and density are not significant under such conditions as evidenced by the insensitivity of the safety factor to the imposed root conditions (GEO, 2000). This transient nature of root reinforcement under these conditions is clearly contradictory with the constant values of root reinforcements that are generally used in slope stability assessments.

Some reinforcement may be derived from the confining root mat or from buttressing and arching but their overall effect will be limited given the patchy nature of the semi-natural vegetation cover (Gray, 1995). Consequently, the preferred shearing plane would be expected at the contact of the regolith with the bedrock as is generally observed in the field where percolating water stagnates and root reinforcement absent. Higher in the profile failure is averted by both soil cohesion and root reinforcement.

Conclusions

Detailed studies on root reinforcement were made at two slope transects susceptible to slope stability in SE Spain. These slopes were formerly cultivated but a semi-natural vegetation cover has established itself after abandonment. The studies revealed that:

(1) the semi-natural vegetation consists of a patchy herbaceous vegetation cover with dispersed Aleppo pine trees of different age and size and remnants of crops. Two vegetation types were identified, differing in cover and vegetation composition, which reflect the activity of mass wasting processes on the slopes.

(2) along the slopes, the soil properties show a large local variability but no apparent lateral trends.

(3) *in-situ* direct shear tests indicate that a contribution of root systems to the soil shear strength within the topsoil is present but limited. This contribution is in the order of 0.6 kPa under herbaceous cover but may be as high as 18 kPa for densely rooted soil under pines.

(4) fine root content is a determining factor in the observed root reinforcement and a sensitive parameter in the model. The influence of coarse roots cannot be fully captured, not even by the large-scale direct shear tests employed here. For a more truthful representation of this influence at least a more accurate count of coarse roots is needed.
(5) root counts and consequently root reinforcement decrease rapidly with depth. Roots cannot penetrate in the underlying bedrock and the anchorage is limited. Shear planes coincide generally with the weathering depth of the soil profile where percolating water stagnates and root reinforcement is absent.
(6) simulation of root reinforcement at the hillslope scale on the basis of the vegetation patterns returns failure areas and potential shear planes that coincide with the observed instability in the field. Translation of the *in-situ* direct shear tests to the hillslope scale by means of the model therefore seems appropriate although the simulated root reinforcement for the direct shear tests is optimistic. The results reveal that the failure mechanisms at the hillslope scale are intrinsically different and limited by the pull-out resistance of the roots under saturated conditions. The effect of root reinforcement at the hillslope scale is limited to a small range of hydrological conditions and predominantly occurring after failure.

Acknowledgements

The following persons are kindly acknowledged for their assistance during the preparation of this paper: Ms Janneke Klein and Ms Linda Bannink, M.Sc. students of the Department of Physical Geography of the University of Amsterdam, for their devotion in the field and laboratory; Dr Thom Bogaard of the Department of Physical Geography, Utrecht University, for his support with the laboratory direct shear tests; Dr Alexia Stokes and Ms Marie Genet (LRBB-France) and Mr Jean-François Dumail (Xylomecá) for the testing of the root properties; Dr Lionel Dupuy for provision of the root distribution for pine trees; the anonymous reviewers whose comments contributed greatly to the readability of this paper. This research was supported by the Eco-Slopes project, which was funded by the European Union (QLK5-2001-00289).

References

Abe K and Ziemer R R 1991 Effect of tree roots on shallow-seated landslides USDA Forest Service, Gen. Tech. Rep. PSW-GTR-130, 11–20.

Allison L H 1935 Organic soil carbon reduction of chromic acid. Soil Sci. 40, 311–320.

BS 1990 Methods of test for soils for civil engineering purposes. Shear strength tests (effective stress). BS 1377-8, British Standards Institute.

Cammeraat L H, Van Beek L P H and Kooijman A 2005 Vegetation succession and its consequences for slope stability in SE Spain. Plant Soil 278, 135–147.

Cammeraat L H, Van Beek L P H and Dorren L K 2002 Eco-Slopes Field Protocol. University of Amsterdam, Amsterdam, the Netherlands, 73 pp.

Danjon F, Bert D, Godin C and Trichet P 1999 Structural root architecture of 5-year-old Pinus pinaster measured by 3D digitising and analysed with AMAPmod. Plant Soil 217, 49–63.

Dawson E M, Roth W H and Drescher A 1999 Slope stability analysis by strength reduction. Géotechnique 49, 835–840.

Drexhage M and Gruber F 1999 Above- and below-stump relationships for *Picea Abies*: Estimating root system biomass from breast-height diameters. Scand. J. Forest. Res. 14, 328–333.

FAO 1990 Guidelines for Soil Profile Description, 2nd edition. Soil Survey and Fertility Branch Land and Water Development Division. FAO, Rome.

Fourcaud T, Danjon F and Dupuy L 2003 Numerical analysis of the anchorage of maritime pine trees in connection with root structure *In* B Ruck et al. (eds), Wind Effects on Trees. Int. Conf., Karlsruhe, pp. 323–329.

Genet M 2004 Le rôle de la cellulose dans le résistance a la traction des racines DEA Sciences du Bois LRBB. 28 pp.

GEO 2000 Geotechnical Manual for Slopes. 4th reprint. Geotechnical Engineering Office. The Government of the Hong Kong Special Administrative Region, Hong Kong.

Gray D H 1995 Influence of vegetation on the stability of slopes. *In* DH Barker (ed), Vegetation and Slopes, pp. 2–25. Institution of Civil Engineers, London.

Itasca 2002 FLAC 4.0 User Manual.

La Roca-Cervigón N and Calvo-Cases A 1988 Slope evolution by mass movements and surface wash (Valls d'Alcoi, Alicante, Spain). *In* A Imeson and M Sala (eds), Geomorphic Processes. Vol. I, pp 95–102. Hillslope Processes. Catena Verlag, Cremlingen.

La Roca-Cervigón N 1991 Untersuchungen zur räumlichen und zeitlichen Variabilität der Massenbewegungen in Einzuchsgebiet des Riu d'Alcoi (Alicante, Ostspanien). Erde 122, 221–236.

Lamas F, Irigaray C and Chacón J 2002 Geotechnical characterisation of carbonate marlsfor the construction of impermeable dam cores. Eng. Geol. 66, 283–294.

LI-COR 1990 LAI-2000 Plant Canopy Analyzer. Instruction Manual. LI-COR, Lincoln, NE.

MacDonald D, Crabtree J R, Wiesinger G, Dax T, Stamou N, Fleury P, Gutierrez-Lazpita J and Gibon A 2000 Agricultural abandonment in mountain areas of Europe: Environmental consequences and policy response. J. Environ. Manage. 59, 47–69.

Mulder H F M 1991 Assesment of landslide hazard. Doct. thesis. Utrecht University, Utrecht, the Netherlands, 150 pp.

O'Loughlin C L 1974 The effect of timber removal on the stability of forest soils. N. Z. J. Hydrol. 12, 121–134.

Riestenberg M M 1994. Anchoring of Thin Colluvium by Roots of Sugar Maple and White Ash on Hillslopes in Cincinnati. U.S Geological Survey, Bulletin 2059-E. United States Government Printing Office, Washington DC, 25 pp.

Schmidt K M, Roering J J, Stock J D, Dietrich W E, Montgomery D R and Schaub T 2001 The variability of root cohesion as an influence on shallow landslide susceptibility in the Oregon Coast Range. Can. Geotech. J. 38, 995–1024.

Sidle R C, Pearce A J and O'Loughlin C O 1985 Hillslope Stability and Land Use. AGU Water Resources Monograph 11.

Skempton A W 1985 Residual strength of clays in landslides, folded strata and the laboratory. Géotechnique 35, 3–18.

Tsakumoto Y and Kusakabe O 1984 Vegetative influences on debris slide occurrences on steep slopes in Japan. *In* Proc. Symp. Effects Forest Land Use on Erosion and Slope Stability. Environment and Policy Institute, Honolulu.

USACE 1983 Engineering Properties of Soil and Rock. United States Army Corps of Engineers, TM 5-818-1/ AFM 88-3 Ch. 7, Chapter 3, Washington, 27 pp.

Van Beek L P H 2002 Assessment of the Influence of Changes in Land Use and Climate on Landslide Activity in a Mediterranean Environment. Netherlands Geographical Studies 294, KNAG, Utrecht, 364 pp.

Vidal H, 1966. La terre armée. Annales Institut Technique Batim., Paris, No. 223–229: 888–939.

Waldron L J 1977 The shear resistance of root-permeated homogeneous and stratified soil. Soil Sci. Soc. Am. J. 41, 843–848.

Wu T H 1984 Effect of Vegetation on Slope Stability. Transportation Research Record 965. Transportation Research Board, Washington, DC, 3746 pp.

Wu T H, McKinnel III W P and Swanston D N 1979 Strength of tree roots and landslides on Prince of Wales Island, Alaska. Can. Geotech. J. 16, 19–33.

Wu T H, Beal P E and Chinchun L 1988 *In-situ* shear test of soil–root systems. J. Geotech. Eng.-ASCE 114, 1377–1393.

Wu T H 1995 Slope stabilization. *In* RPC Morgan. & RJ Rickson (eds), Slope Stabilization and Erosion Control, pp. 221–264. Spon, London.

Wesemael J C 1955 De bepaling van het calciumcarbonaatgehalte van gronden. Chemisch Weekblad 51, 35–36.

Yarbrough L D 2000 Channel bank stability and design-considering the effects of riparian vegetation and root reinforcement. MSc thesis, University of Mississippi, Oxford, 139 pp.

Slope stabilisation by perennial 'gramineae' in Southern Italy: plant growth and temporal performance

D. Cazzuffi[1,3], A. Corneo[2] & E. Crippa[1]
[1]*CESI S.p.A., via Rubattino, 54, I-20134, Milan, Italy.* [2]*Vetivaria s.r.l., via Spallanzani, 6, I-20129, Milan, Italy.* [3]*Corresponding author**

Key words: eco-engineering, erosion control, monitoring programme, perennial 'gramineae', revegetation, roots, slope stabilisation

Abstract

In the territory of Altomonte, a village located in Calabria, in the Southern part of Italy, a new thermoelectrical station is under construction. This work involved major earthworks which regarded new excavated slopes. In order to protect soil from erosion due to rainfall and runoff and also in order to prevent superficial soil instability, it was decided to plant four different species of perennial "gramineae" plants (Eragrass, Elygrass, Pangrass and Vetiver) characterised by deep roots. Works began in November 2002 and ended in May 2003, a period marked by very different climate and meteorological conditions, varying from exceptionally rainy and cold winter to warm and dry spring months. The paper describes the different stages of the project and the monitoring programme for the following months. The extension of the work and the use of four different kinds of vegetation made periodic inspections of the entire site appropriate. Two in situ investigations, respectively performed in August 2003 and in November 2003, are outlined. The aim of these surveys was to confirm the success of the work by verifying the growth of the plants and roots. The principal monitored parameters were the percentage of sprouted plants, the height of the foliage and the depth of roots. The investigations showed good results, keeping in mind the very difficult climatic conditions and the extreme poor fertility of the topsoil laid down upon the clay layer: in particular, high survival rate were measured over the entire area of the works and the root systems have developed sufficiently to grow through the upper topsoil layer (0.2–0.3 m) into the underlying clay layer. In March 2004, a sampling programme was undertaken on the same site. Direct shear tests were carried out in the laboratory in order to evaluate the increase in shear strength of the rooted soil mass. The research involved the recovery of three undisturbed samples of soil with roots for each of the four types of 'gramineae' plants and three undisturbed samples constituted only of soil, from the surface to a depth of 1.0 m. The tests were performed in a large direct shear apparatus on 200 mm diameter samples. The test results allowed to evaluate the roots' contribution of the different gramineous species and to underline the direct correlation between the increase in soil shear strength and the root tensile strengths. In particular, an increase in cohesion ranging between 2 and 15 kPa was recorded, according to the different species: the maximum values of increase in shear strength were reached by Vetiver roots, which are also characterised by the highest tensile strength.

Introduction

The paper describes a study of the effect of revegetation actions on slope stabilisation in an area where a power station is under construction. The site is located in the territory of Altomonte, in Calabria, a region in the South of Italy (Figure 1). The construction of the power station involved a major earth moving operation which required new excavated slopes (Figure 2). The power station, composed of two principal turbines and one auxiliary turbine, will achieve energy production of 800 MW. The fuel is provided by gas that comes directly from Libya by a specific pipeline.

In order to protect soil from erosion due to rainfall and runoff, it was decided to revegetate the site

* E-mail: cazzuffi@cesi.it

This article has been previously published in the following journal—Journal of Geotechnical and Geological Engineering, 24, 3, 429–447.
Stokes et al. (eds), Eco- and Ground Bio-Engineering: The Use of Vegetation to Improve Slope Stability, 111–124.

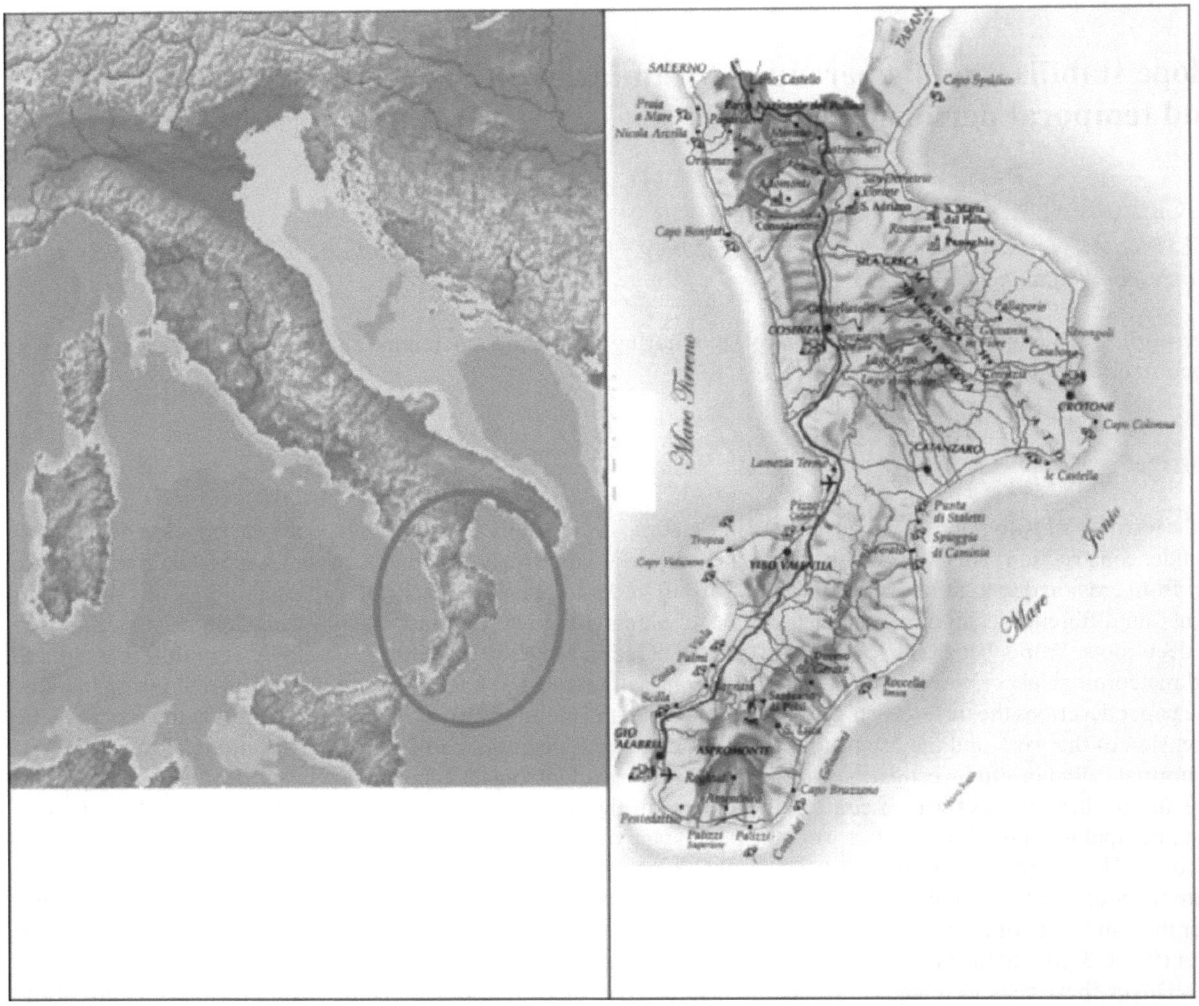

Figure 1. Location of the Altomonte territory in Calabria (Italy).

with ground bioengineering techniques, which permit installation of a vegetation cover on degraded soils (Coppin and Richards, 1990; Morgan and Rickson, 1995; Schiechtl and Stern, 1996).

The territory of Altomonte is particularly characterised by mountain areas with plain zones on only 10% of the total surface of the territory. Concerning the general meteorological conditions in Altomonte area, the average temperature ranges from the minimum value of 2°C in January to the maximum value of 24°C in July, while the precipitations range from the minimum value of 22 mm in July to the maximum value of 146 mm in December.

Focussing attention on the area of interest for the project, it is possible to distinguish in the subsoil only clayey–sandy horizons of Pliocene–Pleistocene age. In particular, the following distinctions could be noted:

Topsoil: $0 \div 0.4/0.7$ *m*: Constituted of sandy silt and clayey silt with organic elements; brown colour with local grey shades. Presence of gravel and rare cobbles, that in some zones could be found more frequently.

Level A: $0.4/0.7 \div 1.7/2.3$ *m*: Soil horizon prevalently constituted of clayey silt with sand of brown colour, natural transition of the upper topsoil. Consistent level, locally with white shades caused by calcium carbonate cementation and with dark red shades.

Level B: $1.7/2.3 \div 10/14$ *m*: Clayey silt with fine sand in traces of hazel-brown colour. Very consistent level and rich in fossil fragments with red shades. Local

Figure 2. Excavation of the hills for the preparation of the site where the power station is under construction.

horizons of coarser material constituted by salt crystals and fine white sand.

Level C: $10/14 \div 46$ *m*: Level represented by blue clay varying from consistent to very consistent and characterised by plastic behaviour.

The topsoil, before excavation, was used for the planting of wheat and for this reason it was yearly subjected to superficial works and chemical–organic manuring. The soil is not rich in nutrients or particularly fertile, but it is adequate for the works on the slopes described later. The plants established have similar behaviour to wheat. A further consideration was the potential presence of weed seeds in the topsoil; perhaps in the last period before the construction of the power station, the fields, from which this soil was removed, were left uncultivated, and for this reason the growth of this vegetation was remarkable, in particular in the autumn–winter season.

It should be underlined that the original design involved the use of topsoil on every part of the slopes, but this did not happen, may be because of the problems encountered in the storage of all this material during the excavation phase. In a project of this scale, as often happens, the agronomic aspects are of a secondary importance in comparison with practical engineering and economical requirements. Thus, the composition of the slopes is not homogeneous, and over the blue clay layer, soils of different types can be found. In many cases the topsoil is comparable to the soil identified as level A.

The slopes are inclined at 33–35% for an inclined length of 35 m, interrupted by berms 4–5 m width. The sub-layer is constituted almost exclusively of blue clay. The excavated slopes in the clayey sub-layer was considered stable after a geotechnical analysis, while the topsoil could be eroded by precipitation and runoff and could translate to a potential failure surface between vegetated soil and clay layer.

This paper aims at evaluating the different bioengineering techniques used on site, the design and the related monitoring surveys; the scope of these surveys is to confirm the success of the work, verifying the growth of the species and of their roots. Quantitative results on the contribution of the different types of gramineous species to the increase in soil shear strength are discussed. The most obvious way in which plants stabilise soils is by root reinforcement, the root tending to bind the soil and to increase its shear strength (Bache and MacAskill, 1984; Wu et al., 1979). For this effect, in recent years, the use of vegetation in civil engineering and landscape works has grown in importance, but specific design standards are still under discussion within the use of vegetation for slope stabilisation.

Materials and methods

Revegetation works

The revegetation works of the slopes were designed and developed by Vetivaria of Milan, a company specialising in bioengineering.

Species

The system developed was a technique of bioengineering applicable for erosion control and superficial soil reinforcement. This technology involved the use of perennial 'gramineae' plants, established with a density of 2–4 plants/m^2: these plants, in one or two seasons, are able to enhance soil shear strength, to modify surface water regime and even to reduce pollutants, if they are present. These plants have the function of pioneer species, which stabilise the soil surface and can be associated with autochthonous shrubs and trees; in this way the site should evolve naturally with low maintenance. The use of perennial "gramineae" species was only recently introduced in Mediterranean areas (Pease, 2000), whilst it is already practised widely in Asia and also in relation to slope stabilisation (Hengchaovanich, 2003).

The revegetation works for the excavated slopes of the power station allowed a gradual process of erosion protection, soil reinforcement, the ability to improve the soil geotechnical properties and the soil structure and to add organic matter to the surface with the annual renewal of vegetation. It was planned to establish four different plants of perennial 'gramineae', with no stolons. These plants have a radial vegetative growth and a fascicle root system, and are able to reach great depths even after the first vegetative seasons. Once having considered the characteristics of the site, the following perennial gramineae were chosen: Vetiver, Pangrass, Eragrass, Elygrass. A brief description of each species is reported here.

Vetiver (Vetiveria zizanioides L.). Vetiver (Figure 3a), relatively unknown until the 1980s, is used for the extraction of the essential oil from roots, for slope stability and water resources conservation. The promotion of the World Bank made this plant known as the model plant at world level for works of 'green engineering'. This plant is a bushy perennial 'gramineae' with rhizomes producing many culms and with aerial development up to 1.60 m in optimal conditions. It presents linear leaves with light green plane lamina and root system with numerous fibrous and cylindrical roots, able to reach 4–5 m of depth. It does not produce spikes because a sterile clone is established that was selected by genetic research for root development. Vetiver, thanks to its reduced number of stoma and to the deep growth of roots, resists very well both drought and immersion in water, tolerating conditions of root asphyxia. It likes to be exposed to full sun. It is also able to adapt itself to a variety of soil conditions, from sand to clay. Plants are able to grow in both acid and alkaline ($4 < pH < 11$) soils, in dry and moisty soils, in peat or in soils with poor organic substances. It resists very high concentration of pollutants and heavy metals, i.e., Cd, Pb, As, Al, Sn, Zn, Hg, Sb, Cr, Ni, etc., present in the soil and which are extracted and retained inside the plant. The plants start growing at soil temperature above 15°C and in this condition, roots are able to grow up to 2 m within the first two years, but they are able to reach greater depths of 5 m, growing vertically and not in a radial direction.

Pangrass (Panicum virgatum L.). This species originated in North America (Figure 3b) and shows a wide genetic variability for a range of climatic conditions. This plant is a bushy perennial 'gramineae' with short and dark rhizome, producing many culms up to 2 m high and with aerial development of 0.6–0.8 m in optimal conditions. It presents persistent leaves, plane

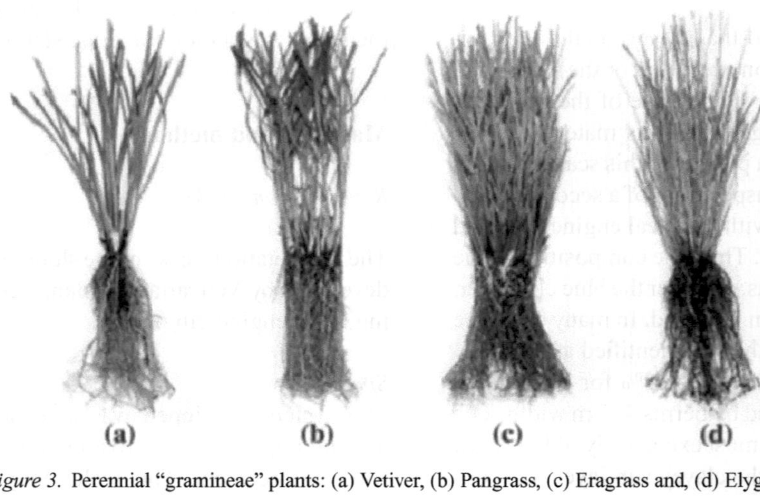

Figure 3. Perennial "gramineae" plants: (a) Vetiver, (b) Pangrass, (c) Eragrass and, (d) Elygrass.

and glabrous lamina up to 0.6 m long and 2 cm wide, but sometimes lightly pubescent at the base. The root system is characterised by numerous cylindrical roots that are able to reach depths of 3–4 m. The flowers mature in autumn and are constituted of spikes, more compacted at the apex and very open at the base, up to 0.5 m long. Pangrass, due to its reduced number of stoma and to the deep growth of roots, resists very well both drought and immersion in water, tolerating conditions of root asphyxia. The plants start growing at soil temperature above 15°C and in this condition, roots are able to grow up to 0.6 m within the first 2 years. The species requires exposition to the sun, adapts itself even to rocky soils with optimum development in the summer and it flourishes with frequent precipitation. The plants are hardy and have a discrete fodder value. They die during winter and germinate again in spring. The plant is able to reproduce both in pots and then with planting on site, giving optimum vegetation within a few months, as long as humidity and temperature are both adequate.

Eragrass (Eragrostis curvala Nees). This species originated from South Africa (Figure 3c). This is the most commonly planted species among the *Eragrostis* family and also has ornamental scope. It is a perennial 'gramineae' with crown producing many culms and with aerial development up to 1.20 m. It has lamina up to 0.6 m long, rough on the upper side and of dark green colour. Root system is characterised by numerous fibrous, very thin roots, able to reach depths of 3–4 m. Dark olive green spikes mature in summer, are open and are 0.3 m long. Eragrass prefers exposed locations, with warm, dry climate and sandy soil. The plants should be established at soil temperature above 15°C and in this condition roots are able to grow up to 0.6 m within the first 2 years. The species requires exposure to sun, adapts itself even to rocky soils with optimum development in the summer and it particularly appreciates frequent precipitation. The plant dies during winter and germinates again in spring. The best propagation method for this plant is in a greenhouse because of the limited dimensions of the seeds. The species has got a discrete fodder value and could be propagated by culm division.

Elygrass (Elytrigia elongata L.). This species originated in southeastern Europe but is now diffused in every continent and is very vigorous and evergreen, even in winter time (Figure 3d). It is a perennial 'gramineae' with crown producing many culms and with aerial development up to 0.6–0.8 m. Its leaves are long, resistant and erect, with culms that could reach the height of 1.0–1.5 m. Root system presents numerous cylindrical and very thin roots, able to reach depths of 2 m. Elygrass prefers an exposed location, with a warm, dry climate and sandy soil. Roots are able to grow up to 2 m within the first 2 years. It has got a high resistance to salinity and can flower twice a year, both in spring and in autumn.

Technical operations of planting

After the excavated areas had been constructed, a cultivation surface was carried out prior to planting. Furrows of approximately 0.25–0.30 m deep were excavated using a caterpillar tractor with trench cutter machine.

Plants were established with a density of 4 plants/m^2 in parallel rows and mature manure was added locally, in quantities of 40 g/m^2, i.e., 10 g for each single plant, with a high nutrient content, especially nitrogen. This was advisable, as the nutrients are largely in a slow-release form and become available as the organic matter decomposes. The works involved an area of 46,400 m^2 with a consequent placing of more than 200,000 plants. In Table 1 the number of established plants, subdivided among different species, is indicated.

A jute geotextile (mass per unit area: 700 g/m^2), completely biodegradable, was placed as mulch to help and support the vegetative growth. It enhances microclimatic conditions (like temperature and soil moisture) and organic matter-levels in the soil (Som et al., 2001). It was fixed with timber stakes linked with a nylon rope in order to resist the action of wind and runoff.

The period of time covered by the execution of the works (17 November 2002–31 May 2003) was characterised by very different climatic and meteorological conditions, varying from exceptionally rainy and cold winter months to a dry and warm spring period. These different conditions made it more difficult for the plants to establish and they decreased their growth rate.

Table 1. Number of plants for each species

Planted species	Number of plants
Vetiver	39,901
Elygrass	44,392
Eragrass	57,352
Pangrass	59,024

At the end of the planting period, maintenance was carried out with localised works, hand weeding and considerable quantities of irrigation, because of the high temperature and exceptional drought for that period. The irrigation was carried out in the warmest period between June and August, in double working shifts from 6 a.m. to 6 p.m. with the daily dispersal of about 100,000 litres of water by a 'jet-gun'. The water was taken from a nearby stream and its quality was particularly good.

Monitoring programme

The considerable extension of the works and the use of different vegetation species made periodic monitoring of the entire site appropriate. Technical inspections on site were carried out every 3 months for the first year and once a year for the following 2 years, with qualitative evaluations and with the measurements of appropriate parameters.

The principal monitored parameters were the percentage of live plants, the height of the foliage and the depth of the roots. The percentage of live plants is the result of a numerical evaluation on a sample of 50 plants for each considered sector of the site. The height of the foliage is the average of five measurements on plants in the same sector, while the depth of the roots is obtained from the direct measurement of plant roots extracted from soil. This last measurement is affected by the extraction method of plants. Roots present a more branched structure towards the apex with their diameters diminishing rapidly. Thus, root systems are difficult to extract without breaking their tips. During the extraction operations, carried out with agricultural tools, a certain number of roots were broken and their total length reduced by approximately 30%–40%. Comparing these results with others obtained employing a more complex technique with water jet pressure, these higher total lengths of roots could be measured and quantified as 33% more than that obtained with the traditional method. For this reason, the lengths of the roots measured on extracted plants were considered as equal to 66% of the real extension and consequently increased in order to provide a more representative result.

Contribution of roots to soil shear strength

While the effects related to the presence of roots are very well known from a theoretical point of view, the research did not yet come to define a sufficiently consolidated methodology for their quantification (Gray and Sotir, 1996). Only few references were found in the literature concerning shear strength evaluation of rooted soil in laboratory (Goldsmith, 1998; Operstein and Frydman, 2000) and on site (Wu and Watson, 1998). In order to quantify the contribution of roots to soil shear strength, direct shear tests both on soil and on root reinforced soil, respectively, were carried out. These tests were performed on undisturbed samples collected from the site of Altomonte.

All these samples were specimens already prepared to be tested in laboratory at different depths (respectively, 0.2, 0.4 and 0.6 m), without placing additional loads on the surface but leaving only acting the weight of the overlying soil. This configuration reproduced the situation on site of superficial soil movement.

The approach to perform tests on undisturbed samples containing roots was absolutely new and it has allowed tests to be performed where plant growth and root development have not been influenced by other factors. Operstein and Frydman (2000) tested soil samples reinforced with root vegetation, but the plant grew in pots in laboratory controlled conditions. Moreover, it was considered the demand to define experimental methods which could be repeated and could be employed to verify the effective stabilisation action of roots. This result would have been reached only by adopting experimental methods, common for the geotechnical engineering and conveniently adapted to the case under study. For this reason it was thought to carry out direct shear tests on large samples (200 mm in diameter), in order to allow the complete development of the root resistance mechanism.

The type of soil on the site of Altomonte allowed the extraction of undisturbed samples of such dimension. For this reason the first metre of soil was sampled in order to carry out directly the shear tests on these samples at three different depths. It was also decided to collect three undisturbed soil samples, each containing a root system for all the four considered 'gramineae' species. To effectively quantify the increase in shear strength, it was thought to retrieve also three samples of soil alone from the same zone: this procedure allowed to obtain samples almost homogeneous, considering the nature of the soil. Fifteen undisturbed samples were collected.

The steel samplers of 200 mm in diameter, having a thickness of 1 mm and a height of 1 m, were prepared. The sampling programme was carried out in March 2004 on an accessible area to the sampling equipment.

After having chosen the three specimens for each of the four 'gramineae' plant species, their aerial parts were cut to allow a better positioning and a perpendicular penetration of the sampler. The entire system was pushed under the soil surface, after having positioned the equipment and after having fixed the sampler with screw bolts. This procedure was repeated for each sampling.

The collected undisturbed samples were sealed with wax application to keep constant moisture content and were transported to the CESI Geo-department in Milan.

In March 2004, the steel samplers were opened just before carrying out the direct shear tests, taking particular care not to damage them during this operation. The samples were then wrapped in a transparent film to be moved and then their heights and weights were measured.

Direct shear tests were carried out in a large direct shear device, designed to allow single shear and opportunely modified to perform this series of tests on rooted soils. In fact, in order to perform tests on the soil column about 1 m high, it was necessary to realise a particular steel support, able to sustain and fix the sample during the test, in such a way not to bend the sample itself (Figure 4).

The shear test apparatus is constituted of two parts: the lower part has the function to fix the sample at the base, while the upper part is able to move in the plane and was assembled with the steel support. Samples were inserted vertically and shear planes were localised respectively at depths (z) of 0.2, 0.4 and 0.6 m below the soil surface.

The vertical stress normal to the shear plane was provided by the weight of the overlying soil, without applying additional loads on the surface. In correspondence of the shear plane, a free distance of 10 mm was left between the upper part and the lower part of the shear test device. The transparent film, in which the sample had been wrapped, was removed in the lower part while it was left on the upper part to eliminate friction between the sample and the steel support during test performing.

Direct shear tests were executed to a maximum displacement of 33 mm, which represents the limit of the apparatus, imposing a constant shear displacement ratio of 0.2 mm/min, allowing to complete mobilise root contribution to shear strength. The shear displacement ratio adopted for this experimental programme was the same as in the tests illustrated by Operstein and Frydman (2000). Each test was completed in about six

Figure 4. Carrying out of direct shear test and positioning of the sample inside the steel support specifically created.

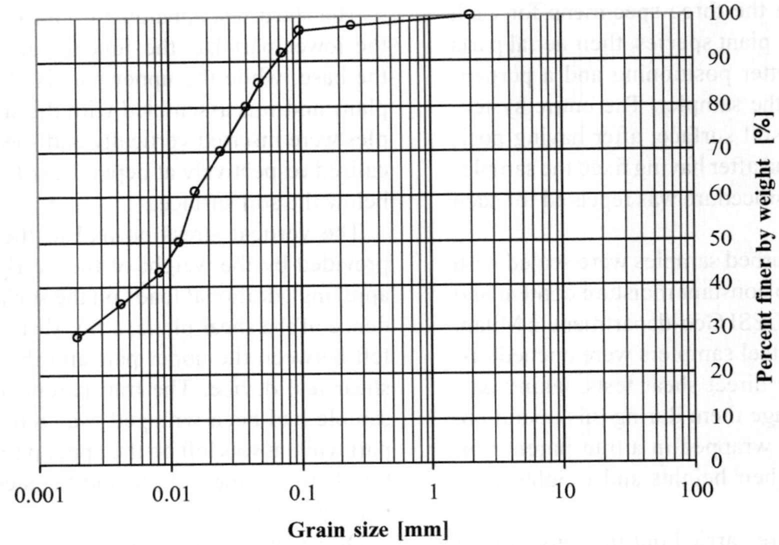

Figure 5. Particle size distribution of an undisturbed soil sample.

working hours and the execution of all the scheduled tests took about 1 month.

After shearing, the moisture content of the soil was measured and values between 25% and 47% were registered, while the particle size distribution was determined for one sample (Figure 5). Originally, it was planned to measure the rooted area (defined as the ratio between the area of the roots and the total section area) on the direct shear plane. However, the species under test were herbaceous plants and were characterised by very fine roots, thus this measure could not be executed because it was not considered significant, as the roots were either translated or torn.

Results and discussion

Revegetation works

August 2003

The variability of the environmental and climatic conditions affected the first phase of plant growth. The worst affected periods seem to be winter (December–January), characterised by continuous rainfall which created asphyxiated soil conditions; in spring (May), characterised by very high temperatures, which was a cause of stress in the planting period and in the first phases of sprouting, even though the plants were irrigated.

Considering the different species, the best results were observed for Pangrass and Eragrass, in particular for the zones where the planting period was carried out in spring (February–April), which had an exceptional growth, both for the aerial vegetation and for roots. Elygrass reflected its characteristic of autumn–winter growth. Good results were achieved for winter plantings, while the plants established during the already very warm period of May were not so well developed. Vetiver was slow to develop, but this is typical of this species.

The sectors of slopes oriented towards north-west revealed a high presence of weeds determined by a great quantity of seeds in the topsoil. For this reason manual removal operations were necessary to avoid competition with established perennial 'gramineae' plants.

The average percentage of sprouted plants for each single species was 81% for Vetiver, 88% for Elygrass, 89% for Eragrass and 90% for Pangrass. The lowest results, as can be observed from Figure 6, are related to the establishment periods of December 2002 and January 2003, due to the adverse meteorological conditions of that period. The percentage of sprouted plants for Vetiver is the lowest among all the species. It was too early to establish definitive conclusions about height of foliage and depth of roots because all the plants were still in the active part of their growth cycle.

Finally, it could be concluded that the works achieved good results for the first phase, considering the growth

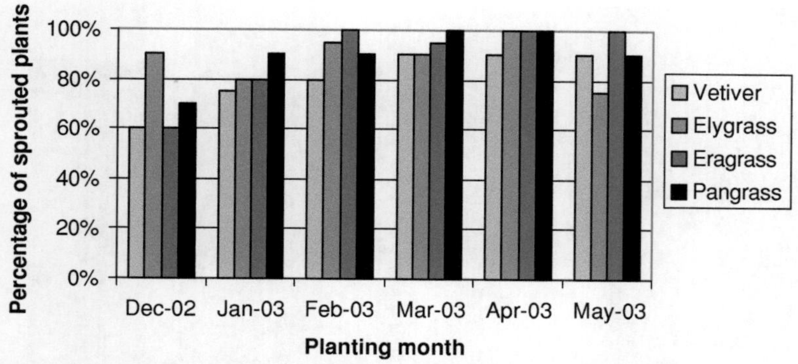

Figure 6. August 2003 monitoring results: percentage of sprouted plants.

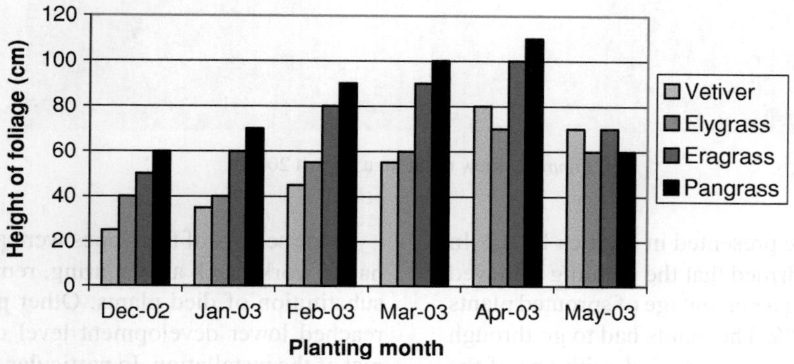

Figure 7. August 2003 monitoring results: height of foliage.

of plants taking into account the difficult climatic conditions and the scarce fertility of the topsoil on the clay sub-layer. In spite of a particularly warm and dry period, plants had established well and showed a vigorous first growth. The nutrient-irrigation interventions allowed the plants to establish without suffering withering from exceptional heat. The results of this survey are presented in Figures 6–8. A picture of the works in August 2003 is illustrated in Figure 9.

November 2003

The growing season could be considered to end in November and thus was the appropriate time to evaluate the definitive growth of plants and the eventual need to replace those which have failed. The measured parameters and the gathered data during the monitoring carried out in November were the same as the previous one in August. The analysed plants were chosen in a random manner in the same zones monitored in

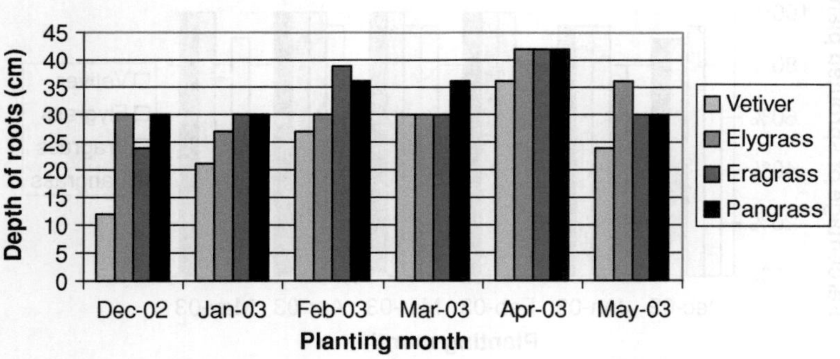

Figure 8. August 2003 monitoring results: depth of roots.

Figure 9. View of the area August 2003.

August. The results are presented in Figures 10–12. In general, it could be affirmed that the planting achieved a good result, obtaining a percentage of sprouted plants between 90% and 100%. The plants had to go through a particularly difficult summer period, with one of the warmest summers among the last 50 years, with temperatures that reached values of 38°C. The plants completed their biological cycle, with the exception of Elygrass which normally carries out its cycle during the autumn–winter period. The jute geotextile was very effective and remained fixed to the soil, keeping it wet for all the summer period and avoiding the propagation of weeds.

Some sectors of the slopes were affected by maintenance works such as manuring, removal of weeds and substitution of died plants. Other parts of the project reached lower development level compared with the rest of the installation. In particular, the sectors planted during May, when the temperature was already too high and the emergency irrigation operations were not sufficient, show a low development. The plants situated in these sectors had a survival rate of 95%–100%, but showed weak growth and yellowing of the leaves due to burning and to lack of nutrition. Anyway, this phenomenon involved only a very small part of the whole area (3%–5%).

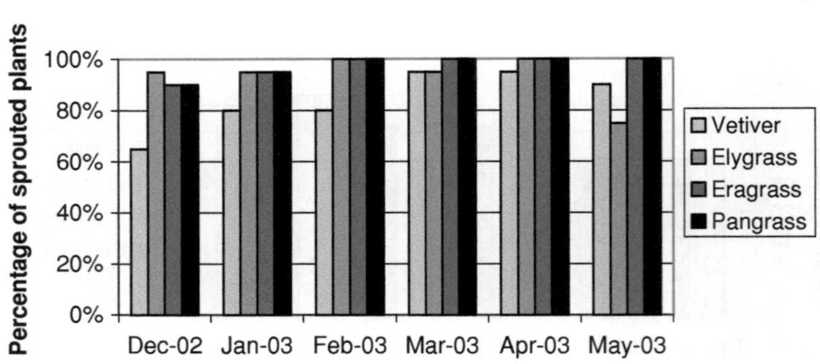

Figure 10. November 2003 monitoring results: percentage of sprouted plants.

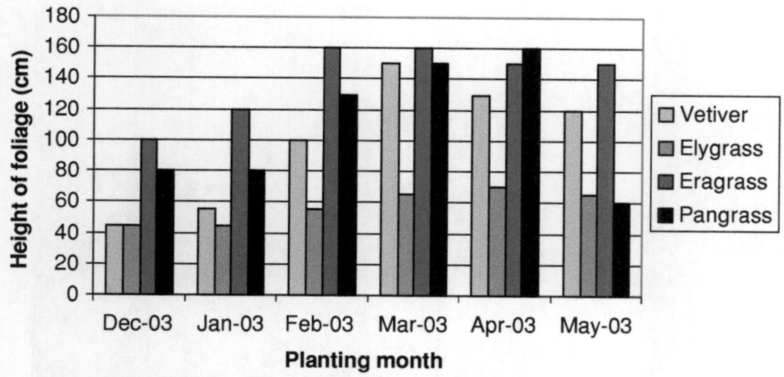

Figure 11. November 2003 monitoring results: height of foliage.

The extraction of the plants for the measurements of the roots was complicated by the growth of roots that had reached the hard clay layer. The effective length of roots was estimated using the same method adopted for the first monitoring campaign (i.e., in August 2003).

The percentage of sprouted plants increased: the averages for each single species had risen from 81% to 84% for Vetiver, from 88% to 93% for Elygrass, from 89% to 98% for Eragrass and from 90% to 98% for Pangrass. The lowest results, as can be observed from Figure 10, are related to the establishment periods of December 2002 and January 2003, when plants were subject to adverse meteorological conditions. The percentage of sprouted plants for Vetiver is the lowest among all the species; although Vetiver was initially affected by difficulties in the establishment period, it showed good results for the height of foliage and for the depth achieved by roots (Figures 11 and 12).

Considering the different species, the best results for root growth were achieved by Vetiver (roots average length 545 mm), Pangrass (500 mm) and Eragrass (475 mm), while for Elygrass (395 mm) it was too early to establish definitive results because it had not commenced its growth cycle which takes place during winter months. From the physiologic point of view, this is reasonably ascribed to the quality of the topsoil, with the addition of nutrients and manure, as a nutritive resource and of the sterile clay layer as a water source, able to maintain its capacity of water retention.

The results showed conclusively that within 5–6 months of establishment, the root systems have developed sufficiently to grow through the upper topsoil layer (0.2–0.3 m) into the underlying clay layer. A picture of the revegetated slopes in November 2003 is illustrated in Figure 13.

Contribution of roots to soil shear strength

In Figure 14, the shear strength values, evaluated in correspondence with the maximum shear stress values

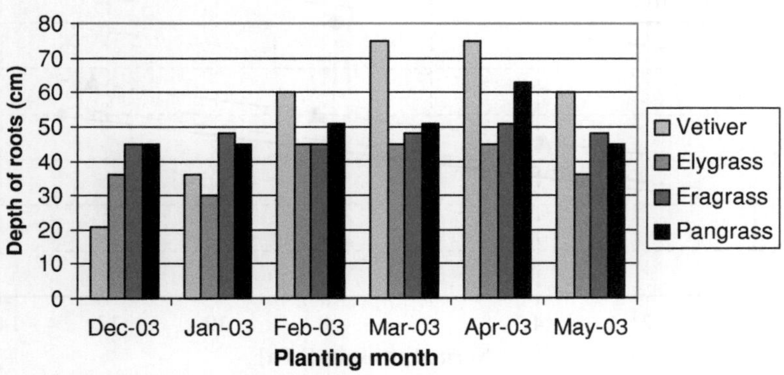

Figure 12. November 2003 monitoring results: depth of roots.

Figure 13. View of the area in November 2003.

registered after the direct shear tests, are presented. These data are expressed in function of the normal stress σ, represented by the weight of the soil over the shear plane and calculated on the base of the weight and of the height measures for each sample. A more detailed description of the results is given by Cazzuffi and Crippa (2005). It should be added that, even if the normal stresses are very low, the measured shear stresses are in very good accordance with the results found in the literature.

In Figure 14, parallel lines are traced, each corresponding to a particular species, that are a reasonable approximation to the trend in the data. In particular, this approach was based on the results of Operstein and Frydman's (2000) study who, from numerous shear tests carried out on plants cultivated in apposite pots,

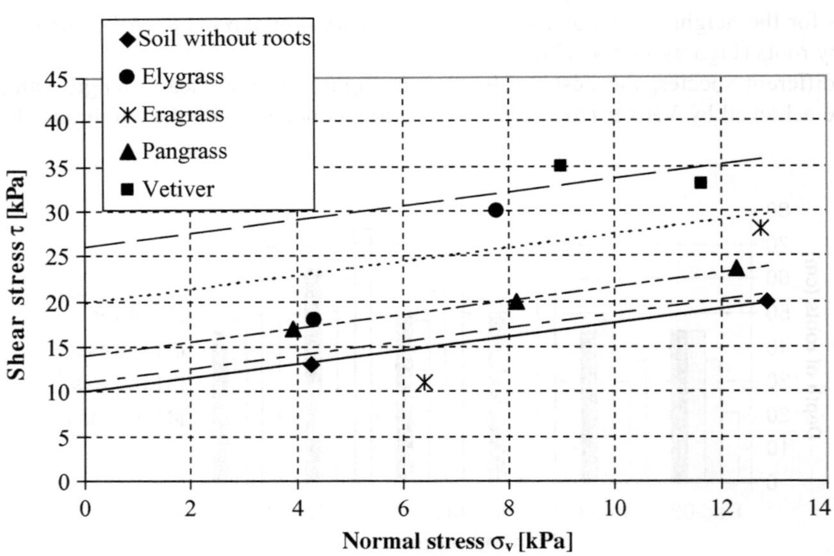

Figure 14. Maximum shear stresses against normal stress registered after direct shear tests.

noticed how, according to different species, trend lines corresponded to parallel lines. These authors concluded that the presence of vegetation roots causes the increase in the soil shear strength and in particular the increase in the cohesion, while the friction angle remains substantially unchanged.

The individuation of the trend lines was realised on the base of the results of the shear tests obtained on the soil samples with Pangrass roots. The traced trend line brought to a cohesion of 14 kPa and a friction angle of about 30°. This result was then extended to the other points on the graph, in order to have five parallel lines. It is effectively difficult to observe a good approximation of the extrapolated trend lines to the data, but it should be considered that data point are not so numerous and a certain variability is more than reasonable for tests on soils incorporating vegetation. Nevertheless, the contribution of roots to the soil shear strength is evident. In each test, the shear strength values of soil samples with roots are always higher than the values obtained from tests on soil samples without roots.

The root tensile strengths of the considered 'gramineae' species were determined by testing in laboratory different root systems sampled on site (Cazzuffi and Tironi, 2003). For the four species considered in this experimental programme, the tensile strength ranges are as follows:

- Elygrass: 25–70 MPa
- Eragrass: 38–55 MPa
- Pangrass: 15–23 MPa
- Vetiver: 25–60 MPa

As it can be noticed from the results of this research, there is a certain influence of the root tensile strength on the increase in soil shear strength. In fact, Elygrass and Vetiver, characterised by the highest root tensile strengths, are the species that were able to offer the highest increase in soil shear strength. On the other hand, the lowest shear strengths corresponded to the soil samples containing Eragrass and Pangrass root systems. Also in this case the variability related to tensile strength intervals should be considered.

With reference to Figure 14, the parallel trend lines are comprised in a range of cohesion of about 15 kPa and in particular the increase in shear strength is ranging between 2 kPa and 15 kPa, which is close to what Operstein and Frydman (2000) found. Moreover, Bache and MacAskill (1984) explained that the increase in soil shear strength due to the roots could vary between 3.4 kPa and 17.2 kPa. Also Belfiore and Urciuoli (2004), on the basis of tensile tests carried out on *Arundo Plinii* and *Poa Pratensis* roots (species belonging to the family of 'gramineae') and after having developed a theoretical model on the root behaviour, obtained a maximum increase in shear strength of about 20 kPa.

Thus, the experimental study presented in this section of the paper seems to confirm that the values obtained for the increase in shear strength represent the actual magnitude of the reinforcement offered by the presence of the root systems. Surely, this increase is a function of the root tensile strength and of the cross-sectional area occupied by the roots, even if this parameter was not considered for this research. On the other hand, authors like Operstein and Frydman (2000) consider it of smaller importance than tensile resistance of roots. It is then necessary to take into account that the influence of the roots to the soil shear strength cannot be a constant value but has to diminish with depth until zero, where roots are not present. Generally, it could be affirmed that the maximum influence limit is about 2–3 m, for species characterised by root systems able to reach high depths.

Conclusion

The paper shows the good results achieved in the Altomonte revegetation works in terms of sprouted plants, aerial growth of plants and root development. In this way, it could be considered that the principle aim of the work to link the topsoil to the clay sub-layer can be achieved, therefore reducing the risk of shallow instability and soil erosion. Moreover, it was demonstrated the importance of monitoring, maintenance and aftercare of bioengineering works, that, in the early years, can be subjected to failures caused by difficult climatic and environmental conditions.

The results obtained from direct shear tests allowed to reveal the influence of roots by direct comparison of tests on soil samples with roots and on soil samples without roots. Following the approach of Operstein and Frydman (2000), it was observed how the increase in soil shear strength can be understood to be the result of an increase in cohesion, which was quantified in a range between 2 and 15 kPa, in optimum accordance with other studies. In particular, it was stressed how the increase in the soil shear strength depends on the considered species and it was also emphasised that the increase is a function mainly of the tensile strengths contributed by the root systems. This conclusion justifies the growing interest on the 'gramineae' species here analysed and in particular on the Vetiver type. These species, in fact, are characterised by very resistant roots and the present study confirms how they, and all the other species with similar properties, could be successfully used with stabilising effects on phenomena like shallow instability.

References

Bache D H and MacAskill I A 1984 Vegetation in Civil and Landscape Engineering. Granada, London, UK.

Belfiore G and Urciuoli G 2004 Analisi del contributo meccanico delle radici alla resistenza del terreno (Interpretation of the root mechanical contribution to the soil shear strength). *In* Proceedings of the Annual Meeting of the Geotechnical Researchers 2004 – IARG 2004, Trento, Italy (CD rom in Italian).

Cazzuffi D and Crippa E 2005 Shear strength behaviour of cohesive soils reinforced with vegetation. *In*: Proceedings of the 16th International Conference on Soil Mechanics and Geotechnical Engineering, Osaka, Japan, pp. 2493–2498.

Cazzuffi D and Tironi F 2003 Contribution of roots to slope stability: an overview of typical results for different plants. *In* Proceedings of the International Conference "Fast slope movements prediction and prevention for risk mitigation," Napoli, Italy, Vol. 1, pp. 101–105.

Coppin N J and Richards I G 1990 Use of Vegetation in Civil Engineering. Butterworths, London, UK.

Goldsmith W 1998 Soil reinforcement by river plants: progress results. *In* Proceedings of the ASCE Conference on wetlands engineering and river restoration, Denver, USA, pp. 301–306.

Gray D H and Sotir R B 1996 Biotechnical and Soil Bioengineering Slope Stabilisation: A Practical Guide for Erosion Control. Wiley, Chichester, USA.

Hengchaovanich D 2003 Vetiver system for slope stabilization. *In* Proceedings of 3rd International Vetiver Conference, Guangzhou, China, pp. 301–309.

Morgan R P C and Rickson R J 1995 *Slope Stabilization and Erosion Control*. A Bioengineering Approach. E & FN Spon, London, UK.

Operstein V and Frydman S 2000 The influence of vegetation on soil strength, Ground Improvement 4, 81–89.

Pease M 2000 The European and Mediterranean Vetiver Network. *In* Proceedings of the Second International Vetiver Conference, Thailand, pp. 185–187.

Schiechtl H M and Stern R 1996 Ground Bioengineering Techniques for Slope Protection and Erosion Control, Blackwell, Oxford, UK.

Som N, Rao P J and Sanyal T 2001 A Manual on Use of Jute Geotextiles in Civil Engineering. Jute Manufacturers Development Council, Hooghly Printing, Kolkata, India.

Wu T H and Watson A 1998 In situ shear tests of soil blocks with roots, *Can. Geotech. J.*, 35, 579–590.

Wu T H, McKinnel W P and Swanston D N 1979 Strength of tree roots and landslides on Prince of Wales Island, Alaska, *Can. Geotech. J.*, 16, 19–33.

Root system morphology and anchorage

Root system asymmetry of Mediterranean pines

P. Ganatsas[1,3] & I. Spanos[2]

[1]*Laboratory of Silviculture, Aristotle University of Thessaloniki, P.O. Box 262,54 124, Thessaloniki, Greece.* [2]*Forest Research Institute, NAGREF, Vassilika, 57006, Thessaloniki, Greece.* [3]*Corresponding author**

Key words: cross-sectional root area, *Pinus brutia*, *Pinus halepensis*, root architecture, root asymmetry

Abstract

Three groups of Mediterranean pines were examined to describe the development of root symmetry on sites characterized by shallow soils and low water availability. Sampling included: (1) 3-year-old planted seedlings of *Pinus halepensis* Mill. taken from Sithonia Halkidiki, northern Greece, (2) 5-year-old natural regenerated seedlings of *Pinus brutia* Ten. taken from Kedrinos Lofos, Thessaloniki and (3) 65-year-old trees of *Pinus brutia* taken from Kedrinos Lofos, Thessaloniki. Root system symmetry was examined by measuring the number, the diameter, the cross-sectional area, the root area index and the length of the lateral roots of each root system, and by analyzing their distribution around the stem. Aboveground plant symmetry was also estimated. The findings of the study indicated that there was an asymmetric root system in all three groups that is characterized by the concentration of the main laterals along the contour lines instead of uphill or downhill; however, the asymmetry was much higher in the young plants. This asymmetry was not correlated with the above-ground plant growth form, which was found to be symmetric. The asymmetric development of root can be attributed to the shallow soil and the high mechanical resistance of the underground bedrock that stopped the taproot growth, restricted the root penetration in the deeper layers and obliged the roots to elongate towards the surface soil layers, where there is more available water.

Introduction

In situ investigations on root systems face many practical difficulties. However, the development of a root system, capable of anchoring the shoot and obtaining water and nutrients, is essential to the terrestrial plants' survival and growth (Bengough et al., 1997; Clark et al., 2003). Since the environment of root systems is highly heterogeneous both in time and space, it appears important that the root systems have the ability to react to that heterogeneity, even at a local level within the root (Stokes et al., 1998); in other words they possess phenotypic plasticity (Fitter, 1991).

Usually, shallow forest soils in combination with high soil consistency affect root architecture since root elongation is permitted only when the root pressure exceeds the soil mechanical impedance. Drought also increases penetration resistance of the soils (Moroni et al., 2003) as soil strength increases with decreasing soil water content (Clark et al., 2003). Pathways of lower mechanical impedance give rise to preferential root growth. As a consequence, the distribution of roots in a soil profile depends on soil depth and the mechanical resistance of the different soil layers. When the root-impeding layers are near the surface, they will slow the downward root growth that results in a shallower root system which finally will be restricted to the upper part 76 of the soil profile (Bennie, 1991; Ehlers et al., 1983).

Root systems of forest trees are often markedly asymmetric and there are many factors that affect asymmetry. However, most of the relevant work has been carried out on species with shallow root systems due to the problems of windthrow (Coutts et al., 1999; Mickovski and Ennos, 2002; Nicoll and Ray, 1996). These findings conclude that the root system of many plant species is often asymmetric. Root systems of trees growing under adverse site conditions such as shade or water stress

* E-mail: pgana@for.auth.gr

This article has been previously published in the following journal—Plant and Soil, Volume 278, 75–83.
Stokes et al. (eds), Eco- and Ground Bio-Engineering: The Use of Vegetation to Improve Slope Stability, 127–134.

may be less symmetrical than normal but there is no information on this (Coutts et al., 1999; Ganatsas and Tsakaldimi, 2003). Concerning trees growing on slopes the reported results are quite controversial; Nicoll et al. (1995) found most roots of Sitka spruce on down-slope side of trees, but, Nicoll and Ray (1996) found for the same species most root mass up-slope; it should be noticed that in both cases root mass was concentrated on the side away form the prevailing wind direction. However, Sundström and Keane (1999) reported that both numbers of roots and root area seemed to concentrate along the contour lines.

The Mediterranean pines *Pinus halepensis* Mill. and *Pinus brutia* Ten. are considered two important tree species for reforestation in the Mediterranean region because they are early successional species, they have a woody deep tap root with vigorous laterals and they are drought tolerant. In deep soils, the diameter of taproot reaches 15–20 cm at a depth of 1 m, while in shallow soils they form a shallow root system (Moulopoulos, 1962).

This research was an *in situ* study of the root architecture of *Pinus halepensis* and *Pinus brutia* grown on sites characterized by shallow soils and low water availability, with the aim of examining to what extent their root system is asymmetric, what the changes are with the tree age and if there is any relationship between root asymmetry and above-ground plant growth.

Materials and methods

Site description

The study was carried out in two areas, the reforestated area of Sithonia Chalkidiki and the artificial periurban forest of Thessaloniki. Both areas are characterized by adverse ecological conditions, namely, shallow soil, high mechanical resistance of the underground bedrock and low water availability (Ganatsas and Tsakaldimi, 2003; Tsitsoni, 2001). The altitude of both studied areas ranges from 100 to 200 m. According to data from the meteorological station of Saint Mamas and the University of Thessaloniki (for the two respective areas), the climate is Mediterranean, with a mean annual precipitation of 420 mm and 397 mm, respectively; the dry period lasts from April or from the middle of May to the end of September. The vegetation of the first area belongs to the *Oleo-lentiscetum* association while the second to the *Ostryo-Carpinion* alliance (Tsitsoni et al., 2004). Geologically, the Sithonia peninsula belongs to the Axios zone and Circum Phodope zone; the rock materials are mainly igneous (granites) and crystalline schists. The area of the peri-urban forest belongs to magmatic series of Chortiatis and consists mainly of green-schists. The soils of both areas are slightly acid up to neutral (pH 5–6.8), and they are characterized by weak structure, low porosity and high percentage of stones and pebbles resulting from soil compaction due to repeated fires and overgrazing. The soil depth ranges from 20 to 30 cm in the first case and from 40 to 50 cm in the second case. Usually, the limiting factor for plant survival and growth in both areas is the low soil water availability during the long dry summer period (Ganatsas and Tsakaldimi, 2003; Radoglou, 1987; Tsitsoni, 2001).

Root sampling

Three groups of trees were sampled for above and below-ground measurements. These were: (1) 3-year-old planted containerized seedlings of *Pinus halepensis* taken from Sithonia Halkidiki, northern Greece (2) 5-year-old natural regenerated seedlings of *Pinus brutia* taken from Kedrinos Lofos, Thessaloniki and (3) 65-year-old trees of *Pinus brutia* taken from Kedrinos Lofos, Thessaloniki. The third group was selected in order to compare the results and to investigate the changes with age from the young individuals to an advanced (mature) stage.

For each of the first two cases 12 randomly selected seedlings or saplings were extracted for 77 root sampling; root excavation was made manually giving special attention to avoid root damage. The third group consisted of 12 trees of *Pinus brutia* that were selected during the Thessaloniki ring-road construction; these trees were cut and their stumps were extracted after the excavation performed during the roadworks. The tree selection included trees with roots that had been least damaged.

The trees were prepared for measurements by removing the litter around the stem, and the root system was revealed by careful removal of the soil (Mickovski and Ennos, 2002). Then, for each single root system of the first two groups of sampling, the following parameters were measured *in situ*, in their original positions: the number of medium sized lateral roots ($d > 1$ mm), the depth of their origin at the tap root, their vertical angle, their orientation using a compass and their length; as it was difficult in many cases to define the

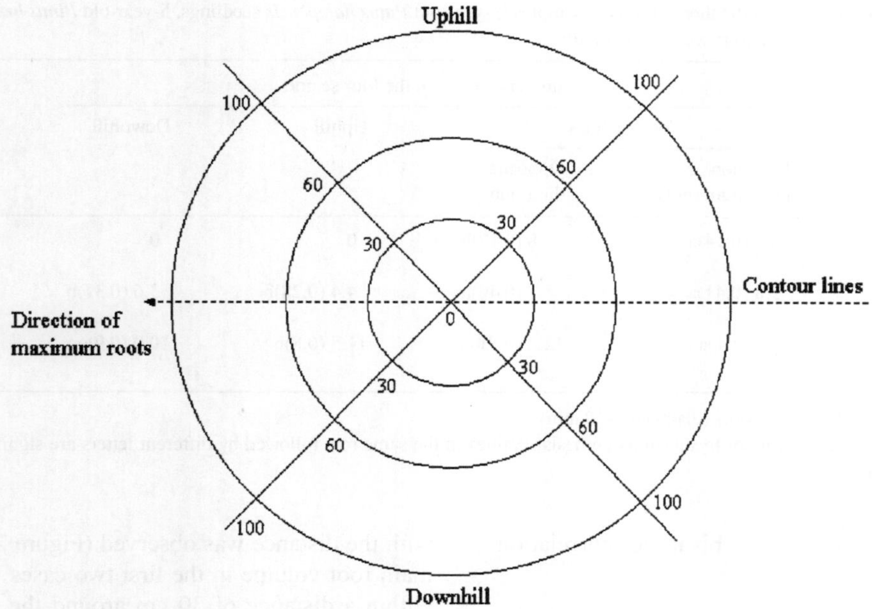

Figure 1. Details on the method of root sampling; the figure shows the distances from the center of root system (in centimeter) to where the CSA was measured and the separation of the four sectors of 90°. The direction of maximum roots was always found to be parallel to the contour lines regardless of the slope orientation.

end of roots, a minimum diameter limit of 0.2 mm was chosen as root ends. Each root system was divided in situ into four sectors of 90° that were based on the four directions of the slope; two directions of contour lines and uphill and downhill (Figure 1). The direction of maximum roots (the thickest and longest laterals) was then determined for each root system, according to the root field measurements. The diameter of each root was recorded at four distances from the stem center (Figure 1), 0 (the edge of the stem), 30, 60 and 100 cm using callipers (Sundström and Keane, 1999). The root diameters were taken in two directions on every occasion to get the cross-sectional area (CSA) of the root. The numbers of all roots and their CSA were totaled for each distance separately and for the entire root system. Number of roots, CSA and total root length were summed up for each of the four sectors and for the entire root system. The root area index (RAI), an index of evenness of root distribution in the four sectors, was used to estimate the root asymmetry (Lindström and Rune, 1999). This index was calculated as the ratio between root area in the sector with the largest root area (maximum roots) and the total root area. For an evenly distributed root system the RAI is 0.25 while higher RAI indices indicate a more asymmetrical root system. However, in the third group only the coarse laterals ($d > 10$ mm) were measured since more detailed measurements were not feasible. For the same reason the CSA at the edge of the stump and the root length measurements were omitted.

Above ground measurements

The above-ground measurements were carried out in all sampled trees. Radial crown width and diameter of three dominant branches were measured in two directions; the direction of the maximum roots and the opposite direction. Based on the branches' diameter, the CSA of the maximum branches was calculated. The stem diameter was measured in the direction of contour lines as well as in the direction perpendicular to them. Finally, the total plant height was measured.

Statistical analysis

The statistical analysis was accomplished by SPSS statistical program. The comparison of means between the four sectors was assessed by one-way ANOVA followed by Waller–Duncan test ($P < 0.05$, Norusis, 2002). Correlations between above and below-ground parameters

Table 1. Number of roots recorded in the three groups of samples (3-year-old *Pinus halepensis* seedlings, 5-year-old *Pinus brutia* saplings and 65-year-old *Pinus brutia* trees) in the four sectors and totally

	Number of roots in the four sectors				
	Contour lines		Uphill	Downhill	Total number of roots
	Direction of maximum roots	Opposite direction			
3-year-old seedlings of *Pinus halepensis*	2.4 (0.14)a	1.8 (0.12)b	0	0	4.2 (0.28)
5-year-old saplings of *Pinus brutia*	5.8 (0.41)a	5.7 (0.40)a	4.4 (0.28)b	3.6 (0.37)b	19.5 (1.31)
65-year-old trees of *Pinus brutia**	15.8 (1.1)a	12.2 (0.8)b	11.5 (0.8)b	10.0 (0.9)c	49.5 (2.3)

*The values concern only the coarse roots (diameter >10 mm).
Values are means and standard errors of mean (in parenthesis). Values in the same row followed by different letters are significantly different ($P < 0.05$, Waller–Duncan test).

were tested with Spearman' bivariate correlation coefficient.

Results

Root characteristics

The number of medium-sized roots was found to be very low in the case of 3-year-old *Pinus halepensis* seedlings; an average of 4.2 lateral roots per seedling were recorded (Table 1); the remaining roots were thinner and usually their development was restricted to within the space occupied by the growing medium. The number of roots was much greater in the case of 5-year-old naturally regenerated saplings of *Pinus brutia*; the average number of roots in this case was 19.5 laterals per individual. Mature trees exhibited an average number of 49.5 coarse roots ($d > 10$ mm) and a much greater number of medium and fine roots (Ganatsas and Tsakaldimi, 2003) that it was unfeasible to record.

The percentage of cross-sectional root area (CSA) of the medium-sized roots near the stem decreased with the age of the plants. It was on average 78% of the total root CSA in the case of 3-year-old *Pinus halepensis* seedlings (Figure 2a), as no roots were recorded further than 100 cm from the centre of the root system, and very few roots were found further than 60 cm from the centre. A lower percentage was found in 5-year-old *Pinus brutia* saplings (65%) as the contribution of the CSA recorded in the other distances increased (Figure 2b). In the case of 65-year-old trees of *Pinus brutia*, a lower decrease rate of the CSA of coarse roots with the distance was observed (Figure 2c). Thus, the main root volume in the first two cases was recorded within a distance of 30 cm around the centre of the stem while in the third case it was observed within the distance of 60 cm from the stump. Roots originating from the upper part of the root system were the thickest and the longest. In the case of young trees, these laterals originated from a depth of 5–15 cm while in the case of mature trees they originated from a depth of 10–30 cm. However, most of the roots originated from the taproot and they developed almost horizontally, parallel to the soil surface (their vertical angle was in almost all cases above 75°).

The total root length was on average 128.7 cm per seedling in the case of 3-year-old *Pinus halepensis* seedlings, while it was found to be much higher (1452.5 cm) in the case of 5-year-old saplings of *Pinus brutia* (Table 2).

Root asymmetry

The general trend observed in both pine species was that almost all the sampled plants developed their main lateral roots concentrated along the contour lines or with small deviation downwards on the slope, instead of uphill or downhill (Table 1); this resulted in an asymmetrical root development in all cases. The direction of maximum roots was always found to be at one of the two sectors along the contour lines, regardless of the slope orientation. The number of roots (Table 1), the CSA (Figure 2a–c) and the sum of root length (Table 2) were asymmetrically distributed around the stems. The RAI index was very high (0.79) in the case of

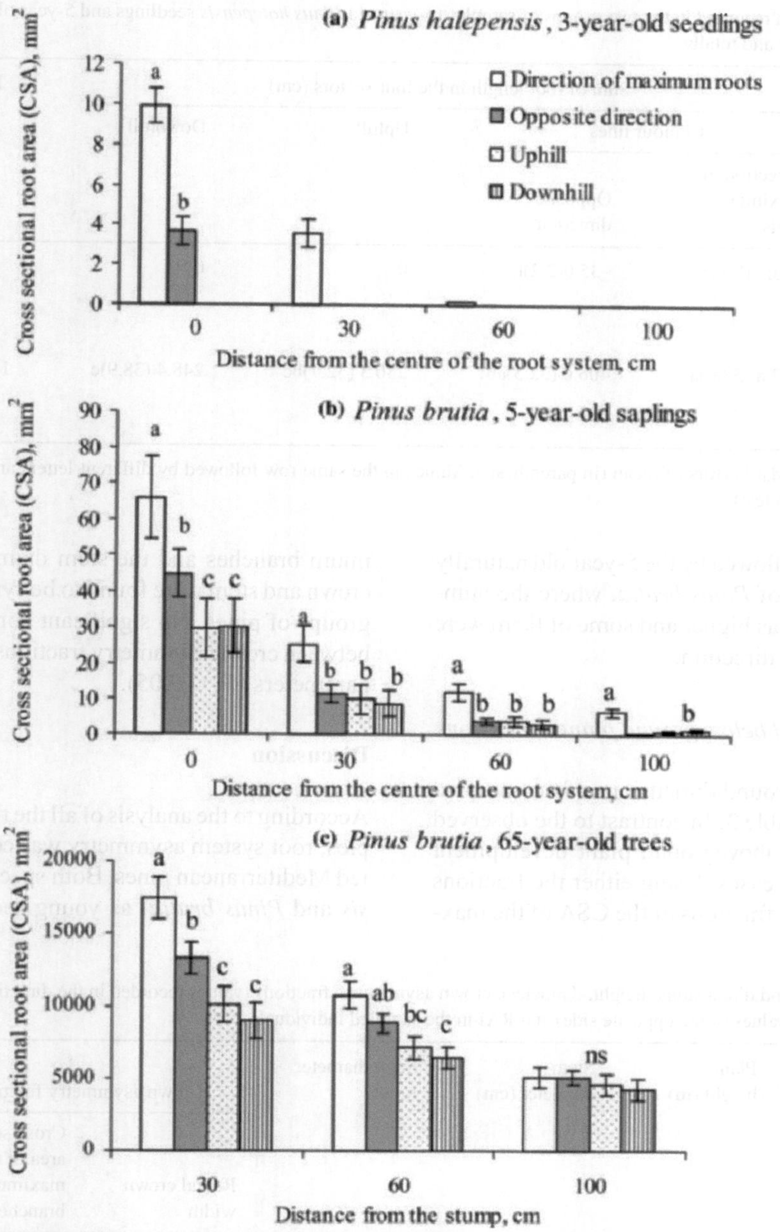

Figure 2. Mean CSA at the four distances from the center of root system and at the four sectors, for the 3-year-old *Pinus halepensis* seedlings (a), for the 5-year-old *Pinus brutia* saplings (b) and for the 65-year-old *Pinus brutia* trees (c). Vertical bars represent the standard error of mean (SE). Values for the same distance followed by different letters are significantly different ($P < 0.05$, Waller–Duncan test). ns = no significant differences.

3-year-old *Pinus halepensis* seedlings, and lower (0.42 and 0.33) in 5-year old *Pinus brutia* saplings and mature *Pinus brutia* trees, respectively (Table 3). The root growth pattern in the case of the planted *Pinus halepensis* seedlings was very characteristic; during the first year after outplanting almost all the roots grew within the space occupied by the growing medium; there were only few main laterals per seedling that elongated to the physical soil, in the same direction parallel to the soil surface, reaching a length of 50–70 cm. The same

Table 2. Sum of root length recorded in the two groups of samples (3-year-old *Pinus halepensis* seedlings and 5-year-old *Pinus brutia* saplings) in the four sectors and totally

	Sum of root length in the four sectors (cm)				Total root length in cm
	Contour lines		Uphill	Downhill	
	Direction of maximum roots	Opposite direction			
3-year-old seedlings of *Pinus halepensis*	93.7 (7.1)a	35.0(2.2)b	0	0	128.7 (8.1)
5-year-old saplings of *Pinus brutia*	517.8 (53.6)a	406.0 (32.3)ab	280.3 (32.7)bc	248.4 (38.9)c	1452.5 (112.7)

Values are means and standard errors of mean (in parenthesis). Values in the same row followed by different letters are significantly different ($P < 0.05$, Waller–Duncan test).

pattern seems to be followed by the 5-year old naturally regenerated saplings of *Pinus brutia*, where the number of main laterals was higher and some of them were found in the opposite direction.

Relation of above and below ground plant dimensions

The average above-ground dimensions of the sampled pines are shown in Table 3. In contrast to the observed root asymmetry, the aboveground plant development was symmetric in all cases. Using either the fractions of crown width or the fractions of the CSA of the maximum branches and the stem diameter ratio, the tree crown and stem were found to be symmetric in all three groups of pines. No significant correlation was found between crown asymmetry fractions and below-ground parameters ($P > 0.05$).

Discussion

According to the analysis of all the three groups of samples, root system asymmetry was common in the studied Mediterranean pines. Both species (*Pinus halepensis* and *Pinus brutia*) as young seedlings and mature

Table 3. Plant above-ground dimensions (height, diameter), crown asymmetric fractions (values recorded in the direction of maximum roots divided by the respective values of the opposite side) and RAI in the sampled individuals

	Plant height (m)	Stem diameter (cm)	Stem diameter ratio[a]	Crown asymmetry fractions		Root area index
				Radial crown width	Cross-sectional area of the maximum branches	RAI[b]
3-year-old seedlings of *Pinus halepensis*	0.33 (0.01)	0.68 (0.03)	1.00	0.98	1.00	0.79
5-year-old saplings of *Pinus brutia*	0.65 (0.02)	2.02 (0.08)	1.00	1.00	1.01	0.42
65-year-old trees of *Pinus brutia*	11.80 (0.29)	29.20 (1.01)	1.02	0.97	1.02	0.33

[a] The value is the fraction of the diameter values recorded in the direction of contour lines divided by the values recorded in the vertical to it axis.
[b] The ratio between the root area in the sectors with the largest root area and total root area. Higher values imply a more asymmetrical root system.
Values are means and standard errors of mean (in parenthesis where appropriate).

trees, planted or natural regenerated, developed their root system asymmetrically, which means that the tree root system is very susceptible to site stress factors during the whole plant life. This root asymmetry confirms that roots react to environmental modifications and show the plasticity of root systems (Fitter, 1991). Both, number of roots and root area, were mostly concentrated along the contour lines instead of uphill or downhill. This pattern of root system seems to allow for efficient water and nutrient uptake from the soil layers. Similar root orientation pattern was observed by Sundström and Keane (1999) for 10-year-old Douglas-fir trees. However, as only a few medium-sized roots were found in the case of 3-year-old *Pinus halepensis* seedlings this can be attributed to the great difficulties that the seedlings face during the first year after outplanting (a crucial period for seedling survival and growth (Tsakaldimi, 2001), combined with the high soil compaction of the area (Tsitsoni, 2001). Number of roots and total root length were found much higher in 5-year old naturally regenerated seedlings of *Pinus brutia*; this may show a greater adaptability of the naturally regenerated seedlings compared to the planting ones (Lindström and Rune, 1999). Also, the site conditions were better in the second case; the soil depth was much greater, 40–50 cm vs. 20–30 cm in the previous case.

The main laterals were observed to originate from the upper part of the taproot, as perhaps they have an advantage over deeper roots because they are the first to receive assimilates from the shoots (Coutts et al., 1999), while roots that originated from the lower part were smaller. Analyzing the changes of root asymmetry with age, it seems that the size of the main laterals increased with the age, while a clear typical taproot system was absent in all cases. However, the asymmetry was much higher in the young seedlings; as the trees grow, the symmetry of the structural root system maybe increasingly influenced by adaptive secondary growth related either to wind sway (Coutts et al., 1999) or to the exploration of microsites for more available water. Thus, young planted seedlings of *Pinus halepensis* had a more asymmetric root development (higher RAI index) than the 5-year old naturally regenerated saplings of *Pinus brutia* while the mature trees generally showed a better and more uniform root distribution. The finding that root distribution is improved by trees age has also been reported earlier (Lindström and Rune, 1999).

The observed root asymmetry was not correlated with the above-ground tree form which was found to be symmetric in all cases. This crown symmetry indicates that wind has a minor effect on tree growth, including probably root-system growth. Furthermore, according to the local meteorological data there is no great risk from wind in the studied areas; the winds are seldom intensive enough to cause problems to tree stands. Taking into account that the risk from wind is very low during the early stages of tree life we concluded that the decisive factor for root development in our case, which results in root asymmetry, is the shallow soil that stopped the growth of the taproot in combination with the soil water scarcity; these probably have a more serious effect on distribution of the roots, than any climatic factor. In contrast to that, many studies reported that the root asymmetry of several tree species, mainly with shallow root systems, is attributed to the requirement for the tree to withstand winds (Coutts et al., 1999; Mickovski and Ennos, 2002, 2003; Nicoll and Ray, 1996). However, Konstandinidou (1998), in Kassandra peninsula North Greece, found that *Pinus halepensis* trees grown on deep soils on marls have a tap root system at the ages of 23 and 48 years and a heart-shaped root system at the ages of 70 and 100 years.

The results obtained from this work suggest that the existence of shallow soil and the mechanical resistance of the underground bedrock in combination with low soil water availability caused a modification of the typical tap rooted pine root system. The taproot growth has stopped, the root penetration in the deeper soil layers was restricted and the laterals were obliged to elongate towards the surface soil layers probably in the direction of the existence of preferential pathways for water infiltration in the surface soil (Builet et al., 2002); this growth pattern results in a root asymmetry. However, this root asymmetry decreases with age and it was not correlated with the above-ground tree form.

Finally, the authors believe that considerations of root system modification of Mediterranean pines could contribute to a better management of stands. It is suggested that more space be provided for each tree along the contour lines rather than perpendicular to them, by thinning methods and planting spacing. However, more studies are needed to improve the knowledge on the tree root modifications under Mediterranean conditions.

Acknowledgments

We are grateful to two anonymous referees for critical reading and helpful suggestions on an earlier version of the manuscript.

References

Bengough A G, Croser C and Pritchard J 1997 A biophysical analysis of root growth under mechanical stress. Plant Soil 189, 155–164.

Bennie A T P 1991 Growth and mechanical impedance. *In* Y Waisel, A Eshel and U Kafkafi, (eds), Plant Roots: The Hidden Half. pp 393–414. Marcel Dekker, New York.

Bouillet J P, Laclau J P, Arnaud M, M'Bou A T, Saint-Andre L and Jourdan C 2002 Changes with age in the spatial distribution of roots of Eucalyptus clone in Congo: impact on water and nutrient uptake. For. Ecol. Manage. 171, 43–57.

Clark L J, Whalley W R and Barraclough P B 2003 How do roots penetrate strong soil? Plant Soil 255, 93–104.

Coutts M P, Nielsen C N and Nicoll B C 1999 The development of symmetry, rigidity and anchorage in the structural root system of conifers. Plant Soil 217, 1–15.

Ehlers W, Kopke U, Hesse F and Bohm W 1983 Penetration resistance and root growth of oats in tilled and untilled loess soil. Soil Till. Res. 3, 261–275.

Fitter A H 1991 Characteristics and functions of root systems. *In* Y Waisel, A Eshel and U Kafkafi (eds), Plant Roots: The Hidden Half. pp 3–25. Marcel Dekker., New York.

Ganatsas P and Tsakaldimi M 2003 Root system modification of *Pinus brutia* Ten. species under adverse ecological conditions. *In* TD Lekkas (ed), Proceedings of the 8th International Conference on Environmental Science and Technology. pp. 216–223, 8–10. Lemnos island, Greece.

Konstandinidou E 1998 Dynamics of root form and biomass in *Pinus halepensis* stand in Kassandra, Chalkidiki, Ph.D thesis, Department of Forestry and Natural Environment, Aristotle University of Thessaloniki, Thessaloniki, 185 p.

Lindstrom A and Rune G 1999 Root deformation in planta tions of container-grown Scots pine trees: effect on root growth, tree stability and stem straightness. Plant Soil 217, 29–37.

Mickovski S B and Ennos A R 2002 A morphological and mechanical study of the root systems of suppressed crown Scots pine *Pinus silvestris*. Trees Struct. Funct. 16, 274–280.

Mickovski S B and Ennos A R 2003 Anchorage and asymmetry in the root system of *Pinus peuce*. Silva. Fenn. 37, 161–173.

Moroni M T, Worledge D and Beadle C L 2003 Root distribution of *Eucalyptus nitens* and *E. globulus* in irrigated and droughted soil. For. Ecol. Manage. 177, 399–407.

Moulopoulos C 1962 Special Applied Silviculture. Thessaloniki (in Greek), Thessaloniki, 362 p.

Nicoll B C and Ray D 1996 Adaptive growth of tree root systems in response to wind action and soil conditions. Tree Physiol. 16, 891–898.

Nicoll B C, Easton E P, Milner A D, Walker C and Coutts M P 1995 Wind stability factors in tree selection: distribution of biomass within root systems of Sitca spruce clones. *In* MP Coutts and J Grace (eds), Wind and Trees, pp. 276–292. Cambridge Univ. Press, Cambridge, UK.

Norusis M J 2002 SPSS 11.0 Guide to Data Analysis, 637 p. Prentice Hall, Englewood Cliff, NJ.

Radoglou K 1987 Site influence on the success of reforestation and on the ecophysiological condition of trees of the Kedrinos Hill Thessaloniki. Ph.D thesis, Department of Forestry and Natural Environment, Aristotle University of Thessaloniki, Thessaloniki, 162 p.

Stokes A, Berthier S, Sacriste S and Martin F 1998 Variations in maturation strains and root shape in root systems of Maritime pine (*Pinus pinaster* Ait.). Trees Struct. Funct. 12, 334–339.

Sundström E and Keane M 1999 Root architecture, early development and basal sweep in containerized and bare-rooted Douglas fir (*Pseudotsuga menziesii*). Plant Soil 217, 65–78.

Tsakaldimi M 2001 Research on the production and quality assessment of the container-planting stock used in the afforestations. Ph.D thesis, Department of Forestry and Natural Environment, Aristotle University of Thessaloniki, Thessaloniki, 198 pp.

Tsitsoni T 2001 Restoration of high-degraded forest ecosystems in *Sithonia peninsula*, North Greece. *In* TD Lekkas (ed). Proceedings of the 7th International Conference on Environmental Science and Technology. Vol B. Hermoupolis, Syros, pp. 886–893, 3–6.

Tsitsoni T, Ganatsas P, Zagas T and Tsakaldimi M 2004 Dynamics of postfire regeneration of *Pinus brutia* Ten. in an artificial forest ecosystem of northern Greece. Plant Ecol. 171, 165–174.

Root morphology, stem growth and field performance of seedlings of two Mediterranean evergreen oak species raised in different container types

M. Tsakaldimi[1], T. Zagas, T. Tsitsoni & P. Ganatsas

Department of Forestry and Natural Environment, Laboratory of Silviculture, Aristotle University of Thessaloniki, University Campus, 54 124, P.O.Box 262, Thessaloniki, Greece. [1]*Corresponding author**

Key words: container seedlings, outplanting performance, *Quercus coccifera*, *Quercus ilex*, root morphology, stem growth

Abstract

Outplanting container-grown oak seedlings with undesirable shoot and root characteristics result in poor establishment and reduced field growth. The objective of this study was to determine the influence of container type on both above-and below-ground nursery growth and field performance of one-year old tap-rooted seedlings *Quercus ilex* L. and *Quercus coccifera* L. The experiment was conducted in an open-air nursery and the seedlings were grown in three container types. At the end of the nursery, growth period seedlings' shoot height, diameter (5 mm above root collar), shoot and root biomass, root surface area, root volume and total root length were assessed. Then the seedlings were planted in the field and their survival and growth were recorded for two growing seasons after outplanting. The results showed a difference between the *Quercus* species in the effect of container type. *Q. ilex* seedlings raised in *paper-pot* had significantly greater height, diameter, shoot and root biomass and root volume than those raised in the other two container types. Similarly, *Q. coccifera* seedlings raised in *paper-pot*, had significantly greater above-and below-ground growth than those raised in the other two container types. Both oak species showed relatively low survival in the field; the mortality was mainly observed the first year after outplanting, especially after the summer dry period. However, 2 years after outplanting, the *paper-pot* seedlings of the two oak species showed better field performance.

Introduction

In ecological studies, the evergreen sclerophylls are regarded as one of most typical components of the Mediterranean type vegetation (Saleo and LoGullo, 1990). Many restoration projects have established plantations of these evergreen resprouting species (Vallejo et al., 2000; Vilagrosa et al., 2003). Despite the great efforts in oak regeneration research, the successful planting of oaks is still fraught with uncertainty (Pope, 1993). Early attempts to introduce broad-leaved resprouting species to the Mediterranean basin (e.g., *Quercus* species) faced high seedling mortality, and until recently, nursery and field techniques were poorly developed for these two species (Pausas et al., 2004).

In eastern Spain as well as in Greece, the field survival and growth of planted Mediterranean oaks are frequently very low (Hatzistathis et al., 1999; Pausas et al., 2004; Tsakaldimi, 2001; Vilagrosa et al., 2003; Villar-Salvador et al., 2004).

The poor development of *Quercus* seedlings plantations, in some cases, could be attributed to the low quality of the planted seedlings. Nursery cultivation regimes can strongly determine the functional characteristics of seedlings and their field performance (Landis et al., 1990; Simpson, 1995; Villar-Salvador et al., 2004). For instance, during the container seedling production, container size, growing density and design characteristics of the containers are important determinants of seedling quality (Landis et al., 1990). The volume of the cavity is one of the most obvious and important characteristics of a container because in general, the larger the container the larger the seedling

* E-mail: marian@for.auth.gr

This article has been previously published in the following journal—Plant and Soil, Volume 278, 85–93.
Stokes et al. (eds), Eco- and Ground Bio-Engineering: The Use of Vegetation to Improve Slope Stability, 135–142.

that can be produced. However, the optimum container size varies according to many different factors, including species, growing density, environmental conditions and length of the growing season. Pine species that are tolerant of crowding, such as loblolly pine, could be produced in small-volume containers with a high growing density. In contrast, broad-leaved species should be produced at lower growing density because their leaves intercept more water and nutrients and generate more shade (Landis et al., 1990). One of the most serious problems in containers, especially in the case of seedlings with tap roots such as oaks, is the tendency of seedling roots to spiral around the inside of the container or to concentrate at the base of the container (Biran and Eliassaf, 1980; Landis et al., 1990). Root spiraling is most serious in round, smooth-walled plastic containers and can seriously reduce seedling quality after outplanting. In contrast, well-developed and well-structured root systems with numerous first order laterals are one of the most essential attributes of high quality oak seedlings (Day and Parker, 1997; Thompson and Schultz, 1995).

However, the influence of container type on seedling quality and the outplanting performance of Mediterranean oak species has received almost no attention and to the best of our knowledge, no study on root morphology of seedlings of these species has been reported.

Thus, the objective of this study was to determine the influence of container type on both above- and below-ground nursery growth and field performance of two tap-rooted seedlings, *Quercus ilex* and *Quercus coccifera*.

Materials and methods

Nursery phase

Experimental treatments

The experiment was conducted in an open-air nursery of Forest Service (N. Chalkidona, North Greece). Acorns of *Quercus ilex* L. and *Quercus coccifera* L. were sown in mid-March. Three container types were selected to provide a wide range in container volumes, density of plants and design characteristics as these have been shown to have a strong influence on the morphology and field performance of seedlings (Jones et al., 2002; Landis et al., 1990; Salonius and Beaton, 1994). The container types used for the tap-rooted seedlings production were: (a) *paper-pot* FS 615; made of biodegradable paper, planted with the seedling, each cavity is hexagonally shaped, bottomless, 482×10^3 mm^3 in volume and 150 mm in depth, and (b) and (c) two rigid reusable plastic containers from which the seedlings are removed before planting: (b) quick pot T18; each cavity is of square shape, tapered from top to bottom, has interior vertical antispiralling ribs and open crossed base and is 650×10^3 mm^3 in volume and 180 mm in depth, and (c) *plantek* 35F; each cavity has similar design features to *quick-pot* but air root pruning is achieved from the sides of the walls and from the base, and is 275×10^3 mm^3 in volume and 130 mm in depth. All cavities were filled with sphagnum Lithuanian peat of medium structure and coarse perlite (3:1, v/v). This potting medium is commonly used in Greek forest nurseries. The potting medium was fertilized with 1.3 kg mixed fertilizer (N:P:K 15:30:15 + micronutrients), 0.6 kg potassium sulfate, 1.0 kg superphosphate (0-20-0), 0.4 kg magnesium sulfate and 2 kg lime (CaO) per m^3 of peat.

The three treatments were arranged in a randomized complete block design with three replications for each of 2 species $\times$ 3 container types. There were 24 seedlings per container type, in each block (total 216 seedlings per species) and all seedlings were identified with a number. All seedlings were irrigated with an overhead irrigation system, as needed.

Growth measurements and destructive sampling

At the end of the growth period in the nursery, on November, the shoot height, the diameter (measured 5 mm above the root collar) of all seedlings were measured with an accuracy of 1 and 0.1 mm, respectively. Twelve randomly selected seedlings per treatment (4 seedlings $\times$ 3 replications) of each species, were collected for destructive sampling and they were transferred to the Laboratory for biomass measurements. From these selected seedlings, five random root samples per treatment were used for the root morphology estimations prior to biomass measurements. The root system was separated from the soil, under a gentle water jet, using a sieve to collect any root fragments detached from the system. Then, each root system was put into a glass box and covered with a white plastic sheet to keep it in a fixed position and improve the contrast of the root image. The box was placed on a scanner (Hewlett Packard, ScanJet 6100C) connected to a computer, and an image analysis system (DT-Scan,

Delta T-Devices) was used to determine the total root length, the root surface area and the total root volume (Barnett and McGilvray, 2001; Fitter et al., 1991). For biomass measurements the seedlings were divided into two parts: shoot (stem + needles) and root system. Both parts were oven-dried at 70°C for 48 h and then they were weighed (Thompson, 1985).

Field experiment

In early December, 8-month-old *Q. ilex* and *Q. coccifera* seedlings were outplanted to the field in 'Kassandra' Peninsula, Chalkidiki (North Greece), which is located 80 km south-east of Thessaloniki at 25°30′ E and 40° N. According to the climatic data (period 1978–1997) from the meteorological station of the Forest Service, the climate of the area is of the Mediterranean type with mild winters and dry hot summers. The mean annual rainfall reaches 581 mm, while the mean annual air temperature goes up to 16.3°C and the mean maximum air temperature of the warmest month (July) is 30.1°C. The dry period begins in the middle of April and lasts until the middle of September (Tsakaldimi, 2001; Tsitsoni, 1997). The vegetation of the area belongs to *Quercetalia illicis* floristic zone.

For each species, 20 seedlings per treatment per replication were planted in a randomized complete block design with three replications; the identity of nursery blocks was maintained in the field. Experimental blocks (500 m^2 each) were located on three independent sites of W, NW and N aspects and of moderate slopes (15–30%) and they were not irrigated. The distance between the sites was approximately 300 m. The soil of the three sites, where the experiment was conducted, is characterized as deep, sandy–clay loam, neutral to moderate alkaline and rich in organic matter at the surface horizons (Tsakaldimi, 2001).

The seedlings being hand planted in pits (0.30 × 0.30 m^2) and they were spaced 2 m apart. The survival was recorded for each seedling for two successive years after planting. Furthermore, 2 years after planting, height and diameter growth of each seedling were assessed (with an accuracy of 1 and 0.1 mm, respectively). The relative growth rates (RGR) for both height and diameter, after a period of 2 years, were calculated as the difference between the natural logarithms of final and initial height or diameter, respectively, divided by time between the beginning and the end of field experiments (in years) (Elvira et al., 2004; Villar-Salvador et al., 2004).

Statistical analysis

All statistics were calculated with SPSS software. Distribution was tested for normality by Kolmogorov–Smirnov criterion and the homogeneity of variances was tested by Levene's test. The percentages were transformed to arsine square root values, before analysis. Significant differences between treatment means were tested using analysis of variance (one-way ANOVA). Wherever treatment effects were significant, the Duncan's Multiple Range Test was carried out to compare the means (Norusis, 1994; Snedecor and Cochran, 1988).

Results

Nursery performance

Both species were affected by the type of container. *Q. ilex* seedlings grown in *paper-pot* were significantly taller, had greater diameter and shoot biomass than seedlings grown in *quick-pot* and *plantek* (Table 1). Also, the root biomass, the shoot/root mass ratio and the total root volume found to be significantly greater in seedlings grown in *paper-pot* than in seedlings grown in *quick-pot* and *plantek*. The total root surface area and root length did not show significant differences among the *paper-pot* and *quick-pot* seedlings, but were significantly greater than those of *plantek* seedlings.

Similar to *Q. ilex* seedlings, *Q. coccifera paper-pot* seedlings exhibited the greatest height, diameter and shoot biomass (Table 2). On the contrary, the seedlings grown in *quick-pot* did not differ from those grown in *plantek* but both were found significantly smaller than *paper-pot* seedlings. The container type significantly affected the root morphology. *Paper-pot* seedlings had a more extended root system; their root surface area and the total root length were significantly greater than that of seedlings raised in plastic containers, and were twice or more greater than those of *plantek* seedlings. The root volume and the root biomass allocation did not differ between *paper-pot* and *quick-pot* seedlings but remained greater than that of *plantek* seedlings. The shoot/root mass ratio was significantly greater in seedlings grown in *paper-pot*.

Table 1. Effects of container type on *Q. ilex* seedling characteristics at the nursery phase

	Container type		
	Paper-pot (FS 615)	Quick-pot (T18)	Plantek (35 F)
Above-ground seedling characteristics			
Shoot height (mm)	401 (12.1)[a]	208 (9.1)[b]	240 (8.1)[b]
Root-collar diameter (mm)	5.1 (0.12)[a]	4.3 (0.09)[b]	4.2 (0.09)[c]
Shoot dry weight (g)	8.3 (0.80)[a]	4.2 (0.44)[b]	3.8 (0.24)[b]
Below-ground seedling characteristics			
Root dry weight (g)	4.6 (0.43)[a]	3.5 (0.30)[b]	2.9 (0.23)[b]
Root surface area (mm^2)	13 168 (1110)[a]	11 806 (1536)[a]	8057 (896)[b]
Root volume (mm^3)	6630 (890)[a]	4220 (570)[b]	4240 (610)[b]
Total root length (mm)	7376 (701)[a]	8144 (863)[a]	4440 (502)[b]
Shoot dry weight/Root dry weight	2.0 (0.1)[a]	1.2 (0.1)[b]	1.3 (0.1)[b]

Values are means ± standard error (in parenthesis). Within a row, means followed by different letters, are significantly different ($P < 0.05$).

Field survival

One year after outplanting (on November), the survival rate presented significant differences among the treated seedlings of *Q. ilex*, and it was negatively affected by the summer drought period (Figure 1a). Seedlings grown in *paper-pot* presented significantly greater survival rate (73.3%) than those grown in *quick-pot* (50.9%) and *plantek* (42.9%). During the second year after outplanting, the summer drought period caused a further reduction of survival rate; 8.3% for *paper-pot* seedlings, 5.3% for *quick-pot* seedlings and 12.5% for *plantek* seedlings.

Q. coccifera seedlings had also difficulties surviving in the field (Figure 1b). The first year after outplanting, the survival rate significantly reduced. The survival of *paper-pot* seedlings was 73.6%, while the survival recorded in the *quick-pot* and *plantek* seedlings reduced at half, and it was 47.9 and 47.7%, respectively. At the end of the second growth period in the field, the survival rate reduced to 71.7% for *paper-pot* seedlings, 41.7% for *quick-pot* seedlings and 45.5% for *plantek* seedlings.

After recording survival rates we excavated five dead seedlings of each treatment and species and found that their roots were restricted to the space of the nursery

Table 2. Effects of container type on *Q. coccifera* seedling characteristics at the nursery phase

	Container type		
	Paper-pot (FS 615)	Quick-pot (T18)	Plantek (35 F)
Above-ground seedling characteristics			
Shoot height (mm)	283 (15.4)[a]	136 (11.5)[b]	139 (9.2)[b]
Root-collar diameter (mm)	4.2 (0.12)[a]	3.1 (0.11)[b]	3.1 (0.09)[b]
Shoot dry weight (g)	4.5 (0.69)[a]	2.1 (0.25)[b]	1.6 (0.11)[b]
Below-ground seedling characteristics			
Root dry weight (g)	3.6 (0.49)[a]	2.8 (0.29)[a]	1.7 (0.14)[b]
Root surface area (mm^2)	12 306 (1996)[a]	7950 (888)[b]	5839 (614)[b]
Root volume (mm^3)	6410 (1290)[a]	4260 (460)[ab]	3300 (480)[b]
Total root length (mm)	6233 (963)[a]	4148 (462)[b]	2973 (353)[b]
Shoot dry weight/Root dry weight	1.3 (0.08)[a]	0.7 (0.05)[c]	0.9 (0.07)[b]

Values are means ± standard error (in parenthesis). Within a row, means followed by different letters, are significantly different ($P < 0.05$).

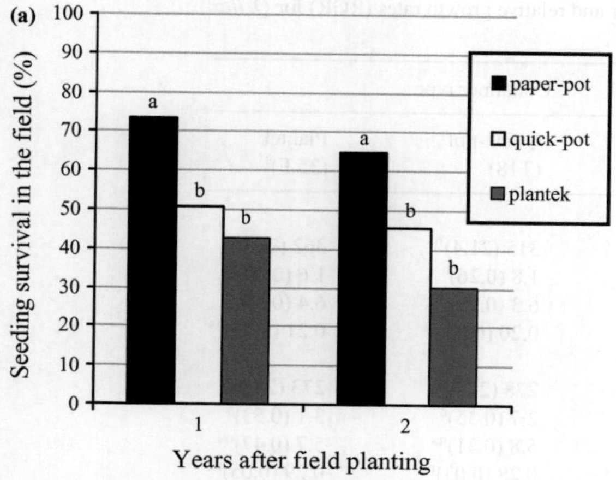

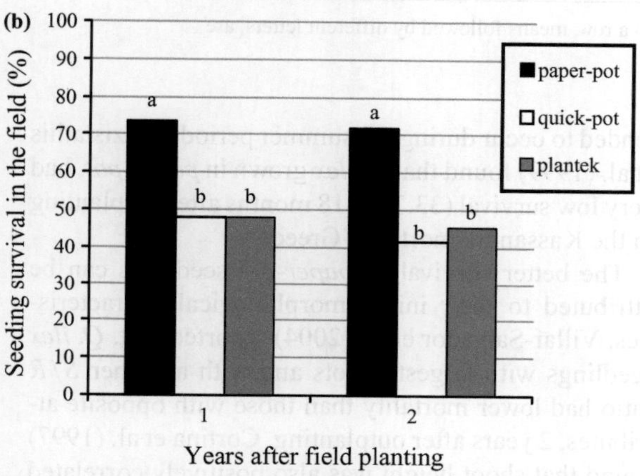

Figure 1. Effect of container type on (a) *Q. ilex* and (b) *Q. coccifera* seedling survival in the field; the first and the second year after outplanting. For the same year, means followed by different letter are significantly different ($P < 0.05$). Error bars are not shown because they are too small.

root plug and none of them had developed new roots out of it.

Growth in the field

At the end of the second year in the field, 23 months after outplanting, the *Q. ilex* seedlings shoot height and diameter presented significant differences among the treatments and followed the same trend as in the nursery (Table 3). The larger and thicker seedlings were those that had grown in *paper-pot* and the smaller seedlings were those that had grown in *quick-pot* and *plantek*. However, *quick-pot* and *plantek* seedlings showed significantly greater height RGR than *paper-pot* seedlings, while the diameter RGR did not differ among treated seedlings.

Similarly, the larger *Q. coccifera* seedlings had grown in *paper-pot* while the smaller ones had grown in *quick-pot* and *plantek*. The *quick-pot* and *plantek* seedlings again showed significantly greater height RGR than the *paper-pot* seedlings while their diameter and diameter RGR did not show significant differences among treated seedlings (Table 3).

Discussion

The seedlings of both oaks produced in the three container types, were healthy, none of them showed root spiraling and all of them approximately reached the appropriate dimensions for planting. According to EU legislation (Council Directive 71/161/EEC, 1971) *Quercus* seedlings, 1 or 2 years old, are considered suitable for planting when their height is 150–250 mm and root-collar diameter is 4 mm. Concerning the Mediterranean oaks, Nardini et al. (2000) found that 2-year-old *Q. ilex* seedlings, raised in containers, were much smaller than those of our study; their stem diameter was only 2.7 mm, height 420 mm, total root dry weight 0.5 g and root surface area was 3680 mm^2. Also, Villar-Salvador et al. (2004), found that 10-month-old *Q. ilex* seedlings, grown in forest-pot 300 containers and fertilized with slow-release fertilizer N:P:K (15:7:15), 1 kg m^{-3} peat, were only 141 mm in height and they allocated shoot dry weight 1.72 g and root dry weight 3.39 g. In our study, seedling dimensions of the two oak species were much differentiated among the container types used. In both oak species, seedlings raised in *paper-pot* were superior to seedlings raised in the other two plastic containers. Contrary to what has been reported for larger containers with lower growing densities (Aphalo and Rikala, 2003; Landis et al., 1990; Tanaka and Timmis, 1974), the *paper-pot* although it is smaller in size and create higher seedling densities than *quick-pot*, it increased shoot height, diameter, biomass allocation and enhanced the root morphology of the oak seedlings. A possible explanation for this is the construction material of the *paper-pot*; the paper is permeable and allow water and soluble salts to move laterally between the cavities of the container. This positively affected the water and nutrient availability for each seedling and thus enhanced the seedlings' growth (Tsakaldimi, 2001). Moreover, although there are no measurements, the *quick-pot* and *plantek*, which are

Table 3. Container type effects on height, root collar diameter and relative growth rates (RGR) for *Q. ilex* and *Q. coccifera* seedlings, 2 years after field planting

Field growth	Container type		
	Paper-pot (FS 615)	Quick-pot (T18)	Plantek (35 F)
Q. ilex			
Height (mm)	473 (22.5)^{a}	315 (21.4)^{b}	362 (26.6)^{b}
Height RGR (mm mm^{-1} year^{-1})	0.9 (0.12)^{b}	1.8 (0.26)^{a}	1.6 (0.39)^{a}
Root-collar diameter (mm)	8.4 (0.37)^{a}	6.3 (0.28)^{b}	6.4 (0.37)^{b}
Diameter RGR (mm mm^{-1} year^{-1})	0.25 (0.02)^{ns}	0.20 (0.03)^{ns}	0.21 (0.03)^{ns}
Q. coccifera			
Height (mm)	367 (23.6)^{a}	278 (28.7)^{b}	273 (24.2)^{b}
Height RGR (mm mm^{-1} year^{-1})	1.6 (0.23)^{b}	2.7 (0.35)^{a}	3.1 (0.53)^{a}
Root-collar diameter (mm)	6.6 (0.32)^{ns}	5.8 (0.31)^{ns}	5.7 (0.47)^{ns}
Diameter RGR (mm mm^{-1} year^{-1})	0.24 (0.03)^{ns}	0.28 (0.03)^{ns}	0.29 (0.05)^{ns}

Values are means ± standard error (in parenthesis). Within a row, means followed by different letters, are significantly different ($P < 0.05$).

plastic and black-colored containers, may absorb more solar radiation which can increase the root temperature. High soil temperatures were especially reported for black plastic containers (Whitcomb, 1989). The high root temperatures can inhibit root growth and may even result in seedling mortality (Landis et al., 1990).

In the field, both oak seedlings had difficulties to survive, but the mortality was much higher in *Q. ilex* seedlings. Similarly, Villar-Salvador et al. (2004) report that *Q. ilex* seedlings have lower survival and growth when compared with other Mediterranean woody species. This indicates that this species is more susceptible to stress factors during its early life stages and especially during the first summer period. In this study also, the mortality was mainly observed at the end of the first year after outplanting and after the summer dry period, and varied considerably among the container seedlings. The survival rate of *paper-pot* seedlings was much greater (73.3% for *Q. ilex* and 73.6% for *Q. coccifera*) than that of the other container seedlings, while the survival rate of the *plantek* seedlings was only 42.9% for *Q. ilex* and 47.7% for *Q. coccifera*. At the end of the second year after outplanting, there was a further reduction of seedling survival. However, the survival of seedlings grown in *paper-pot* remained higher (65 and 71.7% for the *Q. ilex* and *Q. coccifera*, respectively) while the survival recorded for *plantek* seedlings was 45.5% for *Q. coccifera* and only 30.4% for *Q. ilex*. Villar-Salvador et al. (2004) found that 2 years after outplanting, the mortality of *Q. ilex* seedlings reached to 42% and tended to occur during the summer period. Hatzistathis et al. (1999) found that *Q. ilex* grown in *paper-pot*, had very low survival (33.7%), 18 months after outplanting in the Kassandra, northern Greece.

The better survival of *paper-pot* seedlings can be attributed to their initial morphological characteristics. Villar-Salvador et al. (2004) reported that, *Q. ilex* seedlings with largest shoots and with a higher S/R ratio had lower mortality than those with opposite attributes, 2 years after outplanting. Cortina et al. (1997) found that shoot height was also positively correlated with field survival of *Q. ilex* seedlings. Also, in a previous study, Tsakaldimi (2001) found that diameter was a good predictor for field survival of *Q. coccifera* seedlings; the thicker the seedlings the higher the survival. Similarly, in our study, the *paper-pot* seedlings of both oak species, that exhibited the lower mortality, had much greater shoot height, root-collar diameter, shoot dry weight and S/R ratio at the time of planting. The poor performance of smaller seedlings may be due to an unbalanced carbon economy during their establishment phase and the summer period (Villar-Salvador et al., 2004). Root characteristics may also have contributed to the better survival of *paper-pot* seedlings. The greater root volume and root surface area (as well as the greater total root length only in the case of *Q. coccifera*) of *paper-pot* seedlings, may have resulted in a better water and nutrient uptake during their early stages after outplanting and especially during the summer drought. When growth or survival is limited by water (as is observed in the Mediterranean basin) or nutrient availability, immediately after outplanting, roots play a more important role in the performance of container seedlings (Aphalo and Rikala, 2003). Furthermore, it may be important that *paper-pot* seedlings were planted with pots, thus, they had their roots protected not only during the planting work but the whole first year after outplanting until the roots increased and penetrated the soil. In contrast, *quick-pot* and *plantek* seedlings, which were planted without the cavity, had their roots unprotected and moreover their roots had difficulty crossing a textural discontinuity from a light, friable growing medium to natural soil (Tinus, 1986). According to Ruehle and Kormanik (1986), oak seedlings must develop new roots soon after planting if they are to survive and grow. This seems to be confirmed in our study since all the excavated dead seedlings had developed no roots out of the nursery plug.

The differences in seedlings size in the nursery phase, of both oak species, persisted 2 years after the outplanting in the field. *Paper-pot* seedlings remained significantly taller than the other container seedlings, although the height relative growth rate (RGR) was greater in *quick-pot* and *plantek* seedlings. *Q. ilex* seedlings raised in *paper-pot* also had the greatest field diameter, while their diameter RGR did not differ from that of the other container seedlings. In the case of *Q. coccifera*, although *quick-pot* and *plantek* seedlings had smaller diameter at planting they grew as well as the larger *paper-pot* seedlings. However, Villar-Salvador et al. (2004), reported that *Q. ilex* seedlings with larger shoots and with a higher S/R ratio had larger stem volume increase, 2 years after outplanting. In contrast, studies among many other species concluded that, differences in seedlings size at planting disappeared after one or two growth periods in the field (Jones et al., 2002; Simpson, 1995). Actually, growth following outplanting is more complex than mere survival and is related to the planting environment, the genetic potential and the physiological and morphological status of the seedlings, at the time of outplanting (Mexal and Landis, 1990).

Conclusions

The results of this study suggest that the container type has a strong influence on seedling quality and outplanting performance of *Q. ilex* and *Q. coccifera* seedlings. The *paper-pot* contributes to the production of taller, thicker and heavier seedlings with a more extended root system. The better quality of these seedlings in combination with the fact that the *paper-pot* seedlings have their roots protected from transplanting shock, results in better field performance. Also, it is suggested that larger oak seedlings have better survival and they remain greater 2 years after outplanting. Grading criteria for oak seedlings' shoot height and root-collar diameter will be important for sites where environmental stress may be high.

Acknowledgement

We would like to thank the anonymous referees for their comments and suggestions on an earlier version of the manuscript.

References

Aphalo P and Rikala R 2003 Field performance of silver-birch planting stock grown at different spacing and in containers of different volume. New Forest 25, 93–108.

Barnett J P and McGilvray J M 2001 Copper treatment of containers influences root development of Longleaf Pine seedlings. *In* Moorhead D J Tifton (ed), Proceedings of the Longleaf Pine Container Production Workshop. Jan. 16–18, 2001. GA. USDA Forest Service and University of Georgia.

Biran I and Eliassaf A 1980 The effect of container shape on the development of roots and canopy of woody plants. Sci. Hortic. 12, 183–193.

Cortina J, Valdecantos A, Seva J P, Vilagrosa A, Bellot J and Vallejo R 1997 Relación tamaño-supervivencia en plantones de especies arbustivas y arbóreas mediterráneas producidas en vivero. *In* Proceedings of the II Congreso Forestal Español, vol. 3, pp. 159–164. Sociedad Española de Ciencias Forestales and Gobierno de Navarra, Pamplona.

Day D C and Parker W C 1997 Morphological indicators of stock quality and field performance of red oak (*Quercus rubra* L.) seedlings underplanted in a central Ontario shelterwood. New Forest. 14, 145–156.

EEC 1971 Council Directive No 71/161/EEC on external quality standards for forest reproductive material marketed within the community. Official J. Eur. Commun., L 87, 17.4.71, 14 pp.

Elvira S, Bermejo V, Manrique E and Gimeno B S 2004 On the response of two populations of *Quercus coccifera* to ozone and its relationship with ozone uptake. Atmos. Environ. 38, 2305–2311.

Fitter A H, Strickland T R, Harvey M L and Wildson G W 1991 Architectural analysis of plant root systems. 1. Architectural correlates of exploitation efficiency. New Phytol. 118, 375–382.

Hatzistathis A, Zagas T, Ganatsas P and Tsitsoni T 1999 Experimental work on restoration techniques after wildfires in forest ecosystems in Chalkidiki, North Greece. *In* Proceedings of the International Symposium ≪Forest Fires: Needs, Innovations≫ Athens, November 18–19, 1999, pp 310–315.

Jones M D, Kiiskila S and Flanagan A 2002 Field performance of pine stock types: Two-year results of a trial on interior lodgepole pine seedlings grown in Styroblocks, Copper-blocks or Airblocks B.C. J. Ecos. Manage. 2, 1–12.

Landis T D, Tinus R W, McDonald S E and Barnett J P 1990 The Container Tree Nursery Manual, vol. 2. Agriculture Handbook 674. U.S.D.A Forest Service, Washington DC. 88 pp.

Mexal J G and Landis T D 1990 Target seedling concepts: Height and diameter. *In* Proceedings of the Combined Meeting of the Western Forest Nursery Associations, Roseburg, Oregon, August 13–17, 1990, pp. 17–35.

Nardini A, Salleo S, Tyree M T and Vertovec M 2000 Influence of the ectomycorhizas formed by *Tuber melanosporum* Vitt.on hydraulic conductance and water relations of *Quercus ilex* L. seedlings. Ann. For. Sci. 57, 305–312.

Norusis M J 1994 SPSS Professional Statistics. 6.1 SPSS Inc., Chicago, IL.

Pausas J G, Blade C, Valdecantos A, Seva J P, Fuentes D, Alloza J A, Vilagrosa A, Bautista S, Cortina J and Vallejo R 2004 Pines and oaks in the restoration of Mediterranean landscapes of Spain: New perspectives for an old practice-a review. Plant Ecol. 171, 209–220.

Pope P E 1993 A historical perspective of planting and seeding oaks: Progress, problems and status. *In* DL Loftis and CE McGee (eds). Symposium Proceedings: Oak Regeneration, Serious Problems, Practical Recommendations. pp. 224–240. Southeastern Forest Exp. Station, Gen Tech. Rep. SE-84, Asheville, N.C.

Ruehle J L and Kormanik P P 1986 Lateral Root Morphology: A Potential Indicator of Seedling Quality in Northern Red Oak. Southeastern Forest Experiment Station, Research Note SE-344, pp. 1–5. USDA Forest Service, New Orleans, LA.

Saleo S and LoGullo M A 1990 Sclerophylly and plant water relations in three Mediterranean *Quercus* species. Ann. Bot. 65, 259–270.

Salonius P and Beaton K 1994 Effect of root packing on field performance of container seedlings. *In* IUFRO Symposium Abstracts, Making The Grade. Sault Ste. Marie, Ontario, September 1994, p. 29.

Simpson D G 1995 Nursery growing density and container volume affect nursery and field growth of Douglas-fir and Lodgepole pine seedlings. *In* National Proceedings, Forest and Conservation Nursery Associations. Tech Coords T D Landis and R K Dumroese. pp. 105–115. Gen. Tech. Rep. RM-257. USDA Forest Service, Rocky Mountain Forest and Range Experiment Station, Fort Collins, CO.

Snedecor G W and Cochran W G 1988 Statistical Methods. The Iowa State University Press, Ames, IA.

Tanaka Y and Timmis R 1974 Effect of container density on growth and cold hardiness of Douglas-fir seedlings. *In* RW Tinus, WI Stein and W E Balmer, (eds) Proceedings of the North American Containerized Forest Tree Seedling Symposium. Denver, August 26–29, 1974, pp. 181–186.

Thompson B E 1985 Seedling morphological evaluation—What you can tell by looking. *In* ML Duryea (eds), Evaluating Seedling Quality: Principles, Procedures, and Predictive Abilities of Major Tests pp. 59–71. Forest Research Laboratory Oregon State University, Corvallis, OR.

Thompson J R and Schultz R C 1995 Root system morphology of *Quercus rubra* L. planting stock and 3-year field performance in Iowa. New Forest. 9, 225–236.

Tinus R W 1986 Principles of container seedling production. *In* Proceedings of 18th.

IUFRO World Congress, Forest operations and techniques, Division 3, Ljubljana, pp. 292–297.

Tsakaldimi M N 2001 Research on the Production and Quality Assessment of the Container-Planting Stock used in the Afforestations. Ph.D thesis, Department of Forestry and Natural Environment, Aristotle University, Thessaloniki, 198 pp.

Tsitsoni T 1997 Conditions determining natural regeneration after wildfires in *Pinus halepensis* (Miller, 1768) forests of Kassandra Peninsula (North Greece). For. Ecol. Manage. 92, 199–208.

Vallejo V R, Serrasolsas I, Cortina J, Seva J P, Valdecantos A and Vilagrosa A 2000 Restoration strategies and actions in Mediterranean degraded lands. In densertification in Europe: mitigation strategies, land-use and planning. Advanced study course. *In* G Enne, C Zanolla and D Peter (eds), pp. 221–233. Office for Official Publication of the European Communities, Brussels.

Vilagrosa A, Cortina J, Gil-Pelegrin E and Bellot J 2003 Suitability of drought-preconditioning techniques in Mediterranean climate. Restor. Ecol. 11, 208–216.

Villar-Salvador P, Planelles R, Enriquez E and Penuelas-Rubira J 2004 Nursery cultivation regimes, plant functional attributes and field performance relationships in the Mediterranean oak *Quercus ilex* L. Forest Ecol. Manage. 196, 257–266.

Whitcomb C E 1989 Plant Production in Containers, 633 pp. Lacebark Publications, Stillwater.

Stabilising characteristics of New Zealand indigenous riparian colonising plants

M. Marden[1,3], D. Rowan[1] & C. Phillips[2]

[1]*Landcare Research Ltd, P.O. Box 445, Gisborne, New Zealand.* [2]*Landcare Research, Box 69, Lincoln 8152, New Zealand.* [3]*Corresponding author**

Key words: indigenous riparian plants, root growth and morphology, soil and streambank reinforcement

Abstract

This paper presents selected results on the above- and below-ground growth performance of 12 indigenous woody species commonly found growing naturally in unstable riparian slope and/or bank environments throughout New Zealand. This study was needed because little information exists on the effectiveness of New Zealand's indigenous riparian plant species for slope and stream bank stabilisation. By examining the growth performance of selected riparian species during the first 5 years following establishment, we provide valuable insights into the likely strengths and limitations of individual species at maturity and, therefore, into their overall potential usefulness, singularly and/or as mixed plantings, for future riparian stabilisation projects. For all species, their root systems are typically shallow and confined to the uppermost 31 cm of soil. Root spread (mean maximum diameter) increased with increasing age with interspecies differences, by age 5 years, ranging from between ~1 and 2.5 m. At age 5 years the mean root biomass, for all species combined, was 1.2 kg/plant, and averaged ~23% of total plant biomass. Changes in the allocation of biomass for root and shoot growth appear to be species and age dependent. The results of this study indicate that most have above- and below-ground growth attributes well suited to colonising steep and unstable riparian slopes where shallow soil failure is prevalent and/or where stream banks are rocky with skeletal soils. All form part of the early plant succession. Once established, and in the absence of grazing, they are relatively fast growing. The effectiveness of riparian restoration programmes using indigenous species, though potentially high for low-order stream, will be limited by their relatively shallow-rooted habit for bank stabilisation on larger rivers without the prior installation of structural protection works.

Introduction

Since the turn of the 20th century much of New Zealand's indigenous riparian vegetation has been cleared for pastoral use, for the development of an exotic forest industry, and for urban development associated with European settlement. In more recent years, in hill country areas, the regeneration of indigenous species has been suppressed by continual grazing, while remaining stands of riparian vegetation have been further decimated largely through state-assisted land-development encouragement loans or subsidies. The loss of buffering and ecosystem services provided by this riparian vegetation has led to the progressive degradation of waterways through increased sedimentation and nutrient pollution. The consequence has been a loss of in-stream habitat and inferior water quality in many streams and lakes throughout rural New Zealand (Phillips et al., 2001).

Channel widening by bank collapse is now a common occurrence along many kilometers of stream throughout New Zealand. The loss of primary agricultural land and physical property adjacent to eroding stream banks is very costly and the need for their protection against erosion has long been recognised (Acheson, 1968, Eyles, 1983).

* E-mail: mardenm@landcareresearch.co.nz

This article has been previously published in the following journal—Plant and Soil, Volume 278, 95–105.
Stokes et al. (eds), Eco- and Ground Bio-Engineering: The Use of Vegetation to Improve Slope Stability, 143–153.

An increase in awareness of the poor health of New Zealand's water bodies and a genuine willingness to redress this situation have increased the public's desire to become involved in restoring riparian areas by planting indigenous woody rather than exotic species such as willows (*Salix* spp.). Though the role of the latter in improving stream habitat and bank stability and in preventing erosion is well recognised (Van Kraayenoord and Hathaway, 1986), information on the nature, and more importantly on the performance, of New Zealand's indigenous riparian species is generally descriptive, with much of our knowledge anecdotal. In the case of below-ground growth performance and functionality, there are few published studies on root system architecture and biomass of individual tree species (Watson et al., 1995, 1999).

A further need to address this information gap has arisen as a result of the increased risk to many riverbank protection works posed by the introduced willow sawfly (*Nematus oligospilus*), which has caused widespread defoliation and mortality among New Zealand's willow trees (Cowley and Whyte, 1997). Historically, effective structural stream bank protection has been expensive to install and maintain, and as riverbank protection using only willows is no longer practical, other options are needed. This includes combining the proven capability of willows with the untested ability of native species with the view to reducing the longer-term reliance on willows. With public and government pressure to maintain and enhance the indigenous biodiversity of New Zealand, river engineers are seeking to use indigenous plants.

Increasingly, societal considerations have become an integral part of riparian stabilisation projects. These may include incorporating the aspirations of Maori (indigenous people of New Zealand) in plant selection for use in traditional medicine, as fibre for weaving, and for other uses. Other multiple goals may include increasing New Zealand's plant diversification and maximising plant performance for carbon accrediting. New Zealand ratified the Kyoto Protocol (IPCC, 2000) in 1997 (New Zealand Climate Change Office, 2003), and since then the trading of carbon credits by organisations and companies at local, regional and national levels has become an integral part of the country's economy. It is important therefore, that a measure of biomass accumulation, particularly during the early years following the establishment of new plantings of indigenous species, is based on verifiable data.

In this paper we present the results of a trial in which we quantify the, above- and below-ground growth performance of 12, 1- to 5-year-old, indigenous plant species. These species are part of the early plant succession frequently found growing naturally in unstable riparian slope and/or bank environments throughout New Zealand. Despite little documented information on appropriate plant spacing or of an understanding of why certain species appear better adapted to stabilise different riparian environs, e.g., floodplain stream banks, colluvial slopes, landslide scars, many of the trialled species are currently being planted for streamside restoration. By examining their growth performance we aimed to provide valuable insights into the strengths and limitations of individual species and their overall potential usefulness, singularly and/or as mixed plantings, for riparian stabilisation. This trial is the first of its kind to attempt to provide such data in New Zealand.

Methods

It was not considered practical or environmentally acceptable to source plants for destructive sampling, directly from their natural environment. Because of the restrictions planter bags have on root development of containerised plants, it was imperative we secured a source of bare-rooted material at least for plants older than 2 years. We used 2-year-old containerised plants and established a trial site from which plants could be on grown and periodically extracted. To minimise the influence of site variability on growth and to emulate riparian conditions, we chose a streamside location with uniform soils, slope and aspect.

Site details

The trial site was located on a low-lying, even-surfaced alluvial terrace adjacent to the Taraheru River, in Gisborne City, North Island, New Zealand. The soil is free draining, Te Hapara Typic Sandy Brown Soil (Hewitt, 1998) and requires irrigation in summer. The site (50 m by 20 m) was tilled and weed mat was laid down before planting in 1999. The site was subdivided into three blocks. Two-year-old containerised plants were sourced from a local plant nursery and all three blocks were planted in a day. Within each block, plants were arranged so that individuals of the same species were not adjacent. Blocks 1 and 2 (plants to be extracted 1 and 2 years after planting, i.e., at age 3 and

4, respectively) were planted at 1-m spacing, and block 3 (plants to be extracted 3 years after planting, i.e., at age 5) at 1.5–2.0-m spacing. Plants were irrigated for the first 3 months after planting.

Species selection

The species (Allan Herbarium, 2000) chosen (Table 1) are those to appear as part of the natural reversion process on retired pastoral land and on areas of bare ground (mostly landslide scars), or are regarded as suitable for direct planting to restore and stabilise riparian zones.

Biomass and morphology

Plant age was established from the date each species was pricked-out into containers (zero years). For data on 1- and 2-year-old plants, we destructively partitioned the containerised plants. Data for plants 3–5 years old are for bare-rooted plants excavated from the trial site. Using an air lance at 240 kPa, soil surrounding the root systems was removed allowing them to be extracted undamaged and a high percentage of the total root mass to be recovered. We aimed for a sample size of ten plants/species/year. Frost, insect attack and wind-throw accounted for the shortfall in sample size for some of the species, e.g., manuka in year 4. Over the duration of this trial, 554 individual plants were destructively sampled (Table 1).

Above-ground growth parameters measured included height, canopy spread, root collar diameter and diameter at breast height (DBH) (where applicable). Below-ground growth measurements included maximum root depth and lateral root spread. The latter, together with canopy spread, was taken as the average of the maximum diameters measured in two directions. The root system of each plant was photographed before being partitioned into its biomass components. In addition, the above- and below-ground form of a 5-year-old specimen of each species was sketched in detail.

Above-ground biomass was measured by separating the foliage, branches and stem. Below-ground components were partitioned into root bole, tap, lateral and sinker roots. Roots were further partitioned into diameter size classes (<1 mm (fibrous), 1–2, 2–5, 5–10 and 10–20 mm) (Watson and O'Loughlin, 1990), and the total length of roots in each diameter size class (excluding fibrous roots) was measured. Relative to the root bole, a measure of the distribution of root biomass and root length (of roots >1 mm), by diameter size class, was recorded for each 50-cm radius by 50-cm deep concentric disc (similar to growth rings) to the maximum extent of root growth. Each of the above- and below-ground biomass components was oven-dried at 80°C for 24 h then weighed to the nearest 0.1 g.

Root systems and types are described as follows (Phillips and Watson, 1994):

Tap-rooted. The seedling radicle persists and grows into a single or branched massive root (taproot), more or less vertical; it may give rise to planes of lateral roots.

Plate. A shallow spreading root system with abundant surface roots, no taproot.

Heart-rooted. A compact system with many obliquely or vertically descending roots (heart roots)

Table 1. Indigenous riparian species trialled and number of sample trees extracted for each species and year of the trial

		Number of plants extracted/species/year					
Common name	Botanical name	1	2	3	4	5	Species total
Karamu	*Coprosma robusta*	10	10	7	8	10	45
Ribbonwood	*Plagianthus regius*	10	10	10	10	10	50
Kowhai	*Sophora tetraptera*	10	8	8	8	10	44
Lemonwood	*Pittosporum eugenoides*	10	10	10	10	10	50
Kohuhu	*Pittosporum tenuifolium*	10	10	10	9	10	49
Lacebark	*Hoheria populnea*	10	10	10	10	8	48
Mapou	*Myrsine australis*	10	10	10	10	10	50
Fivefinger	*Pseudopanax arboreus*	10	10	10	8	8	46
Cabbage tree	*Cordyline australis*	10	10	10	10	10	50
Rewarewa	*Knightia excelsa*	10	10	10	10	9	49
Manuka	*Leptospermum scoparium*	10	10	5	0	5	30
Tutu	*Coriaria arborea*	10	10	10	8	5	43
Annual totals		120	118	110	101	105	554

rising from or near the root bole and generally replacing the taproot.

Lateral roots. Long and radially spreading from the taproot, or from the root bole, possibly in two or more plates or strata.

Sinker roots. Roots descending more or less vertically from the main laterals at varying distances from the root bole and to depths up to or exceeding that of the taproot.

Statistical analyses

All analyses were carried out in the statistical package GenStat (2002) using the procedure for unbalanced analysis of variance (procedure AUNBALANCED). Analysis on the raw data revealed the residuals (assessed visually) showed significant non-normality (skewed to the right) and an increasing variance with increasing fitted values (heteroscedasticity). All data were $\log_{10}$-transformed to impose normality on the residuals. Throughout this paper, means (back transformed from the $\log_{10}$ scale) are presented ± 1 standard error.

Results

Root system types

Of the trialled species, cabbage tree was the only one to develop a taproot system consisting of a central, frequently branched, tuberous root/rhizome. The remaining species all developed heart-rooted systems. Of note were the lateral roots of ribbonwood, lemonwood and kohuhu, which at an early age developed a few long thin laterals (mean root spread >2 m at age 5). The distribution of their leading lateral roots was highly asymmetric with large areas of soil totally devoid of any roots. Multiple branching was most common at the extremities. Typically, multiple and large-diameter roots developed early and descended obliquely from the base of the root bole. Mapou and kowhai clearly developed sinker roots. These became apparent in year four; they developed between 0.5 and 1 m from the stump and each descended to a depth of ~0.15–0.20 m by age 5 years. Mapou and rewarewa root systems were compact (mean root spread <1 m at age 5 years), highly branched close to the root bole, and had a matted appearance, with the dense cluster of fine roots proving difficult to tease apart.

Root depth

For all species combined there was a significant increase in root depth between years one and five (Figure 1) with interspecies differences being apparent at an early age (Figure 2). Mean root depth for all species combined, at age 5 years, was 0.3 m (Figure 1). The deepest rooted cabbage tree reached a mean depth of 0.4 m, while mapou had the shallowest root system at 0.2 m. At age 5 years the root network was largely confined to the uppermost 31 cm of organic-rich topsoil.

Root spread

For all species combined there was a significant increase in root spread between years one and five (Figure 3). Constriction of the roots in planter bags explains the lack of variation in root spread in years one and two. Lemonwood achieved the greatest mean root spread at 3 m (Figure 4). In contrast, the very compact root systems of mapou and rewarewa barely attained a mean spread of 1 m. For all species trialled, the lateral root spread for plants 3–5 years old generally exceeded crown width. In the trial plot and by year four, the roots of adjacent plants at 2-m spacing were intertwined. The root systems of individual kowhai and tutu extended to a maximum distance twice the diameter of their respective canopy widths.

Root biomass

As for root depth and spread, the relative consistency of root biomass in years one and two for all species combined (Figure 5) was due to the restrictive size of the planting bag. Between years four and five the mean root biomass, for all species combined, more than doubled to 1.2 kg/plant, and averaged ~23% of total plant biomass. At 5 years old, kowhai had the highest percentage of root biomass at 29%, and manuka and cabbage tree the least, at 19%.

All twelve species showed a decrease in fine roots <1 mm, and a corresponding increase in larger diameter roots with increasing age. The fine root fraction of the tap-rooted cabbage tree was consistently lower, at <20% of total root weight, than for the remaining species in each year of the trial. Irrespective of age, the biomass of the root bole (stump) remained relatively constant, at ~25% of total root weight, for all but three species. The small and largely fibrous root systems of manuka and mapou each maintained a comparatively

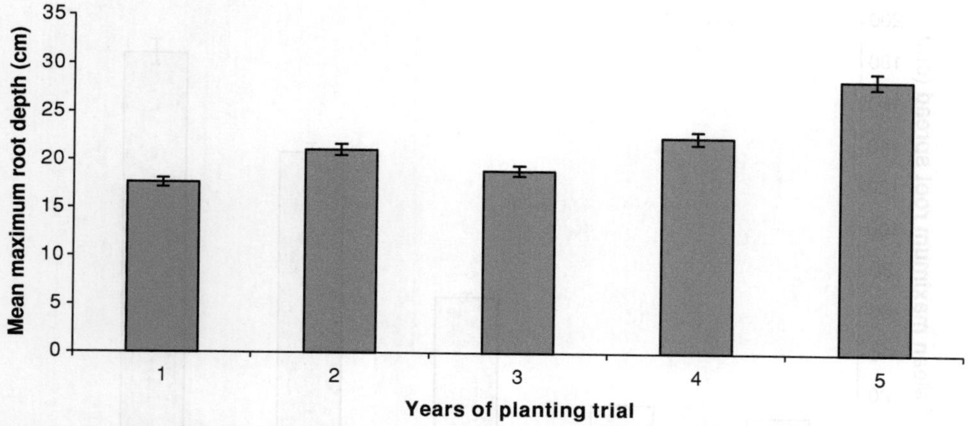

Figure 1. Mean maximum root depth for all 12 riparian species combined, over the 5-year trial period (difference between means is highly significant $F_{4,491} = 43.28$, $P < 0.001$). Error bars are ±1 SE. See Table 1 for the number of plants extracted.

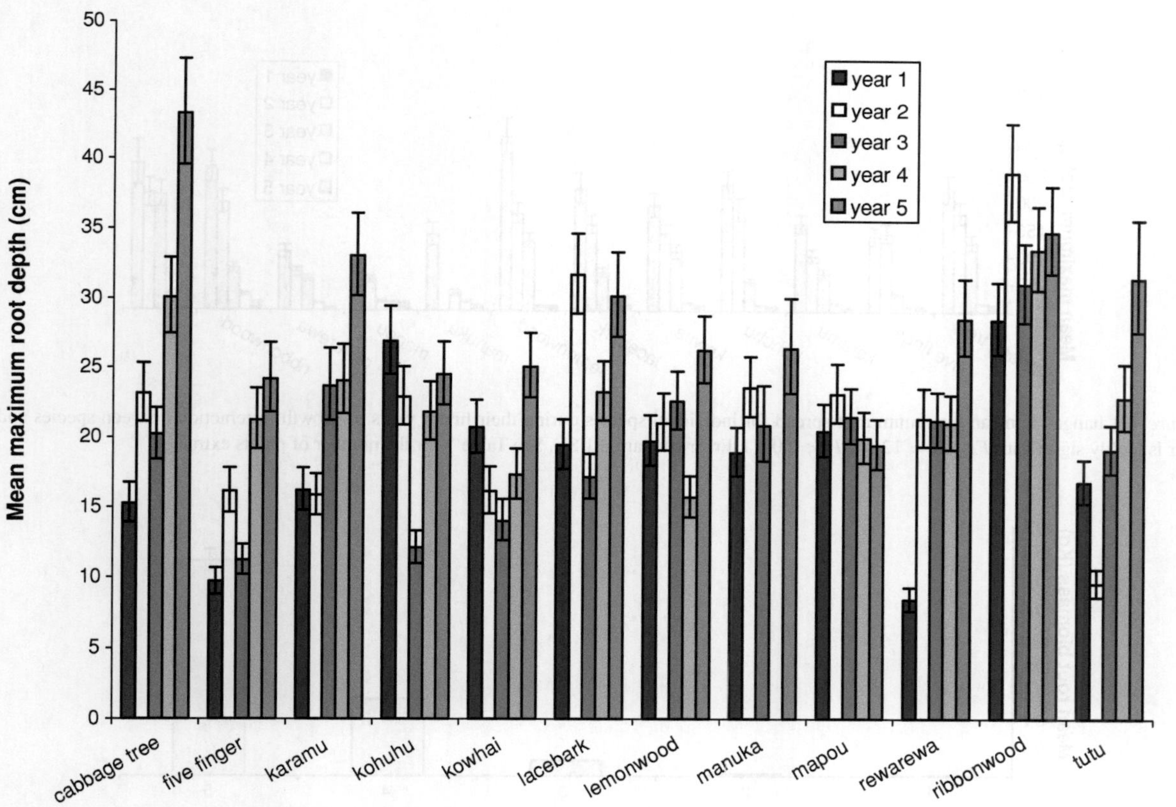

Figure 2. Changes in mean maximum root depth for individual species during their first 5 years of growth (interaction between species and year is highly significant $F_{43,491} = 7.07$, $P < 0.001$). Error bars are ±1 SE. See Table 1 for the number of plants extracted.

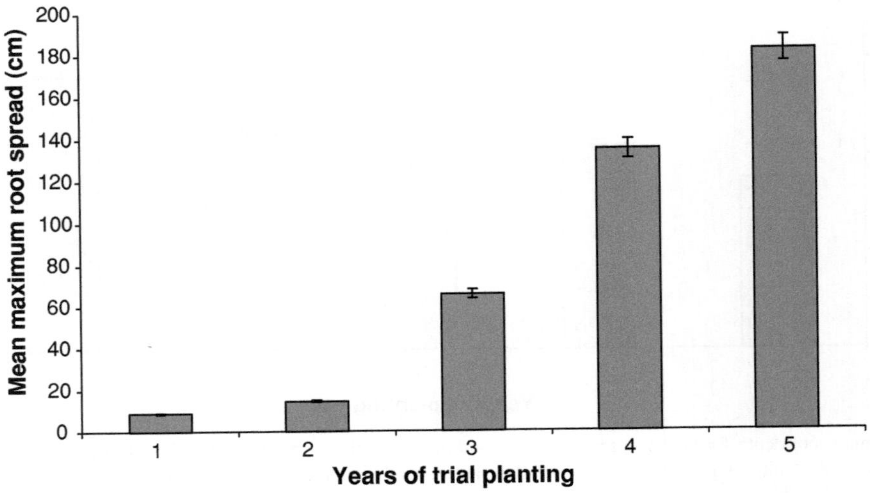

Figure 3. Mean maximum root spread for all 12 riparian species combined, over the 5-year trial period (difference between means is highly significant $F_{4,493} = 1584.8$, $P < 0.001$). Error bars are ±1 SE. See Table 1 for the number of plants extracted.

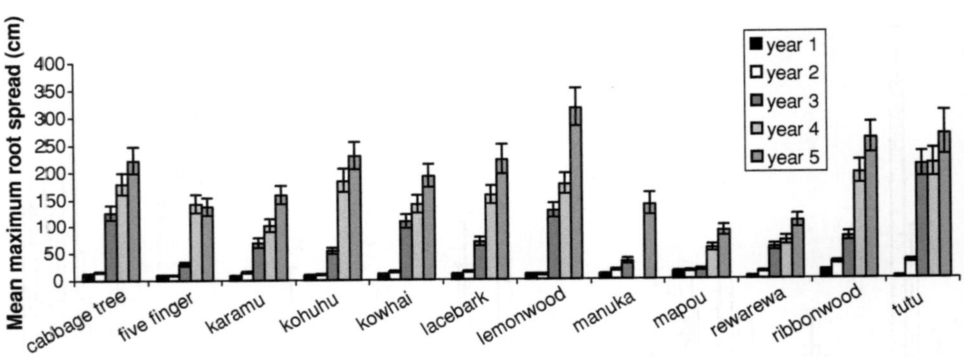

Figure 4. Changes in mean maximum root spread for individual species during their first 5 years of growth (interaction between species and year is highly significant $F_{43,493} = 12.64$, $P < 0.001$). Error bars are ±1 SE. See Table 1 for the number of plants extracted.

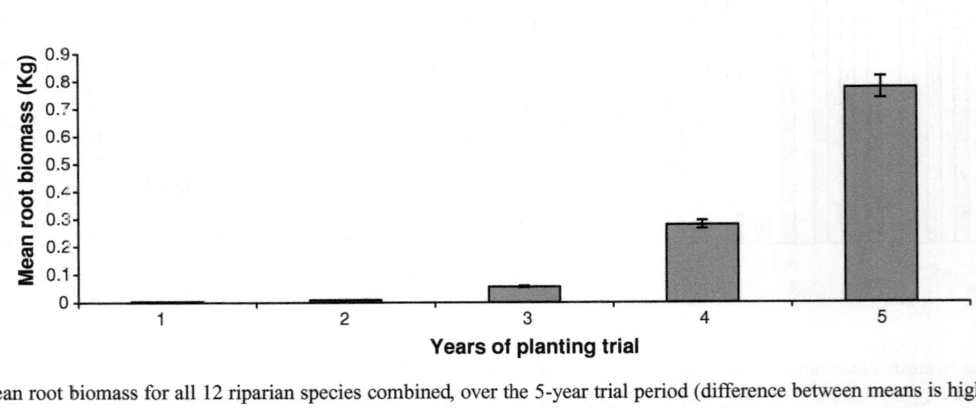

Figure 5. Mean root biomass for all 12 riparian species combined, over the 5-year trial period (difference between means is highly significant $F_{4,494} = 2126.8$, $P < 0.001$). Error bars are ±1 SE. See Table 1 for the number of plants extracted.

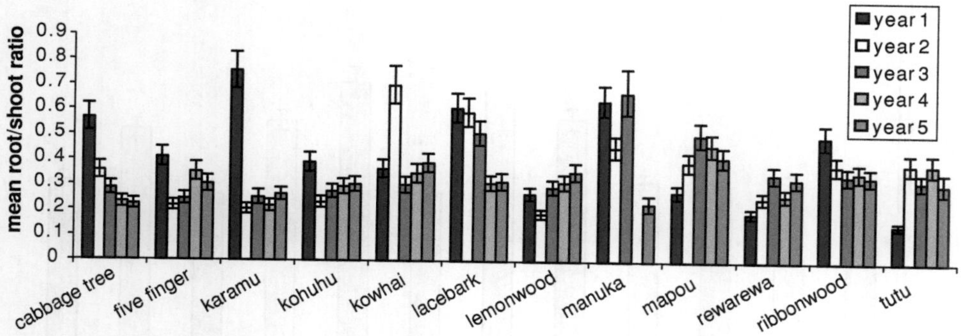

Figure 6. Changes in mean root/shoot ratio for individual species during their first 5 years of growth (interaction between species and year is highly significant $F_{43,494} = 11.22$, $P < 0.001$). Error bars are ±1 SE. See Table 1 for the number of plants extracted.

small root bole for all 5 years of growth, comprising ~10% of total root weight. In contrast, the cabbage tree showed significant growth in its taproot from nothing in year one to ~80% of total root weight in year five.

Root/shoot ratio

Changes in the allocation of biomass for root and shoot growth appear to be species and age dependent. For example, cabbage tree initially invested ~55% of total plant biomass in rhizome growth but by year five this had declined to ~20% (Figure 6). Lacebark showed a similar trend. In year one, the initial high investment in root biomass for some species is apparent and the result of root binding and anchorage strategies. In contrast, fivefinger, kohuhu, rewarewa, ribbonwood and lemonwood showed a steady, but small, annual increment in root biomass relative to shoot biomass. For all species combined, mean root biomass averaged 37% in year one and remained relatively constant between years two and five, at not less than 30% of total plant biomass (Figure 7). For individual species, and over the 5-year duration of the trial, kowhai, lacebark, manuka, mapou and ribbonwood invested significantly more of its total plant biomass as root biomass (~40%) compared with ~30% for the remainder of species trialled (Figure 8).

Discussion

The commonly held belief that New Zealand's indigenous plant species are too slow growing, though true when young, is not the case once plants have become established, with the majority showing exceedingly fast biomass growth within 3 years of planting-out. However, as is common with plants less than 2-year old, factors such as wrenching, root training and planter bag constriction have contributed to the apparent similar rate of growth exhibited during this period (Marden and Phillips, 2002). In years three and four, inter-species differences in overall growth became apparent, with lemonwood, ribbonwood, cabbage tree, karamu,

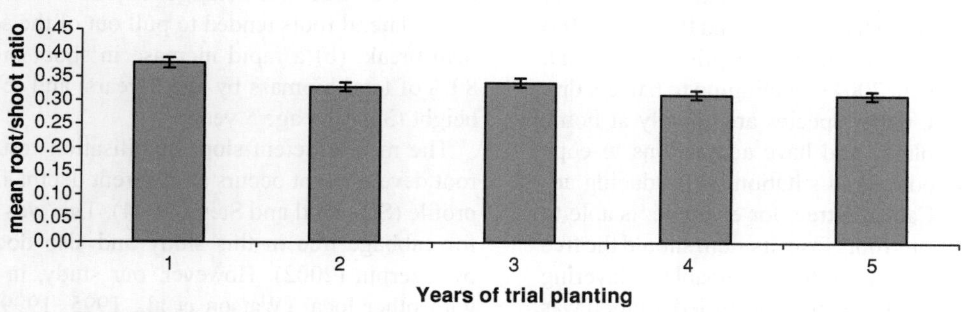

Figure 7. Mean annual root/shoot ratio for all 12 riparian species combined, over the 5-year trial period (difference between means is highly significant $F_{4,494} = 7.27$, $P < 0.001$). Error bars are ±1 SE. See Table 1 for the number of plants extracted.

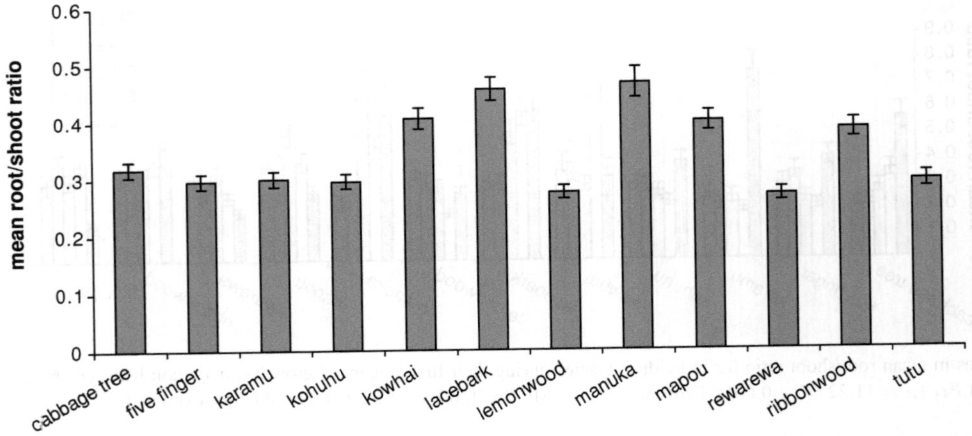

Figure 8. Mean root/shoot ratio for individual species, averaged over the 5-year trial period (difference between means is highly significant $F_{11,494} = 19.57$, $P < 0.001$). Error bars are ±1 SE. See Table 1 for the number of plants extracted.

lacebark and tutu the better performers (Marden and Phillips, 2003, 2004), and this trend continued into the last year of the trial.

Site factors may have contributed to some of the interspecies differences in growth performance. For example, manuka showed significantly slower growth in the fertile soils at the trial site than has previously been measured for similar aged plants growing on less fertile sites more typical of their natural occurrence (Watson et al., 1995). In addition, the openness of the trial site may have affected the early growth performance of the more shade tolerant species including mapou, lemonwood and fivefinger; though once established all proved to be sufficiently hardy to tolerate full sun.

Most are colonising species adapted to harsh growing conditions on steep slopes with skeletal soils deficient in nutrient, and are considered as a nurse crop for emerging shrubby hardwoods. Adaptations include the presence of nitrogen fixing nodules on roots of 4-year-old tutu and kowhai (Marden and Phillips, 2003) and/or roots with a high tensile strength, e.g., kowhai, (Watson and Marden, 2004) for clinging to barren, dry, steep rocky cliffs. Other species are equally at home on alluvial floodplains and have adaptations to cope with periodic flooding and siltation by producing adventitious roots. Cabbage tree, for example, is able to produce adventitious roots from its stem should the tree topple over or break, while tutu is capable of layering should branches break and become buried. Mapou was observed to sprout suckers from lateral surface roots. Foweraker (1929) also highlighted the interesting ability of alluvial communities of some of New Zealand's oldest podocarps to produce a new root system after inundation with river silt.

Potential limitations include susceptibility to frost (e.g., tutu), scale insects (e.g., manuka) and leaf rust (e.g., rewarewa and karamu) (Marden and Phillips, 2003). Some are short lived (e.g., tutu) and together with shade intolerant species will become suppressed and replaced by others. The toxicity of tutu foliage to grazing domestic stock and the potential of contamination of honey produced at the time tutu is in flower, will further limit its application for riparian restoration particularly in urban and unfenced rural settings. Toppling of cabbage trees, whether they are found in isolation or as communities, has previously been documented by Czernin (2002), and was a common occurrence in the latter years of the current study. The prevalence of toppling at our trial site is attributed to a combination of: (a) the structural homogeneity of the alluvial soil where lateral roots tended to pull out of the soil rather than break; (b) a rapid increase in shoot biomass to 80% of total biomass by age 5 years; and (c) tall tree height (3 m), by age 5 years.

The most efficient slope stabilisation results when root development occurs at different depths in the soil profile (Schiechtl and Stern, 1994). This was observed for cabbage tree in this study and also documented by Czernin (2002). However, our study, in common with other local (Watson et al., 1995, 1999) and international (Abernethy and Rutherfurd, 2001; Easson and Yarbrough, 2002) studies on plant root distribution, shows that for most of the riparian species studied here, roots are concentrated only in the upper soil profile. The major vertically and obliquely inclined roots for most of the trialled species were observed to change direction abruptly and strike horizontally at a relatively shallow depth. There was no consistency in the depth at which this occurred, and this may be an adaptation typical of early colonising species. Results also indicate that with increasing depth and distance from the stem there is a rapid decline in roots, that there are interspecies differences in root distribution, and that each species allocates differing proportions of their total biomass to roots at different stages of growth.

Interspecies differences in root distribution has implications for the planting densities required to provide full, near-surface root occupancy of the soil and/or to maximise the root density to the depth requirement at specific sites. The tap-rooted cabbage tree will likely provide a higher level of reinforcement directly under the stand and to a greater depth than heart-rooted species, but the reinforcement quickly tapers off laterally as root density declines away from the stand edge. In contrast, heart-rooted species will likely provide a higher level of near surface reinforcement and to a greater distance from the stem, but reinforcement will rapidly decline at a relatively shallow depth.

Previous research on root depth of older aged riparian species, for example, manuka, shows that the root system of mature trees (13–50 years old) penetrated to a depth of 0.5 m on stony soils and 0.8 m on sandy soils (Watson and O'Loughlin, 1985), and those of another colonising and closely related species of manuka, kanuka (*Kunzia ericoides*), at between 6- and 32-year old reached a maximum depth of between 1.5 and 2.2 m. The latter study concluded that root depth was correlated not to tree age but rather to the stoniness and depth of slope colluvium (Watson et al., 1995). In a detailed investigation of cabbage trees, the root depth of 25-year-old trees growing in alluvial gravel was estimated to be ~2 m (Czernin, 2002). These studies, together with the few published reports on the root depth of some of New Zealand's tallest podocarp forest species (Cameron, 1963), indicate the rooting depth for most of New Zealand 's indigenous species rarely exceeded 2 m. The comparatively shallow rooting depth of New Zealand indigenous riparian species implies most will have a physical limitation to their ability to contribute to deep soil reinforcement.

The time required to achieve appreciable growth for many of these riparian shrub and tree species will vary greatly and depend largely on growth environment. With the exception of the higher altitudes, many of New Zealand's riparian species will achieve appreciable growth 5 years after establishment, but it may require 7–10 years before an effective cover is obtained.

Implications for riparian restoration and management

A significant improvement in riparian slope and bank stability is anticipated for the smaller streams more typical of unmodified, upland stream reaches where current channel form, slope characteristics and hydraulic conditions are better representative of what existed before forest clearance and where the performance of riparian vegetation has proven effective. For these streams, it is not the physical limitations of root system depth, spread and density of individual riparian species to provide effective soil reinforcement that determines the key to successful slope and bank stabilisation but rather the density of plantings and the species mix present. Treatment options that promote the quickest canopy closure and root development at all levels of the soil profile are likely to be the most effective in promoting site stability (Phillips et al., 2001). Where a seed source already exists and if animal stock could be excluded from riparian areas, many would regenerate naturally and at little cost. Excessive vegetative growth may, however, encroach on these channels, and without proper management may create drainage problems by clogging the stream.

As a consequence of their shallow-rooted habit, however, many of New Zealand's indigenous plants will have limited effectiveness in floodplain reaches of higher order streams modified by the building of stopbanks (levees) and where channel hydraulic conditions are likely to undercut stream banks to a very steep and unstable slope ~2 m high. If the potential for bed degradation exists, additional protection in the form of structural materials will be required along the toe of the bank and to some depth below the normal streambed. Similarly, bank materials such as alluvium are prone to undermining, thus riparian plantings must be protected by structural means (e.g., gabion baskets, rip rap, etc.) and/or by bank reshaping until growth is sufficient to achieve effective bank stability.

The limitations of root depth aside, New Zealand's indigenous riparian vegetation is sufficiently diverse to meet most of the requirements for slope and bank restoration, particularly of the lower order streams. The selection of suitable plant materials must take into account both the degree of overbank inundation contemplated and the ability of plant materials to provide year-round protection, have the capacity to become well established under adverse soil conditions, be long lived, develop a root system that will withstand the drag of stream flow on the above ground portion, have multistem and branch characteristics with many stems emerging from the boundary surface, have tough, resilient stems and branches, and require minimum maintenance.

Where stability is required to a known depth, such as to a potential failure plane that lies within the rooting depth of plants being considered to restore stability, the species selection must include those with root systems capable of reaching the specified depth. Failure to meet this goal will undoubtedly be the result of insufficient roots crossing the failure plane as it is below the vertical limit of root growth of the species selected. For many restoration sites the strategy should be to select a mix of species with different rooting habit. To appreciate fully the potential use of indigenous vegetation for the stabilisation of riparian slopes and streambanks in New Zealand, further studies are needed for other riparian plant species that may better meet the slope and bank stability requirements for drainage systems in different soil types, geology and with differing hydraulic characteristics.

Acknowledgements

We acknowledge the support of the Tairawhiti Polytechnic Rural Studies Unit, Gisborne, on whose land the riparian plant trial was located. We thank Alex Watson for reviewing a draft of this paper. Anne Austin is thanked for editing the final version. Statistical analyses were provided by Guy Forester (biometrician), Landcare Research, Lincoln. The Foundation of Research Science and Technology, New Zealand, provided funding under Contract No. COX 0305.

References

Abernethy B and Rutherfurd I D 2001 The distribution and strength of riparian tree roots in relation to riverbank reinforcement. Hydrol. Process. 15, 63–79.

Acheson A R 1968 River control and drainage in New Zealand. Ministry of Works, New Zealand, 296 pp.

Allan Herbarium 2000 New Zealand Plant Names Database, Landcare Research, New Zealand. Available http://nzflora.landcareresearch.co.nz (Accessed 22 November 2004).

Cameron R J 1963 A study of the rooting habits of rimu and tawa in pumice soils. N Z J For. 8, 771–785.

Cowley J M and Whyte C F 1997 Impact assessment for sawfly on Willow, Nematus oligospilus Förster (Hymenoptera: Tenthredinidae). New Zealand Plant Protection Centre, Lynfield, Auckland. Internal Ministry of Agriculture Report, 20 pp.

Czernin A 2002 Above-ground and below-ground characteristics of the New Zealand cabbage tree (Cordyline australis) with particular reference to the root system morphology and root strength. MSc thesis, University of Agricultural Sciences, Vienna, 183 pp.

Easson G and Yarbrough L D 2002 The effects of riparian vegetation on bank stability. Environ. Eng. Geosci. 8, 247–260.

Eyles G O 1983 The distribution and severity of present soil erosion in New Zealand. N.Z. Geog. 39, 12–28.

Foweraker C E, 1929 The podocarp rain forests of Westland, New Zealand. Kahikatea and Totara forests, and their relationships to silting. Te Kura Ngahere. A forestry journal issued by the Forestry Club of the Canterbury College School, Christchurch, New Zealand.

Hewitt A E 1998 New Zealand Soil Classification. 2nd Landcare Research Science Series No. 1 Manaaki Whenua Press, Lincoln, New Zealand.

IPCC, 2000: Land Use, Land-Use Change, and Forestry. A Special report of the IPCC [Watson R T, I R Noble B Bolin N H Ravindranth D J Verardo and D J Dokken, (Eds.)]. Cambridge Univ. Press, Cambridge, UK.

Marden M and Phillips C J 2002 How do native riparian plants grow? Conservation Quorum, 14–15.

Marden M and Phillips C J 2003 Stream bank planting trials. Conservation Quorum, 12–13.

Marden M and Phillips C J 2004 Stream bank planting trial: the final chapter. Conservation Quorum, 7–8.

New Zealand Climate Change Office 2003 Climate change. National Inventory Report for New Zealand. Greenhouse Gas Inventory, 1990–2001. Ministry for the Environment, Wellington, 86 pp.

Phillips C J and Watson A J 1994 Structural Tree Root Research in New Zealand, A Review. Landcare Research Science Series 7. Manaaki Whenua Press, Lincoln.

Phillips C J, Marden M, Rowan D and Ekanayake J C 2001 Stabilising characteristics of native riparian vegetation in New Zealand. *In* Proc. 3rd Australian Stream Management Conference, Brisbane August 2001, 507–512.

Schiechtl H M and Stern R 1994 Handbuch für naturnahen Wasserbau, Eine Anleitung für ingenieurbiologische Bauweisen. Osterreichischer Agrarverlag, Wien

Van Kraayenoord C W S and Hathaway R L 1986 Plant materials handbook for soil conservation, Vol. 2, Introduced Plants. 229. Water and Soil Misscellaneous Publication No. 94. National Water and Soil Conservation Authority, Wellington, New Zealand, ISSN 0110-4705.

Watson A J and O'Loughlin C L 1985 Morphology, strength, and biomass of manuka roots and their influence on slope stability. N Z J. For. Sci. 15, 337–348.

Watson A J and O'Loughlin C L 1990 Structural root morphology and biomass of three age classes of Pinus radiata. N Z J. For. Sci. 20, 97–110.
Watson A J, Marden M and Rowan D 1995 Tree species performance and slope stability. *In* D H Barker (ed), Vegetation and Slopes. pp. 161–171. Thomas Telford Press, London.
Watson A J, Phillips C J and Marden M 1999 Root strength, growth, and rates of decay: root reinforcement changes of two tree species and their contribution to slope stability. Plant Soil 217, 39–47.
Watson A J and Marden M 2004 Live root wood tensile strengths of some common New Zealand indigenous and plantation species. N Z J. For. Sci. 34(3), 344–353.

Mechanical resistance of different tree species to rockfall in the French Alps

Alexia Stokes[1,5], Franck Salin[1], Adzo Dzifa Kokutse[1], Stéphane Berthier[1,2], Henri Jeannin[1], Shaun Mochan[2], Luuk Dorren[3], Nomessi Kokutse[1], Murad Abd. Ghani[1] & Thierry Fourcaud[1,4]

[1]*Laboratoire du Rhéologie du Bois de Bordeaux, (Mixed unit: CNRS/INRA/Université Bordeaux I), 69, rte d'Arcachon, 33612, Cestas cedex, France.* [2]*Forest Research, Northern Research Station, EH25 9SY, Midlothian, Scotland.* [3]*Cemagref Grenoble, 2 rue de la Papeterie, B.P. 76, 38402, St. Martin d'Hères cedex, France.* [4]*AMAP, CIRAD, TA40/PS2, Boulevard de la Lironde, 34398, Montpellier Cedex 5, France.* [5]*Corresponding author**

Key words: protection forest, root anchorage, scar formation, stem breakage, tree stability, uprooting

Abstract

In order to determine the mechanical resistance of several forest tree species to rockfall, an inventory of the type of damage sustained in an active rockfall corridor was carried out in the French Alps. The diameter, spatial position and type of damage incurred were measured in 423 trees. Only 5% of trees had sustained damage above a height of 1.3 m and in damaged trees, 66% of broken or uprooted trees were conifers. Larger trees were more likely to be wounded or dead than smaller trees, although the size of the wounds was relatively smaller in larger trees. The species with the least proportion of damage through stem breakage, uprooting or wounding was European beech (*Fagus sylvatica* L.). Winching tests were carried out on two conifer species, Norway spruce (*Picea abies* L.) and Silver fir (*Abies alba* Mill.), as well as European beech, in order to verify the hypothesis that beech was highly resistant to rockfall and that conifers were more susceptible to uprooting or stem breakage. Nineteen trees were winched downhill and the force necessary to cause failure was measured. The energy (E_{fail}) required to break or uproot a tree was then calculated. Most Silver fir trees failed in the stem and Norway spruce usually failed through uprooting. European beech was either uprooted or broke in the stem and was twice as resistant to failure as Silver fir, and three times more resistant than Norway spruce. E_{fail} was strongly related to stem diameter in European beech only, and was significantly higher in this species compared to Norway spruce. Results suggest that European beech would be a better species to plant with regards to protection against rockfall. Nevertheless, all types of different abiotic stresses on any particular alpine site should be considered by the forest manager, as planting only broadleaf species may compromise the protecting capacity of the forest, e.g., in the case of snow avalanches.

Introduction

The use of protection forests against the impact of natural hazards, e.g., rockfall and snow avalanches is becoming more and more common in Europe (Brang, 2001; Dorren and Berger, 2006; Dorren et al., 2004; Hurand and Berger, 2002; Motta and Haudemand, 2000; Ott, 1996). With the increase in catastrophic events both in the European Alps (Interreg IIIb, 2001; Sauri et al., 2003) and in mountainous regions around the world (Tianchi et al., 2002), research into this phenomenon has accelerated. However, although it is understood that the structure of the forest plays a vital role in determining its effectiveness as a protective barrier (Jahn, 1988; Kräuchi et al., 2000), little information exists concerning the mechanical resistance of different tree species to different types of natural hazards. One particular natural hazard which has been much neglected until recent years is that of rockfall. Not only is the movement of rocks and stones a hazard to both people and infrastructures, but rockfall safety nets are expensive and difficult to install and they deteriorate with time (Dorren, 2003). If further information on the structure of a protection forest, and the most mechanically resistant species to use against rockfall could be obtained, these data could be used as input to models of

* E-mail: stokes@liama.ia.ac.cn

rockfall dynamics (Dorren et al., 2004) and/or fed directly into management and decision support systems (Mickovski, 2005; Stokes et al., 2004).

Even if a tree species is useful as a barrier against one particular type of hazard, the same species may not be suitable in protecting against a different type of hazard e.g. Norway spruce (*Picea abies* L.) is not especially windfirm (Stokes et al., 2000) nor resistant to rockfall (Hurand and Berger, 2002). However, in preventing snow movement, Norway spruce is highly effective in holding in place the snow mantle (Hurand and Berger, 2002). Therefore, it is necessary to determine which species is best suited to a particular function. In the case of rockfall, different types of rockfall exist, including collapsing in mass where the volume displaced is $>5.0\ m^3$. Individual rockfall occurs more often with smaller volumes ($<5.0\ m^3$) displaced (Berger et al., 2002). It is in this latter case that forests can act as a barrier and provide a protective function. When rocks impact against trees, different types of tree failure can occur, including uprooting and stem breakage (Berger et al., 2002). Certain species, particularly angiosperms, appear to be more resistant to failure than others, often sustaining wounds only (Dorren and Berger, 2006). It is not known which species are the most resistant against the impact of rocks, however, foresters have suggested from experience that broadleaf species are more resistant against rockfall impacts, although no particular reasons are given for this hypothesis. Only in the literature concerning wind damage to forests, can comparisons of different species be found with regards to their mechanical resistance (Meunier et al., 2002; Peltola et al., 2000; Stokes et al., 2000). The most common method to compare the likelihood of stem failure or uprooting, is to winch trees sideways until failure occurs (Cucchi et al., 2004; Gardiner et al., 2000; Moore, 2000; Stokes, 1999; Stokes et al., 2000). Certain species, e.g. Sitka spruce (*Picea sitchensis* Bong. Carr) are more susceptible to stem breakage and uprooting than others e.g. European beech (*Fagus sylvatica* L.) (Stokes et al., 2000). Very few data exist concerning broadleaf species, primarily because conifer species are more susceptible to damage during a winter storm, and conifers are economically important timber species.

One of the most important factors governing the ability of a tree to withstand breakage or uprooting during a storm, is the morphology of the root system present (Cucchi et al., 2004; Dupuy et al., 2005; Stokes et al., 2000, 2007) Trees with deep and wide spreading root systems will be better anchored than those with superficial roots only (Stokes, 2002). The shape and size of a root system is influenced by its immediate environment as well as being inherent to a particular species (Köstler et al., 1968). Trees growing on the thin, rocky soils encountered on mountain slopes may therefore possess different rooting types depending on species. The morphology of the root system may also differ to that of the same species growing in a deep soil on flat ground (Köstler et al., 1968). Therefore, a well-anchored species growing in a particular soil type, may become highly unstable in a different environment (Dupuy et al., 2005; Moore, 2000).

In order to identify the type of tree failure which occurs in a forest subjected to rockfall, two studies were carried out on an active rockfall site in the French Alps. Initially, the position, size, species and type of damage sustained to trees growing in a mixed forest were measured. A series of winching tests were then carried out on Norway spruce, Silver fir and European beech, which enabled us to quantify the mechanical resistance of each species. The maximum energy required to cause failure was then estimated, as during an impact between a falling rock and a tree, it is the kinetic energy of the rock which causes tree displacement. Results are discussed with regards to management strategies for protection forests.

Materials and methods

Study site

The study site was situated in the Forêt Domaniale de Vaujany, Vallée de l'Eau d'Olle, Isère (lat 45° 12′, long 6°3′), France, at an altitude of 1350–1600 m. This forest is located on a northwest facing mountain side that can be divided into two areas. First, the rockfall source areas, which are a series of steep cliff faces dissected by some denudation niches occurring on top of each other. The mean slope gradient in the source area is 70° up to vertical cliffs. The second part consists of large post-glacially developed talus cones consisting mainly of rock avalanche deposits, snow avalanche deposits and rockfall scree. These large talus cones were formed after deglaciation of the main valley. The retreat of the glacier resulted in tensional rebound of the oversteepened valley slopes. This retreat led to slope instability and landsliding (mainly rock avalanches),

Figure 1. A digital terrain model of the study area overlain by an orthophoto. This figure shows that the site consists of several talus cones (dashed white lines) on which preferential rockfall and avalanche tracks exist (white arrows; the size indicates the magnitude of processes acting in the preferential track). The study site is depicted by the white rectangle.

which consequently resulted in the build up of the large talus cones (Figure 1). During the Holocene (the last 10,000 years), these talus cones have been colonised by vegetation, eventually resulting in a forest cover. Today, the dominant tree species on the site are Silver fir (*Abies alba* Mill.), Norway spruce (*Picea abies* L.), European beech (*Fagus sylvatica* L.), Sycamore (*Acer pseudoplatanus* L.), European ash (*Fraxinus excelsior* L.) and Common hazel (*Corylus avellana* L.). The forested talus cones have a slope gradient of 38°–42° and act currently as rockfall transit and accumulation zones. Rocks impacting trees can cause damage and are therefore disturbing the forest ecosystem. The other major disturbances are snow creep, snow gliding, snow avalanches, ungulate browsing and wind loading. The effects of the mass movement processes are clearly reflected in the slope relief and in the vegetation as distinct preferential tracks or channels for snow transport and falling rocks. In between the preferential tracks, the forest is dominated by uneven aged Silver fir and Norway spruce and in the preferential tracks the forest is dominated by young European beech, ash and hazel trees. A storm in 1960 resulted in the loss of 2220 m^3 of timber throughout the whole forest, of which the surface area is 818 ha (C. Bazin, personal communication).

Inventory of damage incurred by rockfall

To determine if damage by rockfall was more frequent in certain species compared to others, an inventory of trees with/without damage was carried out. A 200 × 50 m corridor was defined that covered two preferential tracks within the study site, where rockfall appeared to be the most active (Figure 1). All trees >0.1 m diameter were measured within this corridor. For each tree, the species, DBH, diameter at stem base (DSB) and type of damage were noted. Trees were noted as uprooted, broken in the stem or wounded (height and width of wound measured if below DBH).

In order to compare the extent of wound damage in trees of different sizes, the percentage of dysfunctional cambium was calculated using: (width of wound/tree basal circumference) × 100 (Guyette and Stambaugh, 2004). Analysis of variance and χ^2 tests were carried out to determine if the type of damage sustained was influenced by species and size of trees, using size parameters as covariates where necessary.

Winching tests

To quantify the mechanical resistance of a tree to failure by rockfall, bending tests *in situ* were carried out. Trees were winched sideways until failure occurs and the force necessary to cause uprooting or stem breakage measured. Such tests were carried out on 19 adult trees. It was not possible to carry out more tests, as suitable tree material was scarce and winching tests dangerous due to the unstable, steep slope. Three species were chosen for this study, as they appeared to sustain different types of rockfall damage at the site (see results of damage mapping). These species were Silver fir, Norway spruce and European beech. Mean DBH for all trees was 0.23 ± 0.08 m and height was 14.90 ± 0.65 m (Table 1). The system employed was similar to that used by Cucchi et al. (2004), Meunier et al. (2002), Moore (2000), Peltola et al. (2000) and Stokes (1999). A motorised winch (16 kN, Hit-Trac 16B, Habegger, Switzerland) was used to winch trees sideways. For large trees, a pulley was also used which doubled the winch capacity. The winch was attached to the base of an anchoring tree at the longest possible distance to the winched tree, in order to obtain a small angle θ (Figure 2). When pulling trees downhill, a pulley was employed to deviate the force applied, so that the winch user would not be in the pathway of falling rocks

Table 1. Means ± standard error of the parameters measured for each species tested in the winching studies and necessary for the calculation of $TM_{crit,total}$. Mean ± standard error values of the maximum energy (E_{fail}) required to cause failure in each species are also given

Variables	European beech ($n = 7$)	Silver fir ($n = 6$)	Norway spruce ($n = 6$)
Height (m)	17.7 ± 1.8	12.9 ± 0.8	12.7 ± 1.1
DBH (m)	0.23 ± 0.03	0.24 ± 0.02	0.24 ± 0.01
Stem weight (kg)	479.0 ± 129.0	257.5 ± 50.2	235.2 ± 42.1
Crown biomass (kg)	127.2 ± 36.8	180.4 ± 52.9	145.6 ± 28.4
Total biomass (stem + crown) (kg)	606.0 ± 161.0	437.0 ± 92.9	380.8 ± 63.6
Failure energy (kJ)	38.4 ± 12.7	22.7 ± 5.3	15.3 ± 4.7

or branches (Figure 2). The winch cable was attached to the test tree at a height (H_{cable}) of 4.0–8.0 m, as it was physically difficult to attach the cable any higher, due to the high number of branches encountered on most trees, which hindered climbing. The force applied was measured with a load cell (K25H20 kN, Scaime S.A., France) and measured every second using a datalogger (Almemo 2290-8, Ahlborn, Germany). In order to measure the deflection angle, α, of the stem during winching, two inclinometers were nailed to the tree, one at cable height and the second at the stem base (Figure 2), which measured rotation of the root plate (Cucchi et al., 2004). An identical datalogger was used to record α every second during winching. The distance between the test tree and anchorage point was also measured.

Once a tree had been winched to failure, several measurements were carried out on the stem and crown which were necessary for calculations of the total critical bending moment ($TM_{crit,total}$) (Cucchi et al., 2004). The relative crown length i.e. the distance between the first living branch and the stem apex was determined, along with the height of the first living branch. The stem circumference was measured every 1.0 m, avoiding any

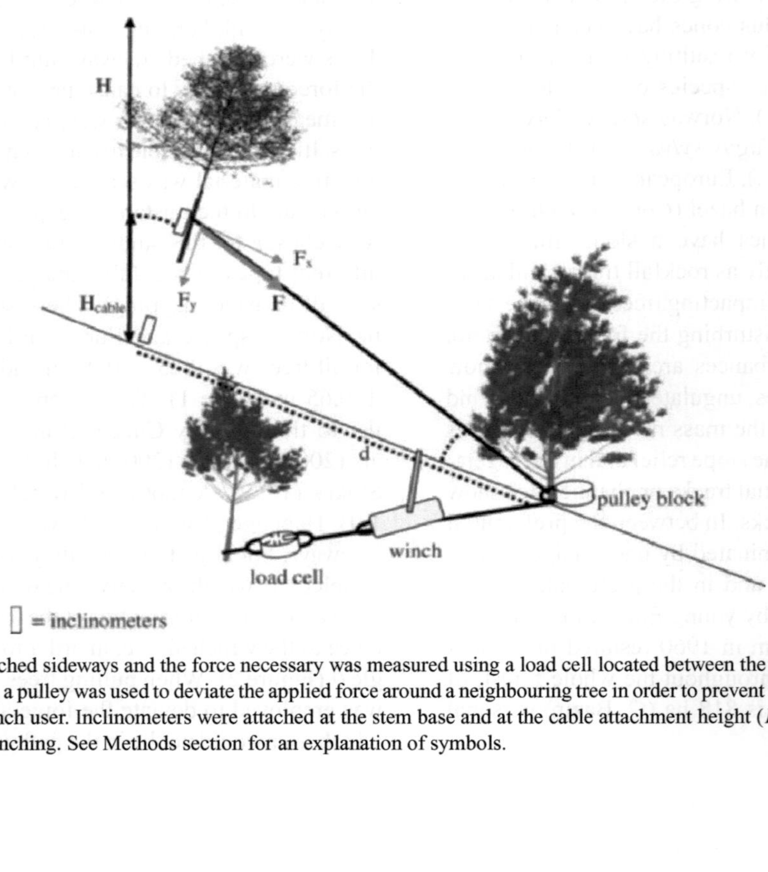

Figure 2. Trees were winched sideways and the force necessary was measured using a load cell located between the winch and anchoring tree. When winching downhill, a pulley was used to deviate the applied force around a neighbouring tree in order to prevent the tested tree or dislodged rocks from hitting the winch user. Inclinometers were attached at the stem base and at the cable attachment height (H_{cable}) in order to measure stem deflection during winching. See Methods section for an explanation of symbols.

bulging whorls or branches. Crown biomass was measured by weighing all the live branches (Table 1). Stem green wood density (wood and bark) was calculated using:

$$\text{Density} = \frac{\text{SectionWeight}}{\text{SectionVolume}}, \qquad (1)$$

where SectionVolume is the volume of a 1.0 m long section of trunk cut from the middle of the stem volume and weighed (SectionWeight). This density was assumed to be constant throughout the stem. Total stem weight was estimated using:

$$\text{StemWeight} = \sum(\text{SectionVolume}) \times \text{Density} \qquad (2)$$

As the local environment around each tree was highly variable, several measurements were made which took into account the immediate vegetation conditions. The number and species of trees within a radius of 5.0 m was recorded, along with the distance to the nearest tree. Soil moisture content was not measured, as it was considered that the large quantity of rocks and stones present would influence uprooting strength more than soil moisture.

Calculation of critical bending moment

The force required to cause failure of a tree was determined from the recorded load cell data, and associated values of σ at the time of uprooting or stem breakage. The critical turning moment applied at the stem base was calculated using the method described in Cucchi et al. (2004):

$$\text{TM}_{\text{crit,applied}} = F_x \times \cos\alpha \times H_{\text{cable}} + F_y \times \sin\alpha \times H_{\text{cable}} \qquad (3)$$

where F_x is the component parallel to the soil surface and F_y the component perpendicular to the soil surface. Both are components of the maximal applied force F(N) and the deflection angle α of the trunk minus the slope angle, when the force was maximal. The tree stem was considered as a rigid cantilever beam and stem curvature was not taken into consideration. The difference between the angle of deflection measured at the stem base and at the height of the cable was therefore considered as negligible. Nevertheless, to account for any stem curvature, we used a mean of the two angles to calculate α. F_x and F_y were deducted from the value of F(N) and the cable angle θ, with regards to the soil surface. θ was derived from the distance d between the tree winched and the anchoring tree (Figure 2), as well as H_{cable}:

$$F_x = F \times \cos\theta \quad \text{and} \quad F_y = F \times \sin\theta \qquad (4)$$

The total critical turning moment $\text{TM}_{\text{crit,total}}$ at the stem base adds the critical turning moment applied by the winch (3) to the critical turning moment $\text{TM}_{\text{crit,weight}}$ due to the force resulting from the overhanging weight of the leaning tree during winching. The weight of the winch and cable were neglected. $\text{TM}_{\text{crit,weight}}$ was calculated by resolving tree weight into stem and crown weights where the tree crown was taken as a whole and the tree stem as 1.0-m long sections stacked on top of each other:

$$\text{TM}_{\text{crit,weight}} = W \times G_x, \qquad (5)$$

where W is the weight in N and G_x is the final horizontal position of the centre of gravity at the middle of the crown or stem section. As tree displacement was small, the horizontal component of G could be assimilated to the height of G on the stem multiplied by the sinus of the leaning angle given by the inclinometers. For the crown and stem section above the cable attachment point, this angle corresponds to the angle measured at this point. Although (3) and (4) assume the stem as a rigid cantilever, we wanted to take into account the stem lean below this point: leaning angle is an evolution in increments between the stem base angle and the attachment point angle. Stem section mass was deducted from the green wood density determined for each tree and the stem section volume assumed to be a truncated cone form. Hence, $\text{TM}_{\text{crit,crownweight}}$ was directly obtained and $\text{TM}_{\text{crit,stemweight}}$ was the sum of all the moments of tree stem sections.

Calculation of energy required to cause failure

The amount of energy (E_{fail}) required to cause tree failure was calculated by integrating numerically the total overturning moment TM, i.e. including crown and trunk weights, over the angle α of stem lean during winching (where $\alpha =$ the deflection angle of the trunk minus the slope angle when the force was maximal.

α was calculated as the mean of the stem two angles measured):

$$E_{\text{fail}} = \frac{1}{2}\sum_{i=1}^{n-1}(\alpha_{i+1} - \alpha_i)(\text{TM}_{i+1} + \text{TM}_i), \quad (6)$$

where n = the number of data recorded just before tree failure, and indices i the ith record in the data base.

Data were analysed using regression analysis and relationships between a given tree parameter and $\text{TM}_{\text{crit,total}}$ or E_{fail} were compared using analysis of covariance.

Results

Inventory of damage incurred by rockfall

423 trees were measured in the active rockfall corridor. The DBH of all trees was 0.31 ± 0.10 m and DSB was 0.41 ± 0.10 m (values are means $\pm$ standard error). The percentage of conifer species on the site was low (23%), with only Norway spruce (5%) and Silver fir (17%) present. Seven broadleaf species were present, with sycamore (24%), European beech (23%) and European ash (23%) being the major species on the site. The remaining four species comprised Common aspen (*Populus tremula* L.) (3%), Silver birch (*Betula pendula* Roth.) (2%), Wych elm (*Ulmus glabra* Huds.) (1%) and European Rowan (*Sorbus aucuparia* L.) (1%). Many trees were damaged or wounded by rockfall, with 50% of broadleaf species and 65% of conifers either wounded or dead. It was not always possible to identify the broken or uprooted conifer species, which were often in an advanced state of decay, therefore both Silver fir and Norway spruce were combined for the statistical analysis. Taking into account only those species present >5% of the total percentage of trees, a higher number of dead conifers was present than either wounded or healthy conifers whereas very few dead broadleaf trees were present. Beech was the species with the highest number of healthy trees whereas sycamore was the most often wounded (Figure 3). Surprisingly, only eight trees were broken in the stem at a height >1.3 m and only 12 trees had sustained wounds above this height.

The size of the tree also appeared to influence its state. Conifers were significantly larger in diameter than broadleaf species, and ash was the species with the smallest DSB ($F_{1,386} = 3.82$, $p = 0.01$, Figure 4). In all species, the largest trees were the most susceptible to being wounded and in broadleaf species, the smallest trees were most likely to be dead ($F_{2,386} = 16.42$, $p < 0.001$, Figure 4). However, a significant negative relationship between the percent dysfunctional cambium and tree DBH existed, even though variability was high ($y = -0.0593x + 39.895$, $R^2 = 0.04$, $p = 0.023$), i.e. although larger trees were more likely to sustain damage, the size of wounds were relatively smaller in larger trees. No significant differences within or between tree species were found.

The results of this inventory permitted us to choose three species for the study of tree winching. Both conifer species were chosen along with beech, in order to compare broadleaf and conifer species. Beech was chosen because it was the species found to have the highest proportion of undamaged trees (Figure 3).

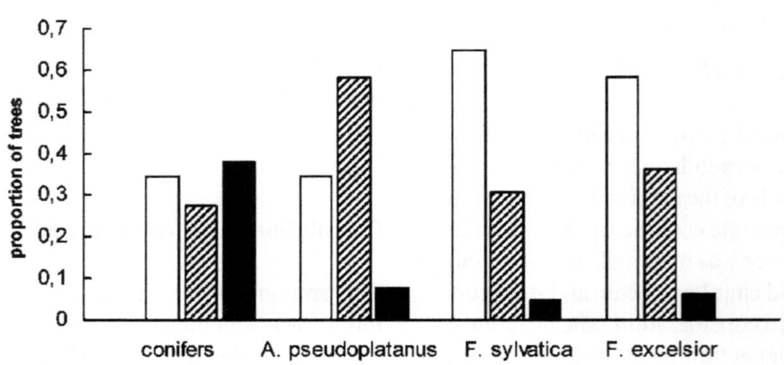

Figure 3. The proportion of healthy, wounded or dead trees were significantly different depending on species ($\chi^2 = 84.4$, $p < 0.001$). Most beech trees were healthy (white bar), whereas a high proportion of sycamore were wounded (hatched bar) and most conifers were either wounded or dead (black bar).

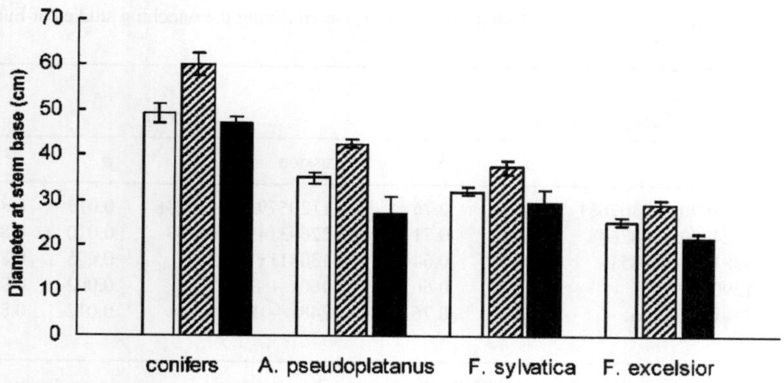

Figure 4. The diameter at stem base (DSB) influenced significantly the state of the tree. Conifers were significantly larger in diameter than broadleaf species, and ash was the species with the smallest DSB. In all species, the largest trees were the most susceptible to being wounded (hatched bar) and in broadleaf species, the smallest trees were most likely to be dead (black bar) when compared to healthy trees (white bar).

Winching tests

Out of the 19 trees winched downhill in this study, 4 European beech, 2 Silver fir and 5 Norway spruce were uprooted, whilst the remaining trees all broke in the stem at a height <1.3 m. European beech failed at a mean height of $5.9 \pm 1.4\%$, and fir at $17.3 \pm 5.0\%$ of the total relative stem length. Out of the beeches that failed in the stem, all trees were found to be growing <0.5 m to a large neighbouring beech tree and root grafting could be seen to occur in superficial roots of neighbouring trees when the topsoil was removed with a trowel. One beech tree was discounted from further analysis, as the base of the tree was found to be associated to a neighbouring tree and may have originated from this tree. Although the Silver fir trees were sometimes growing nearby neighbouring Silver fir and beech, no root grafting could be seen to occur. The Norway spruce which broke failed at the stem base, and was found to contain rotten wood in this area.

Linear regressions carried out between $TM_{crit,total}$ and different tree characteristics showed that the best relationship was obtained for both DBH and total biomass in European beech, and crown biomass in Silver fir (Table 2). No significant relationship was found between $TM_{crit,total}$ and any size parameter for Norway spruce. Results showed European beech was significantly more resistant to failure than Silver fir when $TM_{crit,total}$ was regressed with DBH, DBH_2 and crown biomass (Table 2). Although no significant relationship in $TM_{crit,total}$ and any size parameter was found in Norway spruce, it could be seen that $TM_{crit,total}$ was very low for this species (39.4 ± 7.6 k Nm). No significant differences in $TM_{crit,total}$ were found between uprooted and broken trees, probably due to the low number of trees tested.

Variability in E_{fail} was high both within and between species (Table 1) and significant relationships between E_{fail} and DBH or DBH^2 were found in European beech only, the best being with DBH (Figure 5). E_{fail} in European beech was significantly greater than that in Norway spruce for DBH ($F_{1,8} = 11.55$, $p = 0.009$) and DBH^2 ($F_{1,8} = 7.54$, $p = 0.025$) only.

Discussion

Although we cannot be certain that any particular tree was damaged through falling rocks, the type of damage incurred suggested that rockfall was the main abiotic stress in the corridor chosen. Nevertheless, snow or wind damage may also have resulted in uprooted or broken trees. 66% of uprooted or broken trees were conifers, but it was not always possible to determine the actual species. Only 5% of trees were broken or wounded above DBH, as the average rock rebound height was 1.0 m at this site (Dorren and Berger, 2006). Large trees were more likely to be wounded than smaller trees, which may be due to a decrease in rockfall activity over the last 50 years. However, smaller trees were more likely to die if damage was sustained. Small conifers were less numerous due to over browsing by ungulates in the valley (C. Bazin, personal communications), which will also bias the results in that small, damaged conifers were fewer due to this problem. A large amount of mosses were observed growing on the

Table 2. Significant regression equations for $TM_{crit,total}$ and each parameter measured during the winching studies of European beech and Silver fir

	European beech			Silver fir			Comparison between species	
Variables	Regression	p	R^2	Regression	p	R^2	$F_{2,15}$	p
DBH (m)	$y = 1983690x - 307681$	0.011	0.76	$y = 1120579x - 177738$	0.017	0.80	6.73	0.008
DBH^2 (m^3)	$y = 3824259x - 67648$	0.016	0.71	$y = 2269314x - 44133$	0.010	0.84	5.42	0.017
(H × DBH^2) (m^3)	$y = 5881x + 130851$	0.030	0.64	$y = 130311x - 12751$	0.005	0.88	–	ns
Crown biomass (kg)	$y = 1290x - 2514$	0.027	0.66	$y = 460x + 8856$	0.002	0.92	5.44	0.017
Total biomass (stem + crown) (kg)	$y = 318x - 30875$	0.011	0.76	$y = 248x - 16879$	0.012	0.83	–	ns

Data for Norway spruce were not significant and therefore not included. Regressions between species were compared using analysis of covariance.

rocks along the rockfall corridor, which also suggests a decrease in rockfall activity over recent years.

The increase in the likelihood of damage with stem diameter may not only be due to a decrease in rockfall activity, but the fact that large trees are more likely to be hit by falling rocks. Whereas small trees are more likely to break and therefore die, older species will resist uprooting or breakage and sustain wounds. The larger the tree, the smaller the percentage of cambial damage that will occur. Nevertheless, certain broadleaf species are able to produce scar tissue faster than fir or spruce. This scar tissue will form around the wound and protect it from pathogen attack (Shigo, 1986). Silver fir and Norway spruce will therefore be more susceptible to infection through pathogens, leading to internal stem rot and decay, and ultimately resulting in weakened mechanical resistance to rockfall. Bark thickness was not taken into account, but is known to protect the living cambium from wounding (Guyette and Stambaugh, 2004). A future study should examine the influence of bark thickness on the percentage wound damage to a stem, for different species. Bark is usually thicker in older trees, which may be a further reason why wound size was relatively smaller in large trees.

The winching tests revealed that most Silver fir trees failed in the stem, whereas Norway spruce usually failed through uprooting. Fir trees possessed very few roots, but these roots were large and long and penetrated between the numerous rocks present in the soil (Stokes et al., 2006). Therefore, it can be considered that these

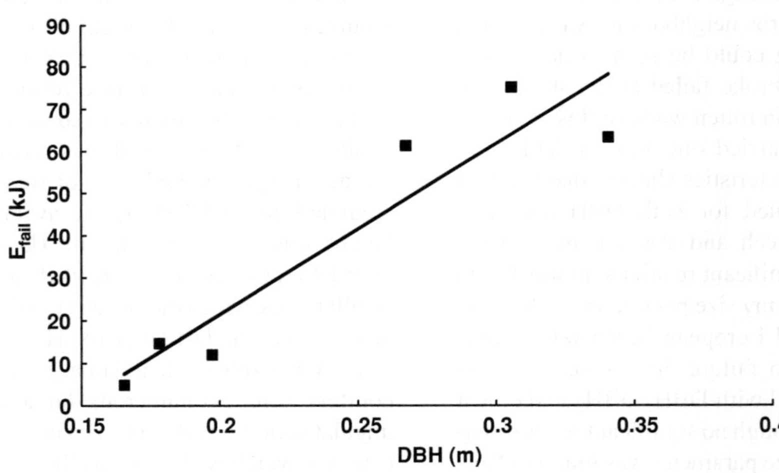

Figure 5. The amount of energy (E_{fail}) required to cause tree failure increased significantly with stem diameter at breast height (DBH) in European beech only ($y = 410.853x - 59.057$, $R^2 = 0.88$, $p = 0.006$).

trees were well anchored, as the moment required to resist overturning was greater than that needed to cause stem failure. Norway spruce possessed more superficial root systems, also with few roots (Stokes et al., 2006), and the only Norway spruce that broke in the stem was rotten at the stem base. Root systems of European beech were highly branched and deeper than spruce (Stokes et al., 2006). European beech failed several times in the stem, but only in trees which were growing nearby other beech trees. A high amount of root grafting could be seen to occur between neighbouring beech trees, which will strongly increase root anchorage. Although spruce and fir can also graft with tree roots of the same species (Bormann and Graham, 1959), no signs of root grafting were visible in the superficial surface roots of these trees.

European beech was the most resistant species to failure, and mean $TM_{crit,total}$ was nearly double that of Silver fir and was three times larger than in Norway spruce. The best correlations between $TM_{crit,total}$ and tree size parameters were found to be DBH and total biomass in beech and crown biomass in fir. In previous studies, the best correlations were usually with stem mass (Gardiner et al., 2000; Meunier et al., 2002) or $H \times DHB^2$ (Cucchi et al., 2004). It is surprising that no significant relationship was found between $TM_{crit,total}$ and any parameter measured in Norway spruce. It would probably be necessary to test a larger number of trees in order to obtain better regressions. The rocky nature of the soil also added to the variability encountered between trees. The value of $TM_{crit,total}$ was high for beech, but comparable to data for other certain species found in the literature, e.g. *Pinus pinaster* Ait. (Cucchi et al., 2004) and *P. radiata* D. Don (Moore, 2000). Nevertheless, $TM_{crit,total}$ is usually much lower, and the values for fir and spruce were similar to those for *Betula* spp. and *P. sylvestris* L. (Peltola et al., 2000). Very few data exist for comparisons with broadleaf species, and none for trees growing in such conditions.

The energy (E_{fail}) required to cause tree failure was calculated from static winching data, and therefore did not take into account crown or whole stem characteristics. The effect of these components on the maximal amount of energy that can be dissipated by a tree (E_{max}) is considerable, as shown by Dorren and Berger (2006). Nevertheless, these authors obtained E_{max} values ranging from 40–115 kJ for Silver fir with a similar DBH to trees from our study. Our results showed that E_{fail} was significantly higher in European beech compared to Norway spruce, and was strongly related to stem diameter in the former species. Therefore, European beech is not only more mechanically resistant to failure, but can also resist rockfall better as stems can deflect more during an impact. These results are coherent with the results obtained by dynamic impact tests as described by Dorren and Berger (2005). Again, the lack of significant relationships between E_{fail} and stem parameters in Silver fir and Norway spruce tested in our experiments, may be due to a lack of data.

These results suggest that beech would be a better species to plant with regards to rockfall protection. Broadleaf species can also regenerate after damage, and produce large quantities of scar tissue if wounded by a falling rock. The disadvantage of broadleaf species is that they do not prevent the formation of homogeneous snow layers due to their reduced canopy surface in the winter. As a result the snow avalanche risk increases in comparison to coniferous forests. Nevertheless, more broadleaf species would need to be tested in order to determine which are the most resistant to rockfall, and thus the most useful species to plant in a rockfall protection forest. The main remaining task for protection forest managers will be to define against which natural hazard the forest has to protect. If both rockfall and snow avalanches are occurring, a mixed forest would be the most effective for protection.

Acknowledgements

Thanks are due to M. Aya, H. Bignalet, G. Cherokee, S. Daviller, M. Genet, N. Harries, B. Issenhuth, H. Lataillade, S. Mickovski, M. Sharman, (INRA), C. Quetel (Cemagref), J. Fannin (University of British Columbia, Canada) and M. Sarnatoro (University of Molise, Italy), for help with fieldwork and F. Danjon (INRA) for his critical comments on the manuscript. We thank C. Bazin (Office National des Forêts), the Marie de Vaujany and F. Berger (Cemagref) for permission to work at the study site, as well as the Hotel Cabri for their help and warm welcome over the last three years. This work was financed by the E.U. project Eco-Slopes (QLK5-2001-00289). A.D. Kokutse was funded by a Marie Curie Bursary.

References

Berger F, Quetel C and Dorren L K A 2002 Forest: a natural protection mean against rockfalls, but with which efficiency? *In* Proceedings of the International Congress. Interpravent 2002 in the

Pacific—Rim. Matsumoto, Japan. Vol. 2, pp. 815–826. Congress Publication.

Brang P 2001 Resistance and elasticity: promising concepts for the management of protection forests in the European Alps. Forest Ecol. Manage. 145, 107–119.

Bormann F H and Graham B F 1959 The occurrence of natural root grafting in eastern white pine *Pinus strobus* and its ecological implications. Ecology 40, 677–691.

Cucchi V, Meredieu C, Stokes A, Berthier S, Bert D and Najar M 2004 Root anchorage of inner and edge trees of Maritime pine (*Pinus pinaster* Ait) growing in different soil podzolic conditions. Trees-Struct. Funct. 18, 460–466.

Dorren L K A 2003 A review of rockfall mechanics and modelling approaches. Prog. Phys. Geog. 27, 69–87.

Dorren L K A, Maier B, Putters U S and Seijmonsbergen A C 2004 Combining field and modelling techniques to assess rockfall dynamics on a protection forest hillslope in the European Alps. Geomorphology 57, 151–167.

Dorren L K A and Berger F 2006. Stem breakage of trees and energy dissipation during rockfall impacts. *Tree Physiol.* 26, 63–71.

Dupuy L, Fourcaud T and Stokes A 2005 A numerical investigation into the influence of soil type and root architecture on tree anchorage. Plant Soil 278, 119–134.

Gardiner B A, Peltola H and Kellomaki S 2000 Comparison of two models for predicting the critical wind speeds required to damage coniferous trees. Ecol. Model. 129, 1–23.

Guyette R P and Stambaugh M C 2004 Post-oak fire scars as a function of diameter, growth, and tree age. For. Ecol. Manage. 198, 183–192.

Hurand A and Berger F 2002 Forêts et risques naturels. Protection contre l'érosion, les mouvements de terrain et les avalanches. La Houille Blanche 3, 64–67.

Interreg III B 2001 Alpine Space Programme. General Document on the Community-Initiative Programme INTERREG III B.

Jahn J 1988 Entwaldung und Steinschlag, Interpraevent, Graz. Tagungspubl Bd.1,185–198.

Köstler J N, Brückner E, Bibelriether H (1968) Die Wurzeln der Waldbäume. Verlag Paul Parey, Hamburg & Berlin, Germany.

Kräuchi N, Brang P and Schönenberger W 2000 Forests of mountainous regions: gaps in knowledge and research needs. For. Ecol. Manage. 132, 73–82.

Meunier S, Ruel J C, Laflamme G and Achim A 2002 Résistance comparéedel–épinette blanche et du sapin baumier au renversement. Can. J. For. Res. 32, 642–652.

Mickovski S B 2007 Decision support systems in eco-engineering: the case of the SDSS. *In* A Stokes, I Spanos, JE Norris and LH Cammeraat(eds), Eco- and Ground Bioengineering: The Use of Vegetation to Improve Slope Stability. Developments in Plant and Soil Sciences, vol. 103, Springer, Dordrecht.

Moore J R 2000 Differences in maximum resistive bending moments of *Pinus radiata* trees grown on a range of soil types. For. Ecol. Manage. 135, 63–71.

Motta R and Haudemand J C 2000 Protective forests and silvicultural stability. An example of planning in the Aosta Valley. Mt Res. Dev. 20, 180–187.

Ott E 1996 Guidelines for the protective role of forests in avalanche formation. Forstwiss Centralbl. 115, 223–230.

Peltola H, Kellomaki S, Hassinen A and Granander M 2000 Mechanical stability of Scots pine, Norway spruce and birch: an analysis of tree-pulling experiments in Finland. For. Ecol. Manage. 135, 143–153.

Saurí D, Milego R, Canalís A, Ripoll A and Kleeschulte S 2003 Mapping the impact of recent natural disasters and technological accidents in Europe. Environmental Issue Report No. 35, European Environment Agency, Copenhagen, Denmark.

Shigo A L 1986 A new tree biology. Shigo & Tree Associates, Durham, New Hampshire.

Stokes 1999 Strain distribution during anchorage failure in root systems of Maritime pine (*Pinus pinaster* Ait.) at different ages and tree growth response to wind-induced root movement. Plant Soil 217, 17–27.

Stokes A 2002 The biomechanics of tree root anchorage. *In* Y Waisel, A Eshel and U Kafkaki (eds), Plant Roots—The Hidden Half, pp. 175–186. Plenum Publishing, New York.

Stokes A, Drexhage M and Guitard D G 2000 A method for predicting the site of failure in trees under mechanical loading. *In* A Stokes (ed), The Structural Roots of Trees and Woody Plants: Form, Function and Physiology, Developments in Plant and Soil Sciences, pp. 279–285. Kluwer, Dordrecht.

Stokes A, AbdGhani M, Salin F, Danjon F, Jeannin H, Berthier S, Kokutse A D and Frochot H 2007 Root morphology and strain distribution during overturning failure of trees on mountain slopes. *In* A Stokes, I Spanos, JE Norris and LH Cammeraat (eds), Eco- and Ground Bio-engineering: The Use of Vegetation to Improve Slope Stability, Developments in Plant and Soil Sciences, vol 103, pp. 165–173. Springer, Dordrecht.

Stokes A, Mickovski S B and Thomas B R 2004 Eco- engineering for the long-term protection of unstable slopes in Europe: developing management strategies for use in legislation. *In* W Lacerda, M Ehrlich, SAB Fontoura, ASF Sayao (eds), Landslides: Evaluation & Stabilization, vol. 2, pp. 1685–1690 A.A. Balkema Publishers, Rotterdam.

Tianchi L, Greminger P, Hofer T (2002) Hazards. Beyond The International Year of Mountains (on-line) URL: http://www.mountains2002.org/files/pdf/factsheets/Hazards-E.pdf.

Root morphology and strain distribution during tree failure on mountain slopes

Alexia Stokes[1,5], Murad Abd.Ghani[1], Franck Salin[1], Frédéric Danjon[2], Henri Jeannin[1], Stéphane Berthier[1,3], Adzo Dzifa Kokutse[1] & Henri Frochot[4]

[1]*Laboratoire du Rhéologie du Bois de Bordeaux (Mixed unit: CNRS/INRA/Université Bordeaux I), 69, rte d'Arcachon, 33612 Cestas cedex, France.* [2]*EPHYSE, 69, rte d'Arcachon, 33612 Cestas cedex, France.* [3]*Current address: Forest Research, Northern Research Station, Roslin, Midlothian, EH25 9SY, Scotland.* [4]*Lerfob, UMR INRA-ENGREF, Equipe Croissance et Production, Nancy 54280 Champenoux, France.* [5]*Corresponding author**

Key words: anchorage, bending tests, French Alps, rockfall, tree stability, windthrow

Abstract

To determine which are the most important characters governing mechanical resistance to rockfall and wind loading, static winching tests were carried out on three tree species: Silver fir (*Abies alba* Mill.),European beech (*Fagus sylvatica* L.) and Norway spruce (*Picea abies* L.) in a mixed forest stand. Trees were winched to an angle of 0.25° at the stem base, both up- and downhill in order to compare how the same individual reacts when tested in two different directions. Trees were then winched to failure. Strain gauges were attached to the stem and one up- and downhill lateral root in order to determine the distribution of strain within the tree during overturning. Root morphology was then measured for all trees which uprooted during failure. No significant differences were found in the force necessary to winch trees up- and downhill in any species, either to an angle of 0.25° or to failure. Strain was significantly higher in lateral roots of Silver fir than in roots of Norway spruce and European beech when winched downhill. Downhill roots of Norway spruce were largely held in tension when trees were pulled downhill, whereas in Silver fir and European beech, they were held in compression. When trees were pulled uphill, no significant differences were found between species, and strain decreased along the lateral root of downhill roots only. European beech possessed a significantly greater number of roots than either Norway spruce or Silver fir. Norway spruce possessed a higher proportion of total root length near the soil surface, whereas European beech had the greatest proportion in the intermediate depth class and Silver fir had the highest maximal root depth. Norway spruce had a significantly lower proportion of oblique roots than the other two species, resulting in a plate-like root system which was less resistant to overturning than Silver fir or European beech.

Introduction

The use of protection forests against rockfall in mountainous regions has begun to be studied in detail over the last few years (Dorren and Berger, 2006; Dorren et al., 2004; Hurand and Berger, 2002; Stokes et al., 2005). Although rockfall results in an isolated impact on a tree, many similarities can be drawn between the tree response to rockfall and other abiotic stresses e.g. wind and snow loading. When trees are subjected to rockfall, they may uproot, break in the stem, or energy may be transferred to the crown, causing it to break (Dorren and Berger, 2006). Although research into the fundamental mechanisms resulting in these three types of failure is little or non-existent, a vast number of studies have been carried out on tree failure through wind loading (Coutts, 1983, 1986; Crook and Ennos, 1996; Cucchi et al., 2004; Peltola et al., 2000; Stokes, 1999), which will provide useful information in the study of tree resistance to rockfall.

* E-mail: stokes@liama.ia.ac.cn

The most common method to determine tree resistance to overturning, is to winch trees sideways and measure the force necessary to cause failure (Coutts, 1983; Cucchi et al., 2004; Peltola et al., 2000; Stokes, 1999). Trees may uproot if poorly anchored, or break in the stem if the moment required to resist overturning is greater than that necessary to break the trunk. Although an important component of root anchorage, root architecture has rarely been measured in studies of overturning resistance. Hypotheses concerning the role of root system shape and morphology have been made by several authors (Coutts, 1983; Cucchi et al., 2004; Mickovski and Ennos, 2002; Nicoll and Ray, 1996), who conclude that root depth, topology, biomass and number are all important factors to consider when examining tree anchorage.

The distribution of strain in root systems during tree winching studies has also been studied using strain gauges (Ennos, 1995; Stokes, 1999; Stokes et al., 2000). Strain gauges convert longitudinal deformations of a metal element into an electrical signal, thus indicating how a material is deformed under loading. Using such gauges, Stokes (1999) and Stokes et al. (2000) determined the mechanical behaviour of several forest species. It was found that in trees which broke in the trunk e.g. Maritime pine (*Pinus pinaster* Ait.), and European beech (*Fagus sylvatica* L.), strain was always found to be higher in the stem than in the roots during winching. However, in Douglas fir (*Pseudotsuga menziesii* Mirb.) and Norway spruce (*Picea abies* L.), which broke at the stem base or uprooted, strain was always highest at the root/stem joint and in the root system, respectively. Stokes (1999) also suggested that wood cells responded to local mechanical stress within the root, which was reflected in the strain values, e.g. leeward roots of wind stressed trees possessed significantly higher values of strain. Therefore, strain measurements in roots and trunks of trees during winching should provide further information about the behaviour of the tree when subjected to mechanical stresses.

A significant amount of information on the structural root architecture of forest trees has been collated in recent years (Coutts et al., 1999; Danjon 1999a,b; Kutschera and Lichtenegger, 2002; Nicoll and Ray, 1996), not only as a response to the need for such data by foresters concerned with the problems of wind storms, but also due to the development of user friendly methods for measuring root topology and geometry (Danjon et al., 1999a,b). Although the classification system used by Köstler et al. (1968), whereby root systems are classified into three shapes, shallow "plate" systems and deeper "heart" and "tap" systems, is still used (Dupuy et al., 2005a; Stokes and Mattheck, 1996), a more accurate description of root architecture is necessary to determine which parameter(s) govern root anchorage. Dupuy et al. (2005b) determined numerically that root topology and biomass were the most important variables influencing root resistance in tension, however, an experimental study on forest trees has not yet tested this hypothesis.

In order to better understand the overturning mechanism of trees when subjected to mechanical loading e.g. rockfall or wind stress, three forest species were winched both up- and downhill on an Alpine slope. Although rockfall can only occur as a downhill abiotic stress, wind loading can occur in any direction, and prevailing winds can often be uphill (Achim et al., 2003). Therefore, results from this study will also be useful to foresters concerned with problems of wind instability of plantations on sloping ground (Achim et al., 2003). Strain in stems and roots was measured during winching and root systems were excavated and root morphology measured in order to determine which parameter best governs root anchorage. Results are discussed with reference to a previous study, whereby trees on the same site were winched to failure downslope only (Stokes et al., in press).

Methods and materials

To determine if tree resistance to bending differed when loaded up- and downhill, static winching tests were carried out on 31 trees. Three species were compared: Silver fir (*Abies alba* Mill., $n = 12$, mean DBH $= 0.23 \pm 0.13$ m), Norway spruce (*Picea abies* L., $n = 10$, DBH $= 0.24 \pm 0.09$ m) and European beech (*Fagus sylvatica* L., $n = 9$, DBH $= 0.23 \pm 0.18$ m) growing at an altitude of 1350–1600 m in the Forêt Domaniale de Vaujany, Vallée de l'Eau d'Olle, Isère, France (means are $\pm$ standard error). Trees were growing on a north-west facing slope with a gradient of 38–42°. Rockfall and windstorms are frequent in the area. Trees were winched sideways (up- or downhill) and the force required to cause failure was measured using a load cell. For a full description of the site, as well as the winching tests carried out, see Stokes et al. (2005). Previous work described only the downhill tests (Stokes et al., in press), however, in this study, each tree was winched uphill as well as downhill. So as to not damage the trees in any way, when a tree was pulled e.g. uphill, it was winched only to a maximal stem basal deflection of 0.25° (Brudi and Wasenaer, 2002), and then released, before being pulled downhill. Half the trees were pulled in the uphill direction first, and then pulled to failure downhill. The remaining half were pulled downhill first, before being uprooted or broken when pulled uphill.

Two lateral roots per tree were excavated, one up- and one downhill (along the winching direction). In order to measure strain in these roots during mechanical loading, plastic backed strain gauges (Kyowa, Japan, KFG-10-120-C1-11, 10 mm gauge length, 120 Ω resistance) were used to estimate longitudinal strains (Stokes, 1999). Strain gauges convert longitudinal deformations of a metal element into an electrical signal and must be connected to a strain indicator (Kyowa, Japan, SD-10) in quarter bridge mode, via a switch and balance unit (Vishay Measurements Group, North Carolina, U.S.A., SB-10), in order for the electrical signal to be converted into micro-deformations (μstrain). Bark was removed with a chisel at DBH and every 0.2 m along the length of a root, starting from the stem-root joint. If it was not possible to attach a gauge every 0.2 m, due to the presence of, e.g. another root branch or a stone, the gauge was glued at the closest possible distance. In the data analysis therefore, classes of distance of 0–0.20 m, 0.21–0.40 m and 0.41–0.60 m along the lateral root were used. Care was taken not to damage the surface fibres of the wood during removal of the bark. Strain gauges were glued to the wood where the bark had been removed, using Loctite 401 multi-usage glue, which took approximately 15 min to dry.

The initial values of each strain gauge were recorded before winching commenced. The trees were then pulled sideways (up- or downhill) using increments of force of 400 N. Strain was measured in each of the gauges after each increment of force had been applied. The tree was winched in one direction up to a maximum deflection of 0.25° at the stem base, during which no plastic deformation or failure appeared to occur (Brudi and Wasenaer, 2002). The same procedure was then repeated in the other direction and the tree winched until failure occurred. The difference in the force necessary to winch the tree in both directions was then calculated for each tree. Therefore, we were able to compare tree resistance to deflection in both directions.

Measurements of root system morphology

Once the winching tests were complete, root systems were extracted by cutting the trunk at the base and winching the root system out of the soil. Only 19 root systems were analysed, as for trees that had failed in the stem, it was too difficult to extract the root system due to the steepness of the slope and its instability (falling rocks were frequent during extraction of the root systems). Root systems were then transported to the laboratory for architectural analysis. Unfortunately, a large number of roots were damaged or lost during the winching and extraction process, therefore most root systems were incomplete.

A topological and geometrical description of the root system was carried out using a low-magnetic field 3D digitiser (3SPACE Fastrak, Polhemus, Long ranger option, www.polhemus.com) driven by the software Diplami (Sinoquet and Rivet, 1997). This device is composed of an electronic unit, a transmitter and a receiver. For each digitised point, $[x, y, z]$ coordinates and diameter (measured manually) were assessed jointly with the topology i.e. how individual roots are connected to each other through branching (Danjon et al., 1999a,b; Tamasi et al., 2005).

Once a tree had been excavated, it was turned upside down and fixed in place for digitising. Root orientation was not taken into consideration as many roots had been broken during the winching and excavation process. Data were saved in files and exported to the software AMAPmod (Godin et al., 1997). In AMAPmod software, root systems are represented by "Multiscale Tree Graphs" (MTG) (Godin and Caraglio, 1998). A MTG is a topological structure in which root data are organised hierarchically in scales. This organisation allows each individual root to be considered as an axis and each axis as a sequence of root segments, a root segment being the part of the root included between two subsequent digitised points. The root length and volume are thus obtained as the sum of each root segment length and volume. A detailed description of the measurement and analysis techniques is given in Danjon et al. (1999 a,b).

Analyses of variance were carried out to determine if the number, size, angle and proportion of roots differed between species and between depth classes of 0.0 – 0.39 m, 0.40 – 0.79 m and >0.80 m. Where proportions were calculated, data were arcsine square root transformed prior to statistical analysis. According to Dupuy et al. (2005b), the best parameters to quantify root anchorage are not single parameters, e.g. root volume or root number alone, but a combination of two parameters. Therefore, we also calculated the following combinations for each tree: total root number × total basal cross-sectional area (CSA) of second order lateral roots (2°Ls), maximal root depth × total root volume, maximal root depth x root number and total root volume × total root number. Regressions were carried out between the total critical overturning moment ($TM_{crit,total}$; Stokes et al., in press) for each tree and certain root architectural parameters, including the combinations of parameters.

Results

In a previous study, it was found that European beech was twice as resistant to mechanical failure as fir and three times more resistant than spruce (Stokes et al., 2005). In this study, the mean force required to winch a tree uphill to a deflection of 0.25° at the stem base was 6231 ± 2133 N and 3340 ± 1075 N when the tree was winched downhill. No significant differences were found in the force necessary to winch trees in both directions for any species.

Strain was usually found to be highest in the region held in tension where the tree failed (Figures 1a,b), but this was only true in 80% of cases. Strain was found to be significantly higher in lateral roots close to the trunk in Silver fir and Norway spruce when winched downhill at a given force of 1500 N (Figure 2). Strain at the root base was similar to that in the trunk for Silver fir only (Figure 2). When trees were winched downhill, strain at the stem base of Silver fir was significantly greater than in either Norway spruce or European beech (Figure 2). In Norway spruce, mean strain at a distance of 0.21 – 0.60 m in the downhill root was found to be in tension whereas the base of the same root was held in compression (Figure 2). When trees were winched uphill, no significant differences between species were found and strain was greatest at the stem base in the downhill root only ($F_{5,108} = 2.57$, $P = 0.031$).

No significant differences were found between species with regards to mean or total volume, or mean or total CSA of 2°Ls, even when divided by the equivalent stem parameter. European beech had a significantly higher number of roots than Silver fir and Norway spruce (Figure 3). The proportion of total root length in each depth class was significantly greater at the most superficial depth for Norway spruce only (Figure 4). European beech had the greatest proportion in the intermediate depth class and fir had similar proportions in all three classes (Figure 5). Mean maximal root depth in Silver fir was twice that found in Norway spruce or European beech (Fig. 6). Norway spruce had a significantly lower proportion of oblique roots (0.25 ± 0.09) compared to European beech (0.55 ± 0.03) and Silver fir (0.52 ± 0.09) ($F_{2,14} = 3.73$, $P = 0.05$). No other significant differences were found with regards to root angle between species.

No significant regressions between $TM_{crit,total}$ and a root architectural variable, including the combinations of parameters, were found for any species when trees were winched to failure either uphill or downhill.

Discussion

In a previous study on the same trees, Stokes et al. (in press) showed that European beech was the most resistant species to uprooting, followed by Silver fir, then Norway spruce. Silver fir tended to break in the stem, whereas Norway spruce and European beech generally uprooted. In the present study, it was surprising that no differences were found in root anchorage when trees were pulled uphill, compared to when they were pulled downhill. According to Achim et al. (2003), Sitka spruce (*Picea sitchensis* Bong. Carr.) required a significantly greater resistance to uproot when winched uphill. The authors attributed this higher resistance to a possible increase in root development on the downslope side of the tree, although root architecture had not been measured. However, variability was high in our results, which was in part due to the heterogeneous and rocky nature of the soil. Although the number of trees tested was low, each tree was tested in both directions, therefore, any differences in anchorage when trees were pulled in different directions, should have been observed. We assume therefore that any root growth asymmetry along the slope direction was not enough to affect the anchorage moment of the trees tested. It was not always possible to predict where a tree would fail from the strain values, as suggested by Stokes et al. (2000), which may also be due to the heterogeneous nature of the rocky soil. However, strain was higher in the region where failure occurred in 80% of the trees tested, whether winched up- or downhill. Therefore, this method of predicting tree failure can still be considered useful, when combined with other techniques (Mattheck and Breloer, 1994).

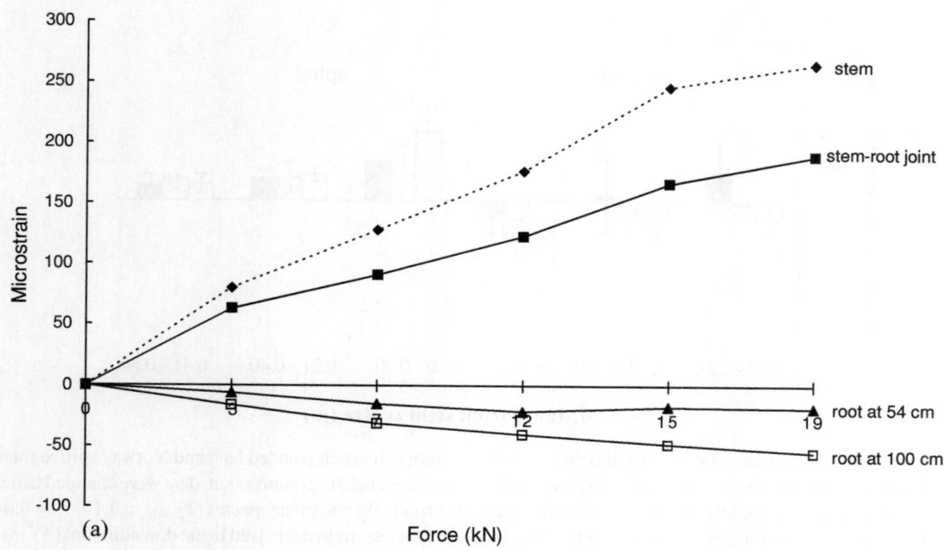

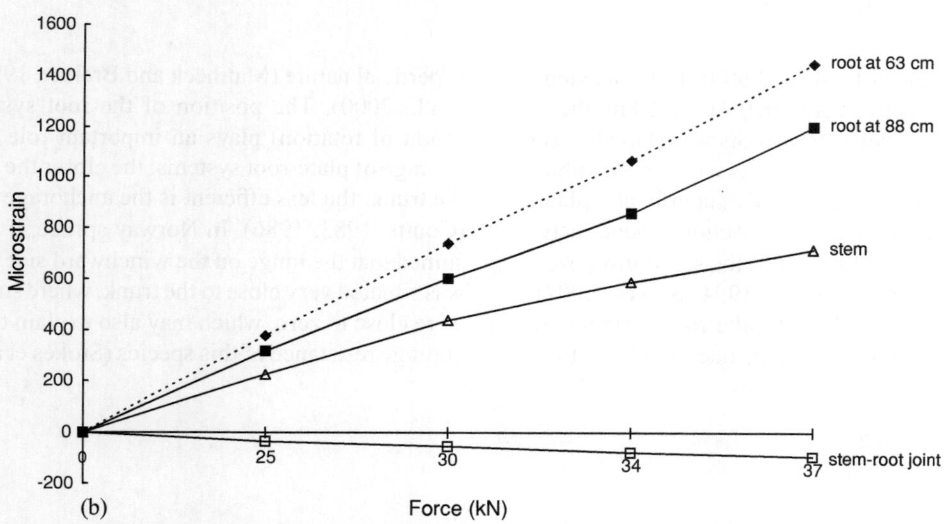

Figure 1. (a) Strain on the counter-winchward (tension) side of Silver fir was significantly greater in the trunk (where failure occurred) as the tree was winched downhill. (b) Failure occurred in the root system of European beech when winched downhill, where highest strain values were observed.

Unfortunately, many roots were broken and lost during the excavation process, therefore, root systems were not intact and a large amount of variability was thus seen in the results. Trees which had broken in the stem (Stokes et al., 2005) were not excavated due to the difficult and dangerous nature of the work, therefore reducing sample size. European beech, the most resistant species to uprooting (Stokes et al., 2005), possessed twice as many roots as Silver fir or Norway spruce. The proportion of roots with regard to length was also greatest at an intermediate depth in European beech, whereas Norway spruce had the highest proportion of superficial roots and lowest proportion of oblique roots. No relationships were found between any root architectural parameter and $TM_{crit,total}$ when all species were considered separately, which may have been due to insufficient data, or the fact that root anchorage is much more complex than the analysis used in this study.

Strain was highest at the stem base in Silver fir and European beech when pulled downhill. In Norway spruce, strain was highest at the stem base in the uphill root only, whereas in the downhill root, strain was

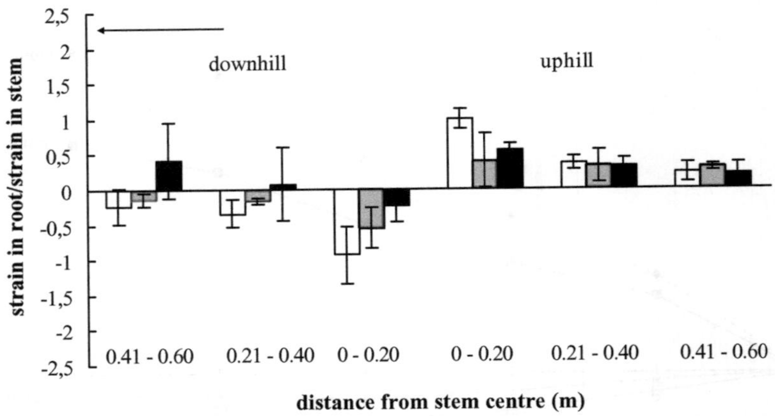

Figure 2. Strain along up- and downhill lateral roots of Silver fir (white bar), European beech (shaded bar) and Norway spruce (black bar) when trees were winched downhill (direction of arrow) at a force of 1500 N. Where roots were held in compression, data were changed to negative values for visual purposes only. Strain was significantly greater in Silver fir close to the trunk, than in other species ($F_{2,109} = 3.19$, $P = 0.045$). In Silver fir and Norway spruce, strain differed significantly along the uphill root and in all species, strain decreased in the downhill root ($F_{5,109} = 2.72$, $P = 0.023$). Data are means ± standard error.

highest at a distance of 0.41 – 0.60 m from the stem. Strain was also positive between 0.21 – 0.60 m, therefore indicating that downslope roots were largely held in tension when Norway spruce was winched downhill. Norway spruce possessed a highly superficial, "plate-like" root system, with very few oblique roots, therefore, the root plate is lifted out of the soil during overturning (Mattheck and Breloer, 1994; Stokes, 2002). It is considered that such plate-like root systems are the least resistant to overturning, due largely to their superficial nature (Mattheck and Breloer, 1994; Stokes et al., 2000). The position of the root system hinge (point of rotation) plays an important role in the anchorage of plate-root systems: the closer the hinge is to the trunk, the less efficient is the anchorage resistance (Coutts, 1983, 1986). In Norway spruce, it can be assumed that the hinge on the winchward side of the tree was situated very close to the trunk, where strain values were close to zero, which may also explain the low anchorage resistance of this species (Stokes et al., 2005).

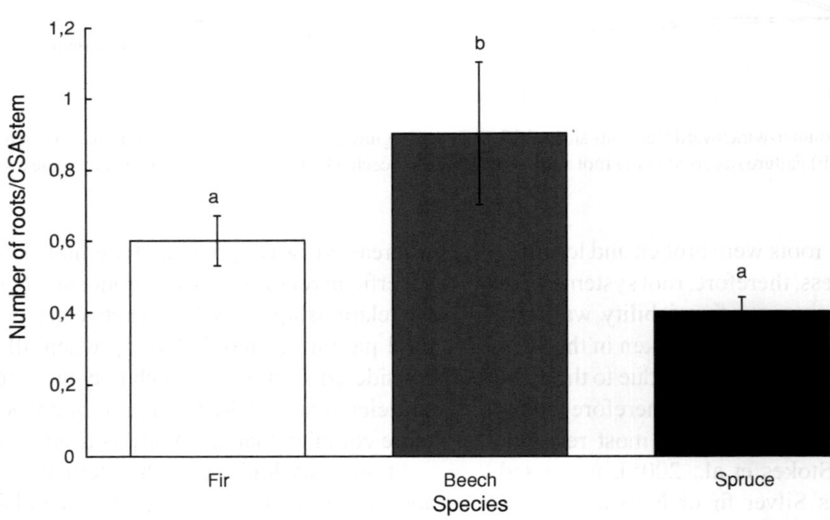

Figure 3. European beech (shaded bar) had significantly more roots/CSAstem than Silver fir (white bar) or Norway spruce (black bar) ($F_{2,16} = 9.38$, $P = 0.002$). Data are means ± standard error. Where superscripts differ, significance <0.05.

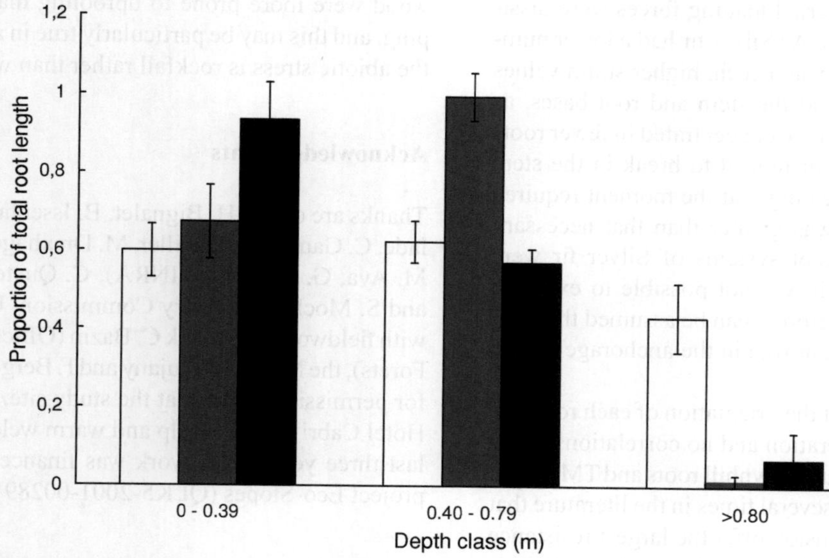

Figure 4. The proportion of total root length in each depth class was significantly greater at the most superficial depth for Norway spruce (black bar) only. European beech (shaded bar) had the greatest proportion in the intermediate depth class and Silver fir (white bar) had similar proportions in all three classes ($F_{2,42} = 23.94$, $P < 0.001$). Data were arcsine square root transformed prior to analysis and are means ± standard error.

Silver fir and European beech possessed deeper rooted systems with a higher proportion of oblique roots. When deeper rooted systems with a large number of branches overturn, e.g. European beech, the root-soil ball slides into the soil and is not lifted out of the soil as in Norway spruce (Mattheck and Breloer, 1994; Stokes, 2002). In Silver fir, however, a tap-rooted system (Kutschera and Lichtenegger, 2002), the long tap root will be pushed into the soil on the counter-winchward side of the tree (Crook and Ennos, 1997). When European beech and Silver fir were pulled downhill, the winchward roots were placed in compression as they were pushed into the soil. European beech roots were highly numerous and strain values were

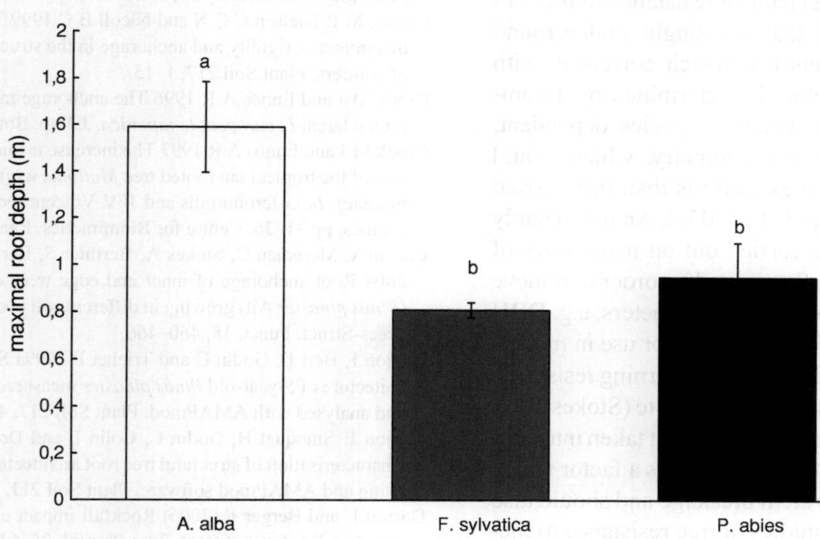

Figure 5. Mean maximal root depth was significantly greater in Silver fir (white bar) than in either European beech (shaded bar) or Norway spruce (black bar) ($F_{2,16} = 5.75$, $P = 0.013$). Data are means ± standard error. Where superscripts differ, significance <0.05.

generally lower, as external loading forces were dissipated through the roots. As Silver fir had a lower number of roots than European beech, higher strain values may be expected around the stem and root bases, as mechanical stresses will be concentrated in fewer roots (Ennos, 1995). Silver fir tended to break in the stem during winching, indicating that the moment required to resist overturning was greater than that necessary to break the trunk. Root systems of Silver fir were very deep, and often it was not possible to excavate the entire system, therefore it can be assumed that root depth plays an important role in the anchorage of this species.

It is unfortunate that the orientation of each root was not taken into consideration and no correlations could be made between up- and downhill roots and $TM_{crit,total}$. It has been suggested several times in the literature that lateral roots held in tension offer the largest resistance to overturning (Coutts, 1983, 1986; Crook and Ennos, 1996; Stokes, 2002), therefore these roots should have been examined more closely. If tap and sinker roots are present, these roots may also contribute significantly to root anchorage (Mickovski and Ennos, 2002). Root topology is also an important factor governing tree anchorage (Dupuy et al., 2005b; Stokes et al., 1995), and should be examined in intact root systems. Nevertheless, results from this study indicate that root architecture differs between forest species, which in turn may influence anchorage strength.

This study shows that different forest species do not possess the same mechanical resistance to rockfall or wind loading, and that no single underground parameter could be identified which correlates with $TM_{crit,total}$. Anchorage may be determined by a combination of parameters which is species dependent, e.g. biomass, topology and geometry, which would necessitate a more complex analysis than that carried out in this study (Dupuy et al., 2005a). An initial study of anchorage should be carried out on many trees of one species growing on flat ground, in order to remove the slope variable. Aboveground parameters, e.g. DBH or stem mass are therefore sufficient for use in models concerned with the prediction of overturning resistance for these species on this type of study site (Stokes et al., 2005). Wood strength of species was not taken into consideration in this study even though it is a factor which will strongly determine stem breakage and should also be examined in future studies of tree resistance to mechanical loading (Dorren and Berger, 2006). Putz et al. (1983) suggested that larger trees with dense, strong wood were more prone to uprooting than stem snapping, and this may be particularly true in a forest where the abiotic stress is rockfall rather than wind loading.

Acknowledgements

Thanks are due to H. Bignalet, B. Issenhuth, H. Lataillade, C. Gauvrit, S. Daviller, M. Drexhage, N. Kokutse, M. Aya, G. Cherokee (INRA), C. Quetel (Cemagref) and S. Mochan (Forestry Commission, U.K.) for help with fieldwork. We thank C. Bazin (Office National des Forêts), the Marie de Vaujany and F. Berger (Cemagref) for permission to work at the study site, as well as the Hotel Cabri for their help and warm welcome over the last three years. This work was financed by the E.U. project Eco-Slopes (QLK5-2001-00289).

References

Achim A, Nicoll B, Mochan S and Gardiner B 2003 Wind stability of trees on slopes. *In* B Ruck, C Kottmeier, C Mattheck, C Quine and G Wilhelm (eds), Wind Effects on Trees, pp. 231–237, University of Karlsruhe, Germany.

Brudi, E and Wassenaer P V 2001 Trees and statics: nondestructive failure analysis. *In* T Smiley (ed), Tree Structure and Mechanics Conference Proceedings, pp 53–69. International Society of Arboriculture.

Coutts M P 1983 Root architecture and tree stability. Plant Soil 71, 171–188.

Coutts M P 1986 Components of tree stability in Sitka spruce on peaty gley soil. Forestry 59, 173–197.

Coutts, M P, Nielsen C C N and Nicoll B C 1999 The development of symmetry, rigidity and anchorage in the structural root system of conifers. Plant Soil 217, 1–15.

Crook M J and Ennos A R 1996 The anchorage mechanics of deep rooted larch, *L. europea*, *L. japonica*. J. Exp. Bot. 47, 1509–1517.

Crook M J and Ennos A R 1997 The increase in anchorage with tree size of the tropical tap rooted tree *Mallotus wrayi*, King (Euphorbiaceae). *In* G Jeronimidis and JFV Vincent (eds), Plant Biomechanics, pp 31–36. Centre for Biomimetics, Reading UK.

Cucchi V, Meredieu C, Stokes A, Berthier S, Bert D and Najar M 2004 Root anchorage of inner and edge trees of Maritime pine (*Pinus pinaster* Ait) growing in different soil podzolic conditions. Trees-Struct. Funct. 18, 460–466.

Danjon F, Bert D, Godin C and Trichet P 1999a Structural root architecture of 5-year-old *Pinus pinaster* measured by 3D digitising and analysed with AMAPmod. Plant Soil 217, 49–63.

Danjon F, Sinoquet H, Godin C, Colin F and Drexhage M 1999b Characterisation of structural tree root architecture using 3D digitising and AMAPmod software. Plant Soil 211, 241–248.

Dorren L and Berger F (2006) Rockfall impact energy dissipation and stem breakage of trees. Tree Physiol. 26, 63–71.

Dorren L K A, Maier B, Putters U S and Seijmonsbergen A C 2004 Combining field and modelling techniques to assess rockfall dynamics on a protection forest hillslope in the European Alps. Geomorphology 57, 151–167.

Dupuy L, Fourcaud T and Stokes A 2005a A numerical investigation into the influence of soil type and root architecture on tree anchorage. Plant Soil. 278, 119–134, DOI: 10.1007/s11104-005-7577-2.

Dupuy L, Fourcaud T and Stokes A 2005b A numerical investigation into factors affecting the anchorage of roots in tension. Eur. J. Soil Sci. 56, 319–327.

Ennos A R 1995 Development of buttresses in rain forest trees: the influence of mechanical stress. *In* MP Coutts and J Grace (eds), Wind and Trees, pp 293–301. Cambridge University Press, Cambridge, UK.

Godin C and Caraglio Y 1998 A multiscale model of plant topological structures. J. Theor. Biol. 191, 1–46.

Godin C, Costes E and Caraglio Y 1997 Exploring plant topological structure with the AMAPmod software: An outline. Silva Fenn. 31, 355–366.

Hurand A and Berger F 2002 Forêts et risques naturels. Protection contre l'érosion, les mouvements de terrain et les avalanches. Houille Blanche 3, 64–67.

Köstler J N, Brückner E and Bibelriether H 1968 Die Wurzeln der Waldbäume. Verlag Paul Parey, Hamburg & Berlin, Germany.

Kutschera L and Lichtenegger E 2002 Wurzelatlas mitteleuropäischer Waldbäume und Sträucher. Leopold Stocker Varlag, Graz.

Mattheck C and Breloer H 1994 The Body Language of Trees. HMSO London.

Mickovski S B and Ennos A R 2002 A morphological and mechanical study of the root systems of suppressed crown Scots pine *Pinus sylvestris*. Trees-Struct. Func. 16, 274–280.

Nicoll B and Ray D 1996 Adaptive growth of tree root systems in response to wind action and site conditions. Tree Physiol. 16, 891–898.

Peltola H, Kellomaki S, Hassinen, A and Granander M 2000 Mechanical stability of Scots pine, Norway spruce and birch, an analysis of tree-pulling experiments in Finland. Forest Ecol. Manag. 135, 143–153.

Putz F E, Coley P D, Lu K, Montalvo A and Aiello A 1983 Uprooting and snapping of trees: structural determinants and ecological consequences. Can. J. Forest Res. 13, 1011–1120.

Sinoquet H and Rivet P 1997 Measurement and visualization of the architecture of an adult tree based on a three dimensional digitising device. Trees-Struct. Func. 11, 265–270.

Stokes A 1999 Strain distribution during anchorage failure of *Pinus pinaster* Ait. at different ages and tree growth response to wind-induced root movement. Plant Soil 217, 17–27.

Stokes A 2002 Biomechanics of tree root anchorage. *In* Y Waisel, A Eshel and U Kafkafi (eds), Plant Roots: The Hidden Half, pp. 175–186. Marcel Dekker, New York.

Stokes A, Fitter A H and Coutts M P 1995 Responses of young trees to wind: effects on root architecture and anchorage strength. J. Exp. Bot. 46, 1139–1146.

Stokes A, Drexhage M and Guitard D G 2000 A method for predicting the site of failure in trees under mechanical loading. *In* A Stokes (ed), Developments in Plant and Soil Sciences, Vol. 87, The Structural Roots of Trees and Woody Plants: Form, Function and Physiology, pp. 279–285, Kluwer, Dordrecht.

Stokes A, Salin F, Kokutse A D, Berthier S, Jeannin H, Mochan S, Kokutse N, Dorren L, Abd.Ghani, M and Fourcaud T (2005) Mechanical resistance of different tree species to rockfall in the French Alps. Plant Soil. 278, 107–117, DOI: 10.1007/s11104-005-3899-3.

Tamasi E, Stokes A, Lasserre B, Danjon F, Berthier S, Fourcaud T and Chiatante D 2005 Influence of wind stress on root system development and architecture in oak seedlings (*Quercus robur* L.). Trees-Struct. Funct. 19, 374–384, Available online DOI: 10.1007/s00468-004-0396x.

A numerical investigation into the influence of soil type and root architecture on tree anchorage

Lionel Dupuy[1,2,4,5], Thierry Fourcaud[1,3] & Alexia Stokes[1]

[1] *Laboratoire de Rhéologie du Bois de Bordeaux, CNRS INRA Université Bordeaux I, Domaine de l'Hermitage, 69 route d'Arcachon, 33612 Cestas Cedex, France.* [2] *Institut pour le Développement Forestier, 23 avenue Bosquet, 75007 Paris, France.* [3] *AMAP CIRAD Département—AMIS, TA 40/PS2 Cedex 5, 34398 Montpellier, France.* [4] *Departement of Plant Sciences, University of Cambridge, Downing Street, CB2 3E4 Cambridge, U.K.* [5] *Corresponding author**

Key words: biomechanics, finite element method, modelling, morphology, SIMUL3R, uprooting

Abstract

The influence of root morphology and soil type on the mechanical behaviour of tree anchorage was investigated through numerical modelling. We developed a simple computer program to construct three-dimensional virtual root architectural patterns. This tool was used to build four schematic patterns: heart-, tap-, herringbone- and plate-like root systems. Each of these rooting types was characterised by specific branching characteristics. However, the total volume (proportional to the wood biomass) and material properties were kept constant. The finite element method was used to calculate the mechanical response of root/soil systems when the stem was subjected to bending forces. The overturning resistance of the four schematic root patterns was determined in four different idealistic soil types. These soils were based on Mohr–Coulomb plasticity models. Results showed that soil internal friction modified the position of the rotation axis during tilting of the root/soil plate. Rooting depth was a determinant parameter in sandy-like soils. Overturning resistance was greatest in heart- and tap-root systems whatever the soil type. However, the heart root system was more resistant on clay-like soil whereas the tap root system was more resistant on sandy-like soil. Herringbone and plate root systems were twice as less resistant on clay soils and 1.5 times less resistant on sandy soils when compared to heart and tap-like structures.

Introduction

The recent wind storms experienced throughout Europe have caused much damage and huge economic losses in both plantation and natural forests (Bergonzini and Laroussinie, 2000; Bouchon, 1987). If predictions that our climate is becoming windier prove to be correct, such disasters are likely to become more frequent in the future. When trees are subjected to strong winds, vari ous types of damage are observed, however, failure usually occurs through uprooting (Cucchi and Bert, 2003; Nicoll et al., 1995).

Traditionally, vulnerability to uprooting has been investigated at the population level, by correlating forest damage to certain stand characteristics i.e. tree species, sylviculture treatment and practice, soil type and wind speed (Putz et al., 1983; Ruel 2000). However, such an empirical analysis of windthrow is incomplete due to the numerous variables influencing tree stability, which are not often taken into account, e.g. root system or soil characteristics and soil hydrology. Experimental studies can provide a certain amount of detailed data concerning tree mechanical stability. Typical tests include winching trees sideways until failure and measuring the force necessary to cause failure (Cucchi et al., 2004; Peltola et al., 2000; Stokes, 1999). However, even if root system architecture is measured, data are often incomplete as roots are usually damaged during the tests (Stokes et al., 2007).

In order to improve tree stability in a plantation, an understanding of the physical processes occurring in the soil is also needed. The root/soil plate of trees is a compound structure, therefore both the physical properties of soils and structural properties of the tree, e.g.

* E-mail: lxd20@cam.ac.uk

This article has been previously published in the following journal—Plant and Soil, Volume 278, 119–134.
Stokes et al. (eds), Eco- and Ground Bio-Engineering: The Use of Vegetation to Improve Slope Stability, 175–189.

root architecture are important in determining the overall anchorage capacity of a tree. The type of failure during a storm depends not only on soil hydrology, but also on soil type. Research has focused on the investigation of fundamental mechanisms involved during uprooting by using mechanical models (Blackwell et al., 1990; Dupuy et al., 2005a,b; England et al., 2000; Ennos, 1990; Mattheck and Breloer, 1994). However, no realistic model yet exists which considers both the soil and the complete root system architecture as distinct elements.

The difficulty of investigating root anchorage is not only due to the complexity of the mechanisms occurring in both roots and soil, but also to their multifactorial aspect (Stokes et al., 2007). A good alternative to difficult and time-consum- ing field experiments is numerical modelling which will also help with the interpretation of experimental data or aid in identifying the parameters needing greatest study. Dupuy et al. (2003) carried out such analyses using the finite element method (FEM), a powerful numerical method used in engineering. These authors developed methods allowing morphological data from real or simulated root systems to be subjected to virtual uprooting tests. Soil mechanical properties could also be changed easily, thereby allowing a rapid assessment of root anchorage in different soils. In a similar numerical investigation, Dupuy et al. (2005a) performed sensitivity analyses on the geometrical, topological and constitutive parameters which affect uprooting. These authors showed that the most important factors governing individual root resistance to pull-out in a given soil, were a combination of root branching parameters and morphology. However, this study was carried out using two- dimensional (2D) models at a very local level, i.e. on small root components approximately 0.1-m long with few branches.

Investigating the anchorage of adult trees is rather more complicated due to the morphological complexity encountered in large tree root systems (Danjon et al., 2005). Several structural elements of importance for mechanical stability can coexist. Coutts (1983, 1986) identified the main components that play a role in the root anchorage of shallowly rooted Sitka spruce (*Picea sitchensis* Bong. Carr). Coutts (1983, 1986) showed that the windward lateral roots held in tension were the most important component of root anchorage, followed by the weight of the root/soil plate, soil cohesion and the bending strength of leeward roots. However, no tap roots were present in these trees. Tap roots are the large, central roots found in root systems of certain species e.g. *Pinus pinaster* Ait or in young trees (Danjon et al., 2005; Hintikka, 1972; Köstler et al., 1968). In winching experiments on different tree species, Stokes et al. (2000, 2005, 2007) found that the most resistant species to overturning were those possessing a 'heart' root system, i.e. with many branched, oblique, horizontal and lateral roots (Köstler et al., 1968). Tap-rooted systems were also tested and were more resistant to overturning when compared with shallowly rooted plate type systems e.g. Norway spruce (*Picea abies* L.). However, root architecture was not accurately measured in these species, therefore, the exact role of either morphology or topology, i.e. the way branches are linked together, is not known.

One further parameter which is rarely measured and understood even less is the relationship between root architecture and potential anchorage, particularly with regards to different soil types. As experimental studies are difficult to carry out and interpret, modelling root architecture and numerical analyses may help to fill this gap in the knowledge (Fourcaud et al., 2003a). It is currently not known whether root architecture or soil type influences uprooting most, and it is likely that each parameter is dependent on the other.

The aim of our study was to determine the anchorage efficiency of four typical root system architectures in theoretical soils representing a wide range of mechanical characteristics. For this purpose, a program was developed to construct simple three-dimensional architectural patterns. The four different root system types were then generated using real data (Dupuy, 2003) as input to the program, considering a small number of parameters. More realistic structures could also be implemented using existing models (Jourdan and Rey, 1997; Pages et al., 2004). Nevertheless, the complexity of these models would not enable us to easily impose given morphological characteristics of root components. Three of the patterns chosen were typical of those found in different forest tree species, and the fourth type, 'herringbone' (Fitter, 1987; Stokes et al., 1996) was an extreme example used to test the model, but which occurs in young plants and mature trees grown from cuttings, e.g. poplar (*Populus* sp.) (Dupuy, 2003). FEM was then used to simulate the mechanism of uprooting for these different root types in idealistic sandy- and clay-like soils of different mechanical properties. The results should enable foresters and tree breeders determine the most important characters of tree stability with regards to root architecture and will also indicate how to manipulate these characteristics depending on soil conditions.

Materials and methods

Generation of schematic root systems

The variability of morphological patterns within root systems, e.g. spatial distribution of branching, branching angle, root length and diameter, needs to be considered carefully when analysing the effect of root architecture on the uprooting process. The use of models is highly beneficial, as they enable deterministic structures to be created, for which the morphological parameters can be controlled. Therefore, we developed a program called SIMUL3R which generates typical root structures using data from real root systems as input. It must be underlined that this program does not simulate dynamic root growth processes, in the sense that chronological events are not taken into consideration, but it can be used to build root systems at a given stage of their development.

Description of the SIMUL3R program

The SIMUL3R program is a stepwise builder that reproduces—in a certain way—the elementary processes of primary elongation and ramification. However, the program does not consider either root mortality or the growth of adventitious roots. The root structure was described as a set of single roots connected together at particular insertion points. The growth of each individual root resulted from the activity at its apex. Root axes were split into unit segments of length Δl and diameter d. The points defined by the segment extremities were called nodes. The distance between two successive nodes bearing lateral ramifications was denoted ΔL (Figure 1).

The program which generates root architecture was based on three main modules (Figure 2): *INITIALISATION*, *EXTENSION* and *VOLUME*.

The *INITIALISATION* stage consisted of defining a preliminary pool of N_0 apexes, i.e. potential roots, which would develop from the stump positioned at the origin (0,0,0). This set of initial apices contained a fixed number of laterals and an optional tap root.

The second module *EXTENSION*, was based on a stepwise procedure where the activity of each apex was computed independently from the others. At each step,

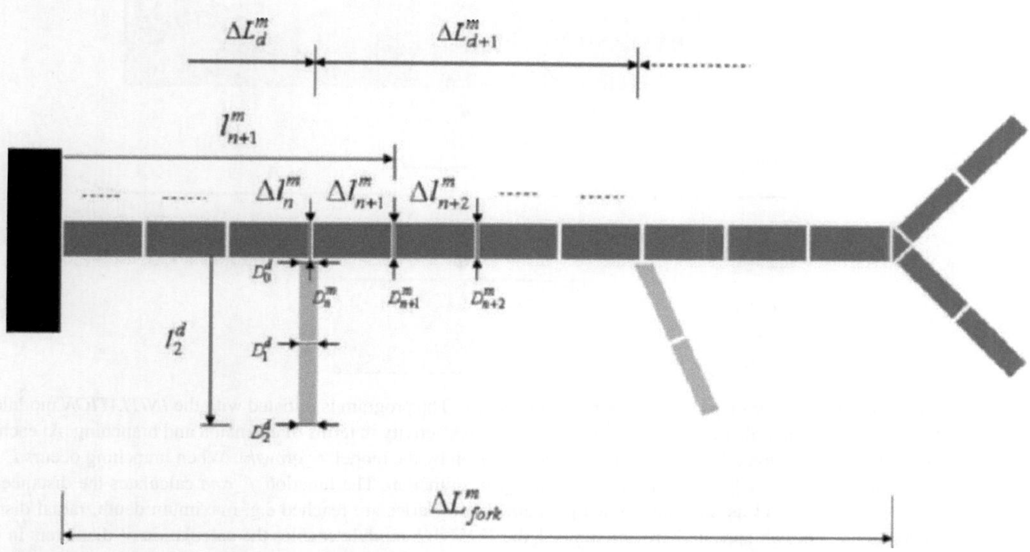

Figure 1. Models of root architecture were used to build typical three-dimensional root structures. Simulated root systems, were described as a set of connected unit segments of length Δl. The nth segment (created at step n of the simulation) of the mth root Δl_n^m is characterised by initial diameter Δ_n^m and final diameter Δ_{n+1}^m. The length of root at step n is written l_n^m. The mth root may initiate lateral branches, i.e. dth root on the figure. The distance between two consecutive branches on the nth root is denoted ΔL_l^m. The mth root may also end with a fork. ΔL_{fork}^m denotes the distance of the fork from the insertion point.

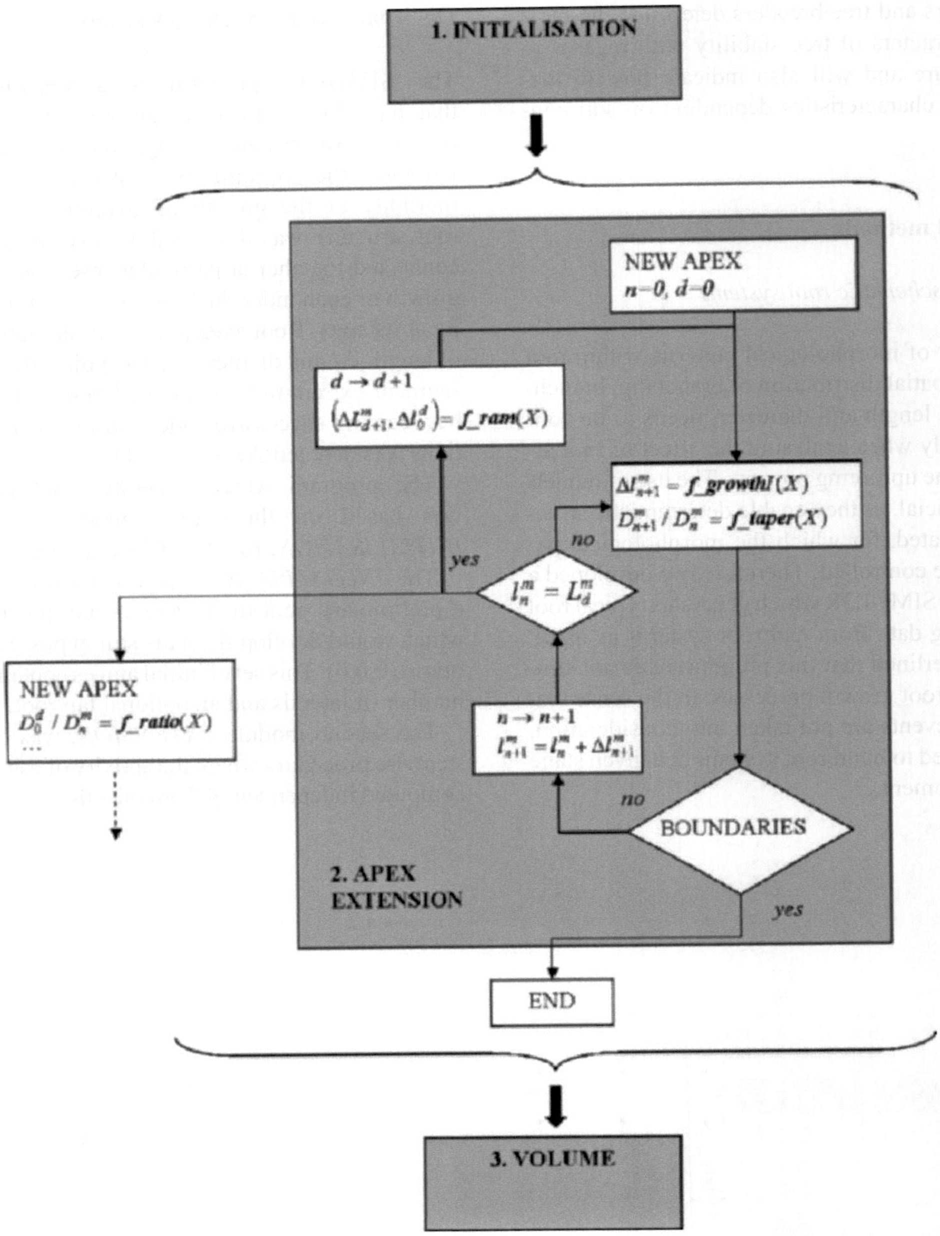

Figure 2. Flowchart of SIMUL3R program for the simulation of root architecture. The program is initiated with the *INITIATION* module which defines a set of apices. The *EXTENSION* module is then used to generate root apex activity in terms of extension and branching. At each step of the procedure, each apex operates an extension of length Δl in the direction given by the model *f_growth*l. When branching occurs $l_n^m = L_I^m$, a new apex is created and its initial diameter D_0^d is calculated from the *f_ratio* function. The function *f_ram* calculates the distance ΔL_{i+1}^m at which the next branch will be inserted and its direction. When predefined boundaries are reached e.g. maximum depth, radial distance or branching order, apex activity stops. After all apex activity has stopped, the *VOLUME* module enables the calculation of diameters in the root system in order to get a fixed total volume of root.

the apex activity corresponded to the creation of a new root segment and any new lateral axes, i.e. new apices. The diameter of each segment was defined using the taper properties of the mother root. These root characteristics were determined at each step using morphological functions: *f_growth*I*(X), f_ram(X), f_ratio(X)* and *f_taper(X)*. The input parameters of these functions could be related to any variable *X*, e.g. time, branching order or spatial position. These functions can be either deterministic or stochastic and could be fitted statistically onto real root systems (Dupuy, 2003), or they could be purely theoretical.

At each calculation step *n*, the current apex created a new root segment. Running of the primary growth function *f_growthI(X)* allowed the root increment Δl_n^m to be determined in length and direction. This apex could also initiate new branches during its extension according to the ramification function *f_ram(X)* that calculated the distance ΔL_d^m between two successive branching nodes, as well as the direction of the new daughter lateral segment ΔL_0^d. Distal roots could also be forked at their extremities. An appearance of such a fork was defined by its distance ΔL_m^{fork} to the root insertion node. The procedure that generated forks was similar to those building lateral branches, except that it caused the death of the bearing root apex. Root diameters were not calculated in the *EXTENSION* procedure. Nevertheless, the *f_ratio(X)* function was used to determine the theoretical coefficient giving the ratio between the initial diameter D_0^d of a root *d* and the diameter D_n^m of the mother bearing segment *m* (Figure 1). Similarly, the function *f_taper(X)* furnished the taper coefficient linking the diameter of successive segments along the root. These ratio coefficients were used by the *VOLUME* module in order to calculate the diameters of all segments at the final stage of the simulation. The building of a single root was stopped when predefined spatial boundaries were reached. The *EXTENSION* module ended when the set of apices to be treated was empty.

At this stage, only the root system skeleton was available. The *VOLUME* module was finally called in order to calculate root diameters when a predefined volume of the whole root system was expected. The diameter of all the root segments was linked together according to the ratio coefficients previously determined during the *EXTENSION* procedure. For that reason, the only variable to be calculated to fit the whole-root volume was the bole diameter D_0^0. This initial diameter was determined using a dichotomy algorithm (a heuristic method of successive calculations) so as to obtain the expected root system volume. Finally, the root structure was recorded under a specific file format allowing three-dimensional visualisation, architectural analysis and mechanical calculations to be performed.

Computation of schematic root patterns

Various types of root structures can be constructed using the SIMUL3R program, by varying the morphological functions *f_growth*I*(X)*, *f_ram(X)*, *f_ratio(X)* and *f_taper(X)*. In our study, simplified structures were used to represent different types of rooting patterns. Therefore, by using such simple representations, we can determine the contribution of certain root morphological characteristics to anchorage efficiency (Ennos, 1990; Niklas et al., 2002). Table 1 shows the characteristics of the morphological functions that were used to generate the schematic heart-, tap-, herringbone- and plate-like root systems. These data were taken from measurements of real root systems of adult trees of different species described elsewhere (Dupuy, 2003). The heart root system was typical of that found in e.g. Red oak (*Quercus rubra* L., Lyford, 1980) with dense branching and a large quantity of forks present; the tap root system was generated using data from Maritime pine (*Pinus pinaster* Ait., Danjon et al., 1999); the herringbone system can be found in mature poplar (*Populus* sp.) trees originating from cuttings (Dupuy, 2003) and the plate-like system was typical of that found in e.g. Norway spruce (*Picea abies* L., Puhe, 2003; Stokes et al., 2007). The final expected volume (proportional to the invested biomass) of each generated structure was fixed to 0.3 m^3. The radius of the root system was fixed to 3 m around the tree, and was assumed to be large enough to consider any root contributing mechanically to tree stability, i.e. roots situated at more than 3 m from the tree are assumed to have minor impact on uprooting.

In the *INITIALISATION* module, the number of main apices initiated at the centre of the bole was varied. Heart-like root systems possessed a maximum of seven main lateral roots whereas plate- and tap-root systems had only five main lateral roots. The herringbone-like structure did not have any main structural lateral roots attached to the bole but showed uniformly distributed finer lateral roots.

Branching characteristics were varied using different *model_ram()* functions. Branching density was

Table 1. Four different types of root systems were simulated using the SIMUL3R program: heart-, tap- herringbone- and platelike root systems

	Types of root system			
	Heart	Tap	Herringbone	Plate
Initiation				
Number of main lateral roots	7	5	0	5
Tap root	Yes	Yes	Yes	No
f_growthl()				
Trajectory	Direction is given at insertion			
f_ram()				
Distance between branching ΔL_i (m)	0.2	0.3	0.3	0.3
Horizontal/vertical	Uniform	40/60	60/40	40/60
Distance of the fork from origin ΔL_{fork} (m)	1.0	–	–	–
Branching on tap root	No	No	Yes	No
f_taper				
Taper equation	$(d_N^m - d_{n+1}^m)/dl = 12 \times (l_n^m)^{-0.6}$			
Taper of the tap root	Yes	Yes	No	–
f_ratio				
Mother/daughter relationship	$d_0^d = 0.45 \times d_n^m$			
Boundaries				
Rooting depth (m)	1.0	1.5	1.0	0.5
Radial distance (m)	3.0	3.0	3.0	3.0
Branching order	<4	<4	<4	<4

Morphological differences were generated by changing the nature of the morphological function. The number of the main apices initiated at the center of the bole was varied in the *INITIALISATION* procedure. Root systems could possess different branching patterns by altering the distances between branches, the direction of roots at the insertion point, or the forking of roots by using the *f_ram* function. Root taper could also have different values in *f_taper*, which emphasised the differences between rooting types. Finally, various types of boundary conditions highlighted differences between the four root morphologies. Data were taken from measurements carried out on root systems of different tree species explained in Dupuy (2003).

higher for the heart root system as the distance between branches was set to 0.2 m, whereas for all the other root systems this distance was set to 0.3 m. Similarly, the growth direction of roots at the insertion point was different in the tap- and plate-root systems, where 40% of roots were horizontal and 60% vertical. In herringbone patterns, 60% of roots were oriented horizontally and 40% vertically, whereas heart root systems possessed a continuous distribution of roots between the two directions. Forking was also present in heart systems, whereas lateral branching was present all along the tap root in the herringbone root systems.

The taper equation in *f_taper* was similar for every root system and was derived from real data from Maritime pine root systems (Danjon et al., 1999; Fourcaud et al., 2003a,b). The only exception concerned the main tap root of the herringbone system, which had no taper, typical of that found in adult poplar trees which have originated from cuttings (Dupuy, 2003). The relationship found between the *f_ratio* of the mother d_n^m and dm daughter d_0^d root diameter at the insertion point was linear and also derived from Maritime pine root system data.

Finally, the set of various types of boundary conditions enabled the differences between the four root morphologies to be appreciated. The tap root system had a maximum rooting depth of 1.5 m, heart and herringbone root systems had a rooting depth of 1.0 m whereas the platelike system was only 0.5-m deep.

Mechanical modelling

Numerical simulations of uprooting were carried out using the finite element software ABAQUS (Hibbitt, Karlsson & Sorensen, Inc., http://www.abaqus.com) and by following the methodology determined by Fourcaud et al. (2003b). The model was composed of two parts; the first being the root system and the second the surrounding soil.

Definition of the domain and meshing for FEM calculation

The root systems were described using a Multi-Tree Graph (MTG) format according to the topological rules given by Godin and Caraglio (1998). These root systems consisted of a set of slender axes connected together at specific branching points. Each axis was split into segments for which basal diameters, as well as spatial coordinates of their extremities were given. The finite element discretisation with three-dimensional linear beam elements (Fourcaud and Lac, 2003) was performed automatically from the MTG file (Dupuy, 2003). The maximum size of beam elements was defined initially by the user.

The root systems were placed in a parallelepipedic block of soil, 7.0 × 7.0 m wide and 2.5-m deep. The surrounding soil domain was composed of one free top surface, four lateral faces for which horizontal displacements where imposed as zero and one bottom face that was blocked in all directions. The whole soil domain was discretized using 20-node quadratic brick elements.

Physical links between the two parts were taken into account considering stiff connections between the beam nodes and the closest brick nodes (Dupuy, 2003), i.e. no slipping or opening was allowed at the interface of the two domains.

Definition of materials

As this study was purely theoretical and parametric, the soil mechanical behaviour chosen was that of a simple, elastic, perfectly plastic model, i.e. without considering hardening. The yield criterion defining the limit of soil failure was given by the classical Mohr–Coulomb model $\tau = c - \tau \tan(\Phi)$, assuming that failure was controlled by the maximum shear stress τ and that this failure shear stress depended on the normal stress σ which is conventionally negative in compression. In this equation, c was the soil cohesion and Φ was the material angle of friction. The Mohr–Coulomb plasticity model provided in ABAQUS (see ABAQUS User's manual for details) was written in terms of three stress invariants in order to apply it to a general state of stress. Mohr–Coulomb yield criterion was classically defined by a straight line in the meridional plane and an irregular hexagonal section in the deviatoric plane. Nevertheless, the non associated flow potential was not a classical hexagonal pyramid, but was defined by a hyperbola in the meridional plane and a smooth deviatoric section (Mené trey and Willam, 1995). The elastic part of the soil constitutive law was assumed to be linear isotropic and governed by the Young's modulus Esoil and Poisson's ratio υ_{soil}.

Table 2. Four different types of idealistic soils were used for the analysis of anchorage efficiency: ideal frictional soil, i.e. dry and saturated sandy sand and ideal cohesive soil i.e. and undrained clay-like soil behaviour

	Types of soil			
	Soft clay	Hard clay	Saturated sand	Dry sand
Young's Modulus (MPa)	20.0	20.0	20.0	20.0
Poisson's ratio	0.49	0.49	0.30	0.30
Cohesion (kPa)	20.0	50.0	2.0	2.0
Friction angle (°)	0.0	0.0	30.0	30.0
Volumetric weight (kN m^3)	20.0	20.0	10.0	20.0

The properties of each of these soil types were modelled using Mohr–Coulombís elasto-plastic model.

The anchorage of tree root systems was analysed in four different types of idealistic soils varying the two parameters c and Φ of the Mohr–Coulomb model (Table 2). These values represent a wide range of soil mechanical behaviours. Ideal cohesive soil, i.e. undrained clay soil, was represented with a null friction angle, a high cohesion value and a Poisson's ratio of 0.49 (Yang and Jeremic, 2002). Cohesion varied from 20.0 to 50.0 kPa and represent variations in clay shear resistance due to different moisture content or geomorphological origins. Young's Modulus (20.0 MPa) remained constant for all soil types. Ideal frictional soils, i.e. drained sandy-like soils, were represented using a friction angle of 30°, a low cohesion value of 2.0 kPa and a Poissonís ratio of 0.3. The distinction between dry and saturated sands, according to effective stress analysis was studied by using a unit weight of 20.0 and 10.0 kN m^{-3} for dry and wet soil, respectively (Whitlow, 1995).

Root wood was considered as an elastic, brittle, perfectly plastic material considering the Mises yield surface with associated flow rule. Deformation prior to yielding was assumed to be linear elastic governed by the Youngís Modulus $E_{roots} = 2.0$ GPa and Poisson's ratio $\upsilon_{roots} = 0.3$. The yield stress σ_{max} was taken to be 15.0 MPa.

Mechanical analysis

An infinitely rigid vertical lever arm was added to the top of the root system bole. This element did not aim to model the tree stem, but was only used in order to impose a rotation of the whole system. An initial calculation step was performed before simulating tree uprooting. This initial step allowed the initial geo-static stresses due to the self-weight of the whole system to be estimated. This phase is particularly important as the initial soil pressure has an impact on the soil shear strength according to the above Mohr–Coulomb criteria. The soil pressure increases linearly with depth; the deeper soil layers being more resistant to shear.

During the following step of analysis, a lateral displacement of 1.5 m was applied incrementally to a point of the upper rigid lever arm situated arbitrarily at a height of 10.0 m in order to apply a rotation to the root structure. The FEM analysis consisted of determining the mechanical response of the whole soil/root system, namely the reaction force calculated at the node where the displacement was imposed, the displacement of the discretisation nodes, the elastic and plastic strains, as well as the stress field in both roots and soil.

Criteria for anchorage analysis

During the stem movement, the increments of reaction forces were recorded at the node where the displacement was imposed. The mechanical response of anchorage was then determined by drawing the force/displacement curve. The maximum force at the end of the lateral displacement was used to compare quantitatively the soil/root anchorage efficiency. Various variables such as displacement or plastic strain fields were monitored in the vertical, symmetrical plane defined by the direction of pulling and which seemed a logical choice for this study. This information was discussed in order to give a qualitative explanation of the uprooting mechanism.

Results

Characteristics of computed root systems

The heart-like root system was typical of that found in Red oak (Lyford, 1980). This structure was densely branched and the secondary lateral roots were oriented randomly between the horizontal and vertical directions. Seven main laterals were axi-symmetrically distributed around the vertical stem axis. The basal diameter of these lateral roots was 0.23 m. The distance between lateral branches *DL* was set to 0.20 m. Forks were generated at a distance of 1.0 m from the stem. The sum of the length of all roots (Σ length) was 160.0 m. The maximum rooting depth was 1.0 m and the root system horizontal diameter was 3.0 m.

The tap-like root system (Figure 3b) was typical of several coniferous species, e.g. Maritime pine (Danjon

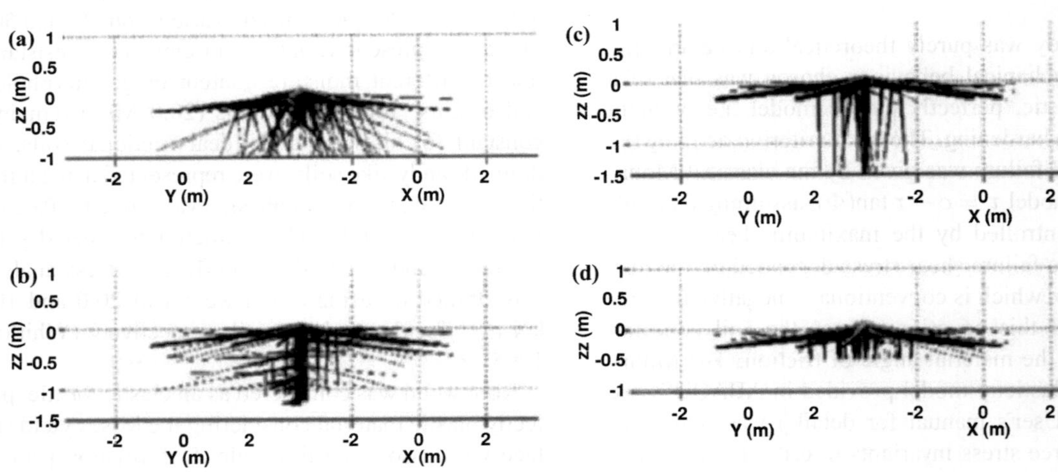

Figure 3. Four types of root systems denoting highly opposed rooting strategies were selected for the simulations: (a) the heart root system was 1.0-m deep and had forks; (b) the tap-root system was 1.5-m deep and constituted a thick first order vertical root and five main lateral roots; (c) the herringbone root system had a main vertical root of constant diameter, uniformly covered by second order lateral roots; (d) the plate like root system had no tap root and was 0.5-m deep.

et al., 1999) and Scots pine (*Pinus sylvestris* L., Köstler, 1968). The main roots were oriented either vertically or horizontally and did not develop forks. Five main laterals were connected to the stump with an axisymmetric distribution. Their basal diameter was 0.20 m. The distance between two lateral branches on a mother root DL was set to 0.30 m (Coutts, 1987). The tap root basal diameter was 0.20 m. The total root length was 1.13 m. The depth of the root system reached a maximum of 1.5 m which corresponded to the length of the tap root. The radius of the whole structure was 3.0 m.

The herringbone-like root system (Figure 3c) possessed layers of horizontal second order lateral roots that were uniformly distributed along the main tap root. The distance between the whorls ΔL was 0.30 m. Each stage was composed of five laterals that were separated from each other by a constant angle. The total length was 170.0 m and the diameter of the main tap root was 0.17 m. The total depth of the root system was equal to 1.0 m and the total horizontal diameter was 3.0 m.

The plate-like root system (Figure 3d) was representative of that found in Sitka spruce (Köstler et al., 1968) or Norway spruce (Puhe, 2003; Stokes et al., 2007). This system did not have a tap root, but five main laterals only, which were attached to the stump. The total length of the root system was 80.0 m. The second order laterals had a basal diameter of 0.27 m and bore sinkers, separated by a constant distance ΔL equal to 0.30 m. The total depth of the root system was limited to 0.50 m, whereas the total diameter was 3.0 m.

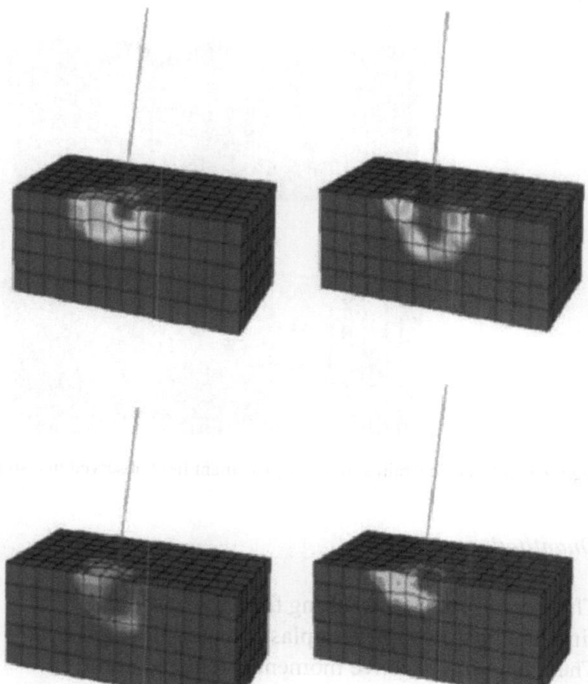

Figure 4. Fields of plastic strain in clay soil (ideal cohesive material with low high resistance): (a) heart-, (b) tap-, (c) herringbone- and (d) plate-like root systems.

Qualitative results

During the uprooting simulations, forces were transmitted by the roots into the soil inducing deformation of the entire root/soil plate. The roots that were located on the leeward side, i.e. the side of the tree held in compression during overturning, tended to plunge downwards, which induced a local increase in soil pressure (Figure 4). Simultaneously, windward roots on the opposite side of the tree moved upwards out of the soil. This movement resulted in a decrease in soil pressure on this side of the root system. After a critical displacement of the stem, the failure yield criterion was reached in certain regions of the soil. These failure areas were highlighted by the field of equivalent plastic strain (Figure 5) and took the form of a slip surface.

Very different uprooting mechanisms were observed in sand and clay soils (Figures 4 and 5). In clay soil, the root/soil system rotated around an axis that was situated straight below the stem. Upward displacements of the windward side were almost equal to downward displacements on the leeward side. The slip surface was symmetric and roughly circular. The diameter of the slip surface was lower in the hard clay than in the soft clay and root plastic strains occurred on large diameter lateral roots close to the bole centre.

In sandy soil, the soil/root system rotated around an axis that was shifted leeward. Permanent strains expanded further away on the windward side, compared to the leeward side. The slip surface was significantly larger than in clay soils.

Plastic strain profiles in the soil were also very different according to the type of root systems considered (Figure 5). Heart type systems showed a rather moderate and homogeneous distribution of permanent strains around the centre of the bole. Tap root systems showed localised slip surfaces extending deeply in the soil. The herringbone root system mobilised only a slight amount of soil during uprooting, indicating that most failure occurred in the roots close to the main vertical root. Soil failure in plate-like root systems was relatively shallow.

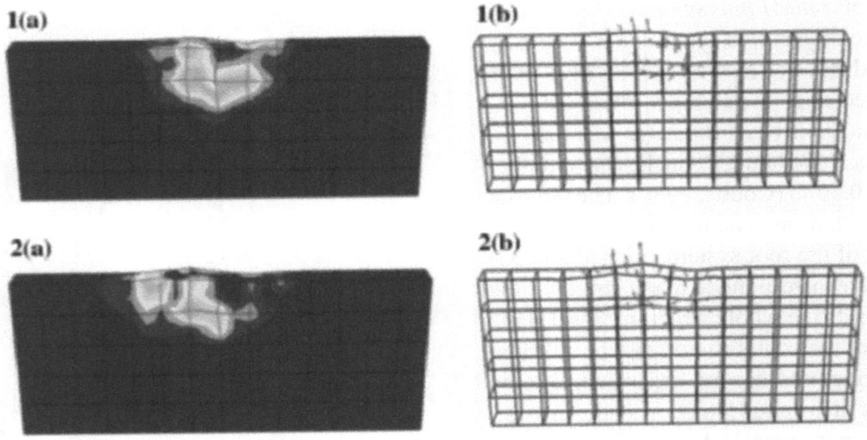

Figure 5. (a) Plastic strain and (b) displacement field observed in a strip of (1) hard clay material (2) dry sand for the plate root system.

Quantitative results

The response curves resulting from the uprooting FEM simulations were typical of plastic materials (Figure 6). The maximum resistive moment calculated at the bole for a range of imposed displacements for all root patterns embedded in both soil types were characteristic of root anchorage (Figure 7).

Resistance to displacement was higher in clay soils (221 ± 37 kN m) compared to sandy soils (127 ± 13 kN m) (Figure 7, means are ± standard deviation). This result was particularly evident in heart (315 kN m in clay and 129 kN m in sand) and tap root systems (292 kN m in clay and 175 kN m in sand). However, differences in the overturning resistance of herringbone and plate root systems were low when clay and sandy

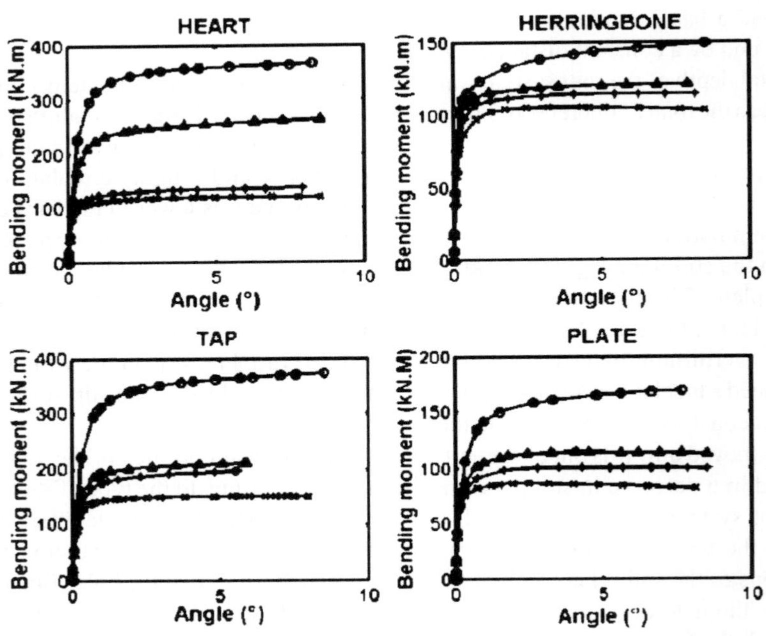

Figure 6. Bending moment vs. rotation angle during the simulation of the four root systems, heart, herringbone, tap and plate, in four different soil types, 'dry sand' (ideal frictional soil in dry conditions) (+), 'saturated sand' (saturated drained behaviour of frictional soils) (x), 'soft clay' (ideal cohesive material with low shear resistance) (Δ) and 'hard clay' (O) (ideal cohesive material with low high resistance). Moment/rotation curves are typical of plastic behaviour.

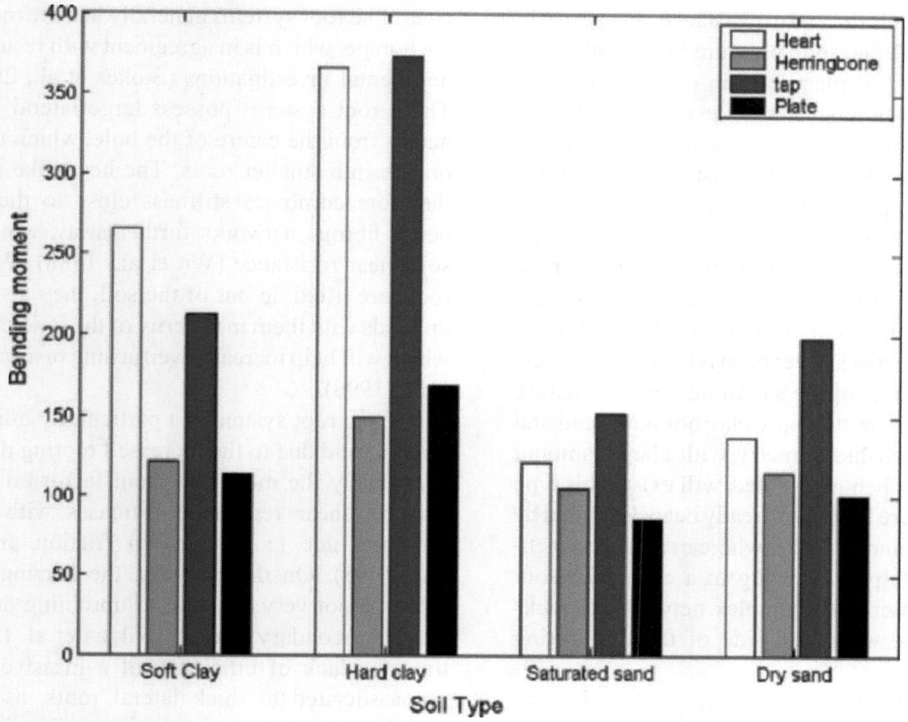

Figure 7. Maximum bending moment of four different root systems in four theoretical soils. The deep rooted tap system was the most resistant to uprooting in sandy-like soil, whereas the heart root system was the most resistant in clay-like soils.

soils were compared (Figure 7). The decrease of shear resistance in clay soils induced a mean decrease of 32% of the global resistance to uprooting. However, water saturation in the sandy soil reduced root anchorage strength by 15%.

Whatever the soil characteristics, heart (222 kN m) and tap (234 kN m) root patterns were the most resistant when compared to her ringbone (122 kN m) and plate (117 kN m) root systems. However, the tap-root system was best anchored in sandy soils (152 kN m in saturated sand and 198 kN m in dry sand) whereas the heart root system was more efficient in soft clay (265 kN m) (Figure 7). The weakest anchorage was found to be that of the plate root system, even though this architectural pattern was slightly more resistant than the herringbone root system embedded in hard clay.

Discussion

The model developed was relatively simple, easy to use and did not require long calculation times to run each simulation. Results for bending moment were realistic and comparable to those found in experimental studies (Cucchi et al., 2004; Stokes et al., 2005). As a tool for studying root anchorage, the model was useful in that branching could be determined by the user, by setting certain architectural parameters. The influence of any particular parameter can then be estimated, as volume, material and soil properties remain constant. The model could be adapted for testing herbaceous species e.g. maize (*Zea mays* L.) which possess a nodal type root system, or even rhizome or bulb species. Similarly, the influence of many different soil types on uprooting could also be tested.

Influence of soil type

The soil type had a strong influence on the resistance to uprooting as well as on the mechanisms occurring during failure. In clay soil, the shear resistance was constant anywhere in the soil and was independent of the isostatic stress due to the chosen null friction angle. Therefore, areas of plastic strain were comparable on both windward and leeward sides of the tree, which thus resulted in a symmetric mechanism (Figure 4a). In sandy soils however, the shear resistance was severely reduced with the decrease in pressure in the soil. Consequently, the field of plastic strain on the windward side of the tree was much more developed, reaching the distal regions of the soil (Figures 4b, 5). On the leeward side of the tree however, the plastic yield remained closer to the stump (Figure 5). The dissymmetry of this type of uprooting mechanism has been highlighted by experimental studies on tap-rooted Maritime pine growing in sandy soils (Cucchi et al., 2004; Stokes, 1999) and also in *Pinus peuce* (Mickovski and Ennos, 2003) even though it has never been related to soil internal friction. For most of the situations found in nature however, the soil is neither pure clay nor pure sand and the presence of such dissymmetry, with a larger amount of windward roots being uprooted, will exist. This type of anchorage failure has been already been described by Mickovski and Ennos (2002), who carried out winching tests on Scots pine growing in a clay loam soil. These authors observed a complex network of cracks in the soil on the windward side of the tree during overturning.

Previous investigations carried out by Crook et al. (1997) showed that uprooting dissymmetry is not only influenced by the soil type but also by rooting depth. This observation is confirmed when observing the qualitative results that were obtained on the tap-root system (Figure 5). This root system was deeper than the others tested and did not exhibit any strong dissymmetry in sandy soils when compared to e.g. the plate root system.

The effect of soil shear resistance tended to decrease the radius of the slip surface and reduce the area where soil is resisting failure through shearing (Mattheck and Breloer, 1994; Yang and Jeremic, 2002). Breaking of roots occurs near the centre of the bole where roots were thicker. The total amount of roots resisting the bending moment was thus increased during uprooting. A similar phenomenon was also observed by Stokes et al. (2000), who carried out winching tests on beech (*Fagus sylvatica* L.), with heart-like root systems growing on a clay loam.

Influence of root morphology

The mode of failure derived from the numerical simulations was strongly determined by root morphology as has also been observed in experimental studies (Coutts, 1986; Crook et al., 1997; Stokes et al., 2000, 2005). Heart like root systems generally had the most efficient anchorage, which is in agreement with results from experimental investigations (Stokes et al., 2000, 2005). These root systems possess large lateral roots originating from the centre of the bole, which then rapidly branch into smaller roots. The heart-like architecture therefore combines stiffness close to the trunk and dense fibrous networks further away, which improve soil shear resistance (Wu et al., 1988). When forked roots are lifted up out of the soil, they also carry soil upwards with them in the crux of the fork, the weight of which will help increase overturning resistance (Stokes et al., 1996).

The tap root system is a particularly efficient structure on sand due to the increased rooting depth. Depth is probably the most important factor on sandy soils because shear resistance increases with the earth's pressure, due to changes in friction angle (Whitlow, 1995). On the contrary, the herringbone structure was not very resistant to uprooting due to a lack of stiff secondary laterals. Niklas et al. (2002) illustrate the lack of efficiency of a massive tap root if not associated to thick lateral roots, as soil is not mobilized around the root system. The insertion angle of lateral roots on the tap root is also important, with an optimal angle around 60°. (Stokes et al., 1996). The influence of this branching angle was also not examined in this study, but the model developed would be a useful tool to investigate these parameters further.

The least resistant rooting type was the platelike root system. Despite possessing massive lateral roots, only a small amount of soil was mobilized during uprooting due to the limited penetration of vertical roots. If there are not many vertical roots embedded in the soil, the soil/root cohesion will be decreased and anchorage strength reduced. Therefore, when plate root systems overturn, the entire soil/root plate is lifted upwards on the windward side (Mattheck and Breloer, 1994). The inefficiency of plate root systems to overturning has also been demonstrated several times through static winching tests (Nicoll et al., 1995; Peltola et al., 2000; Stokes et al., 2000, 2005).

Our studies of root anchorage were carried out on root systems which had the same diameter (6.0 m), but different depths (0.5 m, 1.0 m and 1.5 m). As a consequence, the influence of depth could be easily transposed to the analysis of a shape effect, i.e. the influence of the plate radius/plate depth ratio on root anchorage. However, one can expect a significant scale effect since the root material biomass was fixed to a constant value. This scale effect has not been investigated here, but must be considered in future analyses. Many aspects of the uprooting mechanism still need studying, in particular, the influence of root system volume and diameter. In our study, these parameters were fixed to dimensions assumed large enough not to have an effect on uprooting. Nevertheless, boundary conditions, whether horizontal or vertical, should be taken into consideration. In nature, the boundary conditions which influence uprooting include obstacles in the soil, e.g. stones or bedrock, changes in soil density and waterlogging, resulting in the death of root tips (Cucchi et al., 2004). These parameters translate into complex mechanical and physical phenomena, which influence root/soil frictional resistance, soil plastic flow around the root–soil plate and temporal changes in soil properties. Modelling the influence of such boundary conditions would provide highly useful information in an area where these data are scanty.

The results provided by this study are new qualitative elements for the understanding of uprooting failure in real trees. Phenomena such as stump breakage, root/soil plate asymmetry or overturning of the soil/root plate were explained by physical and geometrical factors. It should be pointed out however, that the root/soil configurations used in our study were theoretical and cannot be found in nature as such. Particularly, root growth is governed by soil type and identical species can produce different root morphologies and volumes depending on local soil conditions (Fitter, 1987). Soil behaviour was also idealistic and uniformly distributed, whereas real soils exhibit strong heterogeneities. Nevertheless, our investigation demonstrated that root architecture and soil type are major factors governing root anchorage, even when volume and material properties are constant between each rooting type.

Although certain simplifications were made in this study, the model developed proved useful for studying different characteristics of root anchorage with regards to root architecture. In the future, certain parameters could be modified to determine the effect of e.g. root wood strength (Stokes and Mattheck, 1996), root shape (Nicoll and Ray, 1996), root asymmetry (Coutts et al., 1999), root slippage through soil, soil type (Moore, 2000), soil hydrology, on tree root anchorage. Not only could these parameters be modified, but the model could also be used to test the effect of cyclic behaviour, e.g. wind gusting on tree anchorage. Further studies should include the comparison of field data with results from the model. However, as it is impossible to simulate field conditions exactly, highly controlled experiments should be set up (Hamza et al., 2006). A considerable improvement in both the hypotheses used for modelling and the understanding of uprooting mechanisms could therefore be obtained.

Acknowledgements

This study was funded by the EU project Eco-slopes (QLK5-2001-00289) and the DERF—GIPECOFOR project Peuplier. L. Dupuy was financed by a French Governmental CIFRE bursary.

References

Bergonzini J C and Laroussinie O 2000 Les écosystèmes forestiers dans les tempêtes. GIP-ECOFOR Ministère de l'agriculture et de la pêche, France.

Blackwell P G, Rennolls K and Coutts M P 1990 A root anchorage model for shallowly rooted Sitka spruce. Forestry 63, 73–91.

Bouchon J 1987 Etat de la recherche relative aux dégâts forestiers dus aux tempêtes. Rev. Forest. Française 4, 301–312.

Coutts M P 1983 Root architecture and tree stability. Plant Soil 71, 171–188.

Coutts M P 1986 Components of tree stability in Sitka spruce on peaty gley soil. Forestry 59, 173–197.

Coutts M P 1987 Developmental processes in tree root systems. Can. J. For. Res. 17, 761–767.

Coutts M P, Nielsen C C N and Nicoll B C 1999 The development of symmetry, rigidity and anchorage in the structural root system of conifers. Plant Soil 217, 1–15.

Crook M J, Ennos A R and Banks J R 1997 The function of buttress roots: a comparative study of the anchorage systems of buttressed (*Aglaia and Nephelium ramboutan* species) and non-buttressed (*Mallotus wrayi*) tropical trees. J. Exp. Bot. 48, 1703–1716.

Cucchi V and Bert D 2003 Wind-firmness in *Pinus pinaster* Ait. stands in Southwest France: influence of stand density, fertilisation and breeding in two experimental stands damaged during the 1999 storm. Ann. For. Sci. 60, 209–226.

Cucchi V, Meredieu C, Stokes A, Berthier S, Bert D, Najar M, Denis A and Lastennet R 2004 Root anchorage of inner and edge trees in stands of Maritime pine (*Pinus pinaster* Ait.) growing in different podzolic soil conditions. Trees—Struct. Funct. 18, 460–466.

Danjon F, Bert D, Godin C and Trichet P 1999 Structural root architecture of 5-year-old *Pinus pinaster* measured by three- dimensional digitising and analysed with AMAPmod. Plant Soil 217, 49–63.

Danjon F, Fourcaud T and Bert D 2005 Root architecture and windfirmness of mature *Pinus pinaster*. New Phytol. 168, 387–400.

Dupuy L 2003 Modélisation de l'ancrage racinaire des arbres forestiers. These de l'UniversitéBordeaux I, France, no. 2788.

Dupuy L, Fourcaud T, Lac P and Stokes A 2003 Modelling the influence of morphological and mechanical properties on the anchorage of root systems. *In* B Ruck, C Kottmeier, C Mattheck, C Quine and G Wilhelm (eds). Proceedings of the International Conference Wind Effects on Trees, September 16–18, 2003. University of Karlsruhe, Germany, pp. 315–322.

Dupuy L, Fourcaud T and Stokes A 2005a A numerical investigation into factors affecting the anchorage of roots in tension. Eur. J. Soil Sci. 56, 319–327. Available online doi: 10.1111/j.1365-2389.2004.00656.x.

Dupuy L, Fourcaud T, Stokes A and Danjon F 2005b A density-based approach for the modelling of root architecture: application to Maritime pine (*Pinus pinaster* Ait.) root systems. J. Theor. Biol., 236, 323–334. Available online doi: 10.1016/j.jtbi.2005.03.013.

England A H, Baker C J and Saunderson S E T 2000 A dynamic analysis of windthrow of trees. Forestry 73, 225–237.

Ennos A R 1990 The anchorage of leek seedlings: the effect of root length and soil strength. Ann. Bot. (London) 65, 409–416.

Fitter A H 1987 An architectural approach to the comparative ecology of plant root systems. New Phytol. 106, 61–77.

Fourcaud T and Lac P 2003 Numerical modelling of shape regulation and growth stresses in trees I. An incremental static finite element formulation. Trees—Struct. Funct. 17, 23–30.

Fourcaud T, Danjon F and Dupuy L 2003a Numerical analysis of the anchorage of Maritime pine trees in connection with root structure. *In* B Ruck, C Kottmeier, C Mattheck, C Quine and G Wilhelm, (eds.), Proceedings of the International Conference Wind Effects on Trees, September 16–18, 2003. University of Karlsruhe, Germany, pp. 323–330.

Fourcaud T, Dupuy L, Sellier D, Ancelin P and Lac P 2003b Application of plant architectural models to biomechanics. *In* BG Hu. and M Jaeger (eds), PMA03—Plant Growth Modeling and Applications. Beijing. pp. 384–400. Tsinghua University press, Springer, Beijing, China.

Godin C and Caraglio Y 1998 A multiscale model of plant topological structures. J. Theor. Biol. 191, 1–46.

Hamza O, Bengough A G, Bransby M F, Davies M C R and Hallett P D 2007 Mechanics of root pullout from soil: a novel image and stress analysis procedure. *In* A Stokes, I Spanos, J E Norris and L H Cammeraat (eds), Eco- and Ground Bio-engineering: The Use of Vegetation to Improve Slope Stability. Developments in Plant and Soil Sciences vol. 103, pp. 213–221. Springer, Dordrecht.

Hintikka V 1972 Wind-induced movements in forest trees. Metsa® ntitkimuslitoksen Julkaisuja 76, 1–56. Jourdan C and Rey H 1997 Modelling and simulation of the architecture and development of the oil-palm Elais guinensis Jacq. root system. Plant Soil 190, 235–246.

Köstler J N, Brückner E and Bibelriether H 1968 Die Wurzeln der Waldbäume. Verlag Paul Parey, Hamburg, Berlin, Germany.

Lyford W H 1980 Development of the root System of Northern Red oak (*Quercus rubra* L.). Harvard University, Harvard forest paper 555.

Mattheck C and Breloer H 1994 The Body Language of Trees. HMSO Dept. of the Environment, London, 240 pp.

Menétrey P and Willam K J 1995 Triaxial failure criterion for concrete and its generalization. ACI Struct. J. 92, 311–318.

Mickovski S B and Ennos A R 2002 A morphological and mechanical study of the root systems of suppressed crown 133 Scots pine *Pinus sylvestris*. Trees—Struct. Funct. 16, 274–280.

Mickovski S B and Ennos A R 2003 Anchorage and asymmetry in the root system of *Pinus peuce*. Silva Fenn. 37, 161–173.

Moore J R 2000 Differences in maximum resistive bending moments of *Pinus radiata* trees grown on a wide range of soil types. For. Ecol. Manage. 135, 63–71.

Nicoll B C, Easton E P, Milner A D, Walker C and Coutts M P 1995 Wind stability factors in tree selection: distribution of biomass within root systems of Sitka spruce clones. *In* MP Coutts. and J Grace (eds), Wind and Trees, pp. 276–292. Cambridge University Press, Cambridge, MA.

Nicoll B C and Ray D 1996 Adaptive growth of Sitka spruce root systems in response to wind action and site conditions. Tree Physiol. 16, 891–898.

Niklas K J, Molina-Freaner F, Tinoco-Ojanguren C and Paolillo D J 2002 The biomechanics of *Pachycereus pringlei* root systems. Am. J. Bot. 89, 12–21.

Pages L, Vercambre G, Drouet J-L, Lecompte F, Collet C and Le Bot J 2004 Root Typ: a generic model to depict and analyse the root system architecture. Plant Soil 258, 103–119.

Peltola H, Kellomaki S, Hassinen A and Granander M 2000 Mechanical stability of Scots pine, Norway spruce and birch: an analysis of tree-pulling experiments in Finland. For. Ecol. Manage. 135, 143–153.

Puhe J 2003 Growth and development of the root system of Norway spruce (*Picea abies*) in forest stands, a review. For. Ecol. Manage. 175, 253–273.

Putz FE, Coley PD, LuK, MontalvoAand AielloA1983 Uprooting and snapping of trees: structural determinants and ecological consequences. Can. J. For. Res. 13, 1011–1020.

Ruel J C 2000 Factors influencing windthrow in balsam fir forests: from landscape studies to individual tree studies. For. Ecol. Manage. 135, 169–178.

Stokes A 1999 Strain distribution during anchorage failure of *Pinus pinaster* Ait. at different ages and tree growth response to wind-induced root movement. Plant Soil 217, 17–27.

Stokes A, Ball J, Fitter A H, Brain P and Coutts M P 1996 An experimental investigation of the resistance of model root systems to uprooting. Ann. Bot.—(London) 78, 415–421.

Stokes A, Drexhage M and Guitard D 2000 A method for predicting the possible site of failure in trees during mechanical loading. *In* A Stokes, (ed) The Supporting Roots of Trees and Woody Plants: Form, Function and Physiology. Developments in Plant and Soil Sciences, vol. 87. pp. 279–285. Kluwer, Doredrecht.

Stokes A, AbdGhani M, Salin F, Danjon F, Jeannin H, Berthier S, Kokutse A D and Frochot H 2007 Root morphology and strain distribution during tree failure on mountain slopes. *In* A Stokes, I Spanos, J E Norris and L H Cammeraat (eds), Eco- and Ground Bio-engineering: The Use of Vegetation to Improve Slope Stability. Developments in Plant and Soil Sciences vol. 103, pp. 165–173. Dordrecht.

Stokes A and Mattheck C 1996 Variation of wood strength in roots of forest trees. J. Exp. Bot. 47, 693–699.

Stokes A, Salin F, Kokutse A D, Berthier S, Jeannin H, Mochan S, Kokutse N, Dorren L, AbdGhani M and Fourcaud T 2005 Mechanical resistance of different tree species to rockfall in the French Alps. Plant Soil, 278: 107–117.

Whitlow R 1995 Basic Soil Mechanics. Addison-Wesley, Reading, MA.

Wu T H, Omber R M, Erb R T and Beal P E 1988 Study of soil–root interaction. J. Geotech. Eng. ASCE 114, 1351–1375.

Yang Z and Jeremic B 2002 Numerical analysis of pile behaviour under lateral loads in layered elastic–plastic soils. Int. J. Numer. Anal. Met. Geomech. 26, 1385–1406.

Methodology applied to eco- and ground bio-engineering

SLIP4EX – A program for routine slope stability analysis to include the effects of vegetation, reinforcement and hydrological changes

John R. Greenwood

School of The Built Environment, Nottingham Trent University, Burton Street, Nottingham, UK, NG1 4BU, England

Key words: hydrology, reinforcement, slopes, stability analysis, vegetation

Abstract

SLIP4EX is a straightforward computer program developed in connection with the EU funded ECOSLOPES project for routine stability analysis and the assessment of the contribution of vegetation to slope stability. The slope section is drawn up and dimensions and parameters are fed in to the Microsoft Excel based program for stability calculations and comparisons of Factors of Safety using different methods of analysis (Bishop, Janbu, Fellenius, Simple, Greenwood). The background and assumptions involved in the derivation of each of the methods is briefly described. The simplicity of the program enables the user to understand the nature of the analysis, explore the parameter assumptions made and compare the different methods of analysis. Soil reinforcement by geosynthetic layers or anchors, and vegetation effects of enhanced cohesion, changed water pressures, mass of vegetation, wind forces and root reinforcement forces are readily included in the analysis. The program is freely available on request from the author.

Introduction

SLIP4EX is a computer program for slope stability analysis, developed in connection with the European Commission funded 'ECOSLOPES' project to help assess the contribution of vegetation to slope stability (http://www.ecoslopes.com). It is based on the earlier SLIP3 'Fortran' program (Greenwood, 1986; Greenwood and Zytynski, 1993). The slope section is drawn up and dimensions and parameters are fed in to the Microsoft Excel based SLIP4EX program for stability calculations and comparisons of Factors of Safety using different methods of limit equilibrium analysis by the method of slices (Bishop, Janbu, Fellenius, Simple, Greenwood).

The simplicity of the program makes it ideal for preliminary problem analysis. It enables the user to understand the nature of the analysis, explore the parameter assumptions made and compare the different methods of analysis. Geosynthetic reinforcement may be included and vegetation effects such as enhanced cohesion, changed water pressures, mass of vegetation, wind forces and root reinforcement forces are readily included in the analysis.

The SLIP4EX program is freely available on request from the author.

The use of the SLIP4EX program is illustrated by an example of a vegetated slope. The notation and the basis for the stability equations used in the spreadsheet are given in the Appendix.

Example application of SLIP4EX to determine the factor of safety of a vegetated slope

Initial calculation without vegetaion

The stability problem is drawn out to scale with the single slip surface defined as shown in Figure 1. Slice dimensions (up to 3 soil layers are permitted) and the angles between the base of each slice and the horizontal, are scaled from the diagram and appropriate soil and water parameters assigned for each slice as indicated in Table 1. The notation used and details of the dimensions are given in the Appendix.

The prepared slice data is then input manually into the SLIP4EX spreadsheet program which calculates the forces acting on each slice of the analysis and the total

This article has been previously published in the following journal—Journal of Geotechnical and Geological Engineering, 24, 3, 449–465.
Stokes et al. (eds), Eco- and Ground Bio-Engineering: The Use of Vegetation to Improve Slope Stability, 193–202.

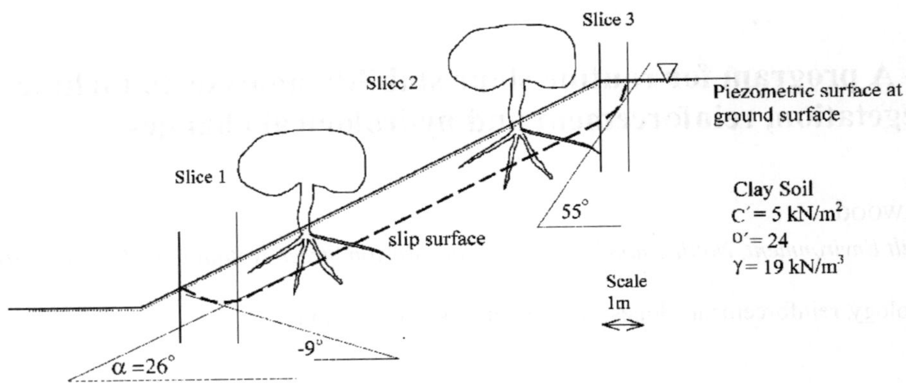

Figure 1. Scale drawing of slope and potential slip surface. Slices selected and parameters assigned.

forces acting on the slip surface. It calculates the Factor of Safety of the slip surface by the different methods commonly used by geotechnical engineers (Table 2). The spreadsheet currently has provision for up to 15 slices to be used.

The Appendix presents a brief review of the different assumptions relating to each method (Greenwood General, Greenwood General with K as input, Simple, Simple with K as input, Fellenius (Swedish), Bishop, and Janbu). The Factor of Safety is calculated both in terms of moment equilibrium and horizontal force equilibrium where appropriate. The iteration for the Bishop and Janbu solutions is done manually in this version by re-inputting the output Factor of Safety until the output value = input value. Automatic iteration can be done on the spreadsheet by addition of more columns.

There is an option in the Greenwood General and Simple methods to assess the additional effects of horizontal earth pressures on the calculated Factor of Safety by assigning an earth pressure coefficient (K value) to each slice. This would be particularly relevant for deeper slip surfaces in overconsolidated soils (Finlayson et al., 1984; Greenwood, 1985; Greenwood et al., 1985). It is conservative to assume $K = 0$.

Once input, as for all spreadsheet work, parameters can readily be changed to demonstrate their sensitivity and influence on the calculated Factor of Safety.

Including the effects of vegetation

The parameters relating to the effects of vegetation (Table A2) may be included in the analysis. Appropriate additional parameters are assigned to each slice as indicated in Table 3 and input to the spreadsheet. In this example an additional tensile root reinforcing force is assumed to act on the base of each slice (in exactly the same way that a geosynthetic layer would contribute to stability). The derivation of this force has been demonstrated (see Note on calculation of Available Root Force, T, Acting on Each Slice). In the example, the fine roots are assumed to have no influence on c′, but the piezometric head is assumed drawn down by 0.1 m under the influence of the vegetation.

Table 1. Slice data prepared from scale drawing ready for input to SLIP4EX spreadsheet

Slice	Avge height of slice	Unit weight of soil	If other soil layers present in slice				Slice width, b	Base angle, α	Cohesion, c' and friction angle, ϕ', at base of slice		Head of water on downslope side, upslope side of slice and average			Earth press coeff.
	Height 1 (m)	Unit wt. 1 (kN/m³)	Height 2 (m)	Unit wt. 2 (kN/m³)	Height 3 (m)	Unit wt. 3 (kN/m³)	Breadth (m)	Alpha (degs)	Cohesion (kN/m²)	ϕ' (degs)	h_{w1} (m)	h_{w2} (m)	h_w (m)	K
1	0.6	19					1.9	−9	5	24	0	1.2	0.6	0.5
2	1.2	19					9.2	26	5	24	1.2	1.2	1.2	0.5
3	0.6	19					0.75	55	5	24	1.2	0	0.6	0.5

Table 2. Input data and output results of SLIP4EX analysis showing calculated forces on each slice of the analysis and comparisons of Factor of Safety calculated by different methods

SLIP4EX - SLOPE STABILITY ANALYSIS (NTU Sept 2004)

Sheet 1 - Comparison of Methods

(See sheet 2, for effects of reinforcement, vegetation and hydrological changes)

PROJECT Thessaloniki conference 2004 DESCRIPTION OF ANALYSIS: 1 in 2 embankment example with vegeation effects

Date:

Enter slice Data

Slice Nr	Height 1 m	Unit wt 1 kN/m^3	Height 2 m	Unit wt 2 kN/m^3	Height 3 m	Unit wt 3 kN/m^3	Breadth m	Alpha degrees	Cohesion* kN/m^2	Phi' degrees	hw1 m	hw2 m	hw m	K
1	0.6	19					1.9	-9	5	24	0	1.2	0.6	0.5
2	1.2	19					9.2	26	5	24	1.2	1.2	1.2	0.5
3	0.6	19					0.75	55	5	24	1.2	0	0.6	0.5
4													0	
5													0	
6													0	
7													0	
8													0	
9													0	
10													0	
11													0	
12													0	
13													0	
14													0	
15													0	

Calculated forces on slices

							Total Resistance - Moment equilibrium						**Total Resistance - Horizontal force equilibr**			
	W	U1	U2	u	Dist force	cohesive res	General	General K'	Simple	Simple K'	Swedish	Bishop	General	General K'	Simple	Simple K'
slice	kN	kN	kN	kN/m2	kN	kN	kN	kN	kN	kN	kN	kN	kN	kN	kN	kN
1	21.66	0.00	7.20	6.00	-3.39	9.62	14.51	14.56	14.13	14.19	14.00	15.22	14.69	14.74	14.31	14.36
2	209.76	7.20	7.20	12.00	91.95	51.18	80.43	85.16	90.94	95.67	80.43	83.85	89.49	94.75	101.18	106.44
3	8.55	7.20	0.00	6.00	7.00	6.54	7.85	8.91	7.57	8.63	5.23	6.14	13.69	15.53	13.20	15.04
4	0.00	0.00	0.00	0.00	0.00	0.00	0.00	0.00	0.00	0.00	0.00	0.00	0.00	0.00	0.00	0.00
5	0.00	0.00	0.00	0.00	0.00	0.00	0.00	0.00	0.00	0.00	0.00	0.00	0.00	0.00	0.00	0.00
6	0.00	0.00	0.00	0.00	0.00	0.00	0.00	0.00	0.00	0.00	0.00	0.00	0.00	0.00	0.00	0.00
7	0.00	0.00	0.00	0.00	0.00	0.00	0.00	0.00	0.00	0.00	0.00	0.00	0.00	0.00	0.00	0.00
8	0.00	0.00	0.00	0.00	0.00	0.00	0.00	0.00	0.00	0.00	0.00	0.00	0.00	0.00	0.00	0.00
9	0.00	0.00	0.00	0.00	0.00	0.00	0.00	0.00	0.00	0.00	0.00	0.00	0.00	0.00	0.00	0.00
10	0.00	0.00	0.00	0.00	0.00	0.00	0.00	0.00	0.00	0.00	0.00	0.00	0.00	0.00	0.00	0.00
11	0.00	0.00	0.00	0.00	0.00	0.00	0.00	0.00	0.00	0.00	0.00	0.00	0.00	0.00	0.00	0.00
12	0.00	0.00	0.00	0.00	0.00	0.00	0.00	0.00	0.00	0.00	0.00	0.00	0.00	0.00	0.00	0.00
13	0.00	0.00	0.00	0.00	0.00	0.00	0.00	0.00	0.00	0.00	0.00	0.00	0.00	0.00	0.00	0.00
14	0.00	0.00	0.00	0.00	0.00	0.00	0.00	0.00	0.00	0.00	0.00	0.00	0.00	0.00	0.00	0.00
15	0.00	0.00	0.00	0.00	0.00	0.00	0.00	0.00	0.00	0.00	0.00	0.00	0.00	0.00	0.00	0.00
				total	**95.57**	67.34	**102.79**	**108.63**	**112.64**	**118.48**	**99.66**	105.20	**117.87**	**125.03**	**128.69**	**135.85**

Factors of Safety (no reinforcement or vegetation)

	Moment equilibrium F_m	Force equilibrium F_f
Greenwood General	**1.08**	**1.06**
Greenwood General (K as input)	**1.14**	**1.13**
Greenwood Simple	**1.18**	**1.16**
Greenwood Simple (K as input)	**1.24**	**1.22**
Swedish	**1.04**	**1.02**
Bishop	1.10	
Janbu (fo =1.05)		1.13

Bishop iteration F initial	F input	F calc	Janbu Iteration F input	F calc
1	1.10	1.10	1.13	1.13

The changes in the Factor of Safety due to the effects of the vegetation (or reinforcement or hydrological changes) are calculated in sheet 2 of the spreadsheet (Table 4). The effects are added to the General, Simple and Swedish equations but not the Bishop and Janbu methods where the iterative process and imposition of the Factor of Safety on to each slice in the stability equations does not permit easy inclusion of the additional forces.

In this example the vegetation has increased the calculated Factor of Safety from 1.08 to 1.21 (General method, Greenwood et al., 2003). It is emphasised that the assumptions made for the vegetation effects and hydraulic changes are to illustrate the application of the stability analysis and should not be applied to particular situations without appropriate investigation and testing.

Note on calculation of available root force, *T*, acting on each slice

Whilst the SLIP4EX spreadsheet is applicable to all stability calculations, it was developed with the intention of including vegetation effects. It may be helpful to describe the way in which a typical available root force

Table 3. Selected parameters to reflect the contribution of vegetation assigned to each slice

	Root force	Root direction	Additional cohesion	Change in water table			Mass of vegetation	Wind force	Wind direction
Slice	T kN (/m)	Theta (deg)	$c'v$ kN/m^2	delta hw1 (m)	delta hw2 (m)	delta hw (m)	Wv kN (/m)	*D* kN (/m)	Beta (deg)
1	0.95	45		0	−0.1	−0.05	0	0	0
2	5	45		−0.1	−0.1	−0.1			
3	0.6	45		−0.1		−0.05			

Table 4. Input 'vegetation' data and output results of SLIP4EX analysis showing calculated 'vegetation' forces on each slice of the analysis and changes to the Factor of Safety calculated by different methods

SLIP4EX - SLOPE STABILITY ANALYSIS (NTU Sept 2004)

Sheet 2 - EFFECTS OF REINFORCEMENT, VEGETATION AND HYDROLOGICAL CHANGES (See sheet 1, for Comparison of Methods)

PROJECT Thessaloniki conference 2004 DESCRIPTION OF ANALYSIS: 1 in 2 embankment example with vegeation effects
Date: 0

Reinforcement, Vegetation and Hydraulic changes
Enter effects for relevant slices

slice	T kN (/m)	Theta deg	c'v kN/m2	delta hw1 m	delta hw2 m	delta hw m	Wv kN (/m)	D kN (/m)	Beta deg.
1	0.95	45		0	-0.1	-0.05	0	0	0
2	5	45		-0.1	-0.1	-0.1			
3	0.6	45		-0.1		-0.05			
4									
5									
6									
7									
8									
9									
10									
11									
12									
13									
14									
15									

Calculated Reinforcement / Vegetation / Hydraulic effects

	Additional disturbing force (to reinf. and veg.)				Additional Restoring Forces									Total additional resista	
	Veg.Weigh	reinf dist fo	Wind dist fo	Total add. dist force	add cohesion	add weight	add u	add U1	add U2	add 'U2-U1	add 'U2-U1	add wind	add T	General	Simple
slice	kN	kN	kN	kN	kN	kN	kN	kN	kN	Gen kN	Simple kN	kN	kN	kN	kN
1	0.00	-0.67	0.00	-0.67	0.00	0.00	-0.43	0.00	-1.15	0.08	0.01	0.00	0.30	0.65	0.72
2	0.00	-3.54	0.00	-3.54	0.00	0.00	-4.56	-1.15	-1.15	0.00	0.88	0.00	1.57	6.13	5.26
3	0.00	-0.42	0.00	-0.42	0.00	0.00	-0.29	-1.15	0.00	0.42	0.20	0.00	0.19	0.06	0.28
4	0.00	0.00	0.00	0.00	0.00	0.00	0.00	0.00	0.00	0.00	0.00	0.00	0.00	0.00	0.00
5	0.00	0.00	0.00	0.00	0.00	0.00	0.00	0.00	0.00	0.00	0.00	0.00	0.00	0.00	0.00
6	0.00	0.00	0.00	0.00	0.00	0.00	0.00	0.00	0.00	0.00	0.00	0.00	0.00	0.00	0.00
7	0.00	0.00	0.00	0.00	0.00	0.00	0.00	0.00	0.00	0.00	0.00	0.00	0.00	0.00	0.00
8	0.00	0.00	0.00	0.00	0.00	0.00	0.00	0.00	0.00	0.00	0.00	0.00	0.00	0.00	0.00
9	0.00	0.00	0.00	0.00	0.00	0.00	0.00	0.00	0.00	0.00	0.00	0.00	0.00	0.00	0.00
10	0.00	0.00	0.00	0.00	0.00	0.00	0.00	0.00	0.00	0.00	0.00	0.00	0.00	0.00	0.00
11	0.00	0.00	0.00	0.00	0.00	0.00	0.00	0.00	0.00	0.00	0.00	0.00	0.00	0.00	0.00
12	0.00	0.00	0.00	0.00	0.00	0.00	0.00	0.00	0.00	0.00	0.00	0.00	0.00	0.00	0.00
13	0.00	0.00	0.00	0.00	0.00	0.00	0.00	0.00	0.00	0.00	0.00	0.00	0.00	0.00	0.00
14	0.00	0.00	0.00	0.00	0.00	0.00	0.00	0.00	0.00	0.00	0.00	0.00	0.00	0.00	0.00
15	0.00	0.00	0.00	0.00	0.00	0.00	0.00	0.00	0.00	0.00	0.00	0.00	0.00	0.00	0.00
	0.00	-4.63	0.00	**-4.63**	0.00	0.00	-5.28	-2.30	-2.30	0.50	1.08	0.00	2.06	**6.84**	**6.26**

Factors of Safety with Reinforcement ,Vegetation and hydraulic changes included

		F_m
Greenwood General	No reinforcement/Veg	**1.08**
	with reinf /veg /water as input	**1.21**
Greenwood General (K as input)	No reinforcement/Veg	**1.14**
	With reinf /veg /water as input	**1.27**
Greenwood Simple	No reinforcement/veg	**1.18**
	With reinf/veg/water as input	**1.31**
Swedish	No reinforcement/veg	**1.04**
	With reinf/veg/water as input	**1.18**

is assigned in the above example following the procedure recommended by Norris and Greenwood (2000) and Greenwood et al. (2001, 2004).

The available root force acting on the base of each slice, T, is calculated by the equation, $T = T_{rd} \times \ell$ where T_{rd} is the available root force per square metre of soil and ℓ is the length of the slip surface.

Typically from observation and tests, assuming 4 roots of 12.5 mm diameter, each having an ultimate pull out resistance of 8 MN/m^2, cross each square metre of soil at 1.2 m depth. The ultimate root force per square metre across the slip plane, T_{ru} would be given by

$$T_{ru} = 4\pi \times 0.0125^2 \times 8 \times 1000/4$$
$$= \text{approx 4 kN per square metre of soil}$$

Applying a partial Factor of Safety of 8 to allow for uncertainty in root distribution and incompatibility of failure strain between the root and the soil (Greenwood et al., 2003), the design root force per square metre, T_{rd}, is given by: $T_{rd} = T_{ru}/8 = 4/8 = 0.5$ kN/m^2.

Root forces, T, for each slice may therefore be calculated as follows:

Slice	T_{rd} kN/m^2	ℓ (approx) m	$T = T_{rd} \times \ell$ kN
1	0.5	1.9	0.95
2	0.5	10.0	5.0
3	0.5	1.2	0.6

The effective angle between the operational roots and the slip surface, θ, is assumed to be 45°. Parametric studies on both geosynthetic and root reinforcement (Greenwood, 1990; Norris and Greenwood, 2003) have indicated that the calculated resistance due to the (root) reinforcement is not particularly sensitive to 0 because as the enhanced normal component acting across the slip surface decreases, the tangential component, will increase.

As more investigation, testing and monitoring of vegetation is carried out, it should be possible to better define the vegetation related parameters and the partial Factor of Safety applicable to root forces for particular sites.

General application of SLIP4EX

SLIP4EX is intended as an easily accessible and available program to help gain an initial understanding of a slope problem and the main influences on stability. The less experienced practitioner can develop a feel for the aspects of the stability analysis and explore different mechanisms of failure before progressing to more sophisticated search programs to find critical slip surfaces. It is valuable as a student learning aid because the engineering process of drawing the slope, deciding on slip surfaces and assigning appropriate parameters is all kept under the userís control. Another application is where a particular slip surface generated by a commercial search program requires an independent check and further study of the significance of the assumed parameters.

Sheet 3 of the SLIP4EX spreadsheet provides opportunity to use the Excel plotting facilities to demonstrate aspects of the calculated output. For example the calculated restoring forces may be displayed for each slice for each of the methods of calculation.

Future developments

Whilst SLIP4EX is particularly valuable to help gain an understanding of the stability problem, it is recognised that the next stage is to set up the full slope model and to run a search program to find the most critical slip surface. An 'automated' version of SLIP4EX (SLIP6EX) in which the problem is set up on the computer, slice dimensions and properties automatically assigned and the critical slip surface (circle) identified, is currently under development in collaboration with Rens Van Beek (personal communication).

Copies of the development version of SLIP4EX together with guidance notes are available by email request to john.greenwood@ntu.ac.uk. As a non commercial package this is provided with no guarantees, backup or support. Any suggestions for improvement or additions will be welcomed by the author.

Acknowledgements

The author is grateful for the debate and support of academic and industrial colleagues and research students who have helped develop and apply this approach to slope stability analysis. The work of Joanne

Norris on the bioengineering applications and in the proof reading of this manuscript is much appreciated. The computer program development work of Rens Van Beek is also much appreciated. The funding of the ECOSLOPES project under the 5th Framework of the European Commission and the support of all partners in the European team is gratefully acknowledged.

Appendix

Notation and equations used in SLIP4EX spreadsheet

The notation used and details of the dimensions are given in Tables A1 and A2, and Figures A1–A3.

The equations used in the SLIP4EX spreadsheet are derived from the basic limit equilibrium stability equation (Lambe and Whitman, 1969):

$$F = \frac{\text{Restoring force (available shear strength)}}{\text{Disturbing force (shear force)}} = \frac{\sum(c'\ell + N' \tan\phi')}{\sum W \sin\alpha}$$

By resolving forces to determine N' (Figure A2), the full stability equation based on effective forces is obtained (See Greenwood, 1987, 1989):

$$F = \frac{\sum(c'\ell + [W\cos\alpha - u\ell - (U_2 - U_1) \times \sin\alpha\,(X_2' - X_1')\cos\alpha - (E_2' - E_1')\sin\alpha]\tan\phi')}{\sum W \sin\alpha} \quad \text{(A.1)}$$

In order to find a solution, assumptions must be made about the 'unknown' interslice forces X' and E'.

Assumption 1

A reasonable assumption is that the resultant of the effective interslice forces is parallel to the base of the slice, i.e. in the direction of movement – a logical assumption as failure progresses.

$$\text{i.e. } (X_2' - X_1')\cos\alpha - (E_2' - E_1')\sin\alpha = 0 \quad \text{(A.2)}$$

This gives the *General equation* (see Greenwood, 1987, 1989; Morrison and Greenwood, 1989)

$$F = \frac{\sum[c'\ell(W\cos\alpha - u\ell - (U_2 - U_1)\sin\alpha)\tan\phi']}{\sum W \sin\alpha}. \quad \text{(A.3)}$$

Assumption 2

An alternative assumption is to ignore vertical interslice forces or at least assume they are equal and opposite (i.e. assume $(X_2' - X_1') = 0$ as Bishop (1955) and others do) – a reasonable assumption when the slip mass is acting as a single unit – and assume that the effective horizontal interslice forces, E_1' and E_2', relate to the horizontal earth pressure, i.e. $\sigma_h' = K\sigma_v'$ where K is the coefficient of lateral earth pressure.

Assuming K is constant with depth and constant water table conditions

$$E_1' = K\gamma' h_1^2/2 \quad \text{and} \quad E_2' = K\gamma' h_2^2/2 \quad \text{(A.4)}$$

$$E_2' - E_1' = K\gamma' h_2^2/2 - K\gamma' h_1^2/2 = K\gamma'/2(h_2^2 - h_1^2) = K(\gamma_b - \gamma_w)(h_2 - h_1)(h_2 + h_1)/2 \quad \text{(A.5)}$$

but for *level ground surface* $h_2 - h_1 = -b\tan\alpha$ and $(h_2 + h_1)/2 = h$ (average height)

$$E_2' - E_1' = -K\tan\alpha(\gamma_b h - \gamma_w h_w b) = -K\tan\alpha(W - ub) \quad \text{(A.6)}$$

and for *sloping ground surface*, parallel to a slip surface, $h_2 - h_1 = 0$ and the term reduces to zero.

It is therefore reasonable to assume the general application of the term (A.6) for $E_2' - E_1'$ and to assign appropriate values of K depending on the location of the slip surface.

The *General equation to include an estimation of the horizontal interslice force based on* 'K' is therefore:

$$F = \frac{\sum(c'\ell + [W\cos\alpha - u\ell - (U_2 - U_1) \times \sin\alpha + K\tan\alpha(W - ub)\sin\alpha]\tan\phi')}{\sum W \sin\alpha} \quad \text{(A.7)}$$

This equation (A.7) is consistent with Mohr–Coulomb retaining wall analysis theory.

For the particular case of horizontal water surface across the slice (static water conditions), from the slice geometry, $U_2 - U_1 = -ub\tan\alpha$, and Eq. (A.7) becomes

$$F = \frac{\sum(c'\ell + [W\cos\alpha - u\ell + ub\tan\alpha\sin\alpha + K\tan\alpha(W - ub)\sin\alpha]\tan\phi')}{\sum W \sin\alpha}$$

which reduces to

$$F = \frac{\sum(c'\ell + [(W - ub)(1 + k\tan^2\alpha)\cos\alpha]\tan\phi')}{\sum W \sin\alpha}. \quad \text{(A.8)}$$

Eq. (A.8) is the *Greenwood Simple equation (K as input)* derived from the *in situ* effective stress state based on Mohr circle/Coulomb criteria (Greenwood, 1983).

The value of K in Eqs. (A.7) and (A.8) may be assigned for a particular situation. For example, a value of $K = 0$ is appropriate where the slip surface is parallel to the slope and a value of $K = K_0$ may be appropriate for slip surfaces passing through the slope foundation (Greenwood, 1985; Greenwood et al., 1985; Finlayson et al., 1984).

Other equations used in SLIP4EX

The *Greenwood Simple* Eq. (A.9), is derived from Eq. (A.3) assuming a consistent horizontal water surface across the slice (Greenwood, 1983; Coppin and Richards, 1990) (i.e. $U_2 - U_1 = -ub\tan\alpha$) or

Table A.1. Notation for slope stability analysis by the method of slices

Term	Units	Description
h	m	Average height of slice
b	m	Width of slice
ℓ	m	Length (chord) along base of slice
R	m	Radius of slip circle
c'	kN/m^2	Effective cohesion at base of slice
ϕ'	degrees	Effective angle of friction at base of slice
γ	kN/m^3	Bulk Unit weight of soil in slice
γ_w	kN/m^3	Unit weight of water (usually taken as 10 kN/m^3)
W	kN	Total weight of soil in slice (for layered soils, with soils 1,2,3 etc. $W = (\gamma_1 h_1 + \gamma_2 h_2 + \gamma_3 h_3 + \text{etc}) \times \text{b}$)
α	degrees	Inclination of base of soil slice to horizontal (negative at toe)
h_{w1}	m	Height of free water surface above left hand (downslope) side of slice
h_{w2}	m	Height of free water surface above right hand (upslope) side of slice
U_1	kN	Water force on left hand (downslope) side of slice (from flow net, seepage calculations or based on h_{w1})
U_2	kN	Water force on right hand (upslope) side of slice (from flow net, seepage calculations or based on h_{w2})
h_w	m	Average piezometric head at the base of the slice. For hydrostatic conditions $h_w = (h_{w1} + h_{w2})/2$
u	kN/m^2	Average water pressure on base of slice ($= \gamma_w \times h_w$)
f_u	kN	Resultant seepage force on slice
τ	kN	Available shear resistance
S or S_f	kN	Shear force ('disturbing' force)
N'	kN	Effective normal force on base of slice
X_1, X_2	kN	Total vertical interslice forces
X'_1, X'_2	kN	Effective vertical interslice forces
E_1, E_2	kN	Total horizontal interslice forces
E'_1, E'_2	kN	Effective horizontal interslice forces
K	ratio	Earth Pressure Coefficient ($\sigma h'/\sigma v'$)
F	ratio	Factor of Safety (usually shear strength/ shear force on slip plane)
F_m	ratio	Factor of Safety in terms of moment equilibrium
F_f	ratio	Factor of Safety in terms of horizontal force equilibrium

from Eq. (A.8) assuming $K = 0$

$$F = \frac{\sum(c'\ell + [(W - ub)\cos\alpha]\tan\phi')}{\sum W \sin\alpha} \tag{A.9}$$

The basic Simple Eq. (A.9) is readily applied and is appropriate for routine analysis where slope, strata and groundwater conditions are not known in any detail. It gives sensible values of the calculated Factor of Safety in most situations.

Table A.2. Notation for additional vegetation, reinforcement and hydrological effects

Vegetation, reinforcement and hydrological effects		
c'_v	kN/m^2	Additional effective cohesion at base of slice (due to vegetation etc.)
W_v	kN	Increase in weight of slice due to vegetation (or surcharge)
T	kN	Tensile root or reinforcement force on slice
θ	degrees	Angle between direction of T and base of slip surface
D_w	kN	Wind force (downslope)
β	degrees	Angle between wind direction and horizontal (often assume equal to slope angle)
Δh_{w1}	m	Increase in height of free water surface above left (downslope) side of slice
Δh_{w2}	m	Increase in height of free water surface above right (upslope) side of slice
ΔU_1	kN	Increase in water force on left hand (downslope) side of slice
ΔU_2	kN	Increase in water force on right hand (upslope) side of slice
Δh_w	m	Increase in average piezometric head at base of slice (due to vegetation)
Δu_v	kN/m^2	Increase in average water pressure at the base of the slice, $= \gamma_w \times \Delta h_w$

The *Swedish equation* (Fellenius, 1936) is derived from the General Eq. (A.3) by making the assumption that the water surface is parallel to the slip surface (Greenwood, 1987; Morrison and Greenwood, 1989) i.e. $U_2 - U_1 = 0$ therefore Eq. (A.3) becomes:–

$$F = \frac{\sum[c'\ell + (W\cos\alpha - u\ell)\tan\phi']}{\sum W \sin\alpha} \tag{A.10}$$

This Eq. (A.10) is shown to give considerable error when steep base angles to the slice are combined with high water pressures (Turnbull and Hvorslev, 1967; Greenwood, 1983). It is generally conservative. It is 'correct' only for the theoretical continuous slope situation with seepage parallel to the slope where it is appropriate to assume $U_2 - U_1 = 0$.

The *Bishop equation* (Bishop, 1955) is

$$F = \frac{\sum\left[\frac{(c'b + (W - ub)\tan\phi')\sec\alpha}{(1 + (1/F_m)\tan\phi'\tan\alpha)}\right]}{\sum W \sin\alpha} \tag{A.11}$$

This equation may be related to the General Eq. (A.3) but in general the assumptions do not correspond with the real distribution of the inter slice pore water forces (Morrison and Greenwood, 1989). The Bishop solution is prone to errors and the equation can become mathematically unstable for high values of α (Turnbull and Hvorslev, 1967; Greenwood, 1983; Krahn, 2001). It may consequently overestimate the Factor of Safety for deep slip surfaces.

The *Janbu* stability equation (Janbu, 1954; Janbu et al., 1956) is identical to Bishop except that the equation is expressed in terms of horizontal force equilibrium (see later), and a compensatory multiplying factor is introduced relating to the geometry of the slip surface

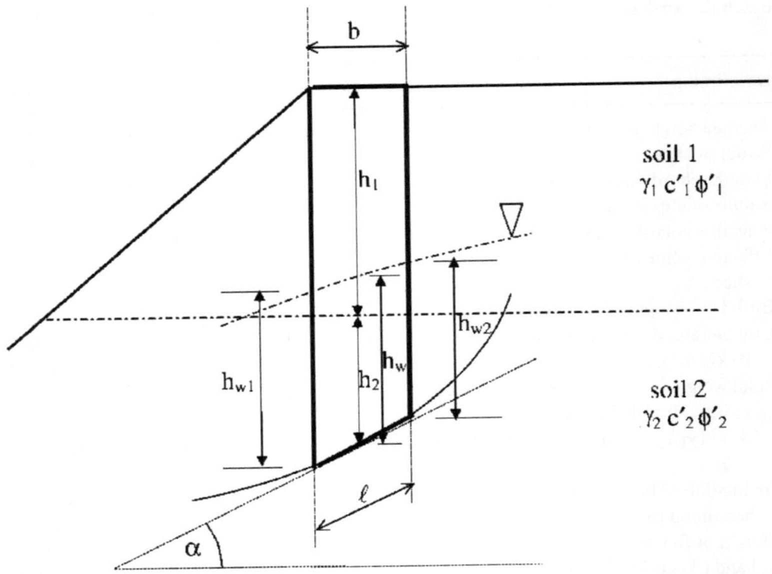

Figure A1. Limit equilibrium slope stability analysis by 'Method of Slices' – Dimensions and parameters assigned for each slice.

(typically $f_0 = 1.05$).

$$F_f = \frac{\sum\left[\frac{(c'b + (W - ub)\tan\phi')\sec\alpha}{(1 + (1/F_t)\tan\phi'\tan\alpha)\cos\alpha}\right]}{\sum W \tan\alpha} \times f_0. \quad (A.12)$$

Horizontal force equilibrium

It is sometimes convenient to express the Factor of Safety in terms of horizontal force equilibrium, for example for slips involving a significant near horizontal movement or to relate to retaining wall design. The equivalent horizontal forces are determined for each slice of the analysis simply by dividing the numerator and denominator of the stability equation by $\cos\alpha$. Eqs. (A.3) General, (A.7) General with K included, (A.8) Greenwood with K included, (A.9) Simple, and (A.10) Swedish, may all be converted to horizontal force equilibrium in the same way as the Bishop equation (A.11) converts to the Janbu equation (A.12).

The notation F or F_m is normally used for moment equilibrium and F_f for horizontal force equilibrium.

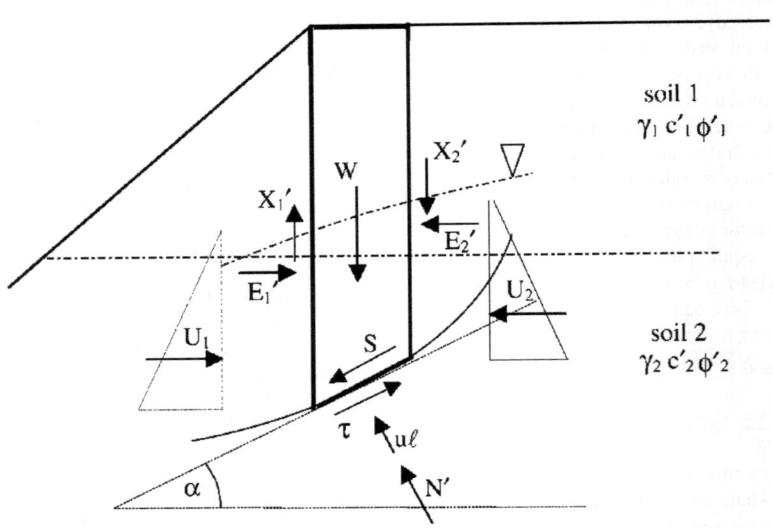

Figure A2. Forces associated with each slice.

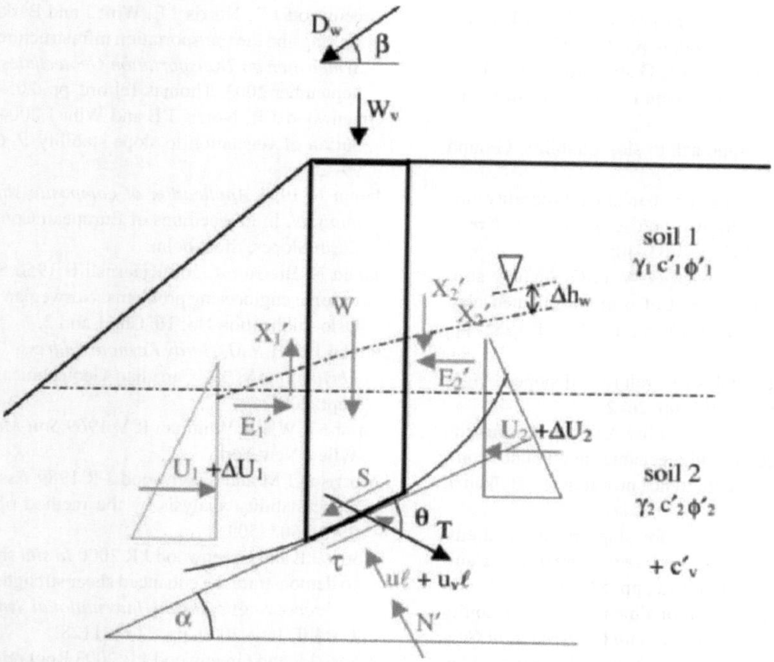

Figure A3. Additional forces due to vegetation, reinforcement and hydrological changes.

Effects of reinforcement, vegetation and hydraulic changes

The simple mathematical form of the Greenwood stability equations with the Factor of Safety simply expressed by a summation of restoring and disturbing moments or forces makes the inclusion of additional forces due to ground reinforcement, anchors or vegetation effects relatively straightforward.

It is not straightforward to add these additional forces in the Bishop, Janbu and other 'sophisticated' published solutions where the global factor of safety is applied to the shear strength parameters for each slice of the analysis resulting in some unrealistic force scenarios for the slices where anchor and reinforcement loads are applied (Krahn, 2001).

The General Eq. (A.3) is adapted for inclusion of the vegetation effects, reinforcement and hydrological changes, (Table A2, Figure A3), as follows (Greenwood et al., 2003, 2004):

$$F = \Big(\sum[(c' + c'_v)\ell + ((W + W_v)\cos\alpha - (u + \Delta u_v)\ell - ((U_1 + \Delta U_{2v}) - (U_1 + \Delta U_{1v}))\sin\alpha - D_w \sin(\alpha - \beta) + T\sin\theta)\tan\phi']\Big) \Big/ \Big(\sum[(W + W_v)\sin\alpha + D_w\cos(\alpha - \beta) - T\cos\theta]\Big). \qquad \text{(A.13)}$$

It is noted that the tangential reinforcement force, $T\cos\theta$, is deducted from the denominator to treat it as a negative disturbing force (shear force) rather than treating it as an additional restoring force. This approach is statically correct in accordance with the force diagram. The calculated value will be identical for a value of Factor of Safety of 1.

The water forces, U_1 and U_2, acting on the downslope and upslope sides of the slice are calculated by the spreadsheet based on an assumed hydrostatic water pressure below the free water surface:

$$\text{i.e., } U_1 = \frac{\gamma_w h_{w1}^2}{2} \qquad \Delta U_1 = \frac{\gamma_w (h_{w1} + \Delta h_{w1})^2}{2} - \frac{\gamma_w h_{w1}^2}{2}$$

Alternatively, values of U1 and U2 may be obtained elsewhere (by flow net or seepage program etc.) and entered directly into the spreadsheet.

The additional reinforcement, vegetation and hydraulic terms are similarly added in to the Greenwood Simple (A.9), Greenwood (*K* as input) (A.8) and Swedish (A.10) equations to provide the new Factor of Safety due to the effects considered.

It is concluded that for routine stability analysis the General equation (A.3) is most appropriate and gives a sensible estimate of the Factor of Safety for all slope and hydrological conditions. Vegetation and reinforcement forces are readily included (A.13).

References

Bishop A W 1955 The use of the slip circle in the stability analysis of earth slopes. Geotechnique, 5(1), 7–17.

Coppin N J and Richards I G (J.R. Greenwood) 1990 Contribution to CIRIA publication on Use of Vegetation in Civil Engineering, pp. 179–182, Butterworths, London.

Fellenius W 1936 *Calculation of the Stability of Earth Dams* 4, Trans. 2nd Int. Cong. Large Dams, Washington, pp. 445–459.

Finlayson D M, Greenwood J R, Cooper C G and Simons N E 1984 Lessons to be learnt from an embankment failure. Proc. Inst. Civ. Engs., 76, 207–220. Part 1.

Greenwood J R 1983 A simple approach to slope stability. Ground Eng., 16(4), 45–48.

Greenwood J R 1985 Wedge analysis of embankment instability during construction of the M25 Motorway, *In ICE Int. Symp. on Earthworks Failures*, 6–7 March 1985, pp. 403–405.

Greenwood J R, Holt D A and Herrick G W 1985 Shallow slips in highway embankments constructed of overconsolidated clay, *In ICE Int. Symp. on Earthworks Failures*, 6–7 March 1985, pp. 69–82.

Greenwood J R 1986 Stability analysis of reinforced slopes. *Highways and Transportation*, Oct. 1986, pp. 26–27.

Greenwood J R 1987 Effective Stress Stability Analysis. Discussion in 9th European Conference on Soil mechanics and Foundations, Dublin, Sept. 1987. Vol 3, post conference proceedings, Balkema, 1989, pp. 1082–1083.

Greenwood J R 1989 Design approach for slope repairs and embankment widening. *Reinforced Embankments Symposium*, Cambridge, Sept. 1989, Thomas Telford Ltd., pp. 51–61.

Greenwood J R 1990 Inclusion of reinforcement forces in stability analysis. Communication in *4th International Conference on Geotextiles Geomembranes and Related Products*, The Hague, May 1990, pp. 114, 997–999.

Greenwood J R and Zytynski M 1993 Stability analysis of reinforced slopes. *Geotextiles and Geomembranes*, 12, 413–424.

Greenwood J R, Vickers A W, Morgan, R P C, Coppin N J and Norris J E 2001 *Bioengineering The Longham Wood Cutting Field Trial*, CIRIA Project Report 81, London.

Greenwood J R, Norris J E, Wint J and Barker D H 2003 Bioengineering and the transportation infrastructure, *In Proceedings of the Symposium on Transportation Geotechnics*, EMGG, Nottingham, September 2003. Thomas Telford, pp. 205–220.

Greenwood J R, Norris J E and Wint J 2004 Assessing the contribution of vegetation to slope stability, *J. Geotech. Eng.*, 157(4), 99–208.

Janbu N 1954 *Application of composite slip surface for stability analysis*, In Proceedings of European Conference on Stability of Earth Slopes, Stockholm.

Janbu N, Bjerrum L and Kjaernsli B 1956 Soil mechanics applied to some engineering problems. Norwegian Geotechnical Institute, Oslo Publication No. 16, Chs 1 and 2.

Krahn J 2001 *R.M. Hardy Keynote Address "The limits of limit equilibrium analysis"*, Canadian Geotechnical Conference, Calgary, Sept. 2001.

Lambe T W and Whitman R V 1969 *Soil Mechanics*, pp. 363–365. Wiley, New York.

Morrison I M and Greenwood J R 1989 Assumptions in simplified slope stability analysis by the method of slices, *Geotechnique*, 39(3), 503–509.

Norris J E and Greenwood J R 2000 *In situ* shear and pull out testing to demonstrate the enhanced shear strength of root reinforced soil. *In Proceedings of the 8th International Symposium on Landslides*, Cardiff, June 2000, pp. 1123–1128.

Norris J E and Greenwood J R 2003 Root reinforcement on unstable slopes in Northern Greece and Central Italy. *In International Conference on Problematic Soils*, Nottingham, July 2003, Nottingham Trent University, pp. 414–418.

Turnbull W J and M J, Hvorslev (1967) Special problems in slope stability. *Proceedings of the ASCE*, 93(SM4), 499–528.

Site investigation for the effects of vegetation on ground stability

John R. Greenwood[1], Joanne E. Norris & Jo Wint
School of the Built Environment, Nottingham Trent University, Burton Street, Nottingham, England, NG1 4BU. [1]*Corresponding author**

Key words: desk study, ground stability, *in-situ* and laboratory testing, site investigation, vegetation

Abstract

The procedure for geotechnical site investigation is well established but little attention is currently given to investigating the potential of vegetation to assist with ground stability. This paper describes how routine investigation procedures may be adapted to consider the effects of the vegetation. It is recommended that the major part of the vegetation investigation is carried out, at relatively low cost, during the preliminary (desk) study phase of the investigation when there is maximum flexibility to take account of findings in the proposed design and construction. The techniques available for investigation of the effects of vegetation are reviewed and references provided for further consideration. As for general geotechnical investigation work, it is important that a balance of effort is maintained in the vegetation investigation between (a) site characterisation (defining and identifying the existing and proposed vegetation to suit the site and ground conditions), (b) testing (*in-situ* and laboratory testing of the vegetation and root systems to provide design parameters) and (c) modelling (to analyse the vegetation effects).

Introduction

The procedures for site investigation before construction and environmental projects and the scope of necessary technical input have been defined by various guidance publications and texts (Site Investigation Steering Group, 1993; Clayton et al., 1995; Simons et al., 2002; Greenwood, 2005; Highways Agency HD22/02). Little attention has been given during routine geotechnical investigation to the part that vegetation might play in contributing to the engineering stability of the existing site or proposed works.

Whilst the potential application of vegetation to assist stability is generally associated with slopes (Barker, 1986; Coppin and Richards, 1990; Gray and Sotir, 1995; MacNeil et al., 2001), it should be noted that vegetation also plays a part in stabilising horizontal surfaces to improve shear resistance. The penalty miss by footballer David Beckham during the European Cup finals of June 2004 (Figure 1) was claimed by Sven Goran Eriksson, the coach, to be due to the fact that 'he slipped with his foot once again because the area around the penalty spot didn't have enough grass'. The significance of ground stability for multi-million pound/euro sporting events should not be underestimated in today's economy which increasingly depends on leisure activities.

The more traditional need for applications of soil bioengineering (or eco-engineering) to sloping ground are illustrated in Figures 2 and 3 where the occurrence of shallow landslides may well have been reduced with appropriate soil bioengineering measures. The investigation of the effects of vegetation is particularly relevant to shallow slope failures, preventative works and erosion control.

Current procedures for geotechnical site investigation

Investigation stages

The investigation work for most construction projects is divided into stages as illustrated in Table 1. The Geotechnical Advisor is normally appointed at the outset of the project and will ensure appropriate geotechnical input at each stage.

* E-mail: john.greenwood@ntu.ac.uk

This article has been previously published in the following journal—Journal of Geotechnical and Geological Engineering, 24, 3, 467–481.
Stokes et al. (eds), Eco- and Ground Bio-Engineering: The Use of Vegetation to Improve Slope Stability, 203–212.

Figure 1. David Beckham misses a crucial penalty in the 2004 European Championships match against Portugal. (Robert Millward/Associated Press Web Site.)

The desk study/preliminary sources study
The desk study, sometimes referred to as the 'initial appraisal' or 'preliminary sources' study is vital for determining a preliminary understanding of the geology of the site and the likely ground behaviour. The term 'desk study' can be misleading because in addition to collection and examination of existing information, it must include a walk-over survey. The study will determine what is already known about the site and how the ground should be investigated.

Before embarking on intrusive ground investigation work, much valuable information may be readily gleaned from existing sources such as geological and Ordnance Survey maps, aerial photographs and archival material. Such documents can yield much about site conditions. The information from these sources is combined with the walkover survey to enable preparation of a geotechnical ('geohazard') plan of the site. A check list of information to be sought in a desk study is given by Perry (1996).

The desk study often represents the most cost effective element of the entire site investigation process revealing facts that cannot be discovered in any other way. The preliminary engineering concepts for the site are prepared and developed at the desk study phase based on the acquired information. The ground investigation in the field is then designed to confirm the conditions

Figure 2. Shallow landslide problems blocking roads and trapping motorists in Scotland, after heavy rains in August 2004 (Times Newspapers).

are as predicted and to provide ground information for the detailed design and project construction.

The walkover survey
The walkover survey is a detailed inspection of the site often done in stages with the initial visit for familiarisation, photography and checking of the current site conditions and with subsequent visits to confirm features noted on historical maps and photographs, etc. Features should be sketched at an appropriate scale on a base plan for inclusion in the desk study report.

The procedural statement
The key to successful site investigation lies in the planning process. If all aspects of the investigation work are considered in advance together with necessary actions

Figure 3. Instability of cutting slopes on the M11 near Loughton. Adjacent vegetated areas appear more stable.

relating to the likely findings, then the outcome is likely to be satisfactory for all parties involved.

A convenient way to bring together and record the proposals for each stage of site and ground investigation is by a 'Procedural Statement' (sometimes referred to as the 'Statement of Intent' or the 'Ground Investigation Brief'). This approach was formally introduced by the Department of Transport/Highways Agency in the 1980s and has now become widely accepted as good practice (Highways Agency HD 22/02). An example of headings and topics covered in a Procedural Statement is given in Table 2. Headings and content will change slightly for each phase of the investigation process as more information is accumulated.

The Procedural Statement is usually prepared by the Geotechnical Engineer/Advisor responsible for the work and should be agreed by all interested parties, and in particular the client, before the investigation proceeds.

The Statement encourages the designer to consider relevant aspects of the proposed investigation and to seek authority to proceed. It forms a valuable document within a quality management system and it becomes a base reference as the investigation proceeds in case changes are needed in the light of the findings.

Addition of the vegetation investigation

The proposed additional sections and notes to consider the effects of vegetation in the Procedural Statement

Table 1. Stages of a geotechnical investigation (Greenwood, 2005)

Construction phase	Investigation work
Definition of project	Appointment of Geotechnical Advisor for advice on likely design issues
Site selection	Preliminary Sources Study (Desk Study) to provide information on relative geotechnical merits of available sites.
Conceptual design	Detailed Preliminary Sources Study (Desk Study) and site inspections to provide expected ground conditions and recommendations for dealing with particular geotechnical design aspects and problems. Plan Ground Investigation (Procedural Statement)
Detailed design	Full Ground Investigation and geotechnical design. (Additional ground investigation if necessary for design changes or for problematic ground conditions)
Construction	Comparison of actual and anticipated ground conditions. Assessment of new risks (Additional ground investigation if necessary)
Performance/maintenance	Monitoring, instrumentation, feedback reporting.

Table 2. Example content of a Procedural Statement to be prepared before the Ground Investigation phase (HD 22/02) (***Suggested additions for vegetation investigation shown in bold italic***)

THE PROCEDURAL STATEMENT—Prepared by the responsible Geotechnical Advisor and agreed by the client and interested parties prior to each investigation phase.

1. SCHEME
Details of Scheme and any alternatives to be investigated; Key location plan.

2. OBJECTIVES
(For example) To provide information to confirm and amplify the geotechnical and geomorphological findings of the desk study as reported separately and to obtain detailed knowledge of the soils encountered and their likely behaviour and acceptability (for earthworks). To ascertain ground water conditions and location of any underground workings ***and nature of existing vegetation and potential for planting to enhance soil stability***. (Work limits to be defined).

3. SPECIAL PROBLEMS TO BE INVESTIGATED
Location of structures. Subsoil conditions below high embankments. Aquifers and likely water-bearing strata affecting the proposed works. Rock stability problems. Man-made features to be encountered. Effects on adjacent properties etc. ***Vegetation problems and benefits***.

4. EXISTING INFORMATION
List of all relevant reports and data. ***Including survey of existing vegetation and its potential contribution to stability. Review of plant suitability guidance***.

5. PROPOSED INVESTIGATION WORK
Fieldwork—Details of exploratory work proposed for specific areas with reasons for choice of investigation methods selected. Proposed sampling to match laboratory testing (***including studies of vegetation and its effects***).
Laboratory work—Details of proposals with reasons for choice of tests and relevance to design (***including root strength assessment***).

6. SITE AND WORKING RESTRICTIONS
Assessment of risk associated with proposals. Site safety, traffic management, difficult access, railway working, ***preservation of existing vegetation, topsoil*** etc.

7. SPECIALIST CONSULTATION
Details of specialist needed to support proposals (***including plant specialists, bioengineers etc***).

8. PROGRAMME, COST AND CONTRACT ARRANGEMENTS
Anticipated start date, work programme, contract arrangements, cost estimates, specification and conditions of contract. Arrangements for work supervision, etc.

9. REPORTING
Responsibility for factual and interpretive reporting. Format of reports and topics to be covered (***including assessment of existing and proposed vegetation***).

are shown in bold italic in Table 2. This will draw the attention of the project team (and funders) to the possible application of the vegetation to assist the engineering performance. It will highlight the need for specialist consultation and help plan the necessary investigation to demonstrate the potential of the vegetation.

Suggested outline procedure for investigation of vegetation

Table 3 outlines the typical factors relating to vegetation which may be considered at each stage of the investigation. It is noted that the major part of the vegetation study can (and should) be completed at the desk study/preliminary stage.

Review of techniques available to help investigate the effects of vegetation

The following paragraphs briefly review the techniques which may be used for investigation of vegetation effects and provide references for further consideration of the various techniques.

Vegetation survey

The extent of a survey of existing vegetation will relate to its relevance to the planned works. There is little point in carrying out detailed surveys of existing vegetation if the proposed works require re-profiling of the ground and removal of vegetation and topsoil. On the other hand, where existing vegetation can be preserved

Table 3. Factors to be considered for inclusion of vegetation effects in stages of routine site investigation

VEGETATION CONSIDERATIONS

Desk study phase

i) Soils

Existing Topsoil—shallow hand dug pits to provide initial information on soil and vegetation Subsoils—likely penetration and distribution of plant roots

Proposed fill materials—possible provision of irrigation/drainage layers to encourage deep root growth

ii) Vegetation

Typical presence and distribution of vegetation (detail depends on project)

Consider use of non invasive techniques (Ground Profiling Radar) to assess root distribution. Identification of indigenous species with potential to assist stability (recognising need for biodiversity)

Grass cover (survey by quadrats—one metre square with 100 mm grid)—Detail to be considered

Plan of vegetation types, trees, etc. across site

List uncertainties re: vegetation (i.e., root distribution, root penetration, tensile strength, pull out resistance, etc.) that may be assessed during main investigation phase

iii) General

Review vegetation influences on adjacent sites

Consider areas of proposed works which might benefit from vegetation to assist stability

Draw up schedule of site zones and information required

Check reference texts and Slope Decision Support Systems for guidance on likely benefit

Check availability of plant / seeds (liaising with specialist plant producers and landscape architect)

Carry out preliminary ground modelling and stability analysis based on assumed properties for soil, hydrology and vegetation.

Main ground investigation

If existing vegetation to be assessed:-

Trial Pits to :-

a) describe topsoil, depth, organic content, standard tests for topsoil classification (BS5930)

b) assess root distribution and carry out *in situ* pull out resistance tests

c) take samples of roots for laboratory tests on tensile strength

d) carry out *in situ* shear tests on root reinforced soils (larger investigations only)

e) compare moisture content profiles in vegetated and non vegetated areas due to different types of vegetation

Possible seasonal monitoring of moisture content profiles by access tube (TDR or Theta Probe technologies)

For future vegetation:-

Assess vegetation growth on adjacent sites

Assess topsoil and subsoil types available and likely vegetation types which can be supported in the region

Analysis

Stability analysis by limit equilibrium methods (numerical methods for ground modelling on larger projects) to assess the influences of the vegetation and help design additional planting and vegetation maintenance schemes

Where little or no existing vegetation is present (regraded slopes etc) analyse benefits/dis- benefits of proposed planting scheme

Construction stage

Monitoring and protection of existing plants and topsoil

Treatment of soils to encourage deeper rooting

Topsoil /subsoil preparation and planting (in association with plant specialist and landscaper) Review conditions on site as found against those predicted—modify design if necessary

Confirm that dependency on vegetation does not introduce inappropriate risks to property and life (If so a 'hard' engineering solution is essential)

Feedback/maintenance

Report on achieved objectives of vegetation and planting and provide programme of necessary on-going maintenance inspections and actions to be taken in light of certain 'foreseen' events

its nature should be recorded and possible contribution to ground stability assessed. The following is recommended:

- All trees and shrubs should be identified and locations recorded with local investigations of root extent where possible.
- The general presence and nature of ground cover (grasses, 'weeds', etc.) should be recorded.
- The maturity and vitality of the vegetation should be recorded.

Where existing (or proposed planted) vegetation is to play a role in engineering stability, more detailed surveys should be carried out as suggested by Cammeraat et al. (2002). The survey is carried out by placing a suitable square grid (quadrat) over the soil and vegetation to record and monitor factors such as the seasonal variation, percentage ground cover and the determination of the mass of vegetation (biomass). The advice of a plant specialist to assist with such surveys is recommended.

Topsoil and subsoil

As the prime growing medium, the available topsoil and subsoils (upper 1.5 m) should be classified in horticultural terms so that existing suitable plants can be encouraged or new plants selected for their engineering contribution.

Consideration might be given to possible treatment of the topsoil and subsoils by aeration and/or fertiliser, to encourage the development of mycorrhizal associations and deeper, healthy root growth (Ryan and Bloniarz, 2000).

Trial pits and boreholes

Shallow trial pits, preferably hand dug, can often be put down with minimal disturbance and provide an excellent means of assessing root distribution and the nature of the topsoil and subsoil layers. As the excavation only represents a snapshot in time, the likely seasonal influences of changing moisture conditions need to be considered (Greenwood et al., 2001).

Root size and distribution may be assessed and recorded by image analysis of the trial pit wall or by manual counting using a 'quadrat' or square grid, typically of 100 mm squares, placed over the vertical sides or horizontal base of the pit (Greenwood et al., 2001).

Boreholes are less valuable than pits for root distribution analysis but horizontal sections through recovered core samples can provide a limited indication of root counts (Greenwood et al., 2001).

Geophysical techniques for root location

Geophysical techniques such as ground penetrating radar have been used with partial success to map tree root systems. The four fundamental factors to consider with any geophysical method are penetration, resolution, signal to noise ratio and contrast in physical properties (McCann et al., 1997). There is a trade off between resolution and penetration depth, penetration may be increased by using a lower frequency but resolution is improved by using a higher frequency (Hruska et al., 1999). However, the attenuation also depends on the conductivity of the soil, therefore, soil type and overall root depth are important factors determining the success of this method. Dobson (1995) and Hruska et al. (1999) have reported successful plan and three dimensional images of roots, but Stokes et al. (2002) reported problems with root crossover and branching, and in determining the location of roots less than 20 mm diameter.

The geophysical techniques are worthy of further consideration to supplement the physical investigations particularly as computer processing power increases to help interpret the geophysical survey results.

Moisture content determination

Moisture content is a fundamental property relating to soil strength and consolidation characteristics. Changes in moisture content will occur primarily due to seasonal effects but also due to the influence of the vegetation. Seasonal comparisons of moisture content profiles in vegetated and non-vegetated areas of the site will be of assistance in considering the vegetation effects.

Physical sampling inevitably involves partial destruction of the site by trial pit or borehole and therefore can only provide a snapshot of conditions at the time of excavation. Moisture profiles at close centres (say 50 or 75 mm) on a vertical profile or as a grid around root networks can provide helpful information. The 'moisture in the bag' technique (Greenwood and Norris, 1999) saves time on sampling and laboratory drying procedures.

Other techniques such as time domain reflectometry (TDR) (Topp and Davis, 1985), Theta probe (Gaskin and Miller, 1996), and Neutron probe (Vickers and Morgan, 1999) permit monitoring of moisture content over extended periods by having either a permanent access tube installed for insertion of a probe or by leaving an instrument buried in the ground to allow continuous real time monitoring. Considerable success is reported with these devices (Greenwood et al., 2001; Vickers and Morgan, 1999) although caution is needed in their calibration which should preferably be done against physical moisture content determination. The remote devices generally record volumetric moisture content (volume of water divided by total volume of specimen) as compared with the gravimetric moisture content (mass of water divided by dry mass of soil specimen) which is more familiar to geotechnical engineers (BS 1377; 1990). Relating the two approaches to moisture content requires the measurement or assumption of the dry density of the soil, i.e.,

$$\text{Gravimetric moisture content} = \text{Volumetric moisture content} \times \frac{\text{Density of water}}{\text{Dry density of soil}}$$

(Greenwood et al. 2001).

Water pressures

Effective stresses which govern the stability of soil slopes are dependent on the pore water pressures present in the soil mass. Traditional monitoring devices of standpipes and piezometers (BS5930, 1999) are valuable for general slope stability monitoring but are unlikely to detect the specific influences of the vegetation (Greenwood et al., 2001). More detailed studies of wetting fronts during rainstorm events (Vickers and Morgan, 1999) and seasonal variation in water pressures are possible by means of tensiometer installations (Greenwood et al., 2001). Tensiometers are considered to be most helpful for assessing water pressures and suctions where the effects of vegetation and other hydrological influences are to be considered in detail (Anderson et al., 1996; Greenwood et al., 2001).

Root strength

For analysis of root reinforced soil an estimate of the contribution of roots to stability is required (see '*stability modelling*'). This may be obtained directly from *in situ* root pull-out tests (Norris and Greenwood, 2003) or from laboratory tests (Coppin and Richards, 1990). Again account needs to be taken of the season at which the testing is completed compared with the most critical 'wet' periods for the site.

Laboratory measurements of root tensile strengths are helpful and should provide root characterisation data to be checked against published results for the particular species (Ecoslopes manual, in preparation).

In situ shear tests can give a direct indication of the shear strength of root reinforced soil but are difficult to interpret in relation to the drained/undrained conditions and the stress distribution within the sample (Greenwood et al., 2004; Norris and Greenwood, 2000a, 2000b, 2003).

Stability modelling

The modelling tools available for analysing the effects of vegetation need to be considered at the outset so that the investigation is designed to provide the required data.

Various methods of limit equilibrium stability analysis are available in commercial packages such as SLOPE/W (Geoslope International Ltd.). Methods based on equilibrium of hydrological forces are shown to be most reliable for estimating the factor of safety and are readily adapted to include the vegetation effects (Greenwood, 2006). The SLIP4EX program based on Microsoft Excel, compares methods for a single slip surface and is freely available (contact: john.greenwood@ntu.ac.uk) for initial exploration of vegetation effects (Greenwood, 2006). Root effects may be represented by radial zones of enhanced soil properties around a single tree or by depth related zones parallel to the slope for general vegetation cover (Greenwood et al., 2003, 2004). Other models for consideration of soil–root interaction are discussed by Wu (1995, 2006) and Operstein and Frydman (2002).

When incorporating vegetation root effects, high partial factors of safety (typically around 8–10) are recommended to take account of the uncertainty of root distribution and anchorage lengths and the large strains necessary to generate the full tensile resistance of the root (Greenwood et al., 2003, 2004).

The power of numerical modelling by finite element or finite difference methods is such that both stress and strain and the generation of water pressures can be modelled for situations of root–soil interaction and ground water infiltration. The problem is that the setting up of accurate models and selection of appropriate parameters is not straightforward. Commercial programs such as Plaxis (Brinkgreve, 2002) and Seep/W (Geoslope International Ltd.) are helpful, particularly for assessing the sensitivity of the analysis to the assumed parameters.

Programs such as Forest Gales (Gardiner et al., 2000) are available to assess specific problems of the vulnerability of trees to wind damage. Other numerical programs are under development to record and model root systems and include their influence in ground models, e.g. Dupuy et al. (2004).

Slope decision support system

One of the key objectives of the EU funded ECOSLOPES project was to provide a slope decision support system (SDSS) to help practitioners to assess their slopes and select appropriate vegetation to help stabilise them. The SDSS may be trialled as a development version (Mickovski and van Beek, 2006; Ecoslopes Manual, in preparation) and it is intended that with the benefit of user feedback its scope will be confirmed to provide the necessary guidance for eco-engineering and soil bioengineering applications.

Discussion

The application of vegetation to assist engineering functions is not always straightforward and expectations as to what might be achieved must be realistic. However the costs are relatively low particularly at the preliminary (desk study) phase and therefore benefit/cost ratios may be high. The linking of the engineering solutions to an improved environment is a satisfactory and rewarding achievement.

Mistakes will inevitably be made and vegetation alone should not be relied on where life and property are directly at risk from resulting landslip.

As experience is gained the checklists and investigation techniques provided in this paper will be reviewed and updated. For all investigation work it has been recognised that there must be a balance of effort between the site and strata definition, the testing and the modelling (Burland, 1989). As vegetation considerations are included, this balance must be maintained with the site characterisation (defining strata, hydrological conditions and vegetation), balanced against the testing (on site and in the laboratory) and modelling (Figure 4). It is pointless carrying out detailed, sophisticated modelling if the strata, hydrology and vegetation properties are not properly defined. Equally, it is pointless doing many tests to determine vegetation characteristics and strengths if the results are not relevant to the site modelling.

Conclusions

Much of the assessment of the potential benefits (and dis-benefits) of vegetation can be efficiently completed at the desk study (preliminary) investigation stage and does not involve large expenditure. Furthermore, vegetation studies at the main ground investigation stage are again relatively low cost involving minimal ground intrusion.

Whilst the application of bioengineering will not be appropriate or relevant for all construction projects, the framework provided should encourage the project team to review the options for preservation or inclusion of vegetation which may enhance the engineering stability in addition to improving the landscape and environment.

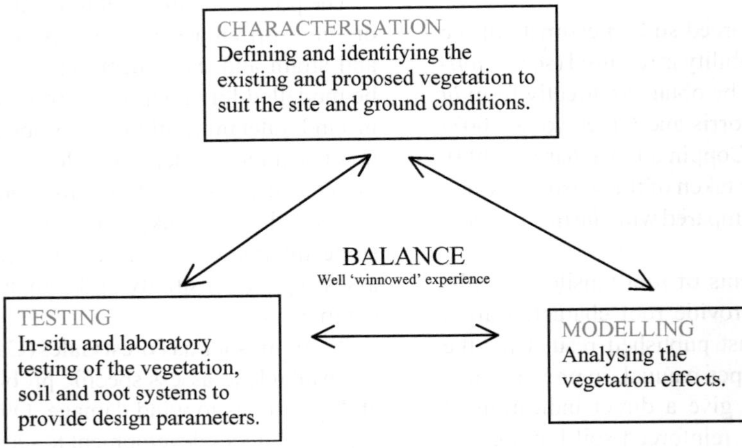

Figure 4. Balance of input into vegetation investigation work (Developed from Burland, 1989).

Acknowledgements

The support, discussion and inspiration provided by European partners during the project on 'Eco-engineering and Conservation of Slopes for Long-term Protection from Erosion, Landslides and Storms' (ECOSLOPES) was much appreciated. The authors are grateful for the funding of this work provided under the 5th Framework of the European Union.

References

Anderson M G, Collison A J C, Hartshorne J, Lloyd D M and Park A 1996 Developments in slope hydrology—Stability modelling for tropical slopes, *In* NG Anderson and SM Brooks (eds), Advances in Hillslope Processes, pp. 799–821. Wiley, Chicester.

Barker D H 1986 Enhancement of slope stability by vegetation, Ground Eng., 19, 11–15.

BS 1377 1990 British Standard Methods of Test for Soils for Civil Engineering, British Standards Institution [Parts 1–9].

BS 5930 1999 Code of Practice for Site Investigation, British Standards Institution.

Brinkgreve R B J 2002 Plaxis 2D—Version 8 Manual, Balkema, Lisse.

Burland J B 1989 Nash lecture: The teaching of soil mechanics—A personal view. *In* Proceedings of 9th European Conference on Soil Mechanics and Foundations, Dublin, September 1987, Balkema, Rotterdam/Boston 3: 1427–1441.

Cammeraat L H, van Beek L P H and Dorren L K A 2002 Ecoslopes Field Protocol. Version 1. University of Amsterdam, Amsterdam, the Netherlands.

Clayton C R I, Matthews M C and Simons N E 1995 Site Investigation: A Handbook for Engineers. Blackwell, Oxford.

Coppin N J and Richards I G 1990 Use of Vegetation in Civil Engineering, Butterworths, London.

Dobson M 1995 Tree root systems. Arboriculture Research and Information Note 130/95/Arb. Parnham Arboricultural Advisory and Information Sytems 1–6.

Dupuy L, Fourcaud T and Stokes A 2005 A numerical investigation into factors affecting the anchorage of roots in tension. Eur. J. Soil Sci., 56(3), 319–327, online doi:10.1111/j.1365-2389.2004.00666.x.

Gardiner B A, Suárez J C, Achim A, Hale S E and Nicoll B C 2000 ForestGALES. A PC-based Wind Risk Model for British Forests. Version 2.0, User's Guide, Forestry Commission, Edinburgh.

Gaskin G J and Miller J D 1996 Measurement of soil water content using a simplified impedance measuring technique, J. Agr. Res., 63, 153–160.

Gray D H and Sotir R B 1995 Biotechnical stabilization of steepened slopes, Transport. Res. Record, 1474, 23–29.

Greenwood J R 2005 Site and ground investigation, *In* Selley, Cocks and Plimer (eds), Encyclopedia of Geology, Vol. 1, pp. 580–594. Elsevier, Oxford.

Greenwood J R 2006 SLIP4EX—Program for routine slope stability analysis to include the effects of vegetation, reinforcement and hydrological changes. J. Geotech. Geol. Eng., 24, 449–465.

Greenwood J R and Norris J E 1999 'Moisture in the Bag', A simplified procedure for the determination of soil moisture content by oven drying, Ground Eng., 32(6), 32–33.

Greenwood J R, Norris J E and Wint J 2004 Assessing the contribution of vegetation to slope stability, J. Geotech. Eng., 157(4), 199–208.

Greenwood J R, Norris J E, Wint J and Barker D H 2003 Bioengineering and the transportation infrastructure. *In* Proceedings of the Symposium on Transportation Geotechnics, Nottingham, September 2003, Thomas Telford, London, pp. 205–220.

Greenwood J R, Vickers A W, Morgan R P C, Coppin N J and Norris J E 2001 Bioengineering—The Longham Wood Cutting field trial. CIRIA Project Report 81, London.

Highways Agency HD22/02 Ground Investigation and Earthworks—Procedure for Geotechnical Certification (DMRB 4.1.2).

Hruska J, Cermák J and Svatopluk S 1999 Mapping tree root systems with ground- penetrating radar, Tree Physiol., 19, 125–130.

MacNeil D J, Steele D P, McMahon W and Carder D R 2001 Vegetation for slope stability. TRL Report 515, Crowthorne, TRL Limited.

McCann D M, Eddlestone M, Fenning P J and Reeves G M 1997 Modern Geophysics in Engineering Geology, Geological Society Special Publication No. 12. Geological Society, London.

Mickovski S B and van Beek L P H 2006 A decision support system for the evaluation of eco-engineering strategies for slope protection. J. Geotech. Geol. Eng., 24, 483–498.

Norris J E and Greenwood J R 2000a Review of in-situ shear tests on root reinforced soil, *In* Stokes (ed), The Supporting Roots of Trees and Woody Plants: Form, Function and Physiology. Developments in Plant and Soil Sciences, Vol. 87, pp. 287–294. Kluwer, Dordrecht.

Norris J E and Greenwood J R 2000b *In situ* shear and pull out testing to demonstrate the enhanced shear strength of root reinforced soil. *In* Proceedings of 8th International Symposium on Landslides, Cardiff, June 2000, Thomas Telford, London, Vol. 3, pp. 1123–1128.

Norris J E and Greenwood J R 2003 *In-situ* shear box and root pull-out apparatus for measuring the reinforcing effects of vegetation. *In* Myrvoll (ed), Field Measurements in Geomechanics, pp. 593–597. Swets and Zeitlinger, Lisse.

Operstein V and Frydman S 2002 The stability of soil slopes stabilised with vegetation, Ground Improvement, 6(4), 163–168.

Perry J 1996 Sources of information for site investigations in Britain. TRL Report 192, Transport Research Laboratory, Crowthorne.

Ryan H D P and Bloniarz D V 2000 Irrigation Systems and Trees. UMASS Extension, University of Massachusetts Amherst, Massachusetts. (http://umassgreeninfo.org/fact_sheets/plantculture.html).

Simons N, Menzies B and Matthews M 2002 A Short Course in Geotechnical Site Investigation, Thomas Telford, London.

Site Investigation Steering Group 1993 Part 1: Without site investigation ground is hazard; Part 2: Planning, procurement and quality management; Part 3: Specification for ground investigation; Part 4: Guidelines for safe investigation by drilling of landfills and contaminated land. Thomas Telford, London.

Stokes A, Fourcard T, Hruska J, Cermak J, Nadyezdhina N, Nadyezhdin V and Praus L 2002 An evaluation of different

methods to investigate root system architecture of urban trees *in situ*: 1. Ground penetrating radar, J. Arboriculture, 28, 1–9.
Topp G C and Davis J L 1985 Measurement of soil water content using time- domain reflectometry (TDR): A field evaluation, Soil Sci. Soc. Am. J., 49, 19–24.
Vickers A W and Morgan R P C 1999 Soil–water monitoring to assess the effectiveness of three bioengineering treatments on an unstable Gault Clay cutting in Southern England. *In* Proceedings of the 2nd International Conference on Landslides, Slope Stability and the Safety of Infra-Structures, Singapore, July 1999, pp. 95–102.
Wu T H 1995 Slope stabilization. *In* RPC Morgan and RJ Rickson (eds), Slope Stabilisation and Erosion Control—A Bioengineering Approach, pp. 221–364. E and FN Spon, London.
Wu T H 2006 Root reinforcement: analysis and experiments. *In* Proceedings of the International Conference on Eco-Engineering: The Use of Vegetation to Improve Slope Stability, Thessaloniki, September 2004 (in press).

Mechanics of root-pullout from soil: A novel image and stress analysis procedure

O. Hamza[1,2], A.G. Bengough[2], M.F. Bransby[1,3], M.C.R. Davies[1] & P.D. Hallett[2]
[1]*Division of Civil Engineering, University of Dundee, Dundee, DD1 4HN, Scotland.* [2]*Plant-Soil Interface Programme, Scottish Crop Research Institute, Invergowrie, Dundee, DD2 5DA, Scotland.* [3]*Corresponding author**

Key words: anchorage, image analysis, plant–soil interaction

Abstract

When plants are loaded by external forces, whether they be above ground e.g. wind or canopy weight, or from within the soil e.g. soil displacement on slopes, the roots will be mechanically loaded. Exactly how the plant roots carry loads during these events is unknown because of their complex morphology and the heterogeneity of the root properties. To gain greater insight into plant root–soil mechanical interactions, a series of tests have been carried out to investigate the mechanical behaviour of roots and rubber root-analogues under tension during pull-out from soil. The results of the mechanical tests are augmented by a novel use of image analysis (specifically Particle Image Velocimetry) of sequential digital photographs taken during loading. This allows root and soil movements to be measured during the tests so that more can be learned about the effects of root morphology on the load distribution and deformation behaviour. The testing methodology and philosophy are presented here together with preliminary results.

Introduction

Plant roots are frequently loaded mechanically. These loads can be applied from external forces acting on the stem e.g. wind or canopy weight causing lodging/blow-down or derived from within the soil e.g. during slope failure. In particular, when a plant root is in a deforming slope, the soil may move past the root, thus loading it laterally (for the case of a well anchored root), or one section of plant may be moved with respect to the other (for a flexible root system) causing the base to be pulled from the soil (e.g. Wu and Watson, 1998). The forces resisting root deformation in the deforming zone will help to stabilise the soil slope and so knowledge of these forces would provide an initial step to the understanding of vegetation stabilised slopes (e.g. Coppin and Richards, 1990; Gray and Sotir, 1996). However, exactly how the plant roots carry loads during these events is unknown because of the complex root system morphology, and the heterogeneity of the root properties. The forces acting on a branched root system during pull-out are illustrated in Figure 1, to indicate the complex loading pattern in even a very simple root system.

This paper reports a series of tests which were carried out in order to gain greater insight into plant root–soil interactions during mechanical loading. The mechanical behaviour of individual root sections under tension, and pull-out behaviour of root systems from soil are studied. The results of the mechanical tests are augmented by the novel use of image analysis, thereby allowing both root and soil movements to be measured during root loading events. It is believed that this approach allows more to be learnt about the effects of root morphology on the load distribution and deformation behaviour of a root system.

To augment the tests, some tests were carried out using defined root analogues. These analogues have the advantage of possessing known, repeatable mechanical properties and can be cast to a selected shape. In this way, the mechanical properties and/or root-analogue architecture could be chosen by the researcher, allowing

* E-mail: m.f.bransby@dundee.ac.uk

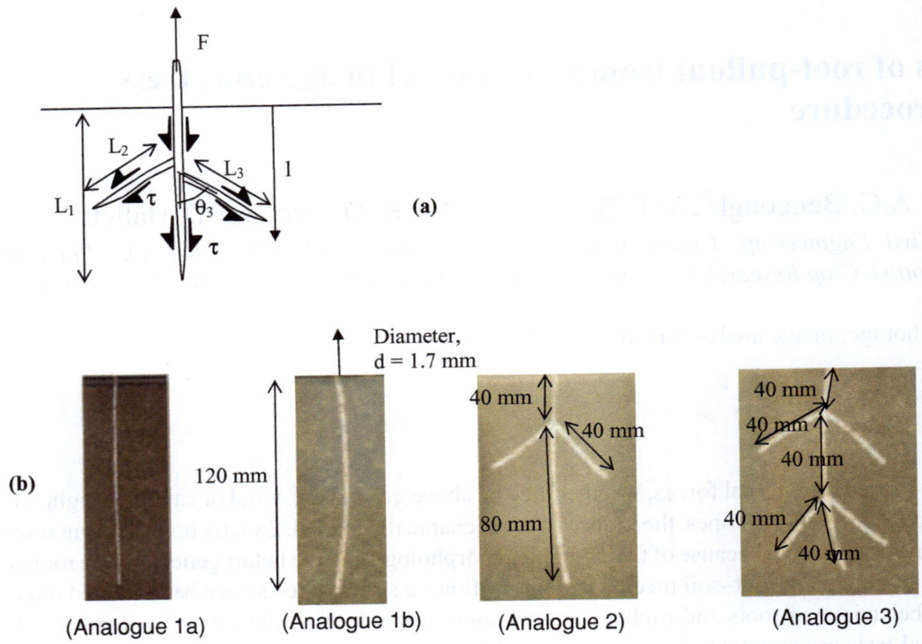

Figure 1. Applied mechanical loadings to plant root systems. (a) Forces resisting pull-out (system with laterals). (b) Vitron-rubber root-analogues used in this series of experiments.

the effects of isolated morphology changes to be investigated. The limitations of this approach will also be considered.

Materials and methods

Apparatus

Anchorage tests were performed in a modified root-box (Bengough et al., 2004) packed with soil, referred to here as a phytoplate apparatus. In the phytoplate apparatus, a 10 mm thick soil layer is packed between two transparent Perspex surfaces of width 210 mm and height 150 mm. Plants are grown or analogue-roots are placed within the soil layer and they can be observed photographically through the Perspex faces during externally applied mechanical loading (Figures 1b and 5c).

Pull-out loads were applied using a mechanical test frame (Model 5544, INSTRON). The cross-head displacement and force transmitted to the loading plate was recorded using INSTRON Merlin software. The cross-head displacement rate was set to 0.05 mm s^{-1}, with displacement accurate to 1 μm. The load cell had a range of 50 N and was accurate to 1% at 1/250 maximum load. The free end of the root was clamped to the displacing end of the test frame using screw-thread grips with hard rubber surfaces, while the soil container was held at the base by the rigid end of the test frame.

Soil preparation

The soil was placed into the phytoplate at a water content of 12 g/100g. A water supply was attached to a base drain and the water table was maintained at a static position below the base of the soil. Measurement of the suction within the soil using tensiometers prior to each loading event and this together with knowledge of the soil bulk unit weight allowed calculation of the effective stress conditions within the soil layer.

Pull-out tests

Two series of tests were performed. In one series, analogue plant roots (with known geometries and mechanical properties) were placed in the soil and pulled out vertically at a rate of 3 mm/min (0.05 mm/s). In the second series, plants were grown in the phytoplates and pulled out vertically by their stems.

Root analogues were made from Viton rubber (circular cross-section $d = 1.7$ mm), divided in half longitudinally. The Viton had a Young's modulus of approximately 7 MN/m^2 which is similar to that of many plant roots (Hamza et al., 2005; Niklas, 1999). The results of the root-analogue tests can be interpreted more thoroughly than for real roots, as their properties are known, homogenous and repeatable.

The analogue plant roots allowed the effects of different branching patterns on pull-out behaviour to be isolated experimentally. Two single unbranched root-analogues were investigated (root-analogues 1a and 1b; analogue 1a is straight and 1b bent) before roots with an additional two (root-analogue 2) or four (root-analogue 3) lateral branches (see Figure 1). All root-analogues were oriented with their flat face against the Perspex of the phytoplate to view the widest point at the line of symmetry. During this study only simple geometries with a small number of secondary roots and homogenous mechanical properties were investigated.

Tests were carried out on Pea (*Pisum sativum* L.) and Maize (*Zea mays* L.) seedlings grown in the soil for 5–14 days at 21°C. Results are presented later for an unbranched primary root of pea (5 d) and a larger maize seedling with a branched primary root.

For all tests carried out using in the phytoplate, sequences of digital photographs of the root and soil system were taken during each loading event using a Nikon D100 digital camera with a Nikkor, 60 mm lens, f2.8. These photographs were later analysed using Particle Image Velocimetry (PIV), which uses a cross-correlation technique to detect the movement of pixels between sequential digital images (Adrian, 1991). PIV has been extended to measure pre-failure strains in geotechnical tests by White et al. (2003), and their program was used in the reported tests. Using this technique, instantaneous axial and radial deformation along the whole or part of a plant root can be quantified (Hamza et al., 2005) alongside soil displacement adjacent to the roots. Computer programs in the Matlab environment (Matlab Guide, 1997) were written to display the measured displacements, and analyse these results further.

Results

Element tests for analogue roots

Results for the analogue element tests revealed that the stress-strain law under monotonic loading was approximated numerically by a power law: $\sigma = a\varepsilon^b$, where σ is the axial stress and ε the axial strain. For the data obtained, $a = 0.104$ N/mm^2 and $b = 0.75$, where strain is given as a percentage.

Straight analogue pull-out (root-analogue 1a)

The loading results of the analogue roots are shown in Figure 2. The axial force measured by the Instron load cell is plotted against the displacement of the Instron grips. Figure 2 also shows the pull-out data after the correction for the shoot extension above the soil surface.

A digital photograph was taken of the straight analogue (root-analogue 1a) taken after a root head displacement of 4 mm. A number of PIV image patches of 50 by 50 pixels along the root length were selected and then the GeoPIV76 program (White et al., 2003) was used to measure the displacements of each of these patches in the vertical (axial) and horizontal (transverse) direction between images (Figure 3). By adding up the displacements measured between pairs of a sequence of images, total displacements of the root could be quantified as the pull-out proceeded.

Figure 3a shows the calculated variation of axial (i.e. vertical) displacement of the root along its length at several instants during the root pull-out. Each profile represents the displacement profile of the root for a given head displacement, z_0 when a photograph was taken and Figure 3a shows only small shoot displacements up to a maximum, $z_0 \approx 5$ mm. It can be seen clearly that for small shoot displacements, only a small section of the root near the soil surface is displaced and as the shoot displacement increases, so does the zone of deformed root beneath it.

The strains within the root can be calculated from the displacement data. Each section of root between the pair of patches j and $j+1$ (numbered sequentially from the patch nearest the root tip) can be examined as an individual root element undergoing constant strain. Each element will have length, $l_j = x_{j+1} - x_j$, where x_j is the axial position of patch j measured from the root tip. Axial and transverse displacements are denoted by z and v respectively and so the PIV analysis measures these two displacement components z_j, v_j and z_{j+1} and v_{j+1}. If the root and displacement axes are coincident, then the axial strain

$$\varepsilon_{ai} = \left(\frac{z_{j+1} - z_j}{x_{j+1} - x_j}\right), \tag{1}$$

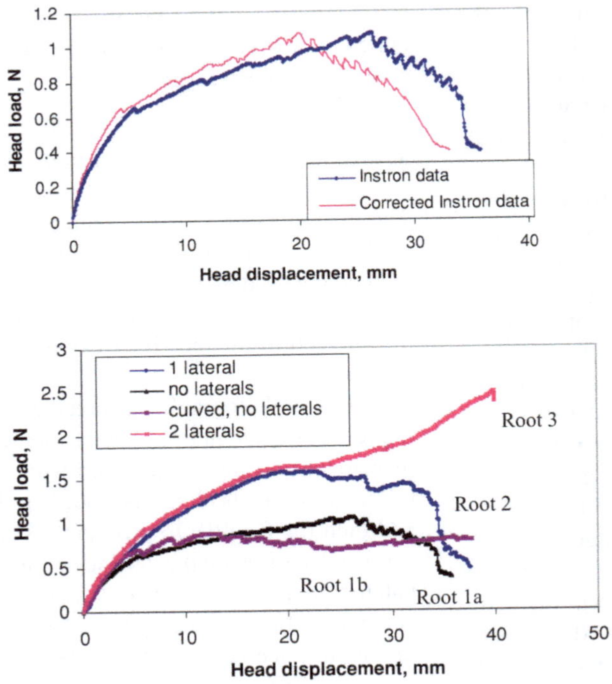

(a) Analogue root 1a

(b) All analogue roots

Figure 2. Head load-displacement responses for root-analogues.

for the root element with the axial position is $(x_{j+1} + x_j)/2$.

The axial strain in the initial stages of the straight analogue test deduced in this way is shown in Figure 3b. For any displacement of the top of the root-analogue, there is strain in the root-analogue near the soil surface, but this reduces to zero beyond a certain effective anchorage length. Because of small displacement measurement errors in the PIV, there is noise in the calculated strain profile, augmented because of the numerical differentiation. However, there is increasing axial strains (concentrated towards the soil surface) and an increasing root straining length with increased shoot displacement. Data for larger head displacements (not shown) show that the root behaves in a similar manner to that described above. Up to a head displacement of 22 mm, there is an increasing length of root which is displacing axially, but the root tip does not displace. However, at head displacements larger than 22 mm (which corresponds closely with the peak head load, Figure 2) the root tip starts to displace and the whole root pulls out of the ground (while the axial load reduces).

The distribution of axial force along the root at any instant can be calculated by combining the root stress-strain law, its cross-sectional area (A) and the instantaneous axial strain, ε_{ai}: $F = Aa\varepsilon_{ai}^{b}$. The deduced value of axial force and displacement from the image analysis data agreed closely with that measured by the loading frame, validating the analysis.

Root–soil interaction can be studied by further image analysis. Figure 4a and 4b show the digital photographs taken before and during root movement (the analogue is pulled from the top-centre of the image). The scales shown are image pixels measured from the base left of the image (note 25 pixels correspond to 1 mm). The light squares represent the position of the measurement (patch) positions in each image. Figure 4c shows displacement vectors for these points. There is negligible soil displacement at any measurement point outside the analogue surface, suggesting the presence of a very thin shear zone at the root–soil interface for these analogue roots. It can also be seen (Figure 4c) that the base of the root is moving a very small amount whereas the root head is displacing significantly (agreeing with the type of mechanism shown in Figure 3).

Branched analogue pull-outs

Figure 2 shows the load-displacement data for the analogue roots with different branching patterns (Figure 1).

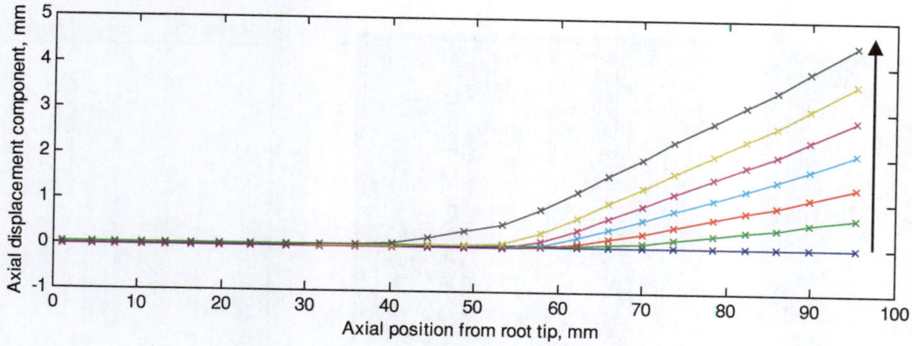

(a) Profile of axial displacement along the root length for a range of root head displacements.

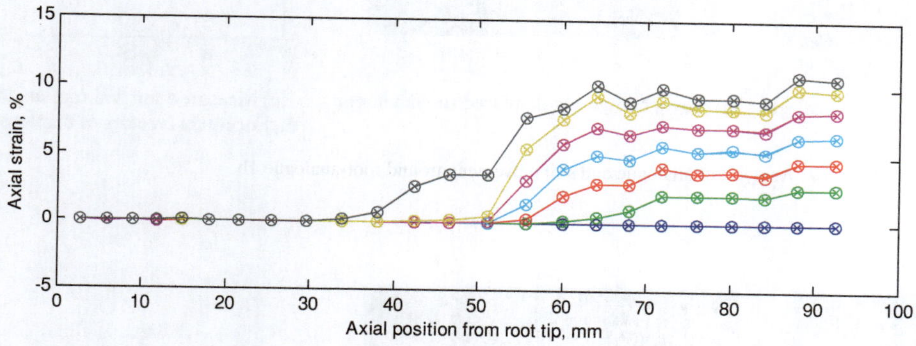

(b) Deduced profile of axial strain along the root length for a range of root head displacements (same displacements as Fig. 3a).

Figure 3. Axial displacement and strain profile along the root length (final head displacement, z_o = 6 mm). The strain profile can be used to calculate directly the force distribution along the root.

As expected, the taproot systems with no laterals have the lowest pull-out capacity (which is achieved at the smallest displacement), whereas the system with two pairs of laterals require the largest force to be pulled out of the ground.

For head displacements of less than about 4 mm there is negligible difference in behaviour between the different root-analogues. This is because the root-analogue deformation is restricted to the unbranched zone above any laterals for small displacements (as shown in Figure 3 for the single root) and so the additional root sections below this are not contributing to the axial force. Figure 3a shows that a 4 mm head displacement deforms only the top 45 mm of the root. This behaviour compares well with that of the single taproot length of 40 mm above the first branch root for all the analogues.

For a displacement larger than 4 mm there is a clear difference between the axial force required to pull the branched root-analogues from the ground compared to the unbranched root-analogue. This is because the higher pair of secondary roots begin to contribute to the axial resistance. However, up to a head displacement of about 20 mm there is no difference between the root-analogue with a single pair of laterals (root-analogue 2) and that with two pairs of laterals (root-analogue 3). Again, this is explained because the root-analogue deformation has not reached the depth on the main taproot where the second laterals branch off. At displacements larger than 20 mm, the root-analogue with two pairs of laterals requires considerably more pull-out force as the second pair of laterals contributes to the capacity.

Figure 5a and 5b show photographs taken during the pull-out of analogue 2. At the displacement level

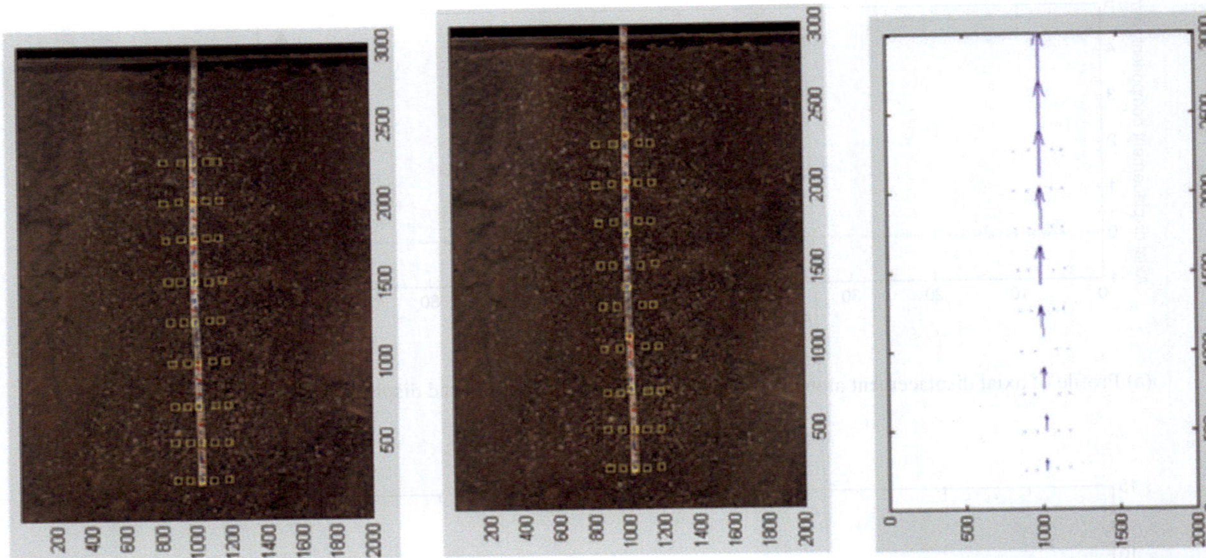

(a) Initial photograph: undeformed root-analogue **(b) Next photograph: displaced root-analogue** **(c) Measured soil and root-analogue displacement (vectors of displacement)**

Figure 4. Analysis of analogue and soil movement around root-analogue 1b.

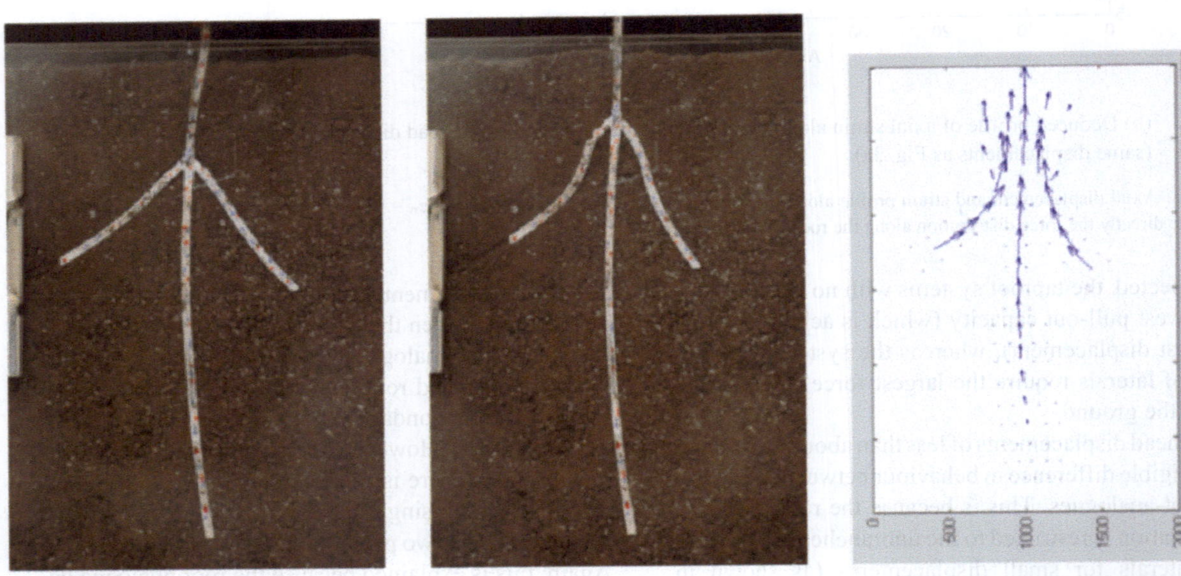

(a) Initial photograph: undeformed root-analogue (b) Deformed analogue after 12 mm of displacement **(c) Measured soil and root-analogue displacement (vectors of displacement)**

Figure 5. Analysis of analogue and soil movement around root-analogue 2.

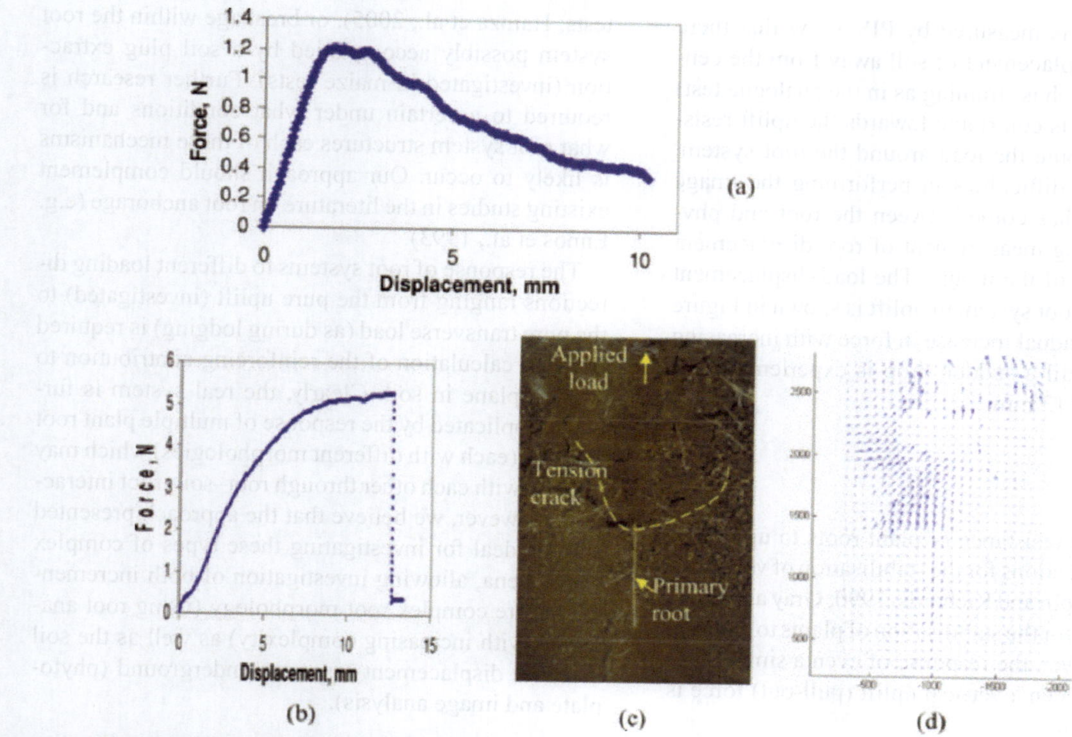

Figure 6. Load-displacement of stem base where the tensile force was applied for (a) an unbranched primary root of pea and (b) a branched primary root of maize. (c) Digital photograph of the maize root system after a large head displacement. (d) Measured soil and root displacements for the maize system.

shown (12 mm) the first pair of laterals is pulling out of the soil and a surface soil plug is lifting above the two laterals. This behaviour can be seen more clearly in the measured vectors of soil displacements at this instant (Figure 5c). However, despite this displacement mode, peak force has not been achieved (Figure 2b) and the tip of the root is still stationary (Figure 5c). Peak force is achieved once the root tip starts to displace.

Pea and Maize seedlings

Figure 6a shows the load-displacement response of a pea seedling (5d) with one single primary root of 120 mm length and 1.5 mm diameter. There were no lateral roots. It can be seen clearly that there is an initial 'elastic' response before a peak tension force, $T_f = 1.2$ N is achieved at a displacement of approximately 2 mm. Tensile force reduces slowly after the peak, so that there is about one third of the peak force, T_f still mobilised at a displacement of 10 mm. Breakage was not observed in the root and so the reduction in pull-out force post-peak must be due to post-peak softening of the interface shear-displacement relationship.

Photographs and image analysis of the pull-out test reveals that two things are different to the response of the unbranched root-analogue: (i) the root does not grow straight and some soil detachment is observed behind laterally moving sections of the root while straightening; (ii) the fine root hairs which can be observed in the photographs affect the deformation field around the root so that a larger body of soil is displaced. These two differences will have opposite effects on the pull-out forces achieved, but should be considered in any correct analysis of the system response.

Figure 6b shows preliminary results for Maize roots pulled from soil by its stem. The root morphology is considerably more complex than that of the analogues with a large number of lateral roots. In addition, as the plant was grown *in situ*, the root–soil interface may be much stronger than for the analogues. The difference in behaviour can be observed in terms of the final photograph after pull-out where there is a clear tension crack in the soil with removal of soil plus a soil plug. The soil displacements measured by PIV show that there is significant displacement of soil away from the central tap root (which is straining as in the analogue test) as the lateral roots contribute towards the uplift resistance and distribute the load around the root system. There are some difficulties in performing the image analysis as soil has come between the root and phytoplate preventing measurement of root displacement in some regions of the image. The load-displacement response of the root system to uplift is shown in Figure 6b. There is a gradual increase in force with increasing displacement until a sudden drop is experienced at a displacement of 13 mm.

Discussion

The mechanical resistance of plant roots to uprooting forces has implications for the stabilisation of vegetated slopes (e.g. Coppin and Richards, 1990; Gray and Sotir, 1996) as well as for the stabilisation of plants to external loadings. However, the response of even a simple root system in soil when a vertical uplift (pull-out) force is applied is complex.

The combination of mechanical testing and image analysis described in this paper allows increased insight into the mechanics of root–soil interactions and the likely root load carrying mechanisms. Load sharing between sections of a root system will depend on the mechanical properties of both the root tissue and the root–soil interaction which are complex (McCully, 1999). Our methodology provides a way of analysing in detail the mechanics of the root–soil interface as affected by factors such as root-hairs and lateral roots (e.g. Bailey et al., 2002). It also provides a way to verify and provide data for theoretical analyses of root system anchorage (e.g. Dupuy et al., 2005).

Extension of the analytical measurement work presented here will lead to identification of strain fields within the soil (including the effects of root 'group' effects), measurement of axial and bending strains in the roots as well as identification and quantification of sliding at the interface. These studies will allow a more detailed mechanical analysis of a root system than previously attempted experimentally and will facilitate understanding and inform modelling hypotheses.

Several mechanisms can occur at the peak uplift force: pull-out of an entire small root system (investigated partly with the analogues), breakage of the root near the stem (investigated in accompanying element tests; Hamza et al., 2005), or breakage within the root system possibly accompanied by a soil plug extraction (investigated in maize tests). Further research is required to ascertain under what conditions and for what root system structures each of these mechanisms is likely to occur. Our approach should complement existing studies in the literature on root anchorage (e.g. Ennos et al., 1993).

The response of root systems to different loading directions ranging from the pure uplift (investigated) to the pure transverse load (as during lodging) is required to allow calculation of the reinforcing contribution to a shear plane in soil. Clearly, the real system is further complicated by the response of multiple plant root systems (each with different morphologies) which may interact with each other through root–soil–root interaction. However, we believe that the approach presented here is ideal for investigating these types of complex phenomena, allowing investigation of both incrementally more complex root morphology (using root analogues with increasing complexity) as well as the soil and root displacement response underground (phytoplate and image analysis).

Acknowledgements

The work reported was funded by the combined UK Medical Research Council, the Biotechnology and Biological Sciences Research Council and the Engineering and Physicial Sciences Research Council's Discipline Hopper scheme. SCRI receives Grant-in-Aid from the Scottish Executive Rural Affairs Department.

References

Adrian R J 1991 Particle-imaging techniques for experimental fluid mechanics. Ann. Rev. Fluid Mech. 23, 261.

Bailey P H J, Currey J D and Fitter A H 2002 The role of root system architecture and root hairs in promoting anchorage against uprooting forces in *Allium cepa* and root mutants of *Arabidopsis thaliana*. J. Exp. Bot. 53, 333–340.

Bengough A G, Gordon D C, Al-Menaie H, Ellis RP, Allan D, Keith R, Thomas W T B and Forster B P 2004 Gel observation chamber for rapid screening of root traits in cereal seedlings. Plant Soil, 262(1-2), 63–70.

Coppin N J and Richards I G 1990 Use of Vegetation in Civil Engineering. Butterworths, London.

Dupuy L, Fourcaud T and Stokes A 2005 A numerical investigation into factors affecting the anchorage of roots in tension. Eur. J. Soil Sci. 56, 319–327.

Dupuy L, Fourcaud T and Stokes A 2005 A numerical investigation into the influence of soil type and root architecture on tree anchorage. Plant Soil. 278(1-2), 119–134.

Ennos A R, Crook M J and Grimshaw C 1993 The anchorage mechanics of maize, *Zea mays*. J. Exp. Bot. 44, 147–153.

Gray D H and Sotir R B 1996 Biotechnical and Soil Bioengineering Slope Stabilization. Wiley, New York.

Hamza O, Bengough A G, Bransby M F, Davies M C R and Hallett P D 2005 Applying geotechnical engineering to plant root biomechanical behaviour. *In* A Stokes, I Spanos, J E Norris and L H Cammeraat (eds), Eco- and Ground Bio-engineering: The Use of Vegetation to Improve Slope Stability. Kluwer, Dordrecht.

Matlab Guide 1997 Image Processing toolbox. For use with Matlab. Computer Solution Europe AB.

McCully M E 1999 Roots in soil: Unearthing the complexities of roots and their rhizospheres. Ann. Rev. Plant Physio. 50, 695–718.

Niklas K J 1999 A mechanical perspective on foliage leaf form and function. New Phytol. 143, 19–31.

White D J, Take W A and Bolton M D 2003 Soil deformation measurement using particle image velocimetry (PIV) and photogrammetry. Geotechnique 53, 7, 619–631.

Wu T H and Watson A, 1998 In situ shear tests of soil blocks with roots. Can. Geotech. J. 35, 579–590.

Dendrogeomorphological observations in a landslide on Tymfristos mountain in Central Greece

A.M. Papadopoulos[1], A. Mertzanis & A. Pantera
Technological Education Institute of Lamia, Forestry Department, 36100 Karpenissi, Greece.
[1]*Corresponding author**

Key words: *Abies borisii regis*, dendrochronology, dendrogeomorphology, mass-movement, Tymfristos mountain

Abstract

This paper presents the effects of a landslide on vegetation located on Tymfristos mountain and more specifically in the basin of the Kati stream, Sperchios river. The landslide areas are described and related to the density and the type of vegetation. The woody vegetation developed on the gliding surface was studied using dendrogeomorphologjcal methods. In particular, the direction and tilting degree of trees as well as the deformation of trunks was compared to trees growing on a neighbouring stable surface. Subsequently, the time of mass movement reactivation was estimated using a visual growth anomaly analysis on dominant fir (*Abies borisii regis* Mattf.) individuals growing on the landslide. From the results it is suggested that the landslide negatively affected fir growth ring width increase at certain times without, however, nullifying it. Certain time periods are identified with intense past sliding phenomena due to increased rainfall that subsequently increased underground water.

Introduction

Dendrogeomorphology is a subfield of dendroecology which utilises dated tree rings to study and date geomorphic processes, e.g. landslides, mass movements and creep (Kaennel and Schweingruber, 1995). Dendrogeomorphology, first introduced by Alestalo (1971), has been used by many researchers (Braam et al., 1987; Fantucci and McCord, 1995; Fantucci and Sorriso-Valvo, 1999; Shroder, 1978) to date landslide events. The trees subjected to stress due to mass-movement, show a tilting and sometimes an S-shape of the stem as well as signs of a sudden decrease in ring growth, or changes in eccentricity (Alestalo, 1971; Fantucci and Sorriso-Valvo, 1999; Goulas, 2003; Lang et al., 1999; Schweingruber, 1996; Shroder, 1978).

* E-mail: ampapadopoulos@teilam.gr

Landslides' dating contributes to understanding the causes, natural or anthropogenic, that provoke them. Severe landslides are noticed in many forested mountainous areas in Greece where flysch is the geological substrate (Kotoulas, 2001). Goulas (2003), studying the biological and biotechnical properties of fir on landslides at Pertouli, a few kilometres north from the present study area, indicated that the species reaction to landslides depends on its biological properties, stand structure and the degree of perturbation imposed to the aboveground and underground tree system. The perturbation degree of each individual tree depends on its location in the area as well as on the technical characteristics of the landslide.

The objectives of the present work were to study the Tymfristos landslide using dendrogeomorphological observations and to date landslide events using dendrochronological techniques and also explore the existing knowledge on the biological and biotechnical behaviour of hybrid fir (*Abies borisii regis* Mattf.) on landslides.

Research area

The research area was located in the upper basin of the Kati stream, (Sperchios river), on the E-SE side of Tymfristos mountain (Figure 1). The area is known as Tymfristos landslide and is known for the problems that it causes to the construction of the Lamia-Karpenissi highway.

The area constitutes an extensive unstable-sliding zone, extending from altitudes of 1300 to 700 m. The upper part of the landslide has a width of approximately 500 m confining, on the lower side, to 100–150 m, whereas its maximum depth reaches 17.5 m below the surface (Paraschoudis, 1999). Hydrogeomorphologically, the landslide shapes a wide territorial depression, with relatively soft relief and platforms, predominated by loose materials and weathered, disturbed flysch parts. These materials are characterised by high permeability, allowing the development of underground flow, supplying the traverse streams running through the region. This landslide undergoes fractures and movement of its territorial materials, depending on the intensity of rainfall–snowfall and other factors that contribute to the erosion of the region (Paraschoudis, 1999).

Geotectonically, the study area is included in the Pindos zone, composed of sedimentary flysch formations (sandstones, siltstones), from the upper Cretaceous—Lower Tertiary era, covered at locations by recently weathered material (Quaternary). The flysch, which is also the geological substrate of the research area, is mainly composed of alternations of sandstones and silts. This formation is sensitive to weathering/erosion and it constitutes, in combination to the intense slope inclination, the rainfall–snowfall and the presence of underground flow, the most important parameter of continuous relief redevelopment, with the creation of landslide phenomena which is very intensive after repeated cycles of saturation and desiccation. The Quaternary sediments constitute mainly the weathered surface mantle and include the loose products from the weathered-eroded substrate.

The region's climate is mountainous Mediterranean with mean annual precipitation of 1400 mm. The vegetation is Oro-mediterranean and characterised by the presence of the species *Abies borisii regis* and in the lower altitudes, *Quercus frainetto* Ten.

Material and methods

In the spring and summer of 1994, the vegetation growing on the landslide was recorded and tree age was estimated both by taking core samples with a Swedish incremental borer and also cutting cross-sections. All woody vegetation higher than 4.50 m was recorded. In total, 119 trees were recorded for (i) the direction towards which the stems were tilting, (ii) the intensity of the tilting (divergence from vertical) and (iii) the amount of deformities per trunk (bending) as a consequence of inclined tree growth.

From the above trees, all fir dominant individuals without wounds and/or fungi and insects attacks were selected for core sampling. In total, 17 trees (Figure 1) growing on the landslide surface were sampled. From each tree, two core samplings at breast-height (BH) were taken, the first core on the side opposite to tilting direction and the second 180° from the first one, making a total of 34 samples. Additionally, six dominant fir individuals were selected from adjacent stand located on a stable area, showing no trace of disturbance. Two core samples were taken from these trees, as mentioned before, one from the northern direction and one from the southern, making a total of 12 samples.

All samples were subsequently treated by the standard methods used in dendrochronology, and specifically, drying and smoothing of their surfaces, cross dating using skeleton plots and ring-width measurement with a 0.01 mm accuracy (Stokes and Smiley, 1968). Tree rings were identified and measured by the Windendro software. A mean chronology was constructed for each tree from its two core samples. For the six sample trees growing outside the landslide (controls), a master chronology was constructed out of the average of the six chronologies, presenting the growth pattern of an undisturbed stand.

A principal component analysis (PCA) using the PPPHALOS software (Guiot, 1990) was applied separately on the dendochronological data from the landslide and the stable area, for the time period from 1972 to 2003, in accordance with the meteorological data. The purpose of this analysis was to compare the annual ring-width variability among trees from the landslide and the stable area, and to study the above-mentioned variability by the available precipitation data. Additionally, all samples from the landslide were analysed using visual growth analysis to identify the abrupt growth changes and specifically, reduction (suppression) of

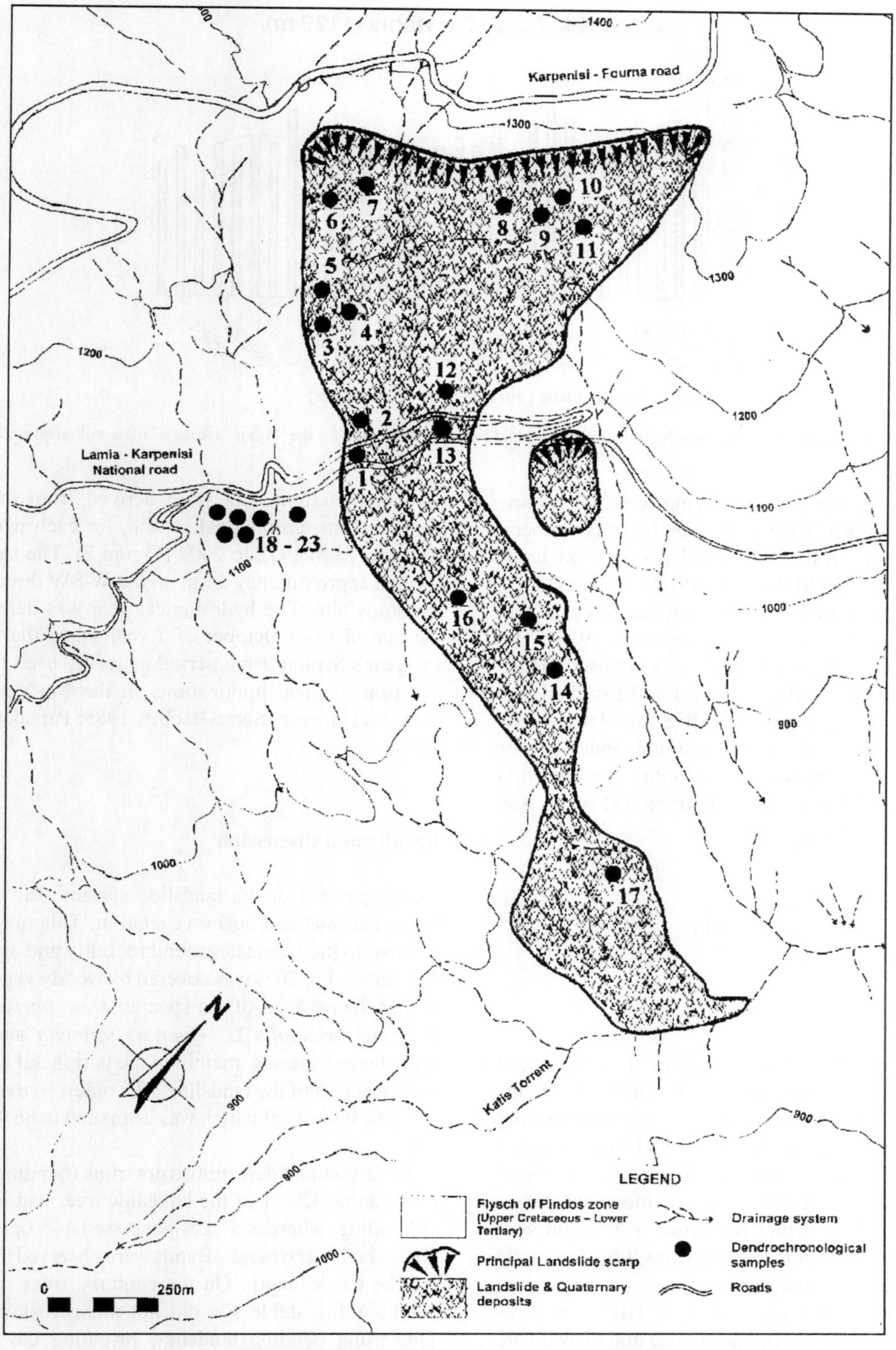

Figure 1. Chart of the Tymfristos landslide study area and location of the dendrochronological samples (1–23).

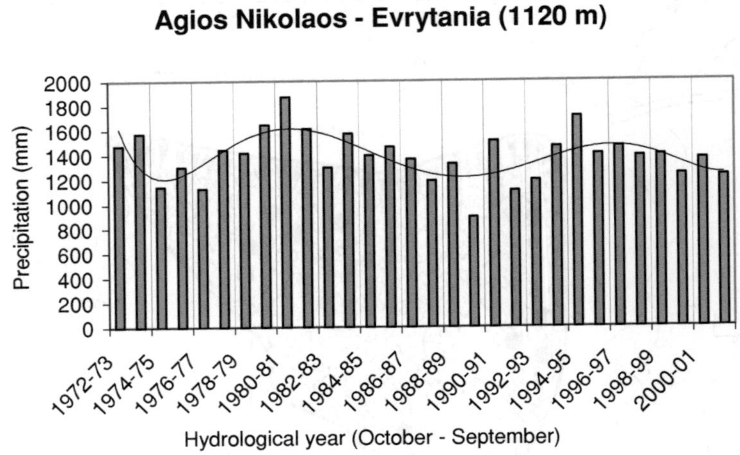

Figure 2. Precipitation in mm for each hydrological year (solid bars) and trend line for the "Agios Nikolaos" meteorological station.

ring width visible in successive ring series (Schweingruber et al., 1990). The suppression had already been noted in dendrogeomorphological research on landslides (Fantucci and McCord, 1995). In sliding slopes, abrupt growth changes in the stem are often an expression of cracker roots (Schweingruber, 1996). The results of the visual growth analysis were used to create a graph of the growth anomalies of the studied site. The graph was based on a modified Shroder formula (Shroder, 1978) to calculate the anomaly index (It) for each of the event–response cases (Fantucci and Sorriso-Valvo, 1999). The index (It), suppression at year t, was calculated as follows:

$$(\mathrm{It}) = \frac{\sum_{t=1}^{n}(\mathrm{Sup}(x)t \times (\mathrm{Fx}))}{\sum_{t=1}^{n}(N_{\mathrm{tot}})t} 100\%$$

where
$\mathrm{Sup}(x)t$ = the number of suppressions of each class (x) of growth anomaly in the year t, Fx = intensity coefficient taking values from 1 to 4 according to increasing intensity of suppression (Fantucci and Sorriso-Valvo (1999) adapted from Schweingruber et al. (1990)): 1 = Slight reduction 40–55%, 2 = moderate reduction 56–70%, 3 = strong reduction >70% and 4 = very strong >70% with missing rings, $(N_{\mathrm{tot}})t$ = total number of samples analysed in year t.

The results of the calculation of (It) were compared and related to precipitation data and to available geomorphological information of the area. The available precipitation data were derived from the "Ag. Nikolaos" meteorological station, for each hydrological year from 1972 to 2002 (Figure 2). The station is located approximately 2 km on the W-SW direction of the study site. The hydrological year was defined for the period from October of a year up to the following year's September, a period generally used for dendroclimatological applications in the Mediterranean basin and Greece (Serre-Bachet, 1985; Papadopoulos, 1993).

Results and discussion

Eighty percent of the landslide surface was covered by herbaceous and bushy vegetation. This may be attributed to the repeated ground mobility and agitation. The remaining 20% was covered by woody vegetation, higher than 4.5 m, of the species *Abies borisii regis, Platanus orientalis* L., *Quercus frainetto* and other broadleaved species, mainly in the peripheral area and the lower part of the landslide. The oldest tree recorded was a fir individual which was estimated to be 98 years old.

The amount of deformities per trunk (bending) study showed that 12.6% of the landslide trees had no signs of bending, whereas 57.2% possessed 1–5 bends, and 30.2% had over 6 bends. Bends were observed throughout the trunk length. On the contrary, trees from the neighbouring stable site did not present such bends. This trunk bending tendency, revealing the leaning tree's ability to straighten up through apical shoot

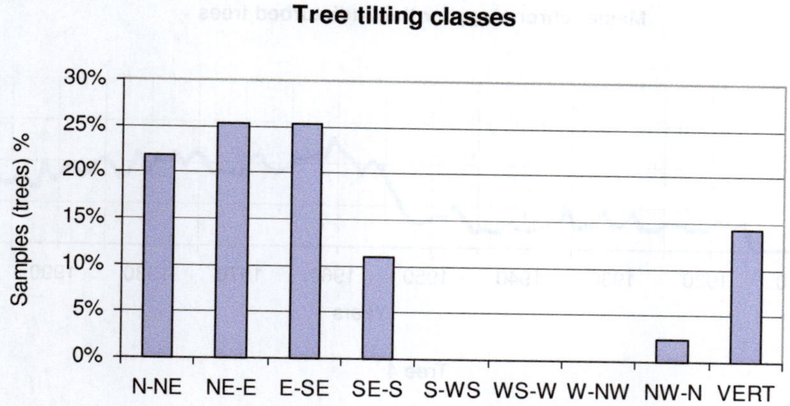

Figure 3. Classes of tree leaning direction in the landslide area.

increase, was more frequent in the broadleaved species. Most broadleaved species demonstrated severe trunk deformities and a high sprouting tendency. Usually, fir had a S-shaped trunk resulting from minor mass movements and tensions applied from neighbouring landslides. According to Goulas (2003), when the landslide is deeper than 1 m, fir trees slide along the moving mass and seldom remain in place.

Concerning the direction and degree of leaning of trees growing on the sliding slope, it appeared that 86% of trees had inclined trunks whereas the remaining 14% had straight trunks. Most of the vertical trees were fir individuals. No trunk tilting was observed in the trees on the adjacent stable site. The major part of trees (72.2%) inclined from N to SE (Figure 3), directions similar to the general slope exposure, but also towards other exposures differing locally by particularities in the landslide's relief. Most of the NW-N leaning trees were concentrated on the upper part of the landslide, showing a backward tilting slumping area.

With regard to the degree of stem lean from the vertical, 15% of trees leaned <10°, 51% were inclined 11°–30°, 20% were leaning >31° and 14% were straight. Severely leaning trees were mainly broadleaved species. This phenomenon of trunk divergence from vertical was more intense in trees growing on the intermediate part of the landslide, indicating the intensity of the sliding phenomena taking place there.

The landslide, in addition to the morphological characteristics, negatively affected width increase of fir trees. This was apparent by comparing the master chronology for the trees of the stable site with the individual ring growth curves for the trees of the landslide. Figure 4 presents some of the most characteristic ring growth curves for the trees growing on the landslide as compared to the master chronology for the trees of the stable site.

The growth pattern for fir trees from the stable stand remained constant, from 1957 onwards and after the trees reached the dominating stand's height. Only a small decrease in growth was observed due to the age of trees. On the contrary, a sudden growth decrease was observed for tree No. 4 from the landslide in 1948 and 1969, for tree No. 15 in 1952, 1973 and 1988 and for tree No. 16 in 1933, 1951, 1964 and 1993. All of the fir trees studied recovered regular growth 4–12 years later. According to Goulas (2003), fir individuals on the fringes of landslides which have >50% of their root system anchored and active inside the stable ground, lose their vitality but usually recover after 3–5 and even 8 years, provided that site conditions remain favourable.

Au: Kindly note that Goul has been changed to Goulas as per the reference list. OK?

From the PCA of the chronologies, it appears that annual variability of ring width in the landslide and stable area do not coincide (Figure 5). The first axis of the PCA, mainly interpreting the common variability due to site factors, expressed 36% of the total variability for the landslide trees whereas, for the trees from the stable site, this percentage reached 63%. This difference indirectly expresses the effect of the landslide on annual tree rings variability.

From the visual growth analysis results (Figure 6), it appears that the larger disturbances in width increase occurred for the period 1933–1935 in 12% of the samples, for the period 1950–1965 in 26–45% of the samples, for the period 1981–1988 in 24–54% of the samples and for the period 1993–2001 in 35–65% of the

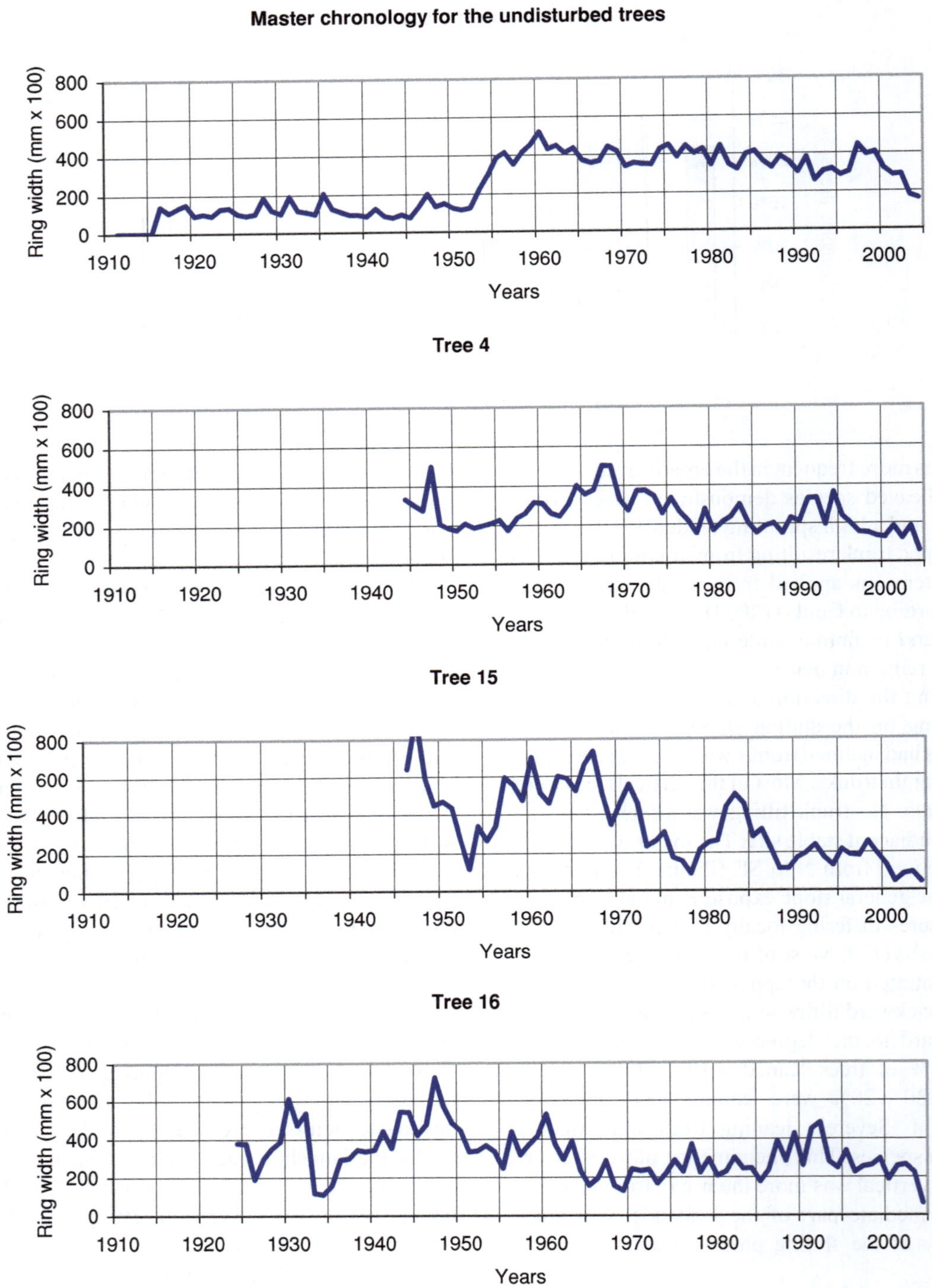

Figure 4. Master chronology curve of the fir trees developed on the stable site and individual ring growth curves of the fir trees developed on the landslide (tree nos. 4, 15 and 16).

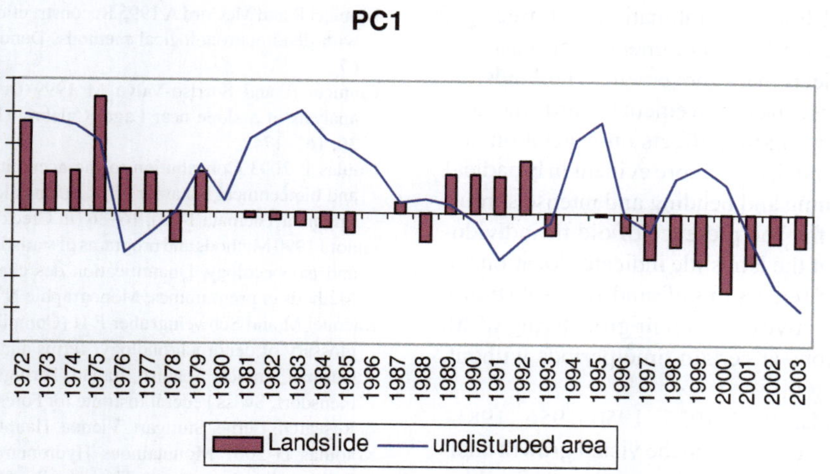

Figure 5. Chronology for the trees from the landslide (solid bars), and from the stable site (line) for the period 1972–2003 on the first PC axis.

samples. From the (It) value it appears that the older the tree, the greater the disturbance is. According to Goulas (2003), fir individuals at the seedling stage and those with thin trunks react more favourably to minor mass movements and, most of the time, soon recover their vitality compared to older individuals.

The comparison of Figure 6 with the precipitation data of the period 1972–2002 from the Agios Nikolaos station (Figure 2) reveals that, generally, the time-periods 1981–1988 and 1993–2001 with growth disturbance (suppression), based on indicator (It), were characterised by increased rainfall tendency. This may be attributed to the part (<50%) destruction of the tree root system because of the mass movement. This root system destruction causes a reduction in width increase, which recovers to normal levels after 3–9 years.

Conclusions

From our study on stem lean direction and degree, as well as from the frequency of trunk bending for trees

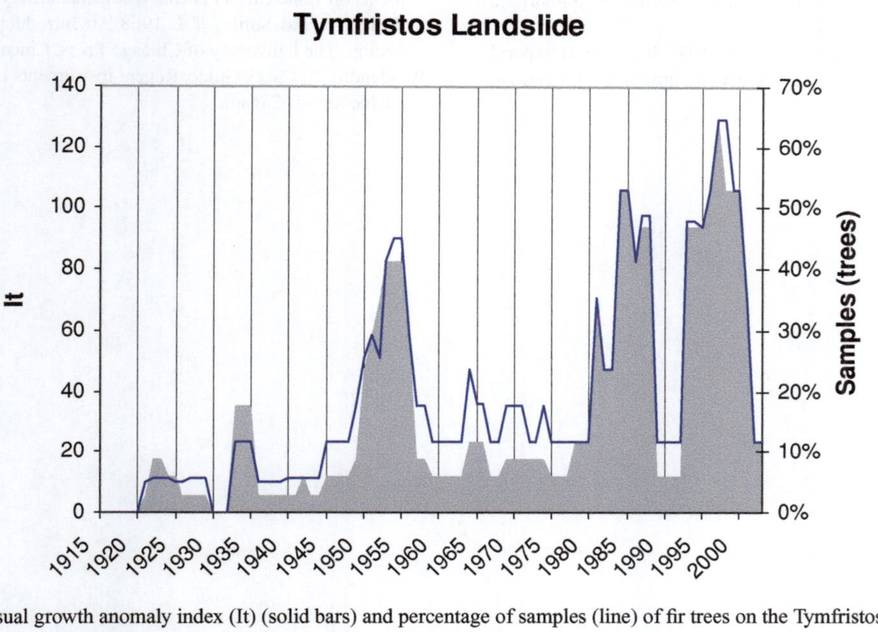

Figure 6. Visual growth anomaly index (It) (solid bars) and percentage of samples (line) of fir trees on the Tymfristos landslide.

growing on a landslide, in combination with tree age, it appears that the landslide commenced at least 100 years ago, via a deep mass movement. The landslide is active with intense mass movement towards the general slope direction. Sliding effects on vegetation are, to a higher extent and degree, more evident in broadleaf species (trunk leaning and bending and intense sprouting) compared to fir. The presence of old fir individuals in the interior of the landslide indicates locations of stable underground (emersions of sandstone substrate). The landslide's negative effect on fir growth ring width increase was obvious at certain time-periods without, however, nullifying it.

The time-periods, 1933–1935, 1950–1956, 1981–1988 and 1993–2001, for which the visual growth analysis and the (It) graph showed abnormalities of sudden growth decreases (suppression), were those years for which intense past sliding phenomena are recorded. Based on the available meteorological data for the two more recent time-periods, this may be attributed to the increased rainfall that subsequently increased the quantity of underground water. Finally, further dendrogeomorphological and hydrogeomorphological data need to be taken for a more detailed dating of the landslide events in the area.

References

Alestalo J 1971 Dendrochronological interpretation of geomorphic processes. Fennia 105, 1–140.

Braam R R, Weiss E E J and Burrough P A 1987 Spatial and temporal analysis of mass movement using dendrochronology. Catena 14, 573–584.

Fantucci R and McCord A 1995 Reconstruction of landslide dynamic with dendrochronological methods. Dendrochronologia 11, 55–67.

Fantucci R and Sorriso-Valvo M 1999 Dendrogeomorphological analysis of a slope near Lago, Calabria (Italy). Geomorphology 30, 165–174.

Goulas P 2003 Contribution to the accreditation of the biological and biotechnical behavior of fir in landslides. Geotechnika Epistimonika Themata 14(3), 16–26 (in Greek).

Guiot J 1990 Methods and programs of statistics for paleoclimatology and paleoecology. Quantification des changements climatiques: Méthods et programmes, Monographie N° 1, INSU, PNEDC.

Kaennel M and Schweingruber F H (Compilers) 1995 Multilingual glossary of dendrochronology. Terms and definitions in English, German, French, Spanish, Italian, Portuguese, and Russian. Birmensdorf, Swiss Federal Institute for Forest, Snow and Landscape Research. Berne, Stuttgart, Vienna, Haupt. 467 pp.

Kotoulas D 2001 Mountainous Hydronomy. Aristotle University Publications Department, Thessaloniki. 682 pp.

Lang A, Moya J, Corominas J, Schrott L and Dicau R 1999 Classic and new methods for assessing the temporal occurrence of mass movements. Geomorphology 30, 33–52.

Papadopoulos A 1993 Dendrochronologie du pin d'Alep en Grèce: contribution aux études climatologiques. Pub. Ass. Inter. Clim. 6, 254–262.

Paraschoudis V 1999 Study on the eastern access of the Tymfristos tunnel. Published by the District of Central Greece, Lamia. 55 pp.

Schweingruber F H 1996 Tree Rings and Environment Dendroecology. WSL/FNP, Haupt, Berne.

Schweingruber F H, Eckstein D, Serre Bachet F and Braker O U 1990 Identification, presentation and interpretation of event years and pointer years in dendrochronology. Dendrochronologia 8, 9–38.

Serre-Bachet F 1985 La dendrochronologie dans le bassin méditerranéen. Dendochronologia 3, 77–92.

Shroder J F 1978 Dendrogeomorphological analysis of mass movements on Table Cliffs Plateau, Utah. Quaternary Res. 9, 168–185.

Stokes M A and Smiley T L 1968 An Introduction to Tree-Ring Dating. The University of Chicago Press, Chicago.

Windendro™. User's Guide, Régent Instruments Inc., Univ. Québec à Chicoutimi, Canada.

Monitoring ground bio-engineering stabilization of landslides in Lazio Region, Italy

Federico Preti[1,3], Chiara Milanese[2]

[1]*Dipartimento Ingegneria Agraria Forestale, Università degli Studi di Firenze, Via San Bonaventura 13—50145 Firenze, Italy.* [2]*Dipartimento GEMINI—Geologia e Ingegneria Meccanica, Naturalistica e Idraulica per il Territorio, Università degli Studi della Tuscia di Viterbo, Via S. Camillo De Lellis, snc—01100 Viterbo, Italy.* [3]*Corresponding author*.*

Key words: bio-engineering, landslide, monitoring form

Abstract

Latium, like most other Italian regions, is subject to numerous landslides and gravitational movements. These landslides have been registered and mapped by the Region of Latium. Sixteen sites were chosen for the monitoring of ground bio-engineering projects which are currently being carried out in collaboration between the Region of Latium and the GEMINI Department of the Tuscia-Viterbo University. The monitoring activity was performed by completing a questionnaire for each site. The main data required, in order to have a complete vision of the site and its problems, concerned descriptions of the morphological, lithological and phytoclimatic aspects of the slope. This information was provided by the planners and contractors, who also gave a description of the ground bio-engineering techniques used to stabilize landslides. The aim of this study was to verify the effectiveness of using ground bio-engineering techniques in the Mediterranean climatic zone. In this zone, rooting of plants is especially complicated, but we can exploit the biotechnical characteristics of plants and adapt ground bio-engineering methods to this area. We can use information from this study to protect the territory, whilst minimizing the environmental impact, thus fostering and protecting the environment in areas where ground bio-engineering is carried out.

Introduction

One of the main objectives of the Latium Region's territorial policy is to orient intervention and soil protection programmes towards criteria that respect the natural equilibrium of the territory in question. Therefore, the Region of Latium has in recent years promoted a series of technical and administrative measures aimed at encouraging the adoption and the spread of ground bio-engineering techniques. Although ground bio-engineering is not a new discipline, it has been re-evaluated in the last 20 years and has been applied in many different parts of the world. The most commonly used definition of ground bio-engineering is the following: Bio-Engineering is a technical discipline that uses living plants in anti-erosive and consolidation operations together with other materials (wood, stones, straw, metallic nets, bio-mat, geo-fabrics, etc. . .) (Schiechtl, 1991).

In Latium, ground bio-engineering began with a decree issued in 1996 by the Region, which defined the criteria governing soil protection projects in this area. These criteria, which were innovative for this territory, were aimed at enhancing the protection of the landscape and environment, with particular attention paid to the restoration of areas where risks related to hydraulics or landslides occur (Regione Lazio, 1999). A fundamental principle to always be applied was that any intervention must have the least possible environmental impact. All organisations and regional offices working for the Latium Region must adhere to this policy. The Region's Environmental Department has organized three

* E-mail: preti@unitus.it

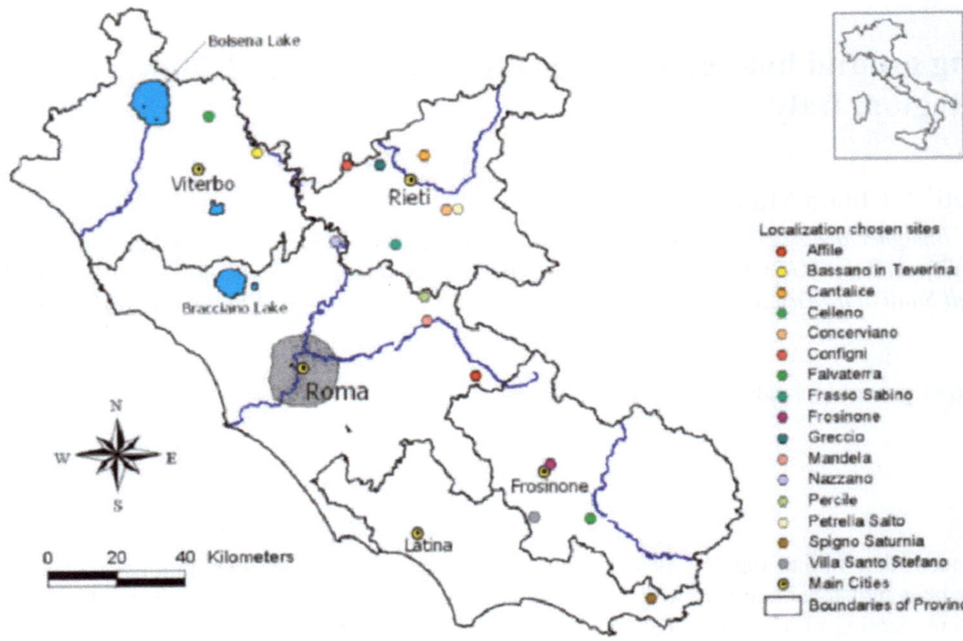

Figure 1. Position of field sites studied.

collaborations to perform studies regarding the application and monitoring of ground bio-engineering in various fields i.e. hydraulic management, quarry and dump sites, revegetation of road embankments, protection of coastal sand dunes and the stabilization of slopes (Regione Lazio, 2003). Our study was carried out with regard to the third convention, between the Region of Latium and the GEMINI Department of the Tuscia-Viterbo University, entitled *Analysis and Applicability of Bio-Engineering Techniques Related to Slope Stabilization Inside the Boundaries of Risk Areas in The territory of the Latium Region* (see Figure 1).

In this research, we present the results from the monitoring of several field sites. These results show the present situation regarding the quality and efficacy of the operations performed on slope stability. The aim of this study was to verify the correct execution of bio-engineering activities, both in technical terms and also of their insertion within the landscape. We also assessed the effectiveness and feasibility of intervening with bio-engineering techniques in a Mediterranean climate and determined which techniques and species are most suitable in this type of climate. As a result of this study, we could identify the sort of training required for ground bio-engineering contractors and planners, and highlight the need to use native plants and inert materials of local origin.

Materials and methods

Site selection

Three operational phases were distinguished for the purposes of monitoring and data: *ante-operam* (before the implementation of works), *in opera* (during the implementation of works), *post-operam* (after implementation of works). Sixteen sites were chosen for the monitoring of ground bio-engineering projects (Figures 2 and 3). A monitoring form was made up to survey slope stability using information provided by a multidisciplinary group of engineers, forestry experts and geologists. Therefore, data should be complete with regard to the different types of problems which could be encountered on a slope.

The field sites were chosen following a series of criteria that would be able to represent all the regional territory, from both the administrative point of view, with sites distributed throughout all the provinces, and also with regard to climate. Having different climate

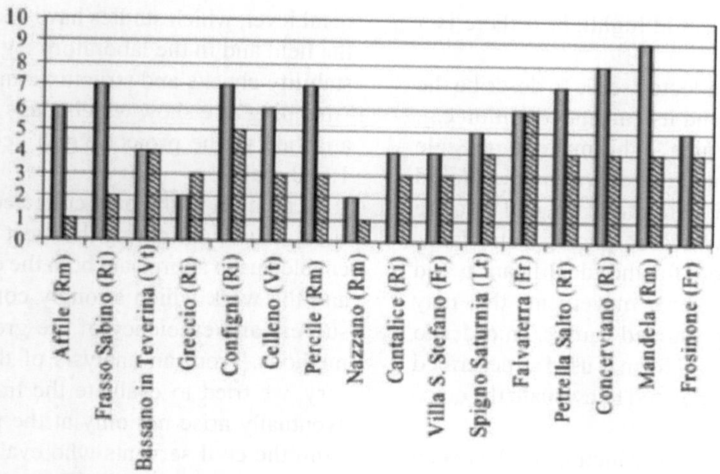

Figure 2. Number of ground bio-engineering works planned (in gray) and implemented (in lines) at the field sites.

types will also result in different types of vegetation. Urban, rural and protected areas were also represented. A wide selection of different ground bio-engineering methods was available and their cost calculated and compared with traditional civil engineering methods.

Monitoring form

A questionnaire or monitoring form was written and consisted of three fundamental parts:

Part 1: This part indicates information related to the position of sites and the identification of the main professional figures involved i.e. the designer, operations manager and contractor. Some of the most important information that we collected in this part falls under the heading "years of activity in the bio-engineering sector". Previous experience has made clear the importance of adequate preparations on the part of contractors for the correct application of techniques in question. Furthermore, such information could provide an indication of state of the art in the ground

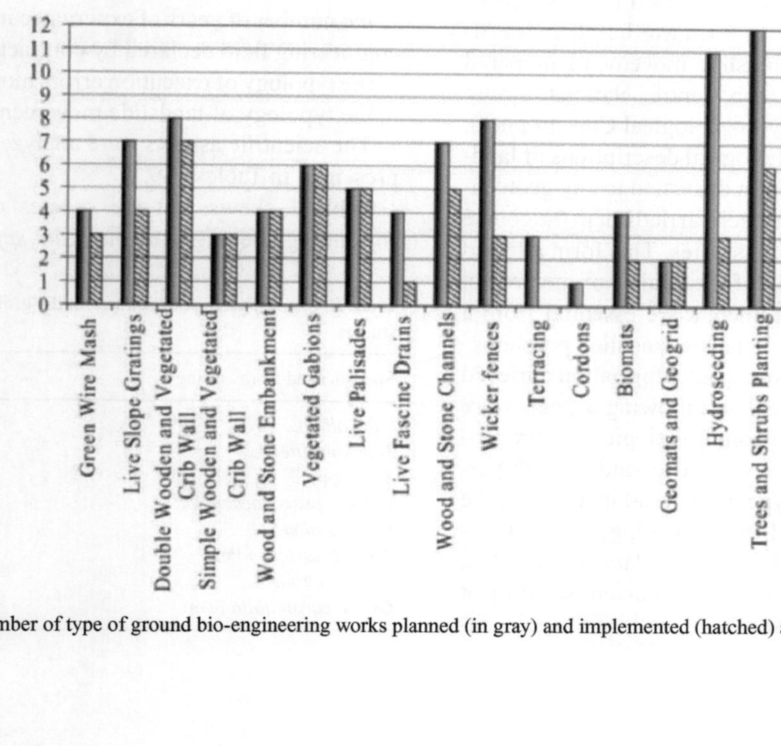

Figure 3. Number of type of ground bio-engineering works planned (in gray) and implemented (hatched) at the field sites.

bio-engineering field and could highlight if there is a need for further training.

Part 2: The data to be collected serve to describe the area under consideration and its current condition e.g. “Vegetation aspects of the area”. This makes it possible to ensure that correct plant varieties will be monitored in the work being planned. Other fundamental data are those related to lithology, geomorphology and landslide typology, which allow for the identification and description of any gravitational movements that may have occurred and their presumed causes. In order to collect such data, a geological form is used as described below. This form will also be used to evaluate the completion of the project.

Part 3: In part 3, the area in which a site has been planned or already set up is described. Therefore, the bio-engineering techniques used and the project scale is determined, with particular attention to the plant species to be used and the criteria for planting, two elements which are fundamental for the success of the operation. Furthermore, costs are reported of all work related to ground bio-engineering techniques, as well as work done with “conventional” civil engineering techniques Some of these techniques remain fundamental for some typologies of landslide phenomena (De Antonis and Molinari, 2003).

In the case of gravitational movements, geological parameters and geotechnical tests are particularly important, therefore, along with the above-mentioned form, a geological form was also used for this type of monitoring. This geological form was based on the official census form for landslide movements prepared by CNR (National Research Centre: National Group for Protection against Hydrogeological Catastrophes), for geological and morphological descriptions of landslides, with a special section dedicated to any geotechnical tests that might have been carried out in the course of preliminary fact-finding studies. This form allowed a more careful evaluation of the quality of the project itself, given that the parameters are essential from a geological point of view to tackle landslide problems.

The geological form contained information retrieved from individual projects. The following aspects were considered: geological, geomorphological, hydrogeological and seismic. Sites were also studied with particular attention to the type of material involved in the landslide movement and to the typology of the landslide movement itself. The recorded data have the dual aim of providing a summary of the current situation at these sites and highlighting, principally at the geotechnical level, which studies have been carried out both in the field and in the laboratory e.g. penetrometric tests, stability checks and structural analysis. In the section “mapping” we show which maps have been drawn and attached to the project e.g. a geomorphological map 1:500 scale.

In relation to the data collected, we designed a fact sheet, thus highlighting the most significant data. This enabled us to appreciate both the quality of the projects and the work which strongly conditions the effective success and efficiency of the ground bio-engineering methods. From an analysis of the parameters necessary, we tried to evaluate the training needs that can eventually arise not only at the project level but also from the civil servants who evaluate the projects and the contractors who actually carry out the operations.

Results and discussion

The parameters that we chose to analyse from the data collected during monitoring were

species type used for cuttings;

species type used for plantations;

planting-out period;

percentage of cuttings and rooted plants which rooted;

planned and implemented seeding;

works planned with live material but carried out with dead material;

the number of years of experience in the ground bio-engineering field declared by contractors;

the typology of execution errors most often recorded

the typology of landslide movements.

The scientific aspects were analyzed and the results presented in Tables 1–9.

Table 1 shows that the species most often used for cuttings belonged to the *Salix spp.* (87.5%). This

Table 1. Species used from cuttings in the ground-bioengineering studies

Species used from cutting	% of use
Salix alba L.	37.50
Salix purpurea L.	37.50
Salix spp.	12.50
Robinia pseudoacacia L.	6.25
Populus alba L.	6.25
Tamarix africana Poir.	6.25
Fraxinus ornus L.	6.25
Ostrya carpinifolia Scop.	6.25

Table 2. Arboreal and shrubby species used in the ground bio-engineering studies

Rooted plant used	% of use
Cornus spp.	37.50
Prunus spinosa L.	25.00
Crataegus monogyna Jacq.	25.00
Quercus ilex L.	18.75
Euonymus europaeus L.	18.75
Spartium junceum L.	12.50
Ostrya carpinifolia Scop.	12.50
Olea europea L.	12.50
Myrtus communis L.	12.50
Rosa spp.	6.25
Pistacia lentiscus L.	6.25
Phillyrea latifolia L	6.25
Corylus avellana L.	6.25
Other species	6.25

was also true in gradient works, where the two main species used were *Salix alba* L. and *Salix purpurea* L. Some attempts to use incorrect alternative species were also evident: for example the use of Locust tree should be avoided as it is not native and is also an invasive species. These considerations highlight the need to try new species as cuttings for gradient works, where conditions for the survival of willows are not favourable. The necessity to use alternative species to willow is shown when we analyse the arboreal and shrubby species used as rooted plants (Table 2). It was observed that a wide variety of species were used and that apart from a few exceptions, these were native plants which, if planted in the proper period and treated with a minimum of care, give excellent results in terms of rooting. Such a result is to be considered positive, and can be implemented if three principal factors are respected:

1. phytoclimatological information is supplied by preliminary field studies (Blasi, 1996);
2. projects are supplied with a botanical report;
3. indications for the period of planting-out are respected, with works being put on hold when necessary.

Consequently, when these factors were respected, higher quality and more careful planning was observed, with live plants beginning to be considered as veritable construction material. Contractors planning to specialize in the ground bio-engineering sector were also more receptive to the suggestions provided during the monitoring phase.

When dealing with living plants, e.g. scion or rooted plants, the field site's needs must be calibrated with suitable periods for planting (Table 3). The summer period is usually excluded, but in certain cases, the planting-out of both cuttings and plants was done in May, when the growing season is fully underway. Such late timing often interferes with the rooting of cuttings, but plants have a higher probability of recovering from transplanting stress if watered. However, irrigation is rarely carried out on the work sites. The latest month for the planting-out of cuttings and plants should be April.

Table 3. Planting-out period of rooted plants (P) and from cutting (C)

	Months											
Sites	Jan	Feb	Mar	Apr	May	Jun	Jul	Aug	Sep	Oct	Nov	Dec
Celleno (VT)												
Bassano in Teverina (VT)						C					P	
Villa S. Stefano (FR)	P	P										P
Falvaterra (FR)										C/P		
Frasso Sabino (RI)		C/P	C/P	C	C							
Nazzano (RM)												
Percile (RM)			C	C								
Mandela (RM)		C	C/P	P								
Spigno Saturnia (LT)	C	C/P									C	C
Frosinone Configni (RI)				C/P	C/P							
Cantalice (RI)												
Concerviano (RI)											C	
Greccio (RI)	P	P	P	P	P						P	P
Affile (RM)					P							
Petrella Salto (RI)										P		

Table 4. Percentage of rooting in species used from cuttings

	Percentage classes of successful rooting				
Species used from cuttings	0–20%	21–40%	41–60%	61–80%	81–100%
Salix alba L.	3	3	0	1	0
Salix purpurea L. + other sp. of willow	3	1	0	2	1
Other species	3	2	0	0	0

Table 5. Percentage of rooting in arboreal and shrubby plant species

	Percentage classes of successful rooting				
Species of rooted plant	0–20%	21–40%	41–60%	61–80%	81–100%
Cornus spp.	2	0	0	0	3
Prunus spinosa L.	1	0	0	0	2
Crataegus monogyna Jacq.	1	0	0	0	2
Euonymus europaeus L.	1	0	0	0	1
Other species	4	0	0	0	2

Another fundamental parameter which should be considered for successful practice of ground bio-engineering, is the percentage of rooting (Tables 4 and 5). This parameter is linked to those above i.e. a higher percentage of rooting will be gained for those plants chosen with regard to the environment in which they will grow and are also planted in the right period. Further factors that influence the percentage of rooting are e.g. seasonal climate (summer 2003 was particularly dry and many plants died that had rooted well at the beginning of the summer season), edaphic conditions, postplanting care, choice of good quality nursery material and the correct size of vegetal material. It is the interaction between all these components that will determine the success of the work. The percentage of successful rooting in cuttings at the field sites was rather low and often only 0–20% successful rooting was observed (Table 4). *S. alba* was the species with the most successful rooting, and sometimes up to 80% success was seen. However, in rooted plants, many plants show either very low rooting ability, or, on the contrary, very high rooting ability (Table 5).

The influence of hydroseeding was also determined (Table 6). Hydroseeding is often successful but is not often used (Regione Lazio, 2002). Table 6 shows that hydroseeding and/or seeding was planned for 12 out of 16 sites, but was only effectively carried out in 25% of the cases, and not done at all at 58% of sites. These data are very important because on denuded slopes, grass coverage is necessary to protect the soil from erosion whilst awaiting the development of tree and shrubby coverage. It is however, worth noting that in most cases, seeding was not carried out because there was an overabundant development of natural grassy vegetation, a condition that can be determined in advance with a local situation study. Therefore, at the design stage, plans could be made in lieu of hydroseeding, for the maintenance of such vegetation. Natural grassy vegetation tends to out-compete cuttings, which usually have to take root in difficult situations, both from a climatic and edaphic point of view. It could be useful to mow this spontaneous vegetation, thereby encouraging the growth of cuttings.

Reinforced terrains are often designed with the intention of planting cuttings (Table 7). However, these cuttings are often not included and are replaced with hydroseeding. Amongst the sites monitored, of the four reinforced terrains which according to plans were to be cultivated with cuttings, none were actually planted

Table 6. Implementation of hydroseeding (Regione Lazio, 2002) and other types of seeding

Hydroseeding and seeding planned	Sites with planned seeding
Implemented	1
Simple seeding	2
To be implemented	2
Not implemented	7

Table 7. Ground bio-engineering works planned with live material but implemented with dead material

Bio-engineering techique (Regione Lazio, 2002)	No. of works planned live	No. of works implemented dead
Green Wire Mash	4	3
Live Slope Gratings	7	3
Double Wooden and Vegetated Crib Wall	8	2
Simple Wooden and Vegetated Crib Wall	3	2
Wood and Stone Embankment	4	3
Vegetated Gabions	6	2
Live Palisades	5	4
Live Fascine Drains	4	0
Wood and Stone Channels	7	2
Wicker fences	8	2
Terracing	3	0
Cordons	1	0
Biomats + seeding	4	2
Geomat + seeding	2	1

with the cuttings and only one of these was hydroseeded (Table 6). These considerations are also important for an efficient management of project's economic resources.

Many errors can occur during the implementation of ground bio-engineering. Some errors can be quite serious and undermine the stability of the work itself

Table 8. Most frequent errors occurring in the implementation of the ground bio-engineering works

Most frequent errors	No. of errors
Work-slope connection	6
Inert material dimensions e.g. pole diameter	6
Works too vertical	5
Disposition of the poles crosspieces	2
Choice of the materials	8
Dimensions of species used from cuttings	9
Wrong joints	6
Wrong nailings	9

(Table 8). The relationship between the number of errors carried out and the years of experience in bio-engineering techniques declared by the contractor was determined (Figure 4). It can be seen that contractors with considerable experience commit as many errors as contractors with little or no experience, which indicates that information was faulty and that the ongoing training of the contractors themselves is necessary. One should therefore arrange for more training through practice work sites.

The final data examined considered the management of landslides (Table 9). When dealing with landslides, it is often necessary to integrate conventional and ground bio-engineering techniques. In general, in the case of slippage, an underground and surface drainage system

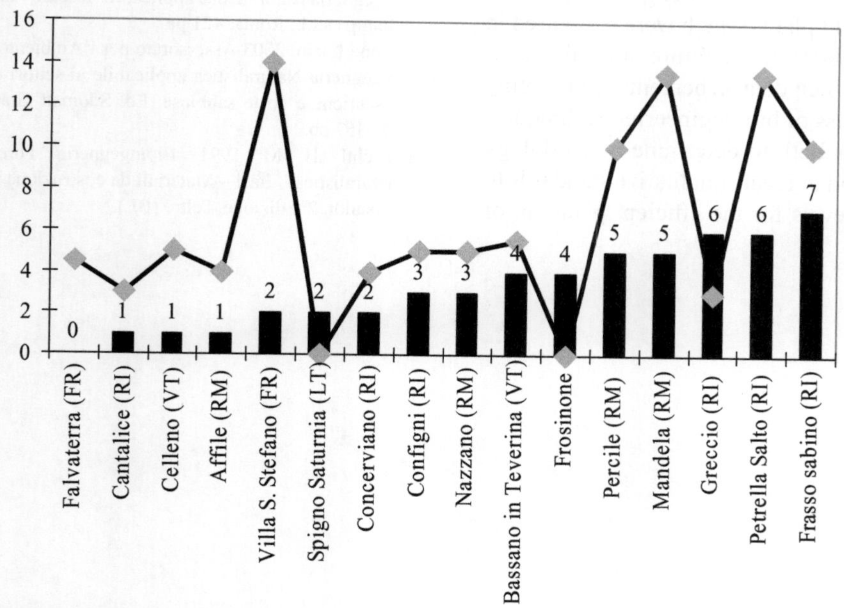

Figure 4. Years of experience (solid line) in ground bio-engineering declared by the contractor compared with occurred errors (black bars).

Table 9. Typology of gravitational movements (Civita, 1982) which occurred in the chosen sites

Typology of gravitational movements	No. of sites
Complex	3
Slump and Slide	8
Fall and Topple	4
Flow	1

is combined with containment and consolidation works carried out with ground bio-engineering techniques.

Conclusion

From the analysis of our monitoring results in Latium, it is clear that many fundamental errors are still being made, especially in the execution stage. These data highlight the need for more training and for a special certification for contractors who have acquired specific qualifications in the field of ground bio-engineering.

The analysis of the species used and the percentage of rooting, indicates the need for further research to identify suitable species within the Mediterranean zone. Another important factor to be considered is the need to encourage local nurseries to furnish this new "construction material". Similar monitoring work has been carried out by the National Park of Vesuvio. For the implementation of ground bio-engineering, managers used rooted wild plants which were not reared in a nursery (Bifulco, 2001). In the future, it would be interesting to compare their data on percentage of rooting and the related success of bio-engineering techniques.

Monitoring serves both to determine methodology for gathering data and to create qualitative standards to serve as reference levels for the efficient planning of slope stability, in both technical and economic terms. For such projects, it is necessary to determine how much to supplement "classic" or "conventional" interventions with bio-engineering, the goal being to protect a territory in such a way as to minimize the environmental impact of these activities.

Acknowledgements

We thank Francesco Gubernale, Simona De Bartoli and Giovanni Falco who managed the Lazio Region grant and Dr. Alexia Stokes for her helpful contribution in the revision of this paper.

References

Bifulco C 2001 Interventi di Ingegneria Naturalistica nel Parco Nazionale del Vesuvio. Ed. Ente Parco, Napoli.

Blasi C 1996, Regione Lazio Assessorato Agricoltura, Foreste, Caccia e Pesca, Usi Civici, Fitoclimatoligia del Lazio.

Civita M 1982 Classificazione tecnica e identificazione pratica dei movimenti franosi. Ed. Levrotto & Bella, Torino.

De Antonis L and Molinari V M 2003 Interventi di sistemazione del territorio con tecniche di Ingegneria Naturalistica. Ed. Regione Piemonte, Torino.

Regione Lazio 1999 Assessorato per le politiche dei trasporti e lavori pubblici, Università di Roma "La Sapienza"—Dipartimento Scienze della Terra Individuazione e perimetrazione delle aree in frana nel territorio della Regione Lazio.

Regione Lazio, 2002 Assessorato per l'Ambiente 2002 Manuale di Ingegneria Naturalistica applicabile al settore idraulico. Ed. Punto Stampa s.r.l., Roma. 421 pp.

Regione Lazio, 2003 Assessorato per l'Ambiente 2003 Manuale di Ingegneria Naturalistica applicabile ai settori delle strade, cave, discariche e coste sabbiose. Ed. Sdomatt Grafica G.S.I, Roma. 18–357 pp.

Schiechtl H M 1991 Bioingegneria Forestale—Biotecnica naturalistica—Basi—Materiali da costruzione vivi—Metodi. Ed. Castaldi, 2ªEdizione, Feltre (BL).

The use of geostatistical techniques applied to soil conservation of low-density woodlands

T. Panagopoulos*
University of Algarve, FERN, Campus de Gambelas, 8000 Faro, Portugal.

Key words: geographic information system, geostatistics, *Quercus suber* L., soil erosion, vegetation management

Abstract

The *Quercus suber* L. woodlands of South Portugal are subjected to numerous attempts of reassessment, in order to control their invasion by shrub vegetation and to prevent fires. Following the intensive soil management techniques, soil is traditionally harrowed in order to establish fodder species, increase aeration and destroy shrubs and other weeds. These practices leave it bare and exposed to erosion for long periods resulting in a poor clay schist soil. In this study, the risk of erosion on *Q. suber* woodland with different vegetation management techniques was examined. The different blocks were: shrub vegetation left undisturbed for at least 11 years, vegetation destroyed by a rotary cutter mower and enrichment with fast growing herbaceous species after tillage. To estimate soil degradation, several soil physical and chemical properties were studied. Graphical interpretation of these soil properties was performed using geostatistics. Ordinary kriging was the geostatistical technique used for the creation of maps, which were then introduced into a geographic information system (GIS) and their values reclassified. Spatial modelling was then applied using the revised universal soil loss equation and the superimposed maps developed a final overlay map for potential soil erosion. On the superimposed final map of erosion risk, the areas with highest risk of erosion were located. In those areas, the erosion risk could be decreased by suggesting simple changes in the vegetation management such as: keeping tillage only in the lowest erosion risk areas and leaving natural vegetation or enriching the soil with fast growing herbaceous species in high-erosion risk areas. The present research shows that geostatistics are useful for sustainable management of extensive agrosilvopastoral woodlands.

Introduction

The *Quercus suber* L. woodlands of Southern Portugal are subjected to numerous attempts to reassessment by linking them with livestock systems, in order to control invasion by shrub vegetation and to prevent fires. Following intensive soil management techniques, soil is traditionally harrowed annually, or every 3 years, in order to establish fodder species, increase aeration and destroy shrubs and other weeds, leaving bare soil exposed to erosion.

Geographic information systems (GIS), modelling and geostatistics are tools becoming progressively more suitable in fields of research like forestry and agriculture (Basso et al., 2001; Bocchi et al., 2000). In recent years, major advancements have been made in the technologies required to implement precision farming practices (Panagopoulos et al., in press; Yalouris et al., 1997). Traditional surveys of soil fertility, together with data from soil survey maps, can be used in combination with geostatistics by decision-makers to support management planning and to predict indicators related to soil quality as a measure of sustainability (Couto et al., 1997).

More specifically, these technologies can enable micro-management techniques on a site-specific basis to account for the natural and human induced variations that exist in agrosilvopastoral woodlands such as

* E-mail: tpanago@ualg.pt

variation in soil type, moisture, topography, chemistry, physical properties, and other factors. These technologies promise the possibility of optimising profit and reducing the adverse environmental impact of forest management (Larson et al., 1997).

Geostatistics provides descriptive tools such as semivariograms to characterize the spatial pattern of continuous and categorical soil properties (Goovaerts, 1999). Various interpolation techniques take advantage of the spatial correlation between observations to predict attribute values at unsampled locations using information related to one or several attributes. Kriging is a statistical interpolation method that uses data from a single attribute to predict values of the same attribute at unsampled locations. Kriging also provides standard errors of the predictions. The kriging interpolator assumes that the distance or direction between sample points reflects spatial correlations that can be used to explain variation in the surface (Chilès and Delfiner, 1999).

An important contribution of geostatistics is the assessment of the uncertainty about unsampled values, which usually takes the form of a map of the probability of exceeding critical values for soil quality (Castrignano et al., 2002). This uncertainty assessment can be combined with expert knowledge for decision making such as a description of contaminated areas where amendment measures should be taken or areas of good soil quality where specific management plans can be developed (Kitanidis, 1997). Ordinary kriging appropriately estimates values in unsampled areas and identifies places where more intensive sampling is required.

The main aim of the present work was to use geostatistical techniques and GIS to identify the risk of erosion on *Q. suber* woodland area where different vegetation management techniques were applied.

Materials and methods

The study area is located in Southeast Portugal in a low-altitude mountainous formation of 'Serra do Caldeirão'. The 65000 people that live in the 2678 km^{2}of the area are mainly farmers, foresters or shepherds with approximately 12 500 goats. Serra do Caldeirão, with a maximum altitude of 589 m, has mostly poor soils with low agricultural potential and moderate slopes. Most of the soils are classified as bedrock or very thin soils derived from clay schist.

Quercus suber L., *Quercus ilex* L., *Pinus pinea* L., *Pinus pinaster* Ait., *Ceratonia siliquia* L., *Eucalyptus* spp., *Cistus* spp., *Olea europaea* L., *Prunus dulcis* Miller, cereals and fodder plants are the main vegetative species that can be found in the area. Soil erosion is one of the major environmental problems of the region.

The climate of the area is continental Mediterranean with very hot and dry summers and mild winters. Algarve is under a strong climatic influence of the Mediterranean, and is therefore characterised by a dry season and a very irregular distribution of rainfall during the year, as well as over the years. Average annual precipitation is between 500 and 800 mm, depending on altitude.

The revised universal soil loss equation (RUSLE) was used to estimate potential erosion. According to Gray and Sotir (1996), RUSLE is defined as $A =$ RKLSCP where $A =$ potential erosion, $R =$ rainfall and runoff factor, $K =$ soil erodibility factor, $LS =$ slope length and gradient factor, $C =$ vegetation cover factor and $P =$ vegetation control practice factor. The rainfall and runoff factor R was estimated from Modified Fournier Index using data from 45 meteorological stations in the study area and surrounding area for 25 years of measurements. The greater the intensity and duration of the rain storm, the higher the erosion potential.

Vegetation cover, management and soil data were collected from an area of 2 $\times$ 2 Km located close to the village Alganduro. In the experimental area, a total of 81 soil samples in a grid scheme of 250 m distance between samples were collected, as suggested for a small heterogeneous areas by Carter (1993). Sampling points were located in the field (Figure 1) using a Global Positioning System (GPS).

Individual soil samples of about 1.0 kg were collected from each sampling position at a depth of 0.1–0.30 m. The mixture of soil and coarse fragments was air-dried, weighed and carefully sieved through a 2 mm screen without breaking up fragile fragments. The fraction passing through the 2 mm sieve was split with a stainless steel riffle and saved for analysis. This fraction was analysed for physical properties (texture, coarse fragments, particle density, specific weight), pH, electrical conductivity, as well as the main macronutrients (available phosphorus, total nitrogen and exchangeable potassium) using standard procedures mentioned from Carter (1993). The soil erodibility K factor was estimated from measurements on organic matter,

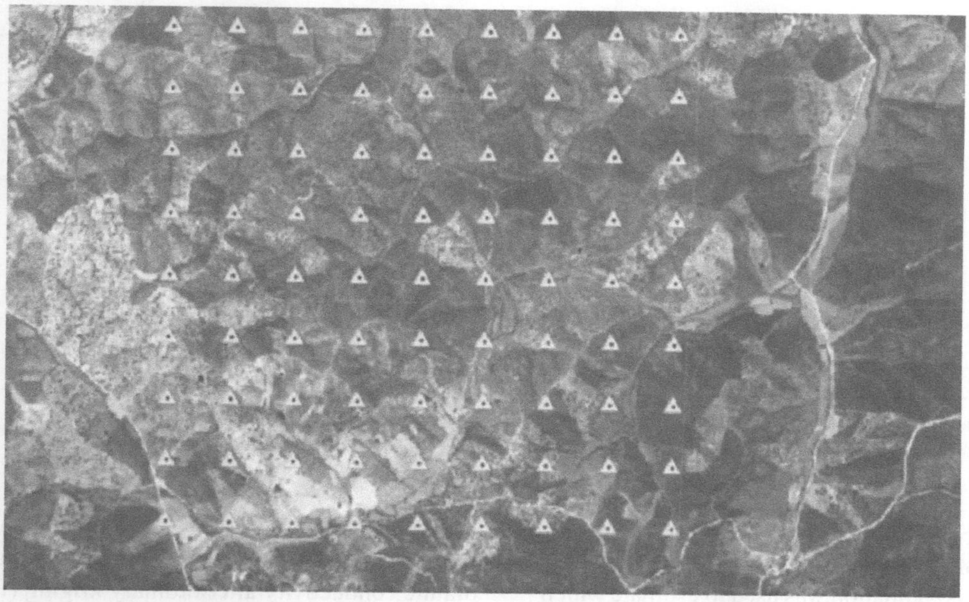

Figure 1. The exact location of the sampling points on the aerial photograph of the 'Alganduro' experimental area.

soil texture, structure and hydraulic conductivity after reclassification of their values. Infiltration was measured using the field-saturated hydraulic conductivity (Kfs) 'Guelf permeameter' method (Reynolds and Elrick, 1985).

In the present study, three types of traditional vegetation management techniques on *Q. suber* woodlands were studied. In the first, the shrub vegetation was left undisturbed for at least 11 years, in the second, vegetation cover was cleared using a rotary cutter mower and in the third, was enriched with fast growing herbaceous species after tillage.

In order to understand the variation of some soil physical and chemical properties, graphical interpretation of these soil properties was performed using geostatistics. A dataset of soil properties and vegetation cover was created with their geo-referenced position in the field by using the Arcview 8 (ESRI). Before creating surface diagrams, the distribution of data was analysed to get a better understanding of trends, directional influences and obvious errors. Ordinary kriging was used for the creation of several maps. Prior to the creation of the maps, semi-variograms were produced for each soil factor. Cross validation was used to compare the prediction performances of the semivariograms. The maps resulting from the interpolation techniques were introduced into a GIS and their values reclassified. After that, spatial modelling was used to develop a final overlay map for potential soil erosion.

Results

A georeferenced dataset of soil properties and vegetation cover was created and the distribution of data was analysed. In most cases the data was not normally distributed, presenting large spread and no symmetry. Transformation and trend removal was performed when necessary to create more accurate prediction maps. Trend removal and logarithmic transformation helped to normalize data distribution. Kriging after trend removal was carried out on the residual data of electric conductivity, texture (clay), organic matter and pH. The prediction map of each factor was calculated and trend was added back to the output surface. Table 1 presents summarily the indicators which helped to choose the most appropriate model of semivariogram for the creation of the prediction map for hydraulic conductivity (one of the soil erosion factors studied). From the cross-validation of the models the mean error (ME), root-mean-square error (RMSE), average

Table 1. Values of model parameters used to find the best semivariogram to predict hydraulic conductivity

Model	Nugget	Partial sill	Major range	Minor range	Direction	ME	RMSE	ASE	RMSSE
Circular	9.42	5.71	1964.9	306.9	357	0.02	4.09	4.57	0.98
Spherical	8.33	6.39	1967.7	306.2	359	0.01	3.60	3.08	0.97
Tetraspherical	10.10	6.49	1966.6	306.5	358	0.02	4.69	4.08	0.98
Pentaspherical	9.65	8.42	1991.1	302.3	354	0.02	4.71	4.09	0.78
Exponential	10.43	6.24	1993.1	306.5	357	0.05	4.75	4.24	0.99
Gaussian	10.51	3.51	1997.0	306.1	351	0.01	5.68	4.91	1.11
Rational Quadr	9.14	4.07	1996.1	306.5	356	0.06	4.83	4.69	1.01
Hole effect	9.53	4.99	1454.5	1131.8	318	0.01	5.73	5.91	0.86
K-Bessel	8.35	8.05	1991.7	306.5	355	0.01	5.74	5.04	0.87
J-Bessel	8.71	6.21	1575.6	1238.2	8	0.04	5.72	5.84	0.95
Stable	9.44	3.90	1910.8	306.5	359	0.04	5.75	5.25	1.06

From the cross-validation of the models the mean error (ME), root-mean-square error (RMSE), average standard error (ASE) and root-mean-square standardized error (RMSSE) were used.

standard error (ASE) and root-mean-square standardised error (RMSSE) were used.

Kriging cross-validation was used to estimate which of the semivariogram models could give the most accurate predictions of the unknown values of the field. The closer the ME was to 0 and the closer the RMSE was to 1, signified that the prediction values were close to measured values (Wackernagel, 1995). When models presented similar values for ME and RMSE, the lowest values of root-mean-square error and average standard error were taken into consideration. The map of hydraulic conductivity after ordinary kriging (Figure 2), demonstrates that the highest permeability occurs at the north-east part of the site, decreasing through the field and reaching the lowest values at the southern part.

The existence of a heavy textured area in the southern part of the field can be seen in the map of texture derived from clay percentage data after ordinary kriging

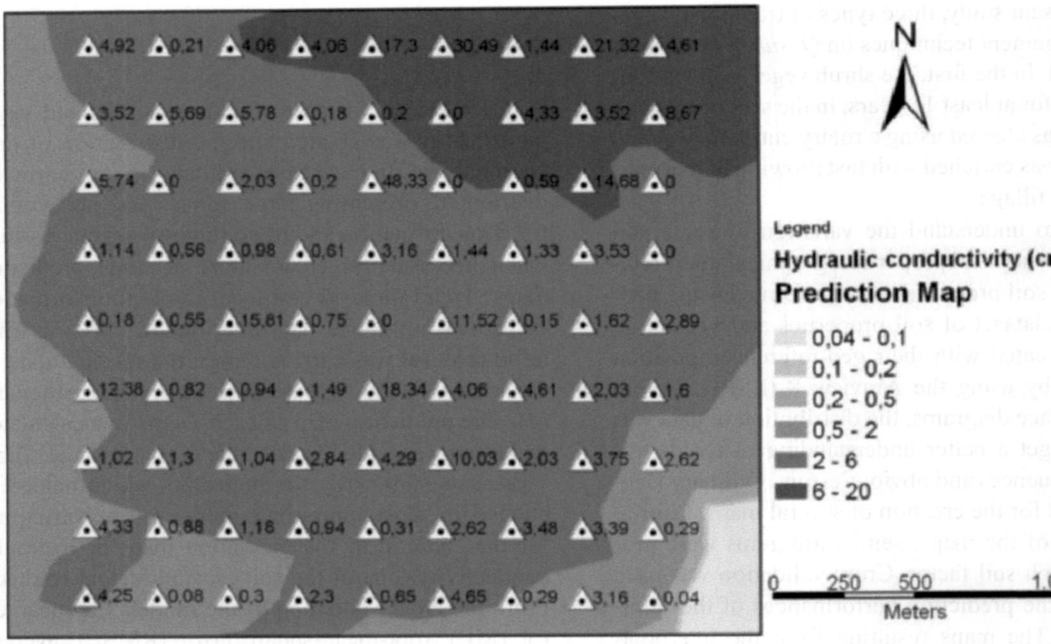

Figure 2. Prediction map of hydraulic conductivity resulted after ordinary kriging (value 0 means missing value).

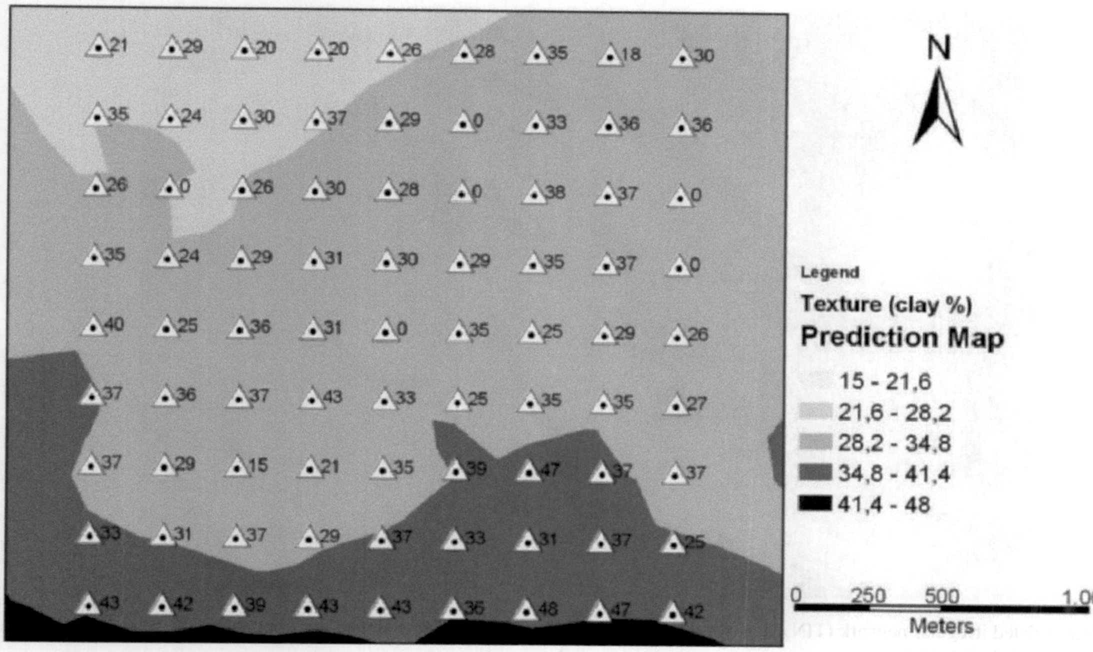

Figure 3. Prediction map of texture resulted after ordinary kriging of the clay percentage (value 0 means missing value).

(Figure 3). The map of hydraulic conductivity (Figure 2) confirms the low permeability of this clay area. The percentage of clay decreased moving to the northern side of the area (Figure 3) whilst at the same location, the percentage of sand increased.

Following the same procedure, maps of the other soil erosion properties based on ordinary kriging were produced. Table 2 presents the final semivariogram model chosen for the prediction map of each parameter analysed using the ME, RMSE, ASE and RMSSE from the cross-validation of the models. Exponential semivariograms and a large nugget effect of some models indicated a high variance. All soil properties showed a positive nugget effect, which can be explained by sampling error, short-range variability, random and inherent variability. Nugget is a parameter of a covariance or semivariogram model that represents independent error, measurement error and microscale variation at spatial scales that are too fine to detect. The nugget effect is a discontinuity at the origin of either the covariance or semivariogram model (Chilès and Delfiner, 1999).

The variable is considered to have a strong spatial dependence if the ratio nugget-to-sill is less than 25%, and has a moderate spatial dependence if the ratio is between 25% and 75%; otherwise, the variable has a weak

Table 2. Semivariogram values used to create the soil erosion prediction maps.

Soil erosion factor	Model	Nugget	Sill	Range	ME	RMSE	ASE	RMSSE	Nugget/sill
PH	Gaussian	0.20	0.49	1997.1	0.02	0.47	0.67	0.86	0.41
Electric cond. (dS/m)	Gaussian	0.51	2.02	1993.0	0.01	0.18	0.11	0.81	0.25
Organic matter (%)	Exponential	0.47	1.52	1943.4	0.01	2.32	2.88	0.91	0.31
Hydraulic cond. (cm/h)	Spherical	8.33	14.72	1967.7	0.01	3.60	3.08	0.97	0.57
Texture (Clay %)	Gaussian	4.83	10.6	1997.8	0.43	1.87	3.94	0.97	0.46
Erodibility factor *K* (t ha/h N)	Exponential	0.002	0.003	1994.5	0.001	0.07	0.08	1.08	0.67
Vegetation cover (%)	Exponential	36	89	1944.5	0.21	4.87	4.54	0.94	0.40
Erosion control practice	Exponential	1	3	1947.5	0.01	0.9	0.9	0.81	0.33

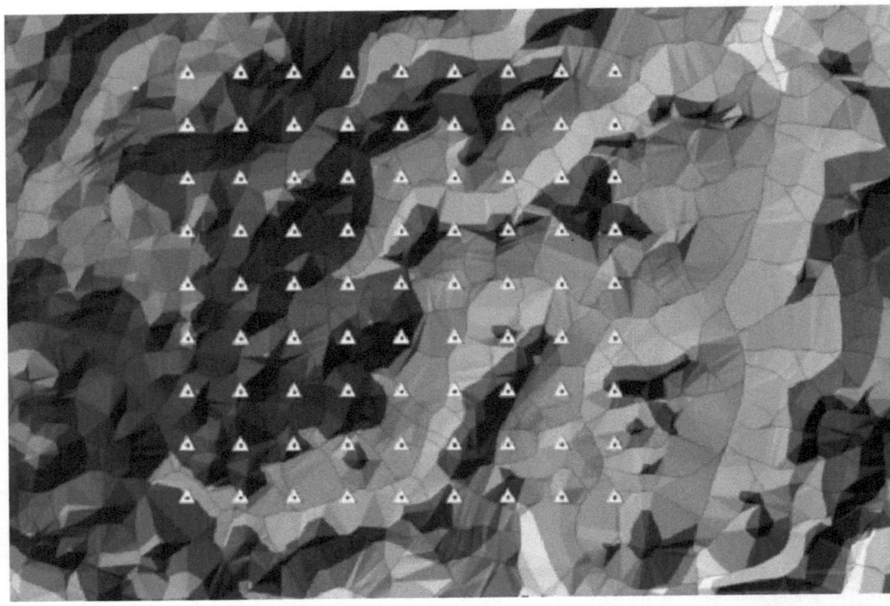

Figure 4. Triangulated irregular network (TIN) layer for three-dimensional surface used to estimate the slope factor of the experimental area. Darker zones indicate higher altitude.

spatial dependence (Cambardella et al., 1994). Nugget-to-sill ratio indicated moderate spatial dependence for all parameters studied (Table 2). Hydraulic conductivity (Kfs) and soil erodibility *K* factor with nugget-to-sill ratio of 57% and 47% respectively, showed the weakest spatial dependence. In the cases that residual spatially uncorrelated nugget was high, block-kriging was done to reduce it, by the use of a different software and importing the values in Arcview 8 (ESRI).

All maps of soil properties were reclassified, weighted and overlaid in Arcview 8 model builder. In Figure 4 it can be seen the three-dimensional surface created as triangulated irregular network (TIN) in Arcview 8, which was the digital terrain model (DTM) for the area of interest. Darker zones indicate higher altitude. The DTM was used to calculate the inclination at the experimental area and to estimate the slope length factor for each of the 81 locations of the grid. The basic idea consisted in dividing the entire watershed into a number of smaller sub-watersheds for each of the 81 locations. Next, the boundaries of the sub-watersheds were detected by the computer program. The slope length and gradient factor (*LS*) were then computed. The vegetation cover factor map was superimposed on the above maps in order to calculate the annual soil loss of the site. The erosion control practice factor map was estimated from values given on each of the 81 sample locations at the moment of sampling and after consulting the shepherds in the area. The final soil loss prediction map resulted according to the RUSLE relationship following map algebra in Arcview 8 (Figure 5). The superimposed final map of erosion risk was created to locate the optimal area for each of the vegetation management techniques.

The rainfall and runoff factor *R* was estimated to be 105.3 for the experimental area. The final soil loss prediction map resulted according to the RUSLE relationship following map algebra in Arcview 8 (Figure 5). This map resulted from the superimposition of vegetation cover factor map, slope length and gradient factor, soil properties and rainfall erodibility maps. In this final map two areas of high soil erosion risk are identified in the eastern and western parts and three smaller spots. The rest of the area has only a moderate risk of erosion.

Discussion

This study showed that simple random sampling and the calculation of an average, usually used as normal procedure for soil sampling in forestry, is not always the best technique for identifying soil erosion risk.

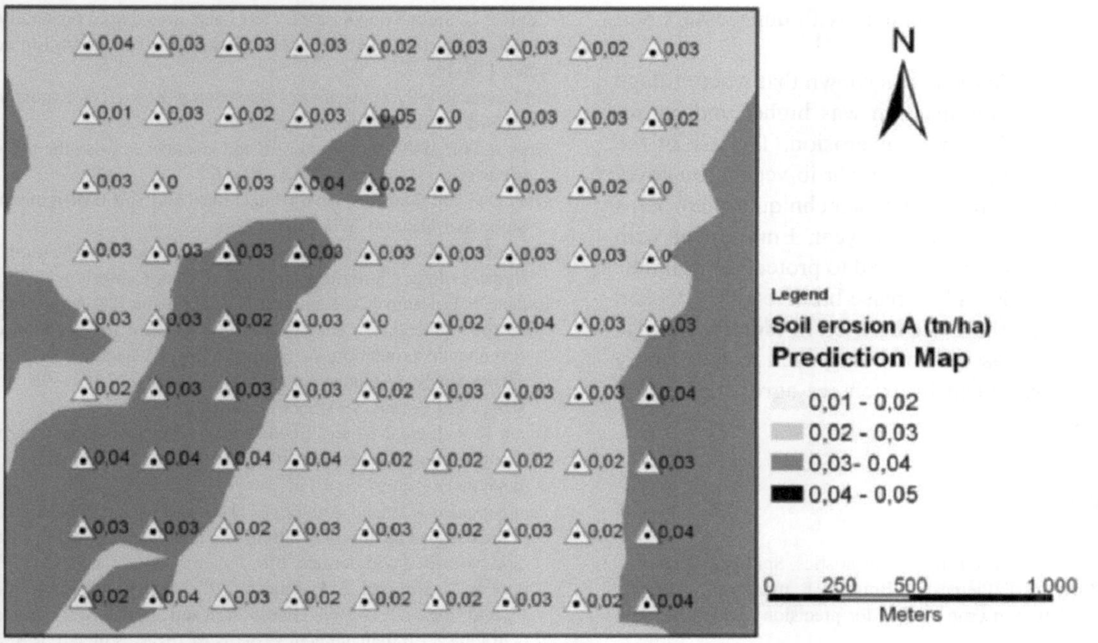

Figure 5. Soil loss prediction map resulted according to the RUSLE relationship following map algebra in Arcview 8 (value 0 means missing value).

Geostatistical methods describe the spatial variability of a site, showing the confidence levels for samples taken. Kriging as a predictor does not require that the data have a normal distribution. The Kriging technique relies on the assumption that all the random errors have zero mean and the covariance between any two random errors depends only on the distance and direction that separates them and not their exact locations (Goovaerts, 1997). From the several parameters that the semivariogram provided, the large nugget effect indicated a high variance at short distance as is mentioned by Armstrong (1998). An additional and powerful advantage of kriging is that the method yields estimates of the errors associated with interpolation.

The maps created demonstrate the existence of a heavy textured area in the southern part of the site which could affect erosion and vegetation management techniques. The higher clay presence in the South was due to continuous tillage during years for cereal production in those areas, which caused erosion, removing the surface sandy horizon and exposing a deeper clay layer. A change in vegetation management of that area could increase organic mater and accelerate soil genesis (Olson et al., 2005; Pottera et al., 1998; Six et al., 2002). Minimum tillage, compost application, the use of cutting mower and enrichment with herbaceous plants can be good practice solutions.

Various, easy to identify, important facts were taken into consideration, like areas of high or low nutrient and organic mater concentrations. Thus, those maps can be important for the estimation of the optimal area for fodder cultivation and help to predict which property is limiting production and where. The most adequate area for fodder cultivation was north-east part of the site, because it had a high-nutrient content and organic matter with low to moderate risk of erosion. The less acceptable area was the south-west.

The understanding of the spatial distribution pattern of soil properties is important to determine soil limitations to plant growth and appropriate management of soil resources in forest areas. Localised problems in soil properties can be solved with simple geographically restricted amendment treatments. The classification techniques of GIS used in the present work to estimate potential erosion were sufficient to determine which areas were the most suitable for the different vegetation management methods and to localize problems with soil properties that could be solved with simple geographically restricted amendment treatments. This approach could help spend less money and stop damaging the environment with unnecessary soil disturbances.

In the present study, it was shown that where tillage was applied, soil compaction was higher and consequently increased the risk of erosion. The use of the rotary cutter mower to destroy shrub vegetation cover can be a vegetation management technique which helps to protect the soil all around year. Enrichment with leguminous species can be used to protect the soil during longer periods and increase biodiversity. GIS and geostatistics can therefore be used to identify the risk of erosion areas as well as being used to help apply precision management in extensive agrosilvopastoral woodlands.

References

Armstrong M 1998 Basic Linear Geostatistics. Springer, Berlin.

Basso B, Ritchie J T, Pierce F J, Braga R P and Jones J W 2001 Spatial validation of crop models for precision agriculture. Agr. Syst. 68, 97–112.

Bocchi S, Castrignano A, Fornaro F and Maggiore T 2000 Application of factorial kriging for mapping soil variation at field scale. Eur. J. Agron. 13, 295–308.

Cambardella C A, Moorman T B, Novak J M, Parkin T B, Karlen D L, Turco R F. Konopka A E, 1994 Field-scale variability of soil properties in Central Iowa soils. Soil Sci. Soc. Am. J. 58, 1501–1511.

Carter M R 1993 Soil Sampling and Methods of Analysis. Canadian Society of Soil Science, Lewis Publishers, Charlottetown, Canada.

Castrignano A, Maiorana M, Fornaro F and Lopez N 2002 3D spatial variability of soil strength and its change over time in a durum wheat field in Southern Italy. Soil Till. Res. 65, 95–108.

Chilès J P and Delfiner P 1999 Geostatistics: Modelling Spatial Uncertainty. Wiley, New York.

Couto E G, Stein A and Klamt E 1997 Large area spatial variability of soil chemical properties in central Brazil. Agr. Ecosyst. Environ. 66, 139–151.

Goovaerts P 1997 Geostatistics for Natural Resources Evaluation. Oxford University Press, New York.

Goovaerts P 1999 Geostatistics in soil science: state-of-the-art and perspectives. Geoderma 89, 1–45.

Gray D H and Sotir R B 1996 Biotechnical and Soil Bioengineering Slope Stabilization. Wiley, New York.

Kitanidis P K 1997 Introduction to Geostatistics: Applications to Hydrogeology. Cambridge University Press, Cambridge, UK.

Larson W E, Lamb J A, Khakural B R, Ferguson R B and Rehm G W 1997 Potential of site-specific management for nonpoint environmental protection. *In* FJ Pierce and EJ Sadler (eds), State of Site-Specific Management for Agriculture, pp. 369–396. ASA Misc. Publ. ASA/CSSA/SSSA, Madison, WI.

Olson K R, Lang J M and Ebelhar S A 2005 Soil organic carbon changes after 12 years of no-tillage and tillage of Grantsburg soils in southern Illinois. Soil Till. Res. 81, 217–225.

Panagopoulos T, Jesus J, Antunes M D C and Beltrão J 2006 Analysis of spatial interpolation for optimising management of a salinized field cultivated with lettuce. Eur. J. Agron. 24, 1–100.

Pottera K N, Torberta H A, Jonesb O R, Matochac J E, Morrison J E and Ungerb P W 1998 Distribution and amount of soil organic C in long-term management systems in Texas. Soil Till. Res. 47, 309–321.

Reynolds W D and Elrick D E 1985 Measurement of field-saturated hydraulic conductivity, sorptivity and the conductivity pressure head relationship using the GUELPH permeameter. *In* Proceedings of National Water Well Association Conference on Characterization and Monitoring of the Vadose (unsaturated) Zone, November 1985, Denver, Colorado.

Six J, Conant, R T, Paul E A. and Paustian K 2002 Stabilization mechanisms of soil organic matter: Implications for C-saturation of soils. Plant Soil 241, 155–176.

Wackernagel H 1995 Multivariate Geostatistics: An Introduction with Applications. Springer, Berlin.

Yalouris K P, Kollias V, Lorentzos N A, Kalivas D and Sideridis A B 1997 An integrated expert geographical information system for soil suitability and soil evaluation. J. Geogr. Inform. Decision Analysis 1, 90–100.

A computer system using two membership functions and T-Norms for the calculation of mountainous watersheds torrential risk: The case of lakes Trixonida and Lisimaxia

F. Maris[1,2] & L. Iliadis[1]

[1] *Department of Forestry & Management of the Environment & Natural Resources, Democritus University of Thrace, 193 Padazidou St., 68200, Nea Orestiada, Greece.* [2] *Corresponding author**

Key words: decision support system, fuzzy sets, lake torrential risk, mountainous watershed, T-Norms, torrential indices

Abstract

This manuscript describes the development and application of a fuzzy computer system for the estimation of a long-term unique torrential risk index, expressing as many parameters as possible. This paper describes the estimation of the torrential risk in the areas of Trixonida and Lisimaxia lakes and it is a part of a bigger project for the torrential risk evaluation of the watersheds, lakes and streams of Greece. The model that was introduced to the computer system works as follows: many torrential (morphometric and hydrographic) parameters are used as input to the System and they are evaluated by Fuzzy functions. This produces separate risk indices with each one associated to a specific parameter. Finally T-Norms are applied in order to unify the risk indices and to produce a unified means of risk measure. The computer system has proven its ability to work more effectively compared to the older (established) methods. From these points of view, this work can be considered as an original and important contribution in the international literature. There are no restrictions in the application-area of the system.

Introduction

As climate changes take place on our planet, the problem of natural disasters grows on an annual basis. Forest fires and large-scale floods occur frequently causing serious erosion problems and destroying settlements and infrastructure. The flooding in central Europe in the summer of 2002 caused considerable damage. The cost in human terms and the damage to infrastructure and the natural and cultural heritage was particularly severe in Germany (EUR 15 billion), Austria (EUR 2 billion), the Czech Republic (EUR 2 to 3 billion) and Slovakia (EUR 35 million). The extent of the damage described above shows the need for an Expert System that will perform an estimation of the degree of torrential risk. The torrential characterisation of each watershed and of each stream will show clearly the way it will behave under extreme weather conditions.

This paper describes the development and the use of a decision support system (DSS) that performs torrential risk evaluation on an annual basis and its output for the areas of Trixonida and Lisimaxia lakes in the region of western Greece. It evaluates both structural and dynamic characteristics of the areas under study in order to identify the areas that are most risky. This will enable the early design of an effective prevention and protection policy. This study is a part of a larger project that aims at estimating the torrential risk in watersheds, streams and lakes of Greece, using concepts of fuzzy algebra. Our research group has already developed two approaches concerning the river Evros and one for the area of Rodopi (Iliadis et al., 2004). Both of our projects have been carried out in Northern Greece. So far, our approaches had a significant success and very encouraging results. These studies used fuzzy concepts that included either triangular or trapezoidal membership functions and fuzzy T-Norm operations (Iliadis et al., 2004).

* E-mail: fmaris@fmenr.duth.gr

The necessity for a new DSS came from the fact that only partial torrential risk estimations of mountainous watersheds were performed by the existing methods. The main drawback of the previous established approaches is that they consider only partial risks and that they apply crisp sets methods that are not capable of describing proper risk linguistics. In our study, it is the first time (globally) that an integrated risk index is produced. Also, it is known that the problem is a composite one and should be seen under different perspectives before the design of a prevention and protection policy. For example, certain areas will be at highest risk under extreme weather phenomena and others will be at highest risk in average situations. The older methods never made such distinctions. The new model performs both a partial risk index estimation (for each factor affecting the problem) and also calculates a unified risk index (URI). The URI estimates the overall risk for each area. This is therefore a pioneer DSS that characterizes each area on *an annual basis* according to its torrential risk, applying many*structural* and*dynamic factors*.

Other systems have been already developed to work on a short-term basis. The EU Commission has developed a flood simulation system (LISFLOOD) providing forecasts between two and ten days in advance and also simulations of the impact. The Commission will provide scientific back-up for a European flood warning system which will contain information about the main European catchment areas and have access to medium-term meteorological forecasts.

The DSS discussed in this paper is called TORRISDESSYS (torrential risk decision support system). It can be unified and work in a parallel way and complimentary with the LISFLOOD System, because they are producing long-term and short-term torrential risk indices respectively.

Materials and methods

Description of the area under study

The research area is located in the region of western Greece. Figure 1 shows the study area of lakes Trixonida and Lisimaxia. Lakes Trixonida and Lisimaxia are very close together and they constitute a very important lake system for the prefecture of Aitoloakarnania (Psilovikos et al., 1998a,b).

Data gathering

The research was based on data gathered from Greek public services who are responsible for meteorological and map data. Also, our research team gathered

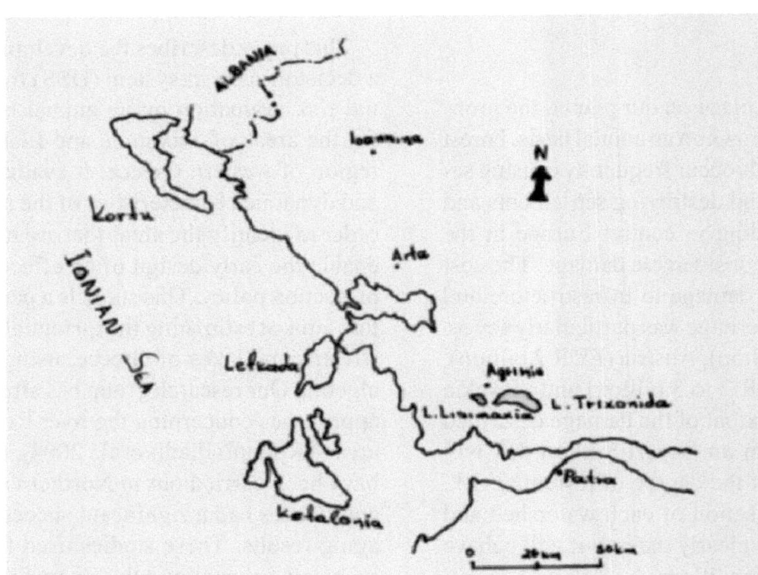

Figure 1. Lakes Trixonida and Lisimaxia.

important data with local visits to the areas. Initially the limits of the research areas were estimated. Maps of the Geographical Army Service (GAS) with a scale of 1:50000 were used for this purpose. The upper and lower limits of the watershed areas are 300 and 2 km^2 respectively (Kotoulas, 1997). For every mountainous watershed area (18 for Trixonida and 12 Lisimaxia) and for each torrential stream the morphometriccharacteristics were specified.

The morphometric characteristics were produced after the process of mapping (scale 1:50000) of the GAS and the accuracy of the data was confirmed by visits of our research teams to the research areas. The results were stored in descending order in matrices. For each research area a matrix was used. The most important morphometric characteristics of the watersheds that influence the torrential risk of an area are the following: the area, the perimeter, the shape of the watershed, the degree of the round shape of the watershed, the maximum altitude, the minimum altitude, the average altitude, the average slope of the watershed, and its maximum altitude (Horton, 1932; Kotoulas, 1969, 1973, 1997; Stefanidis, 1995; Viessman et al., 1989).

Meteorological data was gathered from all of the sources that have meteorological stations in the area of research. The meteorological data come from the Ministry of Agriculture, Ministry of Environment and public works, the Greek Meteorological Service, the Institute of Forest Research, and the National Tobacco Organization. The meteorological data concerned the monthly rain height and the average temperatures. The torrential rock formulations of the area were determined from the 1:50000 scale maps (Kotoulas, 1969). Finally, the determination of land use was performed. This was done using forest maps of scale 1:200000. The land use of the research watershed areas was determined as follows: agricultural lands, range lands, brush lands, thick or thin forests and urban areas.

Fuzzy Algebra concepts applied

Two different algebraic approaches can be applied for the estimation of the torrential risk. One is already established and it uses Crisp sets (Leondes, 1998) and the other is proposed in this paper and it uses fuzzy sets.

Function 1: Definition of a Crisp set

$$\mu_s(X) = \begin{cases} 1, & \text{if } X \in S \\ 0, & \text{if } X \notin S \end{cases}$$

In Crisp sets (Function 1), a function of this type is also called a *characteristic function*. Fuzzy sets can be used to produce the rational and sensible clustering (Kandel, 1992). For fuzzy sets there exists a degree of membership $\mu_s(X)$ that is mapped on [0,1] and every area belongs to the *torrential risky area* fuzzy set with a different degree of membership (Kandel, 1992).

Fuzzy logic models constitute the modelling tools of soft computing. Fuzzy logic is a tool for embedding structured human knowledge into workable algorithms (Kecman, 2001). This project aims at estimating a reliable means of early torrential risk evaluation of mountainous watersheds (TRMW) using fuzzy membership functions and fuzzy relations. The TRMW depends on various risk factors, each one giving a partial degree of risk. Initially a set of indices representing the partial degrees of TRMW for each watershed area was calculated. The torrential factors taken into consideration in the construction of the fuzzy logic model were the following:

1. The average altitude of the watersheds.
2. The average slope of the watersheds.
3. The average annual rain height in the average altitude of each watershed.
4. The percentage of forest-cover of each watershed.
5. The percentage of the compact geological forms of the watershed.

The above characteristics are divided in dynamic (factor number 4) and structural ones (factors 1, 2, 3, 5). Five fuzzy sets were constructed with each one corresponding to a different torrential factor. The five fuzzy sets used by the model are the following:

- Areas with watersheds with high average altitude.
- Areas with watersheds with high slope.
- Areas with watersheds with high average annual rain.
- Areas with high forest cover.
- Areas with high percentage of compact geological forms.

The Degree of membership of each area to each one of the above fuzzy sets is its partial degree of risk for the corresponding torrential factor.

Triangular and TRAPEZOIDAL membership functions were used for the estimation of the degree of

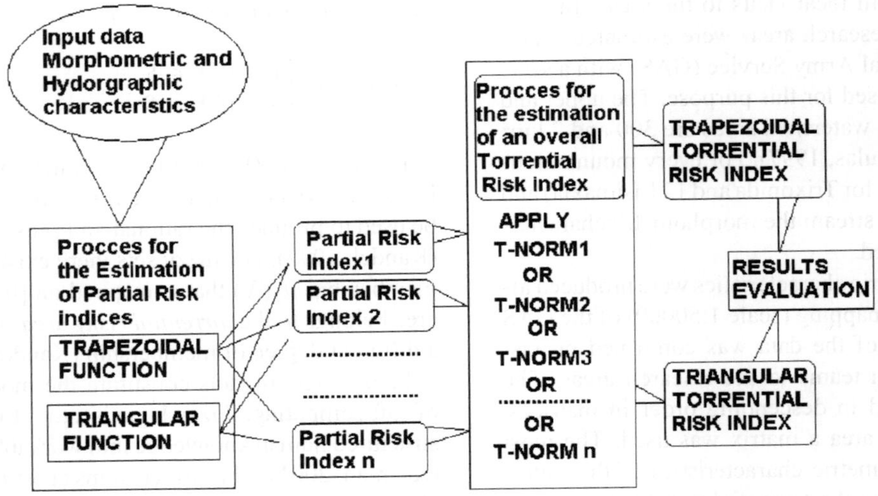

Figure 2. Structure and function of the decision support system.

membership (DOM) of each area to the corresponding fuzzy set. Function 2 is trapezoidal and Function 3 is triangular (Kecman, 2001).

Function 2: A trapezoidal membership function

$$\mu_s(X) = \begin{cases} 0, & \text{if } X \leq a \\ (X-a)/(m-a), & \text{if } X \in (a, m) \\ 1, & \text{if } X \in [m, n] \\ (b-X)/(b-n), & \text{if } X \in (n, b) \\ 0, & \text{if } X \geq b \end{cases}$$

Function 3: A triangular membership function

$$\mu_s(X) = \begin{cases} 0 & \text{if } X < a \\ (X-a)/(c-a) & \text{if } X \in (a, c) \\ (b-X)/(b-c) & \text{if } X \in (c, b) \\ 0 & \text{if } X > b \end{cases}$$

The final target was the unification of all the partial DOM and the estimation of the degree of membership of each area to the final fuzzy set torrential risky area. Special types of fuzzy relations (the fuzzy T-Norms) were applied in order to unify the partial risk indices. The Type 1 function offers a good definition of a fuzzy relation.

A torrential risky area is that which is characterised by the tendency to have high water-flow and to produce large amount of sediments which are the necessary components for a flood to appear. Consequently, the evaluation and ranking of the mountainous watersheds is the most important tool in the hands of the people who are designing torrential prevention and protection policy, and also protection and management measures for the lake ecosystems of Trichonida and Lisimachia.

Let A be an input fuzzy region with element x and membership function μ_A and let B be an output fuzzy region with element $y$and membership function μ_B. The fuzzy relation on the fuzzy product $A \times B$ is a mapping of the following type:

$$\mu_R : A \times B \to [0, 1], \text{ where } \mu_R(X, Y) = \mu_A(X) \Lambda \mu_B(Y)$$

and the fuzzy relation set is defined to be:

Type 1: Definition of a fuzzy relation

$$P = \{((X, Y), \mu_R(X, Y))/(X, Y)) \in A \times B\}.$$

The general aspects concerning the structure and the function of the system can be seen in Figure 2. Table 1 describes the five different cases of T-Norms that were used for the production of the Unified Risk Index (Kecman, 2001).

Implementation of the system

The DSS was developed in MS-Access. It uses a visual environment and applies a modern graphical user interface. It has been designed and implemented to store

Table 1. T-Norms used in the project

1. *Minimum approach*
$URI = MIN(\mu_A(X), \mu_B(X))$
2. *Algebraic product*
$URI = (\mu_A(X) * \mu_B(X)$
3. *Drastic product*
$URI = MIN(\mu_A(X), \mu_B(X))\,..\,\text{if}\,..\,MAX(\mu_A(X), \mu_B(X)) = 1$
otherwise $URI = 0$
4. *Einstein product*
$URI = \mu_A(X) * \mu_B(X)/(2 - (\mu_A(X) + \mu_B(X) - \mu_A(X) * \mu_B(X)))$
5. *Hamacher product*
$URI = \mu_A(X) * \mu_B(X)/(\mu_A(X) + \mu_B(X) - \mu_A(X) * \mu_B(X))$

the data in a relational access database. The retrieval mechanism of the DSS reads the required data from the database. Data is stored in several tables according to the relational philosophy. The main table contains six fields, the primary key and the five fields corresponding to the torrential risk factors. It has been designed to follow the first and second Normal form (Date, 1990).

The DOM to the five fuzzy sets (partial risks) and the URI were calculated by performing SQL (structured query language) operations on the database. Various queries embedding SQL statements were developed. The following statement is an example of a DOM estimation.

Example 1: SQL statement used by the DSS

SELECT fuzzy_help.perioxi, fuzzy_help! help_drastic AS drastic_product FROM fuzzy_help ORDER BY fuzzy_help!help_drastic DESC;

The inference engine of the system is forward chaining and it works in a rather linear manner in order to produce the risk index. Figure 3 shows the friendly graphical user interface of the DSS.

Results

Testing the system

The testing of the DSS was performed using actual data concerning all of the watersheds in the Trixonida and Lisimaxia areas. Table 2 shows the watershed and stream areas under study and their code names. Tables 3 and 4 show the results of the torrential risk estimation for the area of Trixonida and Lisimaxia lakes using the trapezoidal and the triangular membership functions respectively for the partial risk degrees and the Einstein, the algebraic product, the Hamacher Product and Drastic Product T-Norms for the final TRMW estimation. The trapezoidal and the triangular membership functions were applied for the estimation of the partial degrees of risk in Tables 3 and 4 respectively. In both cases the Einstein, the algebraic product, the Hamacher

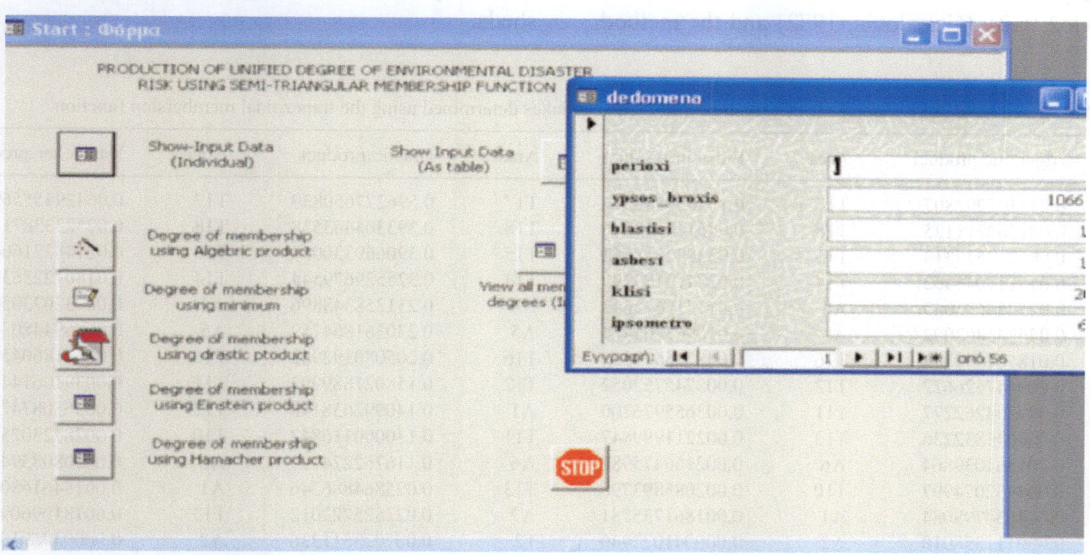

Figure 3. The interface of the decision support system.

Table 2. Mountainous watersheds of Trixonida and Lisimaxia lakes and their code names

Trixonida area		Lisimaxia area
Kato Bloxou (1)	Bathirrema (13)	Aroulas (1)
Kato Tragana (2)	Mpourla (14)	Trianteikon (2)
Koubelorrema (3)	Gavalou (15)	Diamanteikon (3)
Kouforrema (4)	Mperdenikou (16)	Kolosirtis (4)
Palaiokarias (5)	Platanias (17)	Moures (5)
Krinorrema (6)	Tserlis (18)	Katourlis (6)
Mega (7)		Ermitsas (7)
Dixalorrema (8)		Paleobrisis (8)
Sitaralona (9)		Zevgarakiou (9)
Bathirrema (10)		Klisorematon (10)
Palaiochoriou (11)		Anonym 1 (11)
Red Palaiochoriou (12)		Likorema (12)

Product and the Drastic Product T-Norms were applied for the final Unified TRMW calculation.

It is clear that the TORRISDESSYS operates in a totally dynamic way. This means that any change in the data has a direct effect on the characterisation of the area. For example, an area that has high percentage of forest cover (and thus has a low overall torrential risk) can become a highly risky one if the forest cover is reduced dramatically due to e.g. wild fires.

Discussion

The results of the System were compared to the outcome given by the method of Gavrilovic1 (G1), Gavrilovic2 (G2), (Gavrilovic, 1972) and the method of Stiny (St) (Stiny, 1931) which constitute well-established methodologies. The equation of Gavrilovic considers the average annual production of sediments, average annual temperature, average annual rain height of the watershed and its area, the geodeposition, vegetation, erosion of the watershed, and the calculation of the special degradation is very important. The main difference between the G1 and the G2 methods is that G1 refers to the total area under examination and G2 refers to the average per km^2.

The method of Stiny determines the torrential risk based on sediment production with a periodicity of 100 years using the equation of Stiny-Herheulidze (Kotoulas, 1997; Stiny, 1931). This equation was chosen because it fits to the characteristics of the study area. A new model, the LISFLOOD has been developed recently and it has gained wide acceptance (GRDC). The physically based LISFLOOD model has been specifically developed to simulate floods in large European drainage basins. Full basin-scale simulations can be carried out in such a way that influences of land use, spatial variations of soil properties and spatial precipitation differences, e.g. by increased flood frequency through climatic change, are taken into account. Of course this is a system that has a different nature than the one developed here, because the TORRISDESSYS estimates the torrential risk for each single mountainous watershed and it does not deal with each drainage basin as a whole, nor does it deal with urban characteristics of the areas. TORRISDESSYS deals only with risk coming from mountainous watersheds.

Table 3. The fifteen most risky watersheds of Trixonida and Lisimaxia lakes determined using the trapezoidal membership function

Area	Algebraic product	Area	Einstein_product	Area	Drastic_product	Area	Hamacher_product
T17	0.231195292807	T17	0,164654548346	T17	0.596277650839	T17	0.064294555697
T18	0.081562211125	T18	0,046140647048	T18	0.393304463538	T18	0.025233387113
T15	0.065742537313	T15	0,033892318563	T15	0.390669330030	T15	0.021482716045
T14	0.055119046462	T14	0,028413183299	T14	0.325529679544	T14	0.018012236321
Λ4	0.025954635049	Λ4	0,011715625642	Λ4	0.251258348896	Λ4	0.009320730506
Λ5	0.024243402092	Λ5	0,010961765474	Λ5	0.230361864759	Λ5	0.008694480303
T16	0.018501434523	T16	0,008236629522	T16	0.205080192962	T16	0.006718604596
T11	0.009447526622	T12	0,003745752052	T12	0.158021889993	T11	0.003910614430
T12	0.009244262277	T11	0,003655976200	Λ1	0.140992638148	T12	0.003650874770
T10	0.006041832236	T13	0,002211999847	T11	0.130000116842	T10	0.002872302585
Λ6	0.005361039864	Λ6	0,002166475982	Λ6	0.116762274959	Λ6	0.002080839428
T13	0.004932074997	T10	0,002085893798	T13	0.095864886246	Λ1	0.001846168048
Λ1	0.004634798084	Λ1	0,001861735781	Λ2	0.022825722012	T13	0.001810960990
Λ2	0.000112358010	Λ2	0,000041022940	T2	0.007023311236	Λ2	0.000049691897
T2	0.000029960101	T2	0,000012453601	Λ12	0.000100000000	T2	0.000011548602

Table 4. The fifteen most risky watersheds of Trixonida and Lisimaxia lakes determined using triangular membership function

Area	Drastic_product	Area	Hamacher_product	Area	Algebraic product	Area	Einstein_product
T17	0.397518438546	T17	0.025177790957	T17	0.079033455734	T17	0,030300353756
T18	0.262202978761	T18	0.012333997953	T18	0.036313724508	T18	0,006399041601
T15	0.260446223246	T15	0.009290334716	T15	0.025319359726	T15	0,004583969140
Λ9	0.251433928329	Λ7	0.008039198942	Λ7	0.022621866331	T14	0,003800891562
T11	0.232353980280	T14	0.007293829744	T14	0.019865523280	Λ4	0,001517194131
T14	0.217019788903	T1	0.006202160782	Λ8	0.015101534853	Λ5	0,001417577052
Λ4	0.167505567892	Λ8	0.005973638724	T1	0.014613894915	T16	0,001059885603
Λ8	0.157789545729	Λ9	0.005269380251	Λ9	0.013492063402	T12	0,000478559115
Λ5	0.153574578408	T11	0.003976163104	Λ4	0.010009061490	T11	0,000467039708
T16	0.136720130242	Λ4	0.003955572245	T11	0.009043781725	Λ6	0,000282300369
T12	0.105347927967	Λ5	0.003273483631	Λ5	0.008293197916	T13	0,000276492671
Λ1	0.093995093199	T16	0.002841821786	T16	0.007126480810	T10	0,000266214886
Λ6	0.077841517550	T6	0.002196831452	T6	0.004435319509	Λ1	0,000237641435
T13	0.063909924955	T10	0.001555053379	T10	0.003218643638	Λ2	0,000006455648
Λ12	0.052315883723	T12	0.001340464658	T12	0.003160432733	T2	0,000002840775

The comparative study between the TORRISDESSYS and the other existing methods has proven that there exists a good level of compatibility in most of the cases. Table 5 shows clearly the high degree of compatibility between TORRISDESSYS and the existing methodologies. In particular, there exists a high compatibility between the G2 method and TORRISDESSYS that ranges from 73.3% to 80% in the case of triangular membership function and from 60% to 73.3% in the case of the trapezoidal membership function. A high compatibility exists above the average between the G1 method and TORRISDESSYS that ranges from 40 to 60% regardless of the membership function used. Nevertheless, there exists a very low compatibility between the Stiny method and TORRISDESSYS, regardless of the membership function used. This can be easily explained by the fact that the Stiny method does not apply many factors to estimate torrential risk. On the contrary, TORRISDESSYS and G1 and G2 consider many factors affecting directly the problem. It should be mentioned that similar behaviour has been noticed for the Stiny method in the areas of Evros and Rodopi but on a much smaller scale.

Table 5. Compatibility between the results of TORRISDESSYS and Gavrilovicl, Gavrilovic 2, and Stiny methods

	G1		G2		ST	
Torrential risk index for Trixonida and Lisimaxia area with the triangular function						
Algebraic product	9/15	60%	11/15	73.3%	8/15	53.3%
Min	9/15	60%	11/15	73.3%	8/15	53.3%
Drastic	6/15	40%	12/15	80%	4/15	*26.6%*
Einstein	9/15	60%	11/15	73.3%	8/15	53.3%
Hamacher	9/15	60%	11/15	73.3%	8/15	53.3%
Torrential Risk Index for Trixonida and Lisimaxia area with the trapezoidal function						
Algebraic product	6/15	40%	10/15	66.6%	3/15	20%
Min	6/15	40%	10/15	66.6%	3/15	20%
Drastic	5/15	33.3%	11/15	73.3%	1/15	*6.66%*
Einstein	6/15	40%	10/15	66.6%	3/15	20%
Hamacher	6/15	40%	9/15	60%	3/15	20%

Table 5 clearly shows that the triangular membership function is more consistent to the methodologies of Gavrilovic and Stiny. The trapezoidal function tends to normalize the differences and this makes its results less compatible to the ones produced by the existing methods. More specifically, we conclude that the torrential streams with code numbers (17, 18, 15, 14) located in the south part of lake Trichonida and also those located in the north part of lake Lisimachia encoded as 4, 5, 7 and 9, have a very high torrential risk due to their large volume of water-supply and the large amount of sediments that they produce.

Conclusion

One of the very important characteristics of the expert system described here is that it can view the problem from different perspectives. For example, the Drastic Product approach estimates the torrential risky areas *under very extreme* rainfall situations, or in cases that have a very high value in one of the factors thus having a positive effect on the torrential risk. On the other hand, the algebraic product estimates the risky areas under *average rainfall* conditions. The

Einstein T-Norm offers a good approach for an *overall* evaluation.

The local authorities can use our DSS in order to plan an effective protection and prevention policy. It can be applied on a larger scale and for a wider period of time. The more risk factors included in the system's reasoning, the better its performance. It is very important to have an overall risk measure for such a serious problem. The most important aspect of the system is that though it has been tested in Greece, it can be applied on a global basis. There exists no national limitation in its reasoning. The system has been tested so far for the areas of Greek Evros (Northern, Central and Southern Evros) Rodopi, Trixonida and Lisimaxia with remarkable results.

The testing of the TORRISDESSYS must continue for several more years. This of course requires a research team that will gather data on an annual basis. This will enable the system to evaluate the new up to date characteristics of the watersheds and to output the risky areas based on an overall approach, on partial risk factors and on extreme situations formulating a time series of results. Finally, after applying the triangular and the trapezoidal functions, a third type of membership function (the Sigmoid one) can also be used to estimate the partial degrees of risk and then a comparative study (between the three cases) can be performed.

References

Date C J 1990 An Introduction to Database Systems. Addison-Wesley, New York.

Gavrilovic Sl 1972 Inzenjering o bujicnim tovoklima i eroziji. Beograd.

Horton R 1932 Drainage basins characteristics. Transactions, AGU, pp. 350–361.

Iliadis L, Maris F, Marinos D 2004 A decision support system using fuzzy relations for the estimation of long-term torrential risk of mountainous watersheds: The case of river Evros. *In* Proceedings of the 5th International Symposium on Eastern Mediterranean Geology, University of Macedonia, Thessaloniki, Greece, pp. 712–714.

Kandel A 1992 Fuzzy Expert Systems. CRC Press, Boca Raton, FL.

Kecman V 2001 Learning and Soft Computing. MIT Press, London.

Kotoulas D 1969 The streams of N. Greece, their classification in characteristic types and the main principles of their management. Sci. Ann.Agric. Forest., 13 (Appendix) pp. 12–15.

Kotoulas D 1973 The torrential problem in Greece. Report Nr. 47. Laboratory of Silviculture and Mountainous Water Science, School of Agriculture and Forestry, Thessaloniki pp. 8–10.

Kotoulas D 1997 Management of Torrents I. Publications of the University of Thessaloniki.

Leondes C T 1998 Fuzzy Logic and Expert Systems Applications. Academic Press, San Diego.

Psilovikos A, Almpanakis K, Boybalidis K and Palikaridis X 1998a The natural value and the important of lake Trixonida as the biggest natural deposit of clean water of Greece. Greek Geographical Company. *In* 4th National Geographical Conference, 12–14 October 1998.

Psilovikos A, Almpanakis K, Boybalidis K and Palikaridis X 1998b The environmental role of lake Lisimachia as an interjection of a mangrove system in the trench of Agrinio. Greek Geographical Company. *In* 4th National Geographical Conference, 12–14 October 1998.

Stefanidis P 1995 The torrent problems in Mediterranean Areas (example from Greece). *In* Proceedings of the XXIUFRO Congress, Finland.

Stiny J 1931 Die Geologischen Grundlagen der verbanung der Geschiebeherde, Wien.

Valentini C 1930 Sistamatione dei torrenti a dei montani. Milano.

Viessman J W, Levis G L and Knappt J W 1989 Introduction to Hydrology. Harper and Raw, New York.

Comparison between two low cost photogrammetric systems: The analytical instrument Adam ASP2000 and the digital photogrammetric station DVP

Vasilios C. Drosos[1,3], Nikos S. Karantzidis[2] and Vasilios J. Giannoulas[2]
[1]*Department of Forestry and Management of the Environment and Natural Resources, Democritus University of Thrace, Pantazidou 193, 68200 N. Orestiada, Greece.* [2]*Department of Forestry and Natural Environment, Aristotle University of Thessaloniki, 54124 Thessaloniki, Greece.* [3]*Corresponding author*.*

Key words: absolute accuracy, analytical photogrammetric station, digital photogrammetric station, internal accuracy; low cost; mapping

Abstract

The creation of the National Cadastre's map, part of which is the forest cadastre (according to the decision of the Ministry of Agriculture No 99580/506 from 1 July 1999), is based on law 266/1998 'National Cadastre and other provisions' and forecasts the use of photogrammetric methods. The changes in land use, and the stretches of mountainous and semi-mountainous abandoned land are represented on these maps. The photogrammetric community is experiencing a transitional period from the analytical to the digital era. This transition is met with prudent enthusiasm, and care as to which digital systems may replace the analytical ones and under what circumstances this change will take place. This paper endeavours to answer such a question. A complete and fully documented answer lies of course, outside the limitations of this project. Here, however, a comparison was attempted between a contemporary analytical photogrammetric instrument and a digital photogrammetic system, which are both of relatively low cost. This choice was made in order to enable the market in this country to assess the limitations and possibilities of such systems, in view of the large-scale cadastral surveys. Therefore, this comparison is not only confined to accuracy, but touches upon other issues, such as reliability, user friendliness and efficiency. There are many factors, not to mention errors, in the photogrammetric procedure. There main causes are the operator, the instrumentation used and various unpredictable factors. The area of study and stereopairs that were used were chosen so that a reliable terrestrial survey would be available for the purposes of control. A rural forest area and a forest area of Taxiarchi–Vrastamon were selected. For these areas, stereopairs and related diapositives were available at the scale of 1: 20000. They were taken from an E.E.C. GR80-3 ($c = 152.25$ mm) aerial survey camera with an overlap of 65%. The same control points were used for the orientations on both the analytical low cost system and the digital low cost system. The analytical low cost system is a more reliable instrument than the digital low cost system, because both its repeatability and accuracy are better.

Introduction

The evolution in computer technology during the last few years has brought a remarkable increase of computational power, which has drastically affected many applied sciences, including photogrammetry. Photogrammetric mapping has been the most efficient means of producing accurate representations of large, complex areas of land, and technology has radically changed how this process is carried out. We know how to use the knowledge and experience to develop a project while maintaining the projects' goals and budget in a timely manner. The photogrammetric community is still experiencing a transition period from the analytical to digital photogrammetric technologies. Thus, new digital photogrammetric systems are already

* E-mail: vdrosos@fmenr.duth.gr

establishing themselves in the market, offering different prospects and covering a wide cost range.

On the other hand, there are a respectable number of analytical photogrammetric systems that are still in use. The reason for that lies to their high cost, which detains the transition mentioned above.

In Greece the compilation of the National Cadastre's map, part of which is the forest Cadastre (according to the decision of the Ministry of Agriculture No 99580/506 from 1 July 1999) is based on law 266/1998, National Cadastre and other provisions forecasts primarily the use of photogrammetric methods for the production of the relevant cartographic products.

Besides that, photogrammetry can be very helpful for many other aspects in forest science such as forest road construction (Kantartzis et al., 2003), forest inventory (Akca, 2001) and forest management (Gadow, 2000), etc.

The aim of this paper is to compare a contemporary analytical photogrammetric instrument and a digital photogrammetric station, both of relatively low cost, in order to present to the market in Greece the possibilities and limitations of these types of instruments. This comparison focuses mainly on the accuracy of these two instruments, but also examines additional aspects such as reliability, ease of use and user friendliness.

Error Theory

It is considered appropriate to present a few theoretical facts about errors and especially errors occurring during the photogrammetric procedure. This will enable the objective assessment of the practical results later. It is well known that when we referred to the reliability of the result we mean a combination between precision and accuracy. The precision refers to how close the repeated measurements are to each other, while the accuracy, on the other hand, indicates how close the measurements are to the real values.

Errors are categorised according to their qualitative features, their sources and their mathematical properties.

- Regarding their qualitative features, we can have the following errors (Balodimou, 1991; Doukas, 2001): *Gross errors, systematic errors and accidental errors.*
- Concerning their sources, we can have the following errors: *Errors from instruments, errors due to the environmental factors and personal errors.*
- Regarding their mathematical properties, the errors are categorised as follows: *True error, probable error* and *residual error* is the probable error with different prompt.

The main causes for the photogrammetric errors are the operator, various unpredictable factors, the materials, and the instrumentation used.

There are some errors during the photography, which are due to the use of film instead of glass plates and the photographic process itself. Other error sources related to the light sensitive material are focusing and image resolution. The camera itself is a major error source. The camera lens introduces serious radial and asymmetric distortion errors. Additional error sources are earth curvature and atmospheric refraction (Kraus, 1990; Patias, 2001).

The main observation error source is the instrument with which the measurements are performed. In addition, the qualities of the targets and the geometry of the stereopair (base to distance ratio, percentage of overlap etc.) are significant factors affecting observation quality. Finally, the human factor is a source of gross and systematic errors. The errors due to adjustment are caused by the quality, the kind and the distribution of the control points. Another error source is also the cartographic projection used (Hallert, 1960).

Materials and methods

Research area

For the aim mentioned above, a rural forest area and a forest area at the University Forest of Taxiarchi–Vrastamon in the prefecture of Chalkidiki were chosen. For these areas, the related terrestrial survey was carried out during the years 1998–2000 at a scale of 1:5000 (Drosos, 2000). The university forest of Taxiarchi–Vrastamon includes all the characteristics of a typical forest in Greece, as 76.1% of our country's forests lie under the same conditions and is considered as one of the most representative forest areas for the Greek conditions.

The university forest of Taxiarchi–Vrastamon lies on southern and southwestern slopes of the Holomonta Mountain and is situated at a distance of about 70 km from Thessaloniki. It is situated between latitudes 40°23′–40°28′ and longitudes 23°28′–23°34′ and its height is 320–1165 m above sea level.

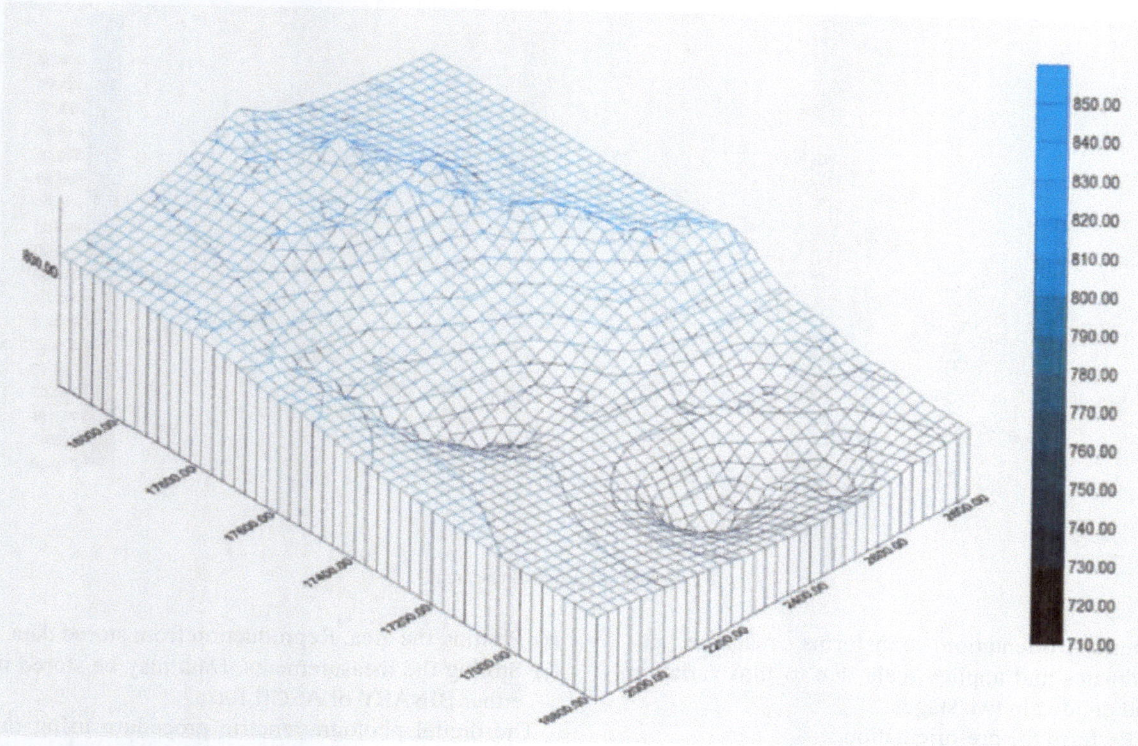

Figure 1. Forest zone of research.

The selected forest area 'Solinaria' of the university forest of Taxiarchi–Vrastamon is situated southwest of the forest office, at the borders of the nationally owned forests of Polygyros–Palaiochora and the university forest of Taxiarchi–Vrastamon.

Broadleaved Oak (*Quercus conferta* Kit) and isolated trees of Fluffy Oak (*Quercus pubescens* Willd), Beech (*Fagus moesiacan* (K. Maly) Srafer), coppice of Trachea Pine (*Pinus brutia* Ten.) and Black Pine (*Pinus nigra* Arnold) cover 96% of the region. The region ranges from 700 to 890 m above sea level and has a SW and NW aspect. The relief is mountainous with inclinations not more than 45% and the trees' height varies between 15 and 25 m, while the road's width ranges from 4 to 5 m and is surrounded by oaks (Figure 1) (AUF, 2002).

The rural forest area 'Livadia' is situated west of the forest office near to the nationally owned forest of Palaiochoras. Broadleaved Oak (*Quercus conferta*), Beech (*Fagus moesiaca*) and Black Pine (*Pinus nigra*) cover 63.52% of the area. The other 36.48% of the area is covered by forests of fir trees (*Abies borisii-regis* MattF). The region lies between 610 and 900 m above sea level and with the following aspects: W (76%), NW (20%), N (10 Ha or 3%) and S (3 Ha or 1%). The slopes do not exceed 25% although in a small and rough area there are slopes that reach 45% (Figure 2) (AUF, 2002).

Instruments

The photogrammetric instruments that were used for the determination and the plotting of the rural forest and forest areas, planimetric and height, were the low cost analytical instrument ADAM ASP2000 and the low cost digital photogrammetric station DVP.

The analytical photogrammetric procedure using the ADAM ASP 2000 follows the next steps:

(a) Setting the diapositives in the carriages.
(b) Setup the job and model of the project.
(c) Filling in ten (10) cards with important information for the process such as the topographic coordinates of the control points, the coordinates of the fiducials, etc.

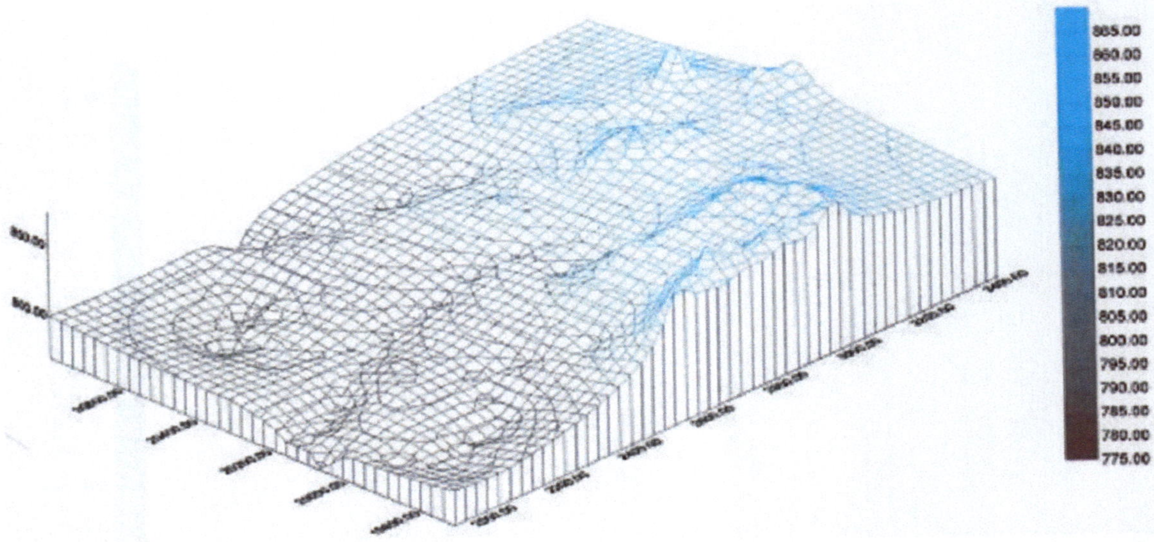

Figure 2. Forest rural zone of research.

(d) Interior orientation transforms machine co-ordinates and applies scale due to film variation and divided in two stages:

Perform the pre-orientation.

Perform the aiming procedure of the fiducial points. The orientation uses an affine transformation and supports error reports and orientation report.

(e) Exterior orientation transforms plate co-ordinates into ground co-ordinates, applies correction for lens distortion, sets the machine parameters to suit model orientation and forms the stereo model. It is divided in two phases:

Relative orientation using the Von Gruber points (These are 6 points at least that, include in the common area of the pair positioning two on the top, two in the middle and the other two at the bottom of the overlap area).

Absolute orientation the routine uses a least squares adjustment to determine the best possible transformation matrix. Generally, only little iteration will be required to bring the model 'down' to absolute, as the corrections are determined from a three-dimensional transformation.

(f) Digitising produces and displays *X*, *Y*, *Z* (Easting, Northing, Height) co-ordinates of observed points in real-time.
(g) Editing of photogrammetric diagram.
(h) Creation and editing of digital terrain model (DTM).
(i) Plotting the area. Reproduction from stored data.
(j) Storing the measurements. Data may be stored in either BINARY or ASCII form.

The digital photogrammetric procedure using the DVP follows the next steps:

(a) Input the digital images.
(b) Determination of job and model of the project.
(c) Interior orientation using the data from the calibration report for each image.
(d) Exterior orientation divided into two phases:

Relative orientation using the Von Gruber points.

Absolute orientation uses a least squares bundle block adjustment algorithm to determine the best possible transformation matrix. Generally, only two iterations were required to bring the model down to absolute, as the corrections are determined from a 3-dimensional transformation. Control points assist with absolute orientation.

(e) Stereo digitising of the area in order to create the Digital Terrain Model (DTM).
(f) Editing of photogrammetric diagram.
(g) Creation of TIN DTM.
(h) Plotting the area. Reproduction from stored data.
(i) Storing the measurements. Data may be stored in DXF (Document eXchange Format) or TXT form.

The control points were chosen in a way that they were easily recognised on the photographs and were spread out all over the overlapping area. The measurements for the calculation of the control points'

Table 1. Technical specifications of the aerial photos and diapositives

Technical specifications	Value
Year	1992
Flying height	3000 m
Scale	1:20000
Focal length	152.25 cm
Size A/P & D/S	23 cm × 23 cm
Film	Black and white

coordinates were carried out with the electro-optical theodolite WILD T2.

A single-engine aircraft did the aerial photography flight. The ADAM software uses pairs of overlapping aerial diapositives that can be obtained from metric cameras. Instead of that, DVP software uses pairs of overlapping aerial images that can be obtained from scanned images.

Materials

The materials that were used for the application of the analytical and digital photogrammetric methods were:

(a) Aerial photo (A/P), photograph of the ground taken from the air by using a flying vehicle (airplane, helicopter etc).

(b) Diapositive (D/S) is the positive aerial photo that has imprinted on a transparent material (film, glass etc.). In our case, the transparent material is film.

The technical specifications of the aerial photos and diapositives, which were used in this research, are included in Table 1.

(c) The calibration report is a special printed matter that contains irreplaceable information for the right carrying out of the analytical and digital photogrammetric processing, such as:
 subject distance
 focal length
 fiducials coordinates
 principal point coordinates
 radial distortion
 asymmetric distortion

Reliability check of the systems

The repeatability or precision (internal accuracy) is determined through repeated pointings on control points and the accuracy or root mean square error (absolute accuracy) through the comparison of these measurements with the geodetic coordinates of the points, which are considered error free.

In order to determine the repeatability, multiple pointings on the control points were performed. Every point was observed three times in such a way that no systematic errors would occur. The equation that was used is

$$\sigma_{V,j} = \left[\frac{\sum_{i=1}^{n_r} \left(v_{i,j} - \bar{v}_j\right)^2}{(n_r - 1)} \right]^{0.5} \tag{1}$$

where v_j is one of the coordinates $(x, y,$ or $z)$ of point j, n_r is the number of repetitions per point (in this case 3) and; $\bar{v}_j$s the numeric mean of the n_r repetitions of v_j, and the value indicated in Table 2 is the arithmetic mean of $\sigma_{V,j}$ calculated for the n control points.

In order to determine the accuracy we have

$$p_V = \left[\frac{\sum_{j=1}^{n} \left(\bar{v}_j - vm_j\right)^2}{n} \right]^{0.5} \tag{2}$$

where $\bar{v}_j$ is the numeric mean of the n_r repetitions of v_j, n is the number of the observations, vm_j is the real value of v_j and the value indicated in the Table 3 is the root mean square error p_V calculated for the n control points.

For the better reliability check of the systems, the following objective criteria were examined:

Table 2. Precision

Research area	Instrument	Number of control points	Precision (m) x	y	z
Livadia	ADAM ASP 2000	30	0.125	0.110	0.035
Solinaria	ADAM ASP 2000	30	0.379	0.409	0.080
Livadia	DVP	30	0.346	0.370	0.093
Solinaria	DVP	30	0.850	0.770	0.297

Table 3. Accuracy

Instrument	Number of control points	Root mean square error (m)			Research area
		x	y	z	
ADAM ASP 2000	30	0.420	0.358	0.084	Livadia
ADAM ASP 2000	30	0.809	0.878	0.271	Solinaria
DVP	30	0.718	0.791	0.142	Livadia
DVP	30	1.693	1.537	0.551	Solinaria

(a) The average of the measurements' divergences by the true value by the formula of the mean absolute error (m_a):

$$m_a = \frac{\sum_{j=1}^{n} \left|\bar{v}_j - vm_j\right|}{n} \quad (3)$$

where $\bar{v}_j$ is the numeric mean of the n_r repetitions of v_j; vm_j is the real value of v_j and n is the number of observations.

(b) The mean square error of the measurements (m_s), which typifies a series of measurements:

$$m_s = \frac{\sum_{j=1}^{n} \left(\bar{v}_j - vm_j\right)^2}{n} \quad (4)$$

The factors are the same as those in Equation 3.

(c) The criterion of the mean square error of average (m_M):

$$m_M = \pm m_s/(n)^{0.5} \quad (5)$$

where m_s is the mean square error of the measurements and n is the number of observations.

The assessment of case of use and user friendliness is achieved with the comparison between the two systems regarding the offer of a great range of hardware and software products easy understandable and learnt by the operators.

Accuracy is the degree of conformity with a standard. Accuracy relates to the quality of a result and is distinguished from precision, which relates to the quality of the operation by which the result is obtained. Both of them depend on the resolution because it helps on how close the repeated measurements are to each other and how close the measurements are to the real values.

Results

From the research and investigation of the data that resulted from the elaboration of the primary data derived from the terrestrial survey with the Wild T2, ADAM ASP2000 and DVP, due to the error theory, the relevant results are presented in Tables 2 and 3.

It is obvious that the repeatability of the analytical system ADAM ASP2000 is better than the one of the digital photogrammetric station DVP, which makes the ADAM ASP2000 more accurate than DVP. Furthermore, it is evident from the data analysis that the accuracy of the analytical system is twice higher than that of the digital one. This is caused by the superior resolution and better stereoscopic vision of the diapositives as compared to the scanned images. The observations performed by the analytical system ASP2000, approach much more the true values in comparison to the observations from the digital photogrammetric station DVP.

The reliability of the two photogrammetric processing methods is presented in Table 4.

The z coordinates are less accurately assessed on forest zone than on forest rural zone because of the high vegetation. It is easier to find control points in forest rural zone than in a forest zone.

ADAM ASP2000 is easy and reliable to use because extensive use of colour throughout to reduce operator error or fatigue. If solutions fail, recovery is easy. In addition, it is more user friendlier because in virtually all situations where a response is required by the operator, the help files are fully accessible and informative. These help files provide hints in non-technical plain English without interrupting any technical process. The time taken to set up and digitise from a model is reduced dramatically. The analytical data is collected, displayed and stored in real coordinates. Data can be stored in ASCII format for transmission to most popularly

Table 4. Results of the reliability check of the systems

Instrument	Research area	Number of control points	Mean numerical error, n_n (m)			Mean square error, n_s (m)			Mean square error of average, m_M (m)		
			x	y	z	x	y	z	x	y	z
ADAM ASP 2000	Livadia	30	0.1411	0.1026	0.0057	0.1764	0.1282	0.0071	0.0322	0.0234	0.0013
ADAM ASP 2000	Solinaria	30	0.5236	0.6167	0.0587	0.6545	0.7709	0.0734	0.1195	0.1407	0.0134
DVP	Livadia	30	0.4124	0.5006	0.0162	0.5155	0.6257	0.0202	0.0941	0.1142	0.0037
DVP	Solinaria	30	2.2930	1.8899	0.2429	2.8663	2.3624	0.3036	0.5233	0.4313	0.0554

available applications packages. The software will obtain data from:

- metric and non-metric cameras,
- virtually any focal length camera,
- aerial or terrestrial photography,
- oblique and convergent camera axes,
- abnormal base/height ratios and
- dissimilar cameras forming a model.

DVP's software will obtain data from:

- digital scanned images or CCD cameras,
- metric and non-metric cameras,
- aerial and close range photos,
- black and white and colour images, with any rotation and
- to support 8/24 bit digital images in all known formats such as TIFF, BMP, TGA etc.

Discussion

It is an undeniable fact that the ASP2000 is a more reliable instrument than the DVP, because both its repeatability and accuracy are better. The major factors, which cause the differences as far as accuracy is concerned, are:

the skill and ability of the operator,

the coordinate measurements of the control points,

the primary quality of the diapositives and

the control points were not pre-marked and significant difficulties have occurred during their precise location in the diapositives, especially in the forest area.

There are additional factors that influence only the digital photogrammetric stations. These are:

pixel size,

the radial distortion was not taken into consideration and

the scanner errors (radiometric and geometric problems).

For comparison of the restitution ability of the two systems, two areas were chosen in the model. One was a typical forest and the other a rural forest area. The errors in both systems were bigger in forest area than the forest rural area. The repeatability, the accuracy, and the objective criteria of the reliability check are better in all cases for the ADAM ASP2000 than the DVP.

Both systems offer the possibility to the user to produce digital vector data in a format, which cooperate easily and direct with other software packages such as AutoCAD and other GIS applications.

DVP includes additionally some very useful possibilities such as the mono-orientation program, the automated production of digital orthophotos or building mosaics, the graphical user interface of the digital photogrammetric station is more friendly than the one of ADAM ASP2000.

The possibility of automation in Digital Video Plotter is better in comparison to the analytical instrument. Nevertheless, the ranges of upgrades available ensure that ADAM is still able to be productive long into the future.

Both systems generated contours as contiguous lines and displayed them as they are formed with the help of a drawing package. The scale of the images used was 1:20000. The final cartographic products were at a scale 1:5000. This is the scale that the forest cadastral maps should be drawn. For this scale, both of the photogrammetric instruments are capable of producing reliable results and products.

The results of the field trial at forest and forest rural areas have indicated that the ADAM Technology software is more accurate, faster, more flexible, easier to use, more robust and provides better quality models, has better software support and requires less training. Based on technical considerations of the trial results, ADAM software provides the best solutions for forest mapping.

References

Administration of University Forests (AUF) 2002 Management Plan of the University Forest of Taxiarchi – Vrastama for the Period 2002–2011, Thessaloniki.

Akca A 2001 Wald Inventur. Sauerlangs' Verlag, Frankfurt.

Balodimou I 1991 Theory of errors and M.E.T. University lecture notes, Department of Surveying Engineering. N.T.U., Athens.

Doukas A K 2001 Topography of Rural and Forest Area, p. 10. Giachoudis Publishing, Thessaloniki.

Drosos V C 2000 Evaluation of Means and Methods Forest Cadastre's Constitution, pp. 183, 185. PhD Thesis, Aristotle University of Thessaloniki, Greece.

Gadow K V 2000 Forst Einrichtung. University lecture notes, University of Goettingen, Germany.

Hallert B 1960 Photogrammetry: Basic Principles and General Survey. McGraw-Hill, New York.

Kantartzis A, Kampouris A, Karantzidis N 2003 Application of digital photogrammetry on forest roads planning. *In* Proceedings of the 11th Greek Forest Conference on Forest Policy: Protection of the Natural Environment, Ancient Olympia, Greece, Vol. 1, pp. 522–529.

Kraus K 1990 Photogrammetrie. Band 1: Grundlange und Standardverfahren, pp. 285–287. Ferd. Duemmler Verlag, Bonn.

Patias P 2001 Introduction in Photogrammetry, pp. 43–45. Ziti Publishing, Thessaloniki.

Applications at the slope level

Vegetative-based technologies for erosion control

R.P.C. Morgan*
National Soil Resources Institute, Cranfield University at Silsoe, Silsoe, Bedfordshire MK45 4DT, United Kingdom.

Key words: erosion control, screening models, vegetation

Abstract

Vegetation is widely used for the control of surface erosion on slopes but different vegetation types vary in their effectiveness and, in some situations, a vegetation cover can have adverse effects and actually increase the rate of erosion. Where climatic or soil constraints exist, there are also concerns about how quickly an effective cover can be obtained. Simple screening models can be used to indicate the likely severity of these issues prior to designing an erosion-control system. As soon as the canopy cover is higher than 0.3 m above the surface, there is a risk that satisfactory protection against soil particle detachment by raindrop impact will not be obtained. With canopies higher than 1.0 m, detachment rates may exceed those from natural rainfall on bare ground. The amount of vegetation needed to prevent soil particle detachment by surface runoff depends upon the steepness of the slope but, for grasses, a stem density of at least 10,000 stems/m^2 is recommended. Uniformity of distribution is important because a clumpy vegetation cover can lead to concentrations of flow between the plants with consequent increases in velocity. A vegetation cover can be used to induce sediment deposition. Where grass is used as a buffer strip, a width of 10–12 m is usually sufficient to trap even the fine sediment. For large areas of the world where water erosion is a problem, it is feasible to establish sufficient grass cover within 1 year.

Introduction

Vegetation is widely used to protect the soil against water erosion. However, vegetation varies widely in its height and density, from grasses through to shrubs and trees, giving rise to differences in the degree of protection that can be attained. Under some circumstances, vegetation can actually enhance the erosion process (Morgan, 2005); for example, where raindrops intercepted by the vegetation fall as leaf drip from tall canopies on to bare soil, they can detach more soil particles than rain impacting directly on the soil surface. Also, where runoff is concentrated in the gaps between clumps of vegetation, it can detach and transport more soil than runoff flowing more uniformly through a more evenly distributed vegetation cover. Before considering vegetative-based technologies, it is therefore helpful to have some simple models to evaluate the beneficial or adverse effects of vegetation and to establish some basic design parameters. Screening models, designed to be indicative rather than precise, can perform this function. In addition, they can be used to determine whether the required level of vegetation cover can be established reasonably quickly under given soil and climatic conditions. This paper describes some simple screening models to address these issues.

Raindrop impacts

The detachment of soil particles by raindrop impact is the first phase of the water erosion process. The detachment rate is generally regarded as a function of either the intensity or the kinetic energy of the rain and a vegetation cover is viewed as a good protector of the soil because it intercepts the rainfall, thereby reducing its intensity and energy at the soil surface. Unfortunately, this view ignores the transfer of a proportion of the intercepted rainfall from the canopy to the ground surface as leaf drainage. For a wide range of vegetation types, the median volume drop diameter of leaf drips

* E-mail: r.morgan@them.org.uk

Stokes et al. (eds), Eco- and Ground Bio-Engineering: The Use of Vegetation to Improve Slope Stability, 265–272.

is 4.8–5.0 mm (Brandt, 1989) compared with 2.0 mm for natural rainfall (Hudson, 1981). Although drops of this size can reach 97% of their terminal velocity when falling from a canopy of 10 m in height (Epema and Riezebos, 1983), assessments of their kinetic energy tend to underpredict their ability to detach soil particles because kinetic energy underplays the importance of drop size. Styczen and Høgh-Schmidt (1988) showed that detachment rates (D_r), as measured in a number of experiments, were better predicted by the following relationship:

$$D_r \propto m^2 v^2 \qquad (1)$$

where m is the mass of the raindrops and v is their impact velocity. More recently, Salles et al. (2000) found that, even for bare soil, the product of $d^4 v^1$, where d is the drop diameter, gave the best predictions of detachment rate, again indicating the importance of the mass or size of the drops in the splash process.

As an example, the following simple calculations are made, using m^2v^2 for a design storm of 50 mm/h intensity and duration of 15 min, giving a total rainfall of 12.5 mm. For screening purposes, calculations are based on the median volume drop diameters of 2.0 mm for the natural rainfall and 5.0 mm for the leaf drainage. It is assumed that the interception capacity of the vegetation cover has been reached and that there is no stemflow, so that all the rainfall reaches the soil surface either as direct throughfall or leaf drainage. The m^2v^2 of the rainfall is then calculated for percentage vegetation covers of 0, 25, 50, 75 and 100% and canopy heights of 0.05, 0.25, 0.5, 1.0, 2.0 and 10.0 m (Table 1). If the values of squared momentum under the vegetation cover are expressed as a ratio of the value of the rainfall on bare ground, it can be seen that once the canopy height exceeds 2.0 m, the potential for soil particle detachment exceeds that from having no vegetation cover at all (Figure 1); further with a 1.0 m canopy height, the detachment rate virtually equates to that on bare ground. The greatest reductions in detachment are obtained from the cover at 0.05 m height. At that height, a percentage cover of at least 75% is needed to reduce detachment to less than 30% of that on bare ground. These results match those of studies of detachment rates under low-growing covers of maize, sugar beet and Brussels sprouts (Finney, 1984; Morgan, 1985) and indicate that, unless the vegetation canopy is in contact with or close to the soil surface, vegetation is likely to have an adverse effect on soil particle detachment by raindrop

Table 1. Values of m^2v^2 for rainfall of 50 mm/h intensity and duration of 15 min for different percentage covers and heights of vegetation

	Percentage cover				
Height (m)	0.00	0.25	0.50	0.75	1.0
0.05	15.63	11.88	8.13	4.38	0.63
0.25	15.63	12.78	9.94	7.09	4.25
0.50	15.63	13.51	11.40	9.28	7.17
1.00	15.63	15.16	14.69	14.22	13.75
2.00	15.63	18.55	21.48	24.41	27.33
10.00	15.63	27.27	38.92	50.57	62.21

Calculations are for: (1) median volume drop diameters of 2.0 mm for natural rainfall and 5.0 mm for leaf drainage, with drop mass calculated assuming a spherical shape; (2) drop velocities of 6.5 m/s for the 2.0 mm drops in natural rainfall and velocities for different fall heights for leaf drainage taken from Epema and Riezebos (1983); (3) rainfall proportioned between natural rain and leaf drainage according to percentage vegetation cover.

impact. It should be noted that under forest covers, the protection of the soil is afforded by the litter layer. Once a decision has been made to establish a low-growing vegetation such as grass to control the first phase of the water erosion process, the question arises as to how well it will control the detachment of soil particles by surface runoff.

Surface runoff

Vegetation reduces the potential of surface runoff to detach soil particles by imparting roughness to the flow

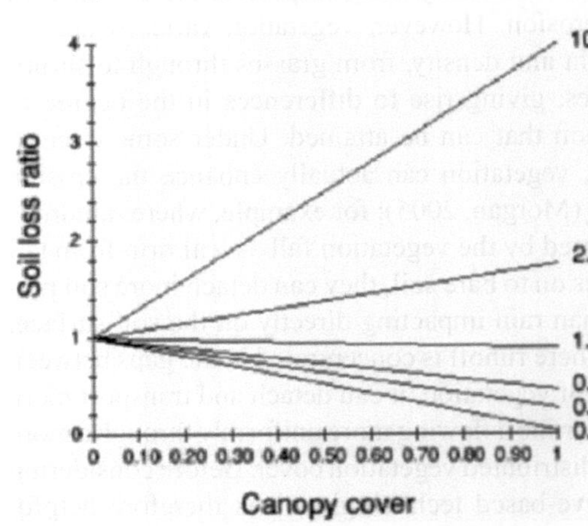

Figure 1. Soil loss ratios for soil particle detachment by raindrop impact based on m^2v^2. The ratio represents the ratio of soil loss with a vegetation cover to the soil loss from bare soil. Canopy cover is the proportion of the soil protected from raindrop impact by the vegetation. Lines denote vegetation at different heights (m).

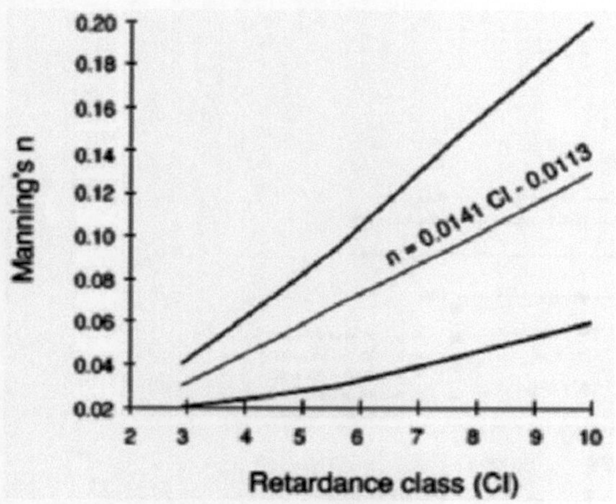

Figure 2. Envelope curves for estimates of Manning's n based on matching descriptions of grass heights to the retardance classes of Temple (1982) and to typical n values (Hudson, 1995) (see Table 10.6 in Morgan, 1995).

and reducing its velocity. Temple (1982) showed that the ability of vegetation to retard flow in this way could be expressed by a retardance class (CI) which is dependant upon the density and height of the plant stems. A simple procedure for assessing the effectiveness of the vegetation is to match its retardance class with a roughness coefficient, such as Manning's n. Figure 2 shows possible relationships for grasses of different heights and densities. Vegetation also changes the hydraulic radius by breaking up the flow into individual flow paths between the stems. Each flow path resembles a rectangular channel with the ground surface as its bed and the stems as its banks. A revised hydraulic radius (r_s) can therefore be calculated based on the spacing and height of the plant stems and defined as:

$$r_s = \frac{ss \cdot h}{2h + ss} \tag{2}$$

where ss is the spacing between the stems and h is the stem height (Tollner et al., 1982). Table 2 shows sample calculations of flow velocity for slopes of 10% (1:10) and 20% (1:5) and flow depths of 0.05 and 0.01 m, for grasses at densities of 0, 500, 1000, 2500, 5000, 7000 and 10,000 stems/m^2 and heights of 0.05 and 0.3 m. The velocities can be compared with typical critical values for particle detachment of 0.3, 0.5, 0.75 and 1.0 m/s for silts, sands and sandy loams, loamy sands and clay loams, and clay soils respectively (Hudson, 1981) (Figure 3). The results indicate that densities of at least 7500 stems/m^2 are required to prevent detachment on the most erodible, silty soils. For sands and sandy loam soils, the density must exceed 2500 stems/m^2. Similar calculations can be made for other slope angles.

Uniformity of stem spacing is important in erosion control. This can be illustrated by considering the case of 0.3 m tall grasses on a 20% slope with a flow depth of 0.05 m. A density of 7500 stems/m^2 will reduce flow velocity to 0.14 m/s compared with 1.50 m/s on bare ground. However, if the density is variable and gaps exist with local densities as low as 500 stems/m^2, flow will be concentrated through the gaps causing local flow depths and velocities to increase. If flow depth doubles to 0.10 m, the flow velocity will increase to 0.56 m/s which is sufficient to detach soil particles on silts, sands and sandy loam soils.

Further research is required to refine the screening model just described. There is a need to determine how the depth of flow changes in relation to different

Table 2. Calculated flow velocities (m/s) for grass covers based on the Manning equation

Plant height (m)	Flow depth (m)	Slope	Stem density (plants/m^2) 0	500	1000	2500	5000	7500	10000
0.05	0.05	0.1	2.12	0.77	0.55	0.35	0.25	0.20	0.18
0.30	0.05	0.1	2.12	0.35	0.26	0.17	0.12	0.10	0.09
0.05	0.05	0.2	3.00	1.08	0.78	0.50	0.35	0.29	0.25
0.30	0.05	0.2	1.50	0.50	0.37	0.24	0.17	0.14	0.12
0.30	0.10	0.2	2.39	0.56	0.40	0.26	0.18	0.15	0.13

Calculations are based on: (1) determining the retardance class (CI) = 2.5 (h vM)$^{1/3}$ (Temple 1982) where h is the height of the plant stems (m) and M is the density of plant stems per m^2; (2) estimating Manning's $n = 0.0141CI - 0.0113$ (Figure 2); (3) using the spacing hydraulic radius = $(ss.h)/(2h + ss)$ where ss is the spacing between the stems and h is the stem height (Tollner et al 1982).

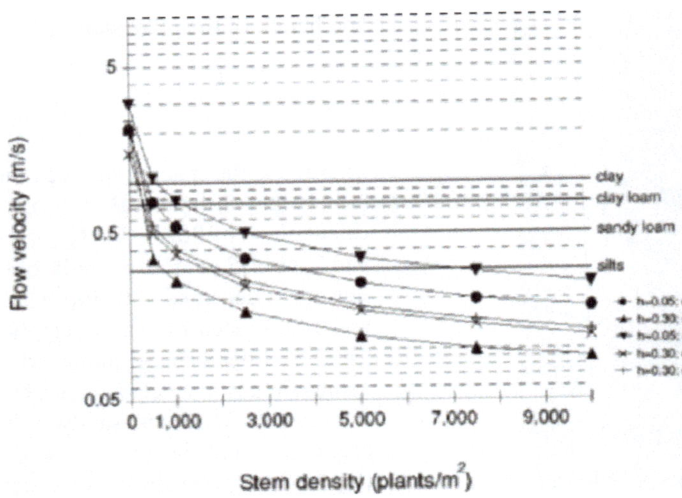

Figure 3. Flow velocity as a function of stem density for grasses: h = grass height; d = flow depth; s = slope. The thick horizontal lines denote the threshold velocities for soil particle detachment by runoff for different soil types.

vegetation types. Flow depths between the plants are likely to reflect the width and thickness of the stems (Kao and Barfield, 1978). Further the effective height of the stems will change as flow depth and velocity increase. Field evidence of flattened grass on slopes after storms suggests that this must occur before the vegetation is submerged since surface runoff on hillsides rarely has depths exceeding the vegetation height. Parameters such as deflected roughness height and the modulus of elasticity of the vegetation are likely to be important for describing the dynamics of changes in roughness during a storm (Kouwen and Li, 1980; Rahmeyer et al., 1996). This research should enable the effects of more woody vegetation types, such as shrubs, with thicker stems and wider spacings, to be evaluated.

Sedimentation

Runoff flowing from bare soil into a vegetated area will have velocity and, therefore, its sediment transport capacity reduced. Figure 4 shows the situation for a grass barrier or buffer aligned across the slope on the contour. In changing to a lower transport capacity, sediment will be deposited both within the vegetation and upslope of the barrier where water ponds. Typical distances of ponding upslope are between 0.1 and 2.3 m, depending on slope steepness (Jin et al., 2000; Melville and Morgan, 2001).

A screening model on the effectiveness of the grass barrier can be developed from the following steps.

1. The transport capacity of the flow (TC) can be evaluated in terms of its stream power, defined as the product of slope and velocity. According to Govers (1990):

$$\mathrm{TC} = a\,(sv - 0.4)^b \tag{3}$$

where TC is in cm^3/cm^3, s is slope (m/m), v is in cm/s and a and b vary with the median grain size of the material (d_{50}). For a sandy soil ($d_{50} = 250$ μm)

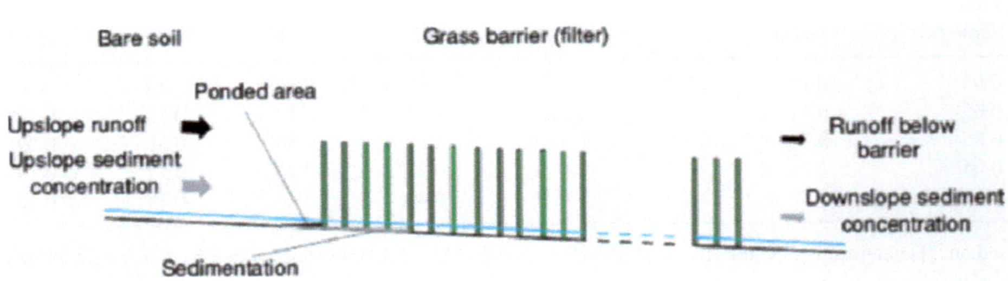

Figure 4. Modelling effect of grass barriers on a slope.

$a = 0.017$ and $b = 0.96$) and for a clay soil ($d_{50} = 50$ μm), $a = 0.063$ and $b = 0.56$.

2. Deposition of sediment in the ponded area upslope is a function of the critical settling velocity (v_c) which can be derived from the ratio between settling depth (d) and residence time (T). Thus:

$$v_c = \frac{d}{T} = \frac{dv}{l} \tag{4}$$

where v is the mean flow velocity and l is the length of the pond (Verstraeten and Poesen, 2001). The rate of deposition in the ponded area $= Cv_c$ where C is the sediment concentration in the flow.

3. Deposition within the grass barrier depends upon the filtering or trapping ability of the vegetation. Since the probability of a particle being trapped increases the more times it falls to the ground surface, Tollner et al. (1976) expressed the trapping ability in terms of a particle fall number (Nf) which is a function of the number of times that a sediment particle can potentially move from the top of the flow to the ground as it passes through the barrier. It is defined as:

$$\text{Nf} = \frac{lv_s}{vd} \tag{5}$$

where l is the length of a section of the barrier, v_s is the settling velocity of the sediment, v is the flow velocity and d is the depth of flow. The percentage of the sediment particles trapped or trapping efficiency (TE; %) can be expressed as a function of Nf by the empirical relationship (Tollner et al., 1976):

$$\text{TE} = 44.1\text{Nf}^{0.29} \tag{6}$$

Table 3 gives some sample calculations for a sandy loam ($v_s = 0.02$ m/s for a d_{50} of 250 μm) and a clay loam ($v_s = 0.0035$ m/s for $d_{50} = 50$ μm) for a flow depth of 0.05 m, 5% and a 0.3 m tall grass with a density of 10,000 stems/m^2. An upslope ponded width of 0.15 m is assumed, giving a ponded depth at the entry point of the barrier of 0.0075 m (ponded width × sin slope). The relevant flow velocities are obtained from Table 2. For the sandy loam soil, it can be seen that the barrier reduces the transport capacity from 0.296 Mg/m^3 to 0.015 Mg/m^3 and that this is achieved within the first 3.0 m of the barrier length. For the clay loam, the respective transport capacities are 0.5008 Mg/m^3 upslope of the barrier, reducing to 0.0714 Mg/m^3 within the barrier within the first 4.0 m of barrier length. The efficiencies of the barriers, based on the percentage reduction in sediment transport capacity, are 95% for the sandy loam and 86% for the clay loam which is comparable to measured efficiencies for barriers of similar density (Melville and Morgan, 2001). Efficiencies of 90–95% were also observed by Wolde and Thomas (1986) for 2.5 m wide grass barriers. Efficiencies obtained from the model after 1 m length are 77% for the sandy loam soil and 47% for the clay loam; these compare well with those of 72–84% measured on sandy loam in laboratory experiments (Lakew and Morgan, 1996). Thus the results obtained with this simple screening model accord reasonably well with measured data and are in line with studies showing that riparian strips of 10–12 m width are more than sufficient to control sediment delivery to water courses (van Dijk et al., 1996; Kronvang et al., 2000).

Table 3. Modelling the effectiveness of grass barriers in erosion control

	Soil type	
	Sandy loam	Clay loam
Flow velocity upslope of barrier (m/s)	1.50	1.50
Sediment transport capacity upslope of barrier (Mg/m^3)	0.2962	0.5008
Effective v_c within pond (m/s)	0.075	0.075
Sediment deposition in ponded area (Mg/m^3)	0.0033	0.0006
Flow velocity within barrier (m/s)	0.06	0.06
Particle fall number (*Nf*)	6.67	1.17
Trapping efficiency of barrier (*TE*; %)	76.5	46.1
Sediment transport capacity within barrier (Mg/m^3)	0.0150	0.0714
Deposition within first 1 m of barrier (Mg/m^3)	0.2241	0.2696
Deposition between 1 and 2 m of barrier (Mg/m^3)	0.0526	0.1453
Deposition between 2 and 3 m of barrier (Mg/m^3)	0.0012	0.0783
Deposition between 3 and 4 m of barrier (Mg/m^3)	0.0000	0.0069
Barrier efficiency (ratio of transport capacity within to that upslope of the barrier)	95%	86%

Calculations are for a 0.3 m tall grass barrier with 10,000 stems/m^2 on a 5% slope with a flow depth of 0.05 m. Sediment transport capacity assumes a particle density of 2.65 g/cm^3. If the runoff (m^3) is known, the quantity of sediment in transport and deposition can be determined.

Vegetation establishment

When using vegetation to control erosion on bare slopes such as embankments, cuttings and pipeline rights-of-way, it is important to have an idea of how quickly a sufficient vegetation cover can be established. The likely success of revegetation can be evaluated in advance using a simple 'screening model' based on climatic and soils data. The rate of vegetation growth can be modelled by the relationship (Biot, 1990):

$$\frac{\partial V}{\partial t} = \frac{V_{\max}}{\left(1 + B \cdot e^{-k\theta}\right)} \tag{7}$$

where V = vegetation biomass (kgDM/ha); t = time; $V_{\max}$ = maximum possible vegetation biomass (kg DM/ha) for the local climatic and soil conditions; θ = available water storage capacity (cm) in the top metre of the soil or a shallower depth if there is impedence to water movement by a compacted layer or hard pan; k = a coefficient related to consumption of water by the plant (for grasses, $k = 0.8$ during the growing season, Biot (1990)); and B = an experimentally derived constant.

According to Biot (1990), reasonable simulations can be obtained if $V_{\max}$ is set at 2550 DMkg/ha and $\partial V = 2500$ for $\theta = 12$ cm and $\partial V = 0$ for $\theta = 0$. With these conditions, $B = 254$.

Whether the available water storage capacity of the soil is filled depends on the quantity of rainfall, the amount of evaporation and the water extracted by the vegetation. A simple method of estimating annual evaporation (E; mm) from annual rainfall (P; mm) and temperature data is to use the following function proposed by Turc (1961):

$$E = \frac{P}{\sqrt{0.9 + \frac{P^2}{L^2}}} \tag{8}$$

where

$$L = 300 + 25t + 0.05t^2 \tag{9}$$

and t = mean annual temperature (°C).

The source of the water in the soil is infiltration. The mean infiltration intensity (i; mm/day) can be calculated from:

$$i = \frac{P - E}{365} \tag{10}$$

Since infiltration will only occur when it is raining, the total infiltration (I) over a year is obtained from:

$$I = i \times N \tag{11}$$

where N = the number of rain days per year.

The ability of the vegetation to extract moisture from the soil depends on the depth of root development. Since most roots are found in the upper layers of the soil, moisture extraction is concentrated there instead of taking place uniformly with depth. As the moisture store is depleted, the level of soil moisture suction increases; once it reaches 100 kPa, the water is held more tenaciously within the soil and it is less easily removed by growing plants. As a general rule of thumb, only 50% of the available water capacity of the soil is usable in supporting plant growth (Withers and Vipond, 1974). The effective available water capacity in respect of plant growth can be therefore be estimated from:

$$\theta_{\text{eff}} = \frac{0.5\,(P - E)\,N}{365} \tag{12}$$

Then,

$$\begin{aligned} \theta &= \theta, && \text{if } \theta_{\text{eff}} \geq \theta \\ \theta &= \theta_{\text{eff}}, && \text{if } \theta_{\text{eff}} < \theta \end{aligned} \tag{13}$$

The procedure can be used to determine the annual growth of vegetative biomass for given values of annual rainfall, mean annual temperature, soil type and vegetation type.

From the level of biomass, it is possible to estimate, in turn, the percentage cover attained by the vegetation and the appropriate C-factor value. Using the procedures proposed by Biot (1990):

$$\%\ \text{cover} = 1.66\left(\frac{V}{0.625}\right)^{0.5} \tag{14}$$

Using data for 1071 climatic stations world-wide (Müller, 1979), Figure 5 has been produced showing, for a sandy loam soil ($\theta = 6$ mm), the predicted vegetation cover after 1 year. The isolines have been interpolated between the data points based on the distribution of climatic regions according to the Köppen classification. As expected, the maps show that in high rainfall tropical and temperate climates, it is possible to establish a cover of 70% or more. In low rainfall areas, it is

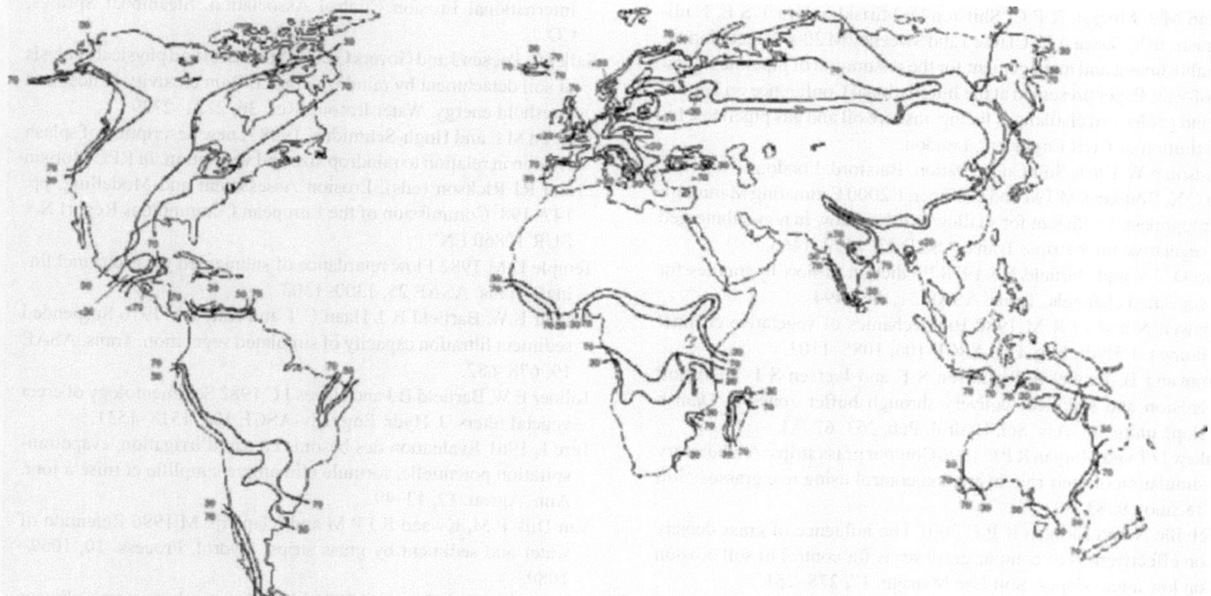

Figure 5. Estimated percentage grass over obtained after 1 year from reseeding a bare sandy loam soil.

not possible to establish a 30% cover. Indeed, in semi-arid and arid areas the cover will be less than 10%. In many areas, the change between achieving 30% and 70% cover occurs over quite short distances. The results broadly conform to experience from land restoration projects along pipeline rights-of-way in Colombia, Georgia, Azerbaijan and northern Canada (Hann et al., 2004).

Conclusions

Vegetation can be used successfully to control water erosion as long as it provides at least a 70% canopy cover with the canopy height preferably 1.0 m or less above the ground. Grasses best meet these criteria. Grasses will also prevent the detachment of soil particles by runoff on hillslopes provided that have a stem density of at least 7500 stems/m^2 and a uniform distribution. By reducing flow velocity, grasses will also encourage deposition of sediment carried in the runoff. When used as barriers or buffers across a low-angled slope, a barrier length of 4.0 m is sufficient to reduce sediment transport to the capacity for the barrier. Sufficient grass vegetation for erosion control can be established within 1 year over large areas of the earth's land surface. Although research is necessary to allow realistic simulation of the effect of non-grass vegetation types on erosion, it is clear that plants with tall canopies, clumpy habitats and wide spacings can either fail to reduce erosion to an acceptable level or even enhance erosion to rates equal to or greater than those on bare ground. It is also clear that, despite their simplifications, for example ignoring the effects of the rooting systems on infiltration and soil strength, simple models can provide a suitable screening process to indicate the likely outcome of introducing vegetative technologies to control surface erosion.

References

Biot Y 1990 THEPROM—an erosion productivity model. *In* J Boardman, IDL Foster and JA Dearing (eds), Soil Erosion on Agricultural Land, pp. 465–479. Wiley, Chichester.

Brandt C J 1989 The size distribution of throughfall drops under vegetation canopies. Catena 16, 507–524.

Epema G F and Riezebos H Th 1983 Fall velocity of waterdrops at different heights as a factor influencing the erosivity of simulated rain. Catena Suppl 4, 1–17.

Finney H J 1984 The effect of crop covers on rainfall characteristics and splash detachment. J. Agr. Eng. Res. 29, 337–343.

Govers G 1990 Empirical relationships on the transporting capacity of overland flow. Int. Ass. Hydrol. Sci. Pub. 189, 45–63.

Hann M J, Morgan R P C, Shilston D, Mirtskhoulava T S E, Nadirashvili V, Gasca A H, Clarke J and Sweeney M 2004 Vegetation establishment and management for the restoration of pipeline rights-of-way. Paper presented at the International Conference on Terrain and geohazard challenges facing onshore oil and gas pipelines. Institution of Civil Engineers, London.

Hudson NW 1981. Soil Conservation. Batsford, London.

Jin C X, Römkens M J M and Griffioen F 2000 Estimating Manning's roughness coefficient for shallow overland flow in non-submerged vegetative filter strips. Trans. ASAE 43, 1459–1466.

Kao D T Y and Barfield B J 1978 Prediction of flow hydraulics for vegetated channels. Trans. ASAE 21, 489–494.

Kouwen N and Li R M 1980 Biomechanics of vegetative channel linings. J. Hydr. Eng. Div-ASCE 106, 1085–1103.

Kronvang B, Laubel A R, Larsen S E and Iversen S L 2000 Soil erosion and sediment delivery through buffer zones in Danish slope units. Int. Ass. Sci. Hydrol. Pub. 263, 67–73.

Lakew D T and Morgan R P C 1996 Contour grass strips: A laboratory simulation of their role in erosion control using live grasses. Soil Technol. 9, 83–89.

Melville N and Morgan R P C 2001 The influence of grass density on effectiveness of contour grass strips for control of soil erosion on low angle slopes. Soil Use Manage. 17, 278–281.

Morgan R P C 1985 Effect of corn and soybean canopy on soil detachment by rainfall. Trans ASAE 28, 1135–1140.

Morgan R P C 2005 Soil Erosion and Conservation. Blackwell, Oxford.

Müller M J 1979 Handbuch ausgewählter Klimastationen der Erde. Forschungsstelle Bodenerosion der Universität Trier.

Rahmeyer W, Werth D Jr, Derrick D and Freeman G E 1996 Predicting flow resistance due to vegetation in flood plains. *In* Erosion Control Technology: Bringing It Home, pp. 431–443. International Erosion Control Association, Steamboat Springs, CO.

Salles C, Poesen J and Govers G 2000 Statistical and physical analysis of soil detachment by raindrop impact: Rain erosivity indices and threshold energy. Water Resour. Res. 36, 2721–2729.

Styczen M E and Høgh-Schmidt K 1988 A new description of splash erosion in relation to raindrop size and vegetation. *In* RPC Morgan and RJ Rickson (eds), Erosion Assessment and Modelling, pp. 147–198. Commission of the European Communities Report No. EUR 10860 EN.

Temple D M 1982 Flow retardance of submerged grass channel linings. Trans. ASAE 25, 1300–1303.

Tollner E W, Barfield B J, Haan C T and Kao T Y 1976 Suspended sediment filtration capacity of simulated vegetation. Trans. ASAE 19, 678–682.

Tollner E W, Barfield B J and Hayes J C 1982 Sedimentology of erect vegetal filters. J. Hydr. Eng. Div-ASCE 108, 1518–1531.

Turc L 1961 Evaluation des besoins en eau d'irrigation, evapotranspiration potentielle, formule climatique simplifie et mise á jour. Ann. Agron. 12, 13–49.

Van Dijk P M, Kwaad F J P M and Klapwijk M 1996 Retention of water and sediment by grass strips. Hydrol. Process. 10, 1069–1080.

Verstraeten G and Poesen J 2001 Modelling the long-term sediment trap efficiency of small ponds. Hydrol. Process. 15, 2797–2819.

Withers B and Vipond S 1974 Irrigation Design and Practice. Batsford, London.

Wolde F T and Thomas D B 1989 The effect of narrow grass strips in reducing soil loss and runoff in a Kabete nitosol, Kenya. *In* DB Thomas, EK Biamah, AM Kilewe, L Lundgren and BO Mochoge, Soil and Water Conservation in Kenya, pp. 176–194. Department of Agricultural Engineering, University of Nairobi and SIDA.

Vegetation succession and its consequences for slope stability in SE Spain

Erik Cammeraat[1,3], Rens van Beek[2] & Annemieke Kooijman[1]

[1]*Institute for Biodiversity and Ecosystem Dynamics-Physical Geography, Nieuwe Achtergracht 166, 1018, Amsterdam, WV, The Netherlands.* [2]*Department of Physical Geography, Utrecht University, P.O. Box 80115, 3508, TC Utrecht, The Netherlands.* [3]*Corresponding author**

Key words: infiltration, land abandonment, macro-pore flow, root–soil anchoring, root strength, shear strength

Abstract

The effect of land abandonment as a result of changing land-use policies is becoming more and more important throughout Europe. In this case study, the role of vegetation succession and landslide activity on steep abandoned slopes was investigated. The influence of vegetation succession on soil properties over time, as well as how developing root systems affect soil reinforcement was determined. The study was carried out in the Alcoy basin in SE Spain, where the marl substratum is prone to landsliding along steep ravines. The bench-terraced slopes have been abandoned progressively over the last 50 years and show various stages of revegetation. The study was carried out at two scales; at the catchment scale long-term evolution of land-use, vegetation succession and slope failure processes were investigated. At a more detailed scale, vegetation cover, soil properties and rooting effects on soil strength were determined.

Results showed that the soil has changed over a period of 50 years with respect to soil properties, vegetation cover and rooting, which is reflected in the activity of geomorphological processes. Vegetation succession progressively limits surface processes (sheet wash and concentrated overland flow) over time, whereas slopes affected by mass wasting processes increase in number.

The spatial heterogeneity of infiltration increases over time, leading to increased macro-pore flow towards the regolith zone, enhancing the potential risk of fast wetting of the regolith directly above the potential plane of failure, as was concluded from rainfall simulations. In situ experiments to determine soil shear strength in relation to rooting indicated that roots contributed to soil strength, but only in the upper 0.4 m of the soil. Most failures however, occur at greater depths (1.0–1.2 m) as anchorage by deeper roots was not effective or absent. The observed initial increase in mass wasting processes after land abandonment can therefore be explained in two ways: (1) the limited contribution of anchorage by root systems at potential slip planes which cannot counterbalance the initial decline of the terrace walls, and (2) the fast transfer of rainfall to the potential slip plane by macro-pores enhancing mass movements. However, after approximately 40 years of abandonment, mass wasting processes decline.

Introduction

Land-use change is an important issue in Mediterranean Europe and is principally driven by socio-economic factors (Geeson et al., 2002; MacDonald et al., 2000). This type of change is dominated by agricultural land abandonment or conversion to agro-industrial land-use, and implies drastic changes in management policies. Sustainable land use can only be achieved when one also assesses changes in process and landscape patterns (Romero-Calcerrada and Perry, 2004). The effect of land abandonment on vegetation, hydrology, soil quality and erosion is not very well known but may have significant consequences (Garcia-Ruiz et al., 1996; Lasanta et al., 2000). A case study is presented where the role of vegetation succession in landslide activity on steep, abandoned slopes was investigated. Land abandonment and vegetation succession affect the mechanical and hydrological properties of soil, especially by the effect of developing

* E-mail: l.h.cammeraat@science.uva.nl

This article has been previously published in the following journal—Plant and Soil, Volume 278, 135–147.
Stokes et al. (eds), Eco- and Ground Bio-Engineering: The Use of Vegetation to Improve Slope Stability, 273–285.

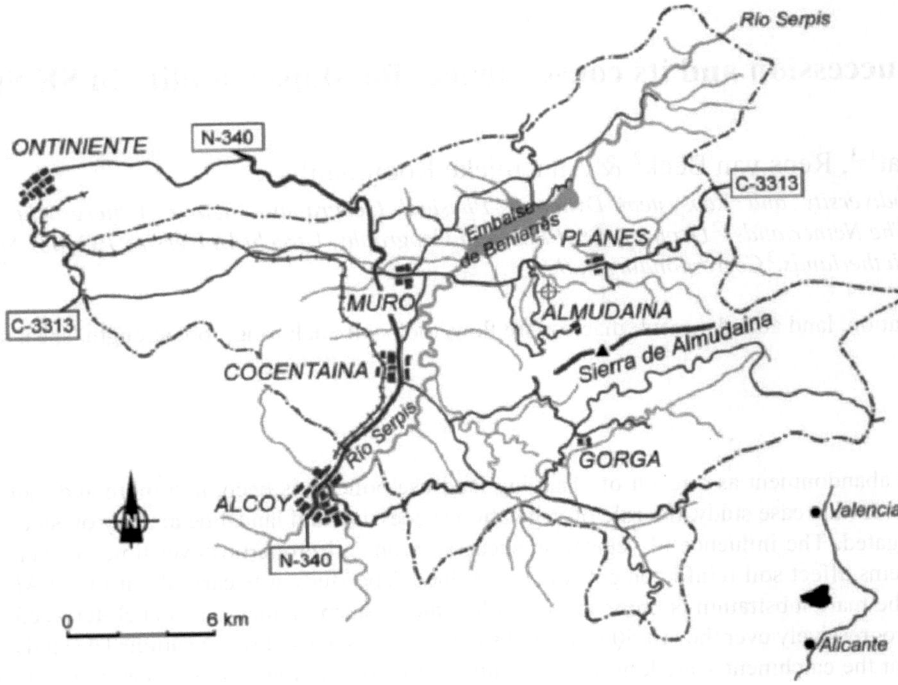

Figure 1. Location of the study area.

root systems. Therefore, measurements were carried out to quantify the possible effect of reinforcement of the soil by roots as well as the effects on soil properties such as infiltration. The effects of land abandonment and vegetation succession on active geomorphological processes related to slope stability have rarely been evaluated.

The study area was located in the Alcoy basin in SE Spain (Figure 1), where the marl substratum is prone to landsliding along steep ravines, which is well described in La Roca-Cervignon and Calvo-Cases (1988). The bench-terraced slopes were abandoned at various times and show various stages of revegetation. A first reconnaissance survey suggested that mass wasting activity increased after land abandonment, which was not fully understood, as vegetation cover appeared greater after abandonment. It might be expected that colonizing plants expanding their roots systems into the soil would enhance slope stability (Coppin and Richards, 1990).

This paper aims at determining the explanation of the processes involved, by exploring the following research questions:

- Do geomorphological processes, on a catchment scale, change in relation to land abandonment and vegetation succession?
- What is the impact of vegetation succession on soil (hydrological) properties and soil mechanical properties and can this effect explain the changes in geomorphological processes observed?

Materials and methods

Field site description

The Valles de Alcoy (Figure 1) are situated within the upper basin of the Serpis river where Cretaceous limestone and Miocene marls of the *Tap* formation dominate the geology. The marls of this formation are very homogeneous in the sense of stratification and composition (IGME, 1975). Colluvial and residual soils have formed on the pediment surface on top of Miocene marl. Deep river incisions with associated steep ravines (barrancos) have developed in the pediment itself. Details of the study area can be found in La Roca-Cervignon and Calvo-Cases (1988) and Van Beek (2002).

On the marl, calcic and haplic regosols (FAO, 1989) have formed with typical depths ranging between 0.6 and 1.2 m (Van Beek, 2002). These profiles are characterised by a relatively well developed root zone over a regolith of weathered marl or colluvium. With depth, the permeability decreases rapidly and at the regolith-bedrock contact, perched water tables may develop (Van Beek, 2002). In combination with the steep slopes, their occurrence makes the barrancos vulnerable to frequent landsliding, especially under extreme rainfall events. Potential planes of failure are typically found between 1.0 and 2.0 m at the contact between the regolith and the unweathered parent material. The climate in the region is sub-humid to humid, with annual rainfall totals between 500 and 1000 mm and a mean annual temperature of 14.5°C.

Land-use and vegetation

On the pediments and marl slopes, bench terraces have been constructed, on which rain-fed perennial crops, mainly cherries (*Prunus ávium* L.), olives (*Olea europaea* L.) and almonds (*Prunus dulcis* (Miller) D.A. Webb), were cultivated. The bench terraces were cut in the marl parent material itself and risers are not reinforced with stones or concrete. As the weathering rate of the marls is very high, the regolith depth in the terraced slopes is also large (1.2–2.0 m). No irrigation or extensive drainage works are provided on these terraces. However, mechanisation and a decline in agriculture have changed cultivation on steep slopes. Bench terraced slopes are abandoned at increasing rates, especially on slopes steeper than 10–15°, and consequently are no longer maintained. Water and nutrient availability do not severely limit the colonisation of the abandoned fields severely and regeneration is prompt. After 1 year, a complete cover with annuals and grasses is present. Common shrubs on longer abandoned fields are Gorse (*Ulex parviflorus* Pourret) and Hawthorn (*Crataegus monogyna* Jacq.), as well as the Aleppo pine tree (*Pinus halepensis* Miller).

Field methods and experimental setup

To answer the research questions different methodologies were applied needing different approaches, set in a nested experimental design.

With regard to the question on the possible changing geomorphological processes in relation to land abandonment, a catchment wide study was carried out based on sequential aerial photo interpretation and a field survey. Within this study area 78 field plots were selected for field validation of the aerial photographs as well as to determine the actual geomorphological processes and vegetation properties (Figure 2). At least five field

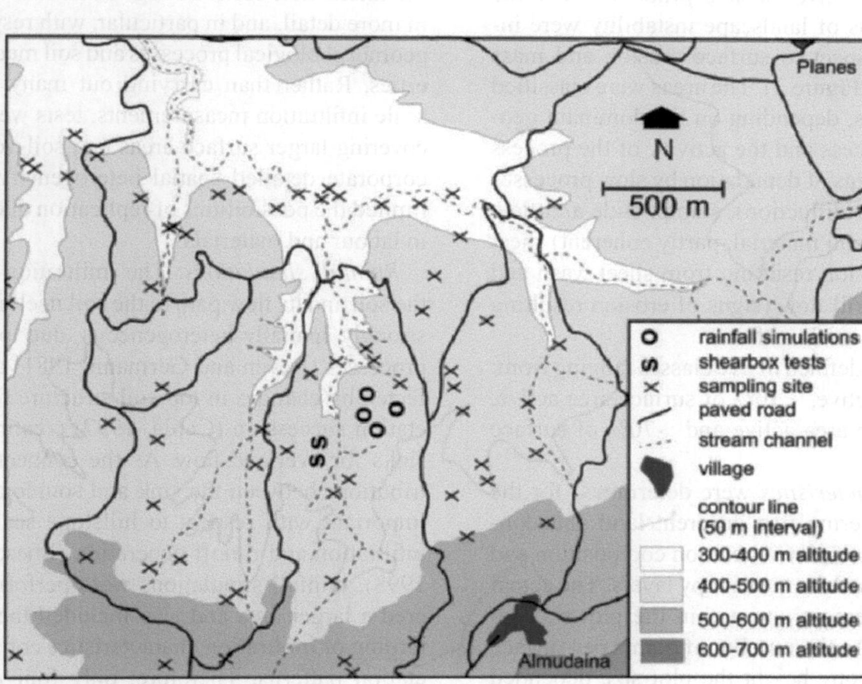

Figure 2. Location of sampling sites and experimental sites within the catchment.

plots were present in each land abandonment class. Space-time substitution (Paine, 1985) was applied to overcome the problem of long term monitoring under the presumption that parent materials do not significantly differ within the area of study.

A further selection of sites was made to gain insight in the vegetation succession related processes observed and to study soil property changes and their effect on geomorphological processes. This selection involved *in situ* detailed scale experiments on infiltration and soil shear strength. As these experiments are time consuming and expensive only a limited amount of these experiments could be carried out.

Catchment scale inventory techniques

Sequential aerial photo interpretation was carried out using a complete set of panchromatic photos from the years 1956, 1965, 1973, 1989 and 1994 at scales ranging from 1:15 000 to 1:30 000. From this interpretation the time of abandonment was inferred and additional information on dates of abandonment was obtained from local farmers. This information resulted in the differentiation into six classes of land abandonment as determined by the time frames between subsequent aerial photographs.

Field survey of active surface processes. For the study area, 78 areas of landscape instability were investigated with respect to surface erosion and mass wasting processes (Figure 2). The areas were classified into activity classes, depending on the dominant geomorphological process and the activity of the process involved: creep (signs of denudation by slow processes such as creep and solifluction), slump, slide and flow (fast movement of soil material, partly coherent) sheet wash (signs of erosion resulting from sheet wash and splash processes), rill flow (signs of erosion resulting from concentrated overland flow).

The activity was defined in five classes ranging from: never affected, inactive, $<10\%$ of surface area active, 10–70% of surface area active and $>70\%$ of surface area active.

Vegetation characteristics were determined for the same 78 plots covering five different land abandonment classes. At each plot vegetation composition and cover was determined for the canopy layers. The actual vegetation measurement site within the plot was selected at random. As the number of plants per surface decreases with canopy height the plot size depended on the type of vegetation. The plot size for herbs was $1\ m^2$, for bushes $25\ m^2$ and for the tree layer $100\ m^2$. The number and cover of each individual plant were determined within the vegetation plots. Standard measurement techniques were applied, using grid frames of $1\ m \times 1\ m$ or measurement tape grids in the case of larger areas (shrubs and trees).

The upper 0.1 m of the soil was sampled and the following standard soil properties were determined; *organic carbon* by the wet oxidation method using $KCrO_6$ (Allison, 1935) and $CaCO_3$ contents using the method of Wesemael (1955), which is based on weight loss on dissolution. *Texture* analysis was carried out by dry sieving and the pipette method (Gee and Bauder, 1990), but without decalcifying the soil, as $CaCO_3$ levels are $>550\ g\ kg^{-1}$. Soils were described and classified following FAO guidelines (FAO, 1989; FAO, 1990).

Detailed scale field and laboratory measurements

The regional characteristics of the land units with different land abandonment histories were derived from the analysis of the sequential aerial photographs. The general characteristics of geomorphological processes and vegetation were studied at the 78 selected plots. Using this information several sites were selected to study the interaction between vegetation and soil properties in more detail, and in particular, with respect to hydro-geomorphological processes and soil mechanical properties. Rather than carrying out many very detailed scale infiltration measurements, tests were carried out covering larger surface areas and soil volumes, to incorporate detailed spatial heterogeneity. This choice limited the possibilities of replication due to high costs in labour and materials.

Rainfall simulations: The infiltration of water into the soil and its flow path in the soil itself is known to be strongly spatially heterogeneous, due to by-pass flow processes (Beven and Germann, 1981) and is also affected by changes in top-soil structure following vegetation succession (Cerdà, 1997) creating source and sinks for overland flow. As the connections and distributions between the sink and source areas are very important with respect to hillslope scale patterns of infiltration and runoff generation (Imeson and Lavee, 1998), rainfall simulations were performed that covered a larger area and also included the detailed patterning of infiltration characteristics connected to vegetation patterns. Therefore, only four *rainfall simulation* experiments were carried out on sites that are considered to be characteristic for the stages of vegetation succession corresponding to the stages following: cultivated, recently abandoned (2001–1989), abandoned (1989–1973), and abandoned before 1973 (see Figure 2 for location). The simulations were carried out using a two nozzle simulator (nozzle type S.S.C.O. Fulljet 3/8 44 SS 27 W), covering an area of 2.75 × 6.5 m. Each experiment consisted of a series of six consecutive artificial rainfall events of 7 mm and 12.2 min duration with a constant rainfall intensity of 34.4 mm h^{-1} spread equally over 2 days, to mimic typical natural rainfall. The rainfall intensity applied is not uncommon and the return period of the total applied rain, based on daily totals was less than one year (Elias-Castillo and Ruiz-Beltran, 1974). Distilled water was used as the soil is sensitive to dispersion (Shainberg et al., 1981). Wetting fronts in the soil profile were measured immediately after all simulations in a trench of 1 m length at the side of the simulated plot. Additionally, colour dyes were applied to accentuate the wetting front (pyranine) and preferential flow paths (rhodamine) as both dyes have different adsorption behaviour.

The *shear strength* of the soil was determined in the field and in the laboratory. Consolidated, drained and strain-controlled direct shear tests were carried out in the laboratory on saturated samples measuring 60 × 60 × 20 mm (e.g., Van Beek, 2002). All 232 samples originated from various soil horizons of the weathered marl throughout the studied catchment. As spatial heterogeneity in shear strength was not known, a pre-investigation was performed using the simpler tor vane test as a first indicator for spatial trends in soil mechanical properties and helped in selecting the proper places for the larger experiments.

Large in situ direct shear tests were carried out on the root permeated soil. The dimensions of the shear box were 0.6 × 0.6 m in plan and the box extended 0.4 m into the soil, comparable to tests as carried out by Wu et al. (1988). These large tests enabled to quantify the possible effects of anchoring roots on soil mechanical strength, which effect is impossible to study in the laboratory or with very small sampling volumes. The tests were limited to two types of cover, a cover type with anchoring taproots (Aleppo pine; *Pinus halepensis* Miller) and a cover with only fine roots (grass, annuals and small herbs, of which the most important were: *Brachypodium distachyon* L. Beauv.; *Tymus vulgaris* L., *Helichrysum stoeachas* (L.) Moench; *Sedum acre* L. and *Sedum album* L. subsp. *album*) representing cover stages in the early stage and in the final stage of vegetation succession after abandonment on directly neighbouring hillslopes (see Figure 2 for location). Shear was applied by means of a jack and the generated shearing resistance was measured with a proving ring. A normal load was applied by means of a dead weight (3.3 and 4.1 kPa) of concrete blocks. The soil was slowly wetted thoroughly prior to testing to eliminate any suction-derived resistance, and the shear rate was kept low (4 mm min^{-1} on average) to avoid the build-up of positive pore pressures. For seven tests, a root count was made at 0.1 m depth intervals while extensive root descriptions were made in soil pits under similar vegetation cover in the direct neighbourhood of the test locations, under the presumption that the local spatial heterogeneity of the soil material was negligible.

Results

Land abandonment and vegetation succession

Aerial photo interpretation, field survey and interviews with farmers resulted in a subdivision in four groups of land units based on date of abandonment: before 1965, 1965–1973, 1973–1989, 1989–2001 and still cultivated. Steep barranco sites were abandoned earlier than less steep areas, and the flat pediment areas are still under cultivation. The change in vegetation cover and the presence of the different types of vegetation is indicated in Figure 3. The vegetation succession evolves from presently cultivated fields to open Mediterranean forest dominated by Aleppo pine. Generally speaking, two trends can be recognised; first of all, with abandonment, the proportion and vitality of the crops declined. Secondly, the bare soil between the fruit trees was colonised by grasses and herbs and finally shrubs (mainly Gorse and Hawthorn) and Aleppo pine trees colonised. During the first ten years after abandonment the fruit trees died off. On land units, which were abandoned before 1965 open natural stands of Aleppo pines are developing. This vegetation was associated with the formation of well-developed and stratified soils on top of the marl regolith and soil. Vegetation succession was found to be independent of aspect and slope angle.

Activity of surface processes

Geomorphological processes were studied in areas representing each phase of abandonment. These processes were divided into surficial (concentrated overlandflow

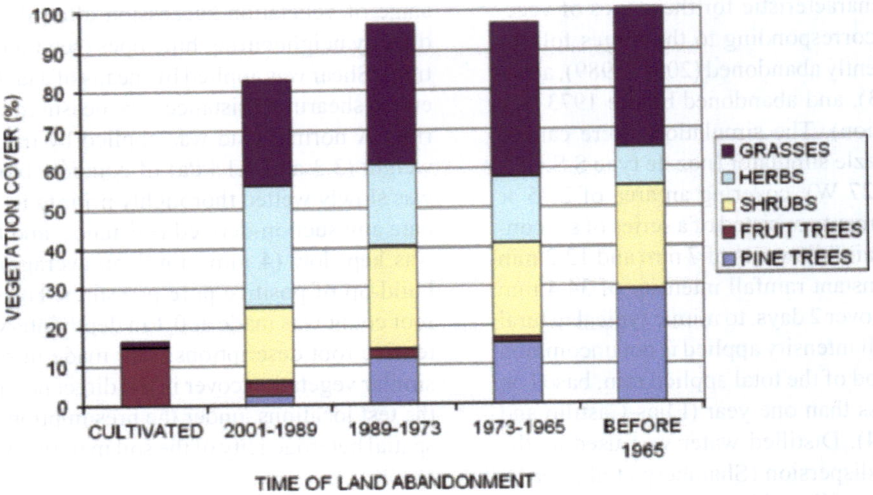

Figure 3. Vegetation cover type and date of abandonment (after Koppel, 2001).

and sheet flow) and shallow mass wasting processes originating from processes in the soil (creep and landsliding). Overland flow and sheet flow were generally present on cultivated fields and clearly diminished over time since abandonment (Figure 4). For the mass wasting processes a reverse trend was observed with at first increasing activity with prolonged abandonment, and a reduction of activity in the fields that were abandoned longest (Figure 4). Bench-terraces are not maintained after abandonment. The increase of overland flow processes directly after abandonment and the lack of vegetation cover, affected the redistribution of water on the terraces. This enhanced local saturation, promoting mass wasting activities and initiation of local disintegration of terraces. However, terraces were still clearly recognisable up to 30–40 years after abandonment.

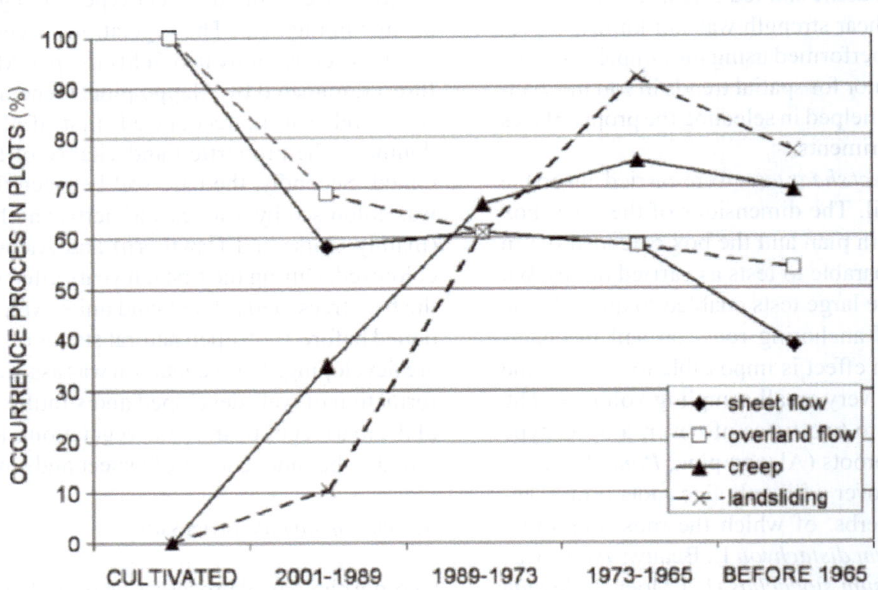

Figure 4. Observed activity of geomorphological processes at the 78 studied sites.

Table 1. Soil properties of a characteristic soil profile in the direct neighbourhood of the area where the *in situ* shear tests were carried out

Horizon	Depth (cm)	Texture Sand ($g\ kg^{-1}$)	Silt ($g\ kg^{-1}$)	Clay ($g\ kg^{-1}$)	Org C ($g\ kg^{-1}$)	$CaCO_3$ ($g\ kg^{-1}$)	Dry bulk density ($g\ cm^{-3}$)	Undrained shear strength (kPa)	Porosity ($m^3\ m^{-3}$)
Ah	0–20/25	20	660	320	220	550	1.03	3.4	0.37
Bw	20/25–30/38	10	650	340	65	570	1.37	3.8	0.32
C1	30/38–56	10	660	330	48	550	1.57	8.0	0.27
C2	>56	20	650	330	35	540	1.61	7.0	0.25

Soil properties

Regolith properties such as bulk density, porosity, $CaCO_3$ content did not change over time. For the organic carbon levels and porosity, there was a clear decrease with depth (Table 1), and a clear increase of the undrained shear strength with depth (Table 2). The upper soil (Ah horizon) showed a gradual but significant increase in organic carbon over time since abandonment (Figure 5) ranging from 14 g kg^{-3} for cultivated fields to 44 g kg^{-3} for fields abandoned before 1965. The amount of $CaCO_3$ is always greater than 500 g kg^{-3} and in some profiles a calcic horizon was present with even higher $CaCO_3$ levels.

Preferential infiltration and soil waterflow

The six large size successive rainfall simulations enabled a study of the progressive evolution of the wetting front through the soil profile (Figure 6). This study revealed that infiltration in the upper soil was quite homogeneous, especially for the cultivated and recently abandoned fields. These areas also showed a regular homogeneous, slightly wavy infiltration after application of larger amounts of precipitation after the third run (>21 mm). The areas which were abandoned for longer showed increasingly irregular wetting fronts with deep (>20 cm) and narrow wet pockets penetrating into the soil. Preferential flow was observed to become greater in the soils abandoned longest, showing deeper flow down vertical root systems and macropores.

Table 2. Shear strength properties (six tests per horizon at three imposed normal loads between 59 and 117 kPa)

Horizon		Depth (m)	c' (kPa)	c_u (kPa)	c_u*kPa	ϕ' (°)
Ah	Top soil	0.00–0.17	4.8	4.5	13.7	34.4
Bw	Weathered soil	0.17–0.26	1.9	4.8	15.9	35.3
C1g	Gleyic horizon	0.29–0.59	10.2	8.1	6.5	31.2
C12	Colluvium	0.59–0.69	9.8	8.2	14.0	31.7
C2	Regolith	0.69–0.92	4.1	6.5	15.6	36.4

c' and ϕ' are the drained shear strength parameters, c_u the average undrained shear strength interpreted as a cohesion ($N = 232$) from field measurements. c_u* is the undrained shear strength of the direct shear samples after consolidation). Depth is the average of the observed layers and indicative only. Note that the horizons C12 and C1g are not necessarily present.

Undrained shear test

The large number ($n = 232$) of consolidated-drained direct shear tests on saturated material revealed that the cohesive component of the shear strength, c', was weak and variable. Statistically, the material strength can be characterised as frictional only with mean friction angles ϕ' between 31° and 36°. Notwithstanding, for the material taken at one of the two sample slopes there seems to be a weak differentiation in cohesion and friction with depth (Table 2). A lower cohesion but a higher friction angle was found in the B horizon below the Ah horizon, while the reverse was observed in colluvial and gleyic horizons above the regolith. There was little variation in the strain at which the maximum shear strength was mobilized (12% or 7 mm on average).

There was a close similarity between the mean tor vane readings and the drained cohesion of the horizons, which suggests that the former can indeed be interpreted as the cohesive strength of the soil. The readings confirmed the variability and trends inferred from the laboratory direct shear tests. The cohesion increased and the differences in cohesion become less marked when the samples were consolidated with the exception of the gleyic layer for which the cohesion decreased.

In situ *shear box testing*

In the case of the *in situ* tests on rooted soil, the strain at failure was more variable than the undrained shear tests, ranging between 6 and 25%, and weakly correlated to

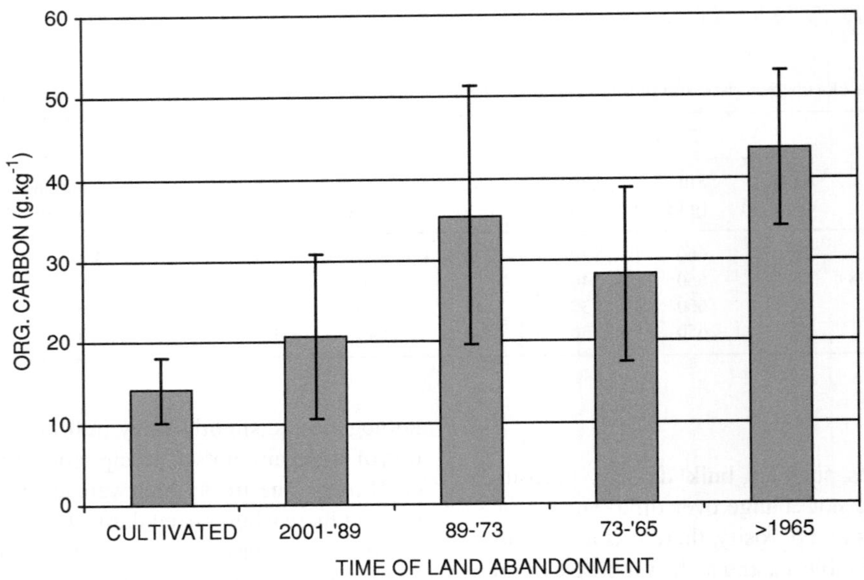

Figure 5. Change of organic carbon after land abandonment. The org. C content of the soil under cultivation and the period 2001–1989 differ significantly from the longest abandoned fields (abandoned before 1965). Means are ±standard deviation (after Hakvoort, 2003).

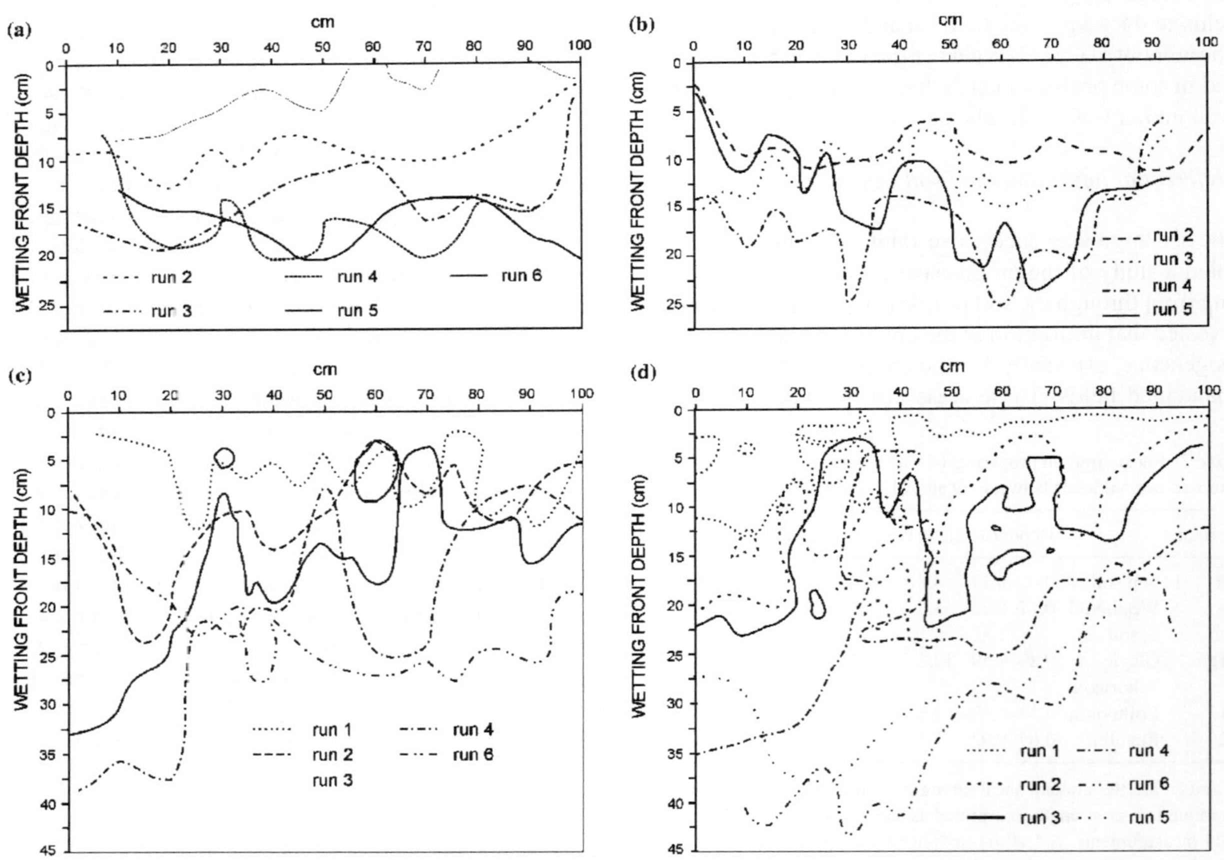

Figure 6. Wetting fronts immediately after four rainfall simulation experiments. (a) Cultivated land in cherry orchard; (b) Abandoned between 2001 and 1989; (c) Abandoned between 1989 and 1973 and (d) Abandoned before 1973.

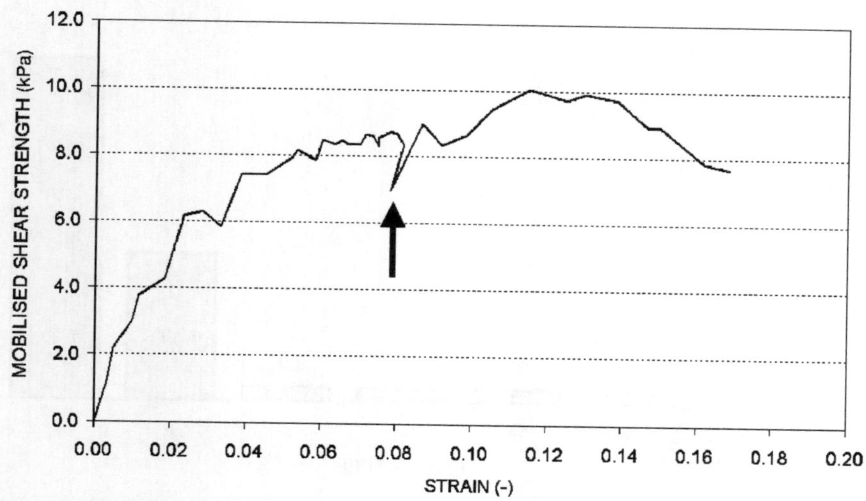

Figure 7. Stress–strain curve of *in situ* direct shear test. Strain is expressed in displacement over length of shear box (-). Arrow indicates snapping of large root.

the root reinforcement of the soil. During the tests roots could be heard snapping and the corresponding drops in the stress-strain curve can be observed (Figure 7). The mean root reinforcement is 3.9 kPa ± 6.3 when the soil shear strength, as calculated from the results of the laboratory direct shear tests, was subtracted. The mean root reinforcement values were 0.6 kPa ± 0.1 for herbs and 5.9 kPa ± 7.5 for Aleppo pine respectively. The root reinforcement ranged from −0.4 kPa (no reinforcement) to 18.2 kPa. Figure 8 shows the strain, reinforcement and fine root content (<1 mm) for seven *in situ* direct shear tests for both herbaceous and Aleppo pine covered soils. The eighth test was not incorporated as no root data were determined from this test. The seventh value was obtained from a test under dense forest cover of *Pinus halepensis* while a test on an 11-year-old sapling of the same species returned a root reinforcement of 7.8 kPa. When the largest reinforcement is excluded, related to large roots, a good correlation between the fine root content (<1 mm) of the slip plane and the root reinforcement is found ($y = 117930x - 1.1309$, $R^2 = 0.96$, $P = 0.004$). Not enough observations have been made to make conclusive remarks on the role of larger roots.

Discussion

The results show that there is a clear effect of land-use change on the development of vegetation cover and type over the last 50 years. Several stages in the development of the vegetation can be recognised, both in the field and from sequential aerial photographs. With the development of biomass over time the incorporation of organic matter into the soil becomes important, as shown by the increasing organic carbon levels with vegetation succession. This is in agreement with studies carried out such as by Martinez-Fernandez et al. (1996), Martínez-Mena et al. (2002) and Ulery et al. (1995), who showed that vegetation removal or development affects the organic carbon levels of the soil. The increase in organic carbon levels over time was also reflected by the development of Ah and ecto-organic horizons in the soil. The high levels of $CaCO_3$ in all soils at all depths hamper the swell and shrinkage properties of clay (Lamas et al., 2002) and reduce the amount of macro-pores present. The porosity showed a clear decrease and bulk density increased with depth. However, in the regolith an inherited higher macro-porosity was found, which is related to the marl shards developing during the first phases of weathering of the marl.

Geomorphological processes such as overland flow and sheet wash acting directly on the surface decreased over time, which can be explained by the increased influence of vegetation cover. However, these processes probably play an important role in the initiation of local mass wasting processes as flow concentration is typical in the first stage of abandonment. The increase in overland flow (sheet wash and rill flow) is related to several factors: Ploughing of the terrace surface will cease with abandonment, favouring soil crust formation and water concentration. The increase of vegetation cover

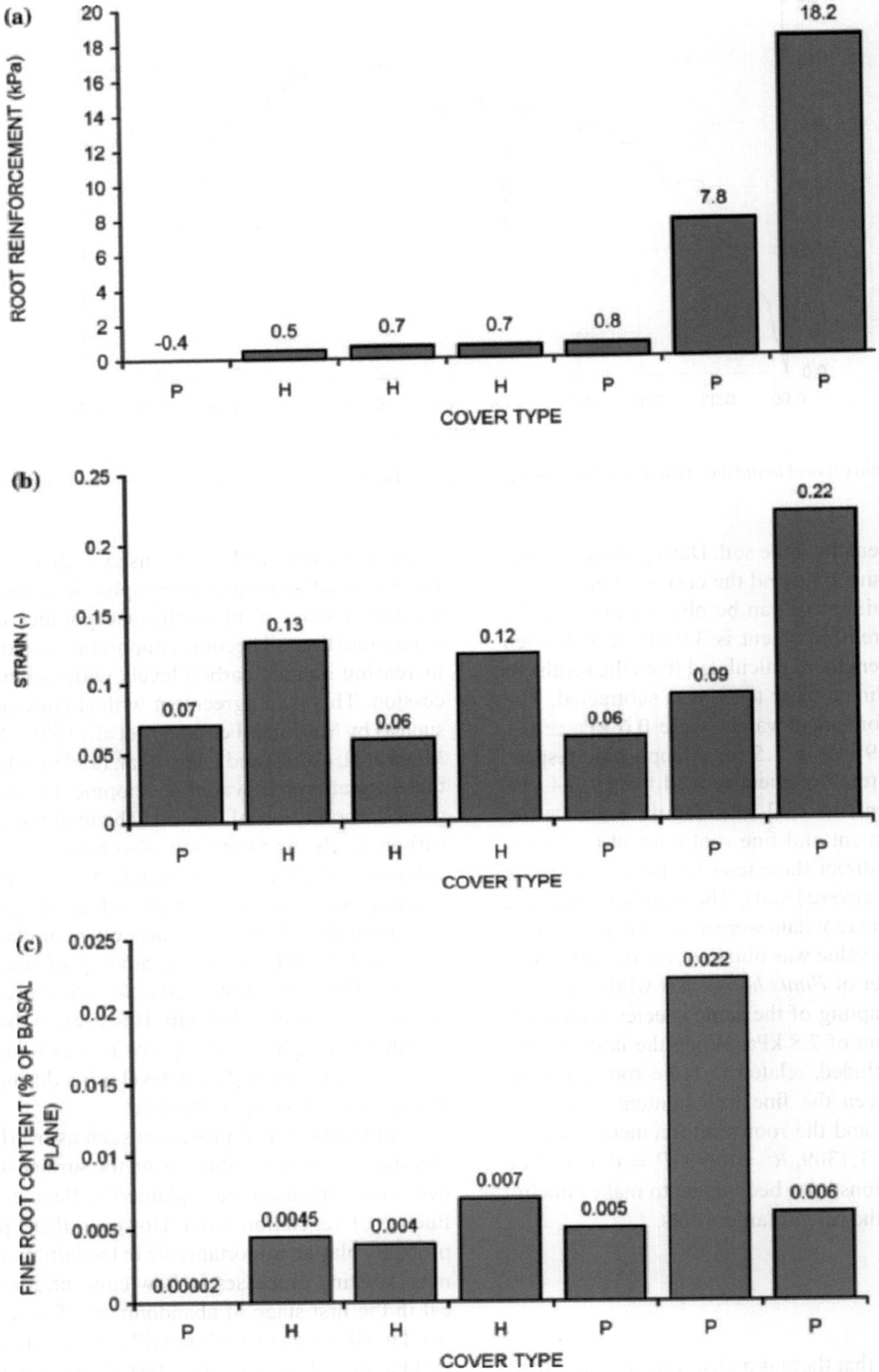

Figure 8a–c. Strain (a), reinforcement (b) and fine root content (<1 mm) (c) for seven *in situ* direct shear tests. H and P refer respectively to herbaceous cover and the presence of Aleppo pine trees.

increases surface roughness, and reduces the size of bare crusted—areas in between the vegetation. Interception by the vegetation reduces the amount of rainfall reaching the soil and the drop impact velocity on the soil (Morgan, 1995). Furthermore, infiltration rates will be higher under the vegetation (Cerdà, 1997; Imeson et al. 1998; Pierson et al., 1994) compared with the surrounding, nonvegetated areas. This higher infiltration is often associated with good soil aggregation and increased macro-porosity (Cerdà et al. 1994; Haynes and Swift, 1990). This latter aspect is clearly visible in the results from the rainfall simulations. The more irregular wetting front and deeper percolation of water into the soil over the 50 years since abandonment are evident. However, the effect of hydrophobic processes might be important, as many studies have shown that these are responsible for heterogeneous infiltration patterns (Doerr et al., 2000) but in this case water repellency tests showed that the effect of water–water repellency could be ignored.

Landsliding and creep processes on the contrary increased over time since abandonment, and became more prominent as the farmers stopped terrace maintenance. Together with the vegetation succession after land abandonment the below ground biomass will increase including larger roots (>5 mm). These roots can have a positive effect on soil strength (Wu et al., 1979) and will be effective in reducing landsliding as long as these roots also contribute to anchoring in the unweathered soil. Where roots increase the soil strength this might have a positive effect on soil creep only. Where this is not the case, there might not be a positive effect from vegetation development on slope stabilisation at all (Gray, 1995). Roots might even induce adverse effects as they uptake water for transpiration and hence desiccate the soil, which can result in large macro-pores (desiccation cracks), which may transport water deep into the soil (Coppin and Richards, 1990).

The *in situ* direct shear tests were performed on samples near pine trees as well as on samples where only grass, herbs or shrubs were present. However, root counts under pine returned higher numbers but the root numbers under both grass and pine decrease exponentially with depth, and virtually no roots, coarse or fine, are present below 0.6 m. The correlation between the fine root content and root reinforcement from the *in situ* shear tests suggests that vegetation type is of minor importance to the actual reinforcement. However, for larger roots such as pine taproots this might be the case when these have a high spatial density. With few trees and under wet conditions on a slope the soil might flow around the large roots as the fine roots do not have enough effect to supply sufficient strength to the soil.

Conclusions

Broad and fine scale research at the Valles de Alcoy revealed that:

(1) mass wasting processes increased over time after abandonment with increasing cover of vegetation following natural succession and only after 40 years mass movement activity was reduced. This is opposite to common literature on root reinforcement on slopes (Coppin and Richards, 1990; Gray, 1995).
(2) organic carbon levels in the upper 0.10 m of the soil increased significantly in the period of study.
(3) rainfall simulations suggested that with increasing time since abandonment, infiltration becomes more heterogeneous and that macropore flow becomes important. This latter process might play an important role in the rapid transfer of water from the surface to potential slip planes, resulting in increased pore pressures at the failure plane, thus enhancing the risk of failure.
(4) the results show that the contribution of the root systems to soil shear strength is present, but limited (~4 kPa). The root system can reinforce the soil up to 40 cm depth, but this does not extend deep enough into the soil to prevent shallow mass wasting processes. These generally develop on potential slip planes at a depth of about 1 m, coinciding with the weathering depth (C1 and C2 horizons) of the soil profile.
(5) vegetation succession does not increase slope stability (no root anchoring) to such a degree that the decline of terraces by mass wasting processes is fully counterbalanced. Instability may be further decreased by developing preferential flow paths feeding potential failure planes in the soil. Eventually a decline in mass wasting processes was observed after 40 years of abandonment.

Acknowledgements

The research was supported by the EU funded Ecoslopes project, under project number QLK5–2001–00289. The support is greatly acknowledged. Research presented was part of several MSc projects at IBED-Physical Geography: Kevin Hakvoort, Zeeger Koppel, Janneke Klein, Casper Cusell, Linda Bannink and Suzanne de Kruif and has also resulted from joint Ecoslope field campaigns with the Trent University Nottingham team (Jo Wint, Joanne Norris and Paul Edwards). The suggestions of the anonymous reviewers have improved the manuscript substantially.

References

Allison L H 1935 Organic soil carbon reduction of chromic acid. Soil Sci. 40, 311–320.

Beven K and Germann P 1981 Water flow in soil macropores II. A combined flow model. J. Soil Sci 32, 15–29.

Cerdà A, García-Alvarez A, Cammeraat L H and Imeson A C 1994 Agregación del suelo en una catena afectada por el abandono del cultivo en la cuenca del Guadalentín (Murcia) I Estabilidad y distribución de los agregados del suelo. *In* JHT García-Ruiz Lasanta (ed), Efectos geomorfológicos del abandono de tierras, pp. 9–19. Sociedad Española de Geomorfología, Zaragoza.

Cerdà A 1997 Erosion and land abandonment in a Semiarid Environment of Southeastern Spain. Arid Soil Res. Rehabil. 11, 163–176.

Coppin N J and Richards I G 1990 Use of Vegetation in Civil Engineering. Construction Industry Research and Information Association, London, 292 pp.

Doerr S H, Shakesby R A and Walsh R P D 2000 Soil water repellency: its causes, characteristics and hydro-geomorpho-logical significance. Earth Sci. Rev. 51, 33–65.

Elias-Castillo F and Ruiz-Beltran L (eds) 1974 Precipitationes maximas en Espana. Servicio de Publicaciones Agrarias, Minesterio de Agricultura, Madrid. 545 pp.

FAO 1989 FAO/Unesco Soil Map of the World Revised Legend World Resources Report 60 FAO, Rome. Reprinted as Technical Paper 20, ISRIC, Wageningen.

FAO 1990 Guidelines for Soil Profile Description. 2nd edition. Soil Survey and Fertility Branch Land and Water Development Division FAO, Rome.

Garcia-Ruiz J M, Ruiz-Flano P and Lasanta T 1996 Soil erosion after farmland abandonment in Submediterranean mountains: A general outlook. *In* JLA Rubio Calvo (eds), Soil Degradation and Desertification in Mediterranean Environments, pp. 165–183. Geoforma Ediciones, Logrono.

Gee G W and Bauder J W 1990 Particle-size analysis. *In* A Klute (ed), Methods of Soil Analysis, part 1, 2nd edition, pp. 383–411. American Society of Agron. and Soil Science Society of American Publisher, Madison.

Geeson N A, Brandt C J and Thornes J B (eds) 2002 Mediterranean Desertification: A Mosaic of Processes and Responses, 440 pp. Wiley, Chichester.

Gray D H 1995 Influence of vegetation on the stability of slopes. *In* DH Barker (ed), Vegetation and Slopes, pp. 2–25. Institution of Civil Engineers, London.

Hakvoort K 2003 Organic Content and Nutrients in the Topsoil as Indicator of Land Degradation. MSc thesis University of Amsterdam, Amsterdam, 63 pp.

Haynes R J and Swift R S 1990 Stability of soil aggregates in relation to organic constituents and soil water content. J. Soil Sci. 41, 73–83.

IGME 1975 Mapa Geologico de España, 1:50.000, Sheet Alcoy (No. 821). Serv. Publ. Min. Industria, Madrid, 31 pp.

Imeson A C, Lavee H, Calvo A and Cerda 1998 The erosional response of calcareous soils along a climatological gradient in Southeast Spain. Geomorphology 24, 3–16.

Imeson A C and Lavee H 1998 Soil erosion and climate change: the transect approach and the influence of scale. Geomorphology 23, 219–227.

Koppel Z 2001 Invloed van vegetatie op landdegradatie. MSc thesis. University of Amsterdam, Amsterdam, 80 pp.

Lamas F, Irigaray C and Chacón J 2002 Geotechnical characterization of carbonate marls for the construction of impermeable dam cores. Eng. Geol. 66, 283–294.

La Roca-Cervignon N and Calvo-Cases A 1988 Slope evolution by mass movements and surface wash (Valls d'Alcoi, alicante, Spain). *In* AC Imeson and M Sala (eds), Geomorphic Processes Vol. I Hillslope Processes, pp. 95–112. Catena-Verlag, Cremlimgen.

Lasanta T J, García-Ruiz M, Pérez-Rontomé C and Sancho-Marcén C 2000 Runoff and sediment yield in a semi-arid environment: the effect of land management after farmland abandonment. Catena 38, 265–278.

MacDonald D, Crabtree D J, Wiesinger G, Dax T, Stamou N, Fleury P, Gutierrez Lazpita J and Gibon A 2000 Agricultural abandonment in mountain areas of Europe: Environmental consequences and policy response. J. Environ. Manage. 59, 47–69.

Martinez-Fernandez J, Martinez-Fernandez J, Lopez-Bermu-dez F, Romero-Díaz M A and Belmonte-Serrato F 1996 Evolution of vegetation and pedological characteristics in fields with different age of abandonment: A case study in Murcia (Spain). *In* JL Rubio and A Calvo-Casis (eds), Soil Degradation and Desertification in Mediterranean Environments, pp. 279–290. Geoforma Ediciones, Logrono.

Martínez-Mena M, Alvarez Rogel J, Castillo V and Albaladejo V 2002 Organic carbon and nitrogen losses influenced by vegetation removal in a semiarid mediterranean soil. Biogeochemistry 61, 309–321.

Morgan R P G 1995 Soil Erosion & Conservation, 290 pp. Longman Group Limited, Essex.

Paine A D M 1985 Ergodic' reasoning in geomorphology: time for a review of the term? Progr. Phys. Geogr. 9, 1–15.

Pierson F B, Van Vactor S S, Blackburn W H and Wood J C (1994). Incorporating small scale spatial variability into predictions of hydrologic response on sagebrush rangelands. *In* WH Blackburn, GE Schuman and FB Pierson (eds), Variability in Rangeland Water Erosion Processes. SSSA Special Publication 38, 23–34. Soil Science Society of America, Madison.

Romero-Calcerrada R and Perry G L W 2004 The role of land abandonment in landscape dynamics in the SPA 'Encinares del río Alberche y Cofio, Central Spain, 1984–1999. Landscape and Urban Planning 66, 217–232.

Shainberg I J, Rhoades D and Prather R J 1981 Effect of low electrolite concentration on soil permeability. Soil Sci. Soc. Am. J. 45, 237–277.

Ulery A L, Graham R C, Chadwick O A and Wood H B 1995 Decade-scale changes of soil carbon, nitrogen and exchangeable cations under chaparral and pine. Geoderma 65, 121–134.

Van Beek L P H 2002 Assessment of the Influence of Changes in Land-Use and Climate on Landslide Activity in a Mediterranean Environment. Netherlands Geographical Studies 294 Utrecht, KNAG. 363 pp.

Wesemael J C 1955 De bepaling van van calciumcarbonaatgehalte van gronden. Chemisch Weekblad 51, 35–36.

Wu T H, McKinnell W P and Swanston D N 1979 Strength of tree roots and landslides on Prince Wales Island, Alaska. Can. Geot. J. 16, 19–33.

Wu T H, Beal P E and Chinchun L 1988 *In-situ* shear test of soil-root systems. J. Geot. Eng. 114, 1377–1393.

Hedge brush layers and live crib walls—stand development and benefits

Rosemarie Stangl[1,2]

[1] *University of Natural Resources and Applied Life Sciences.* [2] *Department of Civil Engineering and Natural Hazards, Institute of Soil Bioengineering & Landscape Construction, Peter-Jordan-Strasse 82, A-1190 Vienna, Austria.*
[2] *Corresponding author**

Key words: adventitious roots, root systems, slope stability, soil bioengineering, soil fixation

Abstract

The long-term performance of soil bioengineering stabilisation methods in torrent catchment areas is difficult to assess due to their dependence on site-specific factors and the interaction between those factors. In order to gain an insight into the dynamic processes of artificially initialised stands and their protection potential, case studies on hedge brush layers and live crib walls aged up to 60 years were carried out. Both construction types were traditional soil bioengineering methods using rooted plants and live plant material for stabilising and revegetation purposes. The plants were horizontally embedded between layers of fill or slope material. Live crib walls additionally included a box-like arranged timber construction which provides further mechanical stability.

The case studies were carried out at 10 different sites in South Tyrol (Italy) and North Tyrol (Austria) and included an inventory of the stands, the excavation of various root systems, an evaluation of the timber constructions, soil chemical analyses and the identification of geological components.

The results of the studies allow a unique and first-time insight into the medium-term development of soil-bioengineering stabilisation methods. Whilst live crib walls maintain their strict linear structures, hedge brush layers develop three-dimensional cover systems with excellent soil armouring effects. With increasing age, the tree species diversity is restricted to only a few species, and the stands are strictly dominated by *Alnus viridis* Vill. and *Alnus incana* L. A change towards natural mixed deciduous forests cannot be expected at this stage.

Using soil bioengineering methods, all of the problematic torrent catchment areas were successfully stabilised, but the geological analyses showed that the risk of further mass movement cannot be eliminated due to the geological site conditions. Hence, mechanical soil reinforcement by means of live plant material, used for both construction types, is of utmost importance for site stability. The excavation of root systems demonstrated the excellent anchoring effect of layer constructions. Dense root systems contribute to the protection against erosion. The former plant inlays function as horizontally oriented main roots which essentially increase slope stability.

Introduction

Case studies on soil bioengineering slope stabilisation

According to Schiechtl and Stern (1992), soil bioengineering is a construction technique that uses biological components for hydraulic engineering and slope stabilisation. In combination with mechanical structures, plants and/or parts of plants are used as living construction elements providing durable stability in the course of their development. Gray and Sotir (1996) use the synonym 'biotechnical', regarding 'soil bioengineering' as a subset of biotechnical stabilisation. Both Gray and Sotir (1996) and Schiechtl and Stern (1992) emphasize soil bioengineering to be a useful

* E-mail: rosemarie.stangl@boku.ac.at

and an essential complementary discipline to conventional technical engineering. Sotir (1990) significantly expresses the biotechnical approach which initialises the establishment of plant communities, hence, supporting the conventional technical techniques to recover damaged land:

> In today's era of high energy costs and depleted resources, soil bioengineering is a practical, cost effective alternative to many land stabilization problems. [...] Soil bioengineering works closely with nature to cause land to become its own self-supporting structure. By offering a low maintenance, rapid recovery system, soil bioengineering produces structures that grow stronger and more beautiful with age and become part of the land. (Sotir, 1990, p. 148)

With age, the artificial structures develop a self-dynamic regulation system and regain their natural stability. A number of site-specific factors influence the success or failure of biotechnical structures and complicate the assessment of the protection effect. Scientists attempt to measure this effect by means of various laboratory tests and field experiments (Järvelä, 2002; Meixner et al., 2002; Stephen, 1999; Visher and Oplatka, 1998), whilst the presented case studies contribute to gain a better understanding of such artificial stands and their underlying dynamic processes. In this multidisciplinary study, focus was put on those site factors which determine the future development of highly sensitive areas. The topics of research involved chemical, physical, mineralogical and clay mineralogical analyses, detailed stand analyses, excavation of root systems and construction and dry rot analyses of timber constructions. This paper discusses the current stand composition and the aspects of the succession tendencies of hedge brush layers and live crib walls. Furthermore the root formation of grey alder (*Alnus incana* L.), green alder (*Alnus viridis* Vill.) and cuttings from bitter willow (*Salix elaeagnos* Scop.) and purpleosier willow (*Salix purpurea* L.) used for layer constructions is introduced and the benefits for stabilisation purposes are highlighted.

Live crib walls for slope stabilisation

Vegetated or live crib walls usually consist of a hollow, box-like arrangement of timber headers and stretchers, filled with slope material and layers of live branch cuttings or rooted plant inlays. According to Schiechtl and Stern (1992), vegetated crib walls function as reinforcement constructions for linear and/or spatial slope stabilisation, whereas Schlüter (1986) also mentions their applicability for selective stabilisation. The instant mechanical stabilisation performance as well as its suitability for toe stabilisation is generally emphasized. Single or double crib walls consisting of timber, concrete, metal or synthetic materials represent technical stabilisation elements, whilst the simultaneous use of live plant material and branch inlays initialises the establishment of the vegetation. Active drainage and the increase of the root systems' armouring effects are the essential benefits of vegetated crib walls. Besides their aesthetic value, Gray and Sotir (1996) point out their flexibility with regards to minor soil movements and settlements.

The durability of the timber construction plays a major role for the medium-term functionality of wooden crib walls. Florineth and Rauch (2001) recommend the use of European larch (*Larix decidua* Mill.), silver fir (*Abies alba* Mill.), pine (*Pinus* L.), oak (*Quercus* L.), European chestnut (*Castanea sativa* Mill.) or black locust (*Robinia pseudoacacia* L.) which are known to be timber species with durable structures. Gray and Sotir (1996) name examples of timber crib walls aged up to 80 years.

The resistance to dry rot and decomposition determines the stability of timber crib walls. Change in colour and structure are caused by destructive mycotic processes, moisture and heat usually fostering the growth of fungi. Noetzli (2002) studied timber constructions used for torrent control in Switzerland and detected significant influences in the change of the water level on the process of timber decomposition. Fully covered with water, timber can be preserved up to several hundred years. Noetzli (2002) showed that the destruction of the driest spruce logs was in the most advanced state of progress, whereas anaerobic conditions caused only minimal dry rot.

Hedge brush layers for slope stabilisation

Hedge brush layers consist of live cut branches and rooted plants placed in layers onto excavated terraces and filled up with soil material. Hedge brush layers are linear structures and, according to Begemann and Schiechtl (1994), have to be completed with additional plantation or seeding. This type of construction

Table 1. Parameters of the sites studied

Site no.	Site	Locality	Geolog. zone	Altitude (m)	Incl. [°]	Type of construction	Age group (years)	Year of construction
1	Tscheiner	Bolzano area	Bolzano porphyry	800	30	live crib walls	1 [2–5]	2000
2	Oberpremer	Bolzano area	Bolzano porphyry	950	31	live crib walls	1 [2–5]	1998
3	Mühlwald	Ahrntal (Pustertal)	old gneiss	1150	34	live crib walls	2 [5–10]	1996
4	Kronengraben	Suldental (Vinschgau)	Vinschgau schist	1200	29	live crib walls	2 [5–10]	1997
5	Widmayer	Bolzano area	Bolzano porphyry	1200	27	live crib walls	3 [10–15]	1991
6	Finsterbach	Gsieser Tal (Pustertal)	old gneiss	1550	25	live crib walls	3 [10–15]	1992
7	Farmahof	Martelltal (Vinschgau)	Vinschgau schist	1700	29	live crib walls	4 [15–20]	1985
8	Holzerlahn	Martelltal (Vinschgau)	Vinschgau schist	1700	31	live crib walls	4 [15–20]	1985
GB	Geroldsbach	Innsbruck area (Stubai Alps)	Ötztal crystalline/ glacial moraines	1800	45	hedge brush layers	5 [50–55]	1950/1959
EB	Enterbach	Innsbruck area (Stubai Alps)	Ötztal crystalline/ glacial moraines	1600	60	hedge brush layers	5 [50–55]	1951–1960

was developed in the 1950s by H.M. Schiechtl (personal communication) and the performance of some examples built by Schiechtl will be evaluated in this paper. At extreme alpine sites, conventional hedge plantation showed only limited success. However, by putting the plant material horizontally onto the terraces, soil stabilisation by Schiechtl (personal communication) was more successful at lower production costs.

Hedge brush layers can be used for deeper soil stabilisation. They are exclusively constructed with live plant materials i.e. long branch cuttings and rooted plants that are capable of adventitious root growth. The length of the plant inlays determines the degree of effectiveness and root growth increases the stabilising effect. Schuppener (2003) associates the pull out resistance and the compound strength as the most relevant parameters with the stabilising effect of layer constructions. Plant inlays can be installed during the procedure of slope filling and have also found to be suitable on very steep slopes.

By integrating both long branch cuttings and rooted plants, Schiechtl (personal communication) intended to control and abbreviate the succession phases, since it is possible to encourage the growth of the species in the successive generation.

Methods and materials

Sites

Hedge brush layers that were used to stabilise former erosion scarps in North Tyrol (Austria) in the 1950s, were investigated in detail in 2001 and 2002. The restored slope failures are situated at the upper course of the torrents Geroldsbach and Enterbach, that flow from the northern parts of the Stubai Alps.

Within the scope of a 5-year research program, biotechnical construction methods applied between the 1980s and 2002 are being evaluated in South Tyrol (Italy). Each year, the focus is put onto a different type of construction. In 2003, the centre of interest was on live crib walls situated across South Tyrol at elevations between 800 and 1500 m (Table 1). The eight investigated sites were categorised into four age groups (1 = 2–5 years, 2 = 6–10 years, 3 = 11–15 years and 4 = 16–20 years) according to the year of construction.

Inventory of the stands

The stand analyses of the former hedgebrush layers of Enterbach and Geroldsbach were based on the sample

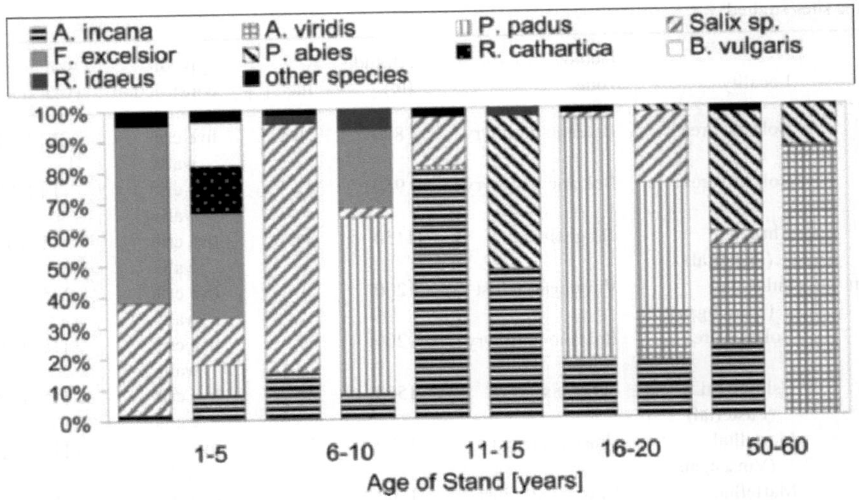

Figure 1. Tree species composition of stands of live crib walls (aged 1–20 years) and hedge brush layers (aged 50–60 years).

plot surveys according to Englisch and Kilian (1999). A full inventory of the woody vegetation, including a general assessment of the local slope area, was carried out at sample areas of 50 m^2. The stand composition of 5% of the total area was registered across a geometric grid and the results extrapolated to per hectare values.

The inventory of the live crib walls had to be adopted according to the linear structures. For each site, three basic topics were addressed separately: (1) general slope situation (adapted according to Grohmann, 2001), (2) rating of the crib walls (construction, timber, plants), (3) single stem inventory at one representative crib wall per site.

Excavation of the root systems

From the hedge brush layers in North Tyrol (site Geroldsbach GB and site Enterbach EB), the root systems of *Alnus incana* L. (one individual) and *Alnus viridis* Vill. (three individuals) were excavated manually up to a distance of 2 m from the stem, by using brushes to remove soil. In South Tyrol, 6 m^3 of the crib walls were excavated at sites 5 and 6 (Table 1). 25 plant individuals of the layer constructions including *A. incana* and cuttings of *Salix eleagnos* Scop. and *Salix purpurea* L. were washed using a water pressure of 2 bar. The diameter, length and branch number of fine, medium, coarse and adventitious roots were then measured for each individual.

Results

Stand analyses of the vegetated crib walls in South Tyrol

The tree species composition of the vegetated crib walls strictly reflects the species that were used for plantation. *Grey alder*, *green alder* and European bird cherry (*Prunus padus* L.) dominate the older sites whereas the younger constructions showed a higher variation in species (Figure 1). Besides various *willows*, European mountain ash (*Sorbus aucuparia* L.), European ash (*Fraxinus excelsior* L.), wayfaringtree (*Viburnum lantana* L.), bloodtwig dogwood (*Cornus sanguinea* L.), maple (*Acer pseudoplatanus* L.), common barberry (*Berberis vulgaris* L.), common buckthorn (*Rhamnus cathartica* L.), and European white birch (*Betula pendula* Roth) increased the diversity in younger sites. Norway spruce (*Picea abies* L.) and American red raspberry (*Rubus idaeus L.*) were found to grow naturally at certain sites, but do not play a major role in soil fixation yet.

The basal diameters of the trees in age groups <10 years were always <50 mm. As most diameters fell into the smallest category (<10 mm), this indicates not only the young age of the stands, but also shows the delay in shoot development after plantation. The trees in stands aged 10–20 years developed basal diameters up to 250 mm, and these stands were dominated by fast growing dendriform species such as grey alder and European

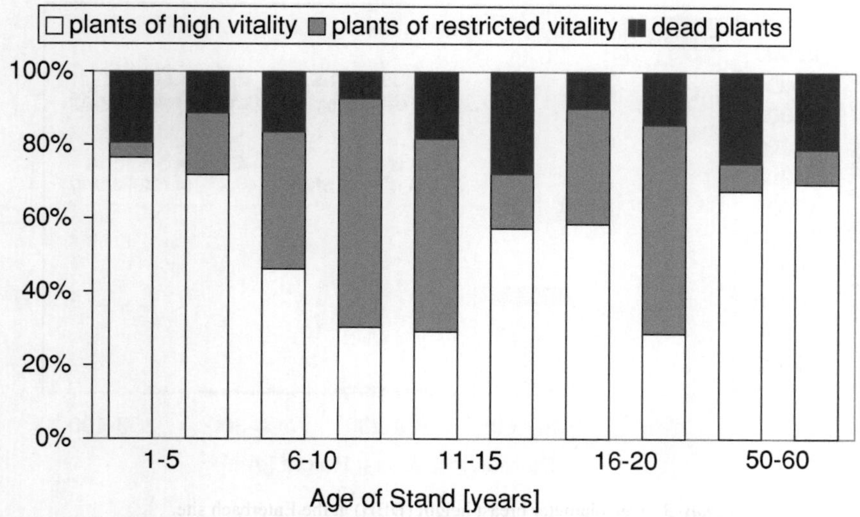

Figure 2. Vitality and mortality rates of stands of live crib walls (aged 1–20 years) and hedge brush layers (aged 50–60 years).

bird cherry. Other species such as green alder, common buckthorn and various willow shrubs, developed basal diameters up to 100 mm. The distribution of basal diameter for trees in older stands was sometimes symmetrical, but usually it was strongly skewed to smaller diameters, thus signifying high shooting and regeneration capacities. There was a good correlation between the diameter distribution and the height distribution. Again, the grey alder and the European bird cherry dominated the upper tree heights, reaching 15 m at the older sites, whilst the remaining species were confined to heights <7 m, and more often <3 m.

In general, tree growth was better at the upper levels of the crib walls. Trees developed larger basal diameters and were taller when growing in the top rows rather than in the lower rows. A higher concentration damage by game animals was found at the base of the crib walls, where trees are more accessible than at the top levels. This damage does not affect the total stand vitality at this stage, as all of the investigated stands were in good condition. Lower vitality rates of some willows and certain grey alder indicated competition between the shoots, but generally the plants showed reasonable to high vitality rates up to 90% (Figure 2).

Stand analyses of the hedge brush layers in North Tyrol

The tree species composition of site Geroldsbach GB was equally dominated by green and grey alder and naturally grown spruce. The Enterbach EB site was dominated by green alders up to 86% (Figure 3), with 13% spruces present and other species e.g. willows, European *larch*, European white birch and Austrian pine (*Pinus nigra* Arnold) representing <1% of the total stands at both sites. The stem numbers of the young stands (diameter breast height, DBH <40 mm) were extraordinarily high at per hectare values of 20 000 (site Geroldsbach) and 24 000 (site Enterbach, Figure 3). The light demanding European larch, European white birch and willows cannot become established in the young stand. They were found in DBH classes <40 mm, but occur more often in the classes 100–200 mm and 200–300 mm as remnants from the previous forest stands before their destruction. The percentage of green and grey alders decreases continuously with increasing DBH classes, they are not represented at classes >200 mm.

The spatial species distribution across the sites was not homogenous. The predominance of a single tree species in certain areas was due to the locally used plants during the construction of the hedge brush layers. At the Enterbach site, local age-related differences were observed across the stand. Due to storm and flood events, the toe of the slope was eroded in the late 1960s and the vegetation cover was completely destroyed. The green alders recovered quickly and soon formed a closed canopy. The tree age is accordingly younger than the average in the upper slope. In the middle slope, the green alders were cut to the base due to increased

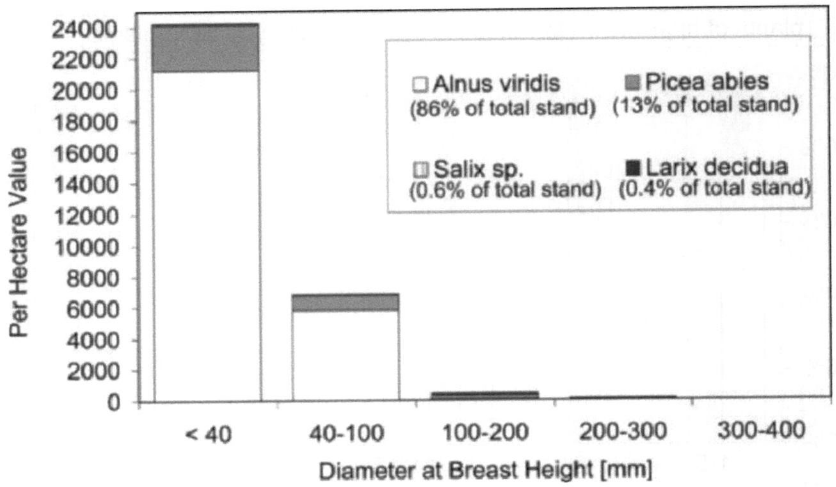

Figure 3. Tree diameter breast height (DBH) at the Enterbach site.

brittleness in 1999. The mortality rate of the pruned stumps was high at 88%. Although their regeneration and sprouting was restricted, the remaining stand is in good condition. The mortality of the green alders was about 20% and the average mortality of the total stand was only 6%. Higher loss in vitality of the young stand at both sites (mortality rates at 21% and 25%) was due to natural selection. As a result of the high stem densities, only the strong individuals can compete. At the Geroldsbach site, the green alder mortality rate was 8%, whereas the mortality of the grey alder was higher at 16%. The average mortality rate of the total stand was 9% (Figure 2).

Root formation

The age of the excavated grey alders ranged from 7 to 48 years (Table 2). All individuals could be identified as plant inlays with an average diameter of 46 mm and an average length of 567 mm. The maximum root length of 2.6 mm was found on a 13-year-old individual. The maximum root length could not be taken from tree number 16 from site Geroldsbach GB (Table 2) since the root system could not be fully isolated. It is assumed that the maximum root length of the 48-year-old grey alder exceeded that from the younger generations. The formation of the roots corresponds to the typical root development of the species. The main tap root was negligible and the coarse roots were of equal size and spread as a uniform cone. Coarse root depth was greater than lateral spread. The development of fine roots was rather low and numerous nitrogen fixing root nodules, as characteristic for alder trees, were found all over the root systems. The adventitious root growth along the former inlays varied widely, and was independent of tree age (Figure 4). There were no adventitious roots found on the oldest individual, whereas the rooting along the newly developed shoot seems to start only after a couple of years after plantation. The grey alders of the live crib wall at site 5 showed intensive root growth within the rows as well as across rows and formed very stable systems.

The age of the green alders was 38, 40 and 42 years with a maximum tree height of 7 m and maximum basal diameter of 100 mm. The length of the plant inlays ranged from 610 to 920 mm with an average diameter of 113 mm (Table 2). The coarse roots spread horizontally close to the surface forming a wide and dense net. These roots were developed from the secondary shoots. Root nodula and fine roots were concentrated in outer regions, with 70% of root diameters <10 mm. There were no adventitious roots found along the former inlays which act as a central anchorage into the slope.

The trees grown from cuttings reached heights up to 3 m with basal diameters ranging from 10 to 40 mm. The lengths of the cuttings varied from 400 to 600 mm with an average diameter of 47 mm. The adventious

Table 2.

species	nr. of tree	site	type of construction	height [cm]	diameter at base [cm]	age of tree	cutting/ plant inlay	diameter of inlay [cm] front/back	length of inlay [cm]	max. root length [cm]	root nodula
Alnus incana	1	5	live crib wall	550	6.5	18	plant inlay	4/3	48	240	8
Alnus incana	10	5	live crib wall	550	9	13	plant inlay	5/3	45	>75	9
Alnus incana	11	5	live crib wall	530	7	13	plant inlay	5.5/2.5	38	135	–
Alnus incana	2	5	live crib wall	480	6	11	plant inlay	5/5	45	100	–
Alnus incana	3	5	live crib wall	470	5.3	12	plant inlay	5/2	60	140	–
Alnus incana	5	5	live crib wall	430	5.3	10	plant inlay	5.3/1.4	45	>114	4
Alnus incana	7	5	live crib wall	580	6	12	plant inlay	6/3	55	>113	1
Alnus incana	8	5	live crib wall	600	7	13	plant inlay	7/2.5	30	>260	–
Alnus incana	18	5	live crib wall	350	3	–	plant inlay	3/1.5	55	>30	5
Alnus incana	19	5	live crib wall	400	3	7	plant inlay	3/1.3	40	>70	19
Alnus incana	21	5	live crib wall	550	5.5	15	plant inlay	6/2	50	>190	15
Alnus incana	21.1	5	live crib wall	k.A	2.3	8	plant inlay	3/2	45	>135	24
Alnus incana	22	5	live crib wall	510	6	12	plant inlay	6/2.5	60	>50	–
Alnus incana	23	5	live crib wall	520	6	11	plant inlay	4/2	36	230	–
Alnus incana	4	6	live crib wall	120	7.5	–	plant inlay	1.9	70	100	–
Alnus incana	10	6	live crib wall	160	2	–	plant inlay	2	–	15	5
Alnus incana	9	6	live crib wall	580	11	9	plant inlay	7	20	165	55
Alnus incana	11	6	live crib wall	–	20	9	plant inlay	4.5	117	160	91
Alnus incana	16	GB	hedge brush layer	1370	27	48	plant inlay	14	150	–	–
Alnus viridis	11	EB	hedge brush layer	650	10	40	plant inlay	12	61	–	–
Alnus viridis	15	EB	hedge brush layer	700	8.5	42	plant inlay	11	92.5	–	–
Alnus viridis	18	GB	hedge brush layer	700	10	38	plant inlay	11	72	–	–
Salix eleagnos	24	5	live crib wall	300	2.5	6	cutting	3.5/3.5	60	>75	–
Salix eleagnos	x	5	live crib wall	60	1	4	cutting	5/5	40	>44	–
Salix eleagnos	13	5	live crib wall	260	3.5	11	cutting	6/5	57	>140	–
Salix eleagnos	20	5	live crib wall	220	3	10	cutting	4.3/5	50	>65	–
Salix eleagnos	4	5	live crib wall	270	4	11	cutting	5/5	40	>135	–
Salix eleagnos	6	5	live crib wall	190	2.5	9	cutting	4.5/3.5	40	>74	–
Salix purpurea	12	5	live crib wall	220.0	2	4	cutting	4	40	>140	–

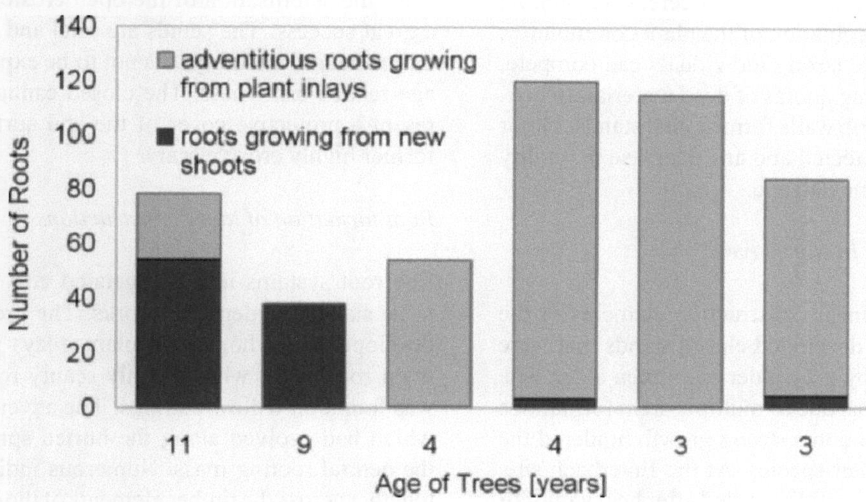

Figure 4. Root growth of plant inlays (*Alnus incana* L.) in live crib walls.

root growth of the willow cuttings differed widely. Both purpleosier willow and bitter willow showed rather weak root development expressed as small root diameters <20 mm and restricted root lengths, although lateral extensions of certain roots up to 1.5 m were observed. The roots were concentrated either at the top or bottom end of the cuttings.

Discussion

Vegetated crib walls in South Tyrol

Within the first 20 years, the vegetation of the stabilised slopes can be characterized by the strict linear structures of the live crib walls. In between the crib walls, vegetation cover depends on the intensity of the tree planting. At the current stage, natural reproduction is of little importance. At the older sites, the zone of brushwood and the lower storey are becoming more dense. Due to root shooting and multiple stem formation, the green and grey alder became dominant species and represent fierce competition for other species.

The single stem inventory of the live crib walls showed homogeneous patterns. The concentration of the small basal diameters reflects the young age of the stands. At most sites, the species diversity was restricted to four to five species, depending on the plants that were used when constructed. The sites of age groups 2 and 3 showed dense tree populations with average distances between plants of only 60 mm. This indicates that the secondary sprout growth seems to reach a maximum after 5 years and then decreases again. At this stage, the self-regulation of the plant communities has started. Only the strong individuals can compete, resulting in increasing quotas of dead material. In general, the vegetated crib walls formed vital stands. Major failures were not detected and any decrease in vitality was due to wild game damage.

Hedge brush layers in North Tyrol

At both sites, the linear construction elements of the hedge brush layers developed closed stands that were strictly dominated by grey alder and green alder. The dense alder brush was due to multiple stem formations and root shooting and this strong growth hindered the establishment of other species. At the Enterbach site, the toe of the slope was fully eroded after heavy rainfall in the 1960s. Due to the high regeneration capacity of the green alders, the stand recovered quickly, whereas the coppice regeneration seemed to be rather low. Due to age-related brittleness, parts of the stands were cut to the base in the late 1990s. Mortality rates of the coppiced stumps (up to 80%) led to the conclusion that the felling was performed too late. These findings contradict Hacker and Paulson (1998) who refer to good coppice regeneration of the green alder. However, the self-regeneration of the remaining stand was very high, and mortality rates of only 5% testify good conditions.

According to the stand inventories, the grade of succession is currently regressive. A static development in the revegetated erosion scarps was reported by Grünwald (1990). On the basis of the recent results it has to be assumed that a change towards mixed deciduous forests according to Raschendorfer's (1954) scenarios of succession cannot be expected, as long as the alders dominate the stands. Due to very dynamic soil processes (Stangl 2003), the alders will maintain their role as important pioneers but also as dominating species. Schwabe-Kratochwil (1998) emphasises the high vegetative regeneration capacity of the grey alder after injuries, as a major advantage to adapt to dynamic processes in slide prone areas. Schiechtl (1998) pointed out the positive qualities of the green alder for vegetation purposes, but he also warned of creating monocultures. By improving simple brush layers to hedge brush layers Schiechtl intended to accelerate the progress of succession (H.M. Schiechtl, personal communication). Although Schiechtl's intentions to establish both initial vegetation as well as species in the following succession phase failed at the Geroldsbach and Enterbach sites, the stabilisation of the open erosion scarps was a great success. The stands are vital and in good condition. Major mortality was not to be expected despite age-related brittleness. The closed canopies also represent a protective cover of the soil surface over the former highly erosive scars.

Root formation of layer constructions

The root systems in the vegetated crib walls proved to be stable and dense networks. The grey alders that developed from horizontal plant inlays produced the main root mass, whereas only scanty root formation was found on willow cuttings. The adventitious roots, which had evolved along the buried sprouts, formed the central rooting mass. Numerous individuals grew together across the timber elements of the cribwall, thus resulting in a higher stability, which could be preserved during the process of excavation and demonstrated the effective mechanical reinforcement of the root collectives. According to Polomski and Kuhn (1998), who detected root grafting in various species, the whole system benefits from nutrient uptake of individual roots. Therefore, within species competition for growth is decreased. As excavated individuals were of a similar size, this hypothesis is supported.

A high amount of roots was observed to grow from the new shoots on two 9 and 11 year old grey alders. The vertical orientation of the coarse roots, which was regularly distributed around the centre, was higher than their lateral orientation, which corresponds to observations of Kutschera and Lichtenegger (2002). The same characteristic was found on a 53-year-old grey alder. Equivalently dominant roots formed a regular cloak around the tree axis, whereas in green alders aged 38 to 42 years, a wide, dense root net close to the surface was found to develop and root growth was significantly higher in the uphill sectors. These findings suggest that uphill root formation is intensified for better mechanical stability in steep slopes.

Both the mature grey and the green alders evolved only a few adventitious roots along the identified former inlays. This leads to the assumption that the adventitious root growth is regressive as soon as the sprout rooting from the new shoot starts.

The root systems clearly demonstrated the anchoring effect of layer constructions such as hedge brush layers and vegetated crib walls. The horizontally placed live branches functioned as mechanical reinforcement elements. According to Schiechtl and Stern (1992), long branch cuttings used for brush layering should have a length of at least 1 m and diameters of more than 50 mm. Gray and Sotir (1996) recommend live branches long enough to reach to the back of the crib boxes or the parental material. Most of the identified former plant inlays only showed maximum lengths up to 1 m. As recommended by Gray and Sotir (1996) and Schiechtl and Stern (1992), using longer branches would support the armouring effect of the root systems significantly.

Conclusions

The grade of succession of the older stands (aged 50–60 years) is currently regressive. Tree species diversity is restricted to the fast growing green and grey alder. Other more light demanding species cannot compete due to high shadowing by the alder canopies. The stands of live crib walls maintain their strict linear characters whereas hedge brush layers form dense and closed canopies across the area. The root systems of layer constructions such as hedge brush layers or live crib walls offer an essential support for slope stabilisation. The former plant inlays act as main roots that are horizontally orientated into the slope and anchor the soil layers efficiently. Adventitious roots form stable networks and protect the upper soil layers from erosion. Dense root mats due to root grafting between trees increase the stabilising effect.

Acknowledgements

Bundesministerium für Landwirtschaft, Forstwirtschaft, Umwelt und Wasserwirtschaft (A); Forsttechnischer Dienst für Wildbach- und Lawinenverbauung, Gebietsbauleitung Mittleres Inntal (A); Autonome Provinz Bozen, Südtirol (I), Abteilung 30 Wasserschutzbauten, Sonderbetrieb für Bodenschutz, Wildbach- und Lawinenverbauung.

References

Begemann W and Schiechtl H M 1994 Ingenieurbiologie. Handbuch zum ökologischen Wasser- und Erdbau. 2. überarb. Aufl., Bauverlag Wiesbaden und Berlin. 154 pp. 203.

Englisch M and Kilian W 1999 Anleitung zur forstlichen Standortskartierung in Österreich. FBVA-Berichte Nr. 104, Schriftenreihe der Forstl. Bundesversuchsanstalt Wien.21 pp. 110.

Florineth F and Rauch H P 2001 Ingenieurbiologie. Studienblätter zur Vorlesung. Eigenverlag Arbeitsbereich Ingenieurbiologie & Landschaftsbau, Universität für Bodenkultur, Wien. 107 pp. 141.

Gray D H and Sotir R B 1996 Biotechnical and Soil Bioengineering Slope Stabilisation. 1, 231 pp. 366. Wiley, New York.

Grohmann J 2001 Untersuchungen zur Wirksamkeit der ingenieurbiologischen Maßnahmen im Wildbacheinzugsgebiet Galina, Gemeinde Nenzing, Vorarlberg. Arbeitsbericht. Eigenverlag Arbeitsbereich Ingenieurbiologie & Landschaftsbau, Universität für Bodenkultur Wien.5 pp. 63.

Grünwald H B 1990 Blaikenbegrünung—Entwicklung und Erfahrungen. Österr. Forstzeitung 11. 101. Jhg. Österr. Agrarverlag Wien, 79–80.

Hacker E and Paulson C 1998 Kurze Übersicht über die Verbreitung der Erlenarten im Exkursionsgebiet. *In* E Hacker and W Pflug (eds), Ingenieurbiologie—Die mitteleuropäischen Erlen, pp. 299–320. Jahrbuch 7 der Gesellschaft für Ingenieurbiologie e.V. Zypresse, Aachen.

Järvelä J 2002 Flow resistance of flexible and stiff vegetation: A flume study with natural plants. J Hydrol. 269, 44–54.

Kutschera L and Lichtenegger E 2002 Wurzelatlas mitteleuropäischer Waldbäume und Sträucher. Leopold Stocker Verlag, Graz. 389, 397, pp. 534.

Meixner H Vollsinger S and Rauch H P 2002 Artificial Flooding of the Soil Bioengineering Test Flume at the Wienfluss River. *In* 2nd International Conference "New Trends in Water and Environmental Engineering for Safety and Life: Eco-compatible Solutions for Aquatic Environments," 24–28, Capri, Italy, pp. 234–247.

Noetzli K P 2002 Ursachen und Dynamik von Fäulen an Holzkonstruktionen im Wildbachverbau. PhD Dissertation, ETH Zürich 46, pp. 128.

Polomski J and Kuhn N 1998 Wurzelsysteme. Birmensdorf, Eidgenösssische Forschungsanstalt für Wald, Schnee und Landschaft. Verlag Haupt, Bern, Stuttgart, Wien. 154 pp. 290.

Raschendorfer I 1954 Blaikentypen in den Tiroler Alpen. Vereinszeitschrift der Diplom-Ingenieure der Wildbachverbauung Band 4-5, Heft 5; Linz, 26–101.

Schiechtl H M 1998 The usefulness of the Alnus-species for bioengineering in the Alps. In E Hacker and W Pflug (eds), Ingenieurbiologie—Die mitteleuropäischen Erlen. Jahrbuch 7 der Gesellschaft für Ingenieurbiologie e.V., pp. 201–211. Zypresse, Aachen.

Schiechtl H M and Stern R 1992 Handbuch für naturnahen Erdbau. Österreichischer Agrarverlag Wien. 90, 108 pp. 153.

Schlüter U 1986 Pflanze als Baustoff. Ingenieurbiologie in Praxis und Umwelt. 2. überarb. Aufl., Patzer Verlag, Berlin. 219 pp. 319.

Schuppener B 2003 Die Geotechnische Bemessung von Böschungssicherungen mit Pflanzen. 4. Österreichische Geotechniktagung, Tagungsbeiträge. Wien: Österr. Ingenieur- und Architekten-Verein, 227–238.

Schwabe-Kratochwil A 1998 Chorological, Sociological and Eological Aspects of Alnus incana. *In* Ingenieurbiologie—Die mitteleuropäischen Erlen. Jahrbuch 7 der Gesellschaft für Ingenieurbiologie e.V. Ed. Hacker E and Pflug W. pp. 299–320. Zypresse, Aachen.

Sotir R 1990 Introduction to soil bioengineering restoration. *In* JJ Berger (ed), Environmental Restoration: Science and Strategies for Restoring the Earth, pp. 146–148. Island Press, Washington, DC.

Stangl R 2003 Wirksamkeit ingenieurbiologischer Maßnahmen zur Sicherung von erosions- und rutschgefährdeten Hängen in Wildbacheinzugsgebieten. Teil A: Heckenbuschlagen zur Blaikensicherung an Enterbach und Geroldsbach. Endbericht vorgelegt dem BMLFUW. Eigenverlag Arbeitsbereich Ingenieurbiologie und Landschaftsbau, Universität für Bodenkultur Wien. A40 pp. 199.

Stephen E D 1999 Effect of riparian vegetation of flow resistance and flood potential. J Hydrol Eng-ASCE 125, 443–454.

Vischer D and Oplatka M 1998 Der Strömungswiderstand eines flexiblen Ufer- und Vorlandbewuchses. Wasserwirtschaft 88/6, 1–5.

Vegetation dynamics on sediment deposits upstream of bioengineering works in mountainous marly gullies in a Mediterranean climate (Southern Alps, France)

F. Rey[1], F. Isselin-Nondedeu & A. Bédécarrats

Cemagref Grenoble UR EM, 2 rue de la Papeterie, BP 76, 38 402, St-Martin-d'Hères cedex, France.
[1]*Corresponding author**

Key words: Bioengineering, erosion, gully, marls, sediment trapping, vegetation dynamics

Abstract

Current erosion-control studies in mountainous catchments emphasise the effectiveness of bioengineering works in constructing vegetation barriers that are designed to trap and permanently retain sediment upstream of such barriers. Plant establishment and succession should result in colonisation of these sediment deposits, thereby improving the trapping capacity of the works. The aim of this study is to evaluate the ability of the natural vegetation to colonise and grow on sediment accumulated upstream of 29 bioengineering works. They were constructed on the channel bottom of two marly gullies in the mountainous Southern French Alps region, which has a Mediterranean climate. We analysed the soil seed bank in sediment deposits after a germination experiment conducted in the laboratory, where soil cores were placed in a non-limited water condition. We also determined the standing vegetation which developed on the sediment deposits on field sites over 2 years of drought (2003 and 2004). The results show that the number of plants was 80/m^2 on average in the samples studied in the laboratory, vs. 31/m^2 in 2003 and 20/m^2 in 2004 on the field sites, with a total diversity of 40 species. Therefore, despite 2 years of drought, natural plant colonisation occurred on the sediment deposits. An improvement in soil water conditions slightly increases the capacity of the sediment deposits to allow seed germination. However, despite the initial success in vegetation colonisation, plant abundance and recovery were rather low, which suggests that vegetation established itself very slowly.

Introduction

Erosion is one of the major problems affecting ecosystem structure and functioning. According to several authors, marly soils, particularly black marls, are one of the most erodible substrates, in particular in the Mediterranean climate (Descroix and Mathys, 2003). A recent study conducted in the Southern French Alps in a badland area devoid of vegetation has shown an erosion rate over 100 m^3 ha^{-1} year^{-1} in marly catchments (Mathys et al., 2003). In gullies, eroded sediment is transported and deposited on the gully floors (Oostwoud Wijdenes and Ergenzinger, 1998). Then it is removed to the gully outlet by concentrated runoff during heavy rainfall events, avoiding soil from developing in eroded gullies.

The vegetation cover can prevent marly soil erosion and trap some of the sediment eroded within a catchment (Bochet et al., 2000; Rey et al., 2004), but erosive conditions in gullies restrict natural colonisation and establishment of vegetation (Cohen and Rey, 2005). However, installing vegetation on eroded lands is possible using afforestation (Toro and Gessel, 1999) and soil bioengineering works (Morgan and Rickson, 1995) which can be made of willow (*Salix*) cuttings (Gray and Sotir, 1996). In gullies, the most effective way to reduce or halt erosion has been to stabilise the gully floor, then revegetate the gully walls to avoid surface erosion

* E-mail: freddy.rey@cemagref.fr

This article has been previously published in the following journal—Plant and Soil, Volume 278, 149–158.
Stokes et al. (eds), Eco- and Ground Bio-Engineering: The Use of Vegetation to Improve Slope Stability, 297–307.

(Yadav and Bhushan, 2002). Successful experiments have been reported from different countries, especially in Europe (e.g., Florineth, 2000; Ternan et al., 1996; Vallauri et al., 2002) and North America (e.g., Li and Eddleman, 2002; Meyer et al., 1997; Pezeshki et al., 2005). In marly gullies, bioengineering works made with brush layers on fascines can be used for gully stabilisation and revegetation (Rey, 2005). These structures can provide vegetative hedges that trap and retain marly sediment going through them, thus creating sediment deposits (or mounds) immediately upslope (Martinez-Turanzas et al., 1997).

These mounds above the brush layers constitute stable ground where the natural vegetation can develop, thus initiating the dissemination of new plant species (Guerrero-Campo and Montserrat-Marti, 2000). Development and recovery of natural vegetation may depend on many factors that can limit recruitment in the plant population. For Eriksson and Ehrlèn (1992), the lack of seed availability and the shortage of microsites are major explanatory factors. Chambers (2000) explains that seeds can be removed and lost by water flow at the soil surface. For Cerdà and Garcia-Fayos (2002), insufficient seed germination and seedling survival seem to be key factors, especially in the Mediterranean climate, which is particularly dry and warm.

If the vegetation cover naturally increases on the sediment deposits created by the bioengineering works, it may subsequently retain more eroded sediment, resulting in the structures trapping even more sediment over time (Bochet et al., 2000). The dynamics involved in this process and the inhibiting factors therefore warranted investigation, in the particular context of mountainous marly gullies in a Mediterranean climate.

The aim of this study is to evaluate the ability of the natural vegetation to colonise and grow on sediment accumulated upstream of bioengineering works, despite the unfavourable climatic and edaphic conditions. We hypothesise that bioengineering works may trap seeds, creating a seed bank, and that soil conditions, especially water availability, govern germination, seedling survival and plant growth. In this study, the relevancy of these hypotheses was tested on 29 bioengineering works built in spring 2002, which had trapped sediment by autumn 2002. We analysed the soil seed bank in sediment deposits after a germination experiment conducted in the laboratory, where soil cores were placed in a non-limited water condition. We also determined flora composition and vegetation dynamics of the vegetation that established on the sediment deposits over 2 years. It was then possible to compare the germinated seed bank of a non-restricted water soil sample and the standing vegetation, in order to determine whether bioengineering works sufficiently improve edaphic water conditions to allow germination.

Materials and methods

Study site and experimental gullies

The observations were carried out over 2 years (2003 and the first 9 months of 2004) in the Saignon catchment, a 400-ha gully catchment on marls (Southern French Alps). Experimental sites are situated in two gullies (gully 1 and gully 2) located on partly eroded black marls. Gully 1 and gully 2 cover 3830 m^2 and 2500 m^2, respectively. Altitude varies from 800 to 905 m. The general exposure of both gullies is to the southwest. Average gully wall slopes range from 100 to 120% and the average gully floor slopes from 35 to 38%. Vegetation cover is 72% in gully 1 and 66% in gully 2. The tree layer is mainly composed of Austrian black pine (*Pinus nigra* Arn. *subsp. Nigra*) and common pine (*Pinus sylvestris*). Whitebeam (*Sorbus aria*), opalus maple (*Acer opalus*) and restharrow (*Ononis fruticosa*) are the principal species of the shrubby layer, and *Achnatherum calamagrostis* dominates the grass layer. Under vegetation, soils are regosoils with mainly fine silt (Vallauri et al., 2002). All the layers are carbonated (with pH varying from 7.8 to 8.1) and poorly structured. However, bio- structuring and biological activities are significant, with large earthworm communities.

The climate is mountainous and Mediterranean (*Vallauri*, 1999). The total average precipitation is 787 mm yr^{-1}. Rainfalls mainly occur within a few months in autumn and spring, with heavy rainfall events in autumn. The average annual temperature is 10.2°C. The average maximum temperature of the warmest month is 28.3°C, whereas the average minimum temperature of the coldest month is −4.2°C. During the observation period of this study, a recording rain gauge was used to measure total rainfall and event characteristics, and a thermometer measured the daily average, minimum and maximum temperatures.

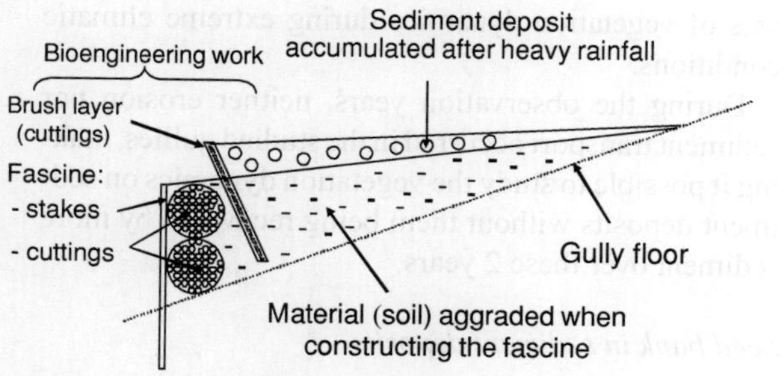

Figure 1. Longitudinal view of a brush layer on a fascine with sediment deposit upstream.

The sediment deposits

We monitored the behaviour of 29 sediment deposits upstream of bioengineering works (BW) made of willow (*Salix*) cuttings, which were brush layers on fascines installed in spring 2002 on the floors of the two experimental marly gullies (Figure 1). Fascines are made of cuttings gathered into bundles and piled up behind stakes. They are installed in gully floors and aid in decreasing erosive and hydrological forces during heavy showers, thus permitting vegetation to develop. Brush layers are installed as rows of cuttings over the aggraded material on top of the fascines. These particular structures were selected for gully restoration because of their proven effectiveness in sediment trapping (Rey, 2005). Thirteen works were built in gully 1 and 16 in gully 2. A set of works was built in each gully, with a single one every 2 m starting from the gully outlet. They were installed along the entire length of gully floors with a slope less than 40%, thus determining the number installed in each gully. They were numbered starting from the gully outlet (for example, BW1 is the first bioengineering work starting from the outlet of the gully). All of them were 1.2 m in width and crossed the entire gully floor. In autumn 2002, heavy rainfall events had led to sediment deposits upstream of the brush layers, an average trapping of 0.06 m^3 sediment per work. The area of each sediment deposit ranged from 1.19 m^2 to 3.2 m^2. As sediment yield at the exit of an eroded gully without bioengineering structures or vegetation is theoretically approximately 100 m^3 ha^{-1} $year^{-1}$ (Mathys et al., 2003), we deduced that a single work trapped 0.06% of this theoretical yield. For all analyses, we divided the longitudinal profiles of the gullies into sectors 10 m in length. For each sector, bioengineering works were grouped. Therefore we determined four groups for gully 1 (BW1–3, 4–6, 7–9, 10–13) and five groups for gully 2 (BW1–4, 5–7, 8–10, 11–13, 14–16).

Measurement and monitoring

Soil seed bank analysis

At the end of the growing season and after the seed release (early spring of year 2004), soil samples were collected using a trowel to cut out and remove 25 × 25-cm blocks of soil. The 25-cm depth was enough to get seeds of the persistent species present in the lower and upper soil layers and to collect transient species only present in the surface soil (Thompson, 1993). Three replicates by group of bioengineering works were collected in the centre of the mounds. This bulked sample procedure was adopted because it is appropriate to avoid standing vegetation damage and side effects due to the small width of certain structures.

A seedling emergence method (Roberts, 1981) was used with these soil samples. This method is simple and appropriate for mid-to large-scale seed bank studies (Gross, 1990). Soil samples were put into cool dry storage, and afterwards stored at 0°C for 4 months. This pre-treatment can break seed dormancy for most of perennial species. Each soil sample was mixed with an equal volume of sterilised potting compost. This mixture was then spread in four, 3-cm-thick sub-sample trails in a greenhouse. Because of different germination requirements, trails were kept under two sets of conditions for 3 months (1.5 months each). In this climate and this type of soil, the maximum seedling emergence occurred from late March to late May (Guàrdia et al., 2000). Early-growing species and annuals that were expected to be found in more mesic soil samples require wet soil as well as mild and fluctuating temperatures in order to germinate. The temperature of the growth chamber was 22°C for 10 h of daylight, then 6°C for 14 h of night, corresponding to the first period of the experiment and to the mean monthly temperatures of the study site. Usually, the species that were expected to be found in sandy soil require hot conditions to germinate (Bakker, 1989). Temperatures were maintained at 25°C for 14 h of daylight and 13°C for the night. To avoid marly-compost mixture dryness, relative air moisture was kept at 75–80%. After 1 month, the position of all samples was reversed in order to limit edge and microclimate effects in the growth chamber. The mixture was also lightly ploughed three times to optimise the seed–light atmosphere interface and consequently ensure a high germination rate (Grime et al., 1988). Seedlings were counted and identified as much as possible (family, genera, species or morph level).

Standing vegetation parameters and dynamics

Adult plants and seedlings were identified and counted on each sediment deposit in spring 2003 and spring 2004, which correspond to the first 2 years of natural colonisation on the mounds. For this study, we retained two vegetation structure indices: plant abundance and species richness, i.e. the number of species. To allow comparisons, plant abundance was expressed in square metres. To study vegetation dynamics, a comparison between years was made based on plant abundance and species richness using non-parametric Wilcoxon tests. The percentage similarity in standing vegetation based on species presence/absence between years was calculated. Therefore, by comparing the results between the standing vegetation and the soil seed bank, we were able to determine the effects of the germination treatment on plant abundance, using a one-way ANOVA.

Results

Meteorological conditions during the observation period

In 2003, total precipitation reached only 586 mm, with 64 mm from 1 May to 31 July (3 months). Heavy rainfall events were rare. The heaviest intensity in 1 h was 24 mm h^{-1}. The average annual temperature was 10.9°C. Average maximum temperatures per month were 28.8°C in June, 29.9°C in July and 31.4°C in August, which was the warmest month of the year. From 1 June to 31 August, the temperatures measured were the warmest ever recorded for approximately 150 years (André et al., 2004), with 53 days when the maximum temperature exceeded 30°C. The total precipitation during the first 9 months of 2004 was only 338 mm, with 15 mm from 1 June to 31 July (2 months). Thus precipitations were particularly low, especially during the beginning of the summer. Few and low rainfall events occurred, the heaviest intensity in 1 h being 11 mm h^{-1}. Average maximum temperatures per month were 25.1°C in June, 27.4°C in July and 26.8°C in August. Thus the years 2003 and 2004 were exceptionally warm and dry, allowing the analysis of vegetation dynamics during extreme climatic conditions.

During the observation years, neither erosion nor sediment transport occurred in the studied gullies, making it possible to study the vegetation dynamics on sediment deposits without them being recovered by more sediment over these 2 years.

Seed bank in sediment deposits

In all the samples, 154 seedlings were observed, 67 in gully 1 and 86 in gully 2 (Table 1). Expressed by area, the number of plants was 80/m^2 on average, with a maximum value of 148/m^2 (BW10–13 in gully 1) and a minimum value of 0/m^2 (BW7–9 in gully 1). Plants germinated in all the samples except in samples corresponding to BW7–9 in gully 1.

The total number of species was 14 in all the samples, 9 in gully 1 and 11 in gully 2. The highest number of species was found above BW8–10 in gully 2 (7 species). Dicotyledons made up the major part of all the plants that germinated. All the observed species were grasses except one shrub species, the restharrow, and one tree species, the black locust (*Robinia pseudacacia*). *Oxalis corniculata*, the Asteraceae, *Robinia pseudacacia, Anthyllis vulneraria, Achnatherum calamagrostis* were the most abundant species.

Vegetation recovery on sediment deposits

Considering the two marly gullies in 2003, we found 559 plants on all the bioengineering works, 203 in gully 1 and 356 in gully 2 (Table 2). Expressed by area, the number of plants was 31/m^2 on average, with a maximum value of 65/m^2 (BW1–4 in gully 2) and a minimum value of 11/m^2 (BW7–9 in gully 1).

In 2004, we found 323 plants over the entire bioengineering works area in both gullies, with 96 plants in gully 1 and 227 in gully 2. Expressed by area, the number of plants was 20/m^2 on average, with a maximum value of 51/m^2 (BW1–4 in gully 2) and a minimum value of 4/m^2 (BW4–6 in gully 1).

For all the bioengineering works, there was a general decline in plant abundance between 2003 and 2004 (Figure 2). Considering the variation at the gully scale (all the bioengineering works), plant abundance slightly decreased between 2003 and 2004 only at the gully scale: for gully 1 ($T = 15$, fd.12, $P = 0.03$) and for gully 2 ($T = 24$, fd.15, $P = 0.02$).

Regarding species richness in 2003 (Table 2), we identified 38 species growing on all the mounds, 19

Table 1. Species list, plant abundance and species richness of plants that germinated from the soil cores of bioengineering works. Unidentified seedlings are listed below like others (the exponent number in brackets refers to different morphs)

	Gully 1				Gully 2				
Bioengineering works (BW) n°	1–3	4–6	7–9	10–13	1–4	5–7	8–10	11–13	14–16
Species									
Achnatherum calamagrostis P. Beauv	2	–	–	1	3	1	–	–	–
Anthyllis vulneraria L.	–	1	–	–	–	1	2	2	3
Aphyllantes monspeliensis L.	–	–	–	–	–	1	1	–	–
Asteraceae	1	3	–	4	2	–	1	2	–
Astragalus semperviens Lamarck	–	–	–	2	–	–	–	–	–
Carex ornithopoides L.	2	–	–	–	–	–	–	–	–
Galium sp.	–	–	–	–	–	–	–	–	1
Hieracium pilosella L.	–	–	–	–	–	–	1	1	4
Ononis fruticosa L.	–	–	–	2	–	1	–	–	–
Oxalis corniculata L.	2	–	–	6	7	1	3	–	5
Potentilla reptans L.	–	–	–	–	–	–	2	–	–
Robinia pseudacacia L.	3	–	–	1	–	–	4	1	4
Trifolium sp.	1	–	–	–	1	–	–	–	–
Others	$12^{(6)}$	$3^{(2)}$	0	$21^{(5)}$	$14^{(4)}$	$6^{(2)}$	$3^{(3)}$	$3^{(1)}$	$5^{(4)}$
Plant abundance (total)	23	7	0	37	27	11	17	9	22
Plant abundance (/m^2)	123	37	0	148	108	59	80	48	117
Species richness	$6^{(12)}$	$2^{(4)}$	0	$6^{(11)}$	$4^{(8)}$	$5^{(7)}$	$7^{(10)}$	$4^{(5)}$	$5^{(9)}$

in gully 1 and 34 in gully 2. BW7–9 in gully 1 and BW14–16 in gully 2 showed the poorest species richness (both with seven species). The highest values were found in BW5–7 in gully 2 (20 species). The Asteraceae are well represented herbaceous species, especially with *Crepis* spp. (62 plants) and *Hieracium* spp. (61 plants). *Achnatherum calamagrostis* (Poaceae) was the most abundant grass species (74 plants) growing in the two gullies. The most abundant shrub was the restharrow (37 plants), and we also found Rosaceae shrubs with *Crataegus monogyna* (12 plants) and *Rosa sempervirens* (four plants). We found three species of tree, mainly the Austrian black pine with 37 plants, the opalus maple and the gean (*Prunus avium*), both with one plant.

In 2004, we found 39 species, 21 in gully 1 and 36 in gully 2. The poorest species richness was found above BW4–6 in gully 1 (five species). The highest number of species was on BW1–4 (24 species). For species richness, there was a general declining tendency between 2003 and 2004, but no significant differences (gully 1: $T = 4.0$, fd.12, $P = 0.34$; gully 2: $T = 16$, fd.15, $P = 0.07$) (Figure 3). The Asteraceae family was still well represented on all the sediment deposits with the genera *Hieracium* (52 plants) and *Crepis* (14 plants), the latter being much less present in 2004 than in 2003. *Achnatherum calamagrostis* was the most abundant species (32 plants), but the number of these plants was halved compared to 2003. The abundance of many species strongly decreased (*Aphyllanthes monspeliensis, Euphorbia cyparissias, Galium* sp., *Sanguisorba minor, Trifolium pratense, Viola hirta*), whereas others became more abundant (*Bromus erectus*) or even appeared for the first time (*Calystegia sepium, Chrysanthemum* sp., *Lactuca perennis*). The most abundant shrub was still the restharrow (25 plants), and we observed quite the same shrub species with similar abundance as in 2003, except two plants of *Hippophae rhamnoides* that installed in 2004. We found one more tree species than in 2003, the downy oak (*Quercus humilis*), with three individuals. Only two plants of Austrian black pine were observed, vs. 37 in 2003. On the contrary, 17 plants of opalus maple were found in 2004 vs. only one in 2003.

When we compared similarity in species composition between 2003 and 2004 (Figure 4), BW1–3 in gully 1, BW8–10 and BW1–4 both in gully 2 showed the highest values of similarity (63.2%, 47.4% and 36.6%, respectively). In the other bioengineering works, less than 30% of species persisted from one year to the next.

Table 2. Species list, plant abundance and species richness of the standing vegetation found on sediment deposits above bioengineering works in 2003 and 2004

	2003									2004								
	Gully 1				Gully 2					Gully 1				Gully 2				
Bioengineering works (BW) n°	1–3	4–6	7–9	10–13	1–4	5–7	8–10	11–13	14–16	1–3	4–6	7–9	10–13	1–4	5–7	8–10	11–13	14–16
Species																		
Acer opalus Miller	1	–	–	–	–	–	–	–	–	1	–	–	–	–	–	–	2	–
Achnatherum calamagrostis P. Beauv	30	5	5	2	2	3	11	16	–	17	3	2	–	–	–	1	2	7
Anthyllis vulneraria L.	8	–	–	1	–	–	1	–	–	4	–	–	–	1	–	–	–	–
Aphyllantes monspeliensis L.	–	–	–	–	11	–	–	–	–	–	–	–	2	–	–	–	–	–
Astragalus monspessulanus L.	1	–	–	–	–	–	–	–	–	1	–	–	–	1	–	–	–	–
Astragalus sempervirens Lamarck	–	–	–	–	–	1	–	4	–	–	–	–	–	–	1	–	1	–
Brachypodium pinnatum L.Beauv.	–	–	–	–	15	–	–	–	–	–	–	–	–	21	–	–	–	–
Bromus erectus Huds.	–	–	–	–	2	–	–	–	–	–	–	–	–	8	3	–	–	–
Calystegia sepium L.R. Br.	–	–	–	–	–	–	–	–	–	–	–	–	–	–	–	–	10	–
Carex ornithopoides L.	–	–	–	–	–	1	–	–	–	–	–	–	–	–	–	–	–	–
Chrysanthemum sp.	–	–	–	–	–	–	–	–	–	–	–	–	–	6	–	–	–	–
Clematis vitalba L.	–	–	–	–	–	–	–	16	–	–	–	–	–	–	–	–	–	–
Cornus sanguinea L.	–	–	–	–	–	–	5	–	–	–	–	–	–	–	–	5	–	–
Crataegus monogyna Jacq.	–	6	1	–	3	1	–	–	1	1	1	–	–	3	3	–	–	–
Crepis pyrenaica (L.)W. Greuter	–	–	–	2	–	–	–	–	17	–	–	–	–	–	–	–	–	–
Crepis sp.	10	1	–	–	14	4	3	10	1	2	–	–	–	6	–	6	–	–
Dactylis glomerata L.	–	–	–	–	–	–	–	–	–	–	–	–	–	–	1	–	–	–
*Dicotyledons**	–	–	–	–	12	1	–	1	–	–	–	–	–	–	–	–	–	–
Euphorbia cyparissias L.	3	–	–	–	–	3	6	–	–	–	–	–	–	1	–	–	–	–
Galium sp.	–	–	–	–	3	4	1	–	17	–	–	–	–	3	2	1	1	–
Genista cinerea Villars (De Candolle)	–	–	–	1	–	2	–	–	–	–	–	–	1	–	2	–	–	1
Geranium robertianum L.	–	–	–	–	–	2	–	–	–	–	–	–	–	–	–	–	–	–
Globularia nudicaulis L.	–	–	–	–	–	1	–	–	–	–	–	–	–	–	–	–	–	–
Hieracium bifidum Hornem	6	–	–	–	13	15	8	7	2	8	2	1	3	5	4	4	8	7
Hieracium pilosella L.	–	–	–	–	10	–	–	–	–	–	–	–	–	10	–	–	–	–
Hippophae rhamnoides L.	–	–	–	–	–	–	–	–	–	–	–	–	–	–	–	–	2	–

Inula conyza DC.	–	–	–	–	–	–	–	–	–	–	–	–	–	1	–	–	–	1
Juniperus communis L.	–	–	–	–	–	1	–	–	–	–	–	–	–	–	–	–	–	–
Lactuca perennis L.	–	–	–	–	–	–	–	–	–	–	–	–	–	3	–	–	–	2
Laserpitium gallicum L.	–	–	–	–	3	–	1	3	–	–	–	–	2	2	1	–	3	1
Lavandula angustifolia Miller	–	–	–	–	–	–	3	2	–	1	–	1	–	–	–	1	2	–
Lotus corniculatus L.	1	–	–	–	2	1	–	–	–	1	2	2	–	2	–	–	–	–
Medicago falcate L.	–	–	–	–	–	–	–	–	–	–	–	1	–	–	–	–	–	–
Ononis fruticosa L.	2	8	3	6	8	1	2	6	1	8	–	3	2	6	2	–	1	3
Ononis rotundifolia L.	2	2	–	–	–	–	–	–	–	3	–	2	–	2	–	–	–	–
Pinus nigra ssp.nigra J.F Arnold	7	9	1	1	9	2	2	3	3	–	–	–	–	1	–	1	–	–
Potentilla reptans L.	–	–	–	–	–	2	–	–	–	1	–	–	–	–	3	–	–	–
Prunus avium L.	–	–	–	–	1	–	–	–	–	–	–	–	–	15	2	–	–	–
Quercus pubescens Willd.	–	–	–	–	–	–	–	–	–	1	–	–	–	1	–	1	–	–
Rosa sempervirens L.	–	–	3	1	–	–	–	–	–	1	–	2	1	–	–	–	–	–
Sanguisorba minor Scopoli.	7	1	–	1	11	6	11	2	–	2	–	1	2	5	1	6	2	–
Teucrium chamaedrys L.	–	–	–	–	–	7	–	–	–	–	–	–	–	–	5	–	–	–
Thymus serpyllum L.	–	–	–	3	4	5	–	–	–	–	–	–	3	5	2	–	–	–
Trifolium pratense L.	22	12	4	4	1	–	–	–	–	4	1	–	–	–	–	–	–	–
Viburnum lantana L.	–	–	–	–	1	–	–	–	–	–	–	–	–	1	–	–	–	–
Viola hirta L.	1	4	5	10	–	–	2	–	–	1	–	–	–	–	–	3	–	–
Plant abundance (total)	101	48	22	32	125	63	56	70	42	57	9	14	16	110	32	29	34	22
Surface area (m^2)	2.55	2.94	1,78	1,19	2,95	2,05	2,56	3,2	1,54	2,55	2,94	1,78	1,19	2,95	2,05	2,56	3,2	1,54
Plant abundance ($/m^2$)	60	19	11	28	65	30	25	22	21	35	4	15	16	51	14	12	14	23
Species richness	14	9	7	11	19	20	13	11	7	17	5	8	8	24	14	10	11	7

*Dicotyledons were unidentified seedlings.

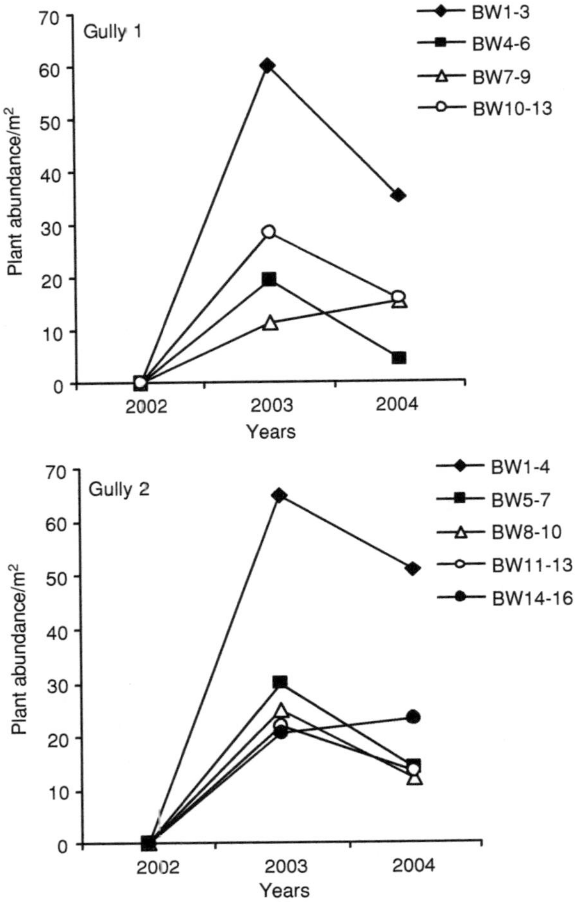

Figure 2. Variation in plant abundance of the standing vegetation since the construction of the bioengineering works in the two marly gullies. Each point represents the mean plant number/m^2 growing on the sediment deposit above the bioengineering works (BW) in 2003 and 2004 (n = number of bioengineering works). No significant differences were shown between the years 2003 and 2004 (standard errors are not represented).

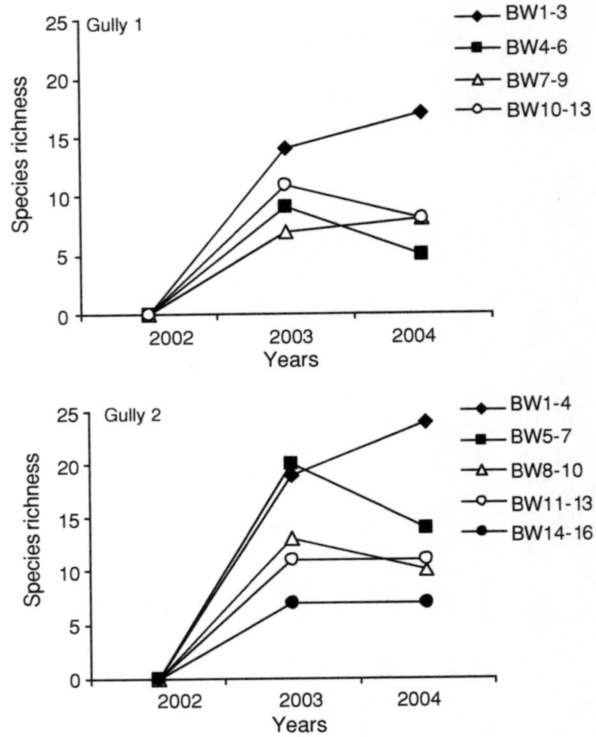

Figure 3. Variation in species richness of the standing vegetation since the construction of the bioengineering works in the two marly gullies. Each point represents the number of species growing on the sediment deposit above the bioengineering works (BW) in 2003 and 2004 (n = number of bioengineering works). No significant differences were shown between the years 2003 and 2004 (Standard errors are not represented).

Effect of water supply improvement on germination

The results concerning the effect of water supply improvement on germination are illustrated in Figure 5. They showed an increasing trend in the mean number of plants by area for nearly all the bioengineering works (except BW7–9 in gully 1). In particular, the increase was statistically significant for BW10–13 in gully 1 ($F = 21.4$, fd.18, $P < 0.05$) and BW14–16 in gully 2 ($F = 9.1$, fd.18, $P < 0.05$). We can note that the black locust and *Oxalis corniculata* were present in the germination experiment (13 and 24 plants, respectively), whereas no plants were observed on the mounds.

Discussion

Natural colonisation by vegetation was observed on sediment deposits, with various plant species developing, thus showing positive vegetation dynamics, even on mineral marly substrates and in a Mediterranean climate in exceptionally dry years. The results show that bioengineering works are able to trap seeds, thereby making a seed bank. This is similar to the results of Urbanska (1997), who observed that structures installed perpendicular to the slope trapped and retained seeds effectively. This also confirms what was established by Guerrero-Campo and Montserrat-Marti (2000), who stated that stable ground can favour vegetation installation and development, whereas erosion processes generally prevent these processes (Chambers, 2000; Cohen and Rey, 2005).

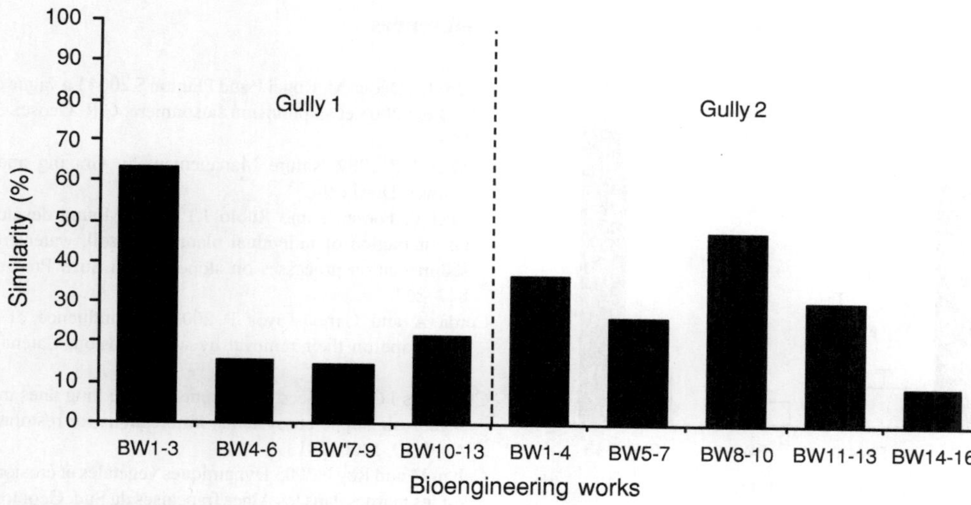

Figure 4. Percentage similarity in standing vegetation based on species presence/absence between 2003 and 2004 on bioengineering works.

However, it is difficult to evaluate the significance of plant abundance based solely on the number of plants. It would have been more useful to estimate the vegetation cover, but plants were not developed enough to measure this. In general, vegetation covers in the mounds were less than 5% after 2 years, a rather low figure. Moreover, plant abundance diminished between spring 2003 and spring 2004, certainly because of the dry summer in 2003. As the summer of 2004 was also very dry, plant abundance may again decrease in 2005. In case of similar climate conditions occurring in the coming years, natural colonisation on mounds should be very slow.

Diversity was rather good, on the mounds as well as in the laboratory. In particular, we observed the development of species that aid in sediment trapping (Rey et al., 2004): grasses, *Achnatherum calamagrostis* and *Aphyllanthes monspeliensis*, and shrubs, *Ononis fruticosa* and *Hippophae rhamnoides*. *Brachypodium pinnatum* and *Bromus erectus*, which have the same morphology as *Achnatherum calamagrostis*, should also act as effective vegetation barriers (Bochet et al., 2000). Investigating how these plants will develop with time would determine whether they lead to the formation of natural vegetation barriers that could trap sediment. Thus, after the use of *Salix* species, the pioneer plants, these colonising plants could make up the post-pioneer vegetation. It appears that the main dissemination factor of these colonising plants is gravity, with superficial micro-landslides bringing with them plants present on gully walls. These results show that plant successions that could lead to sustainable recovering of marly gully floors with vegetation should be considered. Biodiversity should be maintained in the coming years to favour stable and sustainable ecosystems. Trees such as black locust, maple and Austrian black pine can have greater ecological and structural impacts on the bioengineering works.

Germination experiments show that an improvement in soil water conditions slightly increases the capacity of the sediment deposits to allow seed germination. As the two observation years were very dry and warm on the field sites, we surmise, as explained by Cerdà and Garcia-Fayos (2002), that the harsh climatic conditions, and especially the poor water availability, were responsible for the lack of germination, as well as the decreasing number of seedlings between 2003 and 2004. However, the results were not significant for all the bioengineering works. The results are nevertheless quite promising, because the bioengineering works efficiently improved soil conditions to allow germination. Moreover, samples collected in the field were certainly not representative of the whole seed bank present within the mounds. Therefore, it is difficult to assess if adding water and organic matter to the sediment deposits would significantly improve the vegetation dynamics. The high cost of these operations makes their use debatable.

These results do not reflect high effectiveness over the short term of natural vegetation for sediment trapping during future rainfall events. It may therefore be necessary to consider further rehabilitation action.

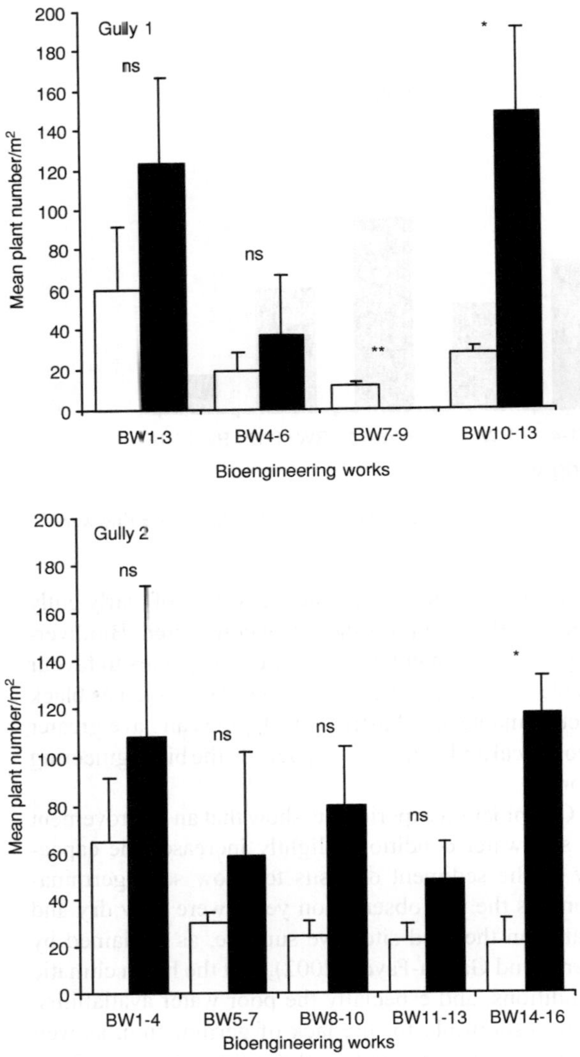

Figure 5. Comparison between standing vegetation and seed bank showing the effects of non-limited water treatment on plant abundance. Mean ($\pm$SE, $n = 3$) plant number expressed by m^2 of sediment deposit above bioengineering works in the two marly gullies (black column: mean number of plants in soil cores; white column: mean number of plant on sediment deposit). Stars indicate a significant effect of the treatment for each bioengineering work, $*P < 0.05$, $**P < 0.01$, $***P < 0.001$, ns: no significant effect.

Brush mats, which are covers of cuttings installed in the same way as brush layers, can be installed directly on the sediment deposits, in order to revegetate them very quickly and effectively, but this will raise the cost of rehabilitation actions.

References

André JC, Déqué M, Rogel P and Planton S 2004 La vague de chaleur de l'été 2003 et sa prévision saisonnière. C.R. Geosci. 336, 491–503.

Bakker J P 1989 Nature Management by Grazing and Cutting. Kluwer, Dordrecht.

Bochet E, Poesen J and Rubio J L 2000 Mound development as an interaction of individual plants with soil, water erosion and sedimentation processes on slopes. Earth Surf. Proc. Land. 25, 847–867.

Cerdà A and Garcia-Fayos P 2002 The influence of seed size and shape on their removal by water erosion. Catena 48, 293–301.

Chambers J C 2000 Seed movements and seedling fates in disturbed sagebrush steppe ecosystems: implications for restoration. Ecol. Appl. 10, 1400–1413.

Cohen M and Rey F 2005 Dynamiques végétales et érosion hydrique sur les marnes dans les Alpes françaises du Sud. Géomorphologie: relief, processus, environnement 1, 31–44.

Descroix L and Mathys N 2003 Processes, spatio-temporal factors and measurements of current erosion in the French Southern Alps: a review. Earth Surf. Proc. Land. 28, 993–1011.

Eriksson O and Ehrlèn J 1992 Seed and microsite limitation of recruitment in plant population. Oecologia 91, 360–364.

Florineth F 2000 Ingenieurbiologische Hangsicherung. Wildbach und Lawinenverbau 142, 29–42.

Gray D H and Sotir R B 1996 Biotechnical and Soil Bioengineering Slope Stabilization: A Practical Guide for Erosion Control. Wiley, New York.

Grime J P, Hodgson J G and Hunt R 1988 Comparative Plant Ecology. Unwin Hyman, London.

Gross K L 1990 A comparison of methods for estimating seed numbers in the soil. J. Ecol. 78, 79–1093.

Guàrdia R, Gallart F and Ninot J M 2000 Soil seed bank and seedling dynamics in badlands of the Upper Llobregat basin (Pyrenees). Catena 40, 189–202.

Guerrero-Campo J and Montserrat-Marti G 2000 Effects of soil erosion on the floristic composition of plant communities on marl in northeast Spain. J. Veg. Sci. 11, 329–336.

Li M H and Eddleman K E 2002 Biotechnical engineering as an alternative to traditional engineering methods. A biotechnical streambank stabilization design approach. Landscape Urban Plan. 60, 225–242.

Martinez-Turanzas G A, Coffin D P and Burke I C 1997 Development of microtopography in a semi-arid grassland: Effects of disturbance size and soil texture. Plant Soil 191, 163–171.

Mathys N, Brochot S, Meunier M and Richard D 2003 Erosion quantification in the small marly experimental catchments of Draix (Alpes de Haute Provence, France). Calibration of the ETC rainfall-runoff-erosion model. Catena 50, 527–548.

Meyer L M, Townsend E C and Thayer G W 1997 Stabilization and erosion control value of Oyster cultch for intertidal marsh. Restor. Ecol. 5, 93–99.

Morgan R P C and Rickson R J 1995 Slope Stabilization and Erosion Control. A Bioengineering Approach, 274 pp. E & FN Spon, London.

Oostwoud Wijdenes D J and Ergenzinger P 1998 Erosion and sediment transport on steep marly hillslopes, Draix, Haute- Provence, France: An experimental field study. Catena 33, 179–200.

Pezeshki S R, Brown C E, Elcan J M and Shields Jr F D 2005 Responses of nondormant Black willow (*Salix nigra*) cuttings to preplanting soaking and soil moisture. Restor. Ecol. 13, 1–7.

Rey F 2005 Efficacité des ouvrages de génie biologique pour le piégeage des sédiments dans des ravines incisées dans des marnes (Alpes du Sud, France). Géomorphologie: relief, processus, environnement 1, 21–30.

Rey F, Ballais J L, Marre A and Rovéra G 2004 Rôle de la végétation dans la protection contre l'érosion hydrique de surface. C.R. Geosci. 336, 991–998.

Roberts H A 1981 Seed banks in soil. Adv. Appl. Biol. 6, 1–56.

Ternan J L, Williams A G, Elmes A and Fitzjohn C 1996 The effectiveness of bench-terracing and afforestation for erosion control on Ranò a sediments in central Spain. Land Degrad. Dev. 7, 337–351.

Thompson K 1993 Seed persistence in soil. *In* GAF Hendry and JP Grime (eds), Methods in Comparative Plant Ecology, pp. 199–202. Chapman and Hall, London.

Toro J and Gessel S P 1999 Radiata pine plantations in Chile. New For. 18, 33–44.

Urbanska K M 1997 Restoration ecology research above the timberline: colonization of safety islands on a machine-graded alpine ski run. Biodiv. Conserv. 6, 1655–1670.

Vallauri D 1999 Quel avenir pour les peuplements RTM de pin noir d'Autriche sur substrats marneux dans les Alpes du sud?. Rev. For. Fr. 51, 612–626.

Vallauri D, Aronson J and Barbero M 2002 An analysis of forest restoration 120 years after reforestation on badlands in the southwestern Alps. Restor. Ecol. 10, 16–26.

Yadav R C and Bhushan L S 2002 Conservation of gullies in susceptible riparian areas of alluvial soil regions. Land Degrad. Dev. 13, 201–219.

Implementation and monitoring of soil bioengineering measures at a landslide in the Middle Mountains of Nepal

Walter Lammeranner[1,3], Hans Peter Rauch[1] & Gregor Laaha[2]
[1]*Institute of Soil Bioengineering and Landscape Construction, University of Natural Resources and Applied Life Sciences, A-1190, Vienna, Austria.* [2]*Institute of Applied Statistics and Computing, University of Natural Resources and Applied Life Sciences, A-1190, Vienna, Austria.* [3]*Corresponding author**

Key words: crib wall, erosion, revegetation, slope stabilisation, soil bioengineering

Abstract

Soil bioengineering is an appropriate approach to deal with erosion problems and shallow seated landslides especially in developing countries such as Nepal. This technique is a cost-effective solution using locally available materials and low-cost labour. Furthermore, this approach allows the involvement of the local population in the management and maintenance aspects. As an example for prevention of soil erosion and increased slope stability by means of vegetation, a landslide in the Middle Mountains of Nepal was selected for an innovative approach of rescheduling the implementation of soil bioengineering stabilisation works into the dry, winter season, testing plants for their suitability and using cut bamboo (*Bambusa nutans* G. C. Wall.) for the construction of vegetated crib walls. This paper deals with investigation, design, construction and monitoring over a period of 32 months. Soil bioengineering solutions for erosion control and shallow slope instabilities in the Middle Mountains proved to be highly successful and all plants, in particular *Salix tetrasperma* Roxb. and *Alnus nepalensis* Don., seemed to be suitable for winter plantation.

Introduction

Soil bioengineering is the use of living plants or cut plant material, either alone or in combination with inert structures, to control soil erosion and the mass movement of land in order to fulfil engineering functions (Howell, 2001). Soil bioengineering can contribute to erosion control and slope stability by achieving the following effects: (1) preventing surface erosion through the soilbinding properties of roots; (2) reducing effects of splash erosion through rainfall interception of vegetation canopy; (3) reducing the incidence of shallow slope instability through the anchoring properties of roots; (4) channelling run-off to alter slope hydrology; and (5) providing support to the base of the slope and trapping material moving down the slope. Live plants and other natural materials have been used for centuries to control erosion problems on slopes and along riverbanks in different parts of the world (Schlüter, 1986). Soil bioengineering may have been used in Nepal in the past, but the methods currently used have been brought into Nepal over the last 30 years (Schaffner, 1987).

The population growth and physiographic limitations of Nepal have led to enormous pressures on the fragile mountain environment and to uncontrolled expansion of farming onto unsuitable marginal lands, deforestation, overgrazing of pasturelands and exploitation of natural resources. In combination with the monsoon climate, very steep slopes and inherently weak geological conditions these factors make Nepal highly susceptible to erosion and landslides (Upreti, 2001). Therefore, it is necessary to develop adequate methods to deal with this erosion and these landslide hazards. Soil bioengineering is a solution to address erosion and slope stability problems. The use of indigenous plant materials, low capital costs compared to civil engineering structures and the possibility for involvement of the local population in management

* E-mail: walter.lammeranner@boku.ac.at

This article has been previously published in the following journal—Plant and Soil, Volume 278, 159–170.
Stokes et al. (eds), Eco- and Ground Bio-Engineering: The Use of Vegetation to Improve Slope Stability, 309–319.

and maintenance are substantial arguments that make soil bioengineering techniques highly appropriate for a developing country such as Nepal. Hence, many soil bioengineering constructions have been carried out in Nepal during the last 30 years and much experience in slope stabilisation works by means of vegetation has been made. Nevertheless, there is still a clear need for scientific research concerning suitability and efficiency of soil bioengineering techniques in Nepal, because the majority of published literature is based on practical experience and not on scientific research and data about conditions and modes of failure (Howell, 2001).

The general practice of the Department of Roads of His Majesty's Government of Nepal, is that all the soil bioengineering stabilisation works are carried out during the pre-monsoon period, mainly in June (Department of Roads, 1997). The advantage of this approach is that the survival rates of cuttings and rooted plants are higher during monsoon. However, not all protection measures are favourable and there are many disadvantages too. Just after construction, the soil surface is vulnerable to erosion and plants cannot provide anchoring and soil binding functions. Due to high precipitation in July and August, there is the possibility of washout of pre-mature plants. Furthermore, local farmers, the main labour source for soil bioengineering works, are busy working on their fields in the pre-monsoon time. A possible solution could be the rescheduling of soil bioengineering works from monsoon to winter. In a subtropical climate, such as exists in the Middle Mountains of Central Nepal, the temperature does not fall below freezing point; hence this should not be the limiting parameter for plant growth in winter. The main limitation of performing soil bioengineering works in winter is dryness, which hinders the growth of plants; therefore it is necessary to find species which are resistant to drought and thus, can be used for soil bioengineering works in the winter period.

Wooden crib walls, a specialized form of gravity-retaining structure, have a long tradition in many parts of the world (Coppin and Stiles, 1995). In Nepal, this technique has not been used so far as the timbers required for the construction are not available in the quality and quantity required. However, bamboo is found almost everywhere in Nepal and is used extensively for various construction purposes. Also, the use of bamboo for erosion control measures has a very long history in Nepal. Therefore, the idea was developed to use bamboo, a cheap, locally available material instead of wood for the construction of vegetated crib walls.

The objective of our research was improvement of soil bioengineering techniques for the use in the Middle Mountains of Nepal, by: (1) rescheduling the implementation of soil bioengineering works from pre-monsoon period into winter season; (2) testing different plant species and rating the suitability for winter plantation; (3) testing the suitability of cut bamboo (*Bambusa nutans* G. C. Wall.) for the construction of vegetated crib walls. The results were discussed with regards to implications for practical use.

Materials and methods

Location, physiography and geology

The project area was near the village of Thankot, 12 km west from Kathmandu along the Tribhuvan Highway. The area lies in the physiographic subdivision of the Middle Mountains, which consist of the southern Himalayan foothills, valleys, and the northern ranges of Mahabharat. The altitude within this subdivision ranges from 800 to 2400 m above sea level and the landscape consists of deep river valleys to high mountains resulting in a wide and complex ecological diversity. For the most part, it is a rugged terrain of deeply dissected mountains and short ranges with sharp crests and steep slopes, which reflect a wide variety of faulted and folded strata (Land Resource Mapping Project, 1986; Wagner, 2000). The very steep slopes and the weak geologic conditions in combination with high precipitation and frequent high-intensity rainfalls make the area very susceptible to landslides. The occurrence of intense weathering phyllites, slates and quartzites add to the instability. Only where the range is made up of rocks like limestone, marble and granite are the slopes more stable (Upreti, 2001).

Climate and vegetation

The Middle Mountains experience a wide range of temperatures depending on elevation and relief. The prevailing subtropical climatic conditions within the project area are subject to excessive rainfalls during the monsoon period from June to September, whereas the winters are rather dry. Whilst there is very little rain in the post-monsoon period from October to December, irregular winter rainfalls of varying intensity can occur from January to March. Many of the smaller streams in this region, e.g. the Gakhcha River below the research site are seasonal, due to the great variation of rainfall throughout the year and steep gradients (Land Resource Mapping Project, 1986; Singh, 1985). The mean annual precipitation for the period 1967–1998 was 2015 mm and varied from 2911 mm in 1978 to 918 mm in 1982. During the monsoon, the mean precipitation was 1581 mm which indicates that about 80% of the average annual rainfall occurred during the period June–September. With an altitude of about 1700 m above sea level, the research site is located in the eastern and central regions of the subtropical zone and hosts a *Schima-Castanopsis* Forest, a subtropical semi-evergreen forest with the dominant trees being *Schima wallichii* DC. Korth. and *Castanopsis indica* Roxb. (in higher regions replaced by *Castanopsis tribuloides* Sm. A. DC. (Department of Forest, 2002; Fleming, 1973). Most of the surrounding area of the research site was young-growth forest or shrubbery with small trees and pioneer vegetation e.g. *Alnus nepalensis* Don. or *Maesa chisia* Buch.-Ham. ex D. Don. Some slopes, mainly those south facing and with dry conditions, were reforested with *Pinus roxburghii* Sarg.—an often-used species for denuded areas (Storrs and Storrs, 1998). Huge areas were covered with *Eupatorium adenophorum* Spreng—a shallow rooting, gregarious weed, which is inedible for livestock and causes serious competition for other plants.

Local co-operation

It has been recognized that co-operation with local land users is essential for the protection of roadside slopes. This co-operation is only possible when both sides can benefit. Local people should for example be allowed to use the trees and grasses planted. The acceptance will be even greater when local groups can profit through income generation. Therefore, our Project had an integrated and participatory approach, strongly involving the local community in all stages of the research project. The research work on the landslide in Thankot was carried out in close cooperation with the women of the Chandragiri Women Forest User Group and the members were actively involved at all stages of the project.

Site assessment

The research site was located in the lower third of a deep-seated landslide which was triggered by heavy monsoon rains in 1996 (Figure 1). The influence of the heavy rains was exacerbated by a number of other factors including: (1) deforestation and lack of vegetation cover; (2) slope undercutting by the Gakhcha River and (3) quarrying of stone at the toe of the landslide. The

Figure 1. View of the research site at a landside near Thankot in the Middle Mountains of Nepal at the beginning of construction works in December 2001.

movement produced a soil with high content of rock fragments in a red coloured, loamy matrix. Soil samples were taken from 0–0.5 m depth and analysed at the Soil Science Division of the Nepal Agricultural Research Council in Lalitpur. The texture was ascertained between loamy (L) and sandy-loamy (sL). The collected soil had a pH value of 7, a high content of stony material and a low organic matter content (1.7 mg kg^{-1}). The mean concentration of N was 0.19 mg kg^{-1}, of P_2O_5 40 kg ha^{-1} and of K_2O 146 kg ha^{-1}. All values related to dry soil.

The exact position of the site was 619097 E and 3063389 N related to the Everest 1830 spheroid at an average altitude of 1700 m above sea level. The slope faced southeast, had an area of about 84000 m^2 (120 × 70 m) and an angle of 30–60°. An attempt to avoid further mass movement was carried out in 1998 by applying civil engineering techniques. Gabion check dams were built into the Gakhcha River to prevent the toe of the slope. Additionally four gabion retaining walls up to 5 m height were arranged throughout the slope.

The types of erosion and slope failures at the site were the following: (1) Surface erosion: The bare unprotected surface was susceptible to erosion and formed rills in the top 0.2 m, which were likely to become deeper if not controlled; (2) Slope undercutting: The toe of the slope was at a risk of undercutting and flooding by the Gakhcha River; (3) Gully erosion: Gullies were established in the slope, continued to develop and had small landslides at their sides; (4) Road protection: The mountainside slope cut above the road was destabilised by seepage erosion.

Soil bioengineering techniques used

To counteract the above hazards, the following types of soil bioengineering techniques were implemented and the suitability of different plants for winter plantation tested.

Hedge-brush layers

Brush layers armor and reinforce the upper soil layers (up to 2 m depth) as the roots anchor and reinforce the soil, furthermore they can help to catch debris (Howell, 1999b). In the project, 17 hedge-brush layers with a total length of 104.7 running meters were applied to strengthen slopes with a steep gradient up to 60° preventing the development of rills and trapping material

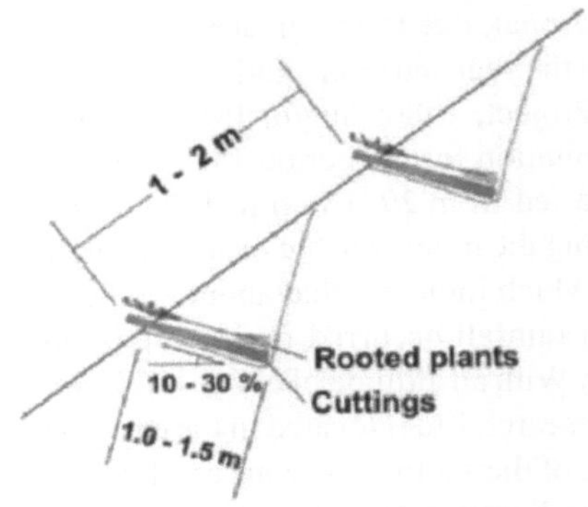

Figure 2. Sketch of hedge-brush layer construction. The hedge-brush layers were applied to strengthen slopes with a steep gradient up to 60° preventing the development of rills and trapping material moving down the slope. Cuttings were laid with rooted plants in alternate into the terraces and covered with soil.

moving down the slope. First 0.8–1.0 m wide terraces inclined at 10% were prepared on the site. Then, cuttings of 20–40 mm in diameter and 0.8–1.2 m in length were laid with rooted plants at alternating intervals of 50 mm. Finally the terraces were covered with soil (Figure 2).

Drainage fascines

Drainage fascines stabilise and drain slopes and can be built into rills or small gullies. Immediately after establishment, draining effects occur (Schiechtl and Stern, 1992). Additionally, this type of fascine enables water-removal due to evaporation of plants (Schiechtl, 1980). In our project, 11 drainage fascines with a total length of 30.7 running meters were used for stabilising the rills which appeared all over the slope. First, appropriate rills were prepared by removing all loose materials and weeds from the site. Cuttings of 1–1.5 m long *Erythrina arborescens* Roxb. were then used as wooden pegs and hammered into the soil at a distance of 1–2 m. Branches were then placed into the rills. The branches were tied together with wire into long bundles of approximately 0.2–0.4 m in diameter and fixed to the pegs. Finally the fascines were buried 20–30 mm with soil (Figure 3).

Palisades

Palisades are generally used for protection of deep, narrow gullies and shallow V-shaped rills. These structures stabilise the gully floor by forming a strong barrier and trapping material moving downwards. The effects increase after shoot development of planted cuttings (Florineth and Rauch, 2001). In the project,

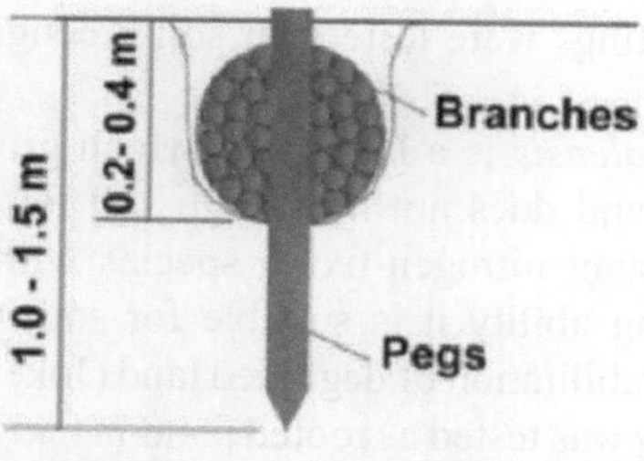

Figure 3. Sketch of drainage fascine construction. The drainage fascines were used for stabilising rills which appeared all over the slope. The branches were tied together with wire into long bundles and fixed to the pegs.

21 palisades with a total length of 26.3 running meters were constructed to stabilise the huge number of gullies, which were often accompanied by soil slippage along their sides. After site selection, holes deep enough to put in vertical hardwood cuttings with a length of 1 m or more were dug out. Then, large hardwood cuttings of *Erythrina arborescens* Roxb. of an adequate length and diameter were fixed horizontally over the gullies. The vertical cuttings were placed at an interval of 0.1 m horizontally into the hole and fastened by wire to the crossbeams to form a wall like structure (Figure 4). Finally the hole was backfilled with soil. Almost 2/3 of the cuttings were covered by soil to keep them moist and to ensure a high survival rate and good shoot development. The construction method suggested by Schiechtl (1980) was modified because species showed rather poor results with the conventional method applied in the first palisades.

Vegetated bamboo crib walls

Crib walls made of timber are a specialised form of gravity-retaining structure using on-site fill material,

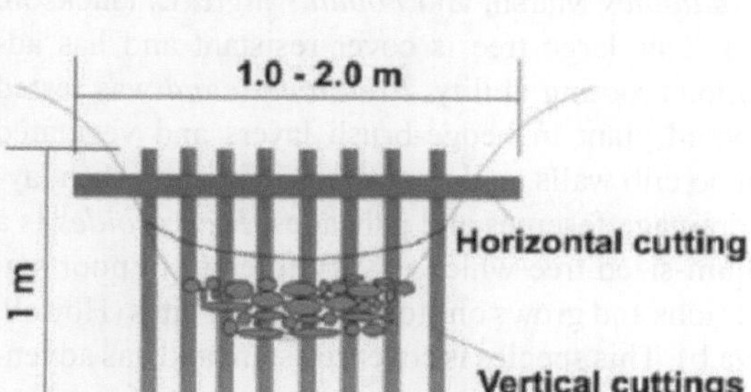

Figure 4. Sketch of palisade construction. The palisades were applied to stabilise the huge number of gullies, which often were accompanied by small landsliding at their sides. The vertical cuttings were placed horizontally into the hole and fastened by wire to the crossbeams to form a wall like structure.

held within a constructed framework (Coppin and Stiles, 1995). Vegetated crib walls provide a combination of the benefits of immediate protection with the long-term advantages of vegetation for stabilisation. Once the plants become established, the subsequent vegetation gradually takes over the structural functions of the wood members (Gray and Sotir, 1996). In the project, the five vegetated bamboo crib walls with a total length of 25 m were used as revetment walls to stabilise the base of the mountainside slope above the road. Bamboos (*Bambusa nutans* G. C. Wall.) with a length of 10 m and an average diameter of 0.1 m were used in the construction. Because of the small diameter of bamboos it was decided to use a bundle of three bamboos for the longitudinal elements fixed together by wire in order to enhance the stability of the wall. The transverse element consisted of only one single bamboo and thus the bamboos with the biggest diameter were used for this purpose. First of all, a terrace approximately 1.5 m wide with a 10% inward slope was prepared and the first course of longitudinal elements were placed parallel to the slope. The longitudinal elements were then fixed into position by interlocking with 1.5 m long header elements that were hammered into the slope. The spacing of the transverse elements was 1.5 m. The interlocking between longitudinal and transversal elements was achieved through cutting notches and fixing with wire. The first horizontal layer of the crib walls was fixed with about 1.5 m long poles of *Erythrina arborescens* Roxb. driven into the soil every 1.5 m. After this procedure the next course of crib elements was placed on top of first layer and again fixed with wire. Then the crib boxes were filled with soil and the plant material was inlayed with an interval of 0.1–0.2 m. This procedure was continued up to the top of the wall (Figure 5).

Plants used

The suitability of a plant for vegetative soil conservation measures depends on the characteristics, requirements and structure of the plant, its usability for certain building systems and its resistance to mechanical forces caused by any form of soil erosion. Species used for soil conservation properties should meet the following criteria: (1) pioneer plant character that means plants grow rapidly on degraded land, landslides, gullies and new road slopes; (2) dense and deep rooting systems which add strength to the surface soil layers and increase the shear strength; (3) adventitious rooting ability and

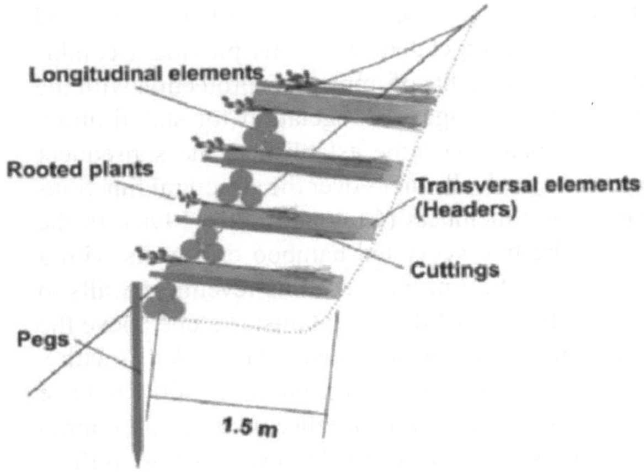

Figure 5. Sketch of vegetated bamboo crib wall construction. The vegetated bamboo crib walls were used as revetment walls to stabilise the base of the mountainside slope above the road. For the longitudinal elements three bamboos were fixed together. The transverse element consisted of only one single bamboo.

coverage resistance and (4) fast and simple propagation such as cuttings and their application in the dormant season (Weigel et al., 1987). Besides these general considerations, the species must fulfil environmental and practical requirements such as: (1) range of altitude; (2) aspects of hill slope; (3) moisture and light requirements; (4) economic value for local population; (5) availability in nurseries and in the test region and (6) preferences of local population.

Most of the used plant species were already well known for their slope stabilising functions and references to that can be found in various publications (Howell, 1999a,b; Howell et al., 1991; Weigel et al., 1987). However, none of these species have been used in winter plantation. Preparatory research work was done in 1999 and tests performed were on the basis of the plant selection. Additionally, in order to learn about local people's preferences regarding species of trees, shrubs and grasses, members of the Chandragiri Women Forest User Group were questioned, using different Participatory Rural Appraisal techniques was carried out (Schönhuth and Kievelitz, 1994). Considering all mentioned parameters it was decided to test the following list of plants: (1) *Alnus nepalensis* Don.; (2) *Callistemon citrinus* (Curtis) Skeels.; (3) *Erythrina arborescens* Roxb.; (4) *Fraxinus floribunda* Wallich.; (5) *Lantana camara* L.; (6) *Morus alba* L.; (7) *Populus x euramericana* (Dode); (8) *Prunus cerasoides* D. Don.; (9) *Salix tetrasperma* Roxb.; (10) *Sapium insigne* (Royle) Trimen. Of each plant species, 100 rooted plants or cuttings were tested per soil bioengineering technique.

Alnus nepalensis is a large tree, which grows well in full light and does not need high soil fertility. As a cover resistant nitrogen-fixing species with adventitious rooting ability it is suitable for soil improvement and rehabilitation of degraded land (Jøker, 2000). *A. nepalensis* was tested as rooted plant in hedge-brush layers and vegetated bamboo crib walls. *C. citrinus* is a small tree, native of Australia, which is cover resistant and has adventitious rooting ability (Howell, 1999b). *C. citrinus* was tested as rooted plant in hedgebrush layers and vegetated bamboo crib walls. *E. arborescens* is a medium sized tree which occurs naturally on dry, rocky slopes and tolerates a wide range of sites (Malla, 1986). This species can be raised from cuttings preferably larger than 50 mm in diameter (Howell, 1999b). *E. arborescens* was tested as cutting in hedgebrush layers, drainage fascines and palisades. *F. floribunda* is a large tree, which is cover resistant, has adventitious rooting ability and is a very good survivor on poor materials (Jackson, 1987). *F. floribunda* was tested as rooted plant in hedge-brush layers and vegetated bamboo crib walls. *L. camara* is a shrub, native of tropical America but now common in many warm-temperate to subtropical regions and known as noxious weed (Polunin and Stainton, 1984). When planted it grows on dry, stony and degraded sites (Howell, 1999a). *L. camara* was tested as cutting in drainage fascines and hedgebrush layers. *M. alba* is a medium-sized tree, which is cover resistant, has adventitious rooting ability and survives on poor and dry sites (Polunin and Stainton, 1984; Storrs and Storrs, 1998). *M. alba* was tested as cutting in hedge-brush layers, drainage fascines and palisades. *P. euramericana* is a hybrid between *Populus deltoides* Marsh. and *Populus nigra* L. (Jackson, 1987). This large tree is cover resistant and has adventitious rooting ability. *P. euramericana* was tested as rooted plant in hedge-brush layers and vegetated bamboo crib walls, and as cuttings in hedge-brush layers, drainage fascines and palisades. *P. cerasoides* is a medium-sized tree which is very tolerant to poor soil conditions and grows on stony and sandy sites (Howell, 1999a,b). This species is cover resistant and has adventitious rooting ability. *P. cerasoides* was tested as rooted plant in hedgebrush layers, and as cutting in vegetated bamboo crib walls. *S. tetrasperma* is a small, light demanding tree, which prefers moist sites but grows also on dry sites when planted (Howell, 1999a). This species is cover resistant and has adventitious rooting ability. *S.*

tetrasperma was tested as rooted plant in hedge-brush layers and vegetated bamboo crib walls, and as cutting in hedge-brush layers, drainage fascines, palisades and vegetated bamboo crib walls. *S. insigne* is a small tree, which grows in varied and dry conditions (Howell, 1999b). *S. insigne* was tested as cutting in hedge-brush layers.

Plant Performance

Features measured were (1) survival rate of plants and (2) length of terminal shoots. The relationship between shoot and root development allows estimation of overall rooting performance and consequently, the anchoring and soil binding functions of the plants. Collected data was used as a basis to report on the survival of rooted plants and cuttings, as well as terminal shoot length of cuttings and rooted plants. Data was collected at months 6, 18 and 32 after planting. Due to infrastructural problems and time limits *C. citrinus*, *L. camara* and *P. euramericana* were measured only 6 and 18 months after implementation.

Statistical analysis

To test the performance of plant species used in different soil bioengineering techniques the following statistical analysis was carried out. Homogeneity of variances were tested using Levine-Test. Because of heterogeneity of variances the Kruskal–Wallis-Test ($P < 0.001$), a nonparametric equivalent to one-way ANOVA was carried out. Length of terminal shoots was used as the dependent variable for testing the influence of: (1) Soil bioengineering techniques on length of terminal shoots, (2) Age of plantation on length of terminal shoots, (3) plant species on length of terminal shoots. After determining differences existing between means, a conservative post-hoc pairwise comparisons test based on a t test, the Temhane T2 was carried out to determine which pair of means differ at an 0.05 level of significance. Because plants for drainage fascines were used as branches and cannot be compared to cuttings or rooted plants they were excluded from statistical analysis.

Results

Cuttings

The survival rate of all cuttings after 6 months was 72%. Subdivided into species there were the following survival rates of cuttings: *E. arborescens* 71%, *L. camara* 60%, *M. alba* 74%, *P. euramericana* 65%, *S. tetrasperma* 100% and *S. insigne* 85%. *S. tetrasperma* had the highest median high growth rate, whereas *S. insigne* had the lowest (Figure 6). A high amount of variability was observed in the response of

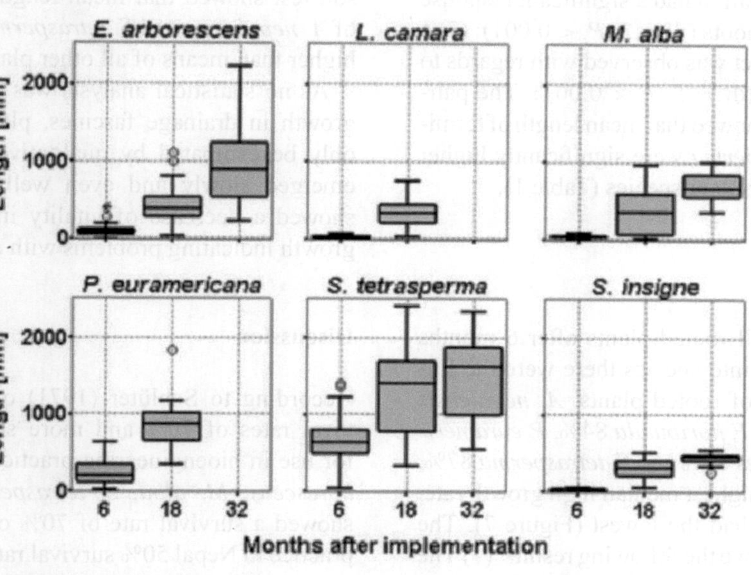

Figure 6. The effect of age (months) on terminal shoot length (mm) for different species of planted cuttings. The boxes represent the interquartile ranges which contain 50% of values. The whiskers are lines that extend from the box to the highest and lowest values, excluding outliers. The lines across the boxes indicate the medians. For *L. camara* and *P. euramericana* no 32-month data available.

Table 1. Mean data for shoot length (mm) of cuttings for different species, analysed using *post-hoc* pairwise comparisons in a Tamhane-Test

Species		Mean difference	Std. error	Sig.
E. arborescens	*L. camara*	210.1	45.7	0.000
	S. tetrasperma	−760.4	54.2	0.000
	S. insigne	165.3	41.7	0.002
L. camara	*E. arborescens*	−210.1	45.7	0.000
	M. alba	−155.1	43.1	0.006
	P. euramericana	−159.1	35.4	0.000
	S. tetrasperma	−970.5	46.6	0.000
M. alba	*L. camara*	155.1	43.1	0.006
	S. tetrasperma	−815.4	52.0	0.000
P. euramericana	*L. camara*	159.1	35.4	0.000
	S. tetrasperma	−811.4	45.9	0.000
	S. insigne	114.3	30.1	0.003
S. tetrasperma	*E. arborescens*	760.4	54.2	0.000
	L. camara	970.5	46.6	0.000
	M. alba	815.4	52.0	0.000
	P. euramericana	811.4	45.9	0.000
	S. insigne	925.7	42.7	0.000
S. insigne	*E. arborescens*	−165.3	41.7	0.002
	P. euramericana	−114.3	30.1	0.003
	S. tetrasperma	−925.7	42.7	0.000

The first two columns show the tested means. The mean difference column displays the difference in length of terminal shoots between pair of means that differ significantly at an 0.05 level of significance. The last two columns display Standard error and Significance.

E. arborescens. The Kruskal–Wallis-Test gave the following results: (1) The soil bioengineering techniques had no significant influence on length of terminal shoots; (2) age of plantation had a significant response to length of terminal shoots (df = 2, $P < 0.001$); (3) a significant species effect was observed with regards to terminal shoot length (df = 5, $P < 0.001$). The pairwise comparison test showed that mean length of terminal shoots of *S. tetrasperma* were significantly higher than means of all other plant species (Table 1).

Rooted plants

The survival rate of all rooted plants after 6 months was 86%. Subdivided into species there were the following survival rates of rooted plants: *A. nepalensis* 86%, *C. citrinus* 97%, *F. floribunda* 84%, *P. euramericana* 77%, *P. cerasoides* 87% and *S. tetrasperma* 87%. *A. nepalensis* had the highest median high growth rate, whereas *F. floribunda* had the lowest (Figure 7). The Kruskal–Wallis-Test gave the following results: (1) The soil bioengineering techniques had no significant response to length of terminal shoots; (2) age of plantation had a significant response to length of terminal shoots (df = 2, $P < 0.001$); (3) a significant species effect was observed with regards to terminal shoot length (df = 5, $P < 0.001$). The pairwise comparison test showed that mean length of terminal shoots of *A. nepalensis* and *S. tetrasperma* were significantly higher than means of all other plant species (Table 2).

As no statistical analysis was carried out for plant growth in drainage fascines, plant performance can only be estimated by qualitative assessment. Plants emerged slowly and even well developed fascines showed a decrease of vitality in their third year of growth indicating problems with dryness.

Discussion

According to Schlüter (1971) only plants with survival rates of 70% and more should be considered for use in bioengineering practice. Cuttings of *E. arborescens*, *M. alba*, *S. tetrasperma* and *S. insigne* showed a survival rate of 70% or higher. In forestry practice in Nepal 50% survival rates are regarded to be enough (Maskey, 1995), accordingly all tested plants can be recommended for soil bioengineering construction works in the winter season. With regards to shoot

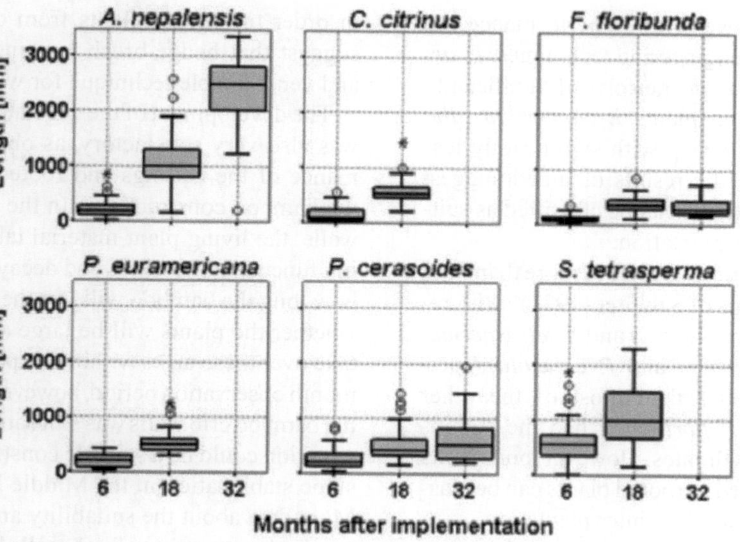

Figure 7. The effect of age (months) on terminal shoot length (mm) for different species of planted rooted plants. The boxes represent the interquartile ranges which contain 50% of values. The whiskers are lines that extend from the box to the highest and lowest values, excluding outliers. The lines across the boxes indicate the medians. For *P. euramericana* and *S. tetrasperma* no 32-month data available.

Table 2. Mean data for shoot length (mm) of rooted plants for different species, analysed using *post-hoc* pairwise comparisons in a Tamhane-Test

Species		Mean difference	Std. error	Sig.
A. nepalensis	*P. euramericana*	652.8	57.5	0.000
	P. cerasoides	548.8	59.2	0.000
	C. citrinus	686.2	57.7	0.000
	F. floribunda	706.7	56.7	0.000
C. citrinus.	*P. cerasoides*	−137.4	27.0	0.000
	S. tetrasperma	−583.2	53.8	0.000
	A. nepalensis	−686.2	57.7	0.000
F. floribunda	*P. cerasoides*	−157.9	24.9	0.000
	S. tetrasperma	−603.7	52.8	0.000
	A. nepalensis	−706.7	56.7	0.000
P. euramericana	*P. cerasoides*	−104.0	26.7	0.002
	S. tetrasperma	−549.8	53.6	0.000
	A. nepalensis	−652.8	57.5	0.000
P. cerasoides	*P. euramericana*	104.0	26.7	0.002
	S. tetrasperma	−445.8	55.4	0.000
	A. nepalensis	−548.8	59.2	0.000
	C. citrinus	137.4	27.0	0.000
	F. floribunda	157.9	24.9	0.000
S. tetrasperma	*P. euramericana*	549.8	53.6	0.000
	P. cerasoides	445.8	55.4	0.000
	C. citrinus	583.2	53.8	0.000
	F. floribunda	603.7	52.8	0.000

The first two columns show the tested means. The mean difference column displays the difference in length of terminal shoots between pair of means that differ significantly at an 0.05 level of significance. The last two columns display Standard error and Significance.

length *S. tetrasperma* showed the best performance and dominated in all soil bioengineering techniques. *E. arborescens* and *P. euramericana* performed significantly better than most of the other plants. *L. camara*, *M. alba* and *S. insigne* formed a group with significantly less shoot growth. However, the results obtained suggest that all plants used as cuttings can be classified as suitable for the use in winter plantation.

All rooted plants had survival rates of more than 70% and fulfil the requirements of Schlüter (1971). With regards to shoot length *A. nepalensis* and *S. tetrasperma*, followed by *P. euramericana* and *P. cerasoides* performed significantly better than most of the other plants. *C. citrinus* and *F. floribunda* had the significantly lowest shoot growth rates. However, our results suggest that all plants used as rooted plants can be classified as suitable for the use in winter plantation.

The qualitative assessment of drainage fascines suggests that drainage fascines should only be used on really damp slopes. The construction of drainage fascines needed a lot of living plant material, which is often difficult to obtain. Florineth (2004) recommends the use of dead branches or stones for the lower third of the fascine, which would be a possible solution to reduce the need of living plant material. However, our results suggests that drainage fascines were not suitable for use in winter plantation and should be replaced by other techniques used for stabilisation of rill erosion.

The modified palisade technique led to satisfactory survival rates and shoot growth of plant species. In contrast to the conventional construction technique (Schiechtl, 1980), the cuttings benefited from additional soil cover, which protected them from drought during winter season. Even without plant growth the palisades gave an immediate mechanical support by catching debris and soil. In Central Europe this old technique is applied seldom (Florineth and Rauch, 2001), but our results and the low demand of plant material and labour suggest that this very cost-efficient technique is suitable for the use in Nepal.

In hedge-brush layer construction the plants grew fast and vigorously. Rooted plants initially had little stabilising effect, as the plants were still very small even after a longer period of growth, but in the long term they will contribute significantly to erosion control and slope stabilisation. Howell (1999b) recommends to cover the plants not more than 50 mm with soil. Florineth (2004) suggests a soil cover of 0.2–0.5 m. According to our results hedge-brush layers implemented in winter plantation should obtain maximum soil cover in order to prevent plants from dryness. Our results suggest that hedge-brush layering is a cost-effective and very suitable technique for winter plantation.

The development of the vegetated bamboo crib walls was also very satisfactory, as observed in the performance of the cuttings and rooted plants inlayed into the bamboo constructions. In the case of wooden crib walls, the living plant material takes over the stabilising functions after the wood decays. As it is not known how long the bamboo will last the question arises as to whether the plants will be large and strong enough to take over these tasks within this period. During the 32-month observation period, however, the development of the bamboo crib walls was still satisfactory and the construction could be a suitable construction technique for slope stabilisation in the Middle Mountains of Nepal. More data about the suitability and durability of bamboo for construction of crib walls has to be attained and verified through long-term research.

In conclusion it can be stated that all plants were suitable for the use in winter plantation. Apart from drainage fascines all techniques showed satisfying results. The use of bamboo for vegetated crib walls was a successful approach in using sustainable resources. Further research on rescheduling of soil bioengineering works from pre-monsoon period into winter season and using bamboo for soil bioengineering constructions is recommended to be the subject of further research in order to improve the soil bioengineering practises in Nepal.

Acknowledgements

The implementation and maintenance of the soil bioengineering techniques by the local population was financed by the Commission for Development Studies at the Austrian Academy of Sciences (Kommission für Entwicklungsfragen bei der Österreichischen Akademie der Wissenschaften), Dr Ignaz Seipel Platz 2, A-1010 Vienna, Austria.

References

Coppin N and Stiles R 1995 Ecological principles for vegetation establishment and maintenance. In RPC Morgan and RJ Rickson (eds), Slope Stabilization and Erosion Control. A Bioengineering Approach. pp. 59–93. E & F N Spon, London.

Department of Forest 2002 Forest and vegetation types of Nepal. Tree Improvement and Silviculture Component (TISC) Document

No. 105, 180 pp. His Majesty's Government of Nepal, Ministry of Forests and Soil Conservation, Kathmandu.
Department of Roads 1997 Use of Bio-engineering in the Road Sector, 120 pp. Guidelines. His Majesty's Government of Nepal, Ministry of Works and Transport, Kathmandu.
Fleming R L 1973 The General Ecology, Flora and Fauna of Midland Nepal, 105 pp. United States Agency for International Development (USAID), Kathmandu.
Florineth F 2004 Pflanzen statt Beton. Handbuch zur Ingenieurbiologie und Vegetationstechnik, 272 pp. Patzer Verlag, Berlin.
Florineth F and Rauch H P 2001 Studienblätter zur Vorlesung Ingenieurbiologie. Institut für Landschaftsbau und Ingenieurbiologie, Universitätfür Bodenkultur, Wien, Österreich, 143 pp.
Gray D H and Sotir R B 1996 Biotechnical and Soil Bioengineering Slope Stabilization. A Practical Guide for Erosion Control, 378 pp. Wiley, New York.
Howell J 1999a Roadside Bio-engineering: Reference Manual, 216 pp. His Majesty's Government of Nepal, Department of Roads, Kathmandu.
Howell J 1999b Roadside Bio-engineering: Site Handbook, 160 pp. His Majesty's Government of Nepal, Department of Roads, Kathmandu.
Howell J 2001 Application of bio-engineering in slope stabilisation: Experience from Nepal. In L Tianchi, S R Chalise and B N Upreti, (eds.). Landslide Hazard Mitigation in the Hindu Kush-Himalayas. pp. 147–161. International Centre for Integrated Mountain Development (ICIMOD), Kathmandu.
Howell J, Clark J, Lawrance C and Sunwar I 1991 Vegetation structures for stabilising highway slopes: A manual for Nepal. UK/Nepal Eastern Region Interim Project of the Maintenance and Rehabilitation Co-ordinating Unit, Department of Roads, Kathmandu, 183 pp.
Jackson J K 1987 Manual of Afforestation in Nepal. United Kingdom Forestry Research Project, Forest Research and Survey Centre, Ministry of Forests and Soil Conservation, Kathmandu, 402 pp.
Jøker D 2000 Seed leaflet No. 8. *Alnus nepalensis* D. Don. Danida Forest Seed Center (DFSC), Hoersholm, Denmark, 2 pp.
Land Resource Mapping Project 1986 Geology Report. His Majesty's Government of Nepal and Government of Canada, Kenting Earth Science Ltd, Ottawa, ON.
Malla S B, Rajbhandari S B, Shrestha T B, Adhikari P M, Adhikari S R and Shakya P R 1986 Flora of Kathmandu Valley. Bulletin of the Department of Medicinal Plants Number 11, 962 pp. His Majesty's Government of Nepal, Ministry of Forests and Soil Conservation, Kathmandu.
Maskey M L 1995 Boöschungsstabilisierung im Mittelgebirge Nepals durch biotechnische Maßnahmen. Dissertation an der Universität für Bodenkultur, Wien, Osterreich, 107 pp.
Polunin O and Stainton A 1984 Flowers of the Himalaya, 580 pp. Oxford University Press, New Delhi.
Schaffner U 1987 Road construction in the Nepal Himalaya. The Experience from the Lamosangu-Jiri Road Project, 67 pp. International Centre for Integrated Mountain Development (ICIMOD), Kathmandu.
Schiechtl H M 1980 Bioengineering for Land Reclamation and Conservation, 404 pp. University of Alberta Press, Edmonton, AB.
Schiechtl H M and Stern R 1992 Handbuch für naturnahen Erdbau. Eine Anleitung für ingenieurbiologische Bauweisen, 153 pp. Österreichischer Agrarverlag, Wien.
Schlüter U 1971 Eignung von Holzarten fürden Busch- und Hecklagenbau, 56 pp. Ulmer, Stuttgart.
Schlüter U 1986 Pflanze als Baustoff. Ingenieurbiologie in Praxis und Umwelt. Patzer Verlag, Berlin, 319 pp.
Schönhuth M and Kievelitz U 1994 Participatory Learning Approaches. Rapid Rural Appraisal. An introductory guide. Gesellschaft für Technische Zusammenarbeit (GTZ). TZ-Verlagsgesellschaft, Eschborn, 183 pp.
Singh I L 1985 Rainfall distribution. In TC Majupuria (eds.) Nepal-Nature's Paradise. Insight into Diverse Facets of Topography, Flora and Ecology. pp. 56–58. White Lotus, Bangkok.
Storrs A and Storrs J 1998 Trees and Shrubs of Nepal and the Himalayas, 367 pp. Book Faith India, Delhi.
Upreti B N 2001 The physiography and geology of Nepal and their bearing on the landslide problem. In L Tianchi, SR Chalise and BN Upreti (eds). Landslide Hazard Mitigation in the Hindu Kush-Himalayas. pp. 31–49. International Centre for Integrated Mountain Development (ICIMOD), Kathmandu.
Wagner M 2000 Slope Stability Mapping of the Chandistan Sub-Watershed, 87 pp. University of Natural Resources and Applied Life Sciences, Vienna, Austria.
Weigel G, Shrestha R B, Meyer W P and Berg C 1987 Vegetative Soil Conservation Measures: A Field Manual. Vol. 1. Soil and Watershed Conservation Section Tansen, Palpa, 153 pp.

Beech coppice short-term hydrological balance for simulated rainfall

P. Trucchi[1] & M.C. Andrenelli

Department of Agriculture and Forestry Engineering, University of Florence, Via S. Bonaventura 13, 50145 Florence, Italy. [1]*Corresponding author**

Key words: *Fagus sylvatica* L., rainfall simulator, runoff, stemflow, throughfall, water balance

Abstract

In an experimental beech (*Fagus sylvatica* L.) coppice plot, throughfall, stemflow and superficial runoff were monitored on a single event basis, in order to verify the short term hydrological balance in different conditions of leaf cover. Rainfall events were simulated using a rainfall simulator, operated at two different intensities. The plot was hydraulically isolated, along its whole perimeter and a small drain, placed in the lowest part of the plot, was used to collect the superficial runoff. Twelve rain gauges were used to survey the throughfall whereas the stemflow was collected through rubber collars. Superficial flows were measured both by sampling the discharges coming out of the plot and by collecting the total water runoff volumes.

The results of the trials underline the very high spatial throughfall variability and the difficulty to evaluate the throughfall itself at the required accuracy. The superficial runoff was very small or practically nil in all the trials and was due to the very high infiltration capacity and hydraulic conductivity of the forest soil.

Introduction

It is well known that in forested basins, total discharges, storm-flow volumes and peak discharges are generally smaller than in adjacent non-forested areas. Forest cover, soil and litter characteristics, interception and evapotranspiration phenomena greatly modify the water balance. For more than 13 years, gross-rainfall (P), throughfall (T), stemflow (S) and superficial runoff (Q) were monitored in a pure beech (*Fagus sylvatica* L.) coppice of medium density and variable age in order to study the forest cover water balance (Giacomin and Trucchi, 1992). During this time, the runoff volumes, cumulated only over the period April–November, in order to avoid snow precipitation, never exceeded 2.5% of the corresponding P which reached an average value of 700 mm/year. Investigation of the relationships between P and Q, on a single event basis, was not possible due to the lack of precipitations of very high intensity (Falciai et al., 2002). For this reason, a rainfall-simulator was built to carry out trials in an experimental plot under controlled conditions of gross-rainfall intensity and different degrees of leaf cover.

Although many researchers have investigated P, T, S and Q relationships in different canopy cover types and under natural rainfall events (Crockford and Richardson, 1990; Ford and Deans, 1978; Giacomin et al., 2003; Johnson, 1990; Loustau et al., 1992; Marukata et al., 1990; Neal et al., 1993; Sinun et al., 1992; Tobon Marin et al., 2000), the research illustrated in this paper aims to study the water balance and the partition of P in T, S and Q as influenced by the leaf cover for single simulated rainfall of very high intensity.

Materials and methods

Experimental plot

The experimental plot was located in the Acquerino forest, Prato province, Italy. The plot had an area of 39.5 m^2, was of rectangular shape (see Figure 1) and its length (10 m) was parallel to the maximum slope (average slope being 60%). The plot was hydraulically

* E-mail: paolo.trucchi@unifi.it

Stokes et al. (eds), Eco- and Ground Bio-Engineering: The Use of Vegetation to Improve Slope Stability, 321–327.

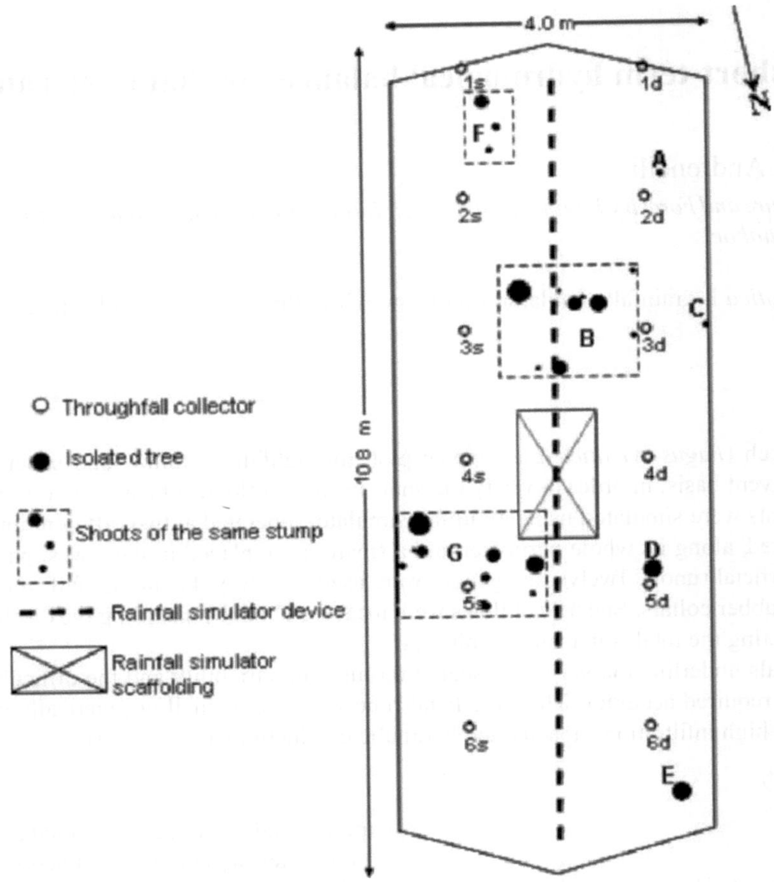

Figure 1. Experimental plot and setup.

isolated, along its whole perimeter, from external surrounding areas, through plate sheets driven 0.1 m into the ground, in order to avoid superficial water exchanges. On the downhill side of the plot, a small drain was placed in order to collect the superficial runoff. As the runoff was to be measured both in discharge terms and in volumetric terms, a joint between the drain and a pipe was made to carry the water to a reservoir. From the geological point of view, the plot was located on the side of a large outcrop of a turbiditic oligocenic formation (Macigno) which was formed by thick banks of quartz-feldspathic and chloritic sandstone alternated with thin silty and argillitic strata.

A soil survey carried out on a profile dug immediately below the plot, highlighted a soil with a considerable coarse fragment content (from 30 to 50%), highly disturbed by the beech root system and characterized by a high macroporosity which promotes water movement. A subangular polyhedric structure, along the whole soil profile, was evidence of its good aggregation and structure stability. The first 0.3 m was identified as an eluvial horizon (A) which is characterized by a loam texture and high permeability. The underlying layer, 0.6 m thick, is an illuvial silty-loam horizon (Bw) of medium permeability. The basal layer (BC), at a depth of 1.3 m, had a clear loam texture, and an increase in the permeability and stoniness was also found.

In order to evaluate the soil infiltration characteristics, infiltration tests were performed outside the plot (the plot was not to be disturbed) by using a single ring infiltrometer (357 mm of diameter), operating with a water head of 50 mm, because of the slope in that area. The infiltration capacity ranged from 600 to 1200 mm h^{-1}. The saturated hydraulic conductivity, as measured in the laboratory, was conducted with a head of water of 50 mm and using stainless steel sampling core rings of 50 mm diameter, reaching 90 mm h^{-1}. The differences

Table 1. Dendrometric survey of the experimental plot

Isolated trees or stumps (label)	Number of shoots or plants	Shoot or plant diameter (mm)	Shoot or plant height (m)
A	1	90	6.5
B	9	250	13.5
		180	11.0
		160	7.5
		150	8.2
		50	5.0
		50	7.0
		40	4.5
		30	3.5
		30	3.5
C	1	60	6.5
D	1	200	17.5
E	1	200	15.0
F	3	140	13.0
		70	7.0
		60	5.0
G	9	240	14.5
		190	15.0
		150	8.0
		120	9.0
		90	4.5
		90	4.0
		60	4.2
		50	3.5
		40	4.2

between field and laboratory results were mainly due to the lack of macro pores within the samples used for the hydraulic conductivity measurements.

A dendrometric survey was performed inside the plot in order to measure the trunk diameters (above 25 mm) and the heights of all the plants (at 0.7 m from ground level), both shoots and isolated trees. The canopy cover, tree crown dimensions, heights and diameters of the plants, were determined on four isolated plants and three stumps (see Figure 1 for tree positions). Heights and diameters of the trees ranged respectively from 3.5 to 13.5 m and from 30 to 250 mm (Table 1). Each tree crown developed vertically at different levels and overlapped and intersected with adjacent ones forming a complex branch texture. Very few areas of the plot remained uncovered and unprotected by the canopy when the leaf cover reached its maximum.

Rainfall simulator

The rainfall simulator consisted of two 6-m joint steel pipes. It was placed at 16 m from the ground, over a scaffolding, at the centre of the experimental plot along its maximum slope (Figure 1). The water was conveyed to the pipes, through a rubber reinforced hose and a high-density polyethylene pipe which was directly connected with the pumping system necessary to ensure the required discharge and pressure. The simulator was formed by 24 full cone nozzles (model 10PCF of Toro Manufacturing, USA), generally utilized for the irrigation of turf, which were selected because of their hydraulic characteristics (pressure-discharge relation, wetted areas, application rate) and for the possibility of mounting a pressure regulating device, useful for the conditions under which the nozzle should have operated (different pressure head values for each nozzle due to the head losses along the pipe and to the elevation the nozzles themselves were placed). The nozzles, with the jet pointed downward, were mounted along the lower face of the two pipes, 0.5 m apart. Varying the pressure between 50 and 150 kPa, the nozzle range reached respectively 3.4 and 3.8 m, which was enough to cover the whole width of the experimental plot, with the requested rainfall intensity.

The performances of the rainfall simulator were evaluated verifying the spatial distribution of the rainfall and its uniformity at the two previous operating pressure heads. These tests were performed employing a rainfall simulator similar to the one described above but only 4 m long, as this length was sufficient to obtain the total overlapping of the wetted area produced by each nozzle in the centre of the simulator pipe. This "test simulator" was provided with 8 nozzles, and in order to ensure the drop falling vertically to the ground, a height of 6 m was estimated sufficient. A different series of measurements was carried out, with all eight nozzles working and a single one, with the aim to verify the interaction between the nozzle jets and to obtain a mathematical relationship between rainfall intensity and distance from the nozzle. According to this relationship, it was possible to draw the rainfall intensity isohyets relative to the experimental plot and to estimate the mean rainfall intensity for the working pressure heads as already mentioned. The average rainfall intensity was 98 mm h^{-1} with 150 kPa and 78 mm $h^{-1}$with 50 kPa.

A uniformity index concerning the spatial water distribution on the plot was also calculated. The uniformity coefficient, obtained using the Christiansen's formula (Christiansen, 1942), reached 90% and 72%, respectively with the highest and the lowest nozzle working pressure values. The rainfall intensity values mentioned above can be obtained taking into account the maximum rainfall recorded at the rain-gauge stations located not far from the experimental site. For a return period of 100 years and according to the Gumbel distribution, the 30-min rainfall height can be set equal to 50 mm ($i = 65\,t^{-0.59}$; i is the rainfall intensity in millimetre per hour and t is the rainfall duration in hours).

Gross-rainfall, throughfall, stemflow and superficial runoff measurement devices

As the rainfall simulator served an area bigger than the plot, the plot gross-rainfall was obtained by subtracting the runoff volume outside the plot (intercepted and collected by a polyethylene sheet laid on the ground outside the plot) from the one distributed by the simulator (measured with a water meter). Twelve rain gauges of 0.01 m^2, placed 0.7 m above ground level, were used to survey T on two alignments arranged along the maximum slope (see Figure 1 for T collectors position). These alignments were set 2 m apart and the collectors 1.6 m one from the other (over an area of 3.3 m^2). Sampling T in different positions (symmetrically arranged on two alignments with respect to the simulator pipe) allowed the evaluation of the T spatial variability mainly due to the canopy cover distribution. A network was not arranged at the beginning of the research to measure S, because the runoff might be influenced by the concentrated inflow of S. After having assessed that S did not influence Q, rubber collars fixed tightly around the trunks were used to collect S on all the shoots of each stump and on all the isolated trees. The runoff was measured both by sampling the discharges coming out of the plot and by collecting the total runoff volumes in a reservoir.

Results and discussion

Having monitored P, T, S and Q, it was possible to verify the variations and the relationships existing between the above mentioned parameters for several trials carried out from October 2001 to May 2003 in different leaf cover conditions. The leaf cover was set at 100% when the leaf index reached the maximum (leaves completely developed). For each trial, the degree of leaf cover was estimated in comparison with the maximum one by counting the number of leaves on several branches taken as a reference.

Figure 2 shows the T patterns surveyed respectively at 50 and 150 kPa operating pressure heads. T showed a pronounced spatial variability among the 12 positions. For each operating pressure head, T patterns were similar (in the performed trials); they were characterized by minima and maxima which generally occur in the same positions and this can be explained with the different local leaf cover and or dissimilar branch slope and architecture among the collecting points.

In the collectors close to an isolated tree or to a stump, where a concentration of rain occurring below the crown was expected, high values of T are not always recorded and low T values do not correspond to maxima of S and vice versa. For example, the T of rain-gauge 2d shows low values as the ones of S, recorded for tree A. In this case T and S values were influenced by stump B which records, on the contrary, high values for both S and T. Moreover, very close to the isolated tree D, T assumes another maximum (collector 5d) as does S, with reference to the same tree. Plant D, was very tall and presents vertically oriented branches but did not have a large crown. In these conditions a T as well as a dramatic S was found.

The medium percentage ratio T/P reached 60% over the plot (T for the entire plot is the average intensity recorded by the 12 gauges which collected the cumulated T volume in each trial), with 30% being the minimum and 70% the maximum. Only in one case (collector 6d on September '02 at a pressure head of 0.5 kPa) a value of T higher than the corresponding P had been recorded ($T = 103$ mm h^{-1}, $P = 87.2$ mm h^{-1}) and it is supposed that the redistribution of P through the canopy is the main factor of the concentration of flow below the tree crown. Nearly always T showed higher values with a higher operating pressure head when compared to the lower one.

S values for the different trials are shown in Figure 3, however, S flows of the isolated trees A and C are not reported because of their negligible values. Due to the low number of trees in the plot, evaluating from a statistical point of view the influence that other vegetative characteristics (tree height, number and disposition of branches etc.) can have on S, is not appropriate. Nevertheless, it can be noticed that trees having the same trunk diameter and/or height may have dissimilar S values due to a different number and architecture of branches as in trees D and E. Therefore, the entity of S, referred in particular to the isolated plants, is more related to their morphology than to the different degree of leaf cover. Consequently, plants with trunks of small diameter and crown have a S which is practically independent from the presence of leaves in contrast to the

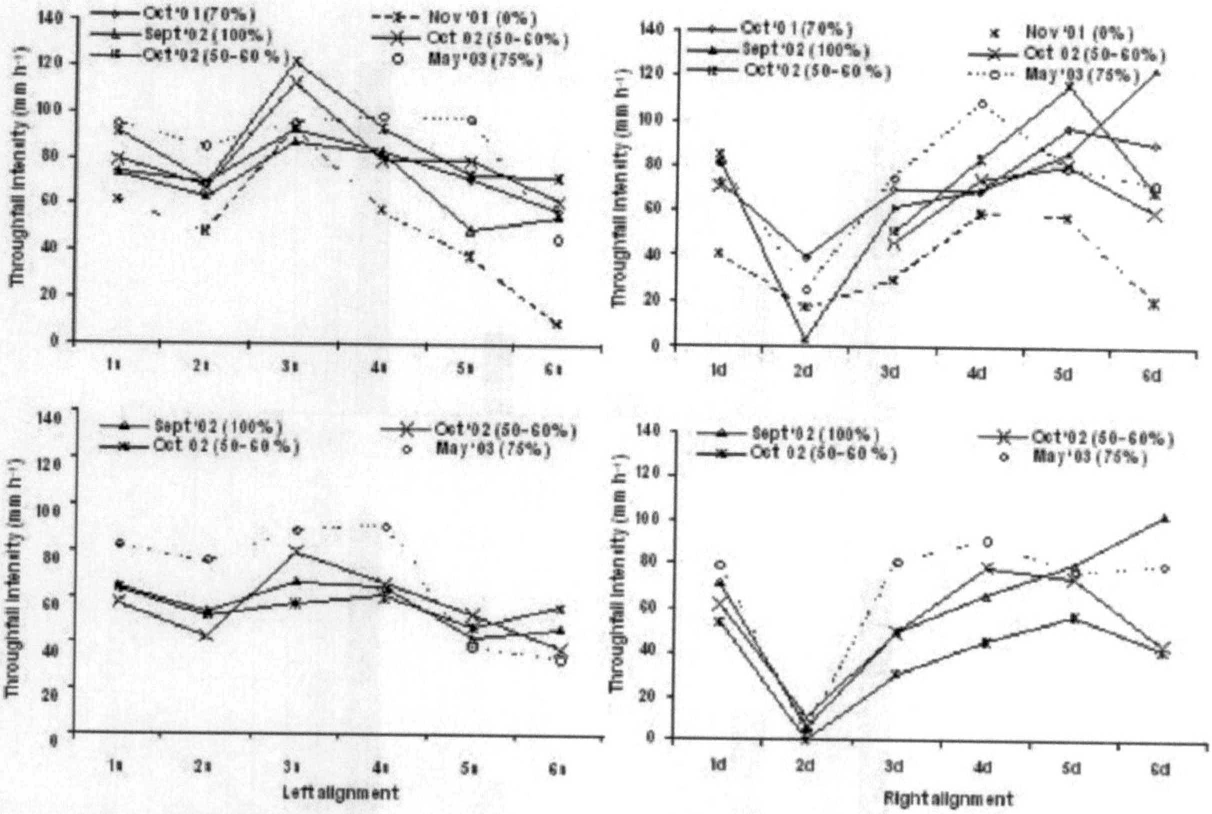

Figure 2. Throughfall patterns surveyed at 150 kPa (above) and at 50 kPa (below) rainfall simulator operating pressure heads for different degrees of leaf cover expressed in percentage.

behavior of the stumps. S varied from a maximum of 8 l min^{-1} (stump G) to a minimum of 0.3 l min^{-1} (isolated tree E) and was practically independent from the simulator operating pressure head. In terms of intensity and taking into account the experimental plot area, the total S varied from a minimum of 17.8 mm h^{-1} (May '03 at 0.5 kPa) to a maximum of 31.8 mm h^{-1} (Sept '02 at 1.5 kPa) and represented 16% and 28% of the corresponding P respectively.

The superficial runoff was always very small or practically nil in all the trials when compared with P. The different degree of leaf cover had no influence on Q. Certainly, the high soil thickness, macroporosity and soil infiltration capacity facilitated both the vertical water movement and the possibility for the soil to hold high water volumes. The water balance referred to in the plot can be written as: $Qi - E = T + S$, where Qi represents the whole water volume distributed by the rain simulator, measured with a water meter, E is the water volume which fell out of the plot on the polyethylene sheet ($Qi - E = P$ and then $P = T + S$), T and S are the already mentioned parameters. The above relationship is correct if the interception is set to be zero. This assumption is quite realistic considering the plot rainfall intensity and the duration of the trials which lasted for more than 30 min.

In Figure 4, T (at plot scale), S and P, for the two rainfall simulator operating pressures of 150 kPa and 50 kPa are reported. The different values of $T + S$ with P depend on how precisely the water balance parameters themselves are gauged. The accuracy of the estimate is related, on the other hand, to the way each parameter is recorded and measured. It must be remembered that Qi is measured through a water meter, E is collected in a reservoir, S is surveyed on all the trees in the plot, T is sampled using twelve collectors. Furthermore, beyond the accuracy of the measurements, the relative weight of each parameter on Qi is relevant. With regards to T and to its high variability, the confidence interval ($p = 0.95$) for each trial can be

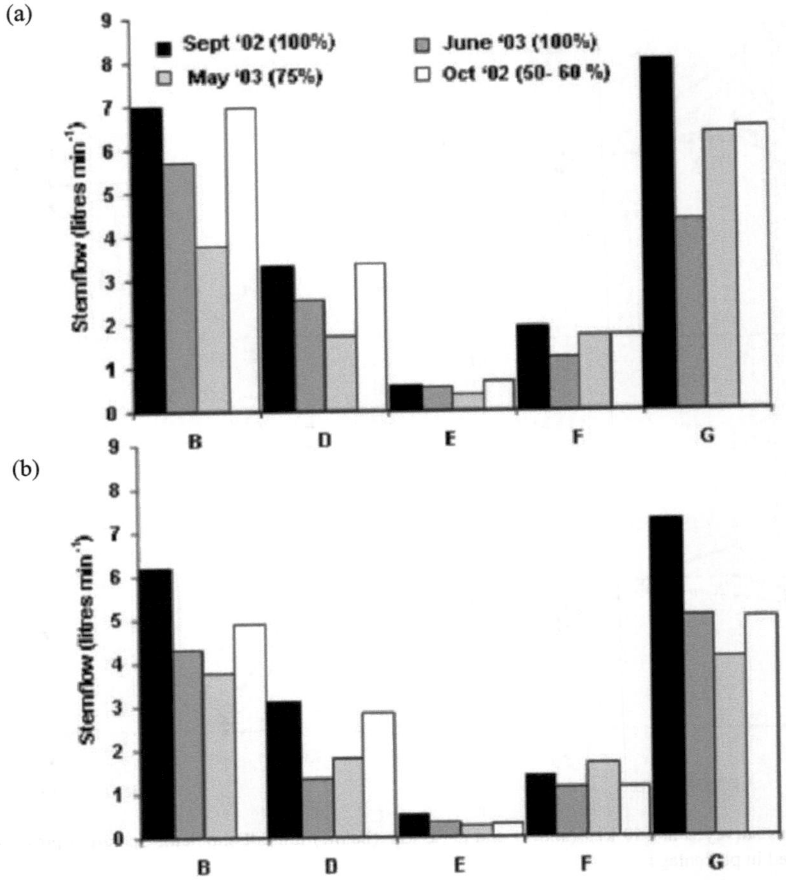

Figure 3. Stemflow values recorded at 150 kPa (above) and at 50 kPa (below) rainfall simulator operating pressure heads for different degrees of leaf cover expressed in percentage.

calculated. With these intervals, the water balance can be re-evaluated. As an underestimation of T is evident in all the trials (Figure 4), the upper confidence limit causes an improvement of the water balance.

In the trials carried out at 150 kPa pressure head, in two cases (Oct '01 and Nov '01) out of five, the sum $T + S$ equals P; in one case (Sept '02) adding the confidence interval to $T + S$ makes P equal; in the remaining cases (Oct '02 and May '03) the relation $T + S = P$ does not balance and this means that an underestimation of T probably occurred or at least, errors in the measurement of the other hydrological parameters were made. With a pressure head of a 50 kPa, in two trials (Sept '02; May '03) out of three, the sum $T + S$ practically equals P; in the remaining case (Oct '02) the confidence interval did not compensate $T + S$ and reach P. If the weight of T is compared with P, it is considerable (60% on average) this is the main reason why an erroneous estimate of T causes errors in the water balance.

Conclusion

The results of the research described above show that in forested areas, even with precipitation of very high intensity, soil characteristics influenced by the forest may prevent or obstruct the formation of a superficial runoff. The forest cover promoted the delayed flow and produced a lengthening of the times of concentration. Since in a forest environment the parameters which characterize the water balance are usually sampled quite generally, a correct estimate of these appears to be crucial (Lloyd et al., 1988; Peterson and Rolfe,

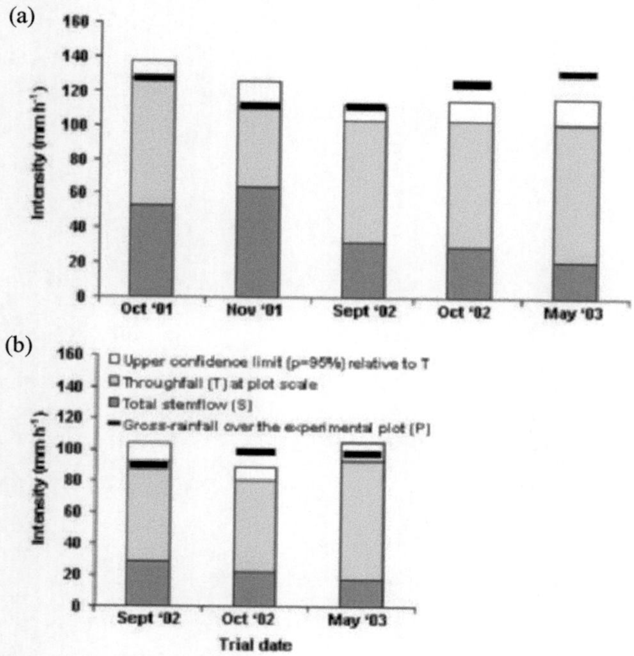

Figure 4. Water balance terms at 150 kPa (above) and at 50 kPa (below) rainfall simulator operating pressure heads.

1979; Puchett, 1990). Nevertheless, from our results, having measured P through a water meter, collected S from all the trees and sampled only T, we show that T is the parameter which is the most difficult to be gauged with the required accuracy.

References

Christiansen J E 1942 Irrigation by Sprinkling. Bull. 670, Agricultural Exp. Station, University of California, Berkley.

Crockford R H and Richardson P 1990 Partitioning of rainfall in a Eucalypt forest and pine plantation in south-eastern Australia: II. Stemflow and factors affecting stemflow in a dry scerophyll eucalypt forest and *Pinus radiata* plantation. Hydrol. Process. 4, 145–155.

Falciai M, Ghinassi G and Trucchi P 2002 Deflussi superficiali stagionali da un bosco ceduo di faggio sottoposto a trattamenti selvicolturali differenziati. Quaderni di Idronomia Montana 18, 95–100.

Ford E D and Deans J D 1978 The effects of canopy structure on stemflow, throughfall and interception loss in a young Sitka Spruce plantation. J. Appl. Ecol. 15, 905–917.

Giacomin A, Trucchi P and Andrenelli M C 2003 Intercettazione e deflussi superficiali in un bosco ceduo di faggio. *In* Proc. on "La difesa della montagna", dicembre 11–12, 2002, Assisi, pp. 490–494.

Giacomin A and Trucchi P 1992 Rainfall interception in a beech coppice (Acquerino, Italy). J. Hydrol. 137, 141–147.

Johnson R C 1990 The interception, throughfall and stemflow in a forest in Highland Scotland and the comparison with other upland forests in the UK. J. Hydrol. 118, 281–287.

Lloyd C R and Marques F de O A 1988 Spatial variability of throughfall and stemflow measurements in Amazonian rain forest. Agr. Forest Meteorol. 42, 63–73.

Loustau D, Berbigier P, Grainier A and El Hadj Moussa F 1992 Interception loss, throughfall and stemflow in a maritime pine stand: I Variability of throughfall and stemflow beneath the pine canopy. J. Hydrol. 138, 449–467.

Marukata H, Ando M and Ogawa H 1990 Throughfall, stemflow and interception of rainwater in a evergreen broadleaved forest. Ecol. Res. 5, 303–316.

Neal C, Robson A J, Bharwaj C L, Conway T, Jeffery H A, Neal M, Ryland G P, Smith C J and Walls J 1993 Relationship between precipitation, stemflow and throughfall for a lowland beech plantation, Black Wood, Hampshire, southern England: findings on interception at a forest edge and effect of storm damage. J. Hydrol. 146, 221–233.

Peterson D L and Rolfe G L 1979 Determining sample size in throughfall studies. For. Sci. 25, 582–584.

Puchett L J 1990 Spatial variability and collector requirements for sampling throughfall volume and chemistry under a mixed-hardwood canopy. Can. J. For. Res. 21, 1581–1588.

Sinun W, Meng W W, Douglas I and Spencer T 1992 Throughfall, stemflow overland flow and throughflow in the Ulu Segama rainforest, Sabah, Malaysia. Philos. T. Roy. Soc. B. 335, 389–395.

Tobon Marin C, Bouten W and Sevink J 2000 Gross-rainfall and its partitioning into throughfall, stemflow and evaporation of intercepted water in four forest ecosystem in western Amazonia. J. Hydrol. 237, 40–57.

Beech coppice leaf cover and gross rainfall quali–quantitative transformation in simulated rainfall events of high intensity

P. Trucchi[1] & M.C. Andrenelli

Department of Agriculture and Forestry Engineering, University of Florence, Via S. Bonaventura 13, 50145 Florence, Italy. [1]*Corresponding author**

Key words: drop-size distribution, forest cover, rainfall simulator, throughfall

Abstract

Rainfall simulators have been developed as tools for evaluating hydrological and erosive processes due to agronomical activities and have been very rarely used in forested areas. When a rainfall simulator is operating under forest cover, the characteristics of this peculiar environment (high trees, irregular topography, steep slopes, difficulties in supplying water) as well as the gross-rainfall transformation and redistribution on soil surface (as rain is filtered through the canopy) have to be taken into account.

The authors illustrate the methods employed during the rainfall simulator calibration procedures and the sprinkling device characteristics. The aspects concerning the uniformity coefficient of the gross-rainfall intensities as well as those regarding the distribution of different classes of raindrop diameters, produced by the rainfall simulator, are examined in detail. The "kinetic" characteristics of the simulated rainfall are also illustrated, though in a forest environment this aspect is not of primary importance because the canopy interception strongly modifies both the diameters and the trajectories of the drops and furthermore, forest litter is able to dissipate the kinetic energy of the splashing raindrops.

A model reproducing the kinematics of the raindrop was developed by the authors: this calculates the impact velocity of a drop at ground level as a function of its diameter, falling height, initial velocity and exit angle of the water jet emitted by the nozzles. In addition, the results obtained in field trials, carried out in an old beech coppice for different seasons, are reported and discussed.

The role played by the tree canopy cover in various vegetative conditions is responsible for the differences occurring between the gross-rainfall characteristics (intensities and drop diameters) and those of the corresponding throughfall. Again, these qualitative-quantitative transformations, operated by the canopy cover, produce diverse erosivity values between gross rainfall and throughfall, as a consequence of different intensities, raindrop diameters and velocities. The results obtained in this research indicate the importance that methods concerning erosion assessment should deal with both leaf cover degree on the whole and vegetative characteristics (arboreal or herbaceous cover).

Introduction

When rain falls on a forest canopy it is redistributed according to a random pattern as it moves toward the forest floor; this phenomenon concerns a twofold transformation which involves qualitative and quantitative aspects. The qualitative modification is related to the drop-size distribution of the throughfall compared with the rain incident, whereas the quantitative transformation is associated with changes in throughfall height distribution at ground level, with regard to the corresponding gross rainfall.

Several attempts were made to understand the qualitative aspect, among them Brandt's hypothesis (Brandt, 1989) which has been employed in many hydrological models (Brandt, 1990; Morgan, 2001). According

* E-mail: paolo.trucchi@unifi.it

to Brandt (1989), the raindrops of the throughfall are composed of two components: clear and intercepted throughfall. The former has the same distribution as the gross rainfall and hence it can be modelled by known laws which relate the intensity of the rain to the drop-size distribution (Hudson, 1981; Laws and Parson, 1943; Zanchi and Torri, 1980). In the latter case, the component is always characterized by a normal distribution with a mean between 4.52 and 4.95 mm and a standard deviation between 0.79 and 1.30 mm. This distribution is related neither to the canopy type nor to the rainfall intensity. No further research with regard to the qualitative features was carried out, whereas the quantitative transformation was investigated in detail by several studies which dealt with sampling design for evaluating hydrological balance (Chen et al., 1995; Giacomin and Trucchi, 1992; Kimmins, 1973). Other studies outlined that partitioning between stemflow and throughfall depends upon rainfall intensities and tree canopy architecture (Li et al., 1997; Qingfu Xiao et al., 2000).

The present research started at the beginning of the eighties with the aim of studying the hydrological balance by monitoring gross rainfall (P), throughfall (T), stemflow (S) and runoff (Q) in an experimental area located in the Acquerino forest (Tuscan Apennines, Italy) from 1981 to 1994 (Giacomin et al., 2003). The forest cover is a pure beech coppice (*Fagus sylvatica* L.) of medium density and variable age (Giacomin and Trucchi, 1992). During these experimental trials, the runoff volumes, cumulated over the months from April to November (in order to avoid snow precipitation), never exceeded 2.5% of the corresponding gross rainfall which reached an average amount of 700 mm year^{-1}. Thus, it was not possible to investigate the relationship between P and Q on a single event basis due to the lack of very high intensity rainfall events (Falciai et al., 2002). Therefore, a rainfall simulator was built to carry out trials under controlled conditions (different degrees of leaf and soil litter cover), which is the purpose of the present paper. This simulator had to simulate rainfall events with intensity up to 100 mm h^{-1} at a constant rate. This intensity value, which was particularly high, corresponds to the maximum rainfall amount of 100 years return period and 30-min duration recordable by a rain gauge located near the experimental site. The high acclivity (more than 60%) and the rough morphology of the experimental site, the canopy density, tree heights and dimensions are the main factors which influenced the type of simulator to be built. Moreover, as the sprinkler system itself had to be placed above the tree crowns, it also had to be easy to assemble, in order not to damage the vegetation and the soil during its installation. Subsequently the calibration procedure of the rainfall simulator is described, as well as field trials performed either to investigate the hydrological balance concerning P, S, T and Q, or to verify the quali-quantitative throughfall transformation depending upon gross-rainfall intensity and leaf cover degree.

Material and methods

Rainfall simulator structure

The rainfall simulator, made up of two 6 m joint steel pipes, was placed on scaffolding at 16 m from the ground, at the centre of the experimental plot along its maximum slope. The surface of 39.5 m^2 had a nearly rectangular shape (Figure 1). The water was conveyed to two pipes through a rubber reinforced hose and a high density polyethylene pipe was directly connected to the pumping system, which ensured the required discharge and pressure. The water distribution device was composed of full cone nozzles (model 10PCF of Toro Manufacturing, USA) which were selected because of their hydraulic characteristics (pressure-discharge relation, wetted areas, application rate) and for the availability of a pressure regulating device, useful to ensure the same performance in all the nozzles. Twenty-four nozzles pointed downward were mounted along the lower face of the pipe, 0.5 m apart, and were subjected to different hydraulic loads due to both pressure losses and different elevations of the nozzles. The number and distance between nozzles were able to guarantee an intensity of about 100 mm h^{-1} at a water pressure head of 150 kPa. Varying the pressure between 50 and 150 kPa, the nozzle jets reached 3.4 and 3.8 m respectively. These values ensured the complete coverage of the experimental area.

Rainfall simulator calibration

The performance of the rainfall simulator at two different operating pressure heads (50 and 150 kPa) was evaluated by verifying the uniformity of the rainfall spatial distribution and determining the drop diameters and impact velocities at ground level. All these tests were performed employing a rainfall simulator,

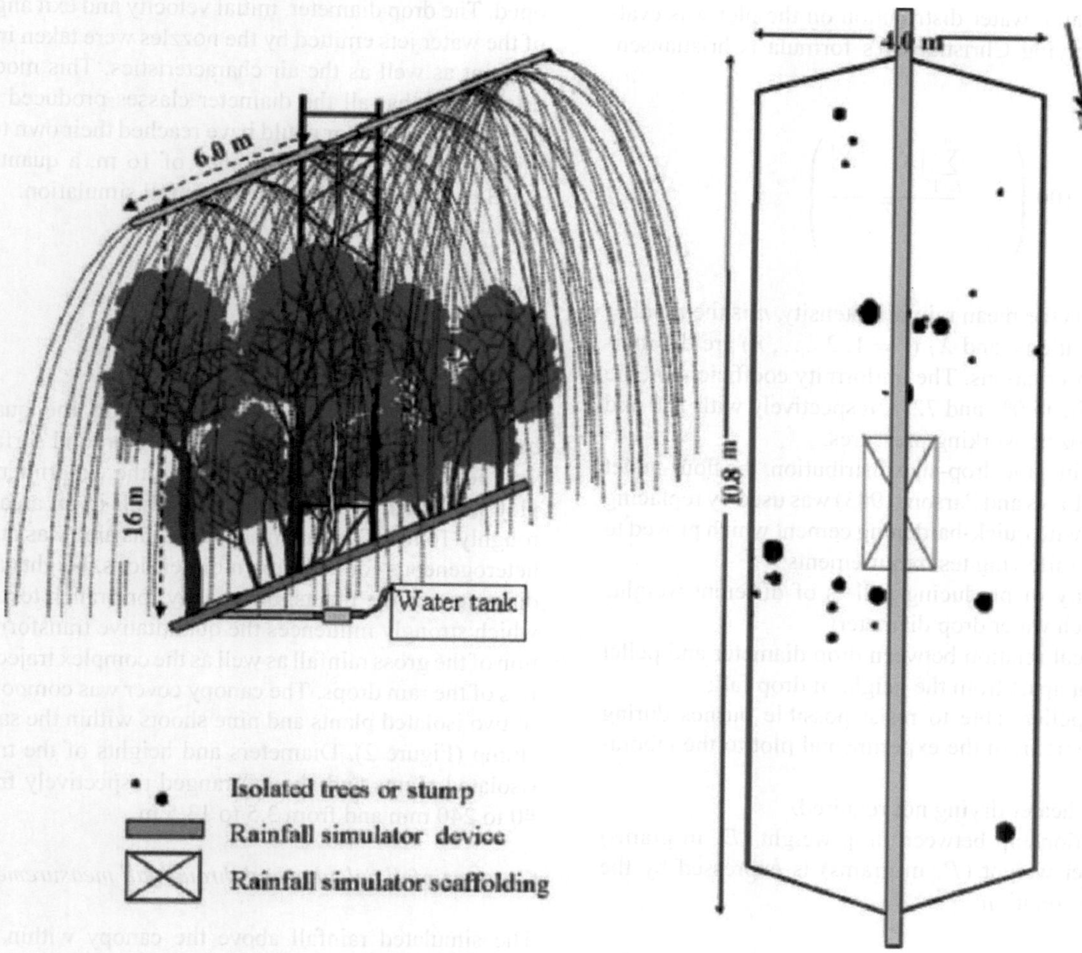

Figure 1. Experimental plot and lay-out of the rainfall simulator.

only 4 m long, similar to that described above, which was able to guarantee the total overlapping of the nozzle jets in the central part of the simulator pipe. This "test device," having only eight nozzles, was placed at a height of 8 m to let the drops reach the ground vertically.

The spatial distribution of the rainfall was evaluated through a sampling grid. Circular rainfall collectors of 870 mm diameter were arranged on the grid interception points. These pluviometers were placed 0.2 m apart along six lines parallel to the simulator and along another line set at right angles to the pipe, between the two central nozzles. Several measurements were carried out, with all eight nozzles working and again with a single nozzle, to verify the interaction among the jets and to obtain a mathematical relationship ($r^2 = 0.999$) between rainfall intensity and distance from the nozzle expressed by the following equation:

$$I = -3.659\ 10^{-11}\mathrm{d}^5 + 3.565\ 10^{-8}\mathrm{d}^4 - 1.082\ 10^{-5}\mathrm{d}^3 + 7.625\ 10^{-4}\mathrm{d}^2 + 0.053\mathrm{d} + 9.815 \qquad (1)$$

where I is the rainfall intensity (millimeter per hour) and d is the distance from the nozzle (in centimeters).

Based on Eq. (1) the rainfall intensity isohyets for the experimental plot (with all 24 nozzles) can be drawn and the mean rainfall intensity for the two working pressure heads estimated. The average rainfall intensity was 98 mm h^{-1} with 150 kPa and 78 mm h^{-1} with 50 kPa.

The spatial water distribution on the plot was evaluated applying Christiansen's formula (Christiansen, 1942)

$$UC = 100\left(1 - \frac{\sum_{i=1}^{n}|X_i - \overline{X}|}{\overline{X}}\right) \qquad (2)$$

where $\bar{X}$ is the mean rainfall intensity, n is the number of observations, and X_i $(i = 1, 2, \ldots, n)$ are the individual observations. The uniformity coefficient on the plot reached 90% and 72%, respectively with 150 and 50 kPa nozzle working pressures.

Regarding the drop-size distribution, the flour-pellet method (Laws and Parson, 1943) was used by replacing the flour with quick-hardening cement which proved to meet the following test requirements:

1. capacity of producing pellets of different weights for each water drop diameter;
2. univocal relation between drop diameter and pellet weight apart from the height of drop fall;
3. hard pellets able to resist possible pushes during transport from the experimental plot to the laboratory;
4. pellet heater drying not required.

The relationship between drop weight (D, in grams) and pellet weight (P, in grams) is expressed by the following relation:

$$D = 0.360P \quad (r^2 = 0.947; n = 15; p < 0.001) \qquad (3)$$

Trays of 0.25 × 0.45 m with a bed of cement 25 mm thick were exposed to simulated rain in different positions on a sampling grid similar to the previous one used for determining the water height distribution. The drop-size distribution on each tray, i.e. for each distance from the simulator test, was evaluated given the frequencies of each diameter class of the drops and the rainfall intensity. The drop diameters ranged from 0.57 to 1.46 mm at a nozzle pressure of 150 kPa and from 0.71 to 3.89 mm at 50 kPa. The median diameters (D_{50}) reached 1.05 and 1.43 mm, respectively for the highest and the lowest pressure value, over the complete overlapping area of the nozzle jets.

To evaluate the terminal velocity of different drops, a mathematical model, based on the Runge-Kutta algorithm for solving differential equations, and able to reproduce the kinematics of the raindrops, was developed. The drop diameter, initial velocity and exit angle of the water jets emitted by the nozzles were taken into account as well as the air characteristics. This model established that all the diameter classes produced by the rainfall simulator could have reached their own terminal velocity within a distance of 16 m, a quantity equal to the height of the field rainfall simulation.

Results and discussion

Characteristics of the experimental plot

The evaluation of the canopy effect on the quali-quantitative redistribution of P on the ground surface was performed on the lower part of the experimental plot subjected to the simulated rainfall, in an area of roughly 16 m^2. The canopy cover in this area was quite heterogeneous for tree crown dimensions, heights and diameters of the plants, but mostly for branch texture which strongly influences the quantitative transformation of the gross rainfall as well as the complex trajectories of the rain drops. The canopy cover was composed of two isolated plants and nine shoots within the same stump (Figure 2). Diameters and heights of the trees (isolated plants and shoots) ranged respectively from 40 to 240 mm and from 3.5 to 13.5 m.

Gross-rainfall calculus and throughfall measurement

The simulated rainfall above the canopy within the smaller experimental area was rebuilt overlaying the contribution of each nozzle previously described from Eq. (1); in addition the throughfall was measured through nine rain gauges of 0.01 m^2, placed 0.7 m above ground level and arranged along three rows, 1.5 m apart. The cross distance between the collectors was 1 m (Figure 3). In the same nine positions, trays with quick-hardening cement beds were placed to collect the drops in order to estimate the drop-size distribution once canopy saturation was reached. Both tests (quantitative and qualitative) were performed in nearly windless conditions.

Field experimental rainfall simulation

In September and October 2002, two simulated rainfall trials were performed with nozzle pressure head values of 50 and 150 kPa. These months indicate different leaf cover conditions: the leaf cover was set at 100% when

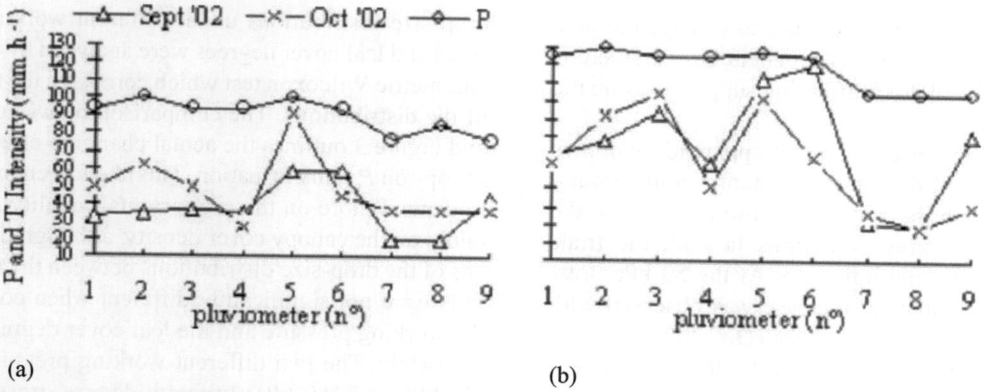

Figure 2. Gross-rainfall and throughfall intensities with two different degrees of leaf cover (Sept. and Oct. 2002) and for two water pressure head values (150 kPa on the right side and 50 kPa on the left side).

the leaf index reached the maximum amount (leaves completely developed) and this condition occurred on the trial performed in September. However, for both trials the degree of leaf cover was estimated by comparing the number of leaves on several sample branches taken as a reference. In Figure 2, P and T intensities for both tests are shown, whereas in Figure 3 the spatial distribution of gross-rainfall intensity at the previous pressure values is plotted. Figure 2 shows that P was always greater than T for both trials, and the different amounts between these hydrological parameters are clearly expressed by S values, according to the hydrological balance relation $P = T + S$. T trends in both tests were similar among the pluviometers, with two maxima in collectors 5 and 6 and as many minima in positions 7 and 8.

The P distribution among the rain collectors was very heterogeneous, especially at 150 kPa along the lengthwise direction, as was clearly stressed by the isohyets pattern (Figure 3). This was due to the boundary position of this experimental area when compared to the position of the simulator pipe. Nevertheless, this

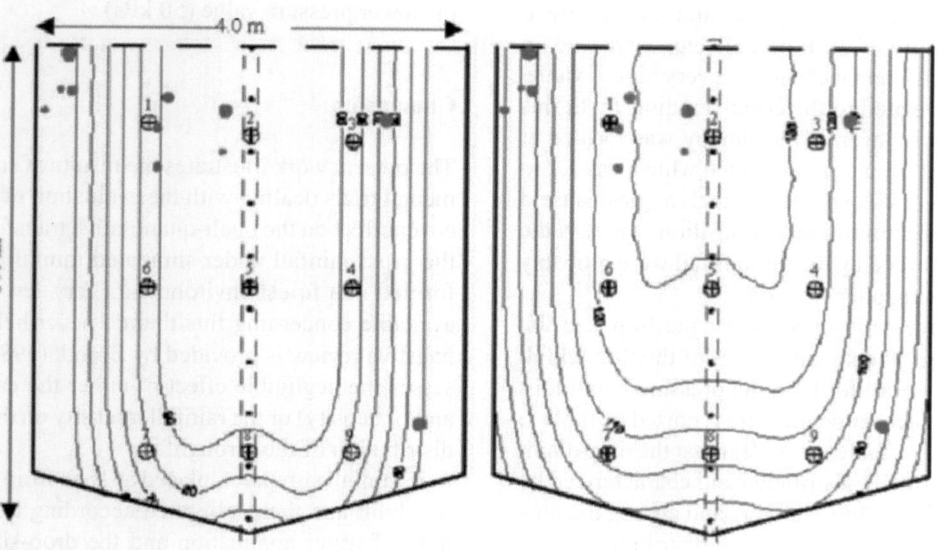

Figure 3. Spatial distribution of simulated gross-rainfall intensities (intensity expressed in millimetre per hour) for nozzle pressure head of 50 kPa (left) and 150 kPa (right). Trunks are represented with a solid circle and pluviometers with an empty crossed circle labelled with a number.

zone was the only available place to arrange the nine pluviometers and to collect the pellets, since numerous trees and the simulator scaffolding support took up the rest of the plot area.

The seasonal leaf cover factor apparently modifies the distribution of the rainfall amounts on the ground surface, although the non exact reproducibility of the two pressure working conditions in both the trials might have had some influence. At the 50 kPa pressure head, a T intensity increase was observed with 50%–60% degree of leaf cover (Oct. '02), in comparison to 100% leaf cover (Sept. '02), as was expected. At 150 kPa, this trend was not remarkably clear as for this pressure condition the jets were probably more easily intercepted by leaves and branches and the canopy degree cover seemed to be of secondary importance. Furthermore, at these pressure values, the wind effect (all the trials were carried out in nearly windless conditions), which might have accidentally and unexpectedly occurred during the trial, could have easily influenced both P heights on the plot and at the interception phenomenon. The larger size of the gross-rainfall drops at 50 kPa probably promotes the clear throughfall compared to those intercepted and lets the drops themselves be less disturbed by the wind effect.

The intensity distribution of the throughfall among the nine gauges was mainly due to the different branch textures above the collecting points and secondarily to the non-uniformity of P. Rain collector 5 showed, in both trials and different leaf covers, a very high T value, apparently not related to the corresponding P. In this position the lower layer of the canopy was located at 12 m high, apart from a small branch which was 1.2 m from the gauge. Evidently, the branch angles cause a sort of rain concentration in this position, whereas the two minima recorded by the throughfall were probably due to an opposing phenomenon.

The descriptive statistics relative to the drop-size distribution amongst the nine positions of the throughfall and the gross rainfall for both the pressure conditions and different leaf cover degrees are reported in Table 1. From data listed in Table 1 it is clear that the drop diameters are not normally distributed and characterized by a great variability within each tray and among the nine trays for all the trials. Generally, all distributions were characterized by a positive skewness, with median values exceeding mean values and by kurtosis coefficients either positive or negative with no evidence of a connection to cover degree and pressure condition. The drop-size distributions under different working pressures and leaf cover degrees were analyzed by the non-parametric Wilcoxon test which compares the medians of the distributions. The comparison between Table 1 and Figure 3 outlines the actual changing effect of the canopy on P transformation. This phenomenon seemed to depend more on the gross-rainfall quality and secondly on the canopy cover density, although the medians of the drop-size distributions between the throughfalls were not significantly different when comparing the working pressure and the leaf cover degree effects separately. The two different working pressure heads (50 kPa and 150 kPa) generate diverse gross-rainfall drop-size distributions: the mean of the drop diameter medians (D_{50}) for the gross rainfall was nearly 0.90 mm at 150 kPa, and reached a mean value of D_{50} around 1.46 mm at 50 kPa. The drop-size distribution of T at the pressure head of 150 kPa was characterised by a mean D_{50} (among the nine trays) ranging from 1.64 mm with 100% leaf cover to 2.50 mm for a leaf cover degree of 50%–60%. In the trial carried out at 50 kPa, the mean D_{50} ranged from 1.58 mm (50%–60% leaf cover) to 1.98 mm (100% leaf cover). A significant increase ($p < 0.01$) of the drop diameters was observed for the throughfall under 100% and 50%–60% of leaf cover degree when compared with the corresponding gross rainfall at 150 kPa. This is due to the interception phenomenon which is however less expressed at the lower pressure value (50 kPa).

Conclusion

The present work illustrates the results of some experimental trials dealing with the evaluation of the canopy cover effect on the quali-quantitative transformation of the gross rainfall under simulated rainfall events performed in a forest environment. Very few studies are available concerning this theme; nevertheless, an exhaustive review is provided by Brandt (1989), who assesses the negligible effect of either the canopy (type and/or density) or the rainfall intensity on the drop-size distribution of the throughfall.

A rainfall simulator, placed at 16 m from the ground, was built and then calibrated according to the uniformity of water application and the drop-size distribution including the kinematics aspect. At pressures of 50 kPa and 150 kPa, the rainfall simulator was able to produce intensities of 78 mm h^{-1} and 98 mm h^{-1} respectively with a uniformity coefficient of 72% and

Table 1. Descriptive statistics of the raindrop diameter (millimeter) distributions of the throughfall and gross rainfall for the nine trays, at 150 and 50 kPa and with the two different leaf cover degrees (50–60% and 100%)

	Throughfall (150 kPa; 100% leaf cover)								
Trays (*n*)	1	2	3	4	5	6	7	8	9
Mean	2.07	2.25	1.84	2.25	2.49	2.77	2.24	2.16	2.10
Standard dev.	1.39	1.71	1.19	1.51	1.75	1.80	1.38	1.43	1.58
Mode	1.46	0.71	1.46	1.46	1.00	1.46	1.46	1.46	1.46
Median	1.46	1.46	1.46	1.46	2.02	2.02	2.02	1.46	1.46
Skewness	1.48	1.28	1.68	1.06	0.88	0.85	0.83	1.32	1.60
Kurtosis	1.85	0.65	2.88	0.34	−0.31	−0.44	−0.44	1.05	1.73
	Throughfall (150 kPa; 50–60% leaf cover)								
Trays (*n*)	1	2	3	4	5	6	7	8	9
Mean	2.40	2.92	3.01	2.85	3.42	3.01	2.89	2.54	3.08
Standard dev.	1.62	1.81	1.92	1.91	2.05	1.85	1.76	1.75	1.91
Mode	1.00	3.89	5.21	1.46	6.71	3.89	3.89	1.46	3.89
Median	2.02	2.74	2.74	2.02	2.74	2.74	2.74	2.02	2.74
Skewness	1.14	0.61	0.56	0.82	0.34	0.57	0.55	0.89	0.48
Kurtosis	0.44	−0.70	−0.97	−0.54	−1.22	−0.83	−0.80	−0.34	−0.96
	Gross rainfall (150 kPa)								
Trays (*n*)	1	2	3	4	5	6	7	8	9
Mean	1.10	1.02	1.10	1.12	0.46	1.12	1.07	0.95	1.07
Standard dev.	0.49	0.46	0.49	0.49	0.42	0.49	0.47	0.41	0.47
Mode	0.71	0.71	0.71	0.71	0.71	0.71	0.71	0.71	0.71
Median	1.10	0.71	1.00	1.00	0.71	1.00	1.00	0.71	1.00
Skewness	1.46	1.76	1.46	1.40	2.32	1.40	1.60	2.18	1.60
Kurtosis	1.76	2.86	1.76	1.55	3.52	1.55	2.24	4.91	2.24
	Throughfall (50 kPa; 100%; leaf cover)								
Trays (*n*)	1	2	3	4	5	6	7	8	9
Mean	2.93	2.71	3.00	2.50	2.81	2.54	2.20	2.12	2.46
Standard dev.	2.05	1.68	2.01	1.82	1.87	1.78	1.43	1.46	1.62
Mode	5.21	3.89	5.21	1.46	2.02	1.46	1.46	1.00	1.46
Median	2.02	2.02	2.74	2.02	2.02	2.02	1.46	2.02	2.02
Skewness	0.77	1.35	0.44	1.08	0.78	1.08	0.85	1.40	0.87
Kurtosis	−1.13	−0.60	−1.28	0.09	−0.61	0.07	−0.49	1.17	−0.44
	Throughfall (50 kPa; 50–60% leaf cover)								
Trays (*n*)	1	2	3	4	5	6	7	8	9
Mean	2.14	2.32	2.19	2.10	2.84	2.54	1.81	1.93	2.09
Standard dev.	1.49	1.69	1.50	1.66	2.05	1.81	1.45	1.36	1.50
Mode	1.46	1.00	1.46	0.71	1.46	1.46	1.46	0.71	1.00
Median	1.46	1.46	1.46	1.46	1.46	2.02	1.46	2.02	1.46
Skewness	1.19	0.43	1.11	1.26	0.74	1.00	1.77	1.33	1.43
Kurtosis	1.54	0.26	0.18	0.36	−0.85	−0.06	2.68	0.72	1.39
	Gross rainfall (50 kPa)								
Trays (*n*)	1	2	3	4	5	6	7	8	9
Mean	1.74	1.59	1.74	1.74	1.74	1.74	1.74	1.59	1.74
Standard dev.	0.93	0.70	0.93	0.93	0.70	0.93	0.93	0.70	0.93
Mode	1.46	2.02	1.46	1.46	2.02	1.46	1.46	2.02	1.46
Median	1.46	1.46	1.46	1.46	1.46	1.46	1.46	1.46	1.46
Skewness	0.99	2.82	0.99	0.98	0.35	0.98	0.98	0.35	0.98
Kurtosis	0.11	−1.15	0.11	0.11	−1.15	0.11	0.11	−1.15	0.11

90%. The trials were performed in September and in October, namely, under two different canopy cover densities of 50%–60% and 100%. As the experimental site was quite a distance from the laboratories and part of the roads were not paved it was necessary to develop a new procedure to collect the drops by replacing flour with quick-hardening cement which was able to produce more resistant pellets.

The results, illustrated in the present work, suggest that the transformations due to the canopy presence depend more upon the intensity of the gross rainfall, namely quality, as was expressed as a drop-size distribution, and therefore the transformations are not consistent with Brandt's data. Moreover, these results suggest that the distribution of the drop diameters was only significantly modified, with an increase of drop dimensions, under a cover density of 50%–60% and a pressure of 150 kPa. For all the other trials, no significant transformation in the drop diameter distribution was observed. Although some variability (not statistically significant) among the trays was shown within each trial, the different arrangement of the branches and density of the vegetation above the nine collectors did not seem to influence the drop-size distribution with a univocal link, valid for all the trials. Evidently, the different working pressure, namely, the quality of the gross rainfall, controls both the drop trajectories and the amount of distribution between real and intercepted throughfall.

At 150 kPa the similar drop-size distributions between *T* with 100% leaf cover and *P* does not mean that no interception factor has taken place; on the contrary, it can be thought that the high density of the canopy greatly controls the trajectories of the drops, and combined with the high intensity of the rainfall, promotes several splashing points and collision among the drops resulting in their breaking up. At 50 kPa the high similarity between the drop-size distributions of the gross rainfall and the throughfall under 50%–60% cover density suggests that the bigger dimensions of the drops produced by the simulator promote the real throughfall. At the same pressure conditions and under a cover density of 100% the drops can preferentially coalesce and collisions with other drops are rarer in contrast to the trial performed at 150 kPa.

References

Brandt C J 1989 The size distribution of the throughfall drops under vegetation canopies. Catena 16, 507–524.

Brandt C J 1990 Simulation of the size distribution and erosivity of raindrops and throughfall drops. Earth Surf. Processes 15, 687–698.

Chen J Y, Hopmans J W and Fogg G E 1995 Sampling design for soil moisture measurements in large field trials. Soil Sci. 159, 155–161.

Christiansen J E 1942 Irrigation by Sprinkling. Bull. 670, Agricultural Exp. Station, University of California, Berkeley, CA.

Falciai M, Ghinassi G and Trucchi P 2002 Deflussi superficiali stagionali da un bosco ceduo di faggio sottoposto a trattamenti selvicolturali differenziati. Quaderni di Idronomia montana. 18, 95–100. Editoriale Bios, Cosenza, Italy.

Giacomin A and Trucchi P 1992 Rainfall interception in a beech coppice (Acquerino, Italy). J. Hydrol. 137, 141–147.

Giacomin A, Trucchi P and Andrenelli M C 2003 Intercettazione e deflussi superficiali in un bosco ceduo di faggio. *In* Proceedings on "La difesa della montagna," dicembre 11–12, 2002, Assisi, 490–494.

Kimmins J P 1973 Some statistical aspects of sampling throughfall precipitation in nutrient cycling studies in British Columbian coastal forests. Ecology 54, 1008–1019.

Laws J O and Parson D A 1943 The relation of rain drop size to intensity. Transaction American Geophysical Union 24, 452–459.

Li Y C, Alva A K, Calvert D V and Zhang M 1997 Stemflow, throughfall and canopy interception of rainfall by citrus tree canopies. Hortscience 32, 1059–1060.

Morgan R P C 2001 A simple approach to soil loss prediction: a revised Morgan-Morgan-Finney model. Catena 44, 305–322.

Qingfu Xiao E, McPherson G, Ustin L S, Grismer M and Simpson R J 2000 Winter rainfall interception by two mature open-grown trees in Davis, CA. Hydrol. Process. 14, 763–784.

Zanchi C and Torri D 1980 Evaluation of rainfall energy in central Italy. *In* M De Boodt and D Gabriels (eds), Assessment of erosion, pp. 133–142. Wiley, New York.

Effect of repeated fire on plant community recovery in Penteli, central Greece

G. Goudelis[1,4], P.P. Ganatsas[2], I. Spanos[3] & A. Karpi[1]

[1]*Technological Education Institute of Lamia, Department of Forestry, 36100 Karpenissi, Greece.* [2]*Aristotle University of Thessaloniki, Department of Silviculture, P. Box. 262, 54124 Thessaloniki, Greece.* [3]*NAGREF, Forest Research Institute, 57006 Vassilika, Thessaloniki, Greece, e-mail: ispanos@fri.gr.* [4]*Corresponding author**

Key words: Aleppo pine, postfire revegetation, regeneration, wildfire

Abstract

In a *Pinus halepensis* Mill. forest of central Greece, we studied the effects of two fires on postfire plant community recovery by comparing a site that was burned once and another adjacent that was burned twice. During the first 15 months after fire, we monitored plant species recruitment, plant density and growth. Lower species richness and plant density were observed in the site burned twice compared to that burned only once. The growth of woody species did not differ between the two treatments and presented high variability even within the same plot. Resprouting plant species appeared earlier than obligatory seeders. Fifteen months after fire, the ecosystem was dominated by the maquis species that existed in the prefire period, mainly *Quercus coccifera* L., *Pistacia lentiscus* L., *Phillyrea latifolia* L., with a low contribution of *Pinus halepensis* seedlings and a greater proportion of *Cistus* species (*Cistus monspeliensis* L., *C. creticus* L., *C. salvifolius* L.).

Introduction

Wildfires comprise a significant factor for ecosystem status in many areas around the world (see Arianoutsou (2001) for a review). However, wildfires play a potentially important role in the Mediterranean basin (Naveh, 1991) where, in combination with human actions such as animal grazing and fuel overexploitation, contribute to long-term land degradation (Thirgood, 1981). According to the data obtained worldwide, some general conclusions have been extracted concerning the postfire process; these can be summarized as follows (Trabaud, 1994): the establishment of the previous communities is usually a rapid phenomenon and in most cases there is no actual succession but an auto-succession process. The plants that persist are those that appear immediately after fire and that existed previously.

However, the above mentioned process seems to be true when fires occur after a long time interval and depends on the type of ecosystem burned (Hatzibiros, 2001; Polakow and Dunne, 1999); in the case of *Pinus halepensis* Mill. Forests, full recovery requires more than 30 years (Arianoutsou and Ne'eman, 2000; Schiller et al., 1997; Trabaud, 2000). However, some land degradation probably takes place (Dafis, 1987; Tsitsoni, 1997). The problem of land degradation increases when grazing of the burned areas occurs after the fire (Spanos, 1992; Spanos et al., 2000). Even though land degradation is more possible in the case of frequent fires repeated within short intervals, there is a scarcity of data concerning the influence of repeated fires on ecosystem recovery and the degree of the degradation process. Such studies can show in real terms if the hypothesis of land degradation after repeated fires is true and to what extent they cause problems to ecosystem resilience with regard to the fire effect.

The aim of this study was to investigate differences in ecosystem recovery after one (area burned in 1998) or two fires (area burned in 1995 and 1998). The findings could result in conclusions about the degradation probability after repeated fires. Thus, postfire community recovery was studied in a *P. halepensis* woodland in Penteli, central Greece, during the early, crucial

* E-mail: ggoudelis@teilam.gr

postfire time of 15 months. Repeated measures were carried out and were focused on plant species regeneration, density and growth. Site characteristics were recorded in order to correctly compare the areas burned once and twice.

Materials and methods

Study area

The study was carried out in Penteli, central Greece, and 30 km from the city of Athens. The region consists of part of the Mount Penteli whose altitude ranges from 100 to 1107 m. Geologically, the area belongs to the Attiko-Cyclades geotectonic zone. The parent rock materials are mainly limestone and schists and a small part is covered by sedimentary formations (Mountrakis, 1985). The slope gradient was 15%–30%. The vegetation of the area belongs to the lower Mediterranean vegetation zone, *Quercetalia ilicis* and particularly to the association *Oleo-lentiscetum*. The climate is characterized as Mediterranean type (Csa) according to Koeppen classification. The annual amount of rainfall is 413 mm and the dry period has an average duration of 5–6 months, lasting from April to September. The ecosystem studied prefire was dominated by *P. halepensis* Mill. with a shrub story of maquis species. The area was burned in July 1995 and a large part of the area was burned again in August in1998. The age of *P. halepensis* stands ranged from 30 to 40 years as shown by tree ring measurements. The entire area is subjected to high human pressure due to animal grazing, illegal claims and trespassing.

Sampling

Sixteen permanent plots of 100 m^2 were established just after the fire in the burned areas in different locations; 11 were established in areas burned twice (B_2), in 1995 and 1998, and five in areas burned once only (B_1) in 1998. The distinction of burned areas was based on maps created by the local forest administration office. Within each of them, five subplots of 1 m^2 were selected systematically within the plots; four in the corner of each plot at a distance of 2 m from the corner, and the fifth in the center of the plot. In each plot, all woody plant species were recorded, as well as their density and height. The monitoring took place during the 15 months after fire, in January 1999 (before spring, to record the early plant regeneration), June 1999 (after spring and before summer) and in October 1999 (after the first postfire summer). Site characteristics were also recorded; these concerned altitude, aspect, topography, angle and postfire management.

Data analysis

Statistical analysis was performed using the SPSS package, version 11.5 for Windows. The differences in stem density, species richness and plant growth between B_1 and B_2 were assessed using t-tests ($P < 0.05$, Norusis, 2002). All figure bars indicated by different letters were statistically different. Error bars represent standard error of the mean. The nomenclature of plant species follows Flora Europea (Tutin et al., 1964–1980).

Results

Effect of repeated fire on woody species richness

Generally, plant species regeneration was delayed and started quite late during the next postfire winter and spring in both B_1 and B_2. Two functional groups, resprouters and seeders were regenerated in the burned areas. The revegetation pattern appeared to be the same in both cases; initially, during the late autumn and winter, only a few woody plant species (resprouters) appeared, whilst most species regenerated later during the spring and summer. The same woody plant species regenerated at both sites and were those which existed before the fire. Therefore, neither of the respouter or seeder species lost the ability to repopulate the area.

Analysis of the monitoring data showed that the species richness was significantly higher in B_1 during the early postfire phase, at all monitoring dates (Figure 1). The species recorded in January, were those regenerating vegetatively: *Quercus coccifera* L., *Pistacia lentiscus* L., *Phillyrea latifolia* L., *Erica arborea* L., *Arbutus unedo* L., *Arbutus andrachne* L., *Nerium oleander* L., *Cotinus coggygria* Scop. Only a few resprouters appeared later (in spring) such as *Pistacia terebibthus* L., *Dittrichia viscose* (L.) W. Greuter *Calicotome villosa* (Poir.) Link and *Lonicera implexa* Ait. Species regeneration continued during the whole period. Woody species richness was low in January, and increased with time; the average number of species

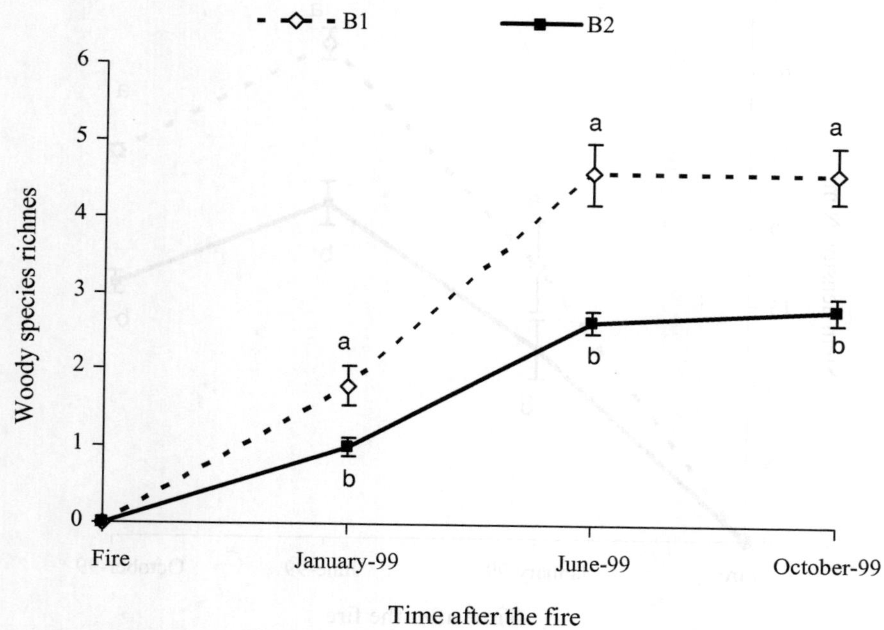

Figure 1. Effect of one or two fires on the postfire number of woody species. Values are the number of species per subplot (1 m^2); values at the same date followed by different letters are significantly different ($p > 0.05$, t- test).

recorded in June was 4.6 species per subplot in B_1 and 2.6 in B_2. Finally, 15 months after the fire, the woody species richness per subplot was significantly greater in B_1 compared to B_2. Repeated fire seems to cause some regeneration problems in certain species e.g. *C. coggygria*, *A. andrachne* and *E. arborea* which were extremely limited in B_2.

Species regenerating from seeds colonized both burned areas much later, mainly during the postfire spring and summer. These species included some seasonal dimorphic sub-shrubs such as the *Cistus* species (*C. monspeliensis* L., *C. creticus* L. and *C. salvifolius* L.) and the tree species *P. halepensis*, as well as many herbaceous species.

Q. coccifera was found in all plots in both B_1 and B_2, whereas *N. oleander* and *A. andrachne* were recorded in only one plot. *L. implexa*, was regenerated in two plots in the spring, but died back during the summer. However, the absence of plant regeneration during the early postfire phase, autumn and early winter, was noticed, and there was a high risk of soil erosion due to the relatively high amount of rainfall that usually happens during this period of the year (average monthly amount of rainfall was 49, 66 and 72 mm for the months October, November and December, respectively).

Effect of repeated fire on stem density

The rate of species regeneration was significantly higher in B_1 than in B_2; thus, stem density was always significantly lower in B_2 during the whole period studied. In October of the next postfire year, regeneration was found to be approximately one third lower in B_2 compared B_1. The average density at this time was 16.1 stems per m^2 and 24.5 stems per m^2 respectively (Figure 2). The stem density appeared to follow the same pattern in both cases, but always with lower values in B_2 due to the differences in the initial plant regeneration. Thus, stem density increased with time until the next spring because of continued plant regeneration; it then decreased due to the summer drought that affected survival of both resprouters and seeders. Specifically, the stem density increased 1.8 times from January to June at both sites; afterwards, it decreased to 78% and 76%, for B_1 and B_2 respectively. Therefore, repeated fire affected early plant density but the pattern of regeneration and survival was unaffected.

Species which had regenerated vegetatively had higher densities, especially during the early postfire months, compared to the seeders (Table 1). These resprouters contributed 83.3% to total density in B_1, and 76.7% in B_2. Seeders contributed only 16.7% and

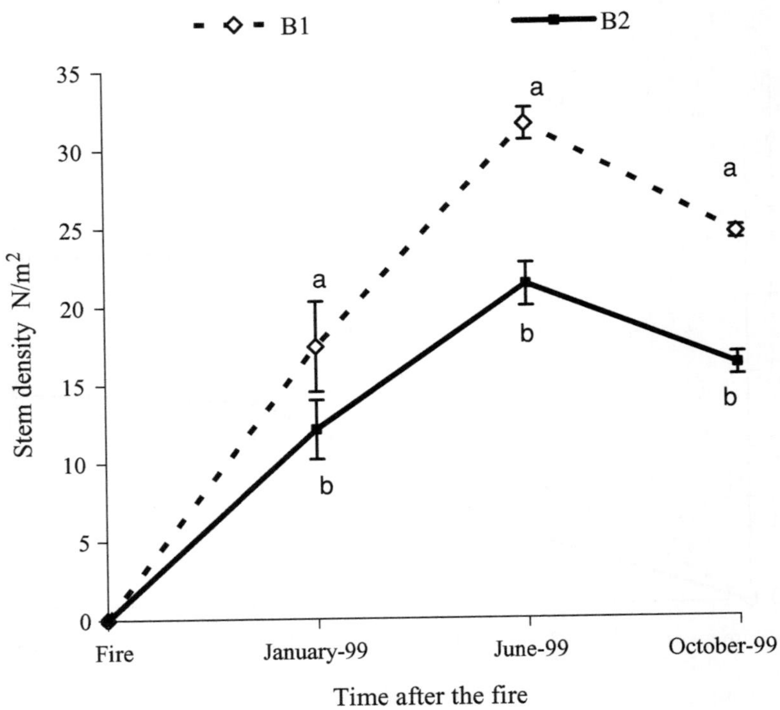

Figure 2. Effects of one or two fires on the postfire vegetation stem density. Values at the same date followed by different letters are significantly different ($P > 0.05$, t-test).

Table 1. Postfire density and growth of woody species, in B_1 and B_2, 15 months after the fire

Species	B_1 areas			B_2 areas		
	Density stems/m^2	Mean height (mm)	Maximum height (mm)	Density stems/m^2	Mean height (mm)	Maximum height (mm)
Resprouters						
Quercus coccifera	12.78a	308ns	520	7.37b	371ns	647
Pistacia lentiscus	0.20*	581ns	600	1.55*	632ns	877
Phillyrea latifolia	0.54*	771a	940	0.68*	392b	610
Arbutus unedo	0.84*	443b	575	1.05*	880a	1400
Arbutus andrachne	1.90*	536*	590	–	–	–
Nerium oleander	–	–	–	0.66*	579*	1170
Calicotome villosa	0.40*	663ns	920	0.51*	633ns	970
Erica arborea	2.12a	244a	360	0.02b	60b	60
Cotinus coggygria	1.24*	267*	455	0.00	–	–
Dittrichia viscose	0.12*	220b	220	0.49*	306a	420
Pistacia terebinthus	0.28*	1190a	1400	0.02*	470b	470
Seeders						
Pinus halepensis	0.40*	78ns	140	0.49*	103ns	160
Cistus monspeliensis	2.58ns	125ns	180	2.24ns	108ns	188
Cistus creticus	1.00*	145ns	210	0.82*	137ns	205
Cistus salviifolius	0.12*	138ns	175	0.20*	133ns	180
Total	24.52a			16.11b		

Values of the same species for the same parameter followed by different letters are significantly different, ns = nonsignificant differences ($p > 0.05$, t-test).
*insufficient data for the test.

23.3% of the total density in B_1 and B_2 respectively. The density of the main seeder species (*Cistus* species) was slightly affected by the repeated fire and density was lower in B_2 than in B_1 (3.26 and 3.70 individuals/m^2 respectively) but with no significant differences. On the contrary, *P. halepensis* density was not affected by the repeated fire. This can be explained from the fact that the fire did not destroy all the forest stands, but some forest patches remained unaffected; thus, these stands produced seeds that colonized the adjacent burned areas. However, the regeneration of *P. halepensis* seedlings was relatively low in both cases (0.40 and 0.49 seedlings per m^2 respectively, 15 months after the fire) and many seedlings died during the summer drought in B_2 (Table 1).

The most abundant species in B_1 were the resprouters *Q. coccifera, E. arborea* and the seeders of the *Cistus* species. This pattern was also found in B_2 (except for *E. arborea*), but the density was significantly lower in these areas. *Q. coccifera* possessed the highest overall resprouting ability (Table 1).

Effect of repeated fire on vegetation growth

Plant growth greatly varied between species and plots as well as within plots for the same species, and did not significantly differ between B_1 and B_2. However, resprouting species were the dominant postfire ecosystem elements due to their high density and greater height (Table 1). On the contrary, the seeders exhibited quite low growth (Table 1). Some species e.g. *P. latifolia, E. arborea* and *P. terebinthus* exhibited significantly greater growth in B_1. On the contrary, other species were taller in B_2 e.g. *A. unedo* and *D. viscosa* (Table 1). *A. andrachne* and *C. coggyria* dominated in some plots of B_1, in terms of height growth, but were scarce in B_2 (Table 1). To sum up the above data, it seems that there was a high variability in plant growth that can mainly be attributed to site variability rather than to the effect of the repeated fire.

Effect of repeated fire on P. halepensis natural regeneration

P. halepensis is an obligatory seeder that was the dominant structural element in the prefire ecosystem. Its natural regeneration was relatively low in both cases, in terms of individual density and growth (Table 1). An average density of 0.40 and 0.49 of 1-year-old seedlings per square meter were found 15 months after fire, in B_1 and B_2. The seedling emergence was delayed in both cases and appeared in the next postfire spring. No seedlings were recorded in January. Seedling growth was low in both cases and did not differ between the two areas. No seedlings died in B_1 during the summer dry period whereas 26.8% of the seedlings died in B_2.

Discussion

The findings of the study indicated that even though there were significant differences in woody species richness and stem density between B_1 and B_2, the community recovery takes place following a similar revegetation pattern. As in other cases (Arianoutsou, 2001; Ganatsas et al., 2004; Trabaud, 1994), an autosuccession process took place and the prefire dominant floristic elements composed the new postfire communities. However, repeated fire affected initial plant regeneration by reducing the early woody species richness and especially the stem density. This resulted in lower values of these plant parameters in B_2 compared to B_1. Species regeneration in both cases started in the late autumn and winter and continued during the entire period studied. Only few species colonized the burned areas during the first postfire months as the January inventory showed. Thus, species richness and density was low. Species richness and stem density then increased until the following spring because of continued plant colonization and then decreased due to the summer drought that affected plant survival. The increase in stem density from January to June as well the decrease from June to October (summer drought effect) was similar in both B_1 and B_2. Finally, the early growth of woody species was highly variable and was similar between the two sites.

Contrary to the results of other studies (Ganatsas et al., 2004; Hatzistathis et al., 1996; Thanos et al., 1996), revegetation took place within a given time delay, i.e. the revegetation process started in the late autumn. Initially, resprouting species occurred and obligatory seeders appeared later, during the next spring and summer. Contrary to the results of other studies (see Arianoutsou and Ne'eman (2000) for a review), which report that seeds germinate after the onset of the rainy season, no seedlings of *P. halepensis* appeared between autumn and January. This can be attributed to the time of fire occurrence, as the fire happened in August, at the end of the growing season; thus, there was not enough time for species to regenerate. This resulted in a fragile, bare land during the first postfire months, a period in which the denuded land was subjected to autumn and winter rainfalls, thus there was a high soil erosion risk. However, this risk was greater in B_2, since in these areas the stem density was much lower than in B_1.

Fifteen months after the fire, the ecosystems were dominated by the maquis species that existed in the prefire period, with a small contribution of the tree species *P. halepensis* and a relatively high proportion of *Cistus* species. The plant community was mainly composed of evergreen broadleaf shrubs: *Q. coccifera*, *P. lentiscus*, *P. latifolia*, *E. arborea*, *A. unedo*, *A. andrachne* and some seasonal dimorphic sub-shrubs e.g. *C. creticus*, *C. salviifolius*, *C. monspeliensis* and *C. villosa*. However, natural regeneration of the prefire dominant tree species *P. halepensis*, was low in terms of stem density and seedling growth was also low. A high percentage of stems in several species died back during the postfire summer due to drought stress.

All the woody species existing before the fires maintained their regeneration ability despite the short interval between the two fires in B_2. Species regenerating vegetatively did not loose their resprouting ability and seeders managed to regenerate either from the soil seed bank or the seeds produced by the new young plants (3 years old in the case of *Cistus* species) or by the aerial seedbank of the adjacent un-burned stands as for *P. halepensis*. Thus, repeated fires in real terms did not affect the floristic elements of the burned areas, even though some species were scarce in B_2; these were few and were also found locally in B_1. In general, both resprouters and seeders kept their regeneration capacity, but this capacity was significantly lower in B_2. What causes this difference seems to be the young age of the plants (3 years old), that affect either their resprouting ability for the species regenerating vegetatively or the fruition of the seeders. In particular, it has been well documented that *P. halepensis* needs at least 7 years before producing mature seeds (Thanos, 2000). However, *Cistus* species are able to produce seeds early; full flowering ability is reached during the third year after a fire (Trabaud, 1987).

Finally, it can be said that despite the short period of this study, the findings indicate that there are differences in the rate of postfire revegetation processes between B_1 and B_2. Repeated fire seems to affect the species regeneration capacity, thus vegetation cover will be lower and the risk of soil erosion greater in the case of repeated fires. In comparison with the postfire process recorded in other ecosystems, usually burned only once (Arianoutsou, 1999; Ganatsas et al., 2004; Trabaud, 1994), the following conclusions can be made; (i) in our case there was a delay in species appearance that can be attributed to the time of fire occurrence (late summer), (ii) average values of species richness and stem density were relatively low, probably due to the dry climate and the influence of grazing animals and (iii) the number of naturally regenerated *P. halepensis* seedlings was low compared to other studies, because of high postfire human pressure at the burned areas.

References

Arianoutsou M 1999 Effects of fire on vegetation demography. *In* Proceedings of the International Symposium "Forest Fires: Needs and Innovations". DELFI Concerted Action, CINAR-EC, DGXII, Athens, pp 265–273.

Arianoutsou M 2001 Landscape changes in Mediterranean ecosystems in Greece: implication for fire and biodiversity. J. Mediterr. Ecol. 2, 165–178.

Arianoutsou M and Ne'eman G 2000 Post-fire regeneration of natural *Pinus halepensis* forests in the East Mediterranean basin. *In* G Ne'eman and L Trabaud (eds), Ecology, Biogeography and Management of *Pinus halepensis* and *P. brutia* Forest Ecosystems in the Mediterranean Basin, pp 269–289. Backhuys Publishers, Leiden, The Netherlands.

Boydak M 2004 Silvicultural characteristics and natural regeneration of *Pinus brutia* Ten.—a review. Plant Ecol. 171, 153–163.

Dafis S 1987 Ecology of *Pinus halepensis* and *P. brutia* forests. *In* Proc. of the 1st Scientific Conference on *Pinus halepensis* and *Pinus brutia* forests. Ed. Hellenic Forestry Society, pp 17–25. Chalkida, Greece.

Daskalakou E N and Thanos C A 1996 Aleppo pine (*Pinus halepensis*) postfire regeneration: the role of canopy and soil seed bank. Int. J. Wildland Fire 6, 59–66.

Ganatsas P, Zagas T, Tsakaldimi M and Tsitsoni T, 2004 Post-fire regeneration dynamics in a Mediterranean type ecosystem in Sithonia, northern Greece: ten years after the fire. *In* Proc. of the 10th MEDECOS Conference, April 25–May 1, 2004, Rhodes, Greece, M Arianoutsou and V Papanastasis (eds), Millpress, Rotterdam.

Hatzibiros K 2001 Setting priorities for wildfire suppression policy in Greece, using a relation between yearly burned areas and recovery time. Global Nest. Int. J. 3, 37–43.

Hatzistathis A, Tsitsoni T, Ganatsas P, Zagas T and Goudelis G 1996 Post-fire vegetation dynamics, in Sithonia peninsula (Chalkidiki, Greece). *In* Proc. of the Second Balkan Scientific Conference on Study, Conservation and Utilisation of Forest Resources, Sofia, pp. 168–173.

Kazanis D and Arianoutsou M 2004 Long-term post-fire vegetation dynamics in *Pinus halepensis* forests of central Greece: a functional-group approach. Plant Ecol. 171, 101–121.

Mountrakis D. 1985 Geology of Greece (in Greek). University Studio Press. Thessaloniki.

Naveh Z 1991 The role of fire in Mediterranean vegetation. Botanical Chronicles 10, 609–615.

Norusis M J 2002 SPSS 11.0 Guide to Data Analysis, 637 p. Prentice Hall, Englewood Cliffs, NJ.

Polakow D A and Dunne T T 1999 Modeling fire-return interval T: stochasticity and censoring in the two-parameter Weibull model. Ecol. Model. 121, 79–102.

Schiller G, Ne'eman G and Korol L 1997 Post-fire vegetation dynamics in a native *Pinus halepensis* forest in Mt Carmel Israel. Israel J. Plant Sci. 45, 297–308.

Spanos I A 1992 Structure analysis and natural regeneration of *Pinus brutia* forests in Thasos island, Ph.D. thesis. Department of Forestrty & Natural Environment, Aristotle University of Thessaloniki, Thessaloniki, Greece.

Spanos I A, Daskalakou E N and Thanos C A 2000 Postfire, natural regeneration of *Pinus brutia* forests in Thasos island, Greece. Acta Oecol. 21, 13–20.

Thanos C A 2000 Ecophysiology of seed germination in *Pinus halepensis* and *P. brutia*. *In* G Ne'eman and L Trabauld (eds) Ecology, Biogeography and Management of*Pinus halepensis* and *P. brutia* Forest Ecosystems in the Mediterranean basin, pp 37–50. Backhuys Publishers, Leiden, The Netherlands.

Thanos C, Daskalakou E N and Nikolaidou S 1996 Early post-fire regeneration of a *Pinus halepensis* forests on Mount Parnis, Greece. J. Veg. Sci. 7, 273–280.

Thirgood J V 1981 Man and the Mediterranean Forest. A History of Forest Depletion, 194 p. Academic Press, New York.

Trabaud L 1987 Fire and survival traits of plants. *In* L Trabaud (ed) The Role of Fire in Ecological Systems, pp 65–89. SPB Academic Publishing, The Hague.

Trabaud L 1994 Postfire plant community dynamics in the Mediterranean Basin. *In* JM Moreno and WC Oechel (eds) The Role of Fire in Mediterranean Type Ecosystems. . Ecological Studies, vol. 107, pp. 1–15. Springer-Verlag, New York.

Trabaud L 2000 Post-fire regeneration of *Pinus halepensis* forests in the West Mediterranean Basin. *In* G Ne'eman and L Trabaud (eds), Ecology, Biogeography and Management of *Pinus halepensis* and *Pinus brutia* Forest Ecosystems in the Mediterranean Bain, pp. 257–268. Backhuys Publishers, Leiden, The Netherlands.

Tsitsoni T 1997 Conditions determining natural regeneration after wildfires in the *Pinus halepensis* (Miller, 1768) forests of Kassandra Peninsula (North Greece). For. Ecol. Manage. 92, 199–208.

Tutin T G, Heywood V H, Burgers N A, Moore D M, Valentine D H, Walters S M and Wedd D A 1964–1980. Flora Europaea I-V. Univ. Press, Cambridge.

Effects of postfire logging on soil and vegetation recovery in a *Pinus halepensis* Mill. forest of Greece

Ioannis Spanos[1,3], Yannis Raftoyannis[2], Gerasimos Goudelis[2], Eleni Xanthopoulou[1], Theano Samara[1] & Alexandros Tsiontsis[1]

[1]*NAGREF, Forest Research Institute, 57006, Vassilika, Thessaloniki, Greece.* [2]*Department of Forestry, TEI Lamias, Karpenisi, Greece.* [3]*Corresponding author**

Key words: log removal, plant regeneration, postfire management, sediment loss, seedling emergence

Abstract

After a wildfire in a *Pinus halepensis* Mill. forest, in northern Greece, the burned trees were logged and the logs were removed either by mechanical or animal traction. The effects of logging and log removal methods on soil and vegetation recovery were evaluated comparing the logged sites with a burned but unlogged site and the unburned forest. Fire and logging did not affect the soil pH and caused only a short-term reduction in organic matter content. Two years after the fire, the highest rates of soil loss were observed in the logged area where mules were used for log removal. Soil moisture showed some differences between treatments during the first year after fire but then values were similar. Logging and particularly the use of skidders for log removal caused an initial increase in the amount of exposed bare ground but later when vegetation cover increased differences were minimized. The main woody species showed a species specific response to the treatments and while seeder species were favoured in the unlogged sites the same was not true for the respouters. In general, the growth and survival of pine seedlings was not affected by treatments.

Introduction

Mediterranean pine forests suffer from wildfires and have adapted to regenerate naturally after a fire (Naveh, 1975; Ne'eman, 1997; Thanos and Doussi, 2000; Trabaud, 1987). Wildfires remove the plant cover and litter layer, which play a major role in the prevention of soil erosion caused by raindrop impact and overland flow, and increase soil vulnerability to erosion processes by altering soil physical and chemical properties (DeBano et al., 1998; Soto et al., 1991).

The environmental effects of postfire management depend on several specific features of burned stands, including the intensity of the burn, slope, soil texture and composition, the presence or building of roads, and postfire weather conditions. Activity effects of logging systems occur within the context of these site-specific factors (McIver and Starr, 2000).

Forest fires have become common in Greece and when productive forests are burned the standard policy is to harvest these trees. However this practice has caused a debate among land managers. Proponents argue that except from the obvious economic return from logging, there are also positive ecological effects such as reduction of erosion due to the scattered logging slash, reduction of fuel loads, reduction of pest populations, increased water infiltration and better conditions for plant regeneration. Opponents of postfire logging argue that the practice increases soil erosion, damages soil processes by compaction and displacement, removes organic matter from the ecosystem, destroys habitats for wildlife and has a negative effect on vegetation recovery.

However, there is little or no scientific information available to support the above arguments and although

* E-mail: ispanos@fri.gr

This article has been previously published in the following journal—Plant and Soil, Volume 278, 171–179.
Stokes et al. (eds), Eco- and Ground Bio-Engineering: The Use of Vegetation to Improve Slope Stability, 345–352.

the effects of fire in Mediterranean pine ecosystems have been the subject of numerous studies few studies focused on postfire management activities and their effects on ecosystem recovery (Ne'eman et al., 1993).

The aim of this study was to assess the effects of postfire logging on plant regeneration and soil erosion in a *Pinus halepensis* forest ecosystem.

Materials and methods

Study area

The study area was located in the forest of "Vozena", 100 km southeast of Thessaloniki, Greece (40°15′ N, 23°37′ E). The climate is typical Mediterranean with mean annual rainfall 420 mm, mean annual air temperature 15°C and a dry season from May to September. The rainfall is irregular, distributed mainly in spring and autumn, with peaks in March and November. The altitude range is 60–200 m and the slope varies from 10% at the hill top to over 50% at the lower parts of the slopes. The unburned forest is dominated by *Pinus halepensis* Mill., with an under storey of shrubs such as *Quercus coccifera* L, *Phillyrea latifolia* L., *Pistacia lentiscus* L. and *Arbutus unedo* L. The stand age is less than 50 years because the present forest has regenerated naturally after a destructive wildfire that happened in 1976. The mean number of trees per hectare is 1050, their mean height is 12 m, the mean diameter at breast height is 15 cm, the wood volume is 95 m^3/ha and the annual wood growth 5.8 m^3/ha. On 6th of September 2001, 1850 ha of mature forest were burned from a high intensity wildfire. Six months after the fire, the burned trees were logged and the logs were removed either by a tractor (of skidder type that pulls timber on the soil with a winch) or animal traction (mules). Ten hectares of the burned forest were not logged in order to be included in our study as a control. In December 2002, 30% of the burned trees in the unlogged site were windthrown and by October 2003 the percentage of collapsed burned trees increased to 70%.

Experimental design and sampling

The effects of postfire logging on vegetation and soil were evaluated in four plots (10 ha each) corresponding to the following treatments: unburned (UB); burned but not logged (UL); burned, logged and use of tractors for log removal (LT); burned, logged and use of mules for log removal (LM). Within each treatment we randomly located four small rectangular plots (30 m in length and 2 m in width) and four large rectangular plots (100 m in length and 3 m in width). We tried to locate the plots in areas with similar stocking densities of trees and avoided areas where green needles remained on the burned trees or where the trees had died but the trunks remained unburned above 5 m to reduce small-scale variation in fire intensity.

The ground cover was characterized along the two 30 m border lines of the small plots, every 10 cm, using a metal pin vertical to the line. The following ground cover categories were used: stones, bare soil, ash, plant debris (dead leaves, needles, stems, twigs, logs), plants (alive herbaceous or woody plants). We recorded groundcover at five dates on May 2002, November 2002, February 2003, October 2003 and March 2004.

Inside the small rectangular plots, we evaluated the regeneration of herbaceous plants by counting the number of individual plants 1 year after the wildfire. In the large plots, we counted the number of woody plants in March 2002, September 2002 and March 2003. Furthermore, we monitored the number of pine seedlings, their height and survival in August 2002, May 2003, August 2003 and July 2004.

The morphological variables and biomass of the main woody species were recorded one year after the fire. Four resprouter (*Quercus coccifera* L., *Phillyrea latifolia* L., *Pistacia lentiscus* L., *Arbutus unedo* L.) and two seeder species (*Cistus creticus* L. and *Pinus halepensis* Mill.) were selected. Twenty plants per species and treatment were randomly selected and either cut at the ground level (resprouters) or carefully uprooted (seeders). We measured the aboveground height, the number of branches, the number of leaves (not for pine), the length of the main root (seeders only), the number of lateral roots (seeders only) and the total dry weight (samples were oven-dried for 5 days at 80°C and then weighted).

The soil organic matter and pH were recorded after the application of treatments (March 2002) and 1 year later, in 6 soil samples per treatment. Soil moisture was sampled monthly by taking three soil samples per treatment, to a depth of 20 cm, at randomly selected points near the sampling plots. Samples were transported to the laboratory in airtight containers, their weight was recorded, oven dried for 24 h at 105°C and then reweighed. Sediment loss was monitored with a network of fashioned sediment traps. Each trap consisted of a plastic square tube (15 × 10 cm cross-section

Table 1. Soil pH and organic matter content of the studied area, recorded at 6 and 18 months after the wildfire

		Treatment			
	Year	LM	LT	UL	UB
pH	2002	7.1 a	7.0 a	7.2 a	7.1 a
	2003	7.2 a	7.1 a	7.0 a	7.2 a
Organic mater (%)	2002	2.9 b	1.9 a	3.3 b	4.2 c
	2003	4.1 a	4.3 a	9.3 b	4.2 a

Means within one soil variable followed by the same letter are not significantly different at $P \leq 0.05$. Treatments: burned, logged and use of mules for log removal (LM); burned, logged and use of tractors for log removal (LT); burned but not logged (UL); unburned (UB).

Table 2. Sediment loss (kg/ha) and rainfall in different periods after fire

		Treatment		
Date	Rainfall (mm)	LM	LT	UL
Jun-Jul/2002	83	6 a	7a	27 b
Aug-Sep/2002	511	323 a	709 b	664 b
Oct-Nov/2002	34	48 b	30 a	30 a
Dec-mar/2003	69	20 a	17 a	24 a
Apr-Jun/2003	55	20 b	21 b	5 a
Jul-Nov/2003	71	114 b	33 a	41 a

Means within the same recording period and species followed by the same letter are not significantly different at $P \leq 0.05$. Treatments: burned, logged and use of mules for log removal (LM); burned, logged and use of tractors for log removal (LT); burned but not logged (UL).

and 40 cm length), nailed to the ground surface, and a durable plastic bag with a carrying capacity of 5 kg, fixed to the lower end of the tube. The dry weight of the sediment inside the traps was recorded six times during the observation period.

Differences between treatments for all variables were analyzed using analysis of variance (ANOVA) and means were compared with Tukey's multiple comparisons test. Percent data did not satisfy the assumptions of ANOVA and subsequently they were subjected to the arcsine transformation for statistical analysis, but actual percentages are given in the tables. All tests for significance were conducted at the $P \leq 0.05$ level.

Results

No effects on soil pH by fire or logging were detected 6 months after the fire and 12 months after the logging

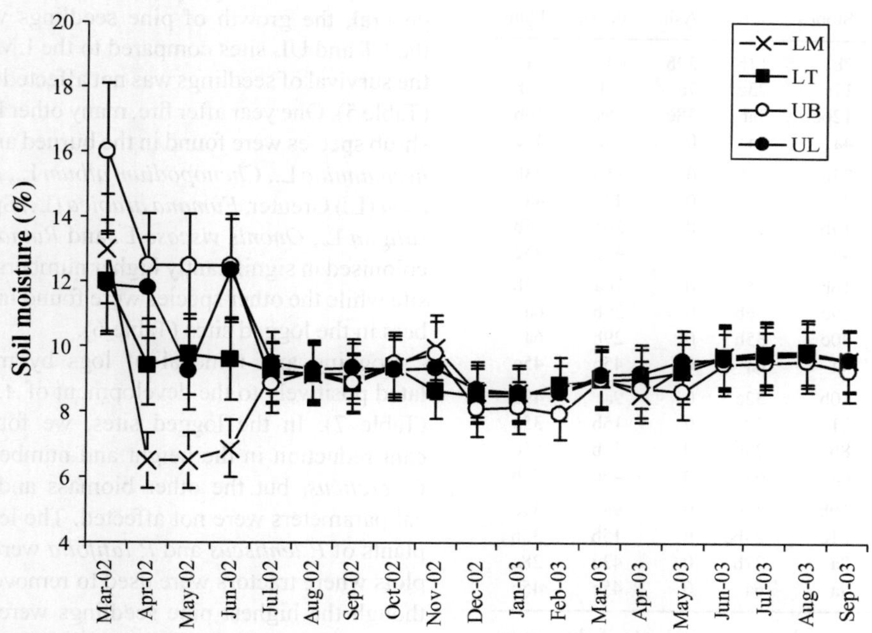

Figure 1. Seasonal courses of soil moisture in the different treatment plots. Unburned (○), burned but not logged (●), burned, logged and use of tractors for log transportation (■) and burned, logged and use of mules for log transportation (×). Symbols represent means ± 1 SD. Treatments: burned, logged and use of mules for log removal (LM); burned, logged and use of tractors for log removal (LT); burned but not logged (UL); unburned (UB).

(Table 1). Six months after the fire, the lowest organic matter content was found in the LT site. A year later, the values of organic matter increased in the logged sites and there were no significant differences from the unburned site while the organic matter increased dramatically in the unlogged burned area.

The highest rates of sediment transfer were recorded after the exceptionally rainy August and September 2002, and particularly in the unlogged site and the LT site (Table 2). Two years after the fire, the highest rates of soil loss were found in the LM area. At the beginning of the recording period and until June 2002, the highest levels of soil moisture were recorded in the unburned site and the lowest in the LM site (Figure 1). After July 2002, soil moisture was similar in all treatments until September 2003.

In May 2002, a significantly higher percentage of exposed bare ground was found in the LT area (Table 3). Two years later, although the percentage of exposed stones and rocks was higher in the LT site, the bare

Table 3. Mean percentage of different groundcover types in each treatment

		Groundcover (%)				
Date	Treatment	Stones	Bare soil	Ash	Plant debris	Plants
May-02	LM	7b	11b	29b	44c	9a
	LT	17c	23c	7a	38b	15b
	UL	12bc	9ab	38c	25a	16b
	UB	4a	6a	0	45c	45c
Nov-02	LM	22b	20bc	0	35b	23b
	LT	37c	17b	0	42c	4a
	UL	19b	25c	0	30a	26b
	UB	4a	6a	0	45c	45c
Feb-03	LM	16b	45c	0	16a	23b
	LT	29c	36b	0	29b	6a
	UL	30c	35b	0	29b	6a
	UB	4a	6a	0	45c	45c
Oct-03	LM	10b	32c	0	9a	49bc
	LT	21c	29c	0	15b	35a
	UL	8b	23b	0	16b	53c
	UB	4a	6a	0	45c	45b
Mar-04	LM	10b	32c	0	9a	49c
	LT	21c	29bc	0	15b	35b
	UL	2a	27b	0	43c	28a
	UB	4a	6a	0	45c	45c

Means within the same groundcover type and date of observation, followed by the same letter are not significantly different at $P \leq 0.05$. Treatments: burned, logged and use of mules for log removal (LM); burned, logged and use of tractors for use of tractors for log removal (LT); burned but not logged(UL); unburned (UB).

soil cover was similar in the logged plots. Plant cover showed significant variations in the first 2 years after fire and a significant reduction of the plant cover in the unlogged site in February 2003 was recorded. After fire, a significant percentage of the ground was covered by ash but later it was removed by wind and especially the high rainfall of the first summer. The percentage of ground covered by dead plant materials decreased with time in all treatments, although a significant increase occurred in the unlogged site after the windthrowing of burned trees.

Six months after fire, the natural regeneration of *A. unedo* was more successful in the unlogged site, but one year later, more plant stems were counted in the LT area (Table 4). *Cistus creticus* seedlings germinated in greater numbers in the UL site but in March 2003 no significant difference between the LM and UL areas could be found while the number of seedlings in the LT area was significantly lower. The highest density of *P. latifolia* stems was found in the LM area, followed by the LT and UL sites. The regeneration of *P. lentiscus* was found to be significantly better in the unlogged site. The highest numbers of *Q. coccifera* stems were found in the LT site. Seedling emergence for *P. halepensis* was higher in the unlogged site, while no significant difference was found among the logged treatments. In general, the growth of pine seedlings was higher in the LT and UL sites compared to the LM site whereas the survival of seedlings was not affected by treatments (Table 5). One year after fire, many other herb and short shrub species were found in the burned areas. *Anthyllis hermanniae* L., *Chenopodium album* L., *Dittrichia viscosa* (L.) Greuter, *Fumana arabica* (L.) Spach, *Lactuca saligna* L., *Ononis viscose* L. and *Rubia peregrina* L. colonised in significantly higher numbers the unlogged site while the other species were found in greater numbers in the logged sites (Table 6).

Logging and removal of logs by mules was related positively to the development of *A. unedo* plants (Table 7). In the logged sites, we found a significant reduction in the height and number of leaves of *C. creticus,* but the other biomass and morphological parameters were not affected. The least developed plants of *P. lentiscus* and *P. latifolia* were found in the plots where tractors were used to remove the logs. Although the highest pine seedlings were found in the unlogged plot, the seedlings in the logged plots developed a bigger root system. With the exception of height, treatments did not affect significantly the growth of *Q. coccifera* plants.

Table 4. Density of the main shurb species (N ha^{-1}) in the different treatment plots measured in March 2002, September 2002 and March 2003

		Species					
Date	Treatment	*Arbutus unedo*	*Cistus creticus*	*Phillyrea latifolia*	*Pistacia lentiscus*	*Quercus coccifera*	*Pinu halepensis*
Mar-02	LM	833b	166667c	916c	167a	454a	458a
	LT	480a	2667b	660b	480b	1167d	640a
	UL	1125c	466667d	498a	541b	625b	1480c
	UB	583a	15a	664b	917c	998c	542c
Sep-02	LM	2166c	101750c	1332c	583a	1498b	1162b
	LT	480a	6250b	660b	480a	3166c	1328bc
	UL	1333b	108000c	522a	996b	992a	3486d
	UB	583a	15a	664b	917c	998a	580a
Mar-03	LM	2166c	101955c	1332c	583a	1498b	1084b
	LT	480a	8460b	660b	480a	3166c	1250bc
	UL	1333b	107950c	522a	996b	992a	3262d
	UB	583a	15a	664b	917c	998a	542a

Means within one species and observation data followed by the same letter are not significantly different at $P \leq 0.05$. Treatments: burned, logged and use of removal (LT); burned, logged and use of tractors for log removal (LT); burned but not logged (UL); unburned (UB).

Discussion

Watersheds severely denuded by fire are often vulnerable to accelerated rates of soil erosion and can yield large but often variable amounts of postfire sediment (DeBano et al., 1998). The use of ground-based logging equipment can cause additional site disturbance and soil compaction (Beschta et al., 2004). High intensity wildfires can consume up to 99% of the organic matter and soil pH usually increases as a result of burning because basic cations released during combustion are deposited on the soil surface (DeBano et al., 1998). However, our results did not detect any significant change of soil pH, 6 months after the wildfire. The high rates of organic matter in the burned unlogged plot could be explained by the large amount of dead tree parts of the wind-thrown trees. In the logged plots, the logging residues increased the soil organic matter content.

Table 5. Height and survival of *Pinus halepensis* seedlings in the burned plots

	Height (cm)			Survival (%)		
Date	LT	LM	UL	LT	LM	UL
Aug-02	7.3a	9.2a	11.8b	100a	100a	97a
May-03	16.2a	14.0a	21.1b	98a	88b	95a
Aug-03	25.3a	20.0a	32.9b	86a	87a	91a
July-04	37.2ab	35.1b	40.2a	82a	87a	89a

Means within the same day followed by the same letter are not significantly different at $P \leq 0.05$. Treatments: burned, logged and use of mules for log removal (LM); burned, logged and use of tractors for log removal (LT); burned but not logged (UL).

Log removal by tractor was related to higher percentages of bare ground compared to the animal traction. Plant growth on skid trails is reduced (Smith and Wass, 1980) and similar results were obtained when skidding was compared to helicopter, skyline, and skidding over snow systems (Klock, 1975).

We did not find any strong evidence that logging and use of mechanical or animal traction for log removal increased the sediment transfer compared to unlogged treatment. In a similar burned *P. halepensis* forest, erosion rates of the clear-cut plot were very similar to those of the non-modified plot (Marques and Mora, 1998). Sediment yield increases after burning and logging have been reported to be short term and equal the unlogged levels in 3–4 years (Van Lear et al., 1985).

However, reports that compare the soil effects of different wood removal methods show that tractor skidding caused significantly higher levels of soil disturbance and compaction compared to other methods such as cable skidding or use of helicopter (Klock, 1975). Also, postfire logging increased sediment transfer (Helvey et al., 1985), while clearcutting and skidding after prescribed fire increased sediment yields in loblolly pine *(Pinus taeda* L.) stands (Van Lear et al., 1985).

One factor that probably minimized the adverse effects of logging was that the chopped woody debris

Table 6. Mean number of herb and small shurb species per hectare (N ha^{-1}), 1 year after the wildfire

Species	Regeneration	Treatment		
		LM	LT	UL
Anthyllis hermanniae L.	Seeder	50b	50b	500a
Asparagus acutifolius L.	Resprouter	250b	85a	80a
Astragalus monspessulanus L.	Seeder	160b	60c	100a
Bituminaria bituminosa (L.) C. H. Stirt.	Seeder	910b	830b	20a
Chenopodium album L.	Seeder	40b	40b	80a
Chrozophora tinctoria (L.) A. Juss.	Seeder	43a	85b	42a
Convolvulus althaeoides L.	Seeder	6580b	1580c	3580a
Dittrichia viscosa (L.) Greuter	Resprouter	30b	20b	80a
Dorycnium hirsutum (L.) ser.	Seeder	960a	160b	1000a
Fumana arabica (L.) Spach	Seeder	45b	80b	410a
Fumana thymifolia (L.) Spach	Seeder	30b	660b	266a
Lactuca saligna L.	Seeder	85b	45c	160a
Ononis viscose L.	Seeder	35b	25b	80a
Portulaca oleracea L.	Seeder	35a	85b	40a
Rubia pergrina L.	Resprouter	35b	15b	160a
Solanum nigrum L.	Seeder	10b	160c	75a
Sonchus sp.	Seeder	15b	75a	85a
Trifolium sp.	Seeder	25b	85c	45a

Means within the same species followed by the same letter are not significantly different at $P \leq 0.05$. Treatments: burned, logged and use of mules for log removal (LM); burned, logged and use of tractors for log removal (LT); burned but not logged (UL).

remained on the site and probably protected the disturbed soils from rainfall and runoff impact. Logging residues have been reported to reduce soil loss up to 95% (Shakesby et al., 1996). Also, 1 year after the fire, in the unlogged plot, burned trees were windthrown and these fallen trees either left a few decimetres of jagged trunk standing where they had broken off or holes in the ground of about 1 m in diameter where they had been uprooted. Nevertheless, this did not increase erosion because it happened when the ground was already protected by vegetation.

Our results did not show a detrimental effect of logging or log removal method on the regeneration of the dominating woody species. The highest stem densities of the resprouting species *A. unedo, P. latifolia* and *Q. coccifera* were recorded in the logged plots although the regeneration of *P. lentiscus* was better in the unlogged site. In many related studies, postfire management affects plant regeneration and can change the vegetation structure. Cutting down the burned trees, removing the twigs or leaving them in the burned area have only a marginal influence on the total species richness or the percentage cover of resprouting species (Ne'eman et al., 1995). Also, postfire logging can encourage colonization of several native ruderal plant species (Abrahamson, 1984; Greenberg et al., 1994). Removal of burned *Pinus pinaster* trees 1 year after fire had a minimal effect on vegetation structure, except for the *Leguminosae,* which increased in the logged sites (Perez and Moreno, 1998). In an unburned *P. halepensis* forest of S. France, clear-cutting and removal of logs with tractors created a plant community similar to that existing prior to the clear cutting (Gondard et al., 2003).

The emergence of pine seedlings was affected negatively by logging but not their survival. Postfire logging affected negatively the establishment of pine seedlings, since the slight cover provided by mature dead trees reduces insolation and, thus, mortality of seedlings during summer (Ne'eman et al., 1993; Saracino and Leone, 1993). However in a similar study postfire cutting of burned trees had no effect on pine seedling recruitment (Ne'eman, 1997) and Martinez-Sanchez et al., (1999) concluded that traditional wood removal practices do not threaten natural postfire *P. halepensis* re-establishment if initial seedling density is large enough. Furthermore, the timing of felling after fire has a significant effect on seedling survival and in our case, logging occurred at the start of seedling and stem emergence. If logging occured after seedling emergence, a high number of seedlings could be killed.

Table 7. Morphological and biomass variables of the main woody plant species 1 year after the fire

Species	Variables	Treatment LM	LT	UL
Arbutus unedo	Dry weight (g)	33a	21b	23b
	Height (cm)	61a	56ab	53b
	Number of branches	3a	2b	2b
	Number of leaves	145a	109b	90b
Cistus creticus	Dry weight (g)	3a	2a	3a
	Height (cm)	17b	12a	25c
	Main root length (cm)	21a	18a	23a
	Number of branches	5a	4a	5a
	Number of lateral roots	12a	14a	15a
	Number of leaves	78b	35a	87c
Philyrea latifolia	Dry weight (g)	25b	20a	24b
	Height (cm)	62a	69a	86b
	Number of branches	21b	12a	16c
	Number of leaves	266b	180a	205a
Pinus halepensis	Dry weight (g)	2a	1a	1a
	Height (cm)	7a	9a	12b
	Main root length (cm)	22ab	25a	18b
	Number of lateral roots	21a	11b	5c
Pistacia lentiscus	Dry weight (g)	35b	22a	30b
	Height (cm)	74b	56a	66b
	Number of branches	4b	2a	5b
	Number of leaves	65b	40a	72b
Quercus coccifera	Dry weight (g)	19b	17a	20a
	Height (cm)	62a	63a	80b
	Number of branches	18a	19a	18a
	Number of leaves	239a	218a	204a

Means within the same variable and species followed by the same letter are not significantly different at $P \leq 0.05$. Treatments: burned, logged and use of mules for log removal (LM); burned and use of tractors for log removal (LT); burned but not logged (UL).

After a fire, *P. halepensis* seedlings emerge, mainly, during the first wet season (Daskalakou and Thanos, 2004). However, in our study, pine seedlings emerged mainly in the period March to June. This shifting in seedling recruitment time was possibly related to the environmental conditions following the fire which are related to the location of our site, the most northern limit of *P. halepensis* in Greece. A similar delayed seedling emergence pattern has been observed for *Pinus brutia* and suggested as a frost avoiding mechanism (Skordilis and Thanos, 1995).

The effect of logging and log retrieval method on plant biomass and growth depended on plant species. In a similar study, Sexton (1994) found that biomass of vegetation produced 1 and 2 years after postfire logging was 38 and 27% of that produced in postfire unlogged stands. Also, skid trails formed in postfire stands can influence productivity of trees growing directly on them (Smith and Wass, 1980). Resprouting species seem to be more resistant to postfire management as has also been observed by Stuart et al. (1993), who found more hardwood cover in their postfire logged and burned treatments, relative to postfire unlogged controls, and that these hardwoods inhibited establishment and growth of Douglas-fir seedlings.

In conclusion, our study showed that early logging after a wildfire was not a detrimental postfire activity in a *Pinus halepensis* forest. Changes in soil pH, organic matter content and soil moisture were only short term. The fast recovery of vegetation compensated for differences among treatments in the initial period after the fire and reduced the danger for ecosystem degradation.

Acknowledgements

The authors wish to thank the European Commission for providing funds to conduct this research, supported by the Fifth Framework Programme, contract no: QLK5-2001-00289, the European partners for their collaboration and Theologis Palatzidis, Anastasios Takouridis and Zoi Pesimitzi for their help in the field and the laboratory.

References

Abrahamson W G 1984 Species responses to fire on Florida Lake Wales Ridge. Am. J. Bot. 71, 35–43.

Beschta R L, Rhodes J J, Kauffman J B, Gresswell R E, Minshall G W, Karr J R, Perry D A, Hauer F R and Frissell C A 2004 Postfire management of forested public lands of the Western United States. Conserv. Biol. 18, 957–967.

Daskalakou E N and Thanos C A 2004 Postfire regeneration of Aleppo pine—the temporal pattern of seedling recruitment. Plant Ecol. 171, 81–89.

DeBano L F, Neary D G and Ffolliott P F 1998 Fire's Effects on Ecosystems. Wiley, New York.

Gondard H, Romane F, Aronson J and Shater Z 2003 Impact of soil surface disturbances on functional group diversity after clear-cutting in Aleppo pine *(Pinus halepensis)* forests in southern France. For. Ecol. Manage. 180, 165–174.

Greenberg C H, Neary D G and Harris L D 1994 Effect of high-intensity wildfire and silvicultural treatments on reptile communities in sand-pine scrub. Conserv. Biol. 8, 1047–1057.

Helvey J D, Tiedemann A R and Anderson T D 1985 Plant nutrient losses by soil erosion and mass movement after wildfire. J. Soil Water Conserv. 40, 168–173.

Klock GO 1975 Impact of five postfire salvage logging systems on soils and vegetation. J. Soil Water Conserv. 30, 78–81.

Marques M A and Mora E 1998 Effects on erosion of two postfire management practices: clear-cutting versus non-intervention. Soil Till. Res. 45, 433–439.

Martínez-Sánchez J, Ferrandis P, De las Heras J and Herranz JM 1999 Effect of burnt wood removal on the natural regeneration of *Pinus halepensis* after fire in a pine forest in Tus valley (SE Spain). For. Ecol. Manage. 123, 1–10.

McIver J D and Starr L 2000 Environmental Effects of Postfire Logging: Literature Review and Annotated Bibliography. Gen. Tech. Rep. PNW-GTR-486.Portland, OR: U.S. Department of Agriculture, Forest Service, Pacific Northwest Research Station. 72 pp.

Naveh Z 1975 The evolutionary significance of fire in the Mediterranean region. Vegetatio 29, 199–208.

Ne'eman G 1997 Regeneration of natural pine forest-Review of work done after the 1987 fire in Mount Carmel, Israel. Int. J. Wildland Fire 7, 295–306.

Ne'eman G, Lahav H and Izhaki I 1993 The resilience of vegetation to fire in an east Mediterranean pine forest on Mount Carmel, Israel: the effects of post-fire management. *In* L Trabaud and R. Prodon (eds), Fire in Mediterranean Ecosystems, pp. 127–141. Commission of the European Communities, Brussels-Luxembourg.

Ne'eman G, Laha v H and Izhaki I1995 Recovery of vegetation in a natural east Mediterranean pine forest on Mount Carmel, Israel as affected by management strategies. For. Ecol. Manage. 75, 17–26.

Perez B and Moreno J M 1998 Fire-type and forestry management effects on the early postfire vegetation dynamics of a *Pinus pinaster* woodland. Plant Ecol. 134, 27–41.

Saracino A and Leone V 1993 Natural regeneration 2 and 4 years after fire of *Pinus halepensis* Miller in dunal environment. *In* L Trabaud and R Prodon (eds), Fire in Mediterranean Ecosystems, pp. 141–150. Commission of the European Communities, Brussels—Luxembourg.

Sexton TO 1994 Ecological Effects of Post-Wildfire Salvage-Logging on Vegetation Diversity, Biomass, and Growth and Survival of *Pinus ponderosa* and *Purshia tridentata*. Corval-lis, OR: Oregon State University, Department of Rangeland Resources, 28 pp.

Shakesby R A, Boakes D J and Coelho C O A 1996 Limiting the soil degradational impacts of wildfire in pine and eucalyptus forests in Portugal: a comparison of alternative post-fire management practices. Appl. Geogr. 16, 337–355.

Skordilis A and Thanos C A 1995 Seed stratification and germination strategy in the Mediterranean pines *Pinus brutia* and *P. halepensis.* Seed Sci. Res. 5, 151–160.

Smith R B and Wass E F 1980 Tree Growth on Skidroads on Steep Slopes Logged after Wildfires in Central and Southeastern British Columbia. BC-R-6. Victoria, BC: Canadian Forestry Service; Pacific Forest Research Centre, 27 pp.

Soto B, Benito E and Díaz-Fierros F 1991 Heat-induced degradation processes in forest soils. Int. J. Wildland Fire 1, 147–152.

Stuart J D, Grifantini M C and Fox L 1993 Early successional pathways following wildfire and subsequent silvicultural treatment in Douglas-fir/hardwood forests, NW California. For. Sci. 39, 561–572.

Thanos C A and Doussi M A 2000 Post-fire regeneration of *Pinus brutia* forests. *In* G Ne'eman and L Trabaud (eds), Ecology, Biogeography and Management of *Pinus halepensis* and *P. Brutia* Forest Ecosystem in the Mediterranean Basin, pp. 291–301. Backheys Publishers, Leiden.

Trabaud L 1987 Fire and survival traits in plants. *In* L. Trabaud (eds), Role of Fire in Ecological Systems, pp. 65–90. BackLuys Publishers, Leiden.

Van Lear D H, Douglass J E, Cox S K and Augspurger M K 1985 Sediment and nutrient export in runoff from burned and harvested pine watersheds in the South Carolina piedmont. J. Environ. Qual. 14, 169.

The contribution of agrotechnical works following a fire to the protection of forest soils and the regeneration of natural forest

M.A. Sapountzis[1], G.S. Efthimiou[2] & P.S. Stefanidis[3]

[1] *University of Thessaly, Department of Agriculture Animal Production and Aquatic Environment, Fitoco st., 38446 N. Ionia Volos, Greece.* [2] *NAGREF, Forest Research Institute, Vassilika, 57006 Thessaloniki, Greece.* [3] *Aristotle University of Thessaloniki, Department of Forestry and Natural Environment, Un. Campus Box 268, 54124 Thessaloniki, Greece.*

Key words: channelling, drift of branches, log erosion barrier, reforestation, soil erosion

Abstract

During 1997 a fire destroyed about 62% of the urban forest in the city of Thessaloniki, northern Greece. Between 50 and 80% of the forest vegetation was burned within the watersheds of the six torrents that flowed across the destroyed area. In this area the forest service constructed agrotechnical, and flood-protection works in the streams and the basins of the torrents to prevent erosion and to protect the city from flooding. The agrotechnical works consisted of channelling, drifts of branches and log erosion barriers. In the present study the effectiveness of the operation of the agrotechnical works in the protection of the erosion phenomena is examined. The success of the agrotechnical works are examined 7 years after their construction and the contribution of these works to the natural regeneration of the forest is evaluated. The erosion prevention and flood protection measures that were constructed contribute significantly to the stability of the forestal soil and the protection of floods. Their contributions in assisting the natural reforestation works were also remarkable.

Introduction

On 6th July 1997 the urban forest of Thessaloniki was set on fire. The fire, which lasted two days, burned down 16.095 km^2 out of 26.095 km^2 of the total urban forested area around the city (Kailidis, 1997; Tourlakidis, 1997). The urban forest covered the slopes of Kedrinos Lofos, Hortiatis massif and was established 65 years ago. It was composed of mainly coniferous species and predominantly by brutia pine (*Pinus brutia* Ten.). The purpose of establishing the urban forest was hydrological, mainly to protect the city of Thessaloniki from floods as well as to protect the slopes from erosion phenomena. In addition, that particular forest served as recreation purposes of the continuously growing city (Gatzogiannis et al., 1996). Immediately after the fire, the forest service constructed agrotechnical and flood-protection works in the streams and the basins of the torrents (rushing streams) to prevent erosion and to protect the city from flooding. An inventory of the work done shows a total construction of 1,500 km of channelling, 663 km of drift branches, 84 km of log erosion barriers, 188 miles of timbered dams and 7 cement dams. These agrotechnical works created the suitable preconditions and prerequisites for the present study. The aim of the present research was to study the effectiveness of the operation of the agrotechnical works in the protection of the erosion phenomena and the success and serviceability of the works, 7 years after their construction. At the same time the contribution of these works to the natural reforestation was evaluated. This study continues and expands on previous studies addressing subjects determining the erosion phenomena that follow after a forested area has been set on fire. Immediately after the fire, experimental plots for studying the erosion (ESE: Experimental Surfaces of Erosion) were established in all the watersheds of the torrents within the urban forest, by employing the method of standard marker-points (Stefanidis et al., 2002).

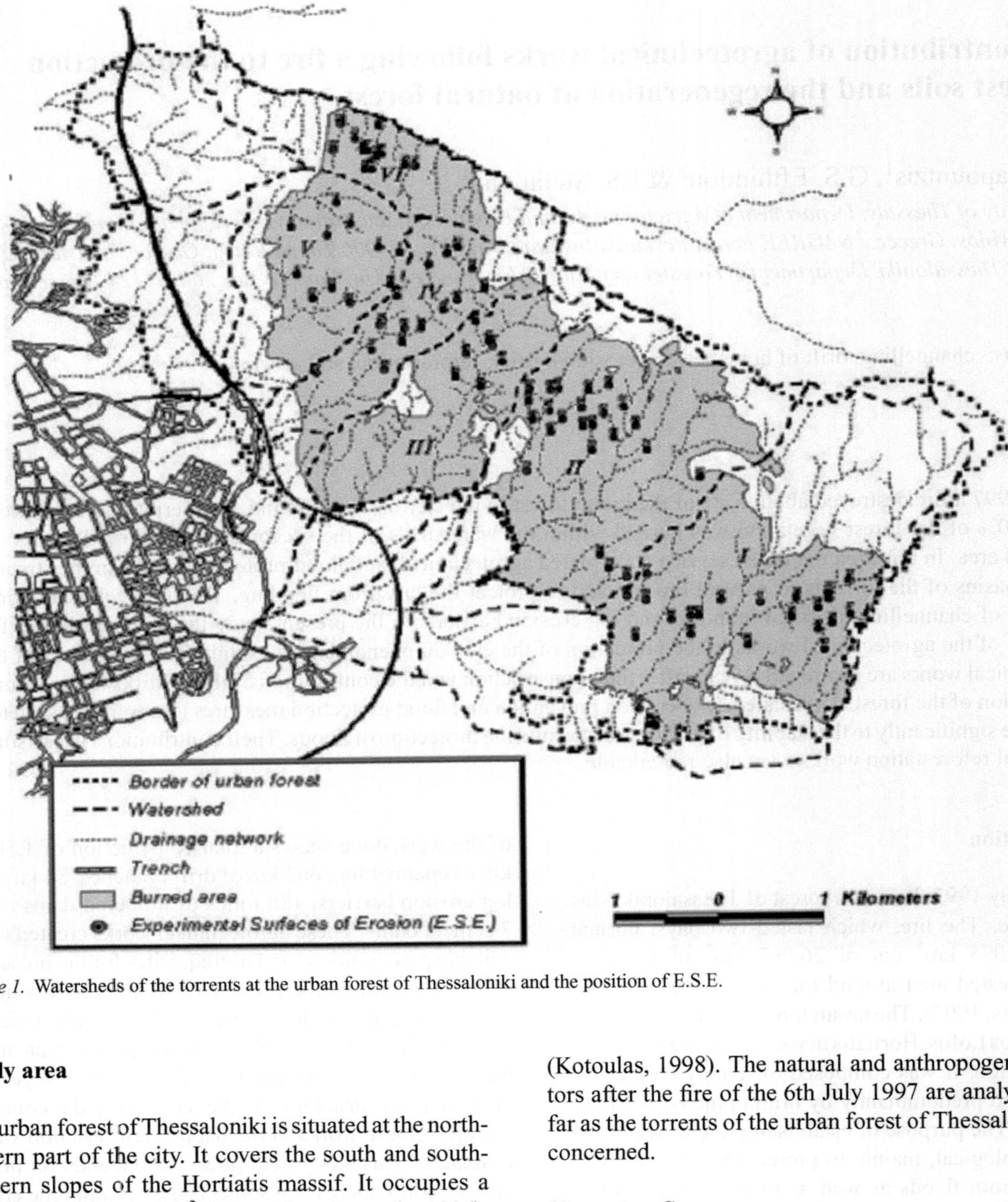

Figure 1. Watersheds of the torrents at the urban forest of Thessaloniki and the position of E.S.E.

Study area

The urban forest of Thessaloniki is situated at the north-eastern part of the city. It covers the south and south-western slopes of the Hortiatis massif. It occupies a total area of 30.19 km^2 (Gatzogiannis et al., 1996), within the following coordinates: Longitude: 22° 57′ E to 23° 04′ E Latitude: 40° 35′ N to 40°39′ N (Figure 1).

The combination of natural and anthropogenic factors in a watershed, determines the way and processes for the movement of the run-off water as well as the appearance and development of degradation phenomena (Kotoulas, 1998). The natural and anthropogenic factors after the fire of the 6th July 1997 are analysed, as far as the torrents of the urban forest of Thessaloniki is concerned.

Vegetation Cover

The wider research area from a phytosociological point of view is part of the Paramediterranean vegetation zone (*Quercetalia pubescentis* Br.-Bl.) and specifically of the growth area of the Kermes Oak (*Coccifero-Carpinetum*) (Athanasiadis, 1978; Dafis,

Table 1. Total burned area and respective number of E.S.E. per basin

Watershed	Total surface (km^2)	Burned area (km^2)	Burned area (%)	Number of E.S.E.
I	8.424	4.237	50.29	25
II	5.757	4.494	78.05	25
III	3.027	2.361	77.99	14
IV	2.865	1.901	66.35	15
V	3.091	2.261	73.15	13
VI	2.931	1.385	47.26	12
Total	26.095	16.095	61.68	104

1973). This artificial forest park is 65 years old and comprises only conifers. Its composition is: 90% *Pinus brutia* with the remaining 10% being *Cupressus sempervirens* L. and *Pinus pinea* L. (Grigoriadis et al., 2003).

After the fire, a major part of the urban forest was destroyed. Table 1 gives the distribution of burned areas in each watershed. As can be seen from Table 1, 61.68% of the forest vegetation in the watersheds was burned during the fire.

Geology

The geological profile of the greater area of Thessaloniki, is as follows:

The plane areas are mainly consisted of sand, conglomerates, clay and terra-rosa. The higher areas are consisted of greenschists, gabbro, limestone phyllites, peritodites, crystalline limestone, limestones and dolomites.

Topography

As far as the relief is concerned, the altitude and the prevailing slopes in the area are of great importance. The altitude is highly correlated to all climatic characteristics. The prevailing slopes in the torrent banks and in the watersheds may accelerate or decelerate the run-off water and thus they have a strong effect on the torrential environment.

The altitude of the research area is between 50 and 700 m and the slopes are in the range of 5–65%. Thus, the relief of the area can be considered as semi-mountainous.

Climate

Among the climatic characteristics of special importance are: the mean annual rainfall and its monthly distribution, the intensity of the rain and the air temperature. A meteorological station that was placed in "Fillipio" (within the experimental area, altitude 367 m) provided data for the needs of the current study. Table 2 gives the monthly rainfall measurements from this station for 3 years (October 1997–September 1998, October 1998–September 1999, October 1999–September 2000).

The rainfall-gauge at "Fillipio" recorded 360.70 mm during the first hydrological year after the fire, the second year 555.00 mm, and during the third year 323.90 mm of rainfall.

Climatic data provided from the Meteorological station of the Forest Research Institute—NAGREF (from 1978 to 2003) suggest that the climate of Thessaloniki and its urban forest could be classified as cold, the bioclimatic strata semiarid with cold winters and an ombrothermic value of $Q = 47.7$. The mean annual rainfall is approximately 435 mm, and the dry season extends from May to October (Figure 2). July is recorded as the hottest month and January the coldest.

Materials and methods

After the fire, the forest service constructed agrotechnical and flood-protection works in the streams and the

Table 2. Meteorological data from the station of Fillipio

	O	N	D	J	F	M	A	M	J	J	A	S	Annual
Fillipio							mm						
1997/98	85	21	67	40	38	7	7	68	3	2	6	16	360
1998/99	27	151	37	10	43	65	31	29	74	21	9	57	555
1999/00	33	81	72	10	47	7	26	22	3	11	0	11	324
Average	48	84	58	20	42	26	21	39	27	11	5	28	413

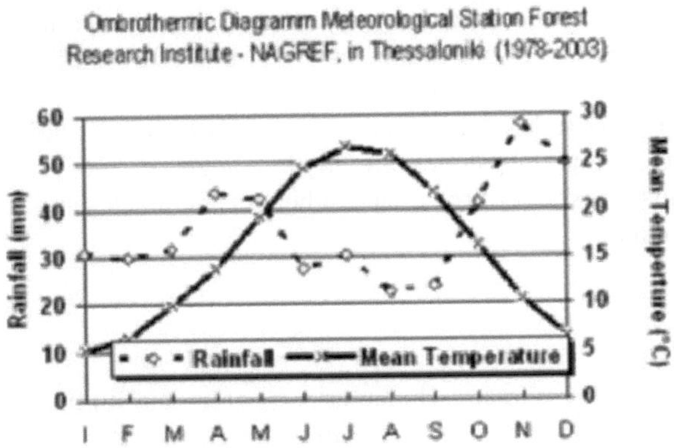

Figure 2. Obrothermic diagram.

basins of torrents for preventing erosion and flooding. A description of the works is as follows.

Agrotechnical works

Drift of branches: All the remains of the logging of the partially burnt trees were stacked along the contours in lines at 0.70 m height and 1.0–1.5 m width. The distance between each contour line was 10 m. Drifts were constructed in places with slopes of 0–30% (Figure 3).

Channelling: The surface between drift branch lines was furrowed with a ripper, forming furrows at a distance of 2.0–2.5 m. The depth of each channel was at least 0.70 m deep. This resulted in a final surface channel of triangular shape.

Log erosion barriers were constructed in places with 30–50% slope, mainly from Pinus and Cupressus logs of about 0.2 m diameters. Log barriers were established

Figure 4. Log erosion barriers across the contours.

across the contours supported by the remains of the burned trees and appropriate poles (Figure 4).

The effectiveness of the works for soil protection and erosion control was determined with the establishment of experimental surfaces (ESE) and the application of the standard point's method (Figure 5). Soil erosion and transferred material were estimated in 104 experimental surfaces from 1998 to 2000 (Stefanidis et al., 2002).

Within each ESE 5 poles were put in a circle area, with a diameter of 4 m. On each pole, the part above the soil surface was painted red, by using a permanent colour, so as to define the starting point (standard marker-point) of possible erosion phenomena in the future.

Two measurements-census of the sheet erosion was carried out in each ESE. The measurements were carried out during September and October 1998, and September and October 1999 (Stefanidis et al., 2002). In each plot, the loss of soil was measured in each pole, by carrying out four measurements, one each side of

Figure 3. Drift of branches on slopes of 0–30%.

Figure 5. Experimental surfaces of erosion (E.S.E.).

the pole, The soil loss was measured at each pole as the mean of the soil loss values from all four sides of the pole:

$$P_i = \frac{a+b+c+d}{4} (mm)$$

The main factor for erosion phenomena, in a certain area, is the rainfall (Kotoulas, 1972; Moench and Fusaro, 2002). For the rainfall monitoring, a rainfall-gauge (type Pentix) with weekly recording, was put in the studied area. The rainfall-gauge established on the 24th of September 1997, at the "Hotel Fillipio" (watershed IV) and is still in use.

For the initiation of sheet erosion, the intensity and the duration of the rain is of major importance. The most important rainfall features that were recorded at the Meteorological Station of Thessaloniki are the following:

- The maximum rainfall intensity (24 h period) was recorded in November 1985 (98 mm in 24 h). During the dates 23–25 November 1985, 114 mm of rain were recorded in a 40 h period.
- On 15th July 1972, 48.7 mm were recorded within 80 min time.

By processing the data from the rainfall-gauge (which was established for the purposes of the present research to the following) resulted:

- The first hydrological year after the fire (October 1997–September 1998) the rainfalls were of minor intensity (4–7.5 mm/ 2 h).
- During the second hydrological year after the fire (October 1998–September 1999) the rainfall were of greater intensity (9–15 mm/h). In addition, rainfalls with intensity of 24 mm/ 17min and 27 mm/ 30min, respectively, were also recorded.

Intense surface erosion occurred after the exponential prevalence of the above fact. The total sediment load of the transferred materials was compared with the volume of the material, which was retained in the constructed works.

The evaluation of the current situation of each agrotechnical construction was performed with *in situ* observation plus weight resistance in the place of the works and all the watersheds of the torrents in the study area.

The contribution of these works to the success of the natural reforestation was carried out with visits to the study area and an inventory of the existing forest species. Records were taken from 5 m long and 1.0 m wide sections of 100 drift branches and a zone of about 1.0 m wide from 100 log barriers.

Results

Natural factors before fire were contributing to a mild torrential environment, while vegetation secured a substantial protective hydrological influence. However, the destruction of nearly 62% of the urban forest increased dramatically the possibility of widespread flooding.

Soil erosion

After completing the second census, the data was processed as follows (Stefanidis et al., 2002): Table 3 gives the values of annual degradation in each watershed. The degradation during the first year after the fire (Nov. 1997–Oct. 1998) was in the range from 0.030 mm/year to 0.180 mm/year, while during the second year after the fire (Nov.1998–Oct.1999) it was from 0.340 mm/year to 1.870 mm/year (Table 3). The degradation in the

Table 3. The annual degradation (sheet erosion) in the burned areas at the Urban forest of Thessaloniki (Stefanidis et al., 2002)

Watershed	Annual degradation (accelerative erosion)		Special annual sediment load (accelerative erosion)		Annual sediment load (accelerative erosion)	
	1998 mm/year	1999 mm/year	1998 m^3/km^2 year	1999 m^3/km^2 year	1998 m^3/year	1999 m^3/year
I	0.060	0.820	60	820	254.2	3,474.3
II	0.180	1.870	180	1,870	808.9	8,403.8
III	0.110	1.740	110	1,740	259.7	4,108.1
IV	0.030	1.530	30	1,530	57.0	2,908.5
V	0.130	1.420	130	1,420	293.9	3,210.6
VI	0.060	0.340	60	340	83.1	470.9

Figure 6. Log erosion barrier after 5–6 years of operation.

total burned area was estimated as the average of the values in all burned areas in the watershed, and it had the value of $Ev_{2000} = 1.357$ mm/year.

The special annual sediment load (per unit of surface) in the total burned area of the urban forest of Thessaloniki was estimated as the average of the values of all the burned areas within the watersheds, and it had the value: $W_{2000} = 1{,}375$ m^3/km^2 year.

Heavy rainfall (4–7.5mm in 2 h during the first year, 9–15 mm in 2 h and strong rainfalls 24.17 mm/30 min and 27 mm/30 min during the second year, June and September) during the experimental period seems to account for the observed soil erosion.

The total sediment load in the total burned area was estimated, namely as the average of the values of the burned areas in each watershed, and it had the value: 22,579.1 m^3 /year. This total quantity was retained by the constructed work (Figure 6).

Evaluation of operation of executed works

Drift of branches: Their effect on preventing erosion was limited because there was insufficient affinity with the ground.

Channelling: It satisfactorily functioned apart from the semi mountainous or green schist areas that resulted rather than the expected results for a natural reforestation.

Log erosion barriers satisfactorily functioned until more recently due to the amount of precipitation in the region (Figure 6).

Current situation of constructions

Drift of branches: Almost all the drifts have dried and constitute inflammable material. Their removal would assist in the protection of future forest fire events (Figure 3).

Log erosion barrier: Trunks of Pinus: 30% show advance decay and 70% show decomposition of rinds.

Trunks of Cupressus: proved particularly durable with 100% showing no signs of decomposition.

Natural reforestation:

The orientation of the slope is a detrimental factor for natural reforestation (Tsitsoni, 1997). In general, sites with northern orientations are more favorable, and therefore, natural reforestation was more pronounced compared with southern sites. This was confirmed after several visits to the experimental urban forest of Thessaloniki.

Table 4 gives the frequency of the occurrence of seedlings that was recorded both on drift of branches and on log erosion barriers. Table 4 shows also that natural reforestation was directly assisted from log contour terraces with the withholding of soil and seeds of the forest species, Pinus (*P. brutia* and *P. pinea*) and Cupressus (*C. sempervirens* and *C. arizonica* Greene) from erosive action of water. Natural reforestation of *Cupressus* is observed everywhere in log places and in less cases in the drift places. Natural reforestation of Pinus is observed less (Figure 7).

Table 4. The frequency of forest species on agrotechnical works

	Forest species							
	Pinus (*P. brutia* and *P. pinea*)				Cupressus (*C. sempervirens* and *C. arizonica*)			
	Frequency							
Agrotechnical works	0	1–2	2–4	>4	0	1–2	2–4	>4
Drift of branches	37	6	2	0	30	17	7	1
Log erosion barrier	8	30	11	1	3	27	16	4

Figure 7. Drift of branches and natural regeneration of forest.

Pinus brutia and *Pinus halepensis*, are considered as fire resistant because they have developed mechanisms for natural reforestation after forest fires (Castri and Mooney, 1973; Dafis, 1983; Spanos et al., 2000, 2001; Thanos et al., 1989; Tsitsoni, 1997; Tsitsoni and Zagas, 1988).

Conclusions

For the area of the urban forest in Thessaloniki before the fire the natural factors determined a wild torrent environment with the existing forestal vegetation providing significant protection from the hydrological effects. After the disaster, the danger of flood events was increased.

The measures against erosion and flood works that were constructed from the Forest Services in the City of Thessaloniki is the objective of current study and contribute significantly to the stability of the forestal soil and the protection of floods. Their contributions to the natural reforestation were also remarkable.

The life span of drift of branches is restricted to 5–6 years, and therefore they cannot provide effective protection from soil erosion. Furthermore, they increase the probability of fires. On the other hand, log erosion barriers continue to be functional, and provide satisfactory protection beyond 5–6 years.

References

Athanasiadis N 1978 Forest Phytosociology. Aristotle University Press, Thessaloniki, Greece (in Greek).

Dafis S 1973 Classification of forest vegetation of Greece. Ass. Tutor at the School of Agronomy-Forestry of AUTH, vol. 15/2, pp. 75–91.

Dafis S 1983 The role of the forest towards the protection of the environment. For. Chronicles T. 9-10, Athens (in Greek).

Gatzogiannis S, Kiriakidis P and Gigis X 1996 Master Plan of Urban Forest of Thessaloniki. Forest Servise—Forest Research Institute, Thessaloniki.

Grigoriadis N, Zigoura V, Efthimiou G and Takos I 2003 Evaluation of the restoration in burned down periurban forest of Thessaloniki. *In* Proceedings of 2nd International Conference on Ecological Protection of the Planet Earth, National Palace of Culture, Sofia Bulgaria, pp. 637–644.

Kailidis D 1997 The great forest fire in the urban forest of Thessaloniki 6-7 July 1997. Thessaloniki (in Greek).

Kotoulas D 1972 Die Wildbache Suddeutschlands und Griechelands, Teil 1. Bericht nr. 25 der versuchsanst. f. Wasserbau d. T.U. München.

Kotoulas D 1998 Mountainous water management and control. Vol. I. Aristotle University of Thessaloniki, Aristotle University Press, Thessaloniki. (in Greek)

Moench R and Fusaro J 2002 Soil erosion control after wildfire. Natural Resources Conservation Service. 4/02. No 6.308.

Stefanidis P, Sapountzis M and Stathis D 2002 Sheet erosion after fire at the urban forest of Thessaloniki (Northern Greece). Silva Balcanica, vol 2 (in press).

Spanos I A, Daskalakou E A and Thanos C A 2000 Postfire, natural regeneration of Pinus brutia forest in Thasos island, Greece. Acta Ökologica 21, 13–20.

Spanos I A, Radoglou K M and Raftoyannis Y 2001 Site quality effects on post-fire regeneration of Pinus brutia on a Greek island. Appl. Vegetation Sci. b4: 229–236.

Thanos A C, Markou S, Christodulakis D and Yannitsanos A 1989 Early post regeneration in *Pinus brutia* forest ecosystems of Samos island (Greece). Acta Ökologica 10, 79–94.

Tourlakidis C 1997 Special study on the construction works, for the protection against erosion and floods, of the burned urban forest of Thessaloniki. Forest Service of Thessaloniki, Thessaloniki. (in Greek).

Tsitsoni T 1997 Conditions determining natural regeneration after wildfires in the *Pinus halepensis* forests of Kassandra Peninsula (North Greece) For. Ecol. Manage. 92, 199–208.

Tsitsoni T and Zagas T 1988 Research of the natural reforestation after the fire in the urban forest of Thessaloniki. Greek Forest Union. D Congress, Larisa 5–7 October 1988, pp. 62–73 (in Greek).

United States Department of Agriculture, NRCS. Natural Resources Conservation Service. http:www.wy.nrcs.usda.gov.

Decision support systems in eco-engineering: the case of the SDSS

Slobodan B. Mickovski[1,2,*] & L. P. H. van Beek[3]

[1] *Laboratoire de Rhéologie du Bois de Bordeaux (CNRS/INRA/Université Bordeaux I), 69, Route d'Arcachon, 33612 Cestas cedex, France.* [2] *Current address: Civil Engineering Department, University of Dundee, Dundee DD1 4HR, Scotland, U.K.* [3] *IBED–Physical Geography, University of Amsterdam, Nieuwe Achtergracht 166, NL 1018, WV Amsterdam, The Netherlands.*

Key words: decision support system, eco-engineering, erosion, expert systems, landslides, windthrow

Abstract

The increased number of catastrophic slope instabilities in Europe in the last decade highlights the need for integrated management of such occurrences. The decision support system presented here incorporates expert knowledge on slope stability and eco-engineering and represents user-friendly solution for help in decision making process on mitigation of slope instability. Based on the results of field studies, experience of the project partners, theory and case studies available in the literature, the slope decision system (SDSS) is a modern decision support system that promotes use of eco-engineering solutions for mitigation of slope stability problems connected with the uppermost soil horizons. The basic considerations, design aims and objectives, together with the advantages and the disadvantages of the approach are discussed in the light of ever-evolving decision making system theory. The future of eco-engineering decision support systems is also discussed with a special emphasis on the learning component for the end-user.

Introduction

In today's Europe we are facing an increased number of catastrophic incidents due to mass instability problems on slopes. The predictions of environmentalists (UNEP, 2002) for the future are not encouraging knowing that climatic change is more likely to produce new occurrences of such disasters and with greater frequency. Torrential rains and floods have lead to soil, debris and mudslides in barren areas, while the extremely hot and dry periods have led to an alarming number of forest fires in Europe, leaving the soil in the aftermath of the events prone to mass wasting during heavy rains. Such instabilities that often have a frequency of 15 to 20 events per year, inflict heavy damages to European agriculture and forestry, especially in the Mediterranean regions of Europe. The cost of such disasters in terms of loss of human life is also staggering with, for example, nearly 6000 deaths in the landslides of the nineties in Italy (Guzetti, 2000). The damage to European forests has also been severe, with frequent loss of trees due to windthrow resulting in large barren areas. Uprooting of a tree usually leaves a void in the ground, which, when left without the reinforcing effect of the roots, is often filled up with water and, in turn, causes further soil erosion and mass instability.

These problems have resulted in an increased research activity of slope instability phenomena. The erosion processes in the affected areas in Europe have been well documented and understood (Rubio and Bochet, 1998; Gimeno et al., 2001); the landslide hazard and occurrences are themes of an increased number of studies (Guzetti, 2000; van Beek, 2002; Kilburn and Pasuto, 2003) whilst the study of tree stability with regards to wind is also being studied intensively in Europe (Gardiner et al., 2000; Mickovski and Ennos, 2002; Cucchi et al., 2004).

In order to be able to deal with the occurrence and the consequences of such phenomena, and also to summarise current knowledge, there was a need for a decision support system that would help end users reach fast and accurate decisions with regards to slope

* E-mail: s.b.mickovski@dundee.ac.uk

instability. The objective of the slope decision support system (SDSS) was to guide agriculturalists, foresters and eco-engineering practitioners to better and more balanced decisions for slope stabilisation, and at the same time providing a learning component that consists of background knowledge available to the authors either from the literature or from experience. This system was developed under the auspices of the EU funded project Ecoslopes: Eco-engineering and Conservation of Slopes for Long-term Protection from Erosion, Landslides and Storms (EU, QLK5-2001-00289, www.ecoslopes.com). This project brought together nine partner scientific institutions from around Europe, and combined their experience and expertise in different scientific disciplines connected to the subject of slope stability.

Creating the SDSS

Knowing that the term 'decision support system' (DSS) is an all-inclusive term for many types of information systems that support decision making at different organizational levels, the idea was to make the SDSS an interactive, computer-based system intended to help users make decisions by assisting in retrieval, summary and analysis of decision relevant factors.

The Ecoslopes partners gained as much knowledge about DSS as possible from the literature available and the expertise of some of the partners in the field of DSS, and then aimed for a DSS design process that involved both the future users of the DSS (the project end-user group committee) and the experts in the fields of slope stability (project partners). The expertise and the knowledge behind the project partnership negotiated the DSS design with the end users about the needs, capabilities, deliverables, outcomes, and what decisions should be supported by the proposed system.

The basis

The premises for the new DSS were that, although multidisciplinary, it had to be simple and intuitive by nature, helping users from different backgrounds to explore the implications of their decisions across the spectrum of the phenomena involved. The DSS was envisaged to provide the user direction and advice in support of decision making by efficiently using the data readily available to the user and correspond to his/hers specific needs and possibilities, at the same time providing a learning component for the user. Finally, among the priorities defined in agreement with all of the project partners, was the unanimous decision to focus on integrated management of slope safety by eco-engineering. This discipline can be defined as the long-term strategy to protect and/or restore a site with regards to natural or man-made hazards (Stokes et al., 2004). For natural slopes within Europe, such hazards include mass movement of soil e.g. landslides, avalanches and rockfall, or erosion e.g. sheet and gully erosion or river bank erosion. The idea of the DSS developers was to effectively manage slopes to minimise the risk of failure by combining ground bio-engineering techniques with long-term solutions. Ground bioengineering methods featured in the DSS would integrate civil engineering techniques with natural materials to obtain fast, effective and economic methods of protecting, restoring and maintaining the environment (Schiechtl, 1980; Coppin and Richards, 1990).

Making the decision on what type of DSS would be chosen as a base for the SDSS was not easy, knowing that the priorities and concerns of each of the project partners were different. These differences, in a way, gave the answer to this question: since all of the partners were experts in different areas of slope stability, it was decided to build the SDSS as an 'expert system', or 'knowledge based system'. The advantages of such a system are that it usually has an explicit knowledge base (KB), analyses data using symbolic logic involving user input parameters, and has the ability to explain the output decision in a way understandable for the user. All of these characteristics coincided with the priorities set up for the design and development of the SDSS. It was therefore decided that the shell to accommodate the SDSS should be the ConFound® software (Toll and Barr, 2001), originally developed for geotechnical foundation design. The basic software package provided an environment in which both the KB could be created and entered hierarchically by the expert (supervisor mode), and where this information can be explored interactively by the user (normal/test mode).

Knowledge base and project specific information

In order to create the KB-basically the scope of the SDSS where the available information on slope stability is stored in pre-designed manner—an assessment of the slope instability problems was needed together with the factors affecting these problems. Identifying the soil

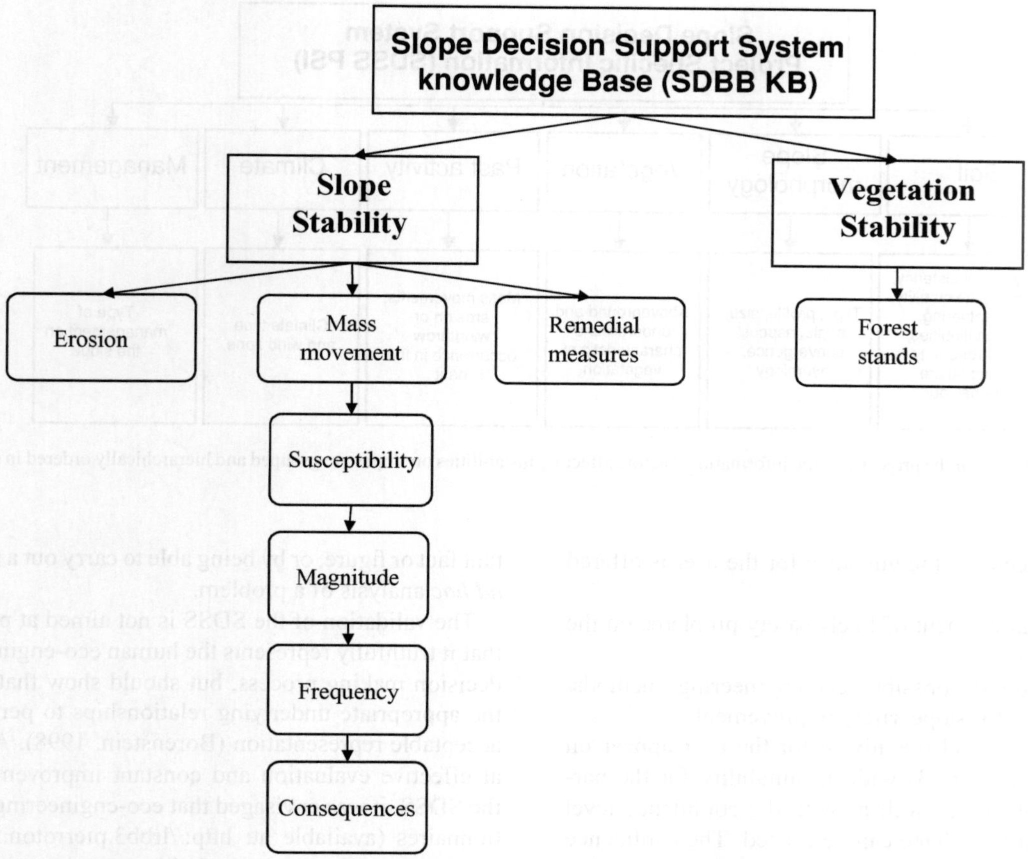

Figure 1. Structure of SDSS knowledge base. Slope stability and vegetation stability are the two main frames, whilst particular instability problems together with the eco-engineering remedial measures are sub-categorised under the main frames.

stability and vegetation stability on slopes as main focus areas for the SDSS (Figure 1) enabled setting the KB frames where, following the envisaged objectives, different instability phenomena together with the eco-engineering remedial measures were sub-categorised.

To solve a particular problem with a DSS the most important issues to be considered are how to elicit the end user's preference information and how to use this information to guide the end user to the best applicable solution (Chen and Lin, 2003). In order to achieve the former, a series of interviews with the project partners and the end users was carried out from which invaluable information from different aspects of slope instability was obtained. In order to achieve the latter, a detailed literature review together with the experimental data from the project investigations were also used in the creation of the KB. This process has proven invaluable not only because it provided the basis for creation of an all-inclusive help file, but also for structuring the factors affecting slope stability as project specific information (PSI). These factors put in order all relevant information that is provided by the user in a predefined form and included environmental, climate, management, morphological and geographical factors influencing instability processes on slopes (Figure 2).

SDSS output

The PSI-structured data entered by the user is evaluated in the software by logic-based rules that reflect the knowledge included in the KB. The rules pertain to the set of frames (objects of interest or themes such as erosion, soil instability or remedial measures, Figure 1), and evaluate the suitability of a selected frame for the PSI that has been entered by the user. Following the

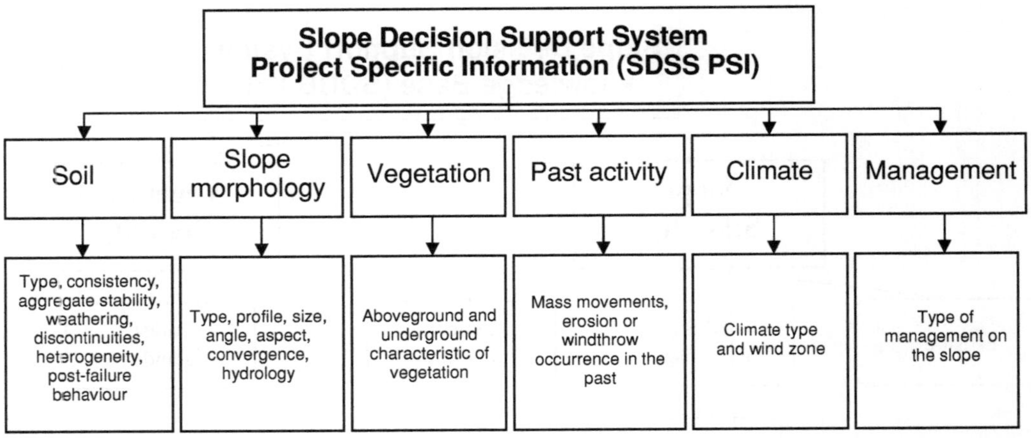

Figure 2. Structure of the project specific information. Factors affecting instabilities on slopes are grouped and hierarchically ordered in separate categories.

design objectives, the guidance for the user is offered on

- hazard assessment of likely safety problems on the slope and
- evaluation of possible eco-engineering methods/strategies for slope safety improvement

The guidance and the advice for the user appear on the screen (Figure 3) with its suitability for the particular problem, as well as with the confidence level with which the advice can be trusted. The confidence level of a particular advice increases with the number of factors (PSI) considered (Stokes et al., 2004).

Any additional data analysis that the user can perform and then return to SDSS with the extra input, could make decision support by the SDSS more unique and appropriate. Ideally, the SDSS should be accompanied by pre- and post-processing that would facilitate the data input and analysis by the user. This pre- and post-processing is suggested and referred to both in the SDSS output and in the help file where it is often presented as additional computer programs, graphs and tables.

Evaluation and validation

Both the DSS developers and the end-users need to assess the potential of the DSS critically. The DSS should not replace the end user and the responsibility for making a decision, or have all the answers for a particular problem. Sometimes, the end-user might benefit from the SDSS usage by rapidly retrieving a single important fact or figure, or by being able to carry out a simple *ad hoc* analysis of a problem.

The validation of the SDSS is not aimed at proving that it truthfully represents the human eco-engineering decision making process, but should show that it has the appropriate underlying relationships to permit an acceptable representation (Borenstein, 1998). Aiming at effective evaluation and constant improvement of the SDSS, it was envisaged that eco-engineering questionnaires (available at: http://lrbb3.pierroton.inra.fr/ecoslopes/conference/quest/bobby-questionnaire.htm) should be sent to relevant institutions and persons who have experience with using eco-engineering for slope protection. Information on (un)successful practices should be put in the SDSS and the output should be checked against the eco-engineering method evaluation from the questionnaire. The benefits of such an action are threefold: the SDSS can be re-evaluated with each new response and any shortcomings of the SDSS can be eliminated whilst at the same time, a database of eco-engineering practice is built from the returned questionnaires. Such a database can be an integral part of the SDSS help file, and will help in providing reference of successful eco-engineering methods for the users who hesitate to use vegetation in slope stabilisation.

Advantages and disadvantages

SDSS incorporates the components of a classic DSS having methods and instruments for dealing with

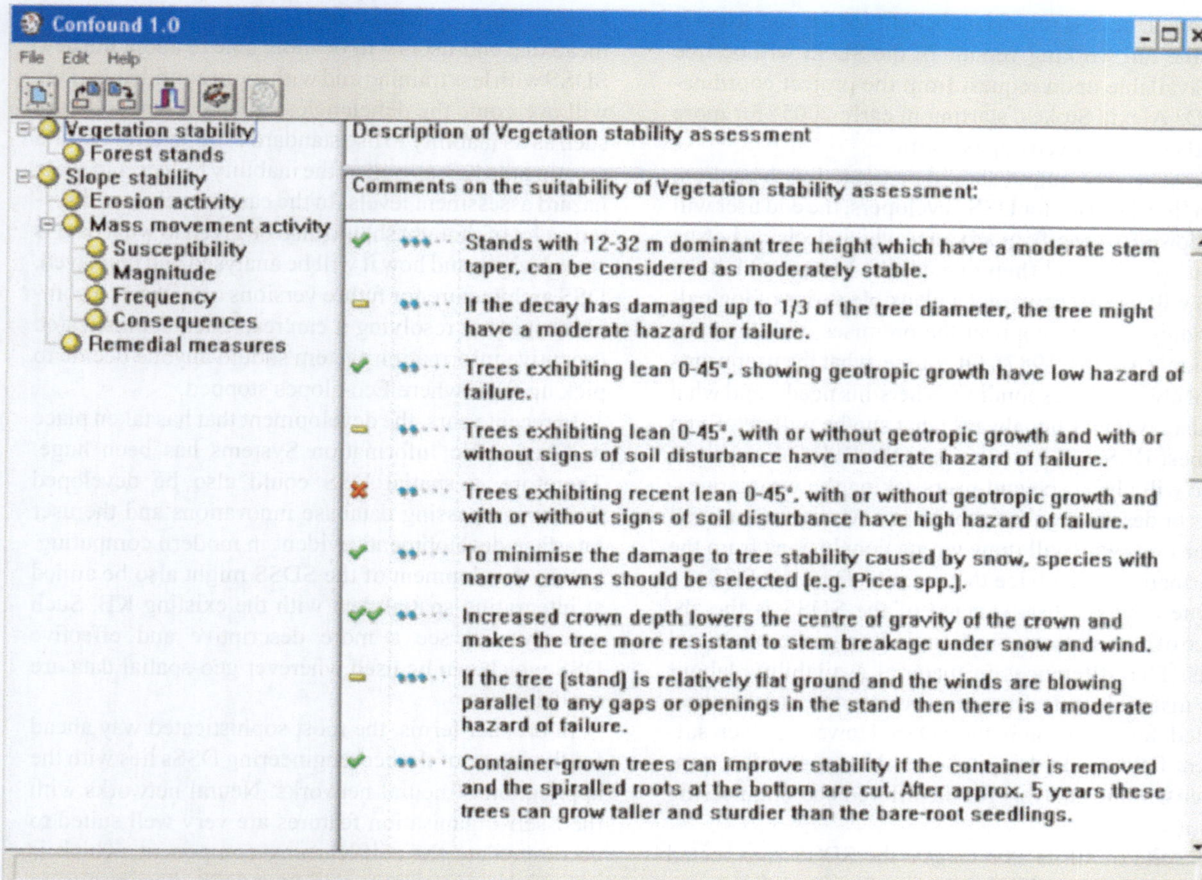

Figure 3. A typical SDSS output screen in Confound®. The window on the left shows the different stability problems that might occur on slopes. The upper window on the right side shows the description of the phenomenon (in this case vegetation stability). The lower window on the right side lists the advice given to the end-user based on already entered project specific information (PSI). The signs before each piece of advice show the suitability of the advice for slope stabilisation and ranges from highly unsuitable to highly suitable, while the number of full dots afterwards reflects the confidence level of each advice (the more full dots the higher the number of PSI were considered and the higher confidence can be placed in this particular piece of advice).

unstructured or semi-structured problems, as well as with incomplete datasets. Being an interactive computer-based system which is more descriptive and user friendly than traditional decision models, it represents a user-oriented system where the modelling and analysis is made to the benefit of the end-user who interacts with the system *via* the dataset entered as PSI.

Using the advantages of the underlying software, allows that the quality of the data that are entered by the user largely defines the outcome. When the user enters no information at all, only general remarks of neutral suitability and unspecified confidence level are returned. This feature is used in the SDSS to help the user to select relevant information from the field, manual, or literature interactively and come back to the SDSS with more information, knowing that the confidence level increases with the inclusion of more types of information. Similarly, the user may follow up the advice presented to a help file through a hyperlink.

SDSS highlights the positive points of the approach used in its design. Representing an 'expert system' it has the potential to remind experienced end-users of the availability of different options to consider and to help a new user reach a complex decision without using optimisation tools and without mastering advanced modelling. One of the greatest benefits to the end-users is that their knowledge and experience can be systematically used in interactive problem solving processes.

The other, no less important benefit for the end-users is that the full working version of the SDSS will be free and available upon request from the project coordinator, Dr Alexia Stokes, starting in early 2005 (for more details see www.ecoslopes.com).

However, the output should be treated with caution. In an 'ideal world' for DSS developers, the end user will be allowed to specify **in advance** which decision he/she wants to make and **then** give all the factors which the DSS will use to 'construct a plausible series of logical-sounding steps to connect the premises with the conclusion' (Adams, 1987). Of course, what the user wants is not always corresponding to hers/his needs, and what the user wants is not always what she/he will get. Even the best DSS will not eliminate 'bad' decisions, since there will always be end-users asking the wrong questions or describing the problem in a wrong way, as well as the ones who will draw wrong conclusions from the information and advice they receive from the DSS.

One obvious disadvantage of the SDSS is the absence of comparison of eco-engineering in terms of real costs. The differences in material availability, labour and material costs from one country to another prevented this inclusion in the SDSS. However, given sufficient field-based data, it is possible to include a simple cost/value analysis for the offered eco-engineering options.

In spite of all its advantages, the SDSS is expected to face 'people problems' arising from cognitive constraints in adopting computer-based systems (Carlsson and Turban, 2001), incomprehension of the advice and guidance obtained as an output which can lead to discarding it in favour of past experience. Reactions such as getting advice and support through oral communication with peers and colleagues when it comes to sensitiveness to the specific preferences and desires of one or several decision makers (Quah et al., 1996) are also not difficult to imagine. However, it is evident that in the future larger groups of eco-engineering practitioners, officers or students who are comfortable with interactive tools and who have hands-on experience with computer software are expected to emerge and to make the tools like SDSS indispensable in day-to-day decision making processes.

Possible improvements and future development

The usability factor of the SDSS can be increased by a shift from the current software package to a web-based system where a standardised interface design would mean that end-users will be more able to adopt the new SDSS with less training and with more confidence. This will overcome the deficiencies of the current software such as its inability to use standard PC operations while creating the KB, as well as the inability to accommodate hazard assessment levels. In the case of web-based system, a lot of thought should also be given to what data is stored where and how it will be analysed and displayed. DSS architecture for future versions of SDSS is a complex topic but resolving it can result in a sophisticated executive information system should anyone decide to pick up from where Eco-slopes stopped.

In recent years, the development that has taken place in Geographic Information Systems has been huge. Therefore, a spatial DSS could also be developed thus encompassing database innovations and the user interface developments evident in modern computing. Future development of the SDSS might also be aimed at integrating spatial data with the existing KB. Such synergy will see a more descriptive and effective DSS which can be used wherever geo-spatial data are available.

In broader terms, the most sophisticated way ahead for the future of the eco-engineering DSSs lies with the application of neural networks. Neural networks with their self-organisation features are very well suited to accommodate the self-learning component, which is one of the cornerstones of each DSS. The ability to adapt to changing circumstances and environmental factors which is unique to neural networks, together with their potential to deal with the fuzziness and bias aspects of human decision making can greatly improve the 'rigid' rule-based systems and transform the existing systems such as the SDSS into an expert network (Quah et al., 1996).

Acknowledgements

The work presented was funded through the Ecoslopes project (EU, QLK5-2001-00289), and the help of all project partners is greatly acknowledged. The author is deeply indebted to Dr Rens van Beek for the cooperation in SDSS creation.

References

Adams D 1987 Dirk Gently's holistic detective agency, p. 78. Heineman. London.

Borenstein D 1998 Towards a practical method to validate decision support systems. Decis. Support Syst. 23, 227–239.

Carlsson C and Turban E 2001 DSS: Directions for the next decade. Decis. Support Syst. 33,105–110.

Chen J and Lin S 2003 An interactive neural network-based approach for solving multiple criteria decision-making problems. Decis. Support Syst. 36, 137–146.

Coppin N J and Richards I G 1990 Use of Vegetation in Civil Engineering, 590 pp. CIRIA/Butterworths, London.

Cucchi V, Meredieu C, Stokes A, Berthier S, Bert D and Najar M 2004 Root anchorage of inner and edge trees of Maritime pine (*Pinus pinaster* Ait) growing in different soil podzolic conditions. Trees – Struct. Funct. 18, 460–466.

Ecoslopes Eco-engineering and conservation of slopes for long-term protection against landslides, erosion and storms. (EU project QLK5-2001-00289). Available at http://www.ecoslopes.com

Gardiner B A, Peltola H and Kellomaki S 2000 Comparison of two models for predicting the critical wind speeds required to damage coniferous trees. Ecol. Model. 129, 1–23.

Gimeno E, Andreu V and Rubio J L 2001 Influence of shrub cover on soil chemical properties in a Mediterranean slope. Commun. Soil Sci. Plan. 32, 1885–1898.

Guzetti F 2000 Landslide fatalities and the evaluation of landslide risk in Italy. Eng. Geol. 58, 89–107.

Kilburn C R J and Pasuto A 2003 Major risk from rapid, large-volume landslides in Europe (EU Project RUNOUT). Geomorphology 54, 3–9.

Mickovski S B and Ennos A R 2002 A morphological and mechanical study of the root systems of suppressed crown Scots pine (*Pinus sylvestris*). Trees – Struct. Funct. 16, 274–280.

Quah T-S, Tan C-L, Raman K S and Srinivasan B 1996 Towards integrating rule-based expert systems and neural networks. Decis. Support Syst. 17, 99–118.

Rubio J L and Bochet E 1998 Desertification indicators as diagnosis criteria for desertification risk assessment in Europe. J. Arid Environ. 39, 113–120.

Schiechtl H 1980 Bioengineering for Land Reclamation and Conservation, 679 pp. The University of Alberta Press, Edmonton, AB, Canada.

Stokes A, Mickovski S B and Thomas B R 2004 Eco-engineering for the long-term protection of unstable slopes in Europe: developing management strategies for use in legislation. *In* W Lacerda, M Ehrlich, SAB Fontoura, ASF Sayao (eds), IX International Society of Landslides Conference, 2004, Rio de Janeiro, Brazil. Landslides: Evaluation & Stabilization, Vol. 2, pp. 1685–1690. A.A. Balkema Publishers, Rotterdam, The Netherland.

Toll D G and Barr R J 2001 A decision support system for geotechnical applications. Comput. Geotech. 28, 575–590.

United Nations Environmental Program (UNEP) 2002 Global Environmental Outlook 3: Past, present and future perspectives.

van Beek L P H 2002 Assessment of the influence of changes in land use and climate on landslide activity in a Mediterranean environment. Tekst. - Proefschrift Universiteit Utrecht, Netherlands.

A decision support system for the evaluation of eco-engineering strategies for slope protection

S.B. Mickovski[1,3,4] & L.P.H. Van Beek[2]

[1]*LRBB, Domaine de l'Hermitage, 69, rte d'Arcachon, 33612, cedex, Cestas, France.* [2]*IBED—Physical Geography, University of Amsterdam, Nieuwe Achtergracht 166, NL 1018, Amsterdam, WV, The Netherlands.* [3]*Civil Engineering Division, University of Dundee, DD1 4HR, Scotland, UK.* [4]*Corresponding author**

Key words: decision support system, erosion, expert systems, knowledge base, landslides, project specific information

Abstract

A decision support system (DSS) has been developed to assist expert and non-expert users in the evaluation and selection of eco-engineering strategies for slope protection. This DSS combines a qualitative hazard assessment of erosion and mass movements with a detailed catalogue of eco-engineering strategies for slope protection of which the suitability is evaluated in relation to the data entered. The slope decision support system is a knowledge-based DSS in which knowledge is stored in frames containing rules that can evaluate the available information for a project, stored as project specific information in a data file. The advantages of such a system are that it accepts incomplete information and that the qualitative nature of the information does not instil the user with a sense of unjustified exactitude. By its multidisciplinary and progressive nature, the DSS will be of value during the initial stages of an eco-engineering project when data collection and the potential of different eco-engineering strategies are considered. The accent of the output of the DSS is on the application of eco-engineering strategies for slope protection as an environmentally friendly solution aiding sustainable development. For its acceptance within the engineering community, the DSS needs to prove its predictive capacity. Therefore, its performance has been benchmarked against successful and unsuccessful cases of slope stabilisation using eco-engineering. The target audience and the areas of application of this DSS are reviewed and the strategies for further development in this area suggested.

Introduction

Recent years have seen an increase in catastrophic incidents due to slope instability problems in Europe. Heavy rains and floods on barren soil are common triggers for mass wasting, while forest fires are devastating the soil under vegetation cover during hot and dry summers, such as the ones in the Mediterranean area, increasing soil exposure to erosive agents (UNEP, 2002). In Alpine areas, rockfalls and avalanches annually cause loss of lives as well as damage worth millions of euros to urban infrastructure, forest and agricultural development.

Knowing that in Europe, catastrophic mass instability events happen every few decades (Kilburn and Pasuto, 2003), and landslides that result in human loss occur with a frequency of 15–20 per year (Guzetti, 2000), it is necessary to focus on accurate hazard assessment and on-slope protection strategies that will be both environmentally-friendly and will provide addition safety for slopes. Using intelligent systems towards reaching decisions on slope stabilisation and the available knowledge on the stabilising and protective benefits of vegetation that depend both on the type of vegetation and type of soil degradation process might be a viable initiative for combating soil degradation and natural resources management. Models that stem from information resulting from data analyses but also include

* E-mail: s.b.mickovski@dundee.ac.uk

This article has been previously published in the following journal—Journal of Geotechnical and Geological Engineering, 24, 3, 483–498.
Stokes et al. (eds), Eco- and Ground Bio-Engineering: The Use of Vegetation to Improve Slope Stability, 369–378.

results of an empirical evaluation and qualitative analysis will be of paramount importance in such cases.

The idea of application of intelligent systems and soft computing in management for problem solving planning and decision making (Carlsson and Turban, 2001) is the basis of decision support system (DSS) design and it can be eficiently applied for creation of methodology for helping hazard assessment and stabilization of slopes. The methodologies used may be analysis or system-oriented, research or case-based or they may be experimentally or empirically focussed. In the DSS creation, field studies should be favoured, especially the ones combining theoretical results with appropriate empirical verification. The DSS presented in this paper combines the knowledge on hazard assessment and remedial measures for soil instability and erosion on slopes which stems from field investigations carried out in the Eco-Slopes Project (QLK5-2001-00289) and is combined with the eco-engineering experience available in the literature. The focus of this DSS is put on eco-engineering as a long-term strategy to protect and/or restore a site with regards to natural or man-made hazards. With a combination of civil engineering techniques and natural materials eco-engineering can be used to provide fast, efiective and economical methods of protecting, restoring and maintaining the environment (Schiechtl, 1980; Coppin and Richards, 1990).

Decision support system for slope stability assessment and remediation

The DSS presented here assesses slope stability from the aspect of difierent scientific disciplines, and suggests eco-engineering measures and strategies for stabilisation and land reclamation. The generic objectives of the DSS presented in this study included:

- providing help for the users in understanding of the eco-engineering concepts involved in slope stability, and assistance in assessing their implementation.
- providing an instrument for measurement of eco-engineering strategy performance parameters.
- considering and evaluating the impact of difierent geo-morphological and environmental parameters in a specific eco-engineering strategy.
- Providing a common language between eco-engineering practitioners, land-use planners, decision-makers, engineers and managers to use in the eco-engineering.

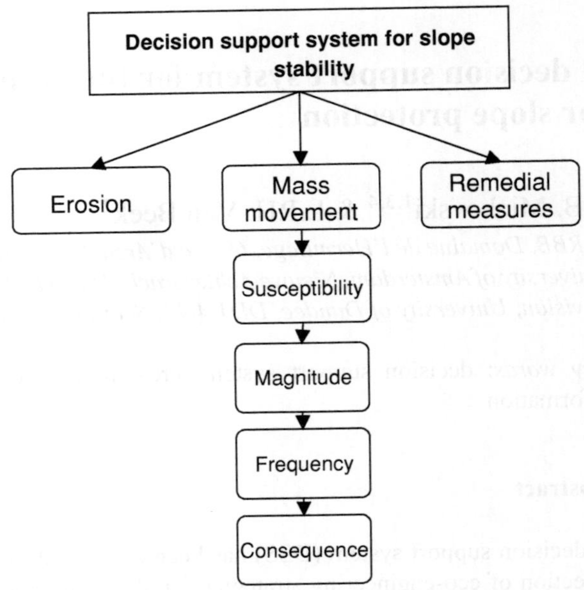

Figure 1. Organogram of the structure of the SDSS KB. Erosion, mass movement and remedial measures for them are the main subjects frames in which the knowledge is classified in difierent subcategories (e.g. susceptibility of a certain site to mass movement, magnitude of possible stability hazards on the site, etc.).

DSS development, structure and operation

DSS development and structure

By combining the available knowledge and experience within the Eco-Slopes project with the experience in the field of slope stability assessment and remediation using eco-engineering strategies available in the literature, the slope decision support system (SDSS) was developed. The SDSS is based on ConFound©, a knowledge base (KB) on foundation design by Toll and Barr (2001). The underlying software package provides both an environment in which knowledge can be entered hierarchically by the expert (supervisor mode) and one in which this information can be explored interactively by the user (normal mode).

Following the structure of ConFound©, the SDSS consists of frames forming the KB, the project specific information (PSI) i.e. information about the factors governing the processes outlined in the KB, and rules which form the connection between the KB and the PSI. The PSI orders all relevant information about a specific slope/site that is provided by the user in a predefined form. This information is evaluated by rules that pertain to a set of frames and that reflect the knowledge that is included in the KB.

Since the SDSS is problem-driven, the structure of the KB consists of two main frames that deal with the knowledge on slope instability and erosion, as the most common stability problems occurring on slopes. Furthermore, each frame consists of knowledge on hazard assessment and eco-engineering strategies for its mitigation classified as shown on Figure 1. For this purpose, the **hazard** is defined as a threat from mass wasting on a slope occurring with given characteristics and probability of occurrence. Every slope is characterised by its **susceptibility** to mass wasting hazards—a factor that is in close relation to the **exposure** of the slope which, in turn, includes the spatial aspects of the hazard (soil type, hydrology, geography, climate, topography). The strategies employed to reduce the mass wasting hazard and to mitigate possible adverse consequences, including anticipation, prevention and remediation, form the **hazard assessment**.

For the SDSS to perform any kind of analysis on a particular site, the information on the level of exposure and susceptibility about a particular site is absolutely necessary. The PSI provide details about a particular site, where the hazards should be assessed and possible mitigation measures suggested. The use of entirely non-numeric (all variables are Boolean or listed types) PSI allowed keeping the PSI-hierarchy as generally applicable as possible. This enables usage of the same piece of information across a broad spectrum of problems concerning soil erosion and mass instability on slopes. This information is variable and it will difier from project to project, i.e. from site to site. The level of detail in entering this information will depend of the level of knowledge the user possesses on it. If little is known on the characteristics above, the SDSS will come out with only general rules for assessment or remediation of any instability, which will bear little confidence.

Hazard assessment and DSS operation

A certain amount of information is requested by the SDSS in order to provide the best possible assessment of mass wasting hazard for the slope. The hazard of mass wasting is assessed based on combination of exposure and susceptibility factors that have major effect on slope stability. These factors can be entered as PSI under the classification shown on Figure 2. Using information available in the literature (Cooke and Doornkamp, 1990; Crozier, 1984; Cruden and Varnes, 1996; Dikau et al., 1996; Rib and Liang, 1978; Sidle et al., 1985), a list of factors and causes known to influence slope stability was chosen. The factors have been classed into descriptions of slope characteristics (Table 1). Amongst the most important criteria to consider when assessing slope stability are slope morphology, including slope gradient, shape and height (Cruden and Varnes, 1996; Varnes, 1978). In this respect, steeper and higher slopes, as well as the ones with irregular profile, are at higher hazard of failure. The type and the stratigraphy of the soil material on slope is also an important characteristic when assessing slope stability. Plastic soils, material sensitive to physical or chemical weathering or heavily fractured or jointed rock are more susceptible to failure, especially when weaker and stronger soil beds or layers with different permeability alternate (Crozier, 1984; Hutchinson, 1988;

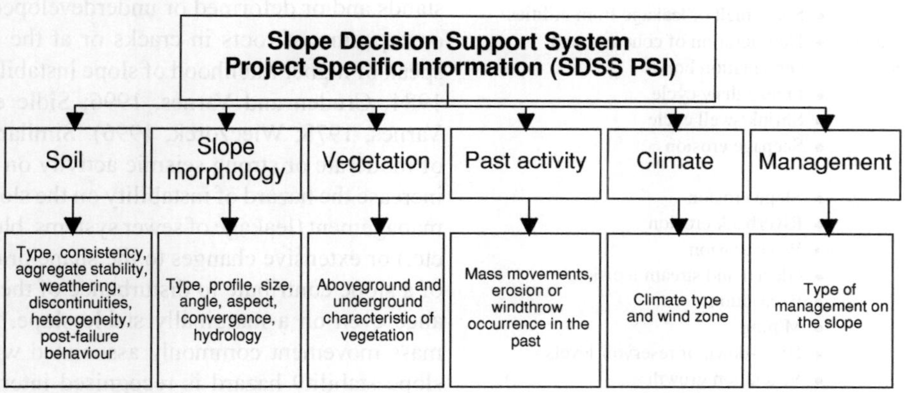

Figure 2. SDSS PSI hierarchy. Factors governing mass instability and erosion processes are analysed in detail and the user is allowed to enter as much information as it is available for a specific slope/site.

Table 1. Site characteristics and slopes with high hazard of slope instability (compiled from Cooke and Doornkamp, 1990; Crozier, 1984; Cruden and Varnes, 1996; Dikau et al., 1996 Rib and Liang, 1978; Sidle et al., 1985)

Site characteristics	Slopes with high hazard of slope instability
Morphology	
Gradient	Moderately steep for landslides (>10°) to extremely steep for falls (>35°). Some flows can maintain momentum even on very gentle slopes
Shape	Convergent or irregular in profile
Height	Short steep slopes for rotational slides, long slopes for translational slides
Material	
Slope material	Plastic soils, material sensitive to physical or chemical weathering or heavily fractured or jointed rock
Stratigraphy	Alternation of weaker and stronger beds, of different permeability
Hydrology	Signs of ponding and springs, presence of gleyic horizons indicating stagnating water in the soil
Drainage	Heavily dissected by ephemeral or permanent streams with signs of undercutting at the base of the slope or signs of disrupted drainage
Climate	Periods of intense or prolonged rainfall or rapid snowmelt; Strong diurnal and seasonal variations in temperature, e.g. freeze-thaw
Seismicity	Evidence of moderately strong to strong earthquakes
Past activity	Signs of previous slope movements (creep, sliding) and/or surface wash
Vegetation	Irregular stands and/or deformed or underdeveloped vegetation; Exposure of roots in cracks or at the surface
Human activity	Evidence of poor site management (leakage of sewer systems, blocked drains etc.) or extensive changes to the shape or composition of a slope. On a marginally stable slope, human intervention can easily upset the critical balance

Varnes, 1978) or where weathering, erosion and progressive failure take place (Table 2). Another factor that affects slope stability is the hydrology, and slopes where signs of ponding and springs can be seen, or where the presence of gleyic horizons indicates stagnating water in the soil are at higher risk of failure. Similarly, slopes where no adequate drainage is provided, or slopes heavily dissected by ephemeral or permanent streams with signs of undercutting at the base of the slope are also prone to instability (Cooke and Doornkamp, 1990; Hutchinson, 1988; Rib and Liang, 1978). Additionally, sites that experience periods of intense or prolonged rainfall, rapid snowmelt or strong diurnal and seasonal variations in temperature, e.g. freeze-thaw cycles pose a higher risk of failure (Dikau et al., 1996; Sidle, 1985; Wieczorek, 1996) due to changes in water regime (Table 2). Signs of loss of support such as previous slope movements (creep, sliding) and/or surface wash manifested *inter alia* as irregular stands and/or deformed or underdeveloped vegetation or exposure of roots in cracks or at the surface will speak of higher likelihood of slope instability (Crozier, 1984; Cruden and Varnes, 1996; Sidle et al., 1985; Varnes, 1978; Wieczorek, 1996). Similarly, evidence of moderate or strong seismic activity on the site will increase the hazard of instability on the slope. Poor site management (leakage of sewer systems, blocked drains etc.) or extensive changes to the shape or composition of a slope contribute to disturbance of the critical balance even on a marginally stable slope. The type of mass movement commonly associated with a certain slope stability hazard is recognised internally by the SDSS according to Table 3, and the user is prompted to be aware of the occurrence of such movement.

Table 2. List of examples of mass movement causes compiled from Varnes (1978), Crozier (1984), Hutchinson (1988), Cruden and Varnes (1996) and Wieczorek (1996)

Mass movement causes	Instability mechanism
Internal	
Changes in water regime	• Pore pressure increase or matric suction decrease upon wetting by rainfall • Snow melt or leakage from utilities
Weathering, erosion and progressive failure	• Deterioration of cohesion and cementation bonds • Freeze/thaw cycle • Shrink/swell cycle • Seepage erosion
External	
Loss of support	• Slope erosion • Riverbank erosion • Wave erosion • Glacial and stream incision • Excavation • Mining • Draw-down of reservoir levels
Increased surcharge	• Vegetation growth • Increasing weight because of wetting

Table 3. Slope features used for recognition of a mass movement by the SDSS

Movement type	Diagnostic features based on PSI
Falls/topples:	• Slope face: Steep to vertical, overhanging or undercut
	• Cracks close to the face
	• A sufficiently large area to unload material from the source area
	• Discontinuities form unfavourable sets projecting out (falls) or running parallel to the slope (topples)
	• Materials sufficiently strong (cohesive soils, rock) to sustain the slope over a period of time
	• Material liable to deterioration: Excavation of more resistant blocks/boulders (soil), physical or chemical weathering (rock) or worn-down or gouge-filled discontinuities
	• The presence of weaker basal layers
	• Environments experiencing periodic freezing and/or large water inflow (snow melt, rainstorms) or dynamic loading (blasting, seismicity)
Slides	• Slopes that are moderately steep and of sufficient height to allow rotational or translational movement
	• Disturbance of the slope by undercutting or surcharges
	• A rise in pore pressures due to undrained loading, changes in the water regime e.g. leaking sewerage, and rainfall
	• Dynamic loading (seismic events, vibrations due to heavy traffic etc.)
	Rotational slides:
	• Uniform deposits of cohesive materials or severely broken down rock
	• The presence of a weaker basal layer, particularly for the formation of multiple rotational slides
	Transitional slides
	• Slopes that are straight or slightly convergent in plan or exhibit a clear break of slope
	• The presence of soil layers of varying or decreasing strength or permeability or the presence of multiple discontinuities in bedrock
Flows	• Steep slopes
	• Availability of loose debris and/or fines
	• Poor drainage as evidenced by high drainage density, impervious substrate or infiltration impeded by permafrost
	• Absence or sparse vegetation cover
	• Intense rainfall or rapid snowmelt
	• Flooding, irrigation or fluctuations in reservoir levels
	• Volcanic eruptions
	• Possibility of earthquakes or vibrations

The quality of the data that are entered by the user thus largely defines the outcome of the DSS. Only general remarks of neutral suitability and unspecified confidence level are returned when the user enters no information at all. This feature largely helps the user to select relevant information from the field, together with the information available in manuals or literature. Finally, the user can follow up the information presented in a specific rule to its source and to the theory behind it in a HTML help file through a hyperlink. It is not expected from the SDSS to make decisions for the user: the SDSS should be used solely as guidance to a spectrum of possible solutions. In the SDSS, the guidance in the form of direction and advice to the user is offered as

- determination of the hazard of slope instability and
- evaluation and suggestion of possible eco-engineering strategies for hazard mitigation.

In the KB, the rules are associated with a suitability indication and a confidence level. In the case of the SDSS the suitability would refer to the slope stabilising or destabilising potential of factor(s) in the PSI. Underlying software package allows the first evaluation to result only in general rules that would appear as default comments without entering detailed PSI. After this, the particular site information entered via the PSI will be evaluated in order to identify one of the possible risks on the slope in question. Subsequently, the feasible management strategies are evaluated and suggested. During all of these processes, hyperlinks to the SDSS help file provide the user with feedback on additional data selection and possible management strategies.

The confidence limits attached to the rules serve as an indication of the trustworthiness of the comments that are returned by the SDSS. The confidence increases as more detailed and better information is entered on the site in question, and more 'trustworthy' rules appear with inclusion of more data in the PSI. Confidence also contains a component of best practice or generally accepted knowledge that is the basis of the SDSS but its provenance and dependability remain hidden from the user. The user can explore the theory and reference to the primary sources following the hyperlinks attached to each rule.

Ideally, the SDSS should be accompanied by pre- and post-processing that will facilitate the data input and analysis by the user. This pre- and post-processing can be performed in conjunction with the SDSS Help File which includes the necessary theoretical and specific knowledge on the processes that govern the erosion and mass instability processes on slopes. An additional data analysis by the user in order to make the decision support by the SDSS more unique and appropriate (i.e. reevaluating the SDSS with more data or additional modelling in the post-processing phase) will be rewarded with more focused and appropriate advice.

DSS Testing

In order to test the SDSS on slopes prone to mass wasting and erosion, four case studies were carried out using data available in the literature (Table 4). The first site (site A) chosen was a north-facing, 192-m wide and 21-m long natural slope of glacial deposits and volcanic ash overlying granitic bedrock in Chelan, Washington, USA (Lewis et al., 2001). The slope received around 277 mm rainfall annually, had no vegetation on it (sparse vegetation cover), and had a history of mass wasting. The second site (site B) was a riverbank with 60° inclination, 5.5 m high, consisting of clay layers in Victoria, Australia (Abernethy and Rutherford, 2000). The groundwater was parallel to the ground surface while streambanks were undercut by the river. The vegetation growing on the slope consisted of predominantly Silver wattle (*Acacia dealbata*) and grass. The third (site C) site was a North-facing bare motorway embankment in Massachusetts, USA (Sotir et al., 1997). The slope angle was 34°, slope length 15 m, slope width 145 m. The slope was made up of 3-m deep glacial till overlying firm bedrock. The surficial soil was saturated with existing groundwater seepage and a seasonal freeze-thaw cycle. The fourth site (site D) was situated in Kataramal (Almora) Himalayas, India (Kumar et al., 2000). The slope was 5-m wide, 20-m long debris dump overlying slightly acidic sandy loam with low water holding capacity. Inclined at 31°, the slope faced South-East and received around 1370 mm of annual rainfall.

The output from the SDSS presents suitable advice for the end-user, together with a confidence level (Table 4). Results from the analysis of the four case studies were different as the SDSS had taken into account the differences in site conditions.

The highest confidence levels of the output were associated with the slope gradient for all of the four case studies—steep slopes pose the highest risk of mass instability and erosion due to high-speed overland flow. Furthermore, for the short steep slope in case study C, the risk of rotational slide is higher than the risk of translational slide, while the slope undercutting by the river on site B increases the risk of falls and topples. In this respect, the user is advised to check the feasibility of sloping, terracing, contouring or regarding measures in order to decrease the risk of mass wasting.

Case studies A, C and D are at risk of sheet erosion combined with rill and/or gully formation because of the high slope gradient. In order to mitigate this risk, planting of vegetation is suggested as a most appropriate measure that will slow the runoff and decrease the risk of rill/gully formation on gentler slopes like in the A and D case studies, while seeding combined with mulching and turfing is suggested for steeper slopes like the one on the site C. Furthermore, to control possible debris erosion/flow as well as to establish immediate soil reinforcement, brushlayer construction is also suggested for the steeper slopes on the sites A and D.

The strong seasonal freeze-thaw cycle on the location of the case study C and the intense and prolonged rainfalls combined with rapid snowmelt on site A, increases the risk of positive pore water pressure build-up and thus decreases the slope stability. Plastic soils, such as the clayey matrix in case study B usually pose higher risk of slope instability due to positive pore water pressure build up, but the presence of vegetation able to decrease them acts as a stabilising factor. Contrariwise, the low water holding capacity of the soil in the case study D is able to dissipate any positive pore pressure and, in this respect, has a low risk of failure. The high ground water table in case study B, together with the fluvial undercutting of the slope pose the highest risk of failure. In order to mitigate these hazards, construction of live fascine drains is suggested for the slopes in the case studies A, B and C. These drains would provide the endangered slopes with extensive surface area drainage and would further decrease the pore water pressures by transpiration after the roots have developed.

Signs of past mass instability and erosion also yielded high confidence levels for slope instability hazard. The series of smaller erosion events and mass movement signs adjacent to the site A, as well as the

Table 4. Input used as PSI and comparative results for the four case studies

SDSS main frames		Site A	Site B	Site C	Site D
Site location		Washington, USA	Victoria, Australia	Massachusetts, USA	Kataramal, India
INPUT (PSI)					
Substrate		Gravely	Clayey	Gravely	Sandy
Soil depth		Intermediate	Intermediate	Intermediate	Shallow
Profile		Weathering	Layered	Uniform	Blanket
Weathering		Slight			
Consistency		Firm	Soft	Firm	Soft
Inclination		<1:2	>2:1	<1:1	<1:2
Profile		Terraced	Rectilinear	Rectilinear	Rectilinear
Length		Intermediate	Long	Long	Intermediate
Aspect		N		N	S
Type		Natural	Embankment	Embankment	Fill
Hydrology		Waterlogged	Waterlogged	Waterlogged	Regional gwt
Vegetation type		No vegetation	Tall trees	No vegetation	No vegetation
Age			Mature		
Stock			Natural regener.		
Height			Medium		
Stem damage			No		
Stem Lean			No		
Stem Taper			<90		
Root depth			Not exceed		
Root system			Plate		
Past mass movements		Shallow		Shallow	
Past erosion		Sheet		Sheet	Sheet
Climate		Alpine	Temp humid	Temp humid	Alpine
Exposition		Fully	Partly	Fully	Fully
OUTPUT					
Hazard/Risk of instability: Causes	Mass stability	High: Debris slides and flows Moderate-High: Shallow slides	High: Falls/topples in undercut slope Moderate-High: Soil slips	High: Soil slips and rotational slides Low: Falls and flows	High: Shallow slides, debris flows Moderate: Mud flows
	Erosion stability	High: Rill and gully erosion	Moderate-Low: Surficial erosion	High: Rill/gully erosion, overland flow	High: Seepage erosion, gully formation Low: Flooding
Improvement—eco-engineering mitigation strategy	Mass stability	Planting Terracing Regarding Wattle fences Drainage: Live fascine drains	Regrading Sloping Gratings	Planting and seeding Live crib walls Live gratings Terracing Drainage: Live fascine drains	Sloping Regarding Terracing Hedge brushlayer
	Erosion stability	Seeding Planting Branch packing Brushlayer Drains	Brush wattles Brushlayer	Planting Sodding and turfing Live fascines Vegetated geogrids Water bars	Planting Branchpacking Wattle fences

mass instability problems with the slope C, speak of increased risk of slope failure. The risk of slope instability incidence in both cases is increased by the absence of vegetation on the slope. Planting and seeding on the slope would decrease the risk of surface erosion and provide immediate soil cover. In order to mitigate the risk of possible debris flows on the site A, live fencing is suggested as an inexpensive and effective mitigation strategy. For the road embankment in the case study C, live crib wall construction might provide both the necessary stability and answer the visual and aesthetic requirements for the site.

Discussion

The SDSS incorporates all components of a classic DSS: it has instruments and methods for dealing with unstructured or semi-structured problems; it has an interactive computer based system which is more descriptive and user friendly than traditional decision models; it represents a user-oriented system where the modelling and analysis is made to the benefit of the end-user who interacts with the system via the dataset entered as PSI.

The implication of these components allows the users to deal easily with complex problems that usually require extensive experience and expert knowledge. Furthermore, it allows the user to make better and more reasoned decisions without using optimisation tools and without mastering advanced modelling. One of the largest benefits to the end-users is that their knowledge and experience can be systematically used in interactive problem solving processes.

SDSS validation is not concerned with proving that it truthfully represents human decision making process—since this is impossible—but with demonstration that it has the appropriate underlying relationships (Borenstein, 1998) to permit an acceptable representation. The validation process for the SDSS consisted of formal validation following cyclic iterative strategy of prototyping methodology development (Borenstein, 1998). The two-stage validation was performed iteratively throughout the SDSS development in a sense that whenever it was changed, modified or expanded (reformulation of rules, redesign of KB or PSI, refinements), it had to be re-evaluated. A slope stability and eco-engineering questionnaire was developed in order to directly compare the SDSS output with an expert assessor's prediction and solution process and sent out to relevant institutions and experts. The data obtained from the returned questionnaires which included (un)successful eco-engineering practices in Europe, was analysed and used as practical basis for addition of new rules if the SDSS underperformed for a particular case described in the questionnaire.

Case study results

The testing of the SDSS against case studies reported in the literature showed the versatility and flexibility of the knowledge base incorporated, while the output coincided in broader terms with the slope stabilization measures undertaken on each site. For example, the slope from the case study A was terraced and seeded in order to decrease the risk of instability, and a bender board fencing was installed on the slope to decrease the risk of flows and falls—strategies suggested by the SDSS output. In spite of several technical difficulties, the undertaken measures proved to really stabilize the soil both against erosion and shallow mass movements (Lewis et al., 2001). The riverbank slope (site B) benefited largely from the vegetation present. Since this site was used as a basis for a theoretical investigation of the effects of tree surcharge on riverbank stability (Abernethy and Rutherford, 2000), no mitigation strategies have been undertaken on the site and no information exists on its current state. However, the SDSS showed that the low risk of erosion actually reflected the beneficial effects of vegetation cover that stabilizes the soil surface, while the slope could have been made even more safe by decreasing the slope inclination. A combination of mitigation strategies was applied to the case study C and the slope was stabilized in accordance to the requirements of the project (Sotir et al., 1997). In general, SDSS output complied with the mitigation measures applied to the slope. The installation of an earthen buttress to stabilize slope toe was not among the output measures suggested by the SDSS. However, the feasibility and the stability of such a construction depend largely on more detailed information about the site which was not reported. Simple planting of native vegetation species on the slope D decreased the mass wasting due to erosion by 50% during high rainfall events in the first year of vegetation establishment (Kumar et al., 2000). Application of the other slope stabilization strategies suggested by the SDSS is likely to further decrease the mass wasting on this slope.

European knowledge base

The non-existence of literature, codes of practice, or laws on eco-engineering in Europe was the main obstacle in the data collection process for the SDSS KB. The widely scattered information of particular studies and reports were combined with the established strategies and experiences worldwide which were adapted for European conditions. However, the main body of information that served as a starting point of the KB creation for the SDSS was derived from the knowledge and the experience of the Eco-Slopes project partners that covered only six European countries. In the course of the project, a series of interviews was carried out with the partner institutions which included theoretical data gathering and practical evaluation of eco-engineering strategies carried out by each institution. This information together with the information and the knowledge accumulated in the returned evaluation questionnaires will be used for creation of an eco-engineering practice database which can be added to the SDSS in a later stage, or published separately as an eco-engineering manual.

Future development of the SDSS

Currently, the SDSS operates in ConFound©, a free software ideally suited for installation on a network, with no system files requiring modification and no write access required. The software used was ideal as a first step of the SDSS development because it offers large possibilities for conversion into another package or system in the future if necessary. However, the slow development of the underlying software might obstruct the development of the SDSS, while the access to the data might become more restricted.

In order to ameliorate the deficiencies of the ConFound©-based SDSS, and with regard to the future of DSSs where eco-engineering management strategies will be included, considerable attention should be given to web-based DSSs. Basing the SDSS on the WWW will strengthen the interactive component of the system and will more easily bring together a variety of disciplines that will jointly handle issues such as mass or vegetation instability both for hazard assessment and remediation. In the same time, the usability factor of the SDSS will also be increased having a standardised interface design which would mean that end-users will be more able to adopt the new SDSS with less training and with more confidence. In this sense *a personalisation* of the SDSS in the future, where the usage patterns of individual users can be identified and automatically modified (Shim et al., 2002) will be a goal to reach for the next generation of DSS in this area.

Another way of improving the efficiency and usability of the SDSS can be the use of neural networks for structuring the logic of the rules. Neural networks with their self-organisation features, are very well suited to accommodate the self-learning component of the SDSS and its potential for parallelism (Quah et al., 1996). The ability to adapt to changing circumstances and environmental factors which is unique to neural networks, together with their potential to deal with the fuzziness and bias aspects of human decision making can greatly improve the 'rigid' rule-based systems and transform the SDSS into an expert network (Quah et al., 1996).

Following the neural network development approach, it might be possible, given sufficient field-based data, to include a simple cost/value analysis for the possible eco-engineering options which will further facilitate decision making process.

Further application and usage of the SDSS can also see collaborative learning in virtual DSS classes where students with different backgrounds will collaborate in solving complex eco-engineering managerial problems as a part of their learning experience. This can also be valid for eco-engineering practitioners who can improve their knowledge on the theory and modern practice in eco-engineering in the same time communicating and consulting with colleagues from different laboratories or different countries. A developmental copy of the software is available for free upon request from the project leader of the Eco-slopes project (http://www.ecoslopes.com).

Acknowledgements

The work on this paper was funded by the Eco-slopes project QLK5—2001-00289), and the authors wish to thank all project partners for their invaluable help and support.

References

Abernethy B and Rutherford I D 2000 Do Riparian Trees Destabilize Riverbanks? Regulated Rivers: Research and Management 16, pp. 565–576, Wiley, New York.

Borenstein D 1998 Towards a practical method to validate decision support systems. Decis. Support Syst., 23, 227–239.

Carlsson C and Turban E 2001 DSS: Directions for the next decade. Decis. Support Syst., 33, 105–110.

Cooke R U and Doornkamp J C 1990 Geomorphology in Environmental Management—A New Introduction. 2nd edition. Oxford Univ. Press, London.

Coppin N J and Richards I G 1990 Use of Vegetation in Civil Engineering, CIRIA, Butterworths, London.

Crozier M J 1984 Field assessment of slope instability. *In* D Brunsden and D B Prior (eds), Slope Instability, pp. 103–143, Wiley, New York.

Cruden D M and Varnes D J 1996 Landslide types and processes. *In* A K Turner and R L Schuster (eds), Landslides–Investigation and Mitigation, National Academy Press, Transportation Research Board Special Report 247, Washington D.C. pp. 36–75.

Dikau R, Brunsden D, Schrott L, and Ibsen M L (eds) 1996 Landslide Recognition, Identification, Movement and Causes. 251 pp. Wiley, Chichester.

Guzetti F 2000 Landslide fatalities and the evaluation of landslide risk in Italy. Eng. Geol., 58, 89–107.

Hutchinson J N 1988 Morphological and geotechnical parameters of landslides in relation to geology and hydrogeology. In Proceedings 5th International Symposium on Landslides, Lausanne, Vol. 1, pp. 3–35.

Kilburn C R J and Pasuto A 2003 Major risk from rapid, large-volume landslides in Europe (EU Project RUNOUT). Geomorphology, 54, 3–9.

Kumar K, Pant D, Panda Y S and Satyal G S 2000 Runofi and soil loss from steep slopes treated with low cost bioengineering measures. The Environmentalist, 22, 133–141.

Lewis L, Hagen S L and Salisbury S L 2001 Soil Bioengineering for Upland Slope Stablization. WSDOT Research Project WA-RD 491.1 WSDOT, Washington, USA.

Quah T-S, Tan C-L, Raman K S and Srinivasan B 1996 Towards integrating rule-based expert systems and neural networks, Decis. Support Syst., 17, 99–118.

Rib H T and Liang T 1978 Recognition and identification: Landslides analysis and control, R L Schuster and R J Krizek (eds), Special Report 176, pp. 34–80.

Schiechtl H 1980 Bioengineering for Land Reclamation and Conservation. The University of Alberta Press, Canada.

Shim J P, Warkentin M, Courtney J F, Power D J, Sharda R and Carlsson C 2002 Past, present and future of decision support technology. Decision Support Syst., 33, 111–126.

Sidle R C 1985 Factors affecting the stability of slopes. *In* Proc. of a Workshop on Slope Stability: Problems and Solutions in Forest Management, U.S.D.A. Forest Service, PNW Research Station, Portland, Oregon.

Sidle, R C, Pearce A J and O'Laughlin C L 1985 Hillslope stability and land use. Water Resources Monograph No. 11, American Geophysical Union, Washington, D.C.

Sotir R B, Difini J T and McKown A F 1997 Soil Bioengineering/biotechnical Stablization of a Slope Failure. Robbin B. Sotir & Associates Inc., Marietta, Georgia, USA. (Available: http://www.sotir.com/pubs/publist/slope_failure.html, assessed July 2004.)

Toll D G and Barr R J 2001 A decision support system for geotechnical applications. Comput. Geotech., 28, 575–590.

United Nations Environmental Program (UNEP). (2002) Global Environmental Outlook 3: Past, Present and Future Perspectives.

Varnes D J 1978 Landslide types and processes. In R L Schuster and R J Krizek (eds), Landslides Analysis and Control, pp. 11–33. Special Report 176.

Wieczorek G F 1996 *Landslide triggering mechanisms*. In A K Turner and R L Schuster (eds), Landslides: Investigation and Mitigation. National Academy Press, Washington, D.C., Special Report 247, pp. 76–90.

Eco- and ground bio-engineering case studies by practitioners

Land restoration for the Olympic Games in Athens, Greece: Case study Lycabettus slopes (cycle road)

Julia Georgi*

Landscape Architect (M.L.A.)-Forester, Hellenic Open University, Ministry of Environment, Planning and Public Works, Athens, Greece.

Key words: land restoration, landscaping, local plants, planting plan

Abstract

A number of landscape projects have been completed around Athens (Greece) for the 2004 Olympic Games. The common purpose of these projects was the promotion of quality in: design, construction and maintenance stages. They also comply with environmental management, which is a method of organising and implementing environmental protection measures.

The paper includes the landscape transportation project—an urban ring road around a main green historical area of Athens, Lycabettus hill slopes, that was used as a cycle road for the Olympic Games. Due to the special natural ecosystem, the technical difficulties (high slopes) and the long history of this site, the development and the landscape design of this project was a highly challenging task and required methods for ecological balance.

The implementation of landscape design in this project, aimed at restoring the landscape affected by construction work, incorporating transport work into the landscape, emphasizing areas of particular interest (e.g. rest areas) while at the same time observing the principles of an efficient operation and safe use of the alignment.

The present application of the landscape restoration aimed to provide, amongst others, for the protection of slopes against surface erosion and the reinforcement of their stability, the improvement of road functionality, the aesthetic improvement of the landscape, adaptation of the road landscape to the greater environment and visual guidance of the user.

The landscape design strategy and landscape applications considered as part of the sustainable development of the road created aesthetic scenery, ensured the security of the transportation and provided environmental protection.

Introduction

On the occasion of the Olympic Games, there was the need for big but also smaller in extent projects, which were included in a more general environmental planning. For this reason new regulations and enactments were established (Georgi, 2003). The region of Attica today is a modern metropolis and at the same time maintains and enhances her historical heritage.

During the planning of the project, there was the effort to combine, (Georgi and Kapnistou, 2004; Kaperoni et al., 2004) the historical memory, the cultural character and the identity of the city of Athens, with the aesthetic upgrade, the environmental sensitivity, as well as the modern needs and the most optimal functionalism of the city (Figure 1).

The purpose of this paper is to point out the present application of the landscape restoration aimed to provide, amongst others for the protection of slopes against surface erosion and the reinforcement of their stability, the improvement of road functionality, the aesthetic improvement of the landscape, adaptation of the road landscape to the greater environment and visual guidance of the user.

* E-mail: jgeorgi@tee.gr

Figure 1. The map of the Olympic works (roads, venues, urban areas)

Materials and method

The method focused on the main planting design and construction concept of the Lycabettus hill ring road that was used as a cycle road during the Olympic Games. The procedure of selection of the plants as well as the estimation of their growth was carried out.

Site description

In the heart of Athens there is a limestone rock reaching almost 227 m of height and 350 acres in extent which is known as Lycabettus. The hill of Lycabettus offers a unique sight of Attica's peninsula with Athens, Piraeus and the Gulf of Saronikos.

The 'circle' road during the Olympic Games crossed the centre of Athens city with its main part crossing the surrounding road of Lycabettus hill. This was a unique chance for improving the factional and the aesthetic quality of the slope, since Lycabettus is one of the main green areas of the city centre of Athens (with Pedion Areos and National Gardens).

Today Lycabettus is vegetated by *Pinus brutia* Ten. and *Cupressus sempervirence* L. These species were

(a) (b)

Figure 2. (a) Lycabettus in 1930 (photo D. Kontsntinou by photo document of Benakion Museum) and (b) Lycabettus today

established by reforestation with a shrub story of maquis species. During the last century, Lycabettus was a bare hill in Athens where several quarries were active and only some natural herbs thrived on the limestone slopes. Different Plans took part in the past for the design of reforestation and restoration of Lycabettus (Dritstas, 2004) (Figure 2).

Finally in 1930 the Greek government decided to reforest Lycabettus with *Pinus* sp. and *Cupressus* sp. At the same time the quarries were closed and the Lycabettus theatre was built. The theatre was redesigned in 1977; Lycabettus theatre hosts various theatrical performances and concerts (Dritstas, 2004) (see Figure 3).

In the following years, the vegetation of Lycabettus was restricted by the urbanization and some pathways was remodelled as roads. Although Lycabettus remained as a place for walking and a natural viewpoint for the Athenians. During the following years, several exotic, non indigenous species (*Agave americana* L., *Aloe arborescence* Mill. (see Figure 4)) have been introduced.

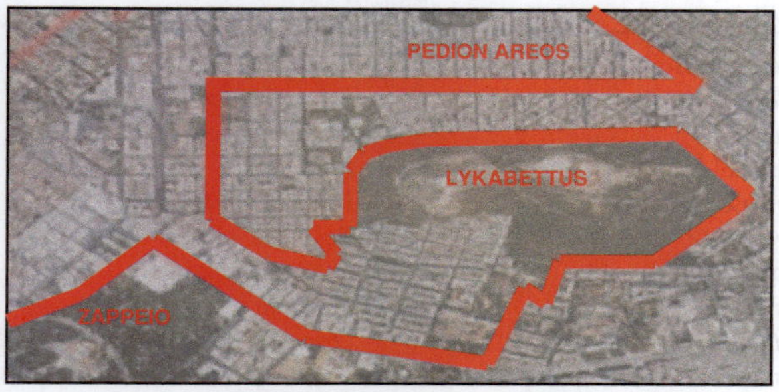

Figure 3. The alignment of the Lycabettus ring road

Results and discussion

The aim of the design of the Lycabettus ring road was:

(1) the integration of the road into the landscape.
(2) the improvement and creation of new landscapes.
(3) uniformity of planting appearance.
(4) group planting with combination of shapes, forms, sizes, height, colours and seasonal colours.
(5) protection of cut slopes.

Therefore planting was used in conjunction with other engineering techniques to stabilize the slope, protect against landslips and surface erosion, as well as to restore negative impacts caused to the landscape by cut slopes. The shrubs were planted in order to retain top soil and absorb horizontal pressure and stress at the cut slopes. The expected growth of these shrubs should develop a rich surface root system in order to create a natural grid to retain the soil.

The surface run-off and erosion expanded constantly on Lycabettus hill except from a small part of the slope that is covered by a stone wall. To reduce and stabilize the large slope gradients, additional works took place such as retaining walls covered by natural material (stones). In order to protect the historical physiological and ecological characteristic of the slope a wall covered by stone in continuity to the old stone wall was built along the ring-road and planting works took place on the slopes of the hill. For this reason research

Figure 4. The non indigenous species that were introduced on the slopes of Lycabettus

took place in order to use the local materials (local stones) and local plants where the local plants that still try to thrive among the *Pinus* and *Cupressus* species and emphasize the focal points, plants and special constructions. The local plants that were found in the area are given in Table 1.

The above survey was followed by a nursery search, since the needs of the Olympic Games increase the cultivation of local plants. Therefore first the plants that were found in the nurseries were identified and secondly assessed for their suitability for planting for aesthetic and functional reasons of reinforcing the slopes with their root systems (Alifragis et al., 1999) and covering the structures (stone wall) to reduce the negative impact to the landscape. The plants selected were: *Rosmarinum officinalis* L., *Rosmarinum officinalis* L. var. *prostrates, Myrtus communis* L., *Lavandula spica* L., *Phillirea* sp., *Cistus incana* L., *Vinca major* L., *Clematis paniculata* Thunb. non Gmel., *Lonicera* sp., *Origanum heracleoticum* L., *Vitex agnus-castus* L.

The plants were transported in their pots from the nurseries to the project site and planted immediately. The work for the installation of plants comprised:

1. The initial weeding of the area before their planting by hand
2. The digging of a pit 0.2 × 0.3 m in difficult planting conditions or 0.3 × 0.3 × 0.5 m in earth or semi-rocky ground for seedlings, nursery plants and bushes, the full removal of stones aids the formation of the planting pit.
3. The carriage of the plant to the planting pit, the extraction from the plant pot, the removal of any dry parts and positioning the plant vertically and at a level of 5 cm lower than the level of the ground

Table 1. Local plants that naturally grow in Lycabettus

Lonicera sp.	*Coronilla emeroides* Bois. & Spr.	*Ramnus alaternus* L.
Clematis paniculata Thunb. non Gmel.	*Arbutus unedo* L.	*Erica arborea* L.
Phlomis fruticosa L.	*Cistus incana* L.	*Origanum heracleoticum* L.
Artemisia absinthium L.	*Cistus salviifolius* L.	*Rhus* sp.
Berberis cretica L.	*Lavandula spica* L.	*Pistacia lentiscus* L.
Vinca major L.	*Vitex agnus-castus* L.	*Rosa canina* L.
Rosmarinum officinalis L.	*Origanum majorana* L.	*Salvia officinalis* L.
Rosmarinum officinalis L. var. *prostratus* hort.	*Myrtus communis* L.	*Phillirea* sp.
Olea oleaster L.	*Ceratonia siliqua* L.	*Quercus coccifera* L.
Iris spp.	*Nerium oleander* L.	
Hedera helix L.	*Buxus sempervirens* L.	

Rosmarinum officinalis

Vitex agnus-castus

Myrtus communis

Cistus incana

Figure 5. We can see the plants that have been planted on the Lycabettus ring road. The plants increased in height by 10–30 cm after 3–4 months growth

surrounding it, the compaction of the soil in the planting pit, the formation of an irrigation basin and the first irrigation application after the installation of the plant. The first irrigation mentioned above performed with at least six litres of water per plant. An irrigation system was installed in order to irrigate the site regularly.

Finally all the plants were successfully established and after 3–4 months the growth was estimated to be between 10 and 30 cm, depending on the species. In general all the plants that were originally 10 cm in height (e.g. *Cistus incana, Vinca major, Origanum heracleoticum*) and those up to 30 cm in height (e.g. *Rosmarinum officinalis, Myrtus communis, Vitex agnus-castus*), now reach between 20 and 45 cm in height (Figure 5). It is now difficult to distinguish which plants grow normally in the area and which have been established recently.

The implementation of landscape design in this project, aimed at restoring the landscape affected by erosion and construction work, incorporating the road into the landscape, emphasizing areas of particular interest (e.g. rest areas, view points), to enhance the historical interest of the site while at the same time observing the principles of an efficient operation and safe use of the alignment. The planting design with local plants aimed to the future restoration of the natural environment of Lycabettus by introducing the rich flora of Attica Landscape and maintaining the special historical memory of the site.

The landscape design strategy and landscape applications considered as part of the sustainable development of the road created an aesthetic scenery, ensuring the security of the transportation and providing the necessary environmental protection.

Finally, the project seems to be successful in its aims as: the landscape restoration and the incorporation of the road into the surrounding various forms of landscape and the particular prevailing local geological, aesthetic and environmental conditions of the area, as well as the aesthetic and the stabilization improvement of the landscape and the safety of the road. The technical difficulties (intensive slopes) and the long history of the site, the development and the landscape design of these projects is a highly challenging task and requires methods for ecological balance (Alifragis et al., 1999; Land Use Consultant, 1999). Also, there was excellent opportunities for all the nurseries to produce indigenous plants in huge quantities and to gain experience in propagation methods.

References

Alifragis D, Chatzistathis A, Kekis G, Ganatsas P, Sotiriadis D, Tsakaldimi M, Valiantza E, Beltsou A, Georgi G N, Stathakopoulos I 1999 Specifications for the production of the suitable plans for the restoration of the slopes of Egnatia Odos. EGNATIA ODOS A.E Thessaloniki, pp. 138.

Dritstas M. 2004 Likabittos & Dexameni, Antropines Roes, Fereniki (in Greek), pp. 155.

Georgi J N 2003 The green spaces in the Olympic works. *In* Proceedings in the Conference: Choice of species for reforestation, improvements of Urban and Suburban and Natural Landscapes. Drama 6/6/2003 (in Greek).

Georgi N and Kapnistou M 2004 Landscaping in Athens due to the Olympic Games in 2004. *In* Proceedings of Central IFLA Symposium, Prague, pp. 35.

Land Use Consultant 1999 Landscape Design Guidelines for Egnatia Odos. S.A., Egnatia Odos S.A., Thessaloniki, pp. 250.

Kaperoni P, Georgi N J, Kapnistou M, Thanasoulia V, Mikros D and Petamidis I 2004 Attic Green areas 2001/2004. Ministry of Environment, Planning and Public Works, Athens (in Greek), pp. 190.

Herbaceous plant cover establishment on highway road sides

Z. Koukoura, A. Kyriazopoulos & I. Karmiris

Aristotle University of Thessaloniki, Faculty of Forestry and Natural Environment, Laboratory of Range Science (236) 54124 Thessaloniki, Greece.

Key words: drought tolerance, plant densities, R/S ratio, survival

Abstract

The ability of different types of plants to become established was studied on the road sides of the Egnatia highway, Thessaloniki, Greece. A mixture of perennial plants (grasses, legumes and forbs) was sown at equal quantities without any preparation of the sowed soil surface. Plant density (individuals/m^2), above and underground dry weight per individual plant (grams/plant), were measured during the growing season. The results showed that drought tolerant species of all plant life forms had high survival percentages and contributed significantly to the vegetation cover at the end of the growing season. Drought tolerance and the existence of rhizomes benefited the establishment ability of grass species. The best adapted species were the grasses *Agropyrum cristatum* L., *Bromus inermis* Leyss., *Dactylis glomerata* L. and *Festuca valesiaca* Schleich, the legume *Medicago sativa* L. and the forb *Sanguisorba minor* Scop.

Introduction

Increased national transportation and subsequent road construction in Greece has resulted in significant impacts to plant communities and soil stability within these areas. Increased soil erosion and sediment decomposition, altered plant community structure, wildlife habitat fragmentation and diminished ecological and aesthetic qualities are evident (Forman and Alexander, 1998; Forman and Delinger, 2000; Harper-Lore, 1996; Petersen et al., 2004). Vegetation removal and the steep slopes of road sides usually increase soil temperature and evapotranspiration as well as decreasing soil moisture. The establishment of vegetation cover is very difficult under these environmental conditions (Johnston et al., 1975).

A wide variety of reclamation methods has been investigated in order to determine the most effective in establishing plants and increasing soil stability on disturbed roadsides. The most cost-effective method to revegetate disturbed areas is considered to be direct seeding (Brown and Bugg, 2001). The implementation of certain revegetation methods e.g. fertilization, tilling of soil and use of mulch, can alter the competitive ability of plant species in different ways (Bugg et al., 1997; Tyser et al., 1998). These techniques may be particularly important in cases where a specific plant species must be protected from direct interspecific competition, e.g. in cases where the seedlings of a species are known to exhibit e.g. low vigor, slow growth rates, or the seeds are expensive and rare (Brown and Bugg, 2001). These techniques increase the cost of establishing sustainable plant communities on highway roadsides. The use of drought tolerant native species could reduce this high cost.

Grass species are suitable for the rehabilitation of disturbed areas because they can reduce soil erosion and protect the soil surface (Lawrence, 1981). Leguminous herbaceous species could also be used for the reclamation of highway roadsides as they can quickly establish vegetation cover (Aronson et al., 1993). Therefore, the objective of this study was to examine the establishment ability of herbaceous plants of different life forms and various water demands on the sides of the newly constructed Egnatia highway, Thessaloniki, Greece.

* E-mail: zoikouk@for.auth.gr, apkyr@for.auth.gr, ikarmiri@for.auth.gr

Materials and methods

The research was conducted on the road sides of the Egnatia highway, on the outskirts of Thessaloniki, Greece at an elevation of about 20 m. The climate is characterized as semiarid, with a mean annual precipitation of 415 mm and mean annual temperature of 15.5°C.

In October 1998, a seed mixture of twelve perennial plants (grasses, legumes and forbs) was sown by hand at equal quantities of 250 g without any previous preparation of the sowing soil surface. The experimental area was watered just once after the seeding. No other practices e.g. fertilization or mulching were used. Each site was 50 m^2 and was replicated three times. Before seeding, a germination test was performed for all plant species. The plant species that were sown and their germination rates are presented in Table 1. Vegetation cover was visually estimated (Cook and Stubbendieck, 1986). Plant species' density (number of plants of each species per square metres) measurements were taken monthly from April to June during the growing season of the first year of establishment with quadrants (0.1×0.1 m^2) applied around six permanent sticks that were randomly dispersed in each replication. On the last sampling date (20 June 1998) all the individuals were removed and the above and underground dry weight per individual plant (grams/plant) was measured. Root/Shoot (R/S) ratio was then calculated. Plant mortality rate (PMR) was calculated using the formula

Table 1. Categories of sown plant species and their germination rates

Turf grasses	Germination rate (%)	Rhizomatous grasses	Germination rate (%)
Agropyrum cristatum L.	88	*Bromus inermis* Leyss	77
Cynosurus cristatus L.	79	*Agrostis tenuis* Sibth.	65
Dactylis glomerata L.	80	Legumes:	
Festuca valesiaca Schleich	85	*Lotus corniculatus* L.	76
Festuca arundinacea Schreb.	77	*Medicago sativa* L. var. *iliki*	82
Lolium perenne L.	70	*Onobrychis sativa* Lam.	62
		Forbs: Sanguisorba minor Scop.	91

PMR = dD/dt where dD is the difference of densities for the period dt.

All data were subjected to two-way analysis of variance through S.P.S.S. statistical package (version 11.0). Plant species and sampling dates (26/03, 30/04, 28/05, 20/06) were treated as factors. Square root transformations were performed on density data in order to correct deviation to normality. To detect differences between plant densities in relation to sampling dates, Tukey's HSD test was employed at $P = 0.05$ level of significance (Steel and Torrie, 1980).

Results

Two-way analysis of variance showed that species density depended on the sampling date with a significant interaction ($F = 2.93$, $P < 0.001$) between these two factors. Significant differences were also detected between species at each sampling date (Figure 1). At the first sampling date, *M. sativa, S. minor, A. cristatum, F. valesiaca* and *D. glomerata* had significantly higher densities. The density of all species decreased significantly over the next sampling dates. The highest mortality rates (PMR) were observed during June. These ranged from 89.8% to 64% for most species. *M. sativa, S. minor* and *B. inermis* had PMR of 6.5%, 8.3% and 23.6% respectively. *S. minor, M. sativa, B. inermis, A. cristatum, F. valesiaca* and *D. glomerata* had significantly higher densities at the end of the growing season.

B. inermis, F. valesiaca, D. glomerata and *A. cristatum* had the highest ratios of R/S dry weight, due to the higher values of their root dry weight. *M. sativa* and *S. minor* had the highest values of root and shoot dry weight despite the lower R/S ratios (Table 2). The total vegetation cover was 60% at the end of the growing season. *M. sativa, S. minor* and *B. inermis* vegetation cover was 12%, 8% and 10% respectively.

Discussion

Road construction alters the environmental characteristics of road sides i.e. soil density and pH, soil and air temperature, intensity of solar radiation and runoff pattern. Hence, the greatest ecological effects on plant communities adjacent to roads can be expected, since a change to physical conditions is usually accompanied by changes in plant communities' characteristics (Bogemans et al., 1989; Farmer, 1993; Fleck

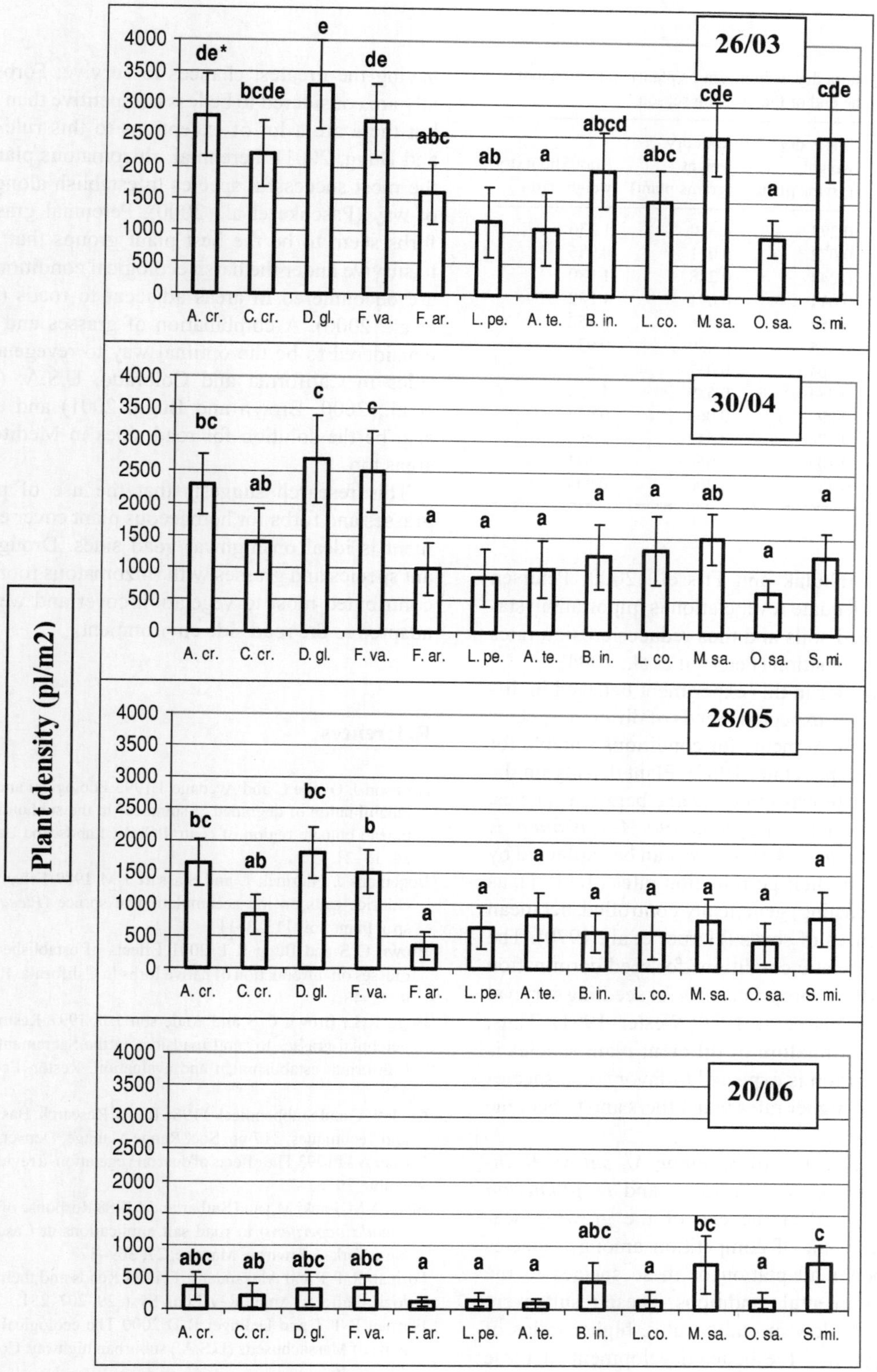

Figure 1. Species density at each sampling date during the growing period. Different letters indicate significant differences of mean plant densities between species on each sampling date, at $P = 0.05$ level of significance, using Tukey HSD test.

Table 2. Root and shoot dry weight (grams/plant) and root/shoot dry weight ratio at the end of the growing season

Plant species	Shoot dry weight (grams/plant)	Root dry weight (grams/plant)	Root/Shoot dry weight ratio
A. cristatum	0.273	0.310	1.136
C. cristatus	0.072	0.071	0.986
D. glomerata	0.244	0.288	1.180
F. valesiaca	0.071	0.091	1.282
F. arundinacea	0.054	0.046	0.852
L. perenne	0.095	0.089	0.937
A. tenuis	0.064	0.057	0.891
B. inermis	0.384	0.589	1.534
L. corniculatus	0.096	0.065	0.677
M. sativa var. iliki	0.570	0.512	0.898
O. sativa	0.103	0.066	0.641
S. minor	0.430	0.406	0.944

et al., 1988;;rombulak and Frissell, 2000; Petersen et al., 2004). Adequate revegetation is important in stabilizing disturbed soils and thus reduces erosion rates and sediment deposition (Petersen et al., 2004).

Different species in this experiment behaved in different ways as was expected. Seeds of different species differ in their requirements for conditions suitable for germination (Harper et al., 1965). Plant density on the first sampling date was quite diverse between species. The higher densities of *D. glomerata, A. cristatum, F. valesiaca, S. minor* and *M. sativa* can be explained by the differences in their germination rates (Table 1), as well as of the existing, genetically controlled, net mean weight differences of seeds (Harper et al., 1970). The best soil and climatic conditions for seed germination are not always the appropriate ones for seedling survival and growth (Harper et al., 1970; Wester, 1991). Thus, in cases where a mixture of different plant species is seeded, competition is expected to favor some species so as to grow at higher rates than others and to become more competitive.

The higher densities of *S. minor, M. sativa, B. inermis, A. cristatum, F. valesiaca* and *D. glomerata* that were observed at the end of the growing season, were the result of competition amongst species and of the better adaptation of these species to the specific environmental conditions. These results were supported by the low mortality rates, higher ratios of R/S dry weight and the better development of these species' biomass per plant during the critical month of June. According to Watt (1981), plant species with large R/S dry weight ratios are the most competitive, having the greatest chances to survive. Forbs generally are considered to be less competitive than grasses, but there are a lot of exceptions to this rule (Brown and Bugg, 2001). Perennial, rhizomatous plants were the most successful species to establish along rights-of-way (Paschke et al., 2000). Perennial grasses and forbs seem to be the best plant groups that are able to survive under the harsh ecological conditions which are encountered in areas adjacent to roads (Paschke et al., 2000). A combination of grasses and forbs is considered to be the optimal way to revegetate roadsides in California and Colorado, U.S.A. (Paschke et al., 2000, Brown and Bugg, 2001) and this may also be the solution for road sides in Mediterranean areas too.

This research suggests that the use of perennial grasses and forbs for herbaceous plant cover establishment is ideal on highway road sides. Drought tolerant species and grasses with rhizomatous root systems contributed most to vegetation cover and were better adapted to the road side environment.

References

Aronson J, Ovalle C and Avedano J 1993 Ecological and economic rehabilitation of degraded "Espinales" in the subhumid Mediterranean climate region of central Chile. Landscape Urban Plann. 24, 15–21.

Bogemans J, Nierinck L and Stassart J M 1989 Effect of deicing chloride salts on ion accumulation in spruce (*Picea abies* (L.) sp.). Plant Soil 113, 3–11.

Brown C S and Bugg R L 2001 Effects of established perennial grasses on introduction of native forbs in California. Restor. Ecol. 9, 38–48.

Bugg R L, Brown C S and Anderson J H 1997 Restoring native perennial grasses to rural roadsides in the Sacramento valley of California: establishment and evaluation. Restor. Ecol. 5, 214–228.

Cook W C and Stubbendieck J 1986 Range Research: Basic Problems and Techniques, 317 pp. Soc. Range Manage, Denver, Colorado.

Farmer A M 1993 The effects of dust on vegetation- a review. Environ. Pollut. 79, 63–75.

Fleck A M, Lacki M J and Sutherland J 1988 Response of white birch (*Betula papyrifera*) to road salt applications at Cascade Lades, New York. J. Environ. Manage. 27, 369–377.

Forman R T T and Alexander L E 1998 Roads and their major ecological effects. Annu. Rev. Ecol. Syst. 29, 207–231.

Forman R T T and Delinger R D 2000 The ecological road-effect zone of Massachusetts (U.S.A.) suburban highway. Conserv. Biol. 14, 36–46.

Harper J L, Williams J T and Sagar G R 1965 The behavior of seeds in soil. I. The heterogeneity of soil surfaces and its role in determining the establishment of plants. J. Ecol. 539, 273–286.

Harper J L, Lovell P H and Moore K G 1970 The shapes and sizes of seeds. Annu. Rev. Ecol. Syst. 1, 327–356.

Harper-Lore B L 1996 Using native plants as problem solvers. Environ. Manage. 20, 827–830.

Johnston R S, Brown R W and Cravens J 1975 Acid mine rehabilitation problems at high elevation. *In* Watershed Management Symposium. Amer. Soc. Civil. Eng, pp. 66–79. Logan, Utah, USA.

Lawrence T C 1981 Intermediate Wheatgrass. Can. J. Plant Sci. 61, 467–469.

Paschke M W, DeLeo C and Redente E F 2000 Revegetation of roadcut slopes in Mesa Verde National Park, U.S.A. Restor. Ecol. 8, 276–282.

Petersen S L, Roundy B A and Bryant R M 2004 Revegetation methods for high-elevation roadsides at Bruce Canyon National Park, Utah. Restor. Ecol. 12, 248–257.

Steel R G D and Torrie J H 1980 Principles and Procedures of Statistics. 2nd edition, 633 pp. McGraw-Hill, New York.

Trombulak S C and Frissell C A 2000 Review of ecological effects of roads on terrestrial and aquatic communities. Conserv. Biol. 14, 18–30.

Tyser R W, Asebrook J M, Potter R W and Kurht L L 1998 Roadside vegetation in Glacier National Park, U.S.A.: effects of herbicide and seeding treatments. Restor. Ecol. 6, 197–206.

Watt L A 1981 Establishment of grasses on cracking clay soils. Seedling morphology characteristics. *In* JA Smith and WV Hays (eds), Proceedings of the XIV International Grassland Congress, pp. 451–453. Westview Press, Boulder, CO.

Wester D B 1991 A summary of range plant seed germination research. International Center for Arid and Semiarid Land Studies (ICASALS) Texas Tech. Univ. Lubbock, T.X. Pub. No 91–92, 112 pp.

Erosion control by application of hydroseeding methods along the Egnatia Motorway (Greece)

M. Katritzidakis[1], A. Liapis, I. Stathakopoulos, E. Pipinis, G. Kekis, E. Ververidou & E. Sevastou

Egnatia Odos S.A., 6th Km Thessaloniki—Thermi, P.O. 30, 57001, Thermi, Greece. [1]*Corresponding author**

Key words: eco-engineering, embankment, erosion, hydroseeding, restoration

Abstract

The Egnatia Motorway is the greatest highway infrastructure project currently being constructed in Southern Europe. The construction of such a large-scale project is expected to cause an environmental and hence landscape disturbance. Landscaping works are, therefore, of a critical importance and are imposed by the Environmental Terms approved for each project. As far as erosion is concerned, the hydroseeding of herbaceous species is a widely used practice for the protection of slopes against surface erosion, which is preferably applied immediately after their final configuration. Undoubtedly, the establishment of woody species immediately after the configuration of the surfaces under restoration (cuts and fills) and at the minimum possible cost constitutes one of the most preferable landscaping solutions in highway construction. In this context, the Green Unit of "Egnatia Odos A.E." carried out an experimental application of woody species hydroseeding, with the purpose of investigating the effectiveness of this method, as far as the establishment of woody vegetation on the Egnatia Motorway slopes in concerned. This application was implemented in two stages. The first one was carried out in cooperation with a group of experts from the Aristotle University of Thessaloniki, at a small scale and after strict experimental planning. The second one was an experimental–pilot application, in order to re-check the positive results of the previous test, to extract more useful conclusions and to check this method in terms of its feasibility as part of the green works performed along the Egnatia Motorway. Another issue, which is closely related to the previous one, is the protection of slopes against erosion by application of various engineering methods using cuttings, apart from hydroseeding and planting. These works, which are compliant with the general environment-friendly philosophy in approaching such issues, are supplementary to the hydroseeding of herbaceous species and are more effective in anti-erosion action when compared to conventional planting works. At the same time, this method serves the need for aesthetic restoration.

Introduction

The Egnatia Motorway is the longest motorway project (680 km) currently under construction in Southern Europe. It crosses five phytosocial zones with a large number of plants (woody perennials and others) of the rich Greek flora. The alternation of mountainous, hilly, flat and peri-urban landscapes, combined with the continuous and rich alternation of ecotopes make up its richness, as well as great ecological, aesthetic, research and education interest. Along the route of the motorway, there are 17 natural habitat areas protected under the European "Natura 2000" Network, 4 wetlands protected under the Ramsar Convention, and 70 wildlife conservation areas (formerly wildlife reserves). Moreover, it crosses most of the longest rivers in Greece (Evros, Nestos, Strymonas, Gallikos, Axios, Loudias, Aliakmonas, Venetikos, Metsovitikos, Arachthos, etc), as well as mountainous areas concerned with intensely and severe stability problems due to poor strength

* E-mail: mkatri@egnatia.gr

rocks (e.g. flysch) or geotechnically downgraded rock-masses.

The construction of such a large-scale project is expected to cause an abrupt and extensive disturbance to the ground. Vegetation is sometimes totally destroyed and broad surfaces become exposed to weather conditions. Environment disturbance and subsequent landscape disturbance when constructing such a large-scale project are due to earthworks carried out during project implementation. These either produce materials that are usually deposited at licensed areas (deposit areas) or require additional geological materials (borrow pits); the construction of service roads opened either as parallel roads or as access roads to worksites; embankments and cuts created by the construction of the mainline; faces created at tunnel portals when driving tunnels; bases of bridges under construction; training works at the main beds of rivers and watercourses, if deemed necessary.

The problem that occurs wherever vegetation is destroyed, e.g. on new slopes, is erosion, including detachment, transfer and deposit of soil material because of the impact of the weather on the ground surface. Each year more than 120,000 km^2 of fertile soil are eroded away globally (Fiedler, 1990). The main factors causing erosion are water, wind and ice, and those affecting the amount of soil lost due to erosion are the climate, ground, topography, and plant cover. The most common erosion factor is water that causes surface erosion and gully erosion as well as problems in deeper strata.

The problem of erosion is more important especially on cut slopes, where the lack of vegetation renders impossible the natural revegetation–restoration of the initial balance by transforming various soil substances to fertile soil. Therefore, fertile soil quantities are continuously reduced with productivity, whereas slope stabilisation problems increase and works' aesthetics are downgraded. According to Megahan and Kidd (1972), 80% of the total erosion is observed on slopes not seeded during the first year after slope construction. Besides, erosion delays or event stops the development of vegetation.

Vegetation can play a very interesting role in surface erosion control (Rey et al., 2004). Vegetation is considered as the main factor for controlling erosion and surface runoff (Thorne, 1980; Gray and Sotir, 1996). Especially by managing thick turf, grass or herbaceous vegetation, soil loss due to water erosion may be reduced 100-fold (USDA Soil Conservation Service, 1978). On seeded slopes, erosion that takes place is 10-fold less compared to uncovered slopes (Barnett et al., 1967). According to Gray and Sotir (1996), the beneficial results of herbaceous vegetation and grass on surface erosion are: interception (foliage and plant residues absorb rainfall energy and prevent the detachment of the soil by raindrops); restraint (root systems bind and restrain soil particles while the above-ground residues filter sediment out of run-off); retardation (above-ground residues (branches, foliage) slow down wind velocity and surface runoff); and infiltration (plants and plant residues help maintain soil porosity and permeability, therefore delaying onset of saturation and increasing water infiltration).

As soon as slopes in an Egnatia Motorway section are finalised, it is a top priority to protect the fine grain surface layer from erosion. Protecting and restoring those surfaces is of paramount importance and is imposed by the environmental terms specific to each project. Because natural restoration is slow or non-existent due to the lack of favourable conditions, we seek to protect slopes immediately after construction by covering them with turf through hydroseeding. Hydroseeding is a method used for revegetation on inaccessible surfaces, e.g. large and steep slopes, where other seeding or planting methods cannot be applied as successfully. Special equipment is used for launching on slopes a mix of seeds, fertiliser, mulches, soil stabiliser, and water. By hydroseeding with the appropriate plant seeds we achieve quick protection of surfaces by installing turf that will stabilise the soil surface with its roots. At the same time, infertile soil is enhanced with materials that will create more favourable conditions for the beginning of microbe activity and will increase its organic matter. The main components of hydroseeding are the seed mix and the auxiliary materials. In addition to the protection of soil against surface erosion, hydroseeding has also certain indirect results, like the immediate aesthetic improvement of the injured landscape, the retaining of organic matter, the achievement of conditions that facilitates and improves soil fertility for future planting, the facilitation of natural revegetation due to the creation of favourable conditions for the migration of flora species from the surrounding areas and the protection of other sections of the project against the debris produced by erosion.

Therefore, the Green Unit of "Egnatia Odos A.E." (EOAE) carried out an experimental application of woody species hydroseeding, with the purpose of investigating the effectiveness of this method, as far as the establishment of woody vegetation on the Egnatia

Table 1. Phytosocial zones along the Egnatia Motorway

	Section Ch.	Length (km)	Altitude (m)	Vegetation zone (development area)
1	0–4.5	4.5	0–200	*Arbuto andrachne* L.-*Quercetum ilicis* Br.-Bl.
2	4.5–29.5	25.0	200–520	*Coccifero-Carpinetum orientalis*
3	29.5–35	5.5	500–640	*Quercetum frainetto* Ten.
4	35–51	16.0	300–600	*Coccifero-Carpinetum orientalis*
5	51–69	18.0	400–790	*Quercetum frainetto*
6	69–116	47.0	480–1000	*Coccifero-Carpinetum orientalis*
7	116–125	9.0	1000–1100	*Pinetum pallasianae*
8	125–136	11.0	800–1100	*Quercetum frainetto*
9	136–137	1.0	900–960	*Pinetum pallasianae*
10	137–141	4.0	820–920	*Quercetum frainetto*
11	141–144	3.0	700–860	*Pinetum pallasianae*
12	144–190	46.0	550–920	*Quercetum frainetto*
13	190–228	38.0	650–760	*Quercetum frainetto* (+ *Q. Trojana* at places)
14	228–251.5	23.5	450–900	*Quercetum frainetto*
15	251.5–259	7.5	100–450	*Coccifero-Carpinetum orientalis*
16	259–323	64.0	10–100	*Coccifero-Carpinetum orientalis* (+ cultivations)
17	323–341	18.0	10–120	*Coccifero-Carpinetum orientalis* (+ settlements)
18	341–359	18.0	90–120	*Coccifero-Carpinetum orientalis* (+ cultivations)
19	359–417	58.0	60–160	*Coccifero-Carpinetum orientalis*
20	417–491	74.0	10–180	*Orno-Quercetum ilicis* (+ cultivations)
21	491–503.5	12.5	40–380	*Coccifero-Carpinetum orientalis*
22	503.5–629	125.5	20–240	*Coccifero-Carpinetum orientalis* (+ cultivations)
23	629–685.25	56.25	10–120	*Orno-Quercetum ilicis* (+ cultivations)

Motorway slopes is concerned. This application was implemented in two stages. The first one was carried out at a small scale in a field, in cooperation with a group of experts from the Aristotle University of Thessaloniki. The second one was an experimental–pilot application, in order to re-check the positive results of the previous test, to extract more useful conclusions and to check this method in terms of its feasibility as part of the green works performed along the Egnatia Motorway.

Materials and methods

Application of hydroseeding on the Egnatia Motorway

Hydroseeding on the Egnatia Motorway for operational purposes, i.e. the immediate protection of the final surface of slopes against surface erosion, has been incorporated in highway construction contracts. To ensure the best possible performance of hydroseeding works by Contractors, EOAE prepared a number of specifications (Article 80, Technical Conditions of Contract) and methods to assess successful results (Article B11, Special Conditions of Contract).

In preparing these specifications, information was extracted from research works conducted by a panel of experts, and from experience, by collecting specifications from various countries as well as by training company officials through international seminars.

The time at which hydroseeding is performed is paramount for the successful installation of turf and woody species on slopes. The most suitable time in the Greek context is October and November. Hydroseeding in areas characterised by heavy winters can be carried out in late February or early March.

The Landscape Guidelines include mixes of seeds that are specific to each phytosocial zone. These mixes are recommended to consultants that prepare landscape restoration studies and to construction contractors. Table 1 refers to phytosocial zones along the Egnatia Motorway. The recommended mix for each phytosocial zone is mentioned in Table 2. Mixes were chosen by a panel of experts (Papanastasis et al., 2000).

For a successful hydroseeding, the quality of seeds and their correct mixing—inter alia—is extremely important. The recommended quantity of seeds is approximately 20 g/m^2 of surface. For the auxiliary material, we use organic and inorganic fertilisers that both enhance the soil with nutrients for the plants and activate

Table 2. Seed mixes for slope turf by phytosocial zone

(a) *Arbuto andrachne-Quercetum ilicis:* Dry and hot environment

Vivacious grasses		Legumes/Others	
Cydonon dactylon	35%	*Phacelia tanacetifolia*	10%
Festuca rubra	15%	*Sangisorba minor*	5%
Agrostis tennuis	5%	*Trifolium subterraneum*	10%
Lolium rigidum	20%		

(b) *Coccifero-Carpinetum ilicis:* Semi-dry–hot environment

Vivacious grasses		Legumes/Others	
Cynodon dactylon	30%	*Medicago sativa*	10%
Festuca arundinacea	20%	*Lotus corniculatus*	10%
Agrostis tennuis	5%		
Poa pratensis (Nudwart)	10%		
Lolium rigidum	15%		

(c) *Orno- Quercetum ilicis:* Semi-dry–cold environment

Grasses		Legumes/Other	
Festuca rubra	30 %	*Medicago sativa*	10%
Festuca arundinacea	10%	*Lotus corniculatus*	10%
Lolium perenne	20%		
Poa pratensis	10%		
Agrostis tennuis	10%		

(d) *Quercetum frainetto:* Subhumid–cold environment

Grasses		Legumes/Other	
Lolium perenne	30%	*Trifolium repens*	10%
Poa pratensis	10%	*Medicago sativa*	10%
Agrostis stolonifera	10%		
Festuca rubra	20%		
Bromus catharticus	10%		

(e) *Pinetum pallasianae:* Humid–cold environment

Grasses		Legumes/Other	
Poa pratensis	20%	*Onobrychis sativa*	10%
Festuca rubra	20%	*Medicago sativa*	10%
Agrostis stolonifera	10%		
Lolium perenne	20%		
Bromus inermis	10%		

microorganisms, and soil stabilisers and bentonite that help seeds stick on the soil. For mulches, i.e. substances that temporarily—until vegetation grows—protect the surface against erosion, natural biodegradable materials are used. Mulches assist the installation of plants because they reduce the adverse impact of extreme temperatures, ensure better humidity conditions for seed germination, reduce the loss of soil humidity through evaporation, as well as protect seeds and subsequently the soil surface against erosion by facilitating the runoff of rainwater. The main type of organic mulches used on the Egnatia Motorway is cellulose. Other mulches used are: properly cut shanks of annual plants, such as wheat, barley, rice etc., suitably smashed wood or bark of trees, suitably cut used paper (old newspapers, etc), wood fibres, and turf.

The improvement and enhancement of hydroseed mixes is being investigated, with the potential change of some species and the introduction of new ones. The introduction of woody species seeds is being tested and the potential combination of the use of seedlings with hydroseeding is being introduced. A proportion of collected seeds of woody species is used for enhancing hydroseed mixes. In parallel, the use of new methods and technologies in hydroseeding is being studied (Papanastasis et al., 2001).

The last few years, the company investigated the installation of woody vegetation on Egnatia slopes with the direct installation of seeds on surfaces to be restored. This investigation concerns mainly rock slopes, where restoration methods are very limited and planting is almost impossible.

All results and indications of innovative applications are evaluated. Special weight is put on unsatisfactory results, so that any negligence is identified, operations are improved and new technologies are utilised. EOSA, in its effort to modernise protection methods for slopes against erosion and restore the disturbed environment, intends to cooperate with research and technological institutes and universities in order to follow up technology.

During previous planting periods trial installations of woody species with hydroseeding were performed at four locations along the Egnatia Motorway (Grevena area—Xanthi area—Kozani area—Prophitis area, Thessaloniki) in autumn and early spring. During these experiments five factors were studied (seed mix, seeding method, type of slope, cover and exposure) in order to study not only the main impact of each one of them, but also the interaction with other factors.

Results

As a strategy, we generally do not provide for any maintenance of hydroseeds, assuming that natural vegetation would gradually install itself. In most cases, not only woody species were not installed naturally, obviously due to unfavourable weather and soil conditions, but hydroseeds themselves showed evidence of weakening. Two years after hydroseeding, the number of species on certain surfaces was reduced (prevalence of one or two species) whereas some gaps appeared on others. The accumulation of biomass on hydroseeded slopes entail risks of fire and inhibits revegetation.

This fact made us revise our initial strategy. We are therefore considering the results of hydroseed maintenance works and dead biomass management works under a green works contract. This is a 2-year term scheme; it started in spring 2004 and will be completed in spring 2006.

Also it is possible to install woody species with hydroseeding on Egnatia slopes under the following conditions:

(a) seeds to be used shall be of high purity and germination ability and shall be treated so that they germinate easily,
(b) seeding shall take place early in autumn, before or in parallel with the beginning of early rainfalls and under no circumstances later than mid-October,
(c) slopes shall be recently formed, i.e. without native vegetation.

Discussion

Based on the previous results, conclusions and recommendations can be proposed and summarised as follows:

1. The proportion of development of woody plant species is reduced when the mix includes also seeds of herbaceous vegetation. Nevertheless, herbaceous vegetation increases the cover percentage that is secured by hydroseeding on slopes.
2. Overlaying with geotextile improves substantially the success of hydroseeding with woody species. On the contrary, adding straw does not seem to contribute to the success of hydroseeded woody species, compared to simple juta overlaying.
3. The application of 0.75–1 kg/acre of woody species seed mix seems to ensure a density of approximately three seedlings per square metre or 750 seedlings/acre, which is considered very satisfactory.
4. It is advisable to apply a mix of woody species in hydroseeding. Nevertheless, the number of species cannot exceed 4–5, and these species should be adapted to the specific conditions of the slope where they shall be planted.
5. Hydroseeding woody species on soily slopes is generally possible, if the above conditions are followed, and it can therefore be an alternative technique for slope revegetation in the Egnatia Motorway. Nevertheless, on difficult slopes, as well as on very steep cuts, the success of woody species hydroseeding is not guaranteed and in such cases it is preferable to plant seedlings produced in a nursery or, alternatively, woody species produced in panels.
6. Hydroseeding of woody species does not ensure high proportions of cover and therefore is not suitable for protecting slopes against surface erosion, especially during the first 1–2 years.
7. On slopes where native herbaceous vegetation is already installed or on slopes where hydroseeding with herbaceous vegetation is applied, the installation of woody species with hydroseeding is impossible and should be avoided. In such cases, it is better to introduce woody species by planting developed seedlings produced in a nursery. Generally, seeding on slopes should immediately follow slope formation so that surface is still rather lost.
8. A necessary condition for successful hydroseeding of woody species is to timely obtain good quality seeds and suitably prepare surfaces. These slope surfaces should have uniform and good quality soil. To this purpose, there should be good coordination between the collection of woody species seeds, the preparation of suitable slopes, and hydroseeding, in order to achieve the best possible results.

The operations that gave the best results (seeding woody plants with mulch and seeding woody plants with straw and overlay of Juta) should be preferred mainly under difficult conditions (steep gradient, long slopes, low permeability), whereas, in the rest of the cases, simple hydroseeding with woody and herbaceous vegetation species seems to produce satisfactory results.

The next step the Company took, after completing this initial stage of investigation, was to proceed to a pilot application of the method in the context of herbaceous vegetation hydroseeding in established highway works contracts.

Experimental application

Upon launching this project, the Green Works Unit was, of course, aware of the fact that simultaneous hydroseeding of woody and shrub species leads to reduced effectiveness of the installation of the former due to competition. This fact could endanger the successful outcome of the whole effort. It is also an undoubted fact that seeding woody species cannot replace herbaceous vegetation in terms of surface protection against erosion and therefore, replacing shrub species with woody species would be useless, whereas their combined application would be the only effective solution. Moreover, it is clear that benefits from simultaneous installation of woody species and herbaceous vegetation (if this can be achieved) are very important. The assessment of both these facts leads to the conclusion that further investigation of the issue would be both useful and justified.

The objective of this second stage was different to that of the first stage. In detail, the targets of this effort were as follows:

- To test the method under real conditions of project execution in the context of a specific contract.
- To apply the method with seed quantities smaller than those used in previous experimental applications (0.75–1 kg/acre) so that the method is practically applicable and financially attractive for large-scale use (note that each year the method of hydroseeding is applied on hundreds of acres of Egnatia Motorway surfaces).
- To test the method under different implementation conditions. Fills under different weather conditions and exposed to different factors, including certain difficult cases, such as rock cuts.
- To assess the method's effectiveness based only on the visible survival of woody species 1 or 2 years after hydroseeding.
- To examine the impact of seed treatment on the effectiveness of the method.

At this stage the company wanted to see whether the combined use of woody species and shrub species would have a practical impact on the installation of woody species and by also calculating the size of the outcome and the cost required (cost of collection, cleaning, storage and treatment of seeds) to conclude on whether this method is suitable to use as supplementary to or alternative of other methods of woody vegetation installation.

Table 3. The treatment processes, as applied to the woody species

Species	Treatment
Spartium junceum, Colutea arborescens, Cercis siliquastrum, Rhus coriaria, Philyrea latifolia, Medicago arborea	Treatment with thick sulphuric acid
Cistus incanus	Treatment with hot water
Anthyllis hermaniae, Paliurus spina cristi, Coronilla emeroides, Nerium oleander	No treatment

The species used were some of those that had been used in a previous trial period and had produced satisfactory results. Specifically a mix of the following was used at all locations: *Spartium junceum* L., *Colutea arborescens* L., *Cercis siliquastrum* L., *Rhus coriaria* L., *Cistus incanus* L., *Phillyrea latifolia* L., *Medicago arborea* L., *Anthyllis hermaniae* L., *Paliurus spina cristi* Mill., *Coronilla emeroides* Bois. & Spr. and *Nerium oleander* L.. The weight distribution of seeds from various species was the same at all positions of application and corresponded to the distribution used at the previous stage.

Three different mix quantities per acre were used (375, 250, 125 g/acre). Certain mix seeds had undergone some treatment (with the exception of *Anthyllis hermaniae, Paliurus spina cristi, Coronilla emeroides, Nerium oleander*) while others no treatment at all (Table 3). Subsequently, six cases were examined (combinations of the above). Each combination of mix quantity per acre and application or no application of treatment was used in four surfaces, in total (one cut and one embankment in West Region, one cut and one embankment in Central Region).

Thus, hydroseeding was applied on 24 different surfaces. Twelve were in West Region, i.e. Baldouma and Metsovo, and 12 in Central Region (9 in the area of Kastania and 3 in Thessaloniki Ring Road). These surfaces include slopes facing north and south, so that each one of the six combinations is applied on slopes with the same orientation (for example, the case of treated seeds and a mix quantity of 375 g/acre was applied in cuts and embankments facing south). Unfortunately, it was not currently possible to use all six combinations at all different cases of slopes (cuts—embankments facing north and south) because that would necessitate 48 different surfaces.

After hydroseeding application, which was carried out in the appropriate period (mid autumn), we visited the areas of application, in order to examine whether seeds had germinated. In several cases, seedlings were found, but densities were not calculated neither in total nor per species. At the last visit, late August (10 months after hydroseeding), no seedlings of the species seeded were found. Probably, a small number of live plants does exist, but due to their small size, these plants could not be detected among dense ground flora. A rule that can be drawn at this stage is that the competition between woody species and herbaceous dramatically affects the survival of the former.

Implementation of plant-engineering works

To stabilise slopes or even restore parts of slopes with intense erosion or localised sliding, we applied eco-engineering methods. Such constructions involve the installation of billets and grids of wooden boards and the installation of layer constructions and fascines. These methods were combined with planting of plants or cuttings and, apart from the protection of slopes against erosion, they also aim at installing vegetation for the aesthetic improvement of disturbed surfaces. These methods are also known abroad, as they have been applied for many years. In Greece, nevertheless, there is no experience in large-scale uses of those methods in similar works. The following is a summary presentation of the implementation of those engineering methods:

Billets are parts of dry tree trunks, 12–20 cm in diameter and up to 4 m in length, that come from various plant species. They were used at parts of slopes in order to install vegetation on very steep slopes or on unstable slopes. They were placed like contour lines, after digging benches 30 m wide with a slight gradient towards the interior of the slope. A billet was placed on the outer side of the berm with a shallow excavation at the interface of the billet and the slope. The billet was stabilised with steel piles. The gap between the billet and the trench towards the interior of the slope was filled in with soil material from the slope and topsoil. Thus, a flat surface, 15–20 cm wide and as long as the billet, suitable for planting plants or cuttings or for seeding, was formed. Billets can be placed in sections or across the entire slope, depending on the needs for protection against erosion. Usually they are placed throughout a length of 15–20 m, with a short interval of 5–10 m and are replaced every 2–5 m in the direction of the slope gradient.

Grid of wooden boards is 3–6 m long, 25–40 cm high—wide and 8–10 mm thick. They are fixed in the ground vertically, following the contour lines, with the assistance of steel piles. Prior to layering the boards, the slope surface is formed by removing all surface abnormalities, surface erosions and local small-scale landslides, especially in the direction of the slope gradient. A small vertical bench 5–8 cm wide is created, to support the placement of boards. Then, the internal gap created by wooden boards towards the slope is backfilled with slope soil material and topsoil. Thus, benches 30–50 cm wide are created in the sense of contour lines on the surface of the slope and are used for planting plants or cuttings as well as for stabilizing, holding surface soil and reducing surface erosions.

Layer constructions can be applied both on minor landslides and where steep slopes may become heavily eroded in the future. Initially, a narrow trench is excavated according to the contour lines, or the disturbed section of the slope is cleared in case of a landslide. Then, green branches of various plant species (*Tamarix sp., Salix sp., Populus nigra* L., etc.) up to 1 m long and 1–5 cm in diameter are fixed to the ground, in the trench, in a double-line arrangement and against the gradient, until they reach the undisturbed part of the slope. The grid density is 60 branches per linear metre. Thirty of the branches are fixed towards one direction and 30 towards the other. Then follows the backfilling of green branches with soil material from the slope and the necessary quantity of topsoil. This forms small benches on the slope surface in the sense of contour lines. This work, depending on the slope needs, extends to disturbed locations or is repeated every 2–5 m in the direction of the slope gradient throughout the slope length.

Fascines are bundles of green branches of species that germinate easily under suitable conditions (*Tamarix sp., Salix sp., Populus nigra, Platanus orientalis* L., etc.). For their construction we open small trenches, vertical to the slope gradient and in the direction of the contour lines. Parts of green branches, consisting of 15–20 or 25–30 branches 60 cm long and 6–40 mm in diameter with propagulums to the same direction, are tied in bundles every 15–30 cm, using a thin wire, and are placed on the trench, on the surface of the slope in the direction of the bundle length and one packet in contact with the next one. The length of the above arrangement of bundles can range between a few metres and the entire length of the slope. Bundles are fixed at their positions with wooden piles fixed to the ground, through the bundles, in a way that all bundles are fixed to the ground by means of the wooden piles. Any gaps are covered with soil material from the slope and topsoil, so that all green branches are in full contact with the soil without being entirely covered by it. Behind the bundles with the green branches and uphill, small benches are formed, where plants or cuttings can be planted or seeds can be seeded. Depending on the needs of the slope, this work can extend to any disturbed areas or be repeated every 2–5 m following the slope gradient throughout the slope length. All constructions mentioned above aim at limiting the movement of rainwater as much as possible. Rainwater creates undesirable grooves on the surface of the slopes. Constructions also aim at retaining the small soil benches, which, due to the steep gradient, tend to slide towards the base of the slope. They also aim at reducing the speed of the rainwater.

The protection of the base of slopes can be achieved in various ways, such as gabions or dry stone walls. Plant engineering methods may also be used. Further to the desirable engineering results they also have aesthetically better results. Such constructions may be carried out by using billets and layer constructions alternatively, at successive levels with intermediate filling with topsoil.

All the above plant-engineering methods were applied during the previous planting period at certain representative cases of Egnatia slopes, in the context of a pilot small-scale scheme. The choice of positions was based on the foreseen need for protection against erosion combined with the foreseen suitability of such interventions (layer constructions, fascines, cuttings with the aforementioned species). The results should justify not only the correct choice of locations for the implementation of these methods but also the effectiveness of plant-engineering methods regarding the targets set when using them.

The authors' company

The company "Egnatia Odos AE" (EOAE) was established in September 1995 by the Greek State in order to manage the design and construction, maintenance, furniture and exploitation of the Egnatia Motorway, the road axes serving the motorway and other works in Greece or abroad. In particular, the Environment & Green Works Unit of EOAE is mainly in charge of issues regarding environmental protection and restoration of landscapes that are disturbed because of construction works. Since 1997, EOAE successfully implements a large-scale scheme for the protection of slopes against surface erosion by installing turf on slopes through hydroseeding. During this period, more than 750 acres have been hydroseeded, 44% of which were cuts and 56% were embankments. As far as the nature and properties of the slopes' geological bedrock are concerned, they involve coarse conglomerates and breccia limestone materials, tertiary period alluvial deposits (marls, lime silt and clay, sandy deposits, sandstone etc), shales – flysch, gneiss and gneiss schist etc.

References

Barnett A P, Diseker E G and Richardson E C 1967 Evaluation of mulching methods for erosion control on newly prepared and seeded highway black-slopes. Agronomy J. 59, 83–85.

Fiedler H J 1990 Bodennutzung und Bodenschutz, EBG. Fischer Verlag, Jena.

Gray D H and Sotir R B 1996 Biotechnical and Soil Bioengineering Slope Stabilization. Wiley, New York.

Megahan W F and Kidd W J 1972 Effect of logging roads on sediment production rates in Idaho Batholithe. USDA Forest Service, Research Paper INT-123, pp. 14.

Papanastasis V, Ispikoudis I, Platis P, Mantzanas K and Iovi K 2000 Reasearch for the improvement of guidelines for hydroseeding in Egnatia Odos Motorway. Thessaloniki.

Papanastasis V, Ispikoudis I, Takos I, Platis P and Brofas G 2001 Reasearch for the installation of woody species by hydroseeding in Egnatia Odos Motorway. Thessaloniki.

Thorne J B 1980 Erosional processes of running water and their spatial and temporal controls: a theoretical viewpoint. *In* MJ Kirkby and RPC Morgan (ed.), Soil Erosion, pp. 129–182. Wiley, Chichester.

USDA Soil Conservation Service, 1978. Predicting Rainfall Erosion Losses: a guide to conservation planning. USDA Handbook No. 537, Washington, DC.

Restoration of slopes disturbed by a motorway company: Egnatia Odos, Greece

M. Katridzidakis[1], E. Pipinis, A. Liapis, I. Stathakopoulos, G. Kekis, E. Ververidou & E. Sevastou

Egnatia Odos S.A., 6th Km Thessaloniki—Thermi, P.O. 30, 57001, Thermi, Greece. [1]*Corresponding author**

Key words: Egnatia Odos S.A., installation of vegetation, plant production, seed collection, seed treatment

Abstract

The Egnatia Motorway is the largest road currently under construction in southern Europe. It is understood that the construction of such a large-scale project causes both environmental and landscape disturbances. Being sensitive to issues concerning environmental protection, Egnatia Odos S.A. has implemented a series of methods aimed at the aesthetic restoration of disturbed surfaces and the prevention of surface erosion. The above-mentioned goal can be achieved through installation of woody vegetation species on the slopes of the Egnatia Motorway. Therefore, the company uses for restoration forestry species of native flora, which are adapted to the environment of the area to be restored. The company policy is to undertake itself the selection of seeds from various forestry species and produce plants in a nursery. For the collection of seeds, suitable biotopes in the wider project area as well as the species that can be massively produced were identified. The seeds collection were transferred to the company laboratory, cleaned, subjected to quality control and stored. Prior to the use of seeds in the nursery, the necessary pretreatment, depending on the species, was carried out in order to terminate the dormancy and enrich the germination. Meanwhile, the production of plants that were considered difficult to germinate was investigated. Seeding took place in QuickPot propagation trays and the filling materials used were: enriched peat, perlite and soil. During production, all the necessary measures were implemented in order to achieve an excellent planting material. After one year, the plants produced from the seeds in the nursery were transferred to the surfaces to be restored and were planted. The plants were planted near the areas from which their seeds were collected. The installation of vegetation on gunite, aiming at the aesthetic restoration of the landscape (apart from the protection of the slope against erosion), has produced very satisfactory results.

Introduction

The Egnatia Motorway is the longest motorway project (680 km) currently under construction in Southern Europe. It crosses five phytosocial zones with a large number of plants (woody, perennials and others) of the rich Greek flora. The alteration of mountainous, hilly, flat and peri-urban landscapes, combined with the continuous and rich alteration of ecotopes make up its richness, as well as great ecological, aesthetic, research and educational interest.

Along the route of the motorway there are

- seventeen Natural Habitat areas protected under the European "Natura 2000" Network;
- four wetlands protected under the Ramsar Convention;
- seventy wildlife conservation areas (formerly wildlife reserves);
- most of the longest rivers in Greece (Evros, Nestos, Strymonas, Gallikos, Axios, Loudias, Aliakmonas, Venetikos, Metsovitikos, Arachthos, etc);
- areas of intensely mountainous terrain and severe stability problems due to poor strength rocks (e.g. flysch) or geotechnically downgraded rockmasses.

The company Egnatia Odos S.A. (EOSA) was established in 1995 for the design, supervision, construction,

* E-mail: mkatri@egnatia.gr

maintenance, furniture and exploitation of the Egnatia Motorway, the road axes serving the motorway, as well as other projects both in Greece and abroad.

The Environment & Green Works Unit of EOSA is mainly in charge of issues regarding environmental protection and restoration of landscapes that are disturbed because of construction works. This unit, having gone through various transformations in the course of the company development, was set up almost since the beginning of the company. In a way that is original for Greek standards, the unit has managed to gather useful know-how. The Environment & Green Works Unit of EOSA has issued (OSAT, 1999):

- Landscape Guidelines
- Guidelines with specifications for collection of seeds and cuttings of native flora
- Guidelines with a record of biotopes for collection of seeds to be used by Contractors
- A search engine (CD form) that supports the search of the plant species used based on their intended characteristics
- Guidelines for seed storage and maintenance.

The Environment & Green Works unit of EOSA with the aid of science and in compliance with the laws of nature:

- collects propagation material
- organizes a bank of native flora plants
- investigates new ways of seed treatment
- improves hydroseeding mixes
- investigates new and more cost effective methods of plant installation
- produces plants that meet the requirements of landscape contracts
- investigates the use of bioengineering methods.

The construction of such a large-scale project is expected to cause an abrupt and extensive disturbance to the ground. Surface vegetation is totally destroyed and broad surfaces become exposed to weather conditions. The restoration of these surfaces is of paramount importance and is imposed by the environmental terms specific to each project.

Environment disturbance, and subsequent landscape disturbance when constructing a large-scale project are due to the following reasons:

- Earthworks carried out during project implementation. These either produce materials that are usually deposited at licensed areas (deposit areas) or require additional geological materials (borrow pits).
- The construction of service roads opened either as parallel roads or as access roads to worksites.
- Embankments and cuts created by the construction of the mainline.
- Faces created at tunnel portals when driving tunnels.
- Bases of bridges under construction.
- Training works at the main beds of rivers and watercourses, if deemed necessary.

EOSA, a company that is particularly sensitive to issues pertaining to environmental protection, has applied several methods and innovations aimed at the aesthetic restoration of disturbed surfaces and the prevention of surficial erosion. In particular

- preparation of designs and aesthetic restoration of tunnel portals
- application of new technologies and materials (reinforced embankments, reinforced earth) for limiting the construction limits
- landscaping by applying cut & cover techniques
- design and installation of Pollution Control Units at sensitive water receivers (pilot application with water plants ponds)
- river bed training with gabions and materials obtained from the area (local rock)
- restoration of the disturbed environment (slopes, deposit areas-borrow pits, junction areas, etc) by applying innovative planting methods
- use of native flora species
- collection of native forest species seeds
- production of plants
- Experimental vegetation installation techniques (small plants, cuttings, direct seeding).

Materials and methods

Environment restoration

General

In order to proceed to an immediate restoration of the disturbed environment, a series of measures are taken, which are considered necessary for the smooth and immediate integration of the project into the surrounding landscape. This objective can be achieved through the installation of vegetation.

After managing to prevent surface erosion, the main company objective is the long-term stabilization and aesthetic restoration of the disturbed slope surfaces thus facilitating the motorway integration into the surrounding landscape. The most appropriate method to achieve this is to install woody vegetation species on the Egnatia Motorway slopes.

EOSA is the first company in Greece to have put an emphasis on the restoration of the disturbed landscape along a motorway project. The know-how it now possesses on environment restoration issues helps improve the traditional vegetation installation techniques and attempt the application new innovating methods.

Principles in selecting plant species

The selection of plant species is based on three key principles:

1. Selection based on ecological criteria: The species used must constitute part of the existing vegetation of the specific area, i.e. they must be ecologically adjusted to the restoration area (Dafis et al., 1997).
2. Selection based on the intended purpose (functional, protective, aesthetic): The purpose of restoration is to smoothly integrate the project (motorway) in the surrounding landscape. Depending on the specific restoration objective, i.e. prevention of surface erosion, stabilization of slopes, audio-visual screening of either adjacent settlements or the project and aesthetic improvement, the appropriate species are used.
3. Engineering and cost considerations: Taking into consideration the above principles and, mainly, the species ecophysiological characteristics and biology, the species selected and proposed are those that can be produced in large numbers.

EOSA is the first company, at least in Greece, using exclusively native flora species for restoration purposes supporting its policy by an integrated mechanism for environmental issues, such as the preparation of Landscape Guidelines.

EOSA has also conducted a phytosocial mapping of the Egnatia Motorway by creating a map indicating the different phytosocial zones of the main axis sections. Five phytosociological zones were recorded along the Egnatia Motorway axis. A list was issued containing the main wood forest species (trees and shrubs) that grow in those five phytosociological zones. This list constitutes the plant species manual that contractors need to consult when preparing landscape designs and a list containing the plants used in vegetation installation contract sites.

Vegetation installation techniques

EOSA also investigates alternative vegetation installation techniques. In particular, innovative methods for the installation of vegetation are mainly attempted at difficult and rocky slopes, where intervention possibilities are limited.

The outcome of this effort was positive. It is now required to work towards improving the methods tested either as it is or in combination with supplementary plantings, in order to achieve the best possible results. Furthermore, it is imperative to enrich the plant inventory with new plant species, the installation of which is attempted on steep slopes.

In addition, test applications of vegetation installation are performed through direct seeding of plant species on disturbed surfaces. This is achieved with various methods, such as installation of seeds in pots, holes or lines, or even seeding by hand. Each method is selected depending upon the specific plant species, as well as upon the type and geological composition of the slope. Further improvement of these methods is required as far as the quantity of seeds used is concerned, in order to achieve the required number of plants (per 0.1 hectare) and eliminate—to the maximum possible extent—erosion phenomena.

These methods continue to be applied under contracts in progress, while their improvement is pursued by adopting new innovative methods, in order to acquire more specialized know-how and help the Green Unit in guiding designers accordingly.

The positive outcome of these experimental applications is also underlined by the successful installation of vegetation with the use of small plants under adverse conditions, e.g. on the bedrock of flysch in the Adriatic-Ionian zone, in the West Region of the Egnatia Motorway. The planting of species such as *Cistus incanus* Rchb. and *Spartium junceum* L. was very successful. This whole effort is still in a trial period and can be further investigated and improved, both with regard to the use of more species under adverse or similar conditions and as concerns finalizing the required number of plants, in order to eliminate—as much as possible—slope erosion. Another aspect that is being investigated is the possible combination of this method (use of small plants) with hydroseeding.

In most cases, however, the optimum vegetation installation technique seems to be the installation of small plants (plants of small dimensions and young age), which require the opening of small holes and, therefore, entail limited intervention works on the slopes.

The results and first indications of those innovative methods are currently in the process of being assessed. Special focus is placed on the negative aspects of their application, in order to indicate possible mistakes, improve operations and use new technologies. In its effort to modernize restoration methods, EOSA aims at cooperations with research and technological institutes, as well as universities, in order to follow up new developments in the field.

In the previous planting period, plants were produced on an experimental level, in order to be used in contracts for the restoration of Egnatia Motorway slopes, from seeds collected under the company responsibility. Certain problems and details have been identified concerning the proper collection, maintenance and storage of seeds, as well as the production procedure itself. In accordance with the strategy adopted, the company is already in the process of proceeding to the organized production of plants of young age, in order to cover its needs.

Seed collection—plant production

General

The majority of the species proposed in EOSA Landscape Guidelines is not produced in Greek nurseries and, therefore, is not available in the market. Two reasons could account for this: either their production is not profitable or their propagation know-how is still not widely known. As regards the species used by EOSA, it is known that no seeds are available in the market coming from seed production units and, therefore, the only way to obtain them is to collect seeds from plant species of the wider project area.

The company has issued a manual on seed collection. Certain areas-biotopes for seed collection have been identified and recorded. This manual records all phenological patterns of each species with special focus on the maturation and seed collection periods (Smiris, 2002).

Forest species, either forming extensive natural populations or scattered, are adjusted to their growing environment because they have been subjected to the long impact of natural selection. The environment affects the genetic composition of the forest species that are developed in a certain area and these species develop mechanisms for their adaptation in the environmental conditions prevailing in the specific area.

The propagation material is collected from the wider project area. The plants produced are planted near the areas from which the material was collected. The use of material coming from an environment different from the planting area means either partial or total failure of plantings.

Successful seed collection requires detailed planning in advance; this includes locating the areas-biotopes in the wider project area where seeds are to be collected, monitoring flowering and fruiting, procuring the appropriate equipment and, finally, accurately predicting the quantity of the seed crop for each species. In case of lack of fruits in a certain species, a different area-biotope is found with similar environmental conditions to the planting area.

It should be noted here that seeds are collected for massive plant production from species that can be propagated by seed, according to the specifications prepared by EOSA. In addition, the use of additional species for restoration purposes and of alternative mass production methods are being investigated.

Collection

Fruits and seeds are collected during the period extending from their full maturation until dispersion from areas-biotopes where they are found, as mentioned above. The time of collection varies from species to species. Even within a species, variations in the ripening time have been observed depending upon the altitude, the current weather conditions, the geographic latitude, the aspect and the special soil conditions. Excellent results can be achieved when seed collection is performed immediately after maturation. However, there are some species the seeds of which can or must be collected prior to their full maturation.

Seed maturation can be detected on the basis of one of the following indices (Boner et al., 1994; Takos and Vergos, 1999):

- *Visual indices*: The colour of fruits and seeds. These indices are based on the experience of the person responsible for the collection and can be immediately applied.
- *Physical indices*: There are three cases as regards the moisture content in seeds during maturation: (1) in dry orthodox seeds and fruits, the moisture content is slowly reduced as seeds mature, (2) in fleshy orthodox fruits, the moisture content is first reduced and then increases, mainly in the pulp and (3) in recalcitrant seeds, the moisture content increases very early and then is slightly reduced.
- *Biochemical indices*: They concern the changes occurring in the seeds, such as, e.g., the reduction of sugar content. Although these are the most reliable indices, they are difficult to use in the field since they require laboratory equipment.

- *Climatic indices*: They concern changes in temperature, especially in the summer, that can affect the seed maturation rate.

Once seed maturity is ascertained, checks are performed, in order to confirm that the majority of seeds is viable and, finally, estimate the required amount of fruits that need to be collected in order to produce the planned quantity of seeds. This is done by calculating the mature viable seeds per fruit in a representative sample of fruits by means of a special sharp knife. When the percentage of non-viable seeds is higher than 50%, then collection is not performed.

Collection methods vary depending upon the type of the specific wood species and the height of trees. The collection of seeds such as *Colutea arborescens* L., *Spartium junceum* L., *Acer sp*., *Fraxinus ornus* L. etc is performed manually directly from trees or shrubs. Heavy fruits-seeds such as *Pyrus amygdaliformis* Violl., *Quercus sp*. etc are collected from the ground after shaking.

The fruits-seeds collected are placed in sacks and transported quickly to a company laboratory, in order to be cleaned. Each sack is appropriately labeled and the following information is recorded: type of species, date of collection, location, and the person responsible for the collection. Until seeds are extracted, fruits are stored in a cool ventilated area, while checks are constantly performed, in order to prevent the fruits' being overheated.

Extraction– cleaning

After seed collection and transportation to a company laboratory, extraction and cleaning of seeds follows. This process includes separating the seed from the rest of the fruit, foliage, branches, etc. Extraction techniques vary depending upon the type of the specific fruit (Smiris, 2002).

- Seed extraction in legumes (*Colutea arborescens* L., *Antyllis hermanniae* L. etc) and capsules (*Cistus sp*., *Erica sp*.) is performed after natural drying and breaking of fruits. Fruits can be broken by rubbing, crushing, etc.
- Seed extraction from fleshy fruits, such as drupes (*Pyrus amygdaliformis* Vill., *Phillyrea latifolia* L. etc) and berries (*Arbutus unedo* L., *Myrtus communis* L.) is performed through de-pulping, i.e. removing the pulp by using water.
- With regard to species of genus *Quercus*, separating the seeds from cups is done manually, since it is very easy and does not require any special handling.

Once seeds are extracted from the fruit, their cleaning follows. It is a procedure comprising the removal of impurities and empty seeds. Cleaning is carried out by means of sieves of various sizes. The first sieve retains larger impurities and the second one retains seeds and removes finer impurities. Empty seeds are removed with water. Floating seeds are removed, after a small sample of them is cut transversely. Furthermore, in certain species, the removal of impurities and empty seeds can be done through blowing.

Quality control

Quality control of each seed-lot includes determining seed purity, moisture content, germination ability and number of seeds per kilogram. Quality control is performed on a relatively small sample and, therefore, the sampling method used plays a decisive role in obtaining a representative sample. When seeds are stored in glass jars, distinct samples are obtained from the upper, middle and lower part of the jar. These distinct samples are thoroughly mixed, in order to have a homogeneous final sample. When seeds are in heaps or spread, distinct samples are obtained from random spots and depths, and then are thoroughly mixed to become a homogeneous final sample. From this final sample, the quantity required for each species is taken. Great care should be taken to commence the quality control tests immediately after obtaining the final sample. In case of delays, the sample is stored in a refrigerator to maintain the seed quality. To determine the above tests, the International Seed Testing Association (I.S.T.A.) guidelines were consulted (ISTA, 1999).

Purity: It involves defining the percentage in weight of seeds of the species being tested. The sample tested is first accurately weighed and then divided into three distinct fractions:

1. clean seeds of the species being tested
2. seeds belonging to other species
3. aggregates.

After separating them, each one of the above fractions is accurately weighed, in order to calculate its weight percentage.

Moisture content: The moisture content of seeds is the most important factor in maintaining their viability. Insects and fungi are activated at certain moisture content levels. Moisture is measured after the extraction-cleaning of seeds and prior to their storage, in accordance with I.S.T.A. specifications. In order to measure moisture content, seed samples are obtained, free from impurities, and, after being weighed at a precision of one milligram (initial weight), placed in an oven at 103 ± 2°C for 17 ± 1 hour. Then, seeds are placed in the desiccation chamber to cool and, finally, accurately re-weighed (final weight). Moisture content is expressed as a percentage of the weight of the original sample on the basis of the following formula:

$$W = (A - T) \times 100/A$$

where W = moisture content, A = the weight of the original sample and T = the weight of the final sample.

Germination ability: Testing the germination ability of seeds can be done either directly by germinating seeds in growth chambers or indirectly by assessing the viability of the embryo (tetrazolium test). The most reliable method is germinating seeds under a controlled environment. For the majority of the species, fluctuating temperatures +30°C for 8 h in light and +20°C for 16 h in the darkness are recommended for the germination of seeds. The lighting used is artificial with cold-light lamps. For each species, 400 seeds are used in 4 replicates and the complete randomized design is used. Seeds are uniformly placed in 9 cm plastic Petri dishes on sand, in such a way so as not to be in contact and fungi that may activate not to spread. The sand used as a substrate is sterilized at 105°C for 24 h. Seeds are regularly watered with de-ionized water, so that sand is kept moist. Germinated seeds are counted every 7 days and the germination criterion used is a 2 mm emergence of the rootlet, in accordance with the I.S.T.A. regulations. This method also indicates whether seeding can be performed immediately or whether seeds need to be subjected to a certain pre-treatment; in additional, germination rate are also determined.

Apart from the direct germination test, which is time-consuming, especially for species in dormancy, germination ability can be indirectly determined by conducting the tetrazolium test. In this test, the seed-coat of the samples being tested (4 replicates of 100 seeds) is removed by application of various methods and then the embryos are immersed in 2,3,5 triphenyl tetrazolium cloride solution of a concentration of 1%. Embryos are completely covered by the solution and should not be exposed to light when in it. The duration of immersion for various species is specified by I.S.T.A.. The main objective of tetrazolium test is to separate viable from non-viable seeds. The embryos of viable seeds are fully stained in red colour. Tetrazolium testing is fast and easy, but rarely the results obtained are reliable.

Number of seeds per kilogram: This parameter is necessary, in order to define the required quantity of seeds. A representative sample is obtained from each seed-lot, which is freed from impurities. Eight replicates of 100 seeds are obtained randomly from the sample. Each replicate are weighed at a precision of one milligram. The average of the 8 replicates is the weight of 100 seeds, which, in turn, determines the number of seeds per kilogram.

Seed storage

In order to maintain seed viability, the factors affecting growing, such as moisture content, ambient temperature, should be kept at levels that limit the physiological functions of seeds (respiration) to a minimum (Boner et al., 1994; ISTA, 1999). Therefore, by reducing the moisture content of seeds (only in the case of orthodox seeds) and the ambient temperature, the longevity of seeds is extended.

Seeds of some species maintain their germination ability under normal storage conditions for a period of 1 year maximum. These seeds are characterized as short-lived. Certain short-lived seeds are also referred to as intermediate or recalcitrant; their viability can be impaired when their moisture content falls to 25–30%. From the species collected, *Aesculus hippocastanum* L. and the species of genus <u>*Quercus*</u> fall into this category. The seeds of these species are collected, spread in a well-ventilated area and, then, regular measurements are performed to monitor their moisture content. When they reach their ideal moisture content (30–35%), seeds are placed in layers with alternating turf layers in container trays and then stored in the refrigerator (+1°C to +4°C). At regular intervals, seeds are stirred and checked, and, when it is deemed necessary, water is added to maintain moisture content rates.

Another wide seed category includes seeds that are desiccation tolerant; such seeds are characterized as orthodox. The moisture content of such species is reduced to a percentage of 5–10% and then seeds are placed in airtight glass jars and in the refrigerator (+1°C to +4°C). Moisture content reduction is achieved by natural means or with special desiccation devices (ovens).

Pre-treatment

Plant germination is affected both by endogenous factors relating to the seed itself (hard seed-coat, insufficient embryo development, etc) and exogenous factors (moisture, temperature, oxygen, light). In many cases, the germination of seeds is achieved without any pre-treatment, as long as seeds are exposed to a favorable combination of factors. Seeds of other species that mature around the end of summer or beginnings of autumn do not germinate immediately but fall into a state of dormancy. Dormancy is due to the impact of one or more of the endogenous factors mentioned above. Seeds in dormancy need to be pre-treated, so that dormancy is broken and germination is activated.

In most of the cases, dormancy is caused by a hard seed coat; this type of dormancy occurs in several species and mainly in leguminous ones. Suspension of germination due to the hardness of seed coat is the result of the following factors:

- Mechanical resistance. They impair the embryo development and rootlet emergence.
- Physical barrier to moisture absorption.
- Physical barrier to gaseous exchange.
- Presence of inhibitors located on the seed-coat.

In nature, such seeds become permeable through mechanical abrasion by the soil, the effect of microorganisms or ingestion by animals. Therefore, a pre-treatment of seeds is required prior to sowing, in order to achieve fast, uniform and high germability. Artificially this can be achieved either mechanically (breaking or scarifying the seed coat) or with immersion of seeds in concentrated sulfuric acid or warm water, resulting in the softening of the hard seed coat.

While germination suspended due to a hard seed coat can be easily restored, germination suspended due to dormancy caused by immature embryos can be restored after a more or less long stratification of seeds. The causes of this type of dormancy are mainly hormonal and is due to a disturbance of the balance between inhibitors and growth regulators. In order to reduce stratification time, seeds are treated with chemical substances. To date, the chemical substances used with successful outcome are KNO_3, ethephon, as well as the growth substances gibberellins and cytokinins.

The pre-treatment of seeds performed for each species prior to sowing was based on bibliographical references. As for the species that are not included in the relevant literature, we pursued and continue to pursue the acquisition of the necessary knowledge as far as the breaking of dormancy is concerned.

Production of plants

Prior to being sown in the nursery, seeds, depending on the species, are subjected to pre-treatment so that dormancy is broken and seeds are rendered suitable for germination. Then, when the time period is suitable, seeding in nurseries takes place. Plants are produced in accordance with the specifications set out by EOSA. The seeding takes place in QuickPot propagation trays and the filling materials used are: turf enriched with almost neutral pH, perlite and soil.

Table 1. Number of seeds per cell of the QuickPot tray according to germination percentage

Germination percentage (%)	Number of seeds per cell
85–100	1
75–85	2
60–75	3

Knowing the number of seeds per kilogram and the percentage of germination of the seeds, we can estimate the approximate quantity of seed to be used for producing a specific number of seedlings. The number of seeds for each cell of the QuickPot tray is decided on the basis of the germination percentage (Table 1).

The factors strongly affecting the quality of seedlings and relating to the production methods are: the size of the QuickPot tray, the mixture used for filling containers, the fertilization method, the irrigation programme as well as the quality of the water used (Alifragis et al., 1997).

The condition of the root system of the seedlings before planting seriously affects their survival and development during the first years following planting. Spiral or deformed roots contribute to the increase of the mortality percentage and the post-planting stress. The volume and, mainly, the height, are some of the cell key-properties affecting the formation of the root system.

The use of QuickPot tray of suitable size, depending on the forest species, with a proper opening at their base to ensure air-root pruning and with impermeable walls, reinforced with vertical, internal grooves to prevent spiral development of the root, contribute to the production of excellent planting material. Continuous research is carried out in order to find out the QuickPot tray type suitable for each species produced.

The filling mixture is where the seedling roots are activated by taking in water and nutrients. Thus, it should have all the physical, chemical and biological properties that contribute to the proper development of plants. The most widely used ingredient of filling mixtures is turf. Apart from turf, QuickPot trays are also filled with mineral soil and perlite. The proportion of the mixture turf-perlite-soil is 30:20:50. The soil used to produce the mixture should be of medium texture, with slightly acid to neutral reaction and with low salt concentrations. First, soil samples are taken (in depths of 0–20 cm and 20–40 cm) from the areas were soil is going to be conducted and the necessary analyses are carried out. The results of the analyses indicate the soil suitability. Prior to being used, the soil is sieved so that skeletal material is removed.

During the mixture production, slow release fertilizer of 8 months, 15-7-15, plus Mg, plus trace elements – 2 kg per cubic meter of turf is used.

Seeds are covered with sand and the QuickPot trays are transferred to parterres under shade shelters. The base of the QuickPot trays stands higher than the ground surface to ensure good aeration and prevent roots from entering the ground.

During germination, the QuickPot trays should be watered frequently, but with small quantities of water, so that the material surrounding the seeds is kept moist. After seed germination, watering is done on a daily basis, also depending on weather conditions and only in the morning. Special attention is given to irrigation, in order to ensure uniform watering of QuickPot trays and watering is carried out up to the point that water starts to flow from their base hole. Irrigation needs to be carefully checked to prevent conditions favoring the development of plant diseases. One month after germination and having overcome the hazard of seedling-rot, we leave one seedling in each container and transplant the others to empty cells. Meanwhile, weeds are removed.

At the beginning of September, seedlings are subjected to hardening, a procedure that involves good watering to wash off as much nitrogen as possible from the substratum and then a gradual decrease of irrigation. Water reduction results in seedling stress, therefore, seedlings cease to grow in height and harden thus adapting to harsher environmental conditions. After the hardening procedure, seedlings are fertilized with a fertilizing solution containing: N (40 mg/l), P (100 mg/l) and K (300 mg/l).

Discussion

Installation of plants

According to the specifications of the Ministry of Public Works, the current planting practice adopted at road works is the installation of two-year plants. Initially, the company adopted these specifications but had the limitation of using only species from the local native flora, produced from external nurseries under supervision of the company. The specifications for the installation of these plants are included in the Technical Specifications of each Contract that involves green works.

The final expenditure of the plant production for the company is lower than the one that would be required for purchasing specific plant species. This enables the company to use the cost benefit for larger-scale restoration works.

Cost-saving is achieved both through the method used to produce plants and during the stage of plant installation and maintenance. The plants installed are not two-year plants but young plants aged up to one year. The method of planting young plants has returned very satisfactory results under difficult installation conditions during previous pilot efforts of the Green Unit in cooperation with a group of experts from the Aristotle University of Thessaloniki.

To reduce the maintenance costs, the use of water-retentive materials placed in the planting pit during plant installation is considered to limit the irrigation needs. Also, a minimum of three different maintenance degrees are considered in order to achieve the optimum result with the lowest cost possible. It is hoped that the maintenance programme to be applied in all future green works of Egnatia motorway will have the best cost-benefit ratio.

Installation of vegetation on gunite

The wider context of activities of the Green Unit of EOSA includes some activities that relate to unique cases but are particularly interesting since they point out the importance of green works in addressing instability, slope erosion and construction problems in highway projects. Such a case is described below:

A gunite structure in the area of Pindos, where the geological materials are extremely unstable and erodable exhibited failure phenomena owing to the accumulation of large quantities of water at the ground level under the structure. There were no water discharge points except for drainage pipes. Thus, the concrete surface was cracked. The problem was quite serious and the company had to select the best among the relevant designs/proposals drawn up either by the contractor or by the company itself.

The Green Unit undertook the initiative of addressing the problem by planting cuttings, native plant species with increased needs for moisture. This method presented three key-advantages compared to other solutions proposed to the company; minimum cost, minimum environmental disturbance and aesthetic improvement of the slope in question. The whole work was carried out by the Green Unit as an experiment on a section of the surface. The experiment was without all the necessary technical means and the conditions ensuring its success, in accordance with the specifications of the Company for the procurement and planting of cuttings.

Hard woodcuttings of the species *Salix purpurea* L., *Salix eleagnus* Scop. and *Platanus orientalis* L. were used. The propagation material was taken from parental plants of an adjacent, riverside area in December. The diameter of the cuttings was 5 to 15 mm and their length 150 to 200 mm. Due to unfavourable weather conditions and other problems, planting of the cuttings was not immediately possible. They were stored in a shady and cool place and were planted in early March, after the opening of 50 to 190 mm diameter holes. Two or three cuttings were placed in each hole. Despite the fact that, due to the long storage, the cuttings were not in the best condition during their planting, the percentage of success was very satisfactory (over 80% of the total number of holes) and so was the plant growth. The average height of the plants at the end of the summer was 0.3–0.4 m and there were no losses during the summer that was expected due to the high moisture levels.

References

Alifragis D, Hadzistathis A and Kekis G 1997 Guidelines of plants' production for the restoration of slopes in Egnatia Odos, E.O.S.A., Thessaloniki.

Boner F T, Vozzo J A, Elam W W and Land S B Jr. 1994 Tree Seed Technology Training Course. U.S. Department of Agriculture, Forest Service, southern Forest Experiment Station, New Orleans, Louisiana, General Technical Report SO-106.

Dafis S, Ganatsas P, Goudelis G, Theodoropoulos K and Georgi N 1997 Selection of appropriate native species for the restoration of landscape disturbances. E.O.S.A., Thessaloniki 1997.

I.S.T.A. (International Seed Testing Association) 1999 International Rules for Seed Testing Rules 1999. Seed Sci. Technol. 27(Supplement), p. 333.

Landscape Guidelines (OSAT). E.O.S.A., Thessaloniki, Greece.

Smiris P 2002 Guidelines for seed collection. E.O.S.A., Thessaloniki, Greece.

Takos I and Vergos S 1999 Guidelines for seed and cutting collection from species of native flora. E.O.S.A., Thessaloniki.

Instability effects of the landforms, in artificially modulated banks along the road Thermopylae–Nafpactos (Ftiotida, Greece)

A. Mertzanis[1,2], A. Papadopoulos[1] & A. Pantera[1]

[1] *Technological Educational Institute of Lamia, Annex of Karpenisi, Department of Forestry, 36100, Karpenisi, Greece.* [2] *Corresponding author**

Key words: environment, environmental impact assessment studies, erosion, geology, geomorphology

Abstract

In this study, we examined the current environmental situation and geological–hydrogeomorphological characteristics, as well as the phenomenon of mass land movement in the area crossed by the National Road Thermopylae–Nafpactos (with an emphasis on the road running between Damasta and Skamnos, Ftiotida) Greece. We also depicted the changes and distortions in the condition of the geological–hydrogeomorphological characteristics due to both roadwork and the operation of the road.

Introduction

The construction and operation of roadworks have a huge impact on the environment and more specifically, on the geomorphological structure, water resources and vegetation. These factors are found to be in direct relation with geological, hydrogeomorphological, soil mechanical and climatic conditions, as well as vegetation prevailing at a site. The type of roadworks, i.e. the size and the configuration of the excavations, cutslopes and bankfills, also have a large effect on the environment (Kallergis, 1999).

During road construction, the changes (impact) to the natural environment are caused mainly by the creation of large pits i.e. excavations and embankments, which in most cases are not replanted subsequently. Special protective construction work and infrastructure e.g. drainpipes and retaining walls also have a large impact on the environment. Therefore, the natural conditions of rainwater drainage are disturbed, which in turn, in combination with the reduction of density and the changes in vegetation, amplify the effects of erosion and soil instability. This leads to the appearance of erosion and landslides on artificial slopes and can even cause the destruction of embankments, which results in economic losses and a decrease in road construction efficiency. The natural environment is also adversely affected (Koukis 1994).

The object of this study was the systematic recording and acknowledgement of the problems relative to the changes and restructuring of the geomorphological structure of the National Road axis Thermopylae–Nafpactos, in the region between Damasta and Skamnos (Fthiotis), Greece. The conclusions with regard to the 'size' and 'reversibility' of the environmental impact caused by roadwork, in connection with decisions taken during the design, planning and implementation of the works are also discussed.

Materials and methods

Research site location and characteristics of roadworks

This research was carried out in two areas located on north-westerly, highly morphological slopes along confluent branches (unnamed stream and Votania stream) of the basin of the river Asopos, Lamia (Figure 1). One site was located at Halvantzeika and the second at the Source of Damasta. These semi-mountainous/mountainous areas were located in a hypsometric area between 0.1 and 0.6 km, through

* E-mail: mertzanis@teilam.gr

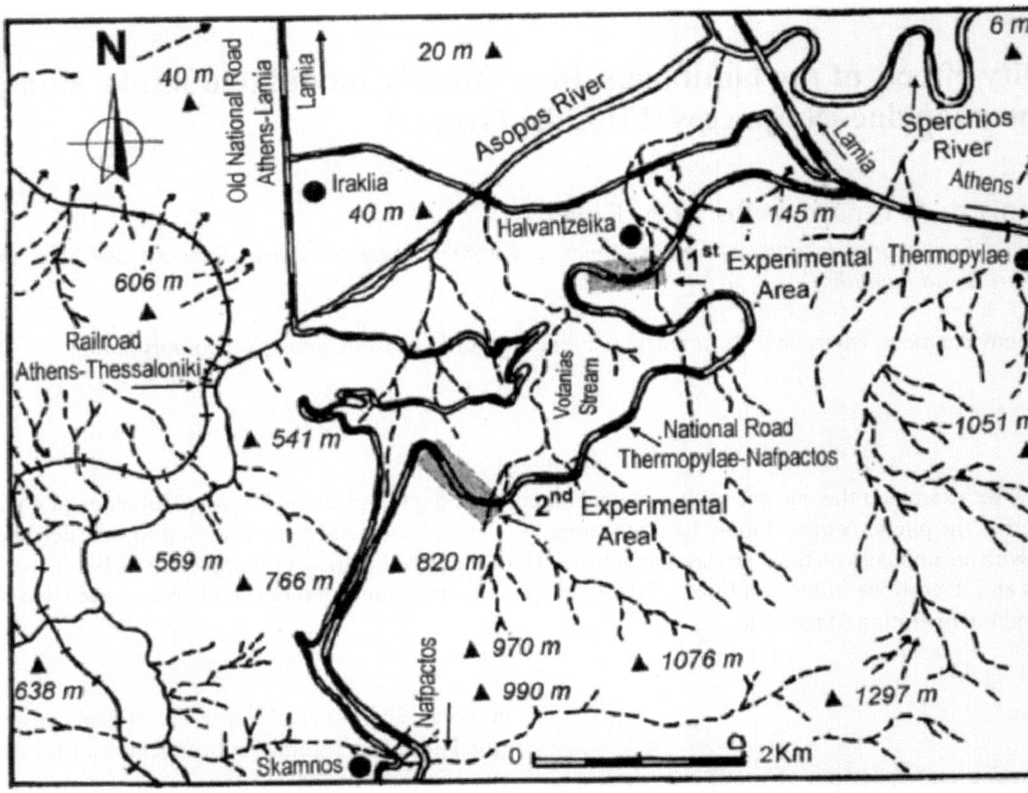

Figure 1. Map of the study area.

which passes the National Road axis, Thermopylae–Nafpactos. The length of the first experimental area (1st E.A.) was 0.5 km, and the second (2nd E.A.) was 0.8 km, with its central longitudinal axis lying alongside that of the existing road. The area of each site was approximately 0.6 km^2 and lay south-west of the crossing with the New National Road Athens, Lamia. The distance of the 1st E.A. from this crossing was 3.2 km and the 2nd E.A. was 10 km. The construction of the national road axis was carried out with many cutslopes and bankfills through the experimental areas, which are covered mostly by pastoral forest areas of maquis vegetation. The cross-section of the road axis, Thermopylae–Nafpactos, in this area has three lanes; one for upward slow moving vehicles and two guidance lanes 1.30 m wide. Total road width was 14 m (Figure 1).

Climatic conditions

The climate of the region was estimated from data gathered at the meteorological stations, Lamia and Lidoriki, available from the Hellenic National Meteorological Service. The mean annual precipitation was between 573 mm (Meteorological Station of Lamia) and 957 mm (Meteorological Station of Lidoriki), with an annual average of 100 days rain and a maximum of 123–154 mm. The mean number of days of snowfall varies from 4.7 to 6.3 days per year.

Vegetation

The vegetation in the area around the experimental sites belongs mainly to the meso-Mediterranean area of vegetation (*Quercion ilicis*) and was characterized by the presence of bushes such as *Quercus coccifera* L. and *Quercus ilex* L. In some alluvial soils, deciduous oak and other broadleaf species existed. In the experimental areas, the vegetation on the lower slopes did not exceed a ground cover of 10% and was composed largely of pioneer species e.g. *Spartium junceum* L., *Rosa sp*, *Rubus sp*, *Cistus sp*. Some species were also present which benefited from the presence of wastewater on the asphalted road surface, e.g. *Fraxinus ornus* L., *Cotinus coggygria* Scop. and *Rhus coriaria* L.

Geomorphology–geology and soil mechanical properties

The morphology of the area surrounding the field sites was mountainous with mild to steep slopes and strong contrasts e.g. gorges, mild valley shapes and alluvial plains, which differentiated according to the nature and site of the geological structures. Geotectonically, the region is considered to belong to the Sub-Pelagonic series and is composed of base sedimentary limestone, from the lower-middle Jurassic, partly covered by continuous Flysch, Pleistocenic and Pliocenic sediments and Quaternary deposits (IGSR 1960, 1967). More specifically, following were the geological composition and soil mechanical properties of the experimental areas: (a) Lower-middle Jurassic limestones, which also constituted the background of the 2nd E.A.. These limestones were compact, dark grey, bituminous, thin layered to medium layered, and at places oolithic. In certain areas they were broken up by crossing rupture surfaces, but satisfactorily maintained their mechanical characteristics. (b) Campanian–Maestrichtian limestone was also present and constituted the geological background of the 1st E.A. This limestone was light coloured and compact and possessed inferior mechanical characteristics as it was highly broken up. (c) The undivided Flysch, which was located in the eastern part of the region, was composed mainly of large particle sandstone, alternating with argillaceous schist and marly sandstone, while at places there were limestone and conglomerate layers. The Flysch had inferior mechanical characteristics and was sensitive to weathering and erosion. This erosion also constitutes the most important parameter in the continuous redevelopment of the relief, with the creation of mass movement phenomena, which was particularly intense after repeated cycles of wetting and drying.

Surface and underground waters

The region is divided by anonymous dendritic sectors of the hydrographic network which drain, via the river Asopos, into the river Sperchios. According to the classification of the hydrographic networks, at a map scale of 1:50000 (Strahler 1957), both 1st and 2nd E.A.s, are run through by 1st class sectors, which present an outflow of waters, after extended rainfalls and snowfalls.

Underground waters in the region are seasonal and limited, with some small local spouts, mainly during the winter period. These spouts are usually located on the axis of flow of the streams Mega and Votanias, which constitute the major local outlets in the region between the 1st and the 2nd experimental areas.

Research method

For the recording of the existing environmental situation, both within and around the experimental sites, we used topographic, geological and vegetation maps. For the determination of changes in the natural environment and particularly the hydrogeomorphological conditions of the region, as well as the trend over time, we used aerial photographs from the Hellenic Military Geographical Service (HMGS), taken in the years 1945, 1960, 1970 and 1986, with scales of 1:42000, 1:30000, 1:15000 and 1:30000, respectively. These data were combined with results from local observations of erosion phenomena and landslides at the artificial cuttings, excavations or embankments. These comparative observations were held diachronically in the month of July, in 1994 and 2004, via six constant points or 'poles' placed in selected places.

For the detailed recognition of individual characteristics of the natural environment (particularly the geomorphological characteristics) and for the evaluation of the size, duration and the possibility of recovery from the impact due to the construction and operation of the road axis Thermopylae–Nafpactos, we recorded all the essential data e.g. geomorphological characteristics of the road axis, vegetation, erosion and landslides, on both sides of the road axis at 0.1-km intervals (Mertzanis et al., 2002). For the evaluation of the extent of the changes in the geomorphological characteristics, as well as the duration and their possibility of recovery, we used approved criteria and indicative magnitudes (Vavizos and Mertzanis 2003). A multi-criteria analysis of the data was then carried out (Figure 2).

Results

Impact on the environment

Following are the main impacts (reversible or not), usually caused to the natural environment, from the

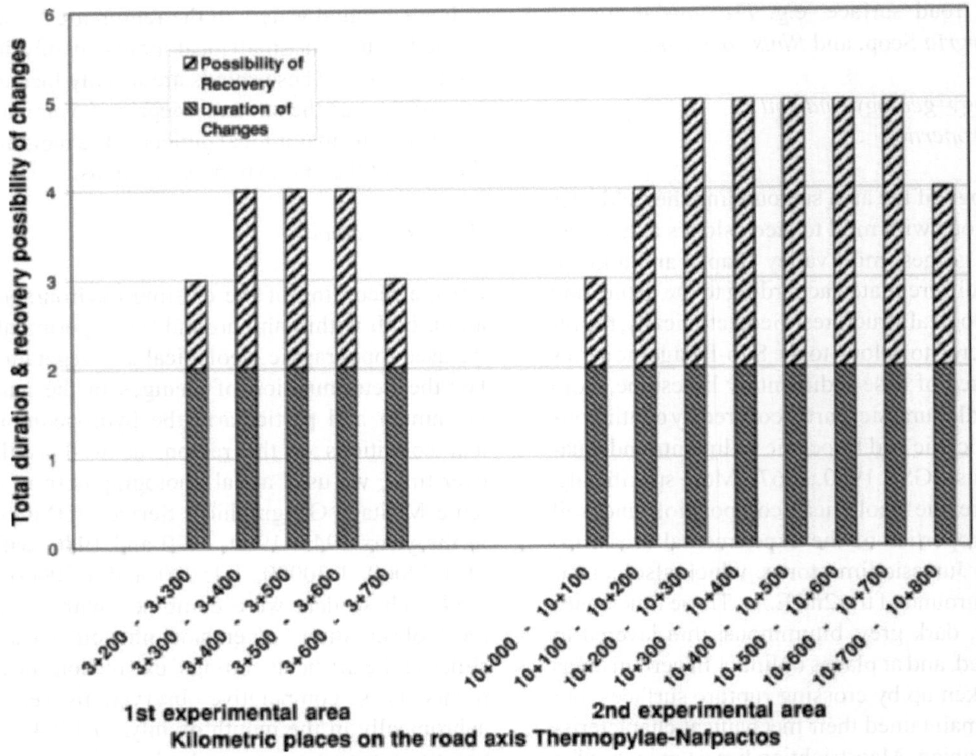

Figure 2. Cumulative assessment of impact on the geological–geomorphological structure of two experimental areas on the road axis, Thermopylae–Nafpactos, with regard to the 'duration' and the 'possibility of recovery'. Duration of changes: 1 stands for short-term changes and 2 stands for long-term changes. Possibility of recovery: 1 stands for reversible changes, 2 for partially reversible changes and 3 for irreversible changes.

construction and operation of roadwork (Tsochos, 1997): (a) Changes in the surface and underground waters, showing as differentiation of their course or direction, their quality and their quantity as well as changes in the rate of absorption of surface and draining waters, or the rate and quantity of eluviation. (b) Changes in the geological–geomorphological structure of the region, showing up mainly as a disturbance of morphology of the relief, changes of the status of erosion–deposition of soil, the creation of unstable situations or changes in the geological layout of the rocks, as well as the splits, shifts, compaction or overlaps of the surface ground layer. (c) Changes in the flora, fauna and the natural ecosystems, the most important being the shrinkage of forests, forestal area and pasture land, because of the creation of a longitudinal area, from which the vegetation is removed and replaced by the road blacktop and raw ground in the cutslopes and embankments slopes. (d) Changes in the landscape, because of differentiation of the appearance of the relief, accompanied by the destruction of natural elements (vegetation, soil, rocks) and by the disappearance of the viewing characteristics of the natural landscape (lines, structure, colour) and their replacement by new man-made viewing characteristics, with intense colours, geometrical lines and figures, differentiated structure and sizes dominating in the landscape, following the removal of large volumes from the trenches-cuttings and the embankments of materials of excavation, for the creation bankfills (Brofas, 1987). (e) Changes in the quality of air and soil, due to the emissions of exhaust gases from the vehicles.

The cumulative evaluation of the extent of changes, in both the short and long term, and the possibility of recovery (reversible, partially reversible, irreversible), at microscale intervals of 0.1 km, in the 1st and 2nd E.A.s, with regard to the above parameters (geological–geomorphological structure) of the multi-variable environmental system, is shown in Figure 2.

Changes to the geological–geomorphological structure and the hydrogeomorphological processes

The changes to the geological–geomorphological and the hydrogeomorphological processes in the area are mainly attributed to the large size bankfills (high embankments) and the high cutslopes, which have been formed in several places and which have caused a longitudinal perturbation to the relief, parallel to and on both sides of the road axis. These changes are described hereafter, and their duration and the possibility of their recovery are shown in Figure 2.

(a) The 'non-reversible' changes (impacts) to the landforms, which are located at the 2nd E.A. (Figure 2), were mainly caused by the construction of
   - the high cutslopes (excavation-cuttings) which were denuded of vegetation and had a height of 0.03–0.06 km. The slope was artificially formed with a gradient of 90–98%, above a limestone bank, southwest of the road axis Thermopylae–Nafpactos, over a length of 0.5 km, which formed at a place called 'Damasta Source (Pigi Damastas)' between the kilometric positions 10 + 200 and 10 + 700. The large size cuts, the artificial formation of vertical slopes and the deforestation, in combination with the construction of the road surface with new gradients in comparison to the pre-existing relief, caused changes to the movement direction and the natural flow conditions of the surface water. These waters flow according to the process of water layer flow (diffuse water flow), in the upstream side of the basin, and by consequence also to the hydrogeomorphological processes related to them. The compaction and water-proofing of the road surface with asphalt, forces the surface draining water and especially the diffuse water flow, to abandon their natural course and follow the direction of the rainwater drain pipe, or to overflow from the road banks. These processes result in intensification of the deep gullying and retrogressing erosion in sections of the hydrological basin, downstream of the road axis and the soil depositions.
   - the high bankfill (height >0.1 km), with an artificial morphological slope of gradient 80–95%, northeast of the road axis, over a length of 0.5 km, which was formed between the kilometric positions 10 + 200 and 10 + 700. More specifically, at the position where a large earth embankment was developed on limestone banking, the pre-existing vegetation has been covered and despite the fact that 30 years have already passed since the road construction, the vegetation has only been replaced by the occurrence of some isolated bush species, not exceeding in coverage 3% of the surface.

(b) The 'reversible' to 'partly reversible' changes, found in both experimental areas (Figure 2), represent changes to the geomorphological processes and the landforms, as in the previous category. However, these were of limited size and were mainly caused by the construction of:
   - cutslopes (excavations) with a height of 0.025–0.04 km, and a morphological slope of gradient 85–95%, and a bankfill height of 0.03–0.06 km, with a morphological slope of 70–85%, formed, respectively, on the limestone bank, south and north of the road axis over a length of 0.5 km, at a place called 'Halvatzeika', between kilometric positions 3 + 200 and 3 + 700. In the same category also belong the sections from kilometric positions 10 + 000 to 10 + 200, and 10 + 700 to 10 + 800, occurring in the 2nd E.A., at a place called Damasta Source (Pigi Damastas). This area mainly represented the transit zone from the high cuttings and embankments, to the natural 'undisturbed' relief.

Discussion

Instability phenomena of landforms on artificial banks, embankment slopes and cuttings

From the monitoring of the evolution over time of instability phenomena in sections of the embankments of the road axis Thermopylae–Nafpactos, we can conclude the following:

(a) The limestone substrate, into which the high and steep roadcuts were constructed, in both experimental areas has contributed to the formation of stable conditions for the cuttings, which showed negligible geomorphologic differentiation from 1994 to 2004.

(b) The broad and high bankfills, constructed at both experimental areas, with steep slopes and low soil mechanical properties, in combination with the lack of vegetation and special protective works, result in instability phenomena e.g. erosion and

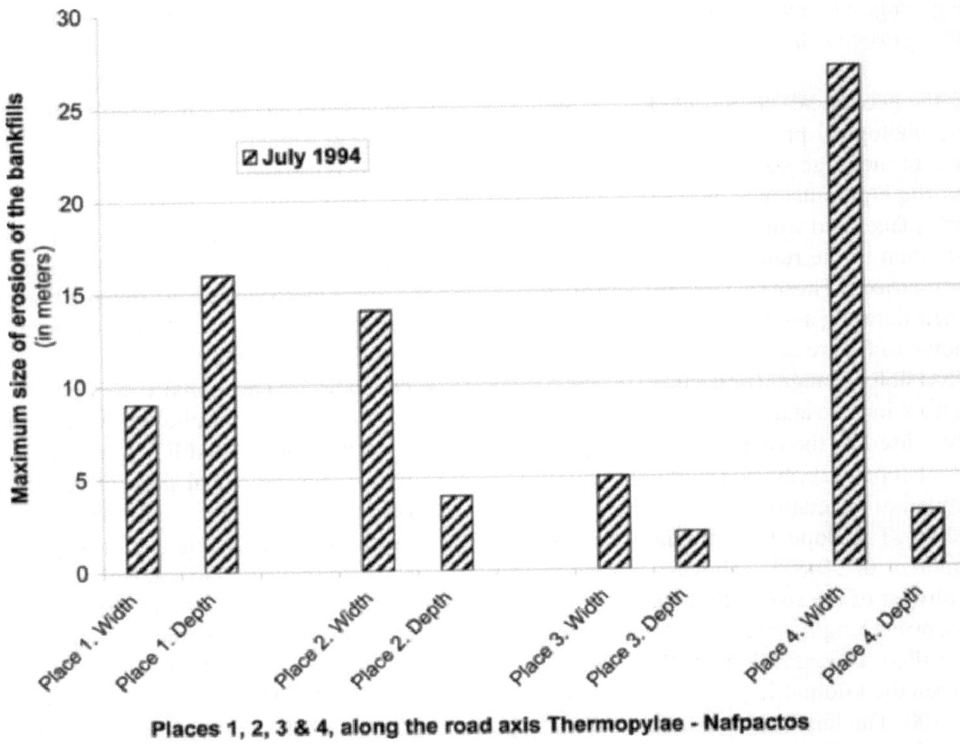

Figure 3. Maximum width and maximum depth of erosion of the bankfills, at positions 1, 2, 3 and 4, in 1994.

landslides, which are clearly visible along several places of the embankments.

In all of the investigated experimental areas, four main places (coded 1, 2, 3 and 4) in the bankfills were recorded, where severe erosion-landslide problems have occurred (before 1994). From the comparison of dimensions relating to the maximum width and the depth of erosion between 1994 and 2004, we can conclude that the erosion fronts have become enlarged at different rates. More specifically, the maximum width of erosion at positions 1, 2, 3 and 4, has increased by 6, 6, 4 and 12 m respectively, and their maximum depth has increased by 4, 2, 3 and 2 m (Figures 3 and 4).

The complete image which is being developed at the place under investigation is attributed to a non-continuous state of erosion- landslide procedure, which shows extreme erosion values, located at specific positions and are connected mainly to the extreme values of long lasting and intense rainfall which damaged the Eastern part of Sterea Hellas (Central Greece) and the area under investigation, between 11 and 13 January 1997. During this storm, the situation was aggravated from the lack of special protective anti-erosion works and vegetation on the artificial bankfills and cutslopes, resulting in an augmentation of the surface flow and the speed of the flowing water running over the road surface.

It is important to note that from studying the aerial photographs taken in previous years (1945, 1960, 1970 and 1986), in both experimental areas, the conditions were mainly stable and there were no visible signs of soil-mass movements due to the presence of limestone rock which is resistant to erosion and was covered with vegetation. Therefore, we can conclude that when we out carry excavation works and bankfills with steep slopes and low soil mechanical properties, without prior suitable measures taken, then we experience the creation of instability effects as well as changes in soil engineering characteristics.

Conclusion

The construction of high-speed roads is imperative for development, but is also a permanent

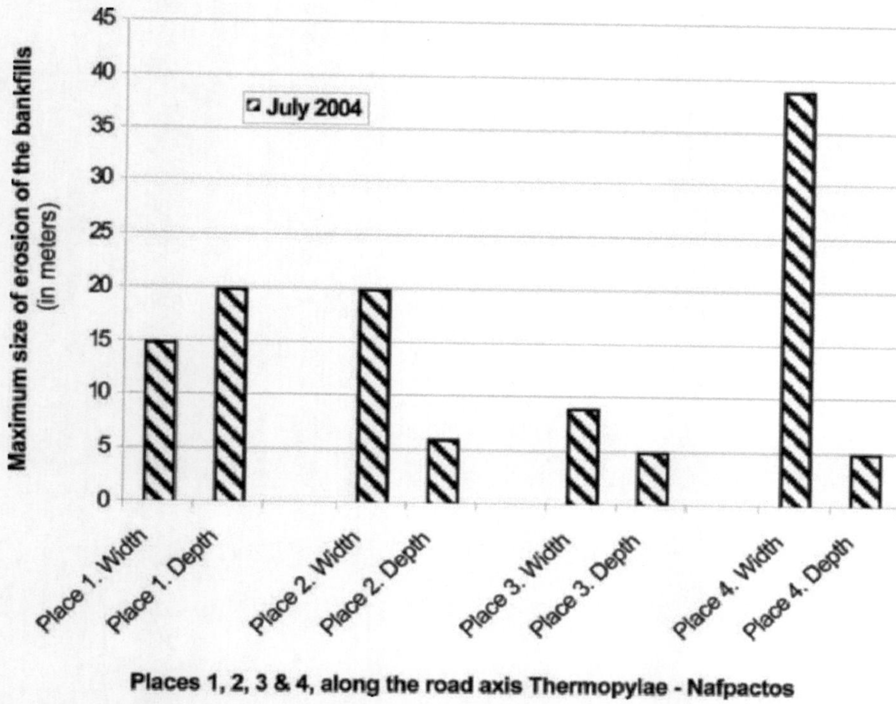

Figure 4. Maximum width and depth of erosion of the bankfills, at positions 1, 2, 3 and 4, in 2004.

intervention to the natural environment of a region. The changes and deformations from roadworks caused to the environment and especially to the geological–hydrogeomorphological conditions, are in direct connection with the laying of the road axis and the size of cutslopes and bankfills, as well as the environmental, hydrogeomorphological and soil mechanical conditions prevailing in the region. The impact to the natural environment is especially important in cases where there are large cuts and bankfills which are not supported by vegetation and/or special protective and infrastructure works, resulting in the occurrence of erosion and instability of the soil mass.

The prompt estimation of the impact on the environment that are expected from the construction of a road axis is, together with the implementation of suitable measures, a basic prerequisite for the economics of the job i.e. total construction, operation and maintenance cost, and also for the possibility of having conditions of 'high acceptance' and 'faster incorporation' into the existing environmental conditions.

References

Brofas G 1987 Research for the recovery of landscape in the quarry of Penteli. Forestal Res. 149–186.

Institute of Geology & Subsurface Research (IGSR) 1967 Geological map of Greece. Lamia sheet.

Institute of Geology & Subsurface Research (IGSR) 1960 Geological map of Greece. Amfissa sheet.

Kallergis G 1999 Applied—Environmental Hydrogeology, pp. 1107. T.E.E. Publications, Athens.

Koukis G 1994 Tank of Dam of Assomati Veroia. Phenomena of Instability in Bankings—Technico—Geological View. Meeting Geology and Dams: Experiences from the Greek space, pp. 79–92. Athens.

Mertzanis A, Marabini F and Papadopoulos A 2002 Changes in the aquatic resources and in the geological-geomorphologic characteristics of mountainous regions from the construction and operation of road work. The case of the road axis Lamia-Karpenisi. 3rd Pan-Hellenic Pasturing Congress Pasturing and Developement of mountainous regions. Hellenic Pasture Range Soc. Proc. 10, 109–116.

Strahler A N 1957 Quantitative analysis of watershed geomorphology. Am. Geophys. Union Trans. 38, 913–920.

Tsochos G 1997 Environmental Road Construction, pp.270. University Studio Press, Thessaloniki.

Vavizos G and Mertzanis A 2003 Environment—Studies of Environmental Impact, pp. 342. Papasotiriou, Athens.

Evaluation of revegetation techniques on mining spoil slopes

G. Brofas[1,3], G. Mantakas[1], K. Tsagari[1] & Ch. Mermiris[2]
[1]*Institute of Mediterranean Forest Ecosystems and Forest Products Technology Terma Alkmanos, 11528 Ilissia, Athens, Greece.* [2]*S & B, Industrial Minerals, S.A. Greece.* [3]*Corresponding author**

Key words: fertilisation, herbaceous species, hydro-seeding, mulching, reclamation, revegetation

Abstract

The effect of hydro-seeding, with different combinations of mulching and fertilisation on the establishment of herbaceous vegetation on a slope of mining spoils, were examined during this work. The field trial consisted of four treatments: (A = hydro-seeding, B = hydro-seeding with cellulose plus the EI-1000 organic glue, C = hydro-seeding with cellulose plus the synthetic glue CURASOL, D = hydro-seeding plus straw mulching plus asphalt fixing) with three replications. Hydro-seeding only, gave initially poor results, but in the following years these results were similar to those of the other treatments. The combined treatments gave better results initially but statistical differences between them were not identified. The vegetation in the third year was dramatically reduced, at approximately a third of the first year. As result of the fertilisation applied during the fourth year, the vegetation recovered, but to a much lower level than in the first and second year. The vegetation composition was dominated by the sown species in the first year but gradually the indigenous species increased up to 40–50% of the composition in the fourth year.

Introduction

The vegetative restoration of disturbed areas—on spoils' slopes—implements functions such as screening, stabilising slopes and erosion control, improving soil conditions and structure, and the visual improvement of landscape characteristics (Coppin and Bradshaw, 1982; FAO, 1985). It usually constitutes first priority activity, on inclined slopes, for their protection from soil erosion (Albaladejo Montoro et al., 2000; Ben-Hur et al., 1989; Jusaitis and Pillman, 1997), though according to Bramble (1952) in spoils with percentage of thin material (<2 mm) smaller than 30%, soil erosion was not observed.

However the establishment of vegetation on spoils slopes, consisting of barren materials from mining exploitations, is usually difficult and this is connected with the unfavourable ecological conditions that prevail in these areas (Brofas, 1979, 1987; Whisenant et al., 1995). Usually, provided that there are no interventions for the improvement of conditions, these areas remain bare for a number of years creating problems of varying severity.

For the improvement of these conditions various materials have been used, while for the establishment of vegetation a number of techniques were employed. Hydro-seeding, mulching and fertilisation have been used extensively to facilitate vegetation establishment on such areas with various results, depending on the conditions and means used. Hydro-seeding with straw mulching and binding with asphalt emulsion, has given very good results (Brofas and Varelides, 2000; Dudeck et al., 1970; Muzzi et al., 1997; Wolf et al., 1984). However this is a laborious three-step process, with problems with the straw mulching, and it is very expensive compared to hydro-seeding plus cellulose (C.E.T.E., 1980; Mermiris and Stefanakis, 2004). In contrast, hydro-seeding with cellulose is simpler to apply, but according to Mermiris and Stefanakis (2004), it is relatively vulnerable to heavy rain.

* E-mail: brofas@fria.gr

Stokes et al. (eds), Eco- and Ground Bio-Engineering: The Use of Vegetation to Improve Slope Stability, 419–425.

Table 1. Extrapolated mean monthly precipitation (millimetre) 1960–2003, of the experimental area[1]

Year	J	F	M	A	M	J	J	A	S	O	N	D	Total	A + M + J
Average 1960–2003	158.1	136.7	126.6	73.3	52.9	29.2	25.9	28.2	42.7	127.0	167.0	208.8	1176.3	155.4

[1] Precipitation figures acquired by extrapolation of Gravia station data, taking altitude difference and precipitation lapse rate into account.

At the same time, however, after successful vegetation establishment and regardless of the means of mulching, it was found that the vegetation on the slopes declines gradually during the years following the first establishment, although temporary recoveries may be observed (Brofas and Varelides, 2000). The result in many cases is a poor vegetation cover and an insufficient protection of soil.

This work aims to compare the hydro-seeding plus straw mulching plus asphalt fixing, to the hydro-seeding plus cellulose; the effectiveness of various means of binding the cellulose, and the effect of fertilisation on the maintenance of vegetation.

Materials and methods

The trial was established at Sideritis location of Giona mountain, near the village of Gravia, in central Greece, at an altitude of 900 m and in an area covered with piles of calcareous spoils derived from bauxite mining. The experimental area lies in the *Abies cephalonica* zone. According to the bio-climatic map of Greece (Mavrommatis, 1980) the area belongs to the sub-humid zone with cold winters and the character of the bioclimate is meso—mediterranean. The annual rainfall exceeds 1000 mm (Table 1) taking into account the data from the nearest meteorological station (Gravia 430 m) and the precipitation lapse rate of the area (Karetsos, 2002; Mariolopoulos and Karapiperis, 1955).

The trial was established on a slope of the spoils heap, which had a steep inclination (78%). The spoils derived from fragmentation of limestone, which lays on top of the bauxite ore in the natural geological strata of the area, and contain coarse fragments of varying size from boulders to thin clay powders. The fragments of less than 2 mm size, command about 20% of the spoil volume. As the fine material (<2 mm) of the spoils is limited, the surface of the experiment was covered with a layer of natural soil (fine earth) about 0.3 m deep. Some of the chemical and mechanical properties of the spoils and the used fine earth, given in Table 2, show that these materials are poor in organic matter and nutrient elements.

The layout of the trial was full-randomised blocks with 4 treatments and 3 replications (blocks); the plot size was 5 m × 18 m. The four treatments included: A = hydro-seeding, B = hydro-seeding with cellulose plus the EI-1000 organic glue, C = hydro-seeding with cellulose plus the synthetic glue CURASOL, D = hydro-seeding plus straw mulching plus asphalt fixing.

The trial was established in the autumn (27 October) 1999 and a seed mixture (the one regularly used by the mining company) was sown at 212 kg/ha (Table 3). The sowing was done with a hydro-seeding machine and the following materials were used depending on the treatment: Cellulose 800 kg/ha, Glue EI-1000 30 kg/ha, Glue Curasol 170 kg/ha, while in all the treatments organic fertiliser 200 kg/ha and fertiliser type 11.15.15, 200 kg/ha was used. For the straw mulching 4000 kg/ha straw were used, and as a binder asphalt solution (9 parts of water to 1 part of asphalt emulsion) 0.5 cubic metres per 0.05 ha. Apart from the fertilisation during hydro-seeding, fertiliser (200 kg/ha) was applied in the spring of 2001, and was repeated in the spring of 2003 (no fertiliser was applied in 2002).

Table 2. Site soil analysis

	pH	Organic matter (%)	N (%)	P ($mg.kg^{-1}$)	Mg ($c.mol_c.kg^{-1}$)	K ($c.mol_c.kg^{-1}$)	Ca ($c.mol_c.kg^{-1}$)	Clay (%)	Silt (%)	Sand (%)
Spoil material	7.8	0.53	0.132	14	5.1	0.16	14.1	9.00	7.00	84.00
Fine earth	6 5	0.37	0.086	4	0.61	0.05	36.4	38.00	30.00	32.00

Table 3. List of species, origin and seed quantities used in the trial

Species	kg/ha	Seed origin	Species	kg/ha	Seed origin
Festuca ovina L.	10	France	*Medicago sativa* L.	20	France
Dactylis glomerata L.	10	France	*Onobrychis sativa* Medik	20	France
Lolium multiflorum Lam.	10	France	*Anthyllis vulneraria* L.	3	France
Lotus corniculatus L.	3	France	*Phacelia tanacetifolia* Bentham.	20	France
Sanguisorba minor Scop.	60	France	*Trifolium subterraneum* L.	20	Australia
Plantago lanceolata L.	3	France	*Melilotus alba* Lam	10	France
Achillea milefolium L.	3	New Zeeland	*Melilotus officinalis* (L.) Pallas.	10	Australia
Medicago lupulina L.	10	France			

At the beginning of July, the number of plants, the dry weight biomass and the percent vegetation canopy cover, were annually estimated for the following 4 years. Five square frames of 0.5 m × 0.5 m in each plot were randomly taken to estimate the number of plants and dry biomass per square metre. Indigenous species in the vegetation composition were also recorded in the plots. The method of points with a 10 needle frame (Cook and Box, 1961; Papanastasis, 1976) was used to estimate the vegetation canopy cover.

The analysis of variance per year was calculated for the data with the plot (treatment per replication) measurements as the basic units, to identify differences between the treatments. All the percentages were transformed with the angular (arcsin) transformation before any statistical treatment; the transformed percentages are shown with brackets in the tables; the comparison of treatment means was done with the least significant difference method.

Results

The results (Table 4) show that in the first year, the simple hydro-seeding (A = hydro-seeding) gave poor results regarding the production of biomass, vegetation cover and number of plants per m^2 in comparison with the combined treatments (B = hydro-seeding with cellulose plus the EI-1000 organic glue, C = hydro-seeding with cellulose plus the synthetic glue CURASOL, D = hydro-seeding plus straw mulching plus asphalt fixing). The combined treatments indicated better results but statistically significant differences between them were not identified. Finally, it should be noted that no differences were found between the treatments with cellulose and different glue material, neither visible phenomena of leaching from rainfall were observed.

The following year (2001) the values for all treatments and all the measured parameters were increased and particularly in the simple hydro-seeding treatment, which reached the values of the other treatments.

During the third year, the dry biomass production, the vegetation cover and the number of plants per square metre were dramatically decreased, while in the next year, after fertiliser application, these parameters were increased again but lower than the levels of the first 2 years.

Concerning biomass production it should be noted that after the first year no statistically significant

Table 4. Biomass production, vegetation cover and number of plants

	Biomass production ($kg \cdot ha^{-1}$) Treatments				Vegetation cover (%) Treatments				Number of plants/square metre Treatments			
Years	A	D	B	C	A	D	B	C	A	D	B	C
2000	2233.3a	4054.7b	3560.0b	3434.7b	52.00 (46.15)a	82.9 (65.5)b	75.6 (60.40b)	72.7 (58.50)b	352.9a	656.1b	687.1b	803.2b
2001	3636.7a	5050.4a	4099.2a	4585.5a	79.8 (63.29)a	92.0 (73.57)a	87.9 (69.66)a	85.0 (67.21)a	711.1a	1007.2b	859.9ab	921.0ab
2002	808.0a	978.1a	1138.9a	979.2a	23.7 (29.15)a	31.2 (33.96)a	33.8 (35.57)a	26.0 (30.63)a	165.3a	252.8b	167.2a	168.3a
2003	1397.0a	1890.9a	1613.8a	1455.5a	43.3 (41.15)a	57.6 (49.37)a	52.7 (46.54)a	54.8 (47.75)a	223.5a	275.0bc	320.0c	261.6ab

Means per row annotated with the same letter do not differ at the $p < 0.05$.
The numbers in brackets represent the angular (arcsin) transformation of percentages.

Table 5. Vegetation composition (mean number of plants m^2 per species and treatments) at the beginning and at the end of the trial period

Treatments	A		B		Γ		E	
Year	2000	2003	2000	2003	2000	2003	2000	2003
Lolium multiflorum	142.3	54.8	237.1	84.6	254.4	86.1	283.4	133.5
Phacelia tanacetifolia	86.3	57.8	118.9	48.0	100.8	26.3	83.6	23.0
Sanguisorba minor	19.9	0.0	77.3	2.3	57.8	2.2	28.3	0.6
Festuca ovina	30.3	0.0	62.9	2.3	84.7	2.2	188.7	0.0
Dactylis glomerata	34.7	2.5	97.1	0.4	86.3	0.4	139.7	0.9
Trifolium subterraneum	13.0	0.0	11.2	0.4	37.0	0.0	45.4	1.9
Onobrychis sativa	3.9	2.5	9.6	3.0	8.1	2.2	5.0	1.2
Plantago lanceolata	6.2	0.0	0.5	3.4	21.0	0.0	7.0	0.0
Medicago sativa	11.5	0.0	0.0	0.0	6.7	0.0	12.6	0.0
Achillea millefolium	0.0	0.0	0.5	0.0	17.6	0.0	0.5	0.0
Medicago lupulina	2.4	0.0	0.0	0.0	0.0	0.0	0.0	0.0
Melilotus officinalis	0.0	0.0	0.0	0.0	0.0	0.0	0.0	0.0
Melilotus albus	0.0	0.0	0.0	0.0	0.0	0.0	0.0	0.0
Lotus corniculatus	0.0	0.0	0.0	0.0	0.0	0.0	0.0	0.0
Anthyllis vulneraria	0.0	0.0	0.0	0.0	0.0	0.0	0.0	0.0
Invaded	2.4 (0.7%)	105.9 (47.4%)	41.0 (6.3%)	130.5 (47.5%)	12.7 (1.8%)	200.6 (62.7%)	9.0 (1.1%)	100.5 (38.4%)

differences between the treatments were found. As for the vegetation cover, in the consecutive years statistically significant differences between the treatments were not apparent. However the number of plants of the straw mulch treatment retained a difference, statistically significant in some cases, from the other treatments.

In the first year, nearly all the sowed species (Table 5) germinated and grew relatively well except the *Anthyllis vulneraria* L., *Lotus corniculatus* L., *Melilotus albus* Lam. and *Melilotus officinalis* (L.) Pallas., which did not appear. In the subsequent years, except 2001, a decline was observed in all sown species. *Phacelia tanacetifolia* Bentham. and *Lolium multiflorum* Lam., which had dominated the vegetation's composition in the first year, were also reduced but conserved their superiority until the end of the trial. In June of the fourth year, two species were mainly found: *L. multiflorum, Ph. tanacetifolia*; and in very limited numbers: *Sanguisorba minor* Scop., *Plantago lanceolata* L., *Onobrychis sativa* Medik, *Festuca ovina* L., *Dactylis glomerata* L. and *Trifolium subterraneum* L. *Ph. tanacetifolia* appeared more numerous in small cavities, furrows and the lower part of the area, where the conditions were better. *L. multiflorum* has considerably declined, and proved inferior for revegetation to *Lolium rigidum* Gaudin, that was used in the past (Brofas and Varelides, 2000) in the same region.

The composition of vegetation in the first year was almost exclusively consisted of sown species (98.2–99.3%) and only in the straw mulching treatment, the number of other species was bigger (6.3%) due obviously to the seed contained in the straw mulch. In the year 2003, the composition of vegetation had been considerably differentiated and the native species constituted 38.4–47.5%, while in the treatment of hydroseeding with cellulose plus the EI-1000 organic glue, the percentage reached 62.7%. The native species, mainly annuals, that dominated the composition of vegetation (Braun-Blanquet climax) were the following: *Vulpia myuros* (L.) C.C. Gmelin *(3)*, *Taeniatherum caput medusae* (L.) Nevski *(2)*, *Avena barbata* Pott ex Link *(2)*, *Lotus sp* L. *(2)*, *Dasypyrum villosum* (L.) Candagy *(1)*, *Bromus sterilis* L. *(1)*, *Daucus guttatus* Sibth.&Sm. *(1)*, *Lactuca seriola* L. *(1)*, *Picnomon acarna* (L.) Cass. *(1)*, *Crepis foetida* L. *(1)*, *Cichorium intibus* L. *(1)*.

Discussion

The treatments with mulching gave good results in all measured parameters, statistically significant in comparison to simple hydro-seeding, and this difference must be attributed to the better conditions created by the mulching (Schuman et al., 1998; Unger, 1978). The results of simple hydro-seeding were inferior. This seems to be related to the lack of mulching, as sufficient rainfall (November 292 mm and December 140 mm) followed the sowing (October 1999) and ensured favourable moisture conditions. Brofas and Varelides (2000) and Muzzi et al. (1997) report poor results with the simple hydro-seeding treatment and according to Bagnaresi et al. (1992, cited by Muzzi et al., 1997) hydro-seeding appears effective in such conditions only when it is combined with tilling and mulching.

Differences were not generally observed between the treatment with straw mulching and the treatments with cellulose. This is probably due to the favourable effect of the cellulose and also to sufficient rainfall that took place after the sowing. Another reason might be the loose soil of the trial that with the natural compaction encloses part of seeds and favours their germination, since it enables better contact between seeds and the soil. However, a rainy season with torrential storms after seeding and before the establishment of an effective vegetation cover could diminish the effectiveness of this treatment. In such a case, if necessary, a full or partial re-application of hydro-seeding should be preferred to that of straw mulching due to its lower cost (1: 2 or 1: 2.5) and easy application.

Compared to the cellulose treatments, the values of biomass production of the straw mulching treatment were higher and in some cases approximated closely to the values that would make them statistically significant. A relative superiority, which became statistically significant during the third year, was also observed in the number of plants of the straw mulching treatment. As for the vegetation cover, significant differences were not observed. Nevertheless, provided that the lack of significance in the measured parameters cannot be attributed to casual facts, it seems that the treatment with cellulose is equally effective to treatment with straw or close behind it. In that case the costly straw mulching may be only applied in sensitive to erosion sites that need immediate protection, or in other areas with marginal ecological conditions.

Concerning the problem with the leaching of cellulose, it is perhaps created in areas prepared in advance and receiving a number of rainfalls before sowing. In this case the soil forms a surface crust, which may prevent the infiltration and facilitate the water runoff, washing away the cellulose with the seeds.

It should also be noted that differences between the treatments with cellulose were not observed. This means that the conjunctive glues are equally effective and the used quantities do not create surface membrane capable of causing problems in the seed germination or preventing rain infiltration in to the soil.

The observed fluctuations in the measured parameters from year to year can be attributed to external factors, such as fertilisation and amount of rainfall. Fluctuations in plant yield are often reported and they are highly correlated with rainfall (Fresquez et al., 1990; Papanastasis, 1982). Richardson and Evans (1986) have attributed the poor establishment and growth of some legumes to the deficiency of N, P and K, while Coppin and Bradshaw (1982) stated that after the initial phase of growth, the vegetation will collapse completely because of lack of nitrogen. Thus the reduced biomass production in the first year (2000) in relation to the second year (2001), taking into account the fertilisation in both years, seems to be related to rainfall which, at the first half of the growing period was much lower (78.5 mm) than the mean (155.4 mm), while in the second year was 242.1 mm. In the third year (2002) the reduction of biomass production seems to be more relevant to the lack of fertilisation, since the rainfall decreased (150.3 mm), but maintained the mean level. In the fourth year (2003) the increase of dry biomass production, and the other parameters appear to be related to repeated fertilisation, since the rainfall was maintained at the same level of the previous year (145.0 mm). If fertilisation had not been applied, the vegetation would at best maintain the previous year level. Consequently, fertilisation can slow down or even reverse the decline process of the established vegetation. However, more research is needed on fertilisation quantities and timing until a permanent vegetation cover is established on these areas.

The improvement of vegetation, in the treatment without mulching after the first year, seems to be relevant with the conditions of mulching, created by the existing vegetation, which dries out and falls on the surface. Thus, with the increase of biomass, the effectiveness of the created mulching is improved and under the same conditions the treatment of simple hydro-seeding can approach the effectiveness of the treatment with mulching after the first year. Consequently, the use or not of mulching is relevant to the climatic conditions following the sowing until the vegetation establishment.

Regarding the vegetation's composition, in the first year it was dominated by the annual species: *L. multiflorum* and *Ph. tanacetifolia*. This was rather expected since annual species have faster emergence rates (Bartolome and Gemmill, 1981), and relatively better growth rates, compared to the perennials (Garnier, 1992; Jackson and Roy, 1986). Albaladejo et al. (2000) and Diaz et al. (1997) report that the initial explosion and predominance of annual plants in soils treated by soil fertilisers is attributed to the pioneer behaviour of these type of plants and whose rapid growth is favoured by the added nutrients.

Perennial species presented much smaller numbers of plants per m^2. Most likely, the fast germination and growth of the above mentioned annual species (*L. multiflorum* and especially *Ph. tanacetifolia*) impeded the development of other species and particularly the perennial ones. According to Bartolome and Gemmill (1981), and to Dyer et al. (1996), annual grasses can have a negative impact on perennial grasses establishment. However, apart from the conditions of competition, some other factors probably contributed to this development of the vegetation. Richardson and Evans (1986) e.g., have attributed the poor establishment and growth of some legumes to the deficiency of N, P, K, while Coppin and Bradshaw (1982) attribute the disappearance of legumes from re-vegetated mining areas to the lack of P. Seedling desiccation and limited nutrient availability, in sand dune and calcareous grassland, was found to have a major impact on seedling establishment (de Jong and Klinkhamer, 1988; Grime and Curtis, 1976). The conditions of germination, which vary between species (Harper et al., 1964; Hilhorst and Toorop, 1997), as well as the weather after seeding (Grantz et al., 1998), can still be some of the factors that influenced the composition and development of the vegetation.

Despite the fact that *L. multiflorum* and especially *Ph. tanacetifolia* may affect negatively the establishment of perennial species, they cover the soil fast and protect it from erosion, while their numbers decrease later permitting the invasion of native species. Based on such performance we believe that these species may be used in simple hydro-seeding or hydro-seeding combined with cellulose mulching, although their percent contribution in the seed mixture should be further studied. On the contrary, they are not necessary in the mixtures with straw mulching, where they may be substituted with perennial species, which after their establishment offer better protection to the soil.

Regarding the native species, they gradually invaded the trial and displaced to a large extent the sown species. According to Scullion (1992), the creation of an initial vegetative cover on such sites enables the introduction or the reintroduction of native species. A first estimate shows that perennial species are few and scattered, in contrast to the annual species that dominated the trial. Especially, *V. myuros* constituted roughly 50% of invaded species and in many places it has created one species populations. This species in combination with the other introduced and native annual plants could be considered as responsible for the displacement of the sown perennial species and the limited invasion of native perennial species. The pure populations of *V. myuros* constitute a good indication of the previous observation. Brown and Rice (2000) also found that native perennial grasses were suppressed by the presence of annual fescue (*V. myuros*), and consider unwise to include *V. myuros* in seed mixtures with the above species. Nevertheless, the performance of this species must be studied as it results in high-density populations, taking also into consideration that extreme fluctuation between years may occur in the annual grass populations (Talbot et al., 1939).

Concerning the seed mixtures, further research is required to investigate the composition and the rates of the species used with emphasis on the perennials, especially the natives ones (established on mining spoils), that can be adapted better to these artificial and steep slopes.

Acknowledgments

The implementation of this experimental set up would not be possible without the valuable help of the S & B, Industrial Minerals, S.A.

References

Albaladejo Montoro J, Alvarez Rogel J, Querejeta J, Diaz E and Castillo V 2000 Three hydro-seeding revegetation techniques for soil erosion control on anthropic steep slopes. Land Degrad. Dev. 11, 315–325.

Bagnaresi U, Ferrari C, Muzzi E and Rossi G 1992 Ricerche per la sistemazione a verde di una cava di argilla. Comune di Carpineti, Regio Emilia. Dipartimento di Colture Arboree, Universita di Bologna, Bologna.

Bartolome J W and Gemmill B 1981 The ecological status of *Stipa pulchra* (Poaceae) in California. Madrono 28, 172–184.

Ben-Hur M, Plaut Z, Shainberg J, Meiri A and Agassi M 1989 Cotton canopy and drying effects in runoff during irrigation with moving sprinkler systems. Agron. J. 81(5), 752–757.

Bramble W 1952 Reforestation of strip-mined bituminous coal land in Pensylvania. J. Forest. 50, 308–314.

Brofas G 1979 Paysages et exploitations minières. Recherche appliquée dans la région de Delphes (Grèce Centrale. Thèse de Docteur—ingénieur Université de Paris VII.

Brofas G 1987 Recherche pour le réaménagement des carrières de marbre dans la région de la montagne de Pendeli. Dasiki Erevna 8(2), 149–186 (in Greek, summary in French).

Brofas G and Varelidis C 2000 Hydro-seeding and mulching for establishing vegetation on mining spoils in Greece. Land Degrad. Dev. 11, 375–382.

Brown C and Rice K 2000 The mark of Zorro effects of the exotic annual grass *vulpia myuros* on California Native Perennial Grasses. Restor. Ecol. 8, 1, 10–17.

C.E.T.E. (Centre d'études Techniques de l'équipement d'Aix-en-Provence) 1980 Revégétalisation des talus en zone Méditerranéenne. Aix-en-Provence, France, pp. 268.

Coppin N J and Bradshaw A D 1982 A guide to quarry reclamation, pp. 79. M.J. Books, London.

de Jong T J and Klinkhamer P G L 1988 Seedling establishment of the biennials *Cirsium vulgare* and *Cynoglossum officinale* in a sand dune area: the importance of growth for differential and growth. J. Ecol. 76, 393–402.

Diaz E, Roldin A, Castillo V and Albaladejo J 1997 Plant colonization and biomass production in a Xeric Torriorthent amended with urban solid refuse. Land Degrad. Dev. 8, 245–255.

Dudeck A E, Swanson N P, Mielke L N and Dedrick A R 1970 Mulches for grass establishment on fill slopes. Agron. J. 62, 810–812.

Dyer A R, Fossum H and Menke J 1996 Emergence and survival of *Nassella pulchra* in a California grassland. Madrono 43, 316–333.

FAO 1985 FAO watershed management field manual-vegetative and soil treatment measures. Conservation Guide 13/1. FAO Rome.

Fresquez P R, Francis R E and Dennis G L 1990 Soil and vegetation responses to sewage sludge on a degraded semiarid broom snakeweed/blue grama plant community. J. Range Manage. 43, 325–331.

Garnier E 1992 Growth analysis of congeneric annual and perennial grass species. J. Ecol. 80, 665–675.

Grantz D A, Vaughn D L, Farber R, Kim B, Zeldin M, VanCuren T and Campbell R 1998 Seeding native plants to restore desert farmland and mitigate fugitive dust and PM10. J. Environ. Qual. 27(5), 1209–1218.

Grime J P and Curtis A V 1976 The interaction of drought and mineral nutrient stress in calcareous grassland. J. Ecol. 64, 975–988.

Harper J L, Williams J T and Sagar G R 1964 The behaviour of seeds in soil. I. The heterogeneity of soil surfaces and its role in determining the establishment of plants from seed. J. Ecol. 53, 273–276.

Hilhorst H W M and Toorop P E 1997 Review of dormancy, germinability, and germination in crop and weed seeds. Adv. Agron. 61, 111–165.

Jusaitis M and Pillman A 1997 Revegetation of waste fly ash lagoons. Plant selection and surface amelioration. Waste Manage. Res. 15, 307–321.

Jackson L E and Roy J 1986 Growth patterns of Mediterranean annual and perennial grasses under simulated rainfall regimes of southern France and California. Acta Ecologica/(Ecologia Plantarum) 7, 191–212.

Karetsos G 2002 Ecology and vegetation study of mount Iti. PhD thesis, University of Patras, Greece, pp. 311+Annexes.

Mariolopoulos H and Karapiperis L 1955 The Precipitations in Greece, pp. 186. National print. Athens, Greece.

Mavrommatis G 1980 The bio-climate of Greece. Dasiki Erevna (Forestry Research), **I**, 1–63 (in Greek, summary in French).

Mermiris Ch and Stefanakis M 2004 Evolution of seeding practice in land reclamation of S&B bauxite mines in Parnasse-Giona massives. *In* Book of Proceedings of First International Conference "Advances in mineral resources management and environmental geotechnology." Chania-Crete, Greece, 7–9 June, 2004, pp. 755–760.

Muzzi E, Roffi F, Sirotti M and Bagnaresi U 1997 Revegatation techniques on clay soil slopes in Northern Italy. Land Degrad. Dev. 8, 127–137.

Papanastasis V 1976 Measuring frequency and cover of vegetation with point-quadrates in a grassland of northern Greece. To Dassos (The Forest), 28, 13–20 (in Greek, summary in English).

Papanastasis V P 1982 Production of natural grasslands in relation to air temperature and precipitation in northern Greece, Dassiki Erevna, 3(III)—Annex (111 p.) (in Greek with English summary).

Richardson J A and Evans M E 1986 Restoration of grassland after magnesian limestone quarrying. J. Appl. Ecol. 23(1), 317–332.

Schuman G E, Booth D T and Cockrell J R 1998 Cultural methods for establishing Wyoming big sagebrush on mined lands. J. Range Manage. 51(2), 223–230.

Scullion J 1992 Re-establishing life in restored topsoils. Land Degrad. Rehab. 3, 161–168.

Talbot M W, Biswell H H and Hormay A L 1939 Fluctuations in the annual vegetation of California. Ecology 20, 394–402.

Unger P W 1978 Straw mulch effects on soil temperatures and sorghum germination and growth. Agron. J. 70, 858–864.

Whisenant S G, Thurow T L and Maranz S J 1995 Initiating autogenic restoration on shallow semiarid sites. Soc. Ecol. Restor. 3, 61–67.

Wolf D D, Blaser R E, Morse R D and Neal J L 1984 Hydro-application of seed and wood-fiber slurries to bind straw mulch. Reclam. Reveg. Res. 3, 101–107.

Revegetation on steep slopes and in subalpine areas using biennial cover plants: A review of Huter's technique

J. Heumader*

Forsttechnischer Dienst für Wildbach- und Lawinenverbauung, Gebietsbauleitung Oberes Inntal, Langgasse 88, A-6460 Imst, Austria

Key words: Revegetation, cover plants, *Secale cereale* L., *Vicia villosa* Roth

Abstract

In the high-mountain region of the Tyrol in Austria, a special method was developed by Huter and Heumader (1981) for the revegetation of soils lacking humus using biennial cover plants. This method shows good results not only under normal conditions, but also on problematic sites such as steep slopes and in subalpine areas. With the exception of fertilizers, no synthetic components, e.g. bitumen emulsions or soil-fixing adhesives are needed. Biennial varieties of rye (*Secale cereale* L.) and sometimes vetch (*Vicia villosa* Roth) are used as cover plants, which produce a lot of organic matter above and below ground. These rapidly growing cover plants prevent surface erosion and protect the slowly growing perennial revegetation plants. After 2 years, the rotting shoots and roots form a green manure, which is important for soil microorganisms. Biennial cover plants, perennial revegetation plants and fertilizer are applied together. On steep terrain small ditches (furrows), running diagonally across the slope, are dug or hoed by hand to control surface runoff and prevent surface erosion and the formation of rills and gullies. This method has proved to be cheap and effective and has the advantage of mostly using natural material.

Introduction

Modern machinery has made large-scale earthworks possible, e.g. quarries or mines. These earthworks can result in large, bare areas that are sometimes difficult to revegetate. Revegetation problems also arise on steep slopes due to the lack of soil organic matter (humus) and through rapid soil erosion, These problems are particularly severe in the harsh environment of subalpine and alpine zones. Natural processes can lead to large disturbances, e.g. landslides, especially in mountainous regions. Natural revegetation of these sites may take decades, sometimes centuries. However, natural revegetation of anthropomorphic disturbances and/or dangerous debris-producing erosion scars is often unacceptably slow and requires human intervention.

* Email: joerg.heumader@wlv.bmlf.gv.at

General remarks

The revegetation of disturbed, bare sites has two main goals:

- Protection against surface erosion and improvement of infiltration and water-holding capacities of the soil.
- Protection of adjacent ecological attributes by adapting the established vegetation cover to the surrounding environment.

Other goals can be the reclamation of grassland for grazing livestock or hay production.

Revegetation using biennial cover plants

The principal idea of a special method developed by Huter and Heumader (1981) in Tyrol/Austria, is the use of biennial varieties of rye *(Secale cereale* L.) as cover plant species. This technique has the following

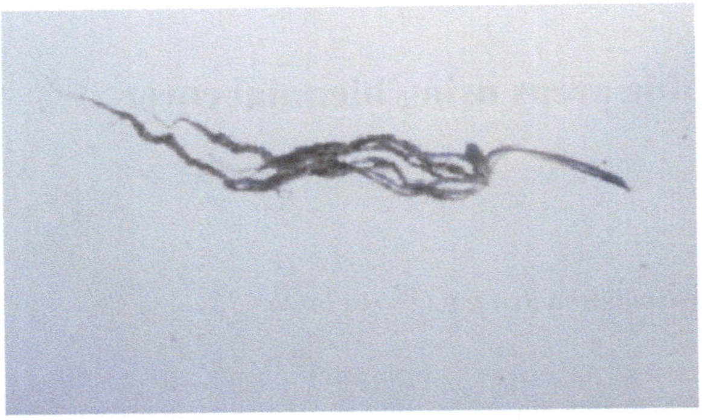

Figure 1. Seedling of biennial rye *(Secale cereale* L) with well-developed root system.

Figure 3. Seedlings of biennial rye in the subalpine zone in spring.

advantages:

- The rapidly growing roots of the cover plants protect the soil from erosion (Figures 1 and 2).
- The rapidly sprouting cover plants shelter the subsequently germinating and slower growing perennial revegetation plants to some extent from negative climatic influences (Figures 3 and 4).
- After 2 years, the cover plants die (Figure 5) and their decomposing stems and leaves remain on the soil's surface as an organic layer whilst their roots enhance soil development and encourage soil microorganism activity.

Revegetation works—practical experiences

The ground bio- and eco-engineer has to find answers to the following questions when planning revegetation at problem sites:

1. How to prevent surface water from eroding bare slopes in the critical period till the vegetation takes over the protective function? In the author's experience, small ditches (furrows) running diagonally across the slope have shown good results on steep terrain. These ditches are dug by hand and should have enough inclination to lead surface water off without overtopping. The higher and steeper the slope, the smaller the distance between these ditches (Figure 6).
2. How to prepare the soil surface for seeding? In case of human caused disturbances, seeding should be performed immediately after the earthworks are finished. In this state there are still many fine soil particles and microsites for the seeds to lodge in, thus enhancing germination. Older, compacted sites can be improved by roughening the surface by hand. The more rough and loose the soil surface, the better the results.

Figure 2. Rye seedlings sprouting rapidly on moraine material lacking humus.

Figure 4. Seedlings of biennial rye in the subalpine zone in autumn (same plot as in Figure 3).

Figure 5. Biennial rye in the autumn of the second year on the crown of an avalanche dam at an altitude of 2300 m.

3. What species of revegetation and cover plants are to be used, in what mixture and in what quantities? The key element of the revegetation method developed by Huter and Heumader (1981) is the use of biennial varieties of rye (*Secale cereale* L.) as cover plants. The rye grains (100–200 kg/ha) can be sown during the whole vegetation period from spring onwards. As biennials, the fruiting culms will grow after the frost period in the second year. The seeded rye needs to be slightly covered with soil to improve germination. There are old varieties of biennial rye available in the Austrian Alps, which have been used successfully as cover plants in the alpine zone up to 2500 m. At high elevations, the rye should be soaked (pre-germinated) in lukewarm water for 24 h. Sometimes biennial vetch (*Vicia villosa* Roth) can be planted with the rye. After sowing the biennial(s), the seeds (100–200 kg/ha) of the perennial revegetation plant species are distributed. The species composition should be determined by experts. Often excess seeding rates are used. Although this may result in a dense green cover in the first 1 or 2 years, after the effects of fertilizers have ceased, competition will result in a rapid decline in plant cover. The seeds of rye and vetch can also be spread mechanically with the perennial plant seeds and fertilizers using a hydro seeder. However, this only works well on sites where the soils have been recently worked and there are enough gaps for the rye and vetch seeds to fall into. Otherwise, the heavy seeds of these species tend to roll off the slope.
4. How much and how often to fertilize? 200–300 kg/ha of synthetic fertilizers can be used with the cover plants. As plants will take up the nutrients rapidly and transform them to decomposing organic matter after 2 years, it is not necessary to use organic fertilizers or manure. The perennial revegetation plants should have limited nutrients so that they develop deep-reaching root systems; too much fertilizing has the opposite effect and only provides a quick visual result.

Figure 6. To prevent surface erosion, small ditches (furrows) are dug by hand, running diagonally across the slope.

Figure 7. Middle reach of a torrent with deeply eroded moraine deposits. Technical countermeasures (concrete consolidation dams) were under construction in 1991.

Figure 8. Middle reach of a torrent in 1993 after scarping, degrading and revegetating the slopes, using the method of Huter and Heumader (1981) (same site as in Figure 7).

Figure 9. Middle reach of a torrent in 2004 after reafforestion (same site as in Figures 7 and 8).

5. What about maintenance? Successful revegetation with the goal of erosion control should need no maintenance after 2 or 3 years. Planting methods should encourage the gradual invasion by neighboring natural vegetation over time.

Concluding remarks

The revegetation method using biennial rye as cover plants, developed by Huter and Heumader (1981), has been applied on many difficult sites such as debris-torrent erosion and avalanche control works in the high-mountain districts of the western Tyrol in Austria (Figs. 7–9). This technique has proven to be effective and relatively inexpensive. Such a method also has the advantage of primarily using natural material; with the exception of seeds and fertilizers, no other organic or synthetic components are needed. Huter and Heumader's (1981) technique does not require mechanical equipment, all work can be done by hand, therefore it would also be a very suitable method for use in developing countries.

References

Huter F and Heumader J 1981 Die Begrünung von Bodenwunden in den Hochlagen mit Hilfe von Winterroggen als Deckfrucht. Wildbach- und Lawinenverbau, Zeitschrift des Vereins der DiplomIngenieure der Wildbach- und Lawinenverbauung, 45. Jahrgang, Heft 2, 57–63.

Comparison of revegetation techniques on alpine slopes prone to avalanches and erosion

T. Schmid[1,4], U. Mueller[2], F. Tognini[3] & J. Meyer[1]
[1]*geoVerde AG, Moserstrasse 8, CH-8300 Schaffhausen, Switzerland.* [2]*Begruenungen Hunn AG, Pilatusstrasse 14, CH-5630 Muri, Switzerland.* [3]*Sezione forestale cantonale, Piazza Nosetto 2, CH-6710 Biasca, Switzerland.* [4]*Corresponding author**.

Key words: Alpine revegetation, hydroseeding, mycorrhiza, sleep seed, straw seeding

Abstract

Ongoing and increasing civil engineering and other construction activities in alpine areas pose a threat to local vegetation, soil cover and finally to slope stability. Most of the alpine vegetation communities are fragile and cannot be restored at all. Vegetation establishment is limited by short summers, heavy rain showers, hail, winds, low temperatures and strong UV radiation. The only persisting plants are specialists that evolved and thrive in these conditions. For revegetation at high altitudes thorough planning respecting the site conditions, use of site-adapted or local plant material and appropriate application techniques are essential.

A case study of a revegetation project at 2000 m altitude on a construction site for the protection of the main transalpine routes (road and railway), power lines and villages against avalanches illustrates possibilities and limits of a selected state of the art revegetation technique. The constructed coffer and diversion dams and barrier hills required classical ground bioengineering practices such as brush and hedge layers and berms for stabilisation.

Key factors for the successful revegetation itself were preliminary drainage system setup in the project area, continuing revegetation work right after construction intervention, the use of a site adapted seed blend, seeding in autumn ('sleep seed') and the application of a straw mulch layer or use of coco nettings on exposed and steep slopes. The vegetation coverage 1 year after seeding was dense enough to protect soil from erosion but still allowing local seeds to establish. The use of limited amount of nurse or pioneer plants such as *Lolium perenne* L., protecting the slower emerging seeded and native plants and disappearing after one or two seasons, proved of value. Uncontrolled pasture poses a threat to plantings and seeded vegetation and should be avoided for several years on these sites.

Results of a field trial on another high altitude site suggests that on sites with severely disturbed or completely lost vegetative soil the inoculation with soil microorganisms helps to re-establish self-sufficient vegetation. Mycorrhizal inoculation performs best with a low nutrient input in organic form.

Introduction

Increasing construction for tourism (winter resorts, cablecars and constructions for artificial snow) as well as the construction of roads, water supply and protective structures damage vegetation cover in alpine sites. Consequences of vegetation loss are increased erosion; primarily the loss of the fine soil matrix which is the basis for plant growth (Rickli, 2001). A slope without vegetation has less shear strength and thus is less stable (Frei et al., 2003). Most of the alpine vegetation communities are fragile and some cannot be restored at all, such as Sphagnion magellanici, Caricion curvulae, Caricion firmae, etc. A damage of these specialist vegetations cannot be restored by artificial means.

High altitudes represent suboptimal conditions for vegetation establishment in several ways. Short summers, low temperatures, rapid temperature changes, heavy rain showers, low humidity, snow, hail, frost, winds and strong radiation are limiting factors. The

* Email: tobias.schmid@geoverde.ch

Stokes et al. (eds), Eco- and Ground Bio-Engineering: The Use of Vegetation to Improve Slope Stability, 433–438.

only surviving plants are the ones that evolved and thrived in the conditions present on alpine sites. Thus, for revegetation at high altitudes knowledge of site conditions, use of site-adapted or local plant material and selected application techniques are essential.

A case study of the revegetation of protective constructions against avalanches at Pizzo Erra (Leventina, Canton Ticino, Switzerland) illustrates possibilities and limits of a selected state-of-the-art revegetation technique. The area was repeatedly affected by avalanche incidences causing death, loss of property and suspension of transport routes. Initial protective structures date back to 1895.

This latest project started in 1986 and ends in 2004. Constructions to protect the main transalpine road and railway routes of the Gotthard pass, power lines and villages against avalanches mainly consisted of a coffer dam, a diversion dam, barrier hills, establishment of protection forest and iron protection structures. The mountainside affected extends between the foot of Pizzo Erra (2416 m altitude) and the village of Anzonico (984 m altitude). Total costs of the project were 15 Mio Euros, financed by state, province, municipality, and the national railway.

The hydroseeding technology employed in this case on the majority of the area is used since the sixties in Switzerland. The main advantages over dry seed are: quick application, immediate superficial erosion control, protection of seed (depending on amount of mulch applied), readily applicable on slopes, quick and regular germination and is cheaper than planting, however it is more expensive than dry seed. Mulching of seeded areas with a layer of straw has proved to be of special use on exposed sites with soils of low water holding capacity, high solar irradiation and on slopes. The mulch cover protects germinating plants from drying, favours microclimate by lowering daily temperature extremes and humidity changes and protects the seed from erosion and rain splash impact destroying soil aggregates.

In a field trial performed at an even higher altitude (2700 m) at Munt da San Murezzan (St. Moritz, Canton Graubünden, Switzerland) different application techniques were compared. At this site, the ski world championship took place in February 2003. Constructional engagements for this event left impacted landscape and offered the chance to install a revegetation trial to test the use of soil microorganism amendments for the revegetation of a harsh high altitude site. A focus was aimed at using mycorrhizal fungi as a key organism group for plant development (Smith and Read, 1997). Most herbaceous plants live in an intricate symbiotic association with arbuscular mycorrhizal (AM) fungi. These fungi play a key role in the exploitation of soil mineral resources and stress tolerance of plants. In herbaceous plant communities, a diverse and efficient AM fungal community contributes greatly to their diversity and ecosystem function (van der Heijden et al., 1998). Mycorrhizal fungi bind the soil in ways that the plants alone cannot, thus promoting the formation of soil aggregation and by this conserving macro-porous soil structure that allows penetration of water and air and prevents erosion (Miller and Jastrow, 1992). In addition, mycorrhized plants may become quicker self-sufficient especially on harsh sites. However, soil disturbance decreases the natural mycorrhizal potential. The reintroduction of mycorrhizal fungi during revegetation may be an appropriate tool in high altitude revegetation (St. John, 2000; Turnau and Haselwandter, 2002).

Materials and methods

Case study at Pizzo Erra

The limiting conditions on the revegetation site at around 2000 m altitude are a short vegetation period of 130 days, average annual temperature of 0–5°C, peak rain showers of up to 120 mm at a time and a 100 year snow maximum 4.6 m. Another constraint is given by the strong winds which together with the southwest position of the slope promote dry conditions.

The original prevailing vegetation community was Larici-Pinetum cembrae – Rhododendro Vaccinion and Nardion. Soils are slightly acidic. Although top soil was secured before intervention it remained scarce for redistribution since the overall surface increased with the construction of the dams and usual losses by soil movement. Pasturing by wild animals is common in the area and not suitable for the initial period after the revegetation. Nevertheless, not all the areas were fenced off.

A network of drainage channels was installed to gather and safely lead away the water that arises quickly with the intense rain showers. The coffer and diversion dams and barrier hills required classical ground bioengineering practices such as brush layers with willow cuttings of local and regional origin (*Salix purpurea* L., *S. nigricans* Sm., *S. daphnoides* Vill., *S. appendiculata* Vill.), hedge layers and plantations with different wooden species (*Larix deciduas* Mill., *Pinus*

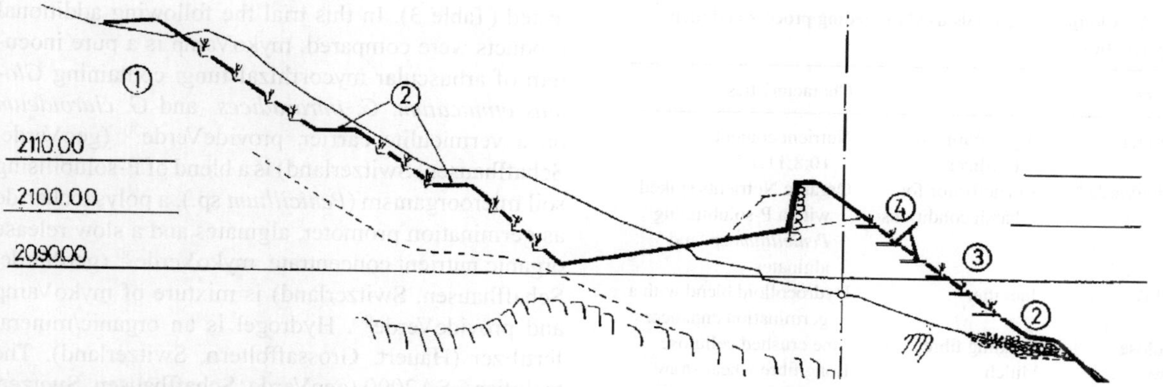

Figure 1. Side view of the coffer dam below Pizzo Erra. 1, Hedge layers with humus; 2, berms; 3, brush layers with humus and wooden sill reinforcement; 4, Snow rack.

mugo Turra, *Pinus cembra* L., *Picea abies* L., *Sorbus aucuparia* L., *Juniperus communis* L.) for stabilisation of the slopes. Berms were formed on the hillside slope above the dam. Brush layers were reinforced with wooden sill. Additional supporting structures against snow movements were installed along the coffer dam (Figure 1)

The revegetation was performed over a period of 5 years totalling 80,000 m^2 on slopes of up to 45° angle. The site adapted seed blend used was Cristallina supplemented with 13 ecotype forbs (Table 1). Products used in the seeding process are described in Table 2. The application was performed with a three-step method straw seed with hydroseeding and power-mulcher equipment:

1 Hydroseeding with seeds (15 g/m^2), soil conditioner provideVerde® (30 g/m^2), organic/mineral fertilizer (10 g/m^2), tackifier geoTak (10 g/m^2) and Cellulose (30 g/m^2).
2 Dry application of a straw mulch cover (300 g/m^2).
3 Bond the straw cover with a dissolved tackifier (20 g/m^2) and Cellulose (60 g/m^2).

The strong winds at the site mean the third step is essential to prevent further losses.

Table 1. Seeds used in the Pizzo Erra revegetation project: Seed blend Cristallina with two high alpine ecotype grasses and the mix of 13 ecotype species of forbs and legumes

	Seed blend Cristallina		Ecotype blend
Grasses	*Festuca nigrescens* Lam.	Herbs	*Anthyllis montana* L.
High Alpine Ecotypes	*Poa alpina* L.		*Lotus alpinus* DC.
			Trifolium alpestre L.
Grasses	*Agrostis stolonifera* L.		*Trifolium badium* Sch.
Useful in high altitudes	*Cynosurus cristatus* L.		*Trifolium nivale* Koch
	Phleum pratense L.		
	Poa pratensis L.	Legumes	*Alchemilla xanthochlora* Rothm.
			Aster alpinus L.
Grasses	*Lolium perenne* L.		*Campanula barbata* L.
For rapid erosion control	*Dactylis glomerata* L.		*Dianthus silvester* Wulfen.
and as nurse plants	*Festuca duriuscula* L.		*Doronicum grandiflorum* Lam.
	Poa trivialis L.		*Hieracium alpinum* L.
	Poa compressa L.		*Myosotis alpestris* Schmidt
			Plantago alpina L.
Legumes	*Trifolium hybridum* L.		*Potentilla aurea* L.
	Lotus corniculatus L.		*Veratrum album* L.
	Anthyllis vulneraria L.		
Herbs	*Achillea millefolium* L.		

Table 2. Additional materials used in seeding process and their characteristics

Product	Use	Characteristics
Hydrogel	Organic/mineral fertilizer	Nutrient content 10:8:11:1.5
provideVerde®	Soil activator for harsh conditions	Organic Nutrients spiked with a P-solubilizing *Penicillium* sp. and alginates
GeoTak	Tackifier	Hydrocolloid blend with a germination enhancer
Cellulose	Bonding fibre	Fine crushed cellulose
Straw	Mulch	Long fibre wheat straw

Most of the area was seeded in autumn (sleep seed) in order to profit from the spring melt waters right away. On steep slopes (>30°) coco nettings were applied to improve protection and top-soil stability. On remote sites such as around the iron protective structures a mobile backpack blower was employed. In this case seed, soil conditioner and tackifier were mixed and applied in a single step onto the coco nettings or the bare soil.

Field trial above St. Moritz

The site lies below Munt da San Murezzan (St. Moritz, Canton Graubünden, Switzerland) on a shallow slope (10–20°) at 2700 m altitude. For the ski world championship water tubes were installed into the ground for an artificial snowing system. During construction the soil was not treated as recommended so that the different layers of A and B horizon were mixed up. The soil was a loamy sand low in organic content and with low nutrient content, especially P (0.3 mg/kg).

The trial was set up and seeded on 17 October 2001 as a sleep seed in dry form by manual means. Five treatments with six replicates each were set up with 20 m^2 per parcel. Seeds were applied already premixed with corresponding additional products to be tested (Table 3). In this trial the following additional products were compared. mykoVamp is a pure inoculum of arbuscular mycorrhizal fungi containing *Glomus etunicatum, G. intraradices* and *G. claroideum* on a vermiculite carrier. provideVerde® (geoVerde, Schaffhausen, Switzerland) is a blend of P-solubilising soil microorganism (*Penicillium* sp.), a polysaccharide as germination promoter, alginates and a slow release organic nutrient concentrate. mykoVerde® (geoVerde, Schaffhausen, Switzerland) is mixture of mykoVamp and provideVerde®. Hydrogel is an organic/mineral fertilizer (Hauert, Grossaffoltern, Switzerland). The tackifier GSA2000 (geoVerde, Schaffhausen, Switzerland) used in this trial in all the treatments to stick seeds and products to the ground is a mix of starch and cellulose. Unfortunately no straw cover could be put on top due to bad weather conditions, though it would be absolutely necessary at this type of site.

Coverage data for grasses, forbs and legumes were determined separately with the quadrat point method (Daget and Poissonet, 1971) in 2002 and 2003. Data were analysed by ANOVA and Duncan's New Multiple Range test to determine significant differences.

Results

Case study Pizzo Erra

No detailed data were collected. However, a rapid coverage of most of the slopes was achieved. Two strong thunderstorms (1998 and 2000) have passed since with no remarkable erosion. Also the vegetation persisted so far with no additional maintenance nor fertilizing. The seed density seemed to be well chosen allowing the colonisation by local plants. The use of nurse or pioneer plants such as *Lolium perenne* L. disappearing after one or two seasons proved of value. Few damages only were observed due to snow movement along hillside. Reinforcement structures against snow movement proved of a value.

Table 3. Treatments and amendments in St. Moritz field trial

Treatments	Amendments	Seed blend
1	*Control* / Cellulose 40 g/m^2 / GSA 2000 5 g/m^2	OH-Piz Alpin 2002 (Alpine ecotype blending)
2	*mykoVamp* 20 g/m^2 / Cellulose 40 g/m^2 / GSA 2000 5 g/m^2	
3	*mykoVerde®* 80 g/m^2 / Cellulose 40 g/m^2/ GSA 2000 5 g/m^2	
4	*provideVerde®* 75 g/m^2 / Cellulose 40 g/m^2/ GSA 2000 5 g/m^2	
5	*Hydrogel* 20 g/m^2/ Cellulose 40 g/m^2/ GSA 2000 5 g/m^2	

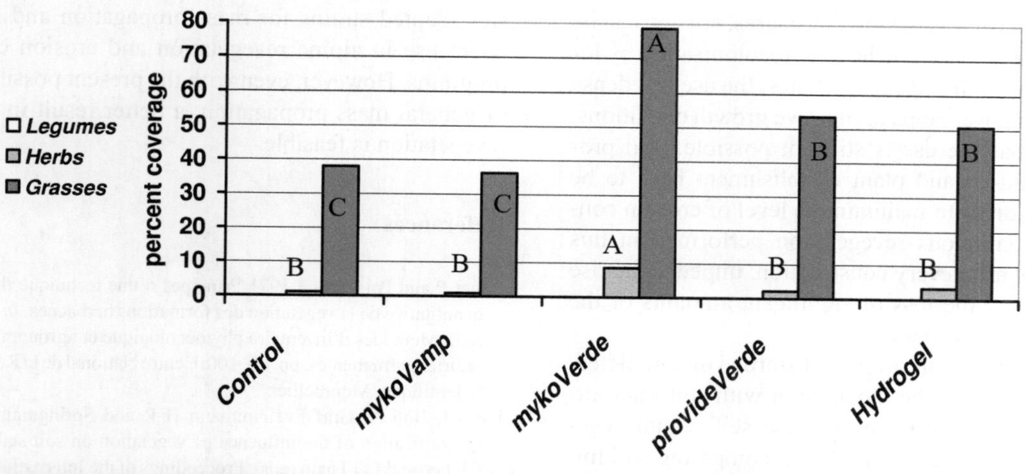

Figure 2. Results of the St. Moritz field trial after 1 year. Significant differences in vegetation coverage are shown with different letters for legumes, herbs and grasses separately (ANOVA, Duncan test $p < 0.05$).

Some plantations were affected by pasturing and did not perform well. Some willows of *S. appendiculata* died off due to the intense grazing, bad rooting and drought. Hillside plantations without snow racks did suffer of snow movement and pressure.

Field trial St. Moritz

Grasses dominated the vegetation especially in the first year in all treatments (Figure 2). On the parcel treated with mykoVerde® a slightly higher forb coverage was observed. No legumes were observed in the first year. Total coverage was best with mykoVerde® (average 78%) significant higher than provideVerde® and Hydrogel, which performed better than the Control and mykoVamp treatment.

In the second year all the four treatments continued and the best were the mykoVerde® treated parcels (Figure 3). The first legumes arose but no significant differences were observed.

Discussion

Any constructional interventions on high altitude sites are a threat to local plant communities and may destroy them irrevocably. The case study of Pizzo Erra demonstrates that by thorough site-adapted planning

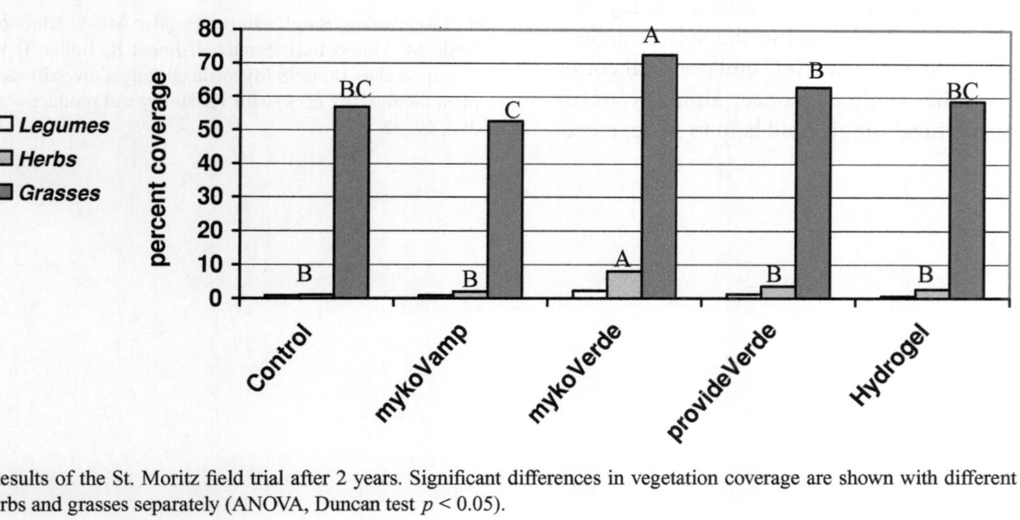

Figure 3. Results of the St. Moritz field trial after 2 years. Significant differences in vegetation coverage are shown with different letters for legumes, herbs and grasses separately (ANOVA, Duncan test $p < 0.05$).

and execution of revegetation measures, erosion can be controlled and a basis be laid for recolonisation by local plants. At high and exposed sites, the use of a dense straw mulch cover helps to improve growth conditions.

Regular pasture use is still not possible. Soil protection, build-up and plant establishment have to be favoured in order to maintain the level of erosion control. The continuous revegetation performed in this project right after every construction, impeded the rise of erosion and the loss of significant amounts of the important fine soil fraction.

On harsh sites with severely disturbed or completely lost vegetative soil the inoculation with soil microorganisms might help to re-establish self reliant vegetation. The field trial at St. Moritz comparing soil microorganism based amendments suggests however, that a mycorrhizal application alone without any nutrient or organic content would not result in a quick enough erosion control effect. The combination of organic bound nutrients, a P-solubilising *Penicillium sp.* and mycorrhizal inoculum (as in mykoVerde®) resulted in quicker vegetation development and performed the necessary 70% vegetation coverage to control erosion within 1 year.

Use of inoculation in erosion control requires a suitable means of introducing both seed and inoculum, and incorporation of a diverse seed mix that includes several good mycorrhizal hosts. The use of fertilizer should be limited to the forms and amounts that approximate the actual needs of the vegetation. Excess fertilization, the rule in erosion control, will discourage formation of the mycorrhizal network and will encourage the growth of weeds (St. John, 2000).

Recently, unexpected diversity of arbuscular mycorrhizal fungi has been discovered in the soil of undisturbed alpine meadows (Oehl, 2005 and presonal communication). Further study of pioneer alpine types of arbuscular mycorrhizal fungi could help to select more site adapted strains for mass propagation and subsequent use in alpine revegetation and erosion control programs. However, even with the present possibilities of general mass propagation, a better result in alpine revegetation is feasible.

References

Daget P and Poissonet J 1971 Principes d'une technique d'analyse quantitative de la vegetation des formations herbacèes. *In* P Daget (ed), Méthodes d'inventaire phytoécologique et agronomique des prairies permanentes, pp. 85–100. Centre National de la Recherche Scientifique, Montpellier.

Frei M, Böll A, Graf F, Heinimann H R and Springman S 2003 Quantification of the influence of vegetation on soil stability. *In* CF Lee and LG Tham (eds), Proceedings of the International Conference on Slope Engineering, pp. 872–877. Department of Civil Engineering, University of Hong Kong.

Miller R M and Jastrow J D 1992 The role of mycorrhizal fungi in soil conservation. *In* GJ Bethlenfalvay and RG Linderman (eds), Mycorrhizas in sustainable agriculture. pp. 29–43. American Society of Agronomy, Madison.

Oehl F 2005 Significance of arbuscular mycorrhizae in the Alpine altitudinal belt discovered. Communication of the Swiss National Science Foundation, The Image of the Month February 2005. www.snf.ch/en/com/imm/imm.asp#feb

Rickli C 2001 Vegetationswirkungen und Rutschungen. Eidg. Forschungsanstalt WSL, Bundesamt für Umwelt, Wald und Landschaft. Birmensdorf, Bern.

Smith S E and Read D J 1997 Mycorrhizal Symbiosis. Academic Press, London.

St. John T 2000 The Instant Expert Guide to Mycorrhiza. www.mycorrhiza.org/EXPERTflat.PDF

Turnau K and Haselwandter K 2002 Arbuscular mycorrhizal fungi, an essential component of soil microflora in ecosystem restoration. *In* S Gianinazzi, H Schüepp, J M Barea and K Haselwandter (eds), Mycorrhizal Technology in Agriculture, pp.137–150.

Birkhäuser Verlag, Basel. van der Heijden MGA, Klironomos JN, Ursic M, Moutoglis P, Streitwolf-Engel R, Boller T, Wiemken A and Sanders IR 1998 Mycorrhizal fungal diversity determines plant biodiversity ecosystem variability and productivity. Nature 396, 69–72.

MIX
Papier aus verantwortungsvollen Quellen
Paper from responsible sources
FSC® C105338

If you have any concerns about our products,
you can contact us on
ProductSafety@springernature.com

In case Publisher is established outside the EU,
the EU authorized representative is:
Springer Nature Customer Service Center GmbH
Europaplatz 3, 69115 Heidelberg, Germany

Printed by Libri Plureos GmbH
in Hamburg, Germany